D1614120

Prefixes for Powers of 10*

Multiple	Prefix	Abbreviation
10^{24}	yotta	Y
10^{21}	zetta	Z
10^{18}	exa	E
10^{15}	peta	P
10^{12}	tera	T
10^{9}	giga	G
10^{6}	mega	M
10^{3}	kilo	k
10^{2}	hecto	h
10^{1}	deka	da
10^{-1}	deci	d
10^{-2}	centi	c
10^{-3}	milli	m
10^{-6}	micro	μ
10^{-9}	nano	n
10^{-12}	pico	p
10^{-15}	femto	f
10^{-18}	atto	a
10^{-21}	zepto	z
10^{-24}	yocto	y

* Commonly used prefixes are in red. All prefixes are pronounced with the accent on the first syllable.

The Greek Alphabet

Alpha	A	α	Nu	N	ν
Beta	B	β	Xi	Ξ	ξ
Gamma	Γ	γ	Omicron	O	o
Delta	Δ	δ	Pi	Π	π
Epsilon	E	ϵ, ε	Rho	P	ρ
Zeta	Z	ζ	Sigma	Σ	σ
Eta	H	η	Tau	T	τ
Theta	Θ	θ	Upsilon	Υ	υ
Iota	I	ι	Phi	Φ	φ
Kappa	K	κ	Chi	X	χ
Lambda	Λ	λ	Psi	Ψ	ψ
Mu	M	μ	Omega	Ω	ω

Terrestrial and Astronomical Data*

Acceleration of gravity at Earth's surface	g	9.81 m/s²
Radius of Earth R_E	R_E	6.38×10^6 m
Mass of Earth	M_E	5.98×10^{24} kg
Mass of the Sun		1.99×10^{30} kg
Mass of the Moon		7.35×10^{22} kg
Escape speed at Earth's surface		11.2 km/s = 6.95 mi/s
Standard temperature and pressure (STP)		0°C = 273.15 K 1 atm = 101.3 kPa
Earth–Moon distance†		3.84×10^8 m = 2.39×10^5 mi
Earth–Sun distance (mean)†		1.50×10^{11} m = 9.30×10^7 mi
Speed of sound in dry air (20°C, 1 atm)		344 m/s
Density of dry air (STP)		1.29 kg/m³
Density of dry air (20°C, 1 atm)		1.20 kg/m³
Density of water (4°C, 1 atm)		1000 kg/m³
Latent heat of fusion of water (0°C, 1 atm)	L_F	334 kJ/kg
Latent heat of vaporization of water (100°C, 1 atm)	L_V	2260 kJ/kg

* Additional data on the solar system can be found in Appendix B and at http://nssdc.gsfc.nasa.gov/planetary/planetfact.html.
† Center to center.

Mathematical Symbols

$=$	is equal to
$\equiv$	is defined by
$\neq$	is not equal to
$\approx$	is approximately equal to
$\sim$	is of the order of
$\propto$	is proportional to
$>$	is greater than
$\geq$	is greater than or equal to
$>>$	is much greater than
$<$	is less than
$\leq$	is less than or equal to
$<<$	is much less than
Δx	change in x
dx	differential change in x
$|x|$	absolute value of x
$n!$	$n(n-1)(n-2)\ldots1$
Σ	sum
lim	limit
$\Delta t \rightarrow 0$	Δt approaches zero
dx/dt	derivative of x with respect to t
$\partial x/\partial t$	partial derivative of x with respect to t
$\int_{x_1}^{x_2} f(x)\, dx$	definite integral $= F(x)\Big|_{x_1}^{x_2} = F(x_2) - F(x_1)$

Abbreviations for Units

A	ampere	h	hour	N	newton
Å	angstrom (10^{-10} m)	Hz	hertz	nm	nanometer (10^{-9} m)
atm	atmosphere	in.	inch	Pa	pascal
BTU	British thermal unit	J	joule	rad	radians
Bq	becquerel	K	kelvin	rev	revolution
C	coulomb	kg	kilogram	R	roentgen
°C	degree Celsius	km	kilometer	Sv	sievert
cal	calorie	keV	kilo-electron volt	s	second
Ci	curie	lb	pound	T	tesla
cm	centimeter	L	liter	u	unified mass unit
eV	electron volt	m	meter	V	volt
°F	degree Fahrenheit	MeV	mega-electron volt	W	watt
fm	femtometer, fermi (10^{-15} m)	mi	mile	Wb	weber
ft	foot	min	minute	y	year
G	gauss	mm	millimeter	μm	micrometer (10^{-6} m)
Gy	gray	mmHg	millimeters of mercury	μs	microsecond
g	gram	mol	mole	μC	microcoulomb
H	henry	ms	millisecond	Ω	ohm

Some Conversion Factors

Length

$1\ m = 39.37\ in. = 3.281\ ft = 1.094\ yard$

$1\ m = 10^{15}\ fm = 10^{10}\ Å = 10^9\ nm$

$1\ km = 0.6214\ mi$

$1\ mi = 5280\ ft = 1.609\ km$

$1\ light\text{-}year = 1\ c \cdot y = 9.461 \times 10^{15}\ m$

$1\ in. = 2.540\ cm$

Volume

$1\ L = 10^3\ cm^3 = 10^{-3}\ m^3 = 1.057\ qt$

Time

$1\ h = 3600\ s = 3.6\ ks$

$1\ y = 365.24\ day = 3.156 \times 10^7\ s$

Speed

$1\ km/h = 0.278\ m/s = 0.6214\ mi/h$

$1\ ft/s = 0.3048\ m/s = 0.6818\ mi/h$

Angle–angular speed

$1\ rev = 2\pi\ rad = 360°$

$1\ rad = 57.30°$

$1\ rev/min\ (rpm) = 0.1047\ rad/s$

Force–pressure

$1\ N = 10^5\ dyn = 0.2248\ lb$

$1\ lb = 4.448\ N$

$1\ atm = 101.3\ kPa = 1.013\ bar = 760\ mmHg = 14.70\ lb/in.^2$

Mass

$1\ u = [(10^{-3}\ mol^{-1})/N_A]\ kg = 1.661 \times 10^{-27}\ kg$

$1\ tonne = 10^3\ kg = 1\ Mg$

$1\ kg = 2.205\ lb$

Energy–power

$1\ J = 10^7\ erg = 0.7376\ ft \cdot lb = 9.869 \times 10^{-3}\ L \cdot atm$

$1\ kW \cdot h = 3.6\ MJ$

$1\ cal = 4.186\ J$

$1\ L \cdot atm = 101.325\ J = 24.22\ cal$

$1\ eV = 1.602 \times 10^{-19}\ J$

$1\ BTU = 778\ ft \cdot lb = 252\ cal = 1054\ J$

$1\ horsepower = 550\ ft \cdot lb/s = 746\ W$

Thermal conductivity

$1\ W/(m \cdot K) = 6.938\ BTU \cdot in./(h \cdot ft^2 \cdot °F)$

Magnetic field

$1\ T = 10^4\ G$

Viscosity

$1\ Pa \cdot s = 10\ poise$

UNIVERSITY PHYSICS

for the Physical and Life Sciences

Volume I

Philip R. Kesten   David L. Tauck

W. H. FREEMAN AND COMPANY
New York

THIS PRELIMINARY EDITION DOES NOT INCLUDE COMPLETE END-OF-CHAPTER PROBLEM SETS OR ANSWERS

Executive Editor: Jessica Fiorillo
Marketing Manager: Alicia Brady
Market Development Manager: Kirsten Watrud
Assistant Market Development Manager: Kerri Russini
Developmental Editors: Kharissia Pettus, Blythe Robbins
Media and Supplements Editor: Dave Quinn
Senior Media Producer: Keri Fowler
Editorial Assistants: Heidi Bamatter, Nicholas Ciani
Marketing Assistant: Joanie Rothschild
Senior Project Editor: Georgia Lee Hadler
Copy Editor: Connie Parks
Photo Editor: Ted Szczepanski
Photo Researcher: Christina Micek
Text and Cover Designer: Blake Logan
Senior Illustration Coordinator: Bill Page
Illustration Coordinator: Janice Donnola
Illustrations: Precision Graphics
Production Coordinator: Paul Rohloff
Composition: Preparé Inc.
Printing and Binding: Quad Graphics

Cover photo by Tim Fitzharris

Library of Congress Control Number: 2011928382

Preliminary Edition Vol. I: 1-4641-1524-9
Preliminary Edition Vol. II: 1-4641-1525-7

Printed in the United States of America
First printing

W. H. Freeman and Company
41 Madison Avenue
New York, NY 10010
Houndmills, Basingstoke RG21 6XS, England
www.whfreeman.com

Contents in Brief

Contents

Biological Applications

These unique and fully integrated physiological and biological applications, are written and carefully explained by a biologist so that the connections are easy for a physicist to teach.

Chapter 1
Lung volumes (unit conversion)
Hair growth (unit conversion)

Chapter 2
Muscle contraction (velocity)
Cheetah running (acceleration)

Chapter 3
Animal navigation (vectors)
Electrocardiograms (vector addition)
Vestibular system (uniform circular motion)

Chapter 4
Microtubules (Newton's second law)
Fish propulsion (Newton's third law)

Chapter 5
Gecko feet (static friction)
Cornea and eyelids (static friction)
Wrist bones (static friction)
Fainting (forces and uniform circular motion)

Chapter 6
Kangaroo hopping (energy transfer)
Rigor mortis (energy transfer)
Actin in horseshoe crab sperm (work–kinetic energy theorem)
ATPsynthesis (work energy)
Weight lifting (potential energy)
Arteries (conservation of energy)
Photosynthesis (energy conversion)

Chapter 7
Squid propulsion (momentum)
Leaping ballet dancer (center of mass)

Chapter 8
Muscle and bone geometry (lever arms)
Alligator death roll (angular momentum)
Falling cat (angular momentum)

Chapter 9
Arterial blood flow (elasticity)
Anterior cruciate ligament (Young's modulus)
Decompression sickness (volume stress and strain)

Properties of lung tissue (bulk modulus)
Endothelial cells in arteries (shear stress and strain)
Connective tissue (elasticity)
Anterior cruciate ligament (tensile stress, failure, yield strength)

Chapter 11
Bones, muscle, and feathers (density)
Cerebrospinal fluid (pressure and depth)
Blood pressure (pressure and depth)
Breathing (pressure differences)
Body fat assessment (Archimedes' principle)
Blood flow (equation of continuity)
Chronic mountain sickness (viscosity)
Blood flow during exercise (laminar flow)
Ischemic heart disease (Poiseuille's law)

Chapter 12
Intracellular Ca^{2+} concentration (oscillations)
Lung volumes during breathing (phase angle)
Eardrum movement (phase angle)
Heart rate (frequency and period)
Swimming dolphin (potential energy to kinetic energy)
Walking (simple pendulum)
Knee-jerk reflex (physical pendulum)
Hip joint (moment of inertia)
Sensory cells and maintaining balance (damped oscillator)
Beating insect wings (resonance)
Buzzing mosquitoes (natural frequency)

Chapter 13
Horned desert viper (surface waves)
Hearing (frequency and period)
Flagellum (wave speed)
Elephant vocalizations (interference)
Bornean tree-hole frog (longitudinal standing waves)
Eardrum (natural frequency and sound level)
Speed of blood flow (Doppler shift)
Bats and dolphins hunting prey (Doppler shift)
Ultrasonic imaging (Doppler shift)

Chapter 14
Fire beetles (infrared receptors)
Body temperature regulation (temperature)
Hypoxia at high altitude (partial pressure)
Water vapor and respiration (partial pressure)
Sweating (heat flow)
Surviving in harsh environments (radiation, convection, conduction)

About the Authors

DR. PHILIP KESTEN

Dr. Philip Kesten, Associate Professor of Physics and Associate Vice Provost for Undergraduate Studies at Santa Clara University, Santa Clara, CA, received a B.S. in physics from the Massachusetts Institute of Technology and received his Ph.D. in high energy particle physics from the University of Michigan. Since joining the Santa Clara faculty in 1990, Dr. Kesten has also served as Chair of Physics, Faculty Director of the ATOM and da Vinci Residential Learning Communities, and Director of the Ricard Memorial Observatory. He has received awards for teaching excellence and curriculum innovation, was Santa Clara's Faculty Development Professor for 2004–2005, and was named the California Professor of the Year in 2005 by the Carnegie Foundation for the Advancement of Education. Dr. Kesten has also served as the Senior Editor for *Modern Dad*, a newsstand magazine, and was co-founder of the Internet software company Docutek, a SirsiDynix Company.

DR. DAVID TAUCK

Dr. David Tauck, Associate Professor of Biology at Santa Clara University, Santa Clara, CA, holds both a B.A. in biology and an M.A. in Spanish from Middlebury College. He earned his Ph.D. in physiology at Duke University and completed post-doctoral fellowships at Stanford University and Harvard University in anesthesia and neuroscience, respectively. Since joining the Santa Clara University faculty in 1987, he has served as Chair of the Biology Department, and as President of the local chapter of Phi Beta Kappa. Dr. Tauck currently serves as the Faculty Director in Residence of the da Vinci Residential Learning Community.

A Letter from the Authors

As faculty members we understand the frustration and hassles caused by errors in a textbook. We also know that no matter how carefully reviewers as well as authors and their editors hunt for mistakes, the first edition of any book is likely to have at least a few typographical errors or other small problems. That's why we convinced our publisher to print a preliminary edition of our book. Distributing a pre-first edition to potential adopters buys us time to send the entire book through another round of reviews by physicists, editors, proof readers, and us. As we approach the publication date of the first edition, it's especially important to us to continue doing every last thing possible to ensure that the book our students use is as close to perfect as possible. So if you come across any problems as you peruse chapters, please alert one of us directly at kestenandtauck@whfreeman.com, or your sales rep if that's easier for you. We would be grateful!

We hope that, by scrutinizing every detail of the preliminary edition one more time, when the full first edition is published it will read much more like a typical second edition. Thank you for considering the preliminary edition of our book when you evaluate textbooks for the upcoming year.

Sincerely,

Philip R. Kesten, Ph.D.

David L. Tauck, Ph.D.

Preface

University Physics for the Physical and Life Sciences

Nearly half of all undergraduate students enrolled in introductory physics courses major in biology and biomedical sciences. Although there are several excellent calculus-based books aimed at physics and engineering students, the same is not true for those in the life sciences. Books for such students are typically either too basic to prepare students for the Medical College Admissions Test (MCAT) or only differ from books geared for physics and engineering students by having a few biomedical examples layered over a traditional university physics text. This book fills that void by being the first calculus-based physics text written specifically for life sciences students, and the first written by both a physicist and a physiologist.

A Seamless Blend of Physics and Biology

Like traditional physics textbooks, this one offers a rigorous presentation of the fundamentals of introductory physics, with reliance on calculus where appropriate. This book integrates biological examples and clinical correlations into the physics discussions; physiological examples are not merely sprinkled throughout as in a standard physics text. Physical principles are often introduced with a brief, qualitative discussion of a biological application. In this way, the book not only encourages students to learn the fundamentals of physics, it also offers them reasons why learning physics should be important to them. The book strives to instill in students an appreciation of physics by showing them how it determines many characteristics of living systems. This is consistent with the recommendation of the American Association for the Advancement of Science to provide students with an education that enables them to integrate knowledge from different fields of science. In addition, the text entices students to see more deeply into the material by using examples and problems from the physics in students' daily lives, including the modern technology they use every day, the natural phenomena they have experienced throughout their lives, and the sports they commonly watch and play.

Just-in-time Presentation of Calculus

In many instances, using calculus and other slightly more advanced mathematical tools ultimately makes physics easier to grasp. For those students with a weaker background in calculus, the text adapts a just-in-time teaching style to explain calculus as needed to meet the Association of American Medical College's (AAMC) recommendation of developing interdisciplinary science curricula and increased scientific rigor at the undergraduate level. While not all medical schools require calculus, many of the best ones do. With this approach, everyone should be able to understand the ideas behind derivations as well as the concepts expressed in the resulting equations, regardless of their mathematical acumen.

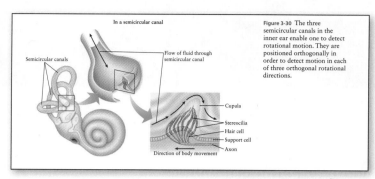

In a semicircular canal

Semicircular canals

Flow of fluid through semicircular canal

Cupula
Stereocilia
Hair cell
Support cell
Axon

Direction of body movement

Figure 3-30 The three semicircular canals in the inner ear enable one to detect rotational motion. They are positioned orthogonally in order to detect motion in each of three orthogonal rotational directions.

Detailed Artwork Designed to Promote Visual Learning

In many textbooks figures contribute little to helping students learn the material. This text uses artwork as a teaching tool. Rather than confining descriptive information in figure legends, as much information as is practical often appears directly on the figures. The result is an annotated figure that reinforces the physics presented in the flow of the text. Moreover, the figures themselves are simple, colorful, and approachable, inviting students to explore them rather than intimidating students into ignoring them.

Text Features

This text includes a problem-solving approach that provides students with all the support they will need, including tools to overcome misconceptions and tackle problems and assessment so students can check for understanding.

Physicist's Toolbox

It's essential for students to have as many tools as possible in their arsenal for tackling the problems found in worked examples and end of chapter problems. **Physicist's Toolboxes** teach the important steps that students can use to successfully set up and solve problems.

Physicist's Toolbox 4-1 Free-Body Diagrams

SET UP

1. Start a free-body diagram by drawing the object or objects on which forces act. Objects can be represented by simple shapes (for example, rectangles or circles) but it is important to get the orientations of the objects approximately correct (Figure 4-9).

2. Identify the origin or agent of any force before adding it to a free-body diagram. In most of the problems that we will encounter, two objects must be touching for one to exert a force on the other. The one exception for now is the force due to gravity. We'll come across some other exceptions later, such as the magnetic force and the force between electric charges.

SOLVE

3. Draw each force acting on an object as an arrow, where the tail end of the arrow starts at the center of the object *on which the force acts*.

4. Point the arrowhead in the direction in which the force acts.

5. Draw the arrow longer or shorter in order to represent the magnitude of a force relative to the other forces.

6. Label each force.

7. Draw and label the coordinate axes.

REFLECT

8. If there is more ... forces in the sam... rate diagram for... the relationship... show the compo... lines for arrows... forces with the f...

9. One way to che... ticular object to... sure that all of t... make sure that f... is at least one fo... the magnitudes o...

Math Box and Math Tutorial

Students may come across a mathematical concept that needs elaboration to allow for deeper understanding of the chapter concepts. **Math Boxes** elaborate on mathematics that is discussed in nearby text. Examples of Math Box topics include the dot product and computing integrals by substitution.

Sometimes it's been a while since students have studied a particular math concept and they need a reminder. Margin notes link concepts in the text to the **Math Tutorial** found in the back of the book. The Math Tutorial serves as a refresher and includes worked examples as well as practice problems. Topics covered by the tutorials include trigonometry, differential calculus, and integrals.

Math Box 4-1 Sigma

Physicists often use mathematical notation in order to represent operations that are either long or of a length not well defined. The capital Greek letter Σ (sigma), for example, is mathematical shorthand for adding together a series of terms. If N is to be found by adding four different values of n or

$$N = n_1 + n_2 + n_3 + n_4$$

the addition can be equivalently written using the summation sign

$$N = \sum_{i=1}^{4} n_i$$

Notice the role of the notation above and below the summation sign; we show the starting value of the variable that counts the terms below it and the final value above. We might not know how many terms will be in the sum. In that case, we write

$$N = $$

We can find that the acceleration vector will point at an angle

$$\phi = \tan^{-1}\left(\frac{a_y}{a_x}\right)$$

counterclockwise from the positive x axis by using Equation 3-4.

 See the Math Tutorial for more information on Trigonometry

Got the Concept?

Most life science students are accustomed to a conceptual approach to learning. Conceptual questions appear as part of the text itself so that students can confirm their readiness to move on to the next topic.

? Got the Concept 6-4
How High?

A block is released from rest at vertical height H on one side of an asymmetric skateboard ramp, as shown in Figure 6-16. The angle from the horizontal of the left ramp is twice the angle of the right ramp. If the ramp is frictionless, how high (vertically) does the block rise on the right side before stopping and sliding back down?

$y_1 = H$

$y = 0$

Figure 6-16 A block is released from rest at vertical height H on one side of an asymmetric skateboard ramp.

Watch Out!

Having taught physics for many years, the authors know which topics are often difficult to learn. Students are often puzzled, for example, by a perception that mirrors reverse images horizontally but not vertically. By tackling such misconceptions directly, the **Watch Out!** feature helps students avoid common pitfalls and gain a deeper understanding of the physics.

! Watch Out
Numeric values can be deceiving.

You might have wondered if there was an error in the two examples of volume in Table 1-2. Could it be that a giant party balloon (Figure 1-2) has a volume of 0.10 m^3 while a typical human head is only 0.005 m^3? The volume of a sphere is given by $\frac{4}{3}\pi r^3$, so the radius of the party balloon is

$$r = \left(\frac{3(0.10 \text{ m}^3)}{4\pi}\right)^{1/3}$$

or about one-third of a meter. That value seems reasonable. How about a head? Try measuring your own—typical values for an adult human head are 17 cm high, 17 cm across, and 21 cm front to back. Taking the shape as box-like, a typical volume is $0.17 \text{ m} \times 0.17 \text{ m} \times 0.21 \text{ m} = 0.006 \text{ m}^3$, so the round number in Table 1-2 is also reasonable.

Problem-Solving Strategy: Set up, Solve, Reflect in Worked Examples

In addition to teaching physics, this text promotes the development of students' problem-solving skills. Too many students would rather memorize equations than comprehend the underlying physics, a mistake that some textbooks encourage in the way they present material. This book models problem-solving skills for students by applying several common steps to all worked example problems. This procedure, summarized by the key phrases "**Set Up**," "**Solve**," and "**Reflect**," mirrors the approach scientists take in attacking problems.

SET UP. The first step in each problem is to determine an overall approach and to gather together the necessary pieces. These might include sketches, equations related to the physics, and concepts. This step relies heavily on diagrams; many sample problems are accompanied by one or more visual aids.

SOLVE. Rather than simply summarizing the mathematical manipulations required to move from first principles to the final answer, the "Solve" section shows many intermediate steps in working out solutions to the sample problems. Other authors too often omit these intermediate steps, either as "an exercise for the student" or perhaps because they appear obvious. Students, however, often do not find the missing steps obvious, and, as a result, simply pass over what might otherwise be a valuable learning experience. In addition, as the authors do during their classroom lectures, the text presents the reasoning and thought process as well as the mathematical steps involved in solving problems. In this way, the text leads students to develop their own problem-solving skills and habits.

REFLECT. An important part of the process of solving a problem is to reflect on the meaning, implications, and validity of the answer. Is it physically reasonable? Do the units make sense? Is there a deeper or wider understanding that can be drawn from the result? This step addresses these and related questions when appropriate. Sometimes the reflection suggests that a second, related problem should be considered; in those cases, a second problem might be tackled or some conceptual questions suggested that allow students to gain greater understanding into the answer.

Associated Practice Problems

Associated **Practice Problems** follow worked examples in the text and reinforce problem-solving skills by asking students to apply the strategies used in the example to a new problem.

Estimate It!

Rough estimations can be a powerful tool in doing science, especially when just starting a new problem. Students will find that it helps them grasp the science at a more intuitive level. The **Estimate It!** feature leads students to develop their estimation skills by modeling the behavior for them.

What's Important

At the end of each chapter section, the **What's Important** feature reminds students about the main points of each section.

Example 6-2 Pushing a Car

Three people each apply a constant force of 200 N horizontally to a car that has run out of gas. To understand the magnitude of a joule, determine the work done on the car by the three people to move it 5 m.

SET UP
The applied force is constant and in the direction in which the car moves, as shown in Figure 6-2, so Equation 6-1 applies. Note that this figure is not a free-body diagram, but instead shows the relevant force and the direction of motion.

SOLVE
We substitute the given values of force and distance into Equation 6-1 to determine the work

$$W = (3)(200\,\text{N})(5\,\text{m}) = 3000\,\text{J}$$

Together the three people do 3000 J of work on the car.

Figure 6-2 Equation 6-1 applies because the applied force is constant and in the direction in which the car moves.

REFLECT
Even if you have never had to push a car that has run out of gas, try to imagine the amount of effort required. Each of the three people in this problem did 1000 J of work on the car.

Practice Problem 6-2 Each of three people apply a constant force of 200 N horizontally to a car. How far will they move it if each does 1 J of work on the car?

Estimate It! 7-1 Enormous Force!

When major league baseball player Derek Jeter hits a home run, the baseball is in contact with his bat for about 0.0007 s. Assuming that the ball leaves Jeter's bat at about the same speed as the pitch (about 95 mph) estimate the average force Jeter's bat exerts on the ball. Major league baseballs weigh between 5.00 and 5.25 oz.

SET UP
We use Equations 7-23 and 7-24 to write the average force as a vector

$$\vec{F}_{\text{avg}} = \frac{\Delta \vec{p}}{\Delta t} = \frac{\vec{p}_{\text{f}} - \vec{p}_{\text{i}}}{\Delta t}$$

So

$$\vec{F}_{\text{avg}} = \frac{m\vec{v}_{\text{f}} - m\vec{v}_{\text{i}}}{\Delta t}$$

The initial and final velocity vectors are in opposite directions, so the velocities have opposite signs. We choose to let v_{f} be positive so v_{i} is negative, and then

$$F_{\text{avg}} = \frac{mv_{\text{f}} + mv_{\text{i}}}{\Delta t}$$

SOLVE
We have assumed that the ball strikes and leaves the bat at the same speed, so $v_{\text{f}} = v_{\text{i}} \approx 95$ mph or 42 m/s. Let m equal 5.125 oz or 0.15 kg and $\Delta t = 0.0007$ s.
So

$$F_{\text{avg}} = \frac{(0.15\,\text{kg})(42\,\text{m/s}) + (0.15\,\text{kg})(42\,\text{m/s})}{0.0007\,\text{s}}$$

$$= 1.8 \times 10^4\,\text{N}$$

REFLECT
The average force exerted on the ball is 1.8×10^4 N—over 2 tons. That's an enormous force to be generated by a human (even a professional baseball player!), but it only lasts for a short time.

⭐ **What's Important 7-4**
The average force that one object exerts on another during a collision is proportional to the change in momentum and inversely proportional to the contact time, which is the length of time objects touch each other during a collision. The difference between the final and initial values of momentum is called impulse.

SUMMARY

Topic	Summary	Equation or Symbol	
acceleration	Acceleration is the rate of change of velocity versus time. In this chapter, we have considered only cases in which acceleration is constant. The SI unit of acceleration is meters per second per second, or meters per second squared (m/s²).	$a = \dfrac{dv}{dt}$	
average acceleration	The average acceleration of an object over time interval $t - t_0$ is the change in velocity $v - v_0$ divided by that time interval. Average acceleration represents the average rate of change in velocity over a time interval.	$a = \dfrac{v - v_0}{t - t_0}$	(2-22)
average speed	The average speed of an object over time interval $t - t_0$ is the net distance the object has traveled divided by that time interval.	$s_{avg} = \dfrac{\text{total distance}}{\Delta t}$	(2-14)
average velocity	The average velocity of an object over time interval $t - t_0$ is the net displacement of the object Δx divided by that time interval. Average velocity equals instantaneous velocity when the velocity of an object is constant. Average velocity represents the average rate of change in position over a time interval.	$v_{avg} = \dfrac{\Delta x}{\Delta t}$	(2-13)

End of Chapter Summary

The end of chapter **Summary** incorporates the key concepts from each chapter as well as any associated equation or symbol into an easy-to-follow table.

End of Chapter Questions and Problems

To reinforce the problem-solving skills taught in the chapters, the end of chapter **Questions and Problems** incorporate conceptual questions and physics problems in the following format:

>   Conceptual Problems
>   Multiple-Choice Problems
>   Estimation/Numerical Analysis
>   Problems by Chapter Section
>   General Problems

Problems include three levels of difficulty: basic, single-concept problems (•), intermediate problems that may require synthesis of concepts and multiple steps (••), and challenging problems (•••).

To help pre-med students prepare for the MCAT® exam, some chapters include problems from actual former MCAT® exams.

Student Ancillary Support

Supplemental learning materials allow students to interact with concepts in a variety of scenarios. By analyzing figures, reinforcing problem-solving methods, and reviewing chapter objectives, students obtain a practical understanding of the core concepts. With that in mind, W. H. Freeman has developed the most comprehensive student learning package available.

Printed Resources

*   ***Problem-Solving Guide with Solutions*** by Timothy A. French, Harvard University

    Volume I (Chapters 1–15) ISBN: 1-4641-0096-9

    Volume II (Chapters 16–28) ISBN: 1-4641-0097-7

    The *Problem-Solving Guide with Solutions* takes a unique approach to promoting students' problem-solving skills by providing detailed and annotated solutions to selected problems. Unlike other solutions manuals, this guide follows the "Set-up," "Solve," and "Reflect" format outlined in the worked examples for worked-out solutions to selected odd-numbered End-of-Chapter problems in the textbook. It also includes integrated media icons which point to selected problem-solving tools that can be accessed.

Free Media Resources

*   **Book Companion Website**

    The *University Physics for the Physical and Life Sciences* Companion Website, www.whfreeman.com/universityphysics1e, provides a range of tools for problem-solving and conceptual support:

    *   Student self-quizzes
    *   Flashcards
    *   Recommended media assets for self-study
    *   Biology Appendix
    *   Access to the Premium Multimedia Resources, which can be purchased for a nominal fee.

MCAT® Section

The section that follows includes material from previously administered MCAT® items and is reprinted with permission of the Association of American Medical Colleges (AAMC).

Passage 13 (81–85)

Tennis balls must pass a rebound test before they can be certified for tournament play. To qualify, balls dropped from a given height must rebound within a specified range of heights. Measuring rebound height can be difficult because the ball is at its maximum height for only a brief time. It is possible to perform a simpler indirect measurement to calculate the height of rebound by measuring how long it takes the ball to rebound and hit the floor again. The diagram below illustrates the experimental setup used to make the measurement.

The ball is dropped from a height of 2.0 m, and it hits the floor and then rebounds to a height h. A microphone detects the sound of the ball each time it hits the floor, and a timer connected to the microphone measures the time (t) between the two impacts. The height of the rebound is $h = gt^2/8$ where $g = 9.8$ m/s² is the acceleration due to gravity. Care must be taken so that the measured times do not contain systematic error. Both the speed of sound and the time of impact of the ball with the floor must be considered.

Four balls were tested using the method. The results are listed in the table below.

Ball	Time (s)
A	1.01
B	1.05
C	0.97
D	1.09

81. If NO air resistance is present, which of the following quantities remains constant while the ball is in the air between the first and second impacts?
A. Kinetic energy of the ball
B. Potential energy of the ball
C. Momentum of the ball
D. Horizontal speed of the ball

Premium Media Resources

The Premium Media Resources can be purchased directly from the Book Companion Website for a small fee, and are also embedded in the multimedia-enhanced eBook and the WebAssign online homework system.

- **Learning Curve** incorporates adaptive question selection, personalized study plans, and state-of-the-art question analysis reports in activities with a game-like feel that keeps students engaged with the material. Integrated eBook sections give students additional exposure to the course text. An innovative scoring system ensures that students who need more help with the material spend more time quizzing themselves than students who are already proficient.

- **P'Casts** are videos that emulate the face-to-face experience of watching an instructor work a problem. Using a virtual whiteboard, the P'Cast tutors demonstrate the steps involved in solving key worked examples, while explaining concepts along the way. The worked examples were chosen with the input of physics students and instructors across the U.S. and Canada. P'Casts can be viewed online or downloaded to portable media devices.

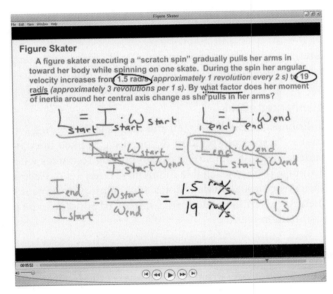

- **Interactive Exercises** are active learning, problem-solving activities. Each Interactive Exercise consists of a parent problem accompanied by a Socratic-dialog "help" sequence designed to encourage critical thinking as users do a guided conceptual analysis before attempting the mathematics. Immediate feedback for both correct and incorrect responses is provided through each problem-solving step.

 Go to Interactive Exercise X.Y for more practice

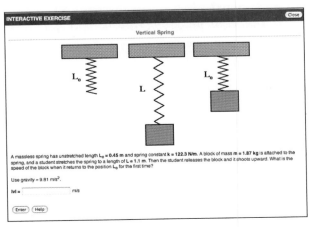

Go to Picture It X.Y
for more practice

- **Picture Its** help bring static figures from the text to life. By manipulating variables within each animated figure, students will be able to visualize and better understand a variety of physics concepts.

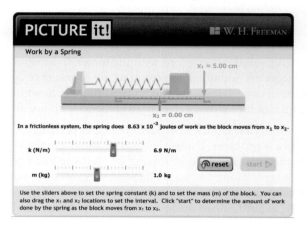

eBook Options

For students interested in digital textbooks, W. H. Freeman offers the complete *University Physics for the Physical and Life Sciences* in two easy-to-use formats.

The Multimedia-Enhanced eBook

The **Multimedia-Enhanced eBook** contains the complete text with a wealth of helpful interactive functions. All student Premium Resources are linked directly from the eBook pages. Students are thus able to access supporting resources when they need them, taking advantage of the "teachable moment" as they read. Customization functions include instructor and student notes, highlighting, document linking, and editing capabilities. Access to the Multimedia-Enhanced eBook can be purchased from the book companion website.

The CourseSmart eTextbook

Though it does not include any Premium Resources, the **CourseSmart eTextbook** does provide the full digital text, along with tools to take notes, search, and highlight passages. A free app allows access to CourseSmart eTextbooks on Android and Apple devices, such as the iPad. They can also be downloaded to your computer and accessed without an Internet connection, removing any limitations for students when it comes to reading digital text. The CourseSmart eTextbook can be purchased at www.coursesmart.com.

Instructor Ancillary Support

For instructors using *University Physics for the Physical and Life Sciences*, W. H. Freeman provides a complete suite of assessment tools and course materials for the taking.

Computerized Test Bank by Don Franklin, Mercer University

ISBN: 1-4641-0098-5

The **Computerized Test Bank** offers over 2,000 multiple-choice questions, tackling both core physics concepts and various life science applications. While the Test Bank is also available in downloadable Word files on the book companion website, the easy-to-use CD includes Windows and Macintosh versions of the widely used Diploma test generation software, allowing instructors to add, edit, and sequence questions to suit their testing needs.

Electronic Instructor Resources

Instructors can access valuable teaching tools through www.whfreeman.com/universityphysics1e. These password-protected resources are designed to enhance lecture presentations, and include Textbook Images (available in .JPEG and PowerPoint format), Clicker Questions, Lecture PowerPoints, Instructor Solutions, and more.

Course Management System Cartridges

W. H. Freeman provides seamless integration of resources in your Course Management Systems. Four cartridges are available (Blackboard, WebCT, Desire2Learn, and Angel), and other select cartridges (Moodle, Sakai, etc.) can be produced upon request.

Online Learning Environments

W. H. Freeman offers the widest variety of online homework options on the market.

WebAssign Premium

For instructors interested in online homework management, **WebAssign Premium** features a time-tested secure online environment already used by millions of students worldwide. Featuring algorithmic problem generation and supported by a wealth of physics-specific learning tools, WebAssign Premium for *University Physics for the Physical and Life Sciences* presents instructors with a powerful assignment manager and student environment. WebAssign Premium provides the following resources:

- Algorithmically generated problems: Students receive homework problems containing unique values for computation, encouraging them to work out the problems on their own.
- Complete access to the interactive eBook is available from a live table of contents, as well as from relevant problem statements.
- Links to the Premium Multimedia Resources are provided as hints and feedback to ensure a clearer understanding of the problems and the concepts they reinforce.

Sapling Learning

Sapling Learning provides highly effective interactive homework and instruction that improve student learning outcomes for the problem-solving disciplines. They offer an enjoyable teaching and effective learning experience that is distinctive in three important ways:

- Ease of Use: Sapling Learning's easy to use interface keeps students engaged in problem-solving, not struggling with the software.
- Targeted Instructional Content: Sapling Learning increases student engagement and comprehension by delivering immediate feedback and targeted instructional content.
- Unsurpassed Service and Support: Sapling Learning makes teaching more enjoyable by providing a dedicated Masters or Ph.D. level colleague to service instructors' unique needs throughout the course, including content customization.

Acknowledgments

Creating a first-edition textbook requires the coordinated effort of an enormous number of talented professionals. We are grateful for the dedicated support of our in-house team at W. H. Freeman; thank you for transforming our concept into a beautiful book.

We especially want to thank our highly-talented developmental editor, Kharissia Pettus, for guiding us through the process of creating a textbook and for sharing her understanding of publishing, editorial prowess, and expertise in thermodynamics. We would also like to thank executive editor Jessica Fiorillo for encouraging us and leading our editorial team, developmental editor Blythe Robbins for coordinating the work of many people and organizing a complicated schedule, media and supplements editor Dave Quinn for producing gorgeous ancillaries and contributing significantly to the design of the book and cover, and of course, we thank editorial assistants Heidi Bamatter and Nick Ciani for their enthusiasm, and hard work on the project. Special thanks also go to our skilled in-house production team, Georgia Hadler, Paul Rohloff, Ted Szczepanski, Bill Page, Janice Donnola, and Blake Logan, for their patience, dedication, and attention to detail.

We are particularly grateful for the substantial contributions of Todd Ruskell, as well as copy editor Connie Parks, accuracy checker Valerie Walters, problems editor Mark Hollabaugh, solutions manual author Tim French, and test bank author Don Franklin. Francesca Monaco of Preparé Inc. also deserves a special thank you for her remarkable attention to details.

Finally, we are grateful to Clancy Marshall both for finding us before we signed with another publisher and for convincing us that we belong on the W. H. Freeman team.

Friends and Family

One of us (PRK) would like to acknowledge valuable and insightful conversations on physics and physics teaching with Richard Barber, John Birmingham, and J. Patrick Dishaw of Santa Clara University, and to offer these colleagues my gratitude. Finally, I offer my gratitude to my wife Kathy and my children Sam and Chloe, for their unflagging support during the arduous process that led to the book you hold in your hands.

One of us (DLT) thanks his family and friends for accommodating my tight schedule during the years that writing this book consumed. I especially want to thank my parents, Bill and Jean, for their boundless encouragement and support, and for teaching me everything I've ever really needed to know; they've shown me how to live a good life, be happy and age gently. I greatly appreciate my sister for encouraging me not to abandon a healthy lifestyle just to write a book. I also want to thank my non-biological family, Holly and Geoff, for leading me to Sonoma County and for making Sebastopol feel like home.

Advisory Board

We sincerely appreciate the following physicists who graciously provided their creative input to the development of the text by serving as members of our Advisory Board:

Timothy A. French, *Harvard University*
Andrew Pelling, *University of Ottawa*
Ryan Snow, *University of California, Davis*
Raluca Teodorescu, *Massachusetts Institute of Technology*
Brian Woodahl, *Indiana University/Purdue University*

Accuracy Review

We know that instructors have particular concerns about using first edition texts because they are prone to errors. We have done everything in our power to alleviate this concern in using our text by submitting chapters and End of Chapter problem sets to several rounds of accuracy reviews and detailed error-checking, by the following:

Wayne R. Anderson
Sacramento City College

Linghong Li
The University of Tennessee at Martin

Marisa Bauza Roman
Drexel University

Todd G. Ruskell
Colorado School of Mines

Kevin W. Cooper
Ohio University

Luke Somers
University of Pennsylvania

Mark Hollabaugh
Normandale Community College

Valerie A. Walters
V Walters Consulting, LLC

Guy Letteer
*Sacred Heart Preparatory and
Santa Clara University*

Class Testers

We thank the faculty below and their 2000+ students for class testing the text.

Neil Alberding, *Simon Fraser University*
George Alexandrakis, *University of Miami*
Philip Backman, *University of New Brunswick*
Philip Blanco, *Grossmont College*
Ethan Dolle, *Northern Arizona University*
Diana Driscoll, *Case Western Reserve University*
John Evans, *Georgia State University*
Melissa Franklin, *Harvard University*
Timothy A. French, *Harvard University*
Logan McCarty, *Harvard University*
Ed Meyer, *Baldwin Wallace College*
Andrew Pelling, *University of Ottawa*
Greg Thompson, *Adrian College*
Carolyne Van Vliet, *University of Miami*
Ruqian Wu, *University of California, Irvine*
Joan Zinn, *Cuyahoga Community College*

Reviewers

We would also like to thank the many colleagues who carefully reviewed chapters for us. Their insightful comments significantly improved our book.

Victor O. Aimiuwu, *Central State University*
Bijaya Aryal, *University of Minnesota*
Quentin G. Bailey, *Embry-Riddle Aeronautical University*
Pradip Bandyopadhyay, *Pennsylvania State University, Berks Campus*
Arun Bansil, *Northeastern University*
David Baxter, *Indiana University*
Bereket Berhane, *Embry-Riddle Aeronautical University*
Luca Bertello, *University of California, Los Angeles*
Nancy Beverly, *Mercy College*
Angela Biselli, *Fairfield University*
Arie Bodek, *University of Rochester*
Robert Boivin, *Auburn University*
Suzanne White Brahmia, *Rutgers University–Busch Campus*
John Broadhurst, *University of Minnesota*

Yorke Brown, *Dartmouth College*
Terry Buehler, *University of California, Berkeley*
James J. Butler, *Pacific University*
Frank Capozzi, *Bradley University, Bridgewater State University*
Duncan Carlsmith, *University of Wisconsin–Madison*
B. Ross Carroll, *Arkansas State University*
Tom Carter, *College of DuPage*
Ryan Case, *Elizabethtown Community and Technical College*
Paola M. Cereghetti, *Lehigh University*
John Cerne, *State University of New York–Buffalo*
Prem Chapagain, *Florida International University*
John S. Conway, *University of California, Davis*
Susan Coppersmith, *University of Wisconsin–Madison*
Nasser Demir, *Duke University*
Margaret Dobrowolska, *University of Note Dame*
Rodney Dunning, *Longwood University*
Michael Eads, *University of St. Francis*
Nayer Eradat, *San Jose State University*
Milton W. Ferguson, *Norfolk State University*
Jane Flood, *Muhlenberg College*
Lewis Ford, *Texas A&M University*
Ted Forringer, *LeTourneau University*
Tim French, *Harvard University*
Lev Gasparov, *University of North Florida*
Vadas Gintautas, *Chatham University*
Yvonne Glanville, *Penn State Worthington*
Benjamin Grinstein, *University of California, San Diego*
Puru Gujrati, *University of Akron*
Ajawad (A.J.) Haija, *Indiana University of Pennsylvania*
Katrina Hay, *Pacific Lutheran University*
Jim Henriques, *Cerritos College*
Andrew S. Hirsch, *Purdue University*
John T. Ho, *University at Buffalo*
Mark Hollabaugh, *Normandale Community College*
John D. Hopkins, *Penn State University*
David Hough, *Trinity University*
Bill Ingham, *James Madison University*
David C. Ingram, *Ohio University*
Sai Iyer, *Washington University*
Bob Jacobsen, *University of California, Berkeley*
Mark C. James, *Northern Arizona University*
Pu Chun Ke, *Clemson University*
Ed Kearns, *Boston University*
Peter Kernan, *Case Western Reserve University*
Kevin R. Kimberlin, *Bradley University*
Jeremy King, *Clemson University*
Yury Kolomensky, *University of California, Berkeley*
J.K. Krebs, *Franklin and Marshall College*
Andrew Kunz, *Marquette University*
David Lamp, *Texas Tech University*
Mark Lattery, *University of Wisconsin–Oshkosh*
Shelly R. Lesher, *University of Wisconsin–La Crosse*
Judah Levine, *University of Colorado*
Hong Lin, *Bates College*
Ramon E. Lopez, *University of Texas at Arlington*
Robert C. Mania Jr., *Kentucky State University*

Muhammed Maqbool, *Mount Olive College*
Eric C. Martell, *Millikin University*
Dario Martinez, *Rice University*
Donald Mathewson, *Kwantlen University College*
Mark Matlin, *Bryn Mawr College*
Dan Mazilu, *Washington and Lee University*
Jeffrey McGuirk, *Simon Fraser University*
James G. McLean, *SUNY Geneseo*
Edwin F Meyer, *Baldwin Wallace College*
Mark Morgan-Tracy, *Central New Mexico Community College*
Jeffrey S. Olafsen, *Baylor University*
John Parsons, *Columbia University*
Andrew E. Pelling, *University of Ottawa*
Chris Petrie, *Brevard Community College*
Jason Pinkney, *Ohio Northern University*
Amy Pope, *Clemson University*
Gurcharan S. Rahi, *Fayetteville State University*
Roberto Ramos, *Indiana Wesleyan University*
Paul Rider, *Grand View University*
Stephen Robinson, *Belmont University*
S. Clark Rowland, *Andrews University*
Mark Rupright, *Birmingham-Southern College*
Mehmet Alper Sahiner, *Seton Hall University*
Stiliana Savin, *Barnard College*
George T. Shubeita, *University of Texas, Austin*
Kanwal Singh, *Sarah Lawrence College*
Ryan Snow, *University of California, Davis*
Michael Sobel, *Brooklyn College*
C.E. Sosolik, *Clemson University*
Achilles Speliotopoulos, *University of California, Berkeley*
Zbigniew M. Stadnik, *University of Ottawa*
J. Scott Steckenrider, *Illinois College*
Jason Stevens, *Deerfield Academy*
Oleg Tchernyshyov, *Johns Hopkins University*
Raluca E. Teodorescu, *Massachusetts Institute of Technology*
Beth A. Thacker, *Texas Tech University*
Gregory B. Thompson, *Adrian College*
Christos Velissaris, *University of Central Florida*
E. Prasad Venugopal, *University of Detroit, Mercy*
Chuji Wang, *Mississippi State University*
Gary A. Williams, *University of California, Los Angeles*
Shannon Willoughby, *Montana State University*
Jeff Allen Winger, *Mississippi State University*
Scott W. Wissink, *Indiana University–Bloomington*
Carey Witkov, *Broward College*
Gregory G. Wood, *CSU Channel Islands*
Brian Woodahl, *Indiana University–Purdue University Indianapolis*
Ruqian Wu, *UC Irvine*
Alexander Wurm, *Western New England University*
Chadwick Young, *Nicholls State University*
Tanya Zelevinsky, *Columbia University*
Ulrich Zurcher, *Cleveland State University*

1 Physics: An Introduction

(Masterfile)

You might be surprised at how much physics you already know or have at least experienced, just because you've lived in the world for a while! You've probably watched a ball roll down a slope or caught the reflection of something (maybe a hummingbird) off the shiny surface of a pond. Perhaps you've ridden a roller coaster that does a vertical 360 or gone bungee jumping. And maybe you've wondered how your brain detects when your body accelerates or how a detailed image of your internal organs can be generated noninvasively. Like the iridescent blue of the hummingbird's feathers, physics lurks in much of what we find in the world around us, just waiting for you to explore.

Physics is the science of how the universe works. Physics helps us understand the tiniest particles that make up atoms and faraway galaxies. In many ways, physics is the most fundamental of the sciences. Chemists apply knowledge gleaned by physicists to understand the properties of molecules and biologists apply the work of chemists to living systems. Yet no matter how complex a living organism, the laws of physics underlie the processes that characterize life. An understanding of physics, therefore, facilitates an understanding of life.

1-1 Speaking Physics

You didn't learn to speak by memorizing millions of phrases for the countless situations that require you to communicate. Instead you gradually acquired the basic rules of grammar and a vocabulary of useful words. Using those basic tools, you learned to express an enormous variety of ideas, even ones no one else has had before, by combining words according to the rules of grammar.

See the Math Tutorial for more information on Equations.

The same principle is true of many complicated skills. A jazz musician, for example, doesn't memorize long sequences of motions required to perform each individual note in every piece of music she performs. Instead she learns how to play individual and small groups of notes and the rules by which they can be combined to improvise riffs that are pleasing to hear.

At the outset it may seem like physics is a long compilation of rules, equations, definitions, and concepts, similar to the long list of words in a phrase book you might take when visiting a foreign country. We will help you discover, however, that learning physics is no different than learning to speak a language or to play jazz. From the long list of physics "phrases" we will distill out the vocabulary and the rules of physics grammar, so that you will learn to *speak* physics rather than read it from a physics phrase book!

The vocabulary of physics is expressed in terms of variables that represent important quantities, and the rules are defined by the fundamental equations which relate them to one another. Expressing the relationship between physical quantities by writing equations is a powerful way to represent complex processes in simple, concise ways. Like most physics books, this one contains many equations. But not all equations express fundamental physical concepts or rules. A much smaller set of basic equations can be used to find relationships of interest to any specific problem. By starting from the fundamental equations and combining them through algebraic manipulation, we can simplify most problems in order to express a quantity of interest in terms of others that we already know.

As you can imagine, the fundamental equations can lead to the generation of many secondary equations. In general, we encourage you not to try to memorize all the myriad variations. The equations are the sentences of physics, not the vocabulary words; memorizing them will not ultimately help you learn the language of physics.

It may be tempting to view the equations in this book, even the fundamental ones, as the science itself and to reach for them first when trying to solve a problem. From the start, we exhort you not to do this! Doing physics is not about picking equations; reaching for a convenient equation that has the "right" variables will usually be counterproductive except in the simplest cases. Studying the form of an equation and the way it relates both to concepts and to other equations is the best way to learn physics.

Calculus is part of the language of physics. Often, for example, calculus offers a natural way to work with very small elements of an object, providing tools and insights that make physics easier to understand.

> **! Watch Out**
>
> **Equations are not the science.**
>
> Equations are compact, convenient ways to represent a scientific concept or relationship. Knowing equations, however, is not a substitute for understanding the concepts that motivate the equations. We will help you to be successful in your study of physics by taking you beyond the equations and showing you the relationships between equations and their applications.

> **✱ What's Important 1-1**
>
> Physics is not a list of definitions and equations to memorize. However, equations help us understand physics by representing complex processes in simple, concise ways.

1-2 Physical Quantities and Units

If you're planning a road trip, you probably know how far it is to your destination and how long it will take to get there. If you're a cook, you probably know how long it takes to hard-boil an egg. If you're a bodybuilder, you probably know how much weight you can lift. But you can only communicate the value of these quantities to someone else if you both agree on basic units of quantities such as *length, time, temperature,* and *mass.*

Scientists measure all sorts of things in their observations and experiments. Many quantities can be determined by measuring others and then combining the measurements according to the laws of physics. Yet practically all physical processes, characteristics, and phenomena can be expressed in terms of a small number of independent, *fundamental quantities.* For example, length, time, temperature, and mass are considered *fundamental* because they cannot be expressed in terms of other physical quantities. To communicate values of any quantity, including the fundamental ones, the scientific community has adopted the **International System of Units,** commonly known as **SI units,** from the French *Système International d'Unités.* Although the choice of the units is arbitrary in that they have been defined by humans rather than prescribed by nature, SI units are the most widely used system in the world. Table 1-1 lists certain fundamental quantities and their corresponding SI units.

Most of the units in Table 1-1 have been defined in terms of natural phenomena. In the late 1700s, the French Academy of Sciences declared the meter to be a specific fraction (1/10,000,000) of the distance from Earth's equator to the pole. Today the meter is defined by the distance light travels in a vacuum in a tiny fraction (1/299,792,458) of a second. The second, in turn, is defined as the time it takes for 9,192,631,770 periods of the transition between the two split levels of the ground state of the cesium-133 atom. Kelvin, the unit of temperature, is defined in terms of the conditions under which water can exist as ice, liquid, and gas simultaneously. Another scientific unit of temperature, degree Celsius, is based on the temperatures at which water freezes and boils. The kilogram is an exception—it is defined by the mass of a carefully protected prototype block made of platinum and iridium that was manufactured in 1889. The block is stored under three bell jars (**Figure 1-1**) in a vault at the International Bureau of Weights and Measures near Paris, France.

Many quantities are not described by one of the SI units in Table 1-1. For example, a quantity representing how fast an object moves or *speed,* which is central to the discussion of motion we'll begin in Chapter 2, is not listed in the table. But note that the rate at which something moves is determined by the distance it travels during a time interval, and distance (m, meter) and time (s, second) are two of the fundamental quantities. The SI units for speed, then, are meters per second, m/s. We will encounter many other so-called derived quantities as we progress through our exploration of physics, some straightforward including area (m^2) and mass density (kg/m^3), and some not so straightforward such as resistance to electric current, $m^2 \cdot kg/(s^3 \cdot A^2)$. Happily, scientists have named additional (non-fundamental) units for some of the more complex and frequently used derived quantities. For example, we define 1 $m^2 \cdot kg/(s^3 \cdot A^2)$ as 1 ohm (1 Ω), the SI unit of electric resistance.

Table 1-1	Fundamental Quantities and Their SI Units		
Quantity		**Unit**	**Abbreviation**
length		meter	m
time		second	s
mass		kilogram	kg
temperature		kelvin	K
electric current		ampere	A
amount of substance		mole	mol
light intensity		candela	cd

Figure 1-1 The kilogram is defined by the mass of a block made of platinum and iridium, stored under three bell jars in a vault at the International Bureau of Weights and Measures near Paris. *(© BIPM. Reproduced with permission of Bureau International des Poids et Mesures.)*

Physicist's Toolbox 1-1   Back-Pocket Numbers

Many physicists find it helpful to have an intuitive feel for the sizes or magnitudes of physical quantities. This is especially true if you grew up using the English system of units—you might have an intuitive feel for, say, how hard it is to lift a block of ice that weighs 10 pounds, but not how it hard it would be to lift a block that weighs 10 newtons. Here are some sizes and magnitudes that you can put "into your back pocket," to have handy when you need an intuitive feel for a value that arises when doing physics (Table 1-2). You might not be familiar with all of the quantities yet, but don't worry. We discuss them all in the course of exploring physics in this book.

Table 1-2	For Your Back Pocket	
Quantity	**Unit**	**For Your Back Pocket**
weight	newton	A medium size apple weighs about 1 N.
mass	kilogram	The mass of a 1-L bottle of water is 1 kg.
temperature	degree Celsius	The temperature on a cold day in the mountains might be −6 °C. On a hot day at the beach the temperature might be 35 °C.
distance	meter	Your arm is probably about 0.75 m long. The ceiling in your room is likely about 2.5 m high. An American football field is about 90 m long.
distance	kilometer	If you're an average person you can walk 1 km in about 12 min.
speed	meter/second	The speed limit on most U.S. highways is about 30 m/s.
power	watt	Your desk lamp likely radiates 40 W or 60 W.
volume	cubic meter	The volume of a typical human head is about 0.005 m³, or 5000 cm³. The volume of a giant party balloon might be 0.10 m³, or 100,000 cm³.
energy	joule	An apple that falls from a height of 1 m has about 1 J of (kinetic) energy. A typical flashlight battery has about 10,000 J of stored energy.

Figure 1-2 How big is a balloon? *(Corbis)*

❗ Watch Out
Numeric values can be deceiving.

You might have wondered if there was an error in the two examples of volume in Table 1-2. Could it be that a giant party balloon (Figure 1-2) has a volume of 0.10 m³ while a typical human head is only 0.005 m³? The volume of a sphere is given by $\frac{4}{3}\pi r^3$, so the radius of the party balloon is

$$r = \left(\frac{3\left(0.10 \text{ m}^3\right)}{4\pi}\right)^{1/3}$$

or about one-third of a meter. That value seems reasonable. How about a head? Try measuring your own—typical values for an adult human head are 17 cm high, 17 cm across, and 21 cm front to back. Taking the shape as box-like, a typical volume is 0.17 m × 0.17 m × 0.21 m = 0.006 m³, so the round number in Table 1-2 is also reasonable.

We use variables to represent the values of physical quantities and relationships between them. Although the quantity is often clear from the letter or symbol used (for example, we use r for the radius of a circle or sphere and m for mass), there is not always an obvious connection between the name of the quantity and the standard variable. Momentum, a measure that takes both mass and speed into account, is represented by p, and a magnetic field is often denoted as B.

We have far fewer letters and symbols available than the number of quantities required to describe the myriad phenomena that constitute all of physics. Even with lowercase and uppercase letters in our alphabet and the Greek alphabet and with a variety of mathematical and other symbols, a good number of variables must serve multiple purposes. T is both the temperature of a system and the period of time over which a pendulum swings back and forth. The letter i represents the square root of -1 and denotes the current in an electric circuit. You will often need to pay careful attention to the context in which a variable is used to discern the quantity it represents. It's also worth noting that some symbols look alike, such as w and ω (the Greek letter omega) as well as u and μ (the Greek letter mu). Don't fall into the trap of confusing them!

In this book, all variables are italicized, so that you can recognize them easily. Sometimes we'll assign values to variables in the example problems we solve, but operating with the letters and symbols is hardly different from operating with numbers. For example, the area of a table top is the product of its length L and width W. Area is then given by LW. Alternatively, if the length and width are given as 1.5 m and 2 m, respectively, the area is given by (1.5 m)(2 m) or 3 m^2. In either case, the area is simply the product of two quantities.

We often use a lowercase letter to represent a quantity which could change in a problem and a capital letter to represent a quantity that is fixed for a given problem. For example, let's say a problem involves the motion of an object. We might solve the problem in a general way, treating the object's mass as a variable. We would likely use lowercase m to represent its mass. But if we're working on a gravitational problem that involves the mass of Earth, we'll likely represent that quantity by uppercase M, because the value of the quantity is constant. In a similar way, in a problem in which a ball is dropped from the roof of a building, H could represent the fixed height of the building but we'd use h for the changing height of the ball as it falls. There are certainly exceptions to this rule, however. For example, by convention we always use capital V to represent voltage, regardless of whether the voltage is constant or changing.

Subscripts are useful when a problem involves more than one value of a particular quantity. In a problem that involves a number of objects that have different masses, for example, we would use the variables m_1, m_2, m_3, and so on, to represent the mass of each object. Words can be useful as subscripts. For example, in a problem that involves the distances a red car and a blue car travel, we might use d_red and d_blue to represent those distances.

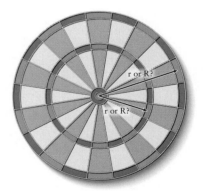

Figure 1-3 Uppercase R suggests a quantity the value of which does not change throughout a problem, while lowercase r suggests a variable or changing quantity.

Got the Concept 1-1
Radius and Radius

A problem involving a dartboard requires you to represent its overall radius and also the distance from the center of the dartboard to the center of each colored region, as in **Figure 1-3**. You want to use the letters r and R to represent quantities of radius. Which is the better choice to represent the (constant) radius of the dartboard and which is the better choice to represent the (changing) radius of each separate colored region?

In addition to the quantities on which the physics of a particular problem or phenomenon depend, a host of physical constants are required to properly describe the laws of physics. A *physical constant* is a quantity with a fixed and universal value. We have no ability to modify the value of a physical constant; its value can be ascertained only by either direct measurement or, in some cases, calculation using established physical laws. Physical constants are nearly always represented by a common letter or symbol, for example, *c*, the speed of light in a vacuum, and *e*, the charge of an electron.

The specific *numerical* value of a physical constant is not fixed in most cases but instead depends on our choice of a system of units. The speed of light in a vacuum, for example, is about 3×10^8 m/s in the SI system and about 7×10^8 mi/h in the English system. Using the right choice of units, we could make the numerical value of *c* equal anything: Units are conventions invented by humans. To emphasize the distinction between a physical quantity and the units used to report its value, we refer to the quantity as having a *dimension or dimensions*, but the value as having units. For example, the quantity that describes how far it is from **Boston** to New York City has dimensions of distance, but SI units of meters.

To avoid the need to define units, physicists are always on the lookout for dimensionless quantities, that is, quantities that carry no units. For example, the physics of motion at speeds faster than the speed of sound depends on Mach number *M*. *M* is defined as the speed of an object divided by the speed of sound, so the same units appear in both the numerator and denominator; after the units cancel, *M* is dimensionless. A plane traveling at Mach 2 is moving at twice the speed of sound, and there's no need to know whether speed is measured in meters per second, miles per hour, or some other set of units.

? Got the Concept 1-2
Going Fast

A special sled is shot down a long, straight track in the New Mexico desert. Scientists decide to use a curious unit to measure the distance, and they report the speed of sound as 205 smoots/s and the speed of the sled as 1780 smoots/s. What was the Mach number of the sled?

✳ What's Important 1-2

Practically all physical processes, characteristics, and phenomena can be expressed in terms of a few independent, fundamental quantities including length, time, mass, and temperature. SI units allow us to communicate values of both fundamental quantities and other quantities derived from them in terms that everyone understands.

1-3 Prefixes and Conversions

The SI unit of distance is the meter, abbreviated m. A certain Web site lists the distance between Boston and New York City as 301,000 m. The way the number is written might look odd to you. Why put those three zeros at the end? Our convention is to define the **prefix** *kilo* to indicate a multiplicative factor of 1000, so that the distance from Boston to New York City can be written as 301 kilometers. The abbreviation for kilometer is km.

Alternatively, we could have written the distance from Boston to New York City using **scientific notation**, in which numbers are expressed as a coefficient multiplied by a power of 10. In scientific notation, the distance could be written as 3.01×10^5 m or 3.01×10^2 km. In scientific notation, the prefix *kilo* is equal to a factor of 10^3. The standard prefixes listed in **Table 1-3** enable scientists to express a wide range of values in a compact way. The prefixes we will encounter most frequently in this book are micro (10^{-6}, μ), milli (10^{-3}, m), centi (10^{-2}, c), kilo (10^3, k), and mega (10^6, M).

Table 1-3	Prefixes	
Factor	**Prefix**	**Symbol**
10^{-24}	yocto	y
10^{-21}	zepto	z
10^{-18}	atto	a
10^{-15}	femto	f
10^{-12}	pico	p
10^{-9}	nano	n
10^{-6}	micro	μ
10^{-3}	milli	m
10^{-2}	centi	c
10^{-1}	deci	d
10^1	deka	da
10^2	hecto	h
10^3	kilo	k
10^6	mega	M
10^9	giga	G
10^{12}	tera	T
10^{15}	peta	P
10^{18}	exa	E
10^{21}	zetta	Z
10^{24}	yotta	Y

 Watch Out

Don't drop the prefixes.

In your excitement about getting a numeric answer to a problem, you might overlook the prefix attached to the units, perhaps using 700 in a calculation instead of 700×10^{-9} when the length given is 700 nm. Wrong answers will abound!

 Watch Out

Beware of prefixes hiding large ranges.

The mass of one object is given as 500 μg, the mass of another is given as 0.5 kg. The difference could be deceiving because the two values have a similar look. Yet 0.5 kg is one million times more massive than 500 μg.

 Got the Concept 1-3

Smallest to Biggest

Each of these numbers represents the size of something. Arrange them from smallest to largest, and try to identify something that would be about that size. (You may need to use Table 1-3 to determine the amount represented by some of the prefixes.)

(a) 0.1 mm
(b) 7 μm
(c) 6380 km
(d) 165 cm
(e) 200 nm

 Got the Concept 1-4

Comparing Sizes

For each pair, determine which quantity is bigger and by how much.

(a) 1 mg, 1 kg
(b) 1 mm, 1 cm
(c) 1 MW, 1 kW
(d) 10^{-10} m, 10^{-14} m
(e) 10^{10} m, 10^{14} m

Prefixes are a convenient and conventional way to express large and small numbers, but use them with care. Regardless of the way that a particular value is written, you must ensure that the values you use are consistent. You are guaranteed consistency when all of the numbers you enter into a calculation are in SI units. To use the distance from Boston to New York City in a calculation, for example, you should not use the value expressed in kilometers (km), but only in the SI unit meter (m). It may feel unnatural to express a large number like that, 301,000 m, but doing so provides assurance that all of the values you use will be consistent. You will need to recognize when a given value must be converted to the proper SI units.

Say a guitar string is to be used in an experiment. Its length is reported on the package as 1140 mm. To make a computation, however, all values should be put into SI units. The value of the length should therefore be converted to meters. By using Table 1-3, 1 mm is equal to 10^{-3} m; that is, the ratio of 10^{-3} m to 1 mm is equal to 1

$$\frac{10^{-3} \text{ m}}{1 \text{ mm}} = 1$$

Notice that the statement can only be correct when the units are included; without units, $10^{-3}/1 = 1$ is not correct. The value of any number remains the same when it is multiplied by 1, so we can convert 1140 mm to meters by multiplying it by a **conversion factor** that indicates the equivalence of 10^{-3} m and 1 mm

$$1140 \text{ mm} = 1140 \text{ mm} \left(\frac{10^{-3} \text{ m}}{1 \text{ mm}} \right) = 1140 \times 10^{-3} \text{ m} = 1.14 \text{ m}$$

The units *mm* are common to the numerator and denominator in the multiplication and therefore cancel, leaving *m* as the units of the result.

Some conversions cannot be carried out by applying a power of 10, but the process is the same. Sixty seconds are in 1 min, for example, so to convert from minutes to seconds express the value of 1 as

$$\frac{60 \text{ s}}{1 \text{ min}} = 1$$

and then multiply a value in minutes by this representation of the value of 1. To convert 25 min to seconds, we apply our conversion factor

$$25 \text{ min} = 25 \text{ min} \left(\frac{60 \text{ s}}{1 \text{ min}} \right) = 1500 \text{ s}$$

▌ Watch Out
● Write each phase of a conversion separately.

Some conversions cannot be easily carried out in a single step. As an illustration, let's convert a time interval of 1 day to seconds. We apply, separately, the equivalence of 1 day and 24 h, 1 h and 60 min, and 1 min and 60 s

$$1 \text{ day} = 1 \text{ day} \left(\frac{24 \text{ h}}{1 \text{ day}} \right) \left(\frac{60 \text{ min}}{1 \text{ h}} \right) \left(\frac{60 \text{ s}}{1 \text{ min}} \right) = 86{,}400 \text{ s}$$

Notice that we are careful to check the cancellation of each pair of like units as the conversion unfolds.

? Got the Concept 1-5
Canceling Units

To ensure that you've written the conversion factor properly, check that the units cancel as necessary between numerator and denominator. Which conversion of 150 s to minutes is correct?

$$150 \text{ s} = 150 \text{ s}\left(\frac{60 \text{ s}}{1 \text{ min}}\right) = 9000 \text{ min} \ \ or \ \ 150 \text{ s} = 150 \text{ s}\left(\frac{1 \text{ min}}{60 \text{ s}}\right) = 2.5 \text{ min}$$

? Got the Concept 1-6
Volume Conversion

The typical person inhales about 500 cm^3 of air into his lungs with every breath. What is this volume in the SI unit of volume, m^3?

Example 1-1 Hair Growth

The hair on a typical person's head grows at a constant 1.5 cm per month, on average (**Figure 1-4**). How far in meters will the ends of your hair move during a 50-min physics class?

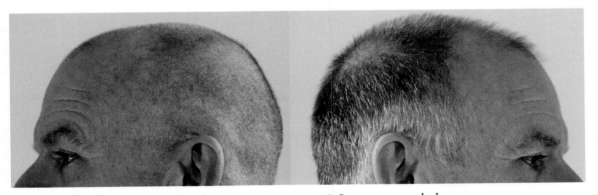

Figure 1-4 If the hair on your head grows at a constant 1.5 cm per month, how far in meters will the ends of your hair move during a 50-min physics class? *(courtesy David Tauck)*

SET UP

Although we haven't yet developed the physics that describes motion, the dimensions of the growth rate, distance per time, suggest that multiplying the rate by a time interval will give the distance of growth in that time. We will therefore multiply the growth rate by the 50-min length of the physics class. Neither the growth rate nor the time interval is given in SI units, so we'll need to convert both before carrying out the computation.

SOLVE

We can always multiply a number by 1 without changing its value. There are 100 cm in 1 m, so

$$100 \text{ cm} = 1 \text{ m} \quad or \quad \frac{100 \text{ cm}}{1 \text{ m}} = 1$$

We use this conversion factor to convert the growth rate from centimeters per month to meters per month

$$1.5 \frac{\cancel{\text{cm}}}{\text{month}} \left(\frac{1 \text{ m}}{100 \cancel{\text{cm}}} \right) = 1.5 \times 10^{-2} \frac{\text{m}}{\text{month}}$$

The units *cm* appear in both the numerator and the denominator and can be canceled to leave us with the desired units (m/month).

Converting months to the SI unit seconds works the same way

$$1.5 \times 10^{-2} \frac{\text{m}}{\text{month}} = 1.5 \times 10^{-2} \frac{\text{m}}{\cancel{\text{month}}} \left(\frac{1 \cancel{\text{month}}}{30 \cancel{\text{day}}} \right) \left(\frac{1 \cancel{\text{day}}}{24 \cancel{\text{h}}} \right) \left(\frac{1 \cancel{\text{h}}}{60 \cancel{\text{min}}} \right) \left(\frac{1 \cancel{\text{min}}}{60 \text{ s}} \right)$$

$$= 5.8 \times 10^{-9} \frac{\text{m}}{\text{s}}$$

Notice that the unit month appears in both numerator and denominator and therefore cancels, as is the case for the intermediate units of days, hours, and minutes. (One month is not exactly 30 days, nor is 1 day exactly 24 h; we are more concerned with the process than in getting an exact answer in this problem.)

The distance the ends of your hair grow in 50 min is found by multiplying the time interval by the growth rate. We'll embed the conversion of 50 min to seconds in Equation 1-1 to find the product of the growth rate (v_{grow}) and time interval (Δt)

$$\text{distance} = v_{grow} \Delta t \qquad\qquad (1\text{-}1)$$

$$= \left(5.8 \times 10^{-9} \frac{\text{m}}{\cancel{\text{s}}} \right) (50 \cancel{\text{min}}) \left(\frac{60 \cancel{\text{s}}}{1 \cancel{\text{min}}} \right) = 1.7 \times 10^{-5} \text{ m}$$

REFLECT

Does the answer make sense? It's certainly a small length of hair, but it adds up. At that rate your hair would grow about 18 cm a year, which seems reasonable. Note also that we might choose to write the answer using a prefix from Table 1-3, for example,

$$1.7 \times 10^{-5} \text{ m} = 17 \times 10^{-6} \text{ m} = 17 \text{ } \mu\text{m}$$

Expressing the value in this way is a more compact way to present this answer.

Practice Problem 1-1 The space shuttle could reach an orbital speed of 7860 m/s. At this speed, how far did the shuttle travel in 7 days?

★ What's Important 1-3

In scientific notation, numbers are expressed as a coefficient multiplied by a power of 10. Standard prefixes, such as micro, milli, centi, kilo, and mega, enable scientists to express large and small numbers conveniently. When solving problems, it is essential to express all values in the same system of units. We use the SI system of units (m, kg, s) in this book.

1-4 Significant Figures

You use an unsharpened pencil, which you know is 18 cm long, to make rough measurements of the length and width of your desk. You find the length and width to be 99 cm and 66 cm, respectively. What is the area of your desk? Multiplying length and width gives 6534 cm², but surely you aren't justified in claiming such a precise answer based on your crude measurements. **Significant figures** characterize the precision, or level of certainty, of a measurement or the statement of a value. We must take care that the number of significant figures shown in a computed value does not exceed the number of significant figures in the values on which it is based. Because each measurement was made to two significant figures, you would give the area of your desk as 6.5×10^3 cm², which also has two significant figures.

 See the Math Tutorial for more information on Significant Figures.

Each nonzero digit in any numerical value is considered a significant digit. A zero between nonzero digits also contributes to the count of significant figures; for example, 1.109 has four significant figures. Leading zeros are not significant, and trailing zeros are considered significant only in the decimal part of a number. For example, 2.4×10^3 has two significant figures while 2.400×10^3 has four. Writing the number without scientific notation could make the determination of significant figures ambiguous; we prefer to write 2400 (without a decimal point) as a number with two significant figures and 2400. (with a decimal point) as a number with four significant figures.

The number of significant figures shown in a computed value should not exceed the number of significant figures in the values on which it is based. If the values used are of different levels of significance, the most poorly defined value takes precedence. For example, multiplying 0.2 by 5.06 yields 1. The first value, 0.2, has one significant digit so the answer should only be given to one significant digit.

Got the Concept 1-7
Significant Figures

All of these numbers have the same value, but not all are written with the same number of significant figures: 123, 1.230×10^2, 123.000, 1.23×10^2, and 0.1230×10^3. Rank them in order of least to most significant figures.

As you solve a physics problem, when should you adjust the number of significant figures? Although the number of significant figures shown in the final numeric answer to a problem should not exceed the number of significant figures in the values on which it is based, as you work through a problem it's important to keep more significant figures in intermediate values. Consider, for example, calculating the volume of a box. You measure its width, length, and height to be 1.2 m, 3.3 m, and 1.6 m. The volume is the product of these three, or 6.336 m³, which we must state with two significant figures as 6.3 m³. But what if you decided to first find the area of the bottom of the box and then multiply that by the height? The area of the bottom is 1.2 m multiplied by 3.3 m, or 3.96 m². To two significant figures, the area would be 4.0 m², which when multiplied by the height of the box gives 6.4 m³. That is not 6.3 m³, because the *intermediate* result, the area of the bottom of the box, was incorrectly taken to the significance we require for the *final* result.

Physicist's Toolbox 1-2 Significant Figures

Significant figures characterize the precision, or level of certainty, of a measurement or the statement of a value. We use these rules to determine the number of significant figures in a value:

- Each nonzero digit in a number is considered a significant digit. Example: The number 1485 is given to four significant figures.

- A zero between nonzero digits is a significant digit. Example: The number 14085 is given to five significant figures.

- Leading zeros are not significant. This statement is true even when the zeros follow a decimal point. Example: The number 0.00519 carries three significant figures, which is clear if the value were written in scientific notation as 5.19×10^{-3}.

- Trailing zeros are considered significant unless the value is stated without a decimal point. Example: The value 300 is considered to have one significant digit, while 300. has three, and 300.00 has five.

- When multiplying or dividing quantities, the number of significant figures in the final answer is no greater than the number of significant figures in the quantity with the fewest significant figures. Example: The product of 1.23 (three significant figures) and 7.6 (two significant figures) is 9.348. We state the result with two significant figures as 9.3, because the most poorly defined value given, 7.6, has two significant figures.

- When adding or subtracting quantities, the number of decimal places in the answer should match that of the quantity with the fewest decimal places showing. Example: Adding 2.66, 9.95, and 4.3 yields 16.91. The value in the sum with the fewest decimal places is 4.3, which is given to the tenths place, so the sum must be stated to the tenths place as 16.9.

- Exact values have an unlimited number of significant figures. A value determined by counting, for example, is an exact value.

Physicist's Toolbox 1-3 Working with Significant Figures

When carrying out calculations, keep more significant figures in any intermediate step than there are in any separate value, and only round to the number of significant figures required for the answer as the very last step. In other words, don't round intermediate values! Consider this simple example. The product of the values 1.2, 1.3, and 1.5 is 2.3; we have stated the answer to two significant figures because each of the values used to calculate the product has two significant figures. What if we first take the product of 1.2 and 1.3, and round before continuing? In that case, 1.2 times 1.3 is 1.56, or 1.6 to two significant figures. Carrying on, the product of 1.6 and 1.5 is 2.4. Rounding before the last step of the calculation has produced an incorrect result.

For calculations that include a small number of intermediate steps, keep one or two more significant figures than are present in the values used. Because the error associated with rounding, the so-called "round-off" error, can accumulate as the number of intermediate calculations grows, it is reasonable to keep more significant figures for more complicated calculations.

✳ **What's Important 1-4**
Significant figures characterize the precision, or level of certainty, of a measurement or the statement of a value. Each nonzero digit as well as each zero between nonzero digits is considered a significant digit. Leading zeros are not significant, and trailing zeros are considered significant only if they appear to the right of the decimal point in a number.

1-5 Solving Problems

More than just broadening our understanding of the world, the process of learning physics involves learning to solve problems. Physics problems can appear challenging because they usually can't be done simply by selecting an equation and plugging in values. Solving physics problems is easier, however, if you build up a set of tools, techniques, and tricks, and then apply a consistent strategy. Our strategy for solving problems involves three steps that we refer to as Set Up, Solve, and Reflect.

Set Up Often the hardest part of a problem is figuring out how to set it up. Start by asking yourself what the problem is asking, and plan an overall approach. Then gather together the necessary pieces, which might include sketches, equations related to the physics, and concepts.

Draw a picture. When encountering a problem for the first time, a physicist's first inclination is to draw a picture. We rely heavily on diagrams; many sample problems are accompanied by one or more. Compare our sketches in the worked examples to the problem statements. A good problem-solving picture should capture the motion, the process, the geometry, or whatever else defines the problem. You don't need an artistic masterpiece; it's perfectly okay to represent most complex objects as squares or circles or even dots.

Label all variables. You probably agree that it makes sense to label all variables right on the picture you've drawn. But be careful! First, more than one kind of a variable is often given in a problem. For example, if two objects are moving, and you use d as the variable that represents displacement, you'll need to distinguish between the displacements of the two objects. Subscripts are useful for making this distinction, for example, d_1 and d_2. Also, don't hesitate to label distances, velocities, and other quantities even if you haven't been given a value for them. You might be able to figure out the value of one of the variables and use it later in the problem, or you might gain an insight that will guide you to a solution.

Look for connections. Often the best strategy for solving a physics problem is to look for connections between variables. These connections could be relationships between two or more variables or a statement that a certain parameter has the same value for two objects. Furthermore, such connections will often either enable you to eliminate variables or to find a value for an unknown variable. As a result, you sometimes will have an equation that defines the variable of interest only in terms of variables that are known. So we'll always be on the lookout for connections among variables when starting a problem.

Solve Rather than simply summarizing the mathematical manipulations required to move from first principles to the final answer, we show many intermediate steps in working out solutions to the sample problems. In addition, we present our reasoning and thought processes. In this way, we lead you to develop your own problem-solving skills and habits. We'll take care to check our units, to look carefully at conversions that are required, and to carry the same number of significant figures through to the answer as are provided in initial values.

Reflect An important but often overlooked part of problem solving is to reflect on the meaning, implications, and validity of the answer. Does it make sense based on what you know? Do the units make sense? Does the result provide any insights into how the world works? Reflecting on—and questioning—answers is a key to understanding and learning physics, and it is also a good way to avoid "silly" mistakes on exams!

Learning physics will be easier, and the disparate pieces of physics will fit together better, if you study the many worked examples in this book and solve some of the problems at the end of the chapters. Getting in the habit of following an organized problem-solving strategy, such as our Set Up, Solve, and Reflect approach, will help immensely. The Set Up process involves determining exactly what the problem is asking, drawing and labeling a sketch that captures the essence of the problem, and looking for connections between the variables presented in the problem statement. Those steps get us to the point where we can calculate an answer. As we Solve a problem, we usually need to substitute relationships between known variables for unknown ones and rearrange equations algebraically before coming to an equation that will give us the answer. Notice that we typically do not plug numbers into equations until the very last step. Waiting to use numbers given in the problem statement makes it easier to find mistakes and often allows us to understand the underlying physics more deeply. Finally, we Reflect on our answer. Not only will reflecting allow you to be more confident that you did it correctly, but reflecting will also help you connect the physics in the problem to other things you know and will lead to a deeper, longer-lasting understanding of the material.

Very often it's handy to have a sense of the magnitude of the answer to a problem before trying to find an exact answer. A good estimate of the answer can help you determine whether the exact answer you find is correct, or at least reasonable. In making an estimate, scientists commonly use rough rather than exact values, and perhaps ignore small effects or details that make the problem hard but don't affect the *sense* of the final answer. It's not unusual to find two or three physicists (in a place where you might find physicists at all) scribbling on whatever writing surface is handy, say, the back of an envelope or the corner of a chalkboard. We'll do estimates like these using big, round—but reasonable!—numbers throughout the book, and we encourage you to do similar estimations when trying to gain an understanding of a problem.

Estimate It! 1-1 Ping-Pong Balls

About how many ping-pong balls would fit into the office of one of the authors?

SET UP
First, because the ping-pong balls are spheres and not rectangular blocks, the balls cannot fill every possible bit of volume in the room. Visualizing spheres filling a volume, we can reasonably guess that well more than half the total volume is occupied by balls, but probably not more than three-quarters. So as a big, round number, we'll take seven-tenths as the volume of the office filled by ping-pong balls. So, the number of ping-pong balls N_{ppb} that fit is seven-tenths of the volume of the room divided by the volume of a ball or

$$N_{ppb} = \frac{0.7(\text{volume of office})}{\text{volume of ping-pong ball}} = \frac{0.7(H \cdot W \cdot L)}{\frac{4}{3}\pi R^3}$$

where H, W, and L are the height, width, and length, respectively, of the office, and R is the radius of a ping-pong ball.

SOLVE

We estimate the dimensions of the office from our knowledge of a typical office on a college campus: a height of 3.0 m, a width of 2.5 m, and a length of 4.5 m. We estimate the radius of a ping-pong ball from remembering what it feels like to hold one; we'll take the radius to be about 2 cm, so

$$N_{ppb} = \frac{(0.7)(3.0\ m)(2.5\ m)(4.5\ m)}{\left(\frac{4}{3}\pi(0.02\ m)^3\right)} = 7 \times 10^5$$

REFLECT

First, notice that we just estimate the dimensions of a typical office, rather than measure one of our actual offices. That's something you would certainly have to do because you've probably never been in one of ours! We imagine a typical office and feel comfortable coming up with distance estimates correct to within half a meter. We could have looked up the size of a ping-pong ball, but our goal is to get a sense of the answer rather than an exact answer. So we pretend to hold a ping-pong ball and estimate its size. In this way, we estimate that about seven hundred thousand ping-pong balls would fit into one of our offices.

Wait! What about the desk and chairs and bookshelves in our offices? They all take up space that would otherwise be occupied by ping-pong balls. A desk might be a half-meter deep, one and a half-meters long, and three-quarters of a meter high. (We're using rough numbers.) The desk isn't solid, so let's guess that half of the volume outlined by the desk is solid and therefore holds no ping-pong balls. That volume is $(\frac{1}{2}\ m)(\frac{3}{2}\ m)(\frac{3}{4}\ m)(\frac{1}{2})$ or about a quarter of a cubic meter. This volume would displace about 8000 ping-pong balls. The chairs and bookshelves would likely displace fewer, so altogether the items in our office reduce the total number of balls by perhaps 20 or 30 thousand. This doesn't significantly change our *estimate* that about seven hundred thousand balls fit into the office! We are able to get a good estimate by ignoring the small details that would otherwise make the calculation more complicated.

> ✳ **What's Important 1-5**
> Physics problems can appear challenging because they usually can't be done simply by selecting an equation and plugging in values. Solving physics problems is easier, however, if you build up a set of tools, techniques, and tricks, and then apply a consistent strategy.

1-6 Dimensional Analysis

When you're taking a physics exam, you won't have an answer key to check your work, but why not arm yourself with a variety of techniques that can help you quickly determine how likely it is that you've done a problem correctly? One valuable approach is to do a **dimensional analysis**; that is, to check that the dimensions of your algebraic answer match what you expect before you substitute values to compute a numeric result. Dimensions are the seven quantities listed in Table 1-1, considered fundamental because they cannot be expressed in terms of other physical quantities.

In Example 1-1, for example, we multiplied growth rate (v_{grow}) by time interval (Δt) to determine the length a hair grows in that time interval (Equation 1-1). The dimensions of growth rate and time interval are length (growth) divided by time and time, respectively. Using square brackets to indicate that we are considering only the dimensions of growth rate and time interval, we see that

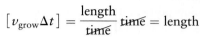

See the Math Tutorial for more information on Direct and Inverse Proportions.

$$\left[v_{\text{grow}}\Delta t\right] = \frac{\text{length}}{\cancel{\text{time}}}\,\cancel{\text{time}} = \text{length}$$

This dimensional analysis convinces us that our answer to the hair growth problem is sensible. However, had we incorrectly found that the distance a hair grows in a certain time interval was given by the product of growth rate and the *square* of the time interval, dimensional analysis would alert us to the error.

$$\left[v_{\text{grow}}\Delta t\right] = \frac{\text{length}}{\text{time}}\,\text{time}^2 = \frac{\text{length}}{\cancel{\text{time}}}\,\cancel{\text{time}}\cdot\text{time} = \text{length}\cdot\text{time}$$

The length of hair grown does not have dimensions of distance multiplied by time, so in this case we would conclude that the relationship we analyzed for hair growth length is incorrect.

Dimensional analysis can tell us if an answer is sensible, but it may not necessarily be correct. Suppose as a result of an error in the hair growth problem we found distance $= 2v_{\text{grow}}\Delta t$. Because the pure number 2 has no dimensions, it wouldn't change the dimensional analysis done above, so a dimensional analysis would not catch the mistake. Performing dimensional analysis can tell you whether an answer is wrong, but not if it's completely correct.

How can you tell the dimensions of a quantity? For some quantities, such as growth rate, the dimensions may be seen directly from the definition of the quantity. For others, the simplest way is to work backward from an equation in which the quantity appears. For example, we will define a quantity K (kinetic energy) as equal to one-half the mass of an object multiplied by the square of its speed ($K = \frac{1}{2}mv^2$). Energy is not one of the fundamental quantities listed in Table 1-1, so we determine the dimensions of K by working backward from the equation. Speed is the distance traveled divided by time interval ($v = d/t$), so

$$\left[K\right] = \left[\frac{1}{2}\right][m][v]^2 = \text{M}\left(\frac{\text{L}}{\text{T}}\right)^2 = \frac{\text{ML}^2}{\text{T}^2}$$

We represent the dimension mass by M, distance by its dimensions of length L, and time by T. Any equation in which a quantity appears can be used to determine its dimensions.

? ## Got the Concept 1-8
Dimensional Analysis

An object of mass m_0 oscillates back and forth on the end of a spring. (Flip forward in the book and look at Figure 12-2.) You determine a relationship for the time it takes for one full cycle T_0, which depends on a constant k that has dimensions of mass divided by time squared, M/T^2

$$T_0 = 2\pi\sqrt{\frac{m_0}{k}} \qquad\qquad (1\text{-}2)$$

Use dimensional analysis to determine whether the result could be correct.

> ### ? Got the Concept 1-9
> #### Working with Units
>
> The relationship between an object's speed v and the time t it has been moving is found to be $v = gt$. The variable g represents a physical constant. What are the SI units of g?

We will often check units rather than dimensions when doing dimensional analysis. Checking units achieves the same goal and can have an added benefit. For example, we checked the dimensions of our answer for the hair growth problem above and found that multiplying growth rate by time interval gave a quantity with the dimensions of distance. In the problem, the value of growth rate was given in cm/month and the time interval in min. Multiplying numbers in those units does not, however, result in a length given in recognizable units

$$v_{grow}\Delta t = \frac{cm}{month}\,min = \frac{cm \cdot min}{month} \neq cm$$

This form of dimensional analysis suggests that we must convert one of the quantities so that its units match that of the other. We could, for example, convert months to minutes. An analysis of the units then shows that

$$v_{grow}\Delta t = \frac{cm}{\cancel{min}}\,\cancel{min} = cm$$

The result is units of length, so we once again conclude that our answer to the hair growth problem is reasonable. Along the way we also recognized that at least one of the numeric values must be converted to match the units of the other.

> ### ✳ What's Important 1-6
>
> Dimensional analysis is the technique of checking the dimensions of your algebraic answer before you substitute values to compute a numeric result; it allows you to make sure that you calculate a sensible answer to a problem.

Answers to Practice Problems

1-1 4.75×10^9 m

Answers to Got the Concept Questions

1-1 Uppercase R suggests a quantity the value of which does not change throughout the problem, whereas lowercase r suggests a variable or changing quantity. In Figure 1-3, the radius of the dartboard is fixed, so R is the better choice. For the radius of each separate colored region, you might use r, or perhaps r_i.

1-2 Mach number is the speed of the object divided by the speed of sound. In this case, divide 1780 smoots/s by 205 smoots/s to get 8.7, so the Mach number M of the sled is 8.7. Notice that the units, smoots per second, cancel in the division.

1-3 From smallest to largest, the order is e < b < a < d < c. Two hundred nanometers (e) is the diameter of a typical virus, 7 μm (b) is the diameter of a human red blood cell, 0.1 mm (a) is the diameter of a typical human hair, 165 cm (d) is an average height of a woman in the United States, and 6380 km (c) is the mean radius of Earth.

1-4 (a) One kilogram is bigger than 1 mg by a factor of 1 million: 1 mg equals 10^{-3} g (0.001 g) and 1 kg equals 10^3 g (1000 g), so the ratio is 10^6 or one million. (b) One centimeter is 10 times bigger than 1 mm: 1 mm equals

0.001 m and 1 cm equals 0.01 m. (c) One megawatt is 1 thousand times bigger than 1 kW: 1 MW equals 1,000,000 W and 1 kW equals 1000 W. (d) The amount 10^{-10} m is 10 thousand times bigger than 10^{-14} m. The negative exponent means that each number is smaller than 1: 10^{-10} m equals 0.0000000001 m (the "1" is in the tenth place after the decimal) and 10^{-14} m equals 0.00000000000001 m (the "1" is in the fourteenth place after the decimal). (e) The amount 10^{14} m is 10 thousand times bigger than 10^{10} m: 10^{14} m is a "1" followed by 14 zeros, while 10^{10} m is a 1 followed by 10 zeros.

1-5 The second answer is correct because the units of seconds cancel. In the first answer, seconds in the numerator does not cancel with minutes in the denominator.

1-6

$$500 \text{ cm}^3 = 500 \text{ cm}^3 \left(\frac{1 \text{ m}}{100 \text{ cm}}\right)^3 = 500 \ \cancel{\text{cm}^3}\left(\frac{1 \text{ m}^3}{10^6 \ \cancel{\text{cm}^3}}\right)$$

$$= 0.0005 \text{ m}^3$$

Notice that because centimeters are cubed in units of the original number (cm^3), the conversion factor must also be cubed.

1-7 The number 123 has three significant figures, as does 1.23×10^2. The number 1.230×10^2 is written

with four significant figures, as is 0.1230×10^3. The number 123.000 has six.

1-8 The dimensions are

$$[T_0] = [2\pi]\sqrt{\frac{[m_0]}{[k]}} = \sqrt{\frac{M}{M/T^2}} = \sqrt{T^2} = T$$

The dimension of T_0 should be time, and because this expression is seen to be correct from Equation 1-2, you can have confidence that the relationship is correct. You can't say, however, that Equation 1-2 is exactly correct because the pure number 2π has no dimensions and therefore does not play a role in the dimensional analysis. Had your answer been, say, $T_0 = 5\sqrt{(m/k)}$, the dimensional analysis would be no different.

1-9 Rearrange the equation as a relationship for g, and then determine the dimensions

$$g = \frac{v}{t}$$

$$[g] = \frac{[v]}{[t]} = \frac{L/T}{T} = \frac{L}{T^2}$$

The SI units of distance and time are meter and second, respectively. The units of g are therefore m/s^2.

SUMMARY

Topic	Summary	Equation or Symbol
conversion factor	All values in scientific computations should be expressed in the same system of units. Although the speed of an object could be found by dividing a distance in furlongs by a time measured in fortnights, it is more meaningful to convert both to a common system of units before carrying out the calculation.	
dimensional analysis	One valuable way to test the validity of an expression determined as part of solving a physics problem is to check that the dimensions of your algebraic answer match those of the desired result. Check your answer by writing out and then simplifying the dimensions of all variables in the expression and then comparing the dimensions to the dimensions of the variable you're trying to find.	
prefix	A prefix can be attached to the units of a variable as a substitute for the power of 10 in scientific notation. For example, 3400 m can be written as 3.4×10^3 m in scientific notation or 3.4 km. The prefix k stands for "kilo," or 10^3.	Some common prefixes: μ micro or 10^{-6} m milli or 10^{-3} c centi or 10^{-2} k kilo or 10^3 M mega or 10^6

scientific notation	Scientific notation is a compact way to express large and small numbers. Numbers expressed in scientific notation are written as a coefficient multiplied by a power of 10; for example, 3400 in scientific notation is 3.4×10^3.
significant figures	Significant figures characterize the precision, or level of certainty, of a measurement or the statement of a value. Each nonzero digit in the answer is considered a significant digit. A zero between nonzero digits also contributes to the count of significant figures. Leading zeros are not significant, and trailing zeros are considered significant only in the decimal part of a number. The number of significant figures shown in a computed value should not exceed the number of significant figures in the values on which it is based. If the values used are of different levels of significance, the most poorly defined value takes precedence.
SI units	SI units (or International System of Units) is the system of units commonly used in physics and in this book. In the SI system, distances are measured in meters (m), time in seconds (s), mass in kilograms (kg), and temperature in kelvins (K).

QUESTIONS AND PROBLEMS

In a few problems, you are given more data than you actually need; in a few other problems, you are required to supply data from your general knowledge, outside sources, or informed estimate.
- Basic, single-concept problem
- • Intermediate-level problem, may require synthesis of concepts and multiple steps
- ••• Challenging problem
SSM *Solution is in Student Solutions Manual*

Conceptual Questions

1. •Define the following basic quantities in physics (length, mass, time, and temperature) and list the appropriate SI unit that is used for each quantity.

2. •Is it possible to define a system of units in which length is not one of the fundamental properties?

3. •Why do physicists and other scientists prefer to use metric units and prefixes (for example, micro, milli, and centi) over American Standard Units (for example, inches and pounds)? SSM

4. •If you use a calculator to divide 3411 by 62.0, you will get something like 55.016129. (Exactly how many digits you get will depend on your calculator.) Of course, you know that all those decimal places aren't significant. How should you write the answer?

5. •What properties should an object, system, or process have for it to be a useful standard of measurement of a physical quantity such as length or time?

6. •If two physical quantities are to be added, must they have the same dimensions? The same units? What if they are to be divided? Explain your answers. If your answers are different, explain why.

7. •All valid equations in physics have consistent units. Are all equations that have consistent units valid? Support your answer with examples. SSM

8. •Consider the number 61,000. What is the least number of significant figures this might have? The greatest number? If the same number is expressed as 6.10×10^4, how many significant figures does it have?

9. •Acceleration has dimensions L/T^2, where L is length and T is time. What are the SI units of acceleration? SSM

Multiple-Choice Questions

10. •Which of the following are fundamental quantities?
 A. mass density
 B. length
 C. area
 D. resistance
 E. all of the above

11. •Which length is the largest?
   A. 10 nm
   B. 10 cm
   C. 10^2 mm
   D. 10^{-2} m
   E. 1 m

12. •One nanosecond is
   A. 10^{-15} s
   B. 10^{-6} s
   C. 10^{-9} s
   D. 10^{-3} s
   E. 10^9 s

13. •How many square centimeters are there in a square meter?
   A. 10
   B. 10^2
   C. 10^4
   D. 10^{-2}
   E. 10^{-4} SSM

14. •How many cubic meters are there in a cubic centimeter?
   A. 10^2
   B. 10^6
   C. 10^{-2}
   D. 10^{-3}
   E. 10^{-6}

15. •Calculate $1.4 + 15 + 7.15 + 8.003$ using the proper number of significant figures.
   A. 31.553
   B. 31.550
   C. 31.55
   D. 31.6
   E. 32

16. •Calculate $0.688/0.28$ using the proper number of significant figures.
   A. 2.4571
   B. 2.457
   C. 2.46
   D. 2.5
   E. 2

17. •Calculate 25.8×70.0 using the proper number of significant figures.
   A. 1806.0
   B. 1806
   C. 1810
   D. 1800
   E. 2000 SSM

18. •Which of the following relationships is dimensionally consistent with a value for acceleration? In these equations, x is distance, t is time, and v is speed.
   A. v^2/t
   B. v/t
   C. v/t^2
   D. v/x^2
   E. v^2/x^2

19. •N and N_0 represent the number of nuclei at different times, so λ in the equation $N = N_0 e^{-\lambda t}$ must
   A. have dimensions of T (time).
   B. have dimensions of $1/T$.
   C. have dimensions of T^2.
   D. have dimensions of $1/T^2$.
   E. be dimensionless. SSM

Estimation/Numerical Analysis

20. •Estimate the distance from your dorm room or apartment to your physics lecture room. Give your answer in meters.

21. •Estimate the mass of Mt. Everest. SSM

22. •Estimate the time required for a baseball to travel from home plate to the center field fence in a major league baseball game.

23. •Estimate the number of laptop computers on your campus.

24. •Estimate the amount (in kg) of nonrecyclable and noncompostable garbage that each student at your school produces each day.

25. •Estimate the volume (in liters) of water used daily by each student at your school. SSM

26. •Estimate the height (in m) of the tallest building on your campus.

27. ••Enrico Fermi once estimated the length of all the sidewalks in Chicago in the 1940s. Estimate the length (in km) of sidewalks in present-day Chicago.

28. •Biology Estimate the volume flow rate (in units of m^3/s) of the air that fills your lungs as you take a deep breath.

29. •Biology Although human cells vary in size, the volume of typical cell is equivalent to the volume of a sphere that has a radius of approximately 10^{-5} m. Estimate the number of cells in your body. *Hint*: Consider your body as a collection of cylinders. SSM

Problems

1-3: Prefixes and Conversions

30. •Write the following numbers in scientific notation:
   A. 237
   B. 0.00223
   C. 45.1
   D. 1115
   E. 14,870
   F. 214.78
   G. 0.00000442
   H. 12,345,678

This page is intentionally left blank.

For complete end of chapter problem sets, please go to
www.whfreeman.com/kestentauck

1-4: Significant Figures

44. •Give the number of significant figures in each of the following numbers:

A. 112.4
B. 10
C. 3.14159
D. 700

E. 1204.0
F. 0.0030
G. 9.33×10^3
H. 0.02240

45. •How many significant figures should be associated with measurements made using a standard meter stick that has *mm* as its smallest division?

46. •How many significant figures should be associated with measurements made using a thermometer that is subdivided into 1 °C increments and can be used for temperatures between 0 and 100 °C?

47. •How many significant figures should be associated with measurements made using a thermometer that is subdivided into 0.1 °C increments and can be used for temperatures between 0 and 10 °C?

48. •Complete the following operations using the correct number of significant figures:

A. $5.36 \times 2.0 = $ _____

B. $\dfrac{14.2}{2} = $ _____

C. $2 \times 3.14159 = $ _____

D. $4.040 \times 5.55 = $ _____

E. $4.444 \times 3.33 = $ _____

F. $\dfrac{1000}{333.3} = $ _____

G. $2.244 \times 88.66 = $ _____

H. $133 \times 2.000 = $ _____

49. •Complete the following operations using the correct number of significant figures: SSM

A.   4.55
    +21.6

B. 80.00
    −112.3

C. 71.1
    +3.70

D. 200
    +33.7

1-5: Solving Problems

50. •(a) What area is cut by a girl who mows a lawn with dimensions shown in the **Figure 1-5**? (b) Determine her rate of pay in dollars per square meter if she earns $125 for mowing the lawn.

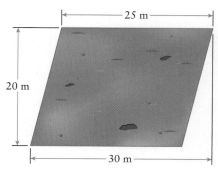

Figure 1-5 Problem 50

51. •One sphere has a radius of 5 cm, and a second has a radius of 10 cm. What is the ratio of the volume of the second to the volume of the first? Why isn't the ratio equal to 2?

52. •An aluminum spacer has a diameter of 12 cm and is 1 cm thick (**Figure 1-6**). The diameter of the hole is 2 cm. Aluminum has density 2700 kg/m^3. Find the mass of the spacer.

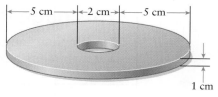

Figure 1-6 Problem 52

53. •**Calc** A string of length S is to be stretched around the perimeter of a rectangular area of width W and length L. Find the ratio of W to L that maximizes the area. SSM

54. •What is the volume of the object shown in **Figure 1-7** that has a height equal to 15 cm, a thickness equal to 8 cm, and a width equal to 5 cm?

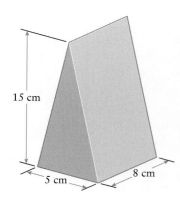

Figure 1-7 Problem 54

55. •Determine the circumference of a path that follows the equator of a spherical planet that has a radius of 5000 km.

56. •Determine the circumference of a path that follows a great arc at a latitude of 45° around a spherical planet that has radius R of 5000 km (**Figure 1-8**).

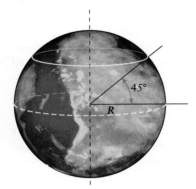

Figure 1-8 Problem 56

1-6: Dimensional Analysis

57. •The equation that describes the motion of an object is $x = vt + x_0$, where x is the position of the object, v is its speed, t is time, and x_0 is the initial position. Show that the dimensions in the equation are consistent. SSM

58. •The equation that describes the motion of an object is $x = \frac{1}{2}at^2 + v_0 t + x_0$, where x is the position of the object, a is the acceleration, t is time, v_0 is the initial speed, and x_0 is the initial position. Show that the dimensions in the equation are consistent.

59. •The equation that describes the kinetic energy of a moving particle is $K = \frac{1}{2}mv^2$, where m is the mass of the particle and v is its speed. Show that 1 joule (J), the SI unit of energy, is equivalent to $1 \text{ kg} \cdot \text{m}^2/\text{s}^2$.

60. •The motion of a vibrating system is described by $y(x, t) = A_0 e^{-\alpha t} \sin(kx - \omega t)$. Find the SI units for k, ω, and α.

61. •Which of the following could be correct based on a dimensional analysis?
  A. The volume flow rate is 64 m^3/s.
  B. The height of the Transamerica building is 332 m^2.
  C. The duration of a fortnight is 66 m/s.
  D. The speed of the train is 9.8 m/s^2.
  E. The weight of a standard kilogram mass is 2.2 lb.
  F. The density of gold is 19.3 kg/m^2. SSM

62. •The period T of a simple pendulum, the time for one complete oscillation, is given by $T = 2\pi\sqrt{(L/g)}$, where L is the length of the pendulum and g is the acceleration due to gravity. Show that the dimensions in the equation are consistent.

General Problems

63. •During a certain experiment, light is found to take 37.1 μs to traverse a measured distance of 11.12 km. Determine the speed of light from the data. Express your answer in SI units and in scientific notation, using the appropriate number of significant figures.

64. •**Medical** The concentration of PSA (prostate-specific antigen) in the blood is sometimes used as a screening test for possible prostate cancer in men. The value is normally reported as the number of nanograms of PSA per milliliter of blood. A PSA of 1.7 (ng/mL) is considered low. Express that value in (a) g/L, (b) standard SI units of kg/m^3, and (c) μg/L.

65. ••**Medical** Express each quantity in the standard SI units requested. (a) An adult should have no more than 2500 mg of sodium per day. What is the limit in kg? (b) A 240-mL cup of whole milk contains 35 mg of cholesterol. Express the cholesterol concentration in the milk in kg/m^3 and in mg/mL. (c) A typical human cell is about 10 μm in diameter, modeled as a sphere. Express its volume in cubic meters. (d) A low-strength aspirin tablet (sometimes called a "baby aspirin") contains 81 mg of the active ingredient. How many kg of the active ingredient does a 100-tablet bottle of baby aspirin contain? (e) The average flow rate of urine out of the kidneys is typically 1.2 mL/min. Express the rate in m^3/s. (f) The density of blood proteins is about 1.4 g/cm^3. Express the density in kg/m^3.

66. ••**Calc** The horizontal range R of a projectile, the distance that a projectile covers horizontally, is given by $R = (v_0^2/g)\sin(2\theta)$, where v_0 is the initial speed of the projectile, g is the projectile's acceleration due to gravity (9.8 m/s^2), and θ is the initial angle that the projectile makes with the horizontal. Find the angle that gives a maximum value of R when v_0 is fixed. (*Hint*: Set the derivative of R with respect to θ equal to zero.)

67. •When a woman who is 2-m tall stands near a flagpole, the shadow of her head and the shadow of the top of the flagpole are superimposed (**Figure 1-9**). The woman's shadow is 10 m long, as shown in the figure. Find the height of the flagpole. SSM

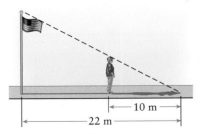

Figure 1-9 Problem 67

68. •**Calc** Evaluate the third derivative with respect to time of the function $x(t) = 4t^4 + 6t^3 + 12t^2 + 5t$ when t equals 4 s.

69. •The acceleration g of a falling object near a planet is given by the following equation $g = GM/R^2$. If the planet's mass M is expressed in kg and the distance of the object from the planet's center R is expressed in meters, find the units of the gravitational constant G. The acceleration g must have units of m/s^2.

70. •**Biology** A typical human cell is approximately $10\,\mu m$ in diameter and enclosed by a membrane that is 5.0 nm thick. (a) What is the volume of the cell? (b) What is the volume of the cell membrane? (c) What percent of the cell volume does its membrane occupy? To simplify the calculations, model the cell as a sphere.

71. •**Medical** At a resting pulse rate of 75 beats per minute, the human heart typically pumps about 70 mL of blood per beat. Blood has a density of $1060\ kg/m^3$. Circulating all of the blood in the body through the heart takes about 1 min in a person at rest. (a) How much blood (in L and m^3) is in the body? (b) On average, what mass of blood (in g and kg) does the heart pump with each beat? SSM

72. •Determine the volume of a cone that has a radius equal to 2.25 m and a height equal to 3.75 m.

73. ••**Medical** A typical prostate gland has a mass of about 20 g and is about the size of a walnut. The gland can be modeled as a sphere 4.50 cm in diameter and of uniform density. (a) What is the density of the prostate? Express your answer in g/cm^3 and in standard SI units. (b) How does the density compare to that of water? (c) During a biopsy of the prostate, a thin needle is used to remove a series of cylindrical tissue samples. If the cylinders have a total length of 28.0 mm and a diameter of 0.100 mm, what is the total mass (in g) of tissue taken? (d) What percentage of the mass of the prostate is removed during the biopsy?

74. ••**Medical** The body mass index (BMI) estimates the amount of fat in a person's body. It is defined as the person's mass m in kg divided by the square of the person's height h in m. (a) Write the formula for BMI in terms of m and h. (b) In the United States, most people measure weight in pounds and height in feet and inches. Show that with weight W in pounds and height h in inches, the BMI formula is $BMI = 703\,W/h^2$. (c) A person with a BMI between 25.0 and 30.0 is considered overweight. If a person is 5'11" tall, for what range of mass will he be considered overweight?

2 Linear Motion

(Steve Romano/Velocity Media Systems.)

As an object moving in a straight line accelerates (that is, as its speed changes), the distance traveled per unit time can change dramatically. Although the three superimposed images of the cheetah were taken at uniform time intervals, the cat moves much farther in the second time interval than the first. The speed of the cheetah is increasing, in other words, it is accelerating. Not only is the cheetah fast, but it is also capable of rapidly accelerating to its top speed.

2-1 Constant Velocity Motion

2-2 Acceleration

2-3 Motion under Constant Acceleration

2-4 Gravity at the Surface of Earth

No land animal runs faster than a cheetah. The felines reach top speeds of over 45 m/s (100 mi/h)! In contrast, moving at a rate of only about 8 cm/h, the banana slug (**Figure 2-1**) must be among the slowest animals on Earth. In the next few sections, we'll lay out our approach and define the variables we'll use to *describe* motion. Later on, we'll address the question of what *causes* motion and what *changes* it. For now, though, descriptions will suffice. We'll answer questions such as these:

1. How fast is an object moving?
2. How far will an object travel in a certain amount of time?
3. How long will it take an object to move a certain distance?

Let's start by asking these three important questions about motion and defining the physics terminology we need to answer them.

How fast is an object moving? Although the words **speed** and **velocity** both describe how fast an object moves, the two terms are not precisely the same. As we'll see more clearly in the next chapter, velocity includes information about the direction an object moves as well as how fast it travels. In contrast, speed only refers to how fast an object moves and tells us nothing about the direction of movement. Speed is, therefore, always

Figure 2-1 The mascot of the University of California at Santa Cruz is the banana slug, a colorful but slow animal that lives in the redwood forests around campus. *(Mav888/Dreamstime.com.)*

25

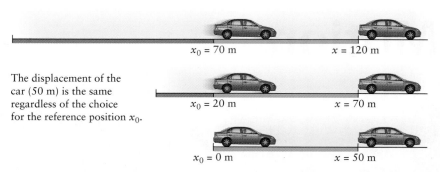

The displacement of the car (50 m) is the same regardless of the choice for the reference position x_0.

Figure 2-2 We describe the change in position of an object, or its displacement, relative to an arbitrary reference position.

a nonnegative number. We'll use the variable s for speed and v for velocity in this chapter.

How far will an object travel in a certain amount of time? The **displacement** of an object is the difference between its positions at two different times. If we use the variable x to describe position, then displacement is $x - x_0$, where x_0 ("ex naught" or "ex zero") is either some previous position of the object or a position selected to serve as a **reference position**. There is no mandatory reference position, so we are always free to choose the one that makes the most sense or makes a particular problem easy to visualize. For example, suppose that a certain problem says an object travels 50 m. If we choose $x_0 = 70$ m, then the final value of x would be 120 m. As you can see in **Figure 2-2**, had we chosen the reference position to be $x_0 = 20$ m, then the final value of x would be 70 m. You might ask, then, why not choose the reference position to be $x_0 = 0$ m, so that the values of $x - x_0$ and x are the same, which would be 50 m in our example? There's no reason not to do that, as long as you remember that if you use x and not $x - x_0$ in a physics equation, you're still describing *displacement* and not *position*. That is, you will be describing "how far" and not "where." "How far," or displacement, is the change in position, so we will sometimes refer to it using the notation Δx, where the capital Greek delta (Δ) means "change in." The change in position is equivalent to displacement or the distance that an object moves.

How long will it take an object to move a certain distance? Here we're talking about the **time interval**, or "elapsed time" in everyday language. We'll use the variable t to represent a time measurement, so that the time interval is $t - t_0$. Similar to displacement, we are free to choose the reference time t_0. Often it will be convenient to set the reference time equal to zero, so that the quantity t can be treated as if it were a time interval. Note that in physics, time is not time of day but rather elapsed time.

2-1 Constant Velocity Motion

We'll start by considering motion with **constant velocity**, that is, motion in one direction along a straight line, at a speed that does not change. Imagine an axon regenerating through a nerve guide or a car traveling in cruise control on a long, straight highway such as Interstate 5 in California. We've declared that in the language of physics, speed (the magnitude of velocity) describes how fast an object moves. That is,

$$\text{speed} = \frac{\text{displacement}}{\text{time interval}}$$

Or, using the variables for speed, displacement, and time interval, we define speed mathematically as

$$s = \frac{|x - x_0|}{t - t_0} \tag{2-1}$$

The absolute value symbols (the vertical bars) guarantee that the numerator of the equation is the magnitude of the displacement, and that the quantity s is nonnegative. If we also want to indicate the direction of motion we remove absolute value symbols and define velocity as

$$v = \frac{x - x_0}{t - t_0} \tag{2-2}$$

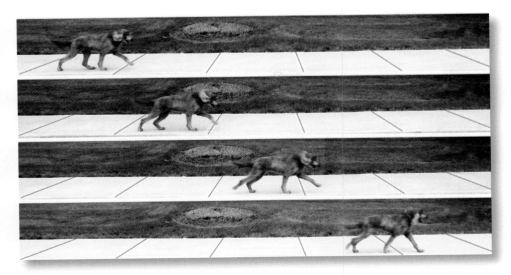

Figure 2-3 The time interval between each of the four images is the same. Notice that from one picture to the next the dog travels approximately the same distance. *(David Hazy/Custom Medical Stock Photography.)*

For example, we often let positive values indicate motion from left to right and negative values indicate motion from right to left.

The definition of velocity in Equation 2-2 only applies to an object moving in a straight line at a constant speed. Physicists often refer to motion along a straight line as **linear motion** and motion at a constant speed as **uniform motion**. Equation 2-2 therefore applies to uniform linear motion. **Figure 2-3** shows four images of a dog walking along a path. The time interval between each is the same, and from one to the next the dog travels the same distance. The dog is traveling in uniform motion.

To make the mathematical description more straightforward, we'll adopt $t_0 = 0$ as the standard reference value for time, remembering that the variable t will then represent a time interval. We can then write the definition of velocity as

$$v = \frac{x - x_0}{t} \qquad (2\text{-}3)$$

How far does an object travel in a straight line at velocity v during a time interval t? Let's rearrange Equation 2-3 to solve for displacement:

$$x - x_0 = vt \qquad (2\text{-}4)$$

How long does it take for an object traveling in a straight line at velocity v to travel a distance $x - x_0$? Let's rearrange Equation 2-3 to solve for the time interval:

$$t = \frac{x - x_0}{v} \qquad (2\text{-}5)$$

Does it look like the number of equations we have is increasing quickly? Not so! Equations 2-2, 2-3, 2-4, and 2-5 are all just variations of the same equation. The definition of velocity (Equation 2-2) involves three variables; rearranging the equation to express one of the three variables in terms of the other two does not result in "new" physics. As the physics we encounter becomes more involved, we'll need to avoid the temptation to think of (or memorize) the many variations of a single equation as if each represents a separate piece of the physics.

Physicist's Toolbox 2-1   The Displacement versus Time Plot

Graphing the displacement of an object as a function of time often helps us understand its motion. In general, when displacement is plotted against time, the slope of the line gives the velocity. Consider Equation 2-4, which we can write as $x = vt$

Figure 2-4 The slope of a plot of displacement versus time gives velocity. (a) When the curve is a straight line, velocity is the slope of the line. (b) When the curve is not a straight line, velocity at any time is given by the slope of the line tangent to the curve at that time.

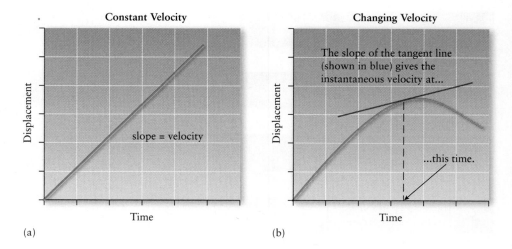

by setting x_0 equal to 0 for simplicity. When velocity is constant, Equation 2-4 generates a straight line, as seen in **Figure 2-4a**. Slope describes the rate at which a line rises or falls, so in this case, slope is equivalent to velocity. For a curve that isn't a straight line, the slope at any point is given by the slope of the line tangent to the curve at that point. Thus, the line tangent to any point on a displacement versus time curve represents the *instantaneous velocity* at that point, as shown in Figure 2-4b.

Example 2-1 Down the Stretch

A thoroughbred racehorse covers the last, straight part of the racecourse, or homestretch, in 21 s. If the homestretch is 420 m long, find the horse's speed assuming it is constant.

SET UP
First, we need to identify exactly what the problem asks us to calculate and to express it in the terminology of physics. We have been asked to determine speed, the magnitude of the velocity, so our goal is to find an expression in which the variable v is isolated on the left side and all variables that have known values are on the right side. In other words, we need an equation which expresses v in terms of known variables.

Next, we need to decide when and where to start and stop our measurement. Because the final answer will not depend on these choices, we are free to pick whatever reference points make the problem easier. For example, if we set the initial position to be $x_0 = 0$ m, then the position x of the horse at the end of the problem would equal 420 m. On the other hand, perhaps the start of the homestretch is 130 m from the barn, so we might set $x_0 = 130$ m. In that case, the end of homestretch would be at $x = 550$ m. Either choice is acceptable. Of course, we could simply set the displacement $x - x_0 = 420$ m. In the same way, we can write the time interval covered by the problem as $t - t_0 = 21$ s.

SOLVE
The problem is straightforward. Equation 2-2 defines the velocity in terms of variables we know, $x - x_0$ and $t - t_0$:

$$v = \frac{x - x_0}{t - t_0} \tag{2-2}$$

Thus, the velocity of the horse is

$$v = \frac{420 \text{ m}}{21 \text{ s}} = 20 \frac{\text{m}}{\text{s}}$$

The horse's speed is the magnitude of the velocity, also 20 m/s.

REFLECT

To decide whether the answer makes sense, compare it to other speeds you know. For example, a good high school sprinter can run 100 m in 10 s or $v = 10$ m/s. A car traveling at highway speed moves at 65 mi/h or $v = 29$ m/s. It seems reasonable that a racehorse moves faster than a person but slower than a car on the freeway. Also, note that although we found the horse's velocity and speed to be the same, it was our choice of position measurements that resulted in a positive displacement and therefore a positive velocity. Another choice could have resulted in velocity being negative, say, because we chose to make "to the right" the positive direction but the horse was running to the left. In that case we would find the horse's velocity to be equal to -20 m/s, but its speed would still be $+20$ m/s.

Practice Problem 2-1 A hockey puck travels 10 m in 0.25 s. Find the speed of the puck, assuming that it is constant.

Example 2-2 What's the Same?

A truck and a car enter opposite ends of a 5.00-km straight section of highway at the same time and are moving toward each other. The truck is traveling at a constant speed of 20.0 km/h and the car is traveling at a constant speed of 105 km/h. How far from where the truck enters the straight section do the truck and car pass each other?

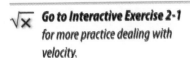

$\sqrt{x}$ **Go to Interactive Exercise 2-1** for more practice dealing with velocity.

SET UP

To make the sketch in **Figure 2-5** we needed to label the distances involved, which also provides a clue about how to proceed. Although it might be fun to make a sketch such as the one shown in Figure 2-5, it would be fine to make a simpler one in which the car and truck are dots or small rectangles. We are given only one value of distance, the length of the section labeled D in the picture, but we also need to show the distance that the car and the truck each travels before passing. We have chosen Δx_{car} and Δx_{truck} to represent these distances. (Because we have to keep track of two distances—one for the car and one for the truck—we use subscripts to

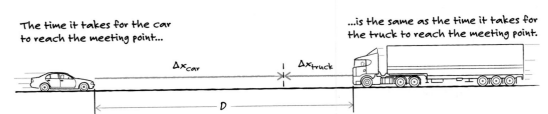

The time it takes for the car to reach the meeting point...   ...is the same as the time it takes for the truck to reach the meeting point.

The sum of distances Δx_{car} and Δx_{truck} must equal the total distance D.

Figure 2-5 A truck and a car enter opposite ends of a 5.00-km straight section of highway at the same time. How far from where the truck enters the straight section do the truck and car pass each other?

avoid confusing the variables.) What clue does the sketch provide? It shows us an important connection between the distances:

$$\Delta x_{\text{truck}} + \Delta x_{\text{car}} = D \tag{2-6}$$

We want to find Δx_{truck} in terms of values we know or can find from the given information; although Δx_{car} is unknown also, we might be able to use the connection between what we want to find and a value we know ($D = 5.00$ km) as a starting point.

What else, if anything, is the same in this problem? Although we don't know how long it takes for the truck and car to pass, it must take the same amount of time for each to reach the passing point! Using Δt to represent the time interval required for the car and truck to pass, we can then write

$$\Delta t_{\text{truck}} = \Delta t_{\text{car}} = \Delta t$$

We can leave off subscripts for the time interval because it is the same for both car and truck.

How can we connect the distances with the time interval? From Equation 2-1, we know that the definition of speed connects distance and time:

$$s_{\text{car}} = \frac{|\Delta x_{\text{car}}|}{\Delta t} \tag{2-7}$$

and

$$s_{\text{truck}} = \frac{|\Delta x_{\text{truck}}|}{\Delta t} \tag{2-8}$$

We now have four relationships containing six variables (Δx_{truck}, Δx_{car}, D, Δt, s_{car}, and s_{truck}), three of which have known values. As a rule, we need at least as many equations as unknown variables in order to fully solve a problem. Here we have three unknowns and four equations, so the problem is solvable.

SOLVE
Let's rearrange both Equations 2-7 and 2-8 to solve for Δt and, again, because the quantity Δt is the same for both the car and truck, set them equal:

$$\Delta t = \frac{\Delta x_{\text{car}}}{s_{\text{car}}} = \frac{\Delta x_{\text{truck}}}{s_{\text{truck}}} \tag{2-9}$$

To simplify the notation, we treat both Δx_{truck} and Δx_{car} as positive numbers, in keeping with the definition of speed as the magnitude of velocity, so we will drop the absolute value signs for the rest of this example.

Our goal is to find Δx_{truck}, so rearrange Equation 2-9 to get

$$\Delta x_{\text{truck}} = \frac{s_{\text{truck}}}{s_{\text{car}}} \Delta x_{\text{car}} \tag{2-10}$$

We're close, but we don't have a value for Δx_{car}. But notice that Equation 2-6 relates Δx_{car} to Δx_{truck} and D, both of which we do know. From Equation 2-6

$$\Delta x_{\text{car}} = D - \Delta x_{\text{truck}}$$

so Equation 2-10 becomes

$$\Delta x_{\text{truck}} = \frac{s_{\text{truck}}}{s_{\text{car}}} (D - \Delta x_{\text{truck}}) \tag{2-11}$$

Although some rearranging is necessary, clearly we're almost done, because the only unknown quantity remaining is Δx_{truck}. We gather together the two Δx_{truck} terms in Equation 2-11:

$$\Delta x_{truck} = \frac{s_{truck}}{s_{car}} D - \frac{s_{truck}}{s_{car}} \Delta x_{truck}$$

$$\left(1 + \frac{s_{truck}}{s_{car}}\right) \Delta x_{truck} = \frac{s_{truck}}{s_{car}} D$$

$$\Delta x_{truck} = \left(\frac{s_{car}}{s_{car} + s_{truck}}\right) \frac{s_{truck}}{s_{car}} D$$

$$\Delta x_{truck} = \frac{s_{truck}}{s_{car} + s_{truck}} D \qquad (2\text{-}12)$$

That's it! We now have an equation for Δx_{truck}, the variable of interest, only in terms of quantities that are known. So now we can now plug in the numbers, and find

$$\Delta x_{truck} = \frac{20.0 \text{ km/h}}{20.0 \text{ km/h} + 105 \text{ km/h}} 5.00 \text{ km} = (0.160)5.00 \text{ km} = 0.800 \text{ km}$$

The truck and the car pass when the truck is 0.800 km from its starting point.

REFLECT
Does the answer make sense? We certainly expect that the distance the truck moves before the truck and car pass must be between 0 and 5.00 km, and our answer is consistent with that expectation. Also, because the truck moves more slowly than the car, we'd expect that the car goes farther to get to the point where the two pass each other. The distance of the truck is $\Delta x_{truck} = 0.800$ km, so Δx_{car} must equal 4.20 km, which also is consistent with our expectation.

Practice Problem 2-2 Mary and Sue are 90 m apart when they start running toward each other along a straight road. If they start running at the same instant and Sue runs twice as fast as Mary, how far will Mary have run when the two meet?

Example 2-3 What's the Same?—Check Your Answer

To check their answers, physicists often assign an extreme value to one of the variables that makes the answer obvious. To check the solution to Example 2-2, substitute extreme values for the speed of the truck. First consider the case in which the speed of the truck is zero, and then let the truck be much faster than the car.

SET UP
In the previous example (Example 2-2), using zero as an extreme value for the speed of the truck, we can easily arrive at the answer. If the truck doesn't move, the distance the truck travels before the car passes must be zero. Does Equation 2-12 predict this result? Another extreme is found by letting the speed of the truck s_{truck} be much faster than the speed of the car s_{car}. For $s_{truck} \gg s_{car}$, we'd guess that the car would hardly move at all while the truck goes nearly all of the distance D before the two pass. Does Equation 2-12 predict $\Delta x_{truck} \approx D$?

SOLVE
In the first case, when the truck does not move, we see that

$$\Delta x_{truck} = \left(\frac{0}{0 + s_{car}}\right) D = 0$$

as we expect.

In the second extreme, when $s_{truck} \gg s_{car}$, we can ignore the s_{car} term in the denominator of Equation 2-12:

$$\Delta x_{truck} = \left(\frac{s_{truck}}{s_{truck} + s_{car}} \right) D \approx \left(\frac{s_{truck}}{s_{truck}} \right) D$$

The fraction in parentheses is equal to 1, so $\Delta x_{truck} \approx D$, again consistent with our expectation.

REFLECT

These two tests of extreme values provide convincing evidence that our result is a good one! You should consider two other extreme values, $s_{car} = 0$ and $s_{car} \gg s_{truck}$. What do you expect under those conditions? Does our equation give answers that make sense?

Watch Out
$\triangle x$ stands for a single quantity that is the change in x.

Although two symbols are required to form the quantity Δx, pronounced "delta x," it is nevertheless a single value. Δx is not a product of Δ and x.

We defined Equations 2-2 through 2-5 as different ways to describe uniform linear motion. For example, Equation 2-2 gives velocity in terms of displacement and a time interval:

$$v = \frac{x - x_0}{t - t_0}$$

However, we stipulated that the velocity of the object in question would not change during the time interval $t - t_0$. If the velocity *is* changing over the time interval, does the equation no longer apply? Let's consider an example.

It's a relatively straight road, 315 km along Interstate 94 from Bismarck to Fargo, North Dakota. You leave Bismarck, travel at constant velocity for 2 h, stop for 30 min at a rest area, then drive 1 h more at your original velocity before reaching Fargo. How can we apply the physical description of motion that we've developed to this situation?

We can certainly determine the velocity at which you traveled while moving. Using $x - x_0 = 315$ km and $t - t_0 = 3$ h, we find

$$v = \frac{x - x_0}{t - t_0} = \frac{315 \text{ km}}{3 \text{ h}} = 105 \text{ km/h}$$

At any instant while moving, your velocity is 105 km/h.

Watch Out
Speed is the magnitude of velocity.

You might be tempted to refer to the value we just determined, 105 km/h, as your "speed." In this case your speed, the magnitude of your velocity, is indeed the same 105 km/h. But although we are unlikely to do so in everyday life, we could have defined a directional sense in which your velocity would be negative. If it were convenient to define the direction from Bismarck to Fargo as the negative direction, then the displacement $x - x_0$ would be negative in Equation 2-2, resulting in a negative velocity. However, even if your velocity were negative, your speed (the magnitude of velocity) would still be positive.

The velocity of 105 km/h is misleading. The entire trip actually took you 3.5 h, not 3 h. Would it mean anything if we did the calculation with $t - t_0 = 3.5$ h? Let's see:

$$v = \frac{x - x_0}{t - t_0} = \frac{315 \text{ km}}{3.5 \text{ h}} = 90 \text{ km/h}$$

Although you were never traveling at 90 km/h, that is the velocity you would have needed to travel in order to make the trip in 3.5 h without a stop. That information would be useful, for example, if you were planning a much longer trip but intended to take a 30-min break every 3 h. Using Equation 2-5, a 900-km trip would take 10 h:

$$t = \frac{x - x_0}{v} = \frac{900 \text{ km}}{90 \text{ km/h}} = 10 \text{ h}$$

Because 90 km/h represents an average of the velocity while moving (in this case 105 km/h) and the velocity while stopped (0 km/h), we call it the **average velocity**. The equation for the average velocity,

$$v_{\text{avg}} = \frac{\Delta x}{\Delta t} \tag{2-13}$$

applies to any time interval, even if the velocity of an object is changing during that time. We are no longer confined to constant velocity!

Motion that combines displacements in opposite directions results in partial, possibly total, cancellation of the displacement Δx in Equation 2-13. In the special case of an object that moves and then returns to its starting point, the total displacement in the numerator of Equation 2-13 is zero, resulting in an average velocity of zero. So if you make a round-trip from Bismarck to Fargo and back in 7 h, your average velocity is 0 km/h because your net displacement is zero. Speed, however, depends on the *distance* traveled in a given time interval; it is sometimes more useful to characterize a motion by **average speed**,

$$s_{\text{avg}} = \frac{\text{total distance}}{\Delta t} \tag{2-14}$$

Your average speed for the round-trip from Bismarck to Fargo is

$$s_{\text{avg}} = \frac{630 \text{ km}}{7 \text{ h}} = 90 \text{ km/h}$$

? Got the Concept 2-1
Average Speed

It takes you 15 min to drive 6 mi to the local hospital to shadow a doctor. It takes 10 min to go the last 3 mi, 2 min to go the last mile, and only 30 s to go the last half mile. What is your average speed for the trip? Can you estimate your instantaneous speed at any point during the trip? If so, what is it?

? Got the Concept 2-2
Average Velocity

Is it possible to determine your average velocity for the trip described in the previous problem? Why or why not?

Table 2-1
Displacement versus Time

Time (s)	Displacement (m)
1	1
2	4
3	9
4	16
5	25
6	36
7	49

In the example of your trip from Bismarck to Fargo, it was easy to find your **instantaneous velocity**, the velocity at any instant of time. Although your velocity was not constant throughout the entire trip because of the rest breaks, we set the problem up so the motion consisted of several segments of constant motion. How could we find the instantaneous velocity for an object not moving at a constant velocity? The answer lies in the definition of average velocity.

Consider a skier zipping in a straight line down a slope. During each second that the skier moves, her displacement also increases, as shown in the table (**Table 2-1**). If she were traveling at constant velocity, dividing the displacement by the corresponding time interval would give her velocity at any instant. Because her velocity is changing, however, the calculation results in a value that is the average over the time interval. For example, her average velocity between $t = 2$ s and $t = 7$ s is

$$v_{2-7} = \frac{x - x_0}{t - t_0} = \frac{49 \text{ m} - 4 \text{ m}}{7 \text{ s} - 2 \text{ s}} = 9 \text{ m/s}$$

where $t_0 = 2$ s and $x_0 = 4$ m have been used as the reference points. We label the velocity v_{2-7} to show that it is the skier's average velocity over the time interval from $t = 2$ s to $t = 7$ s. What if we narrowed the time interval and averaged from $t = 2$ s to $t = 5$ s? You can verify that $v_{2-5} = 7$ m/s, and in the same way $v_{2-3} = 5$ m/s.

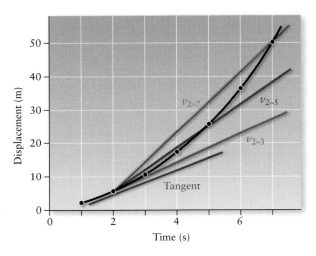

Figure 2-6 The black, curved line in the displacement versus time plot shows the skier's motion. The red, blue, and green straight lines represent the skier's motion as if she were moving at the constant, average velocity computed for different time intervals. As the time interval becomes smaller, the lines that represent those average velocities approach the purple line, which is tangent to the curve at the time of interest.

Figure 2-6 shows the physical meaning these averages represent. The black, curved line in the displacement versus time plot shows the skier's motion. The red, blue, and green straight lines represent the skier's motion as if she were moving at a velocity equal to the average we computed for the time interval ending at the greater time coordinate for each. We got different results for the average velocity depending on the time interval over which we computed it, and each average velocity results in a different straight line in the displacement versus time plot shown in Figure 2-6. Notice that as the time interval becomes smaller and smaller, the lines that represent average velocity approach (in slope) the purple line, which is tangent to the curve at $t = 2$ s.

We saw earlier that the line tangent to a displacement versus time curve at any point represents the velocity of the motion at the time coordinate of that point. For nonuniform motion, then, the tangent line at any point gives the instantaneous velocity, the velocity of an object at one specific instant of time.

Putting all the pieces together, we define the instantaneous velocity as the limit of the average velocity as the time interval over which the average is determined shrinks to smaller and smaller values. The instantaneous velocity is

$$v = \lim_{(t-t_0) \to 0} \frac{x - x_0}{t - t_0} = \lim_{\Delta t \to 0} \frac{\Delta x}{\Delta t} \tag{2-15}$$

which you might recognize as the definition of the derivative of x with respect to t. So, finally, we can write

$$v = \frac{dx}{dt} \tag{2-16}$$

as the mathematical definition of instantaneous velocity. The derivative is just a shorthand way to express the definition of velocity as a change in position taken over an *infinitesimally* small time interval. By using a derivative to define velocity, the time interval is so extremely small that we can treat the velocity of the object as a constant at any point in time. Therefore, the definition of velocity (Equation 2-16) applies at any instant of time without requiring that the velocity is constant for any extended period of time.

Instantaneous speed is the speed of an object at any instant of time.

Math Box 2-1   An Aside on Taking Derivatives

In our study of motion, the most important aspects of derivatives are the two definitions in the preceding discussion. First, the derivative of a function gives the slope of a line tangent to the curve at any point. Second, the derivative expresses how one variable changes with respect to a small change in another variable. In addition to that basic understanding, however, it will be necessary to find the derivatives of some basic functions to solve problems. For that purpose, we will rely on a few operational definitions or rules. To take the derivative of a polynomial, reduce the exponent by 1 and multiply the resulting expression by the original exponent:

$$\frac{d}{dx}x^n = nx^{n-1} \tag{2-17}$$

 See the **Math Tutorial** for more information on Differential Calculus.

The derivative of an exponential results in an expression identical to the original one, but the constant is now also a multiplicative factor:

$$\frac{d}{dx}e^{nx} = ne^{nx} \tag{2-18}$$

Because the value of a constant is, by definition, not changing, the derivative of a constant must be zero:

$$\frac{d}{dx}c = 0 \tag{2-19}$$

The derivative of a combination of functions may be determined by the product rule; if $f(x)$ and $g(x)$ are functions, then

$$\frac{d}{dx}f(x)g(x) = f(x)\frac{d}{dx}g(x) + g(x)\frac{d}{dx}f(x)$$

For example, the product rule allows us to determine the derivative of a polynomial multiplied by a constant:

$$\frac{d}{dx}cx^n = c\frac{d}{dx}x^n + x^n\frac{d}{dx}c = cnx^{n-1} + 0 = cnx^{n-1} \tag{2-20}$$

Rules for taking derivatives of functions other than polynomials and exponentials can be found in the Math Tutorial.

Example 2-4   Hockey Puck

As a hockey puck slides along nearly frictionless ice, its position at any time t relative to one edge of the rink is described by the equation $x(t) = 38t + 7.6$. (a) If x and t are given in SI units, what, if any, units are associated with the numbers 38 and 7.6 in the equation? (b) Use the mathematical definition of velocity to determine the velocity of the puck.

SET UP

The velocity of the puck, which is the rate at which its displacement changes, is the derivative of its position $x(t)$ with respect to t:

$$v = \frac{d}{dt}x(t)$$

We will apply our basic rules about taking a derivative to find v.

SOLVE

(a) The coefficient 38 must carry units of meters per second (m/s), so that the units of $38t$ are meters (m). Similarly the constant 7.6 must carry units of meters (m). In that way, the units of both the left-hand side and the right-hand side of the equation are m.

(b) The equation that describes position has two terms which we'll consider separately. The first term is $38t$, so the first term in the equation for velocity is

$$v_{\text{first}} = \frac{d}{dt}(38t^1)$$

We added the exponent (1) to t in order to make the expression look like the rule for taking the derivative of a polynomial, Equation 2-17. According to the rule,

$$v_{\text{first}} = 38(1)t^{(1-1)}$$

so $v_{\text{first}} = 38$ m/s because t^0 equals 1.

The second term in the equation for position is the constant $+7.6$, so the second term in the equation for speed is

$$v_{\text{second}} = \frac{d}{dt}(+7.6)$$

According to Equation 2-19, the derivative of a constant is zero:

$$v_{\text{second}} = 0 \text{ m/s}$$

Altogether, then, the velocity of the puck is

$$v = v_{\text{first}} + v_{\text{second}} = 38 \text{ m/s} + 0 \text{ m/s} = 38 \text{ m/s}$$

REFLECT

The velocity of the puck is 38 m/s, or 85 mi/h. That value is typical of the speed of a hockey puck when it is shot by a professional hockey player. You may have noticed that, having determined that the number 38 in the position versus time relationship carried units of m/s, it was possible to guess the velocity from the form of the equation. It's a fine idea to use your intuition to guess the answer to a problem, but it's nevertheless important to confirm a guess by applying the physics.

Practice Problem 2-4 As a hockey puck slides along nearly frictionless ice, its position at any time t relative to one edge of the rink is described by the equation $x(t) = 38t - 4t^2 + 7.6$. (The units of 38, 4, and 7.6 are m/s, m/s^2, and m, respectively.) What is the velocity of the puck after 0.50 s has elapsed?

Example 2-5 Hair Growth

In Example 1-1, we found the rate at which hair grows to be $v_{\text{g.r.}} = 5.8 \times 10^{-6}$ m/s. What is the rate at which the *volume* of hair grows? Human hair is cylindrical in shape with a typical diameter of 1.0×10^{-5} m.

SET UP

The volume of a cylinder is given by the length L multiplied by the cross-sectional area A:

$$V = A L$$

The rate at which the volume changes is the derivative of the volume with respect to time:

$$\frac{dV}{dt} = \frac{d}{dt}(A L)$$

SOLVE

The area A is a constant, so according to Equation 2-20, the rate of volume growth is

$$\frac{d}{dt}V = \frac{d}{dt}(A L) = A\frac{dL}{dt}$$

The last term (dL/dt) is just the rate $v_{g.r.}$ at which the hair grows. The cross-sectional area of the cylinder is $A = \pi(D/2)^2$ where D is the diameter of the cylinder, so

$$\frac{d}{dt}V = Av_{g.r.} = \pi\left(\frac{D}{2}\right)^2 v_{g.r.}$$

Using the numbers given in the problem statement,

$$\frac{d}{dt}V = \pi\left(\frac{1.0 \times 10^{-5}\,\text{m}}{2}\right)^2 (5.8 \times 10^{-6}\,\text{m/s}) = 4.6 \times 10^{-16}\,\text{m}^3/\text{s}$$

REFLECT

The increase in volume per second is a small number—less than the volume of a typical atom. In an hour, the volume change is

$$\Delta V = \left(4.6 \times 10^{-16}\,\frac{\text{m}^3}{\text{s}}\right)(3600\,\text{s}) = 1.6 \times 10^{-12}\,\text{m}^3$$

which is about the volume of a few dozen typical human cells.

Practice Problem 2-5 The radius of a spherical balloon grows at a constant rate as it is inflated. Does the volume grow at a constant rate? Why or why not?

Example 2-6 Muscle Contraction

Muscle contraction enables animals to move. Each muscle cell consists of thousands of fibers that are segmented into tiny structures called *sarcomeres*. The sarcomeres in turn contain filaments of the contractile proteins actin and myosin (**Figure 2-7a**). As these proteins slide past one another, the sarcomeres can shorten and lengthen, resulting in the muscle contracting and relaxing. Sarcomeres contract at a relatively constant rate. In the data shown in **Figure 2-7b**, for example, the speed of sarcomere contraction is in the range of a few micrometers per second for the first 0.07 s or so. The change in the position of the end of one of the sarcomeres during contraction can be modeled empirically according to

$$x - x_0 = -a(t - b(1 - e^{-ct})) \tag{2-21}$$

where a, b, and c are constants determined from the data in the figure. In this case $a = 4\ \mu\text{m/s}$, $b = 0.003$ s, and $c = 400$ s^{-1}. The change in length is negative because the sarcomere is getting shorter over time. What is the speed of the end of a sarcomere (relative to the other end) 0.002 s after it has started to contract? After 0.02 s?

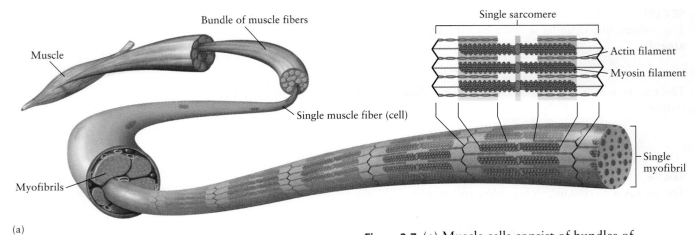

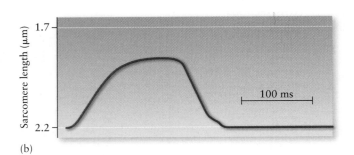

(a)

(b)

Figure 2-7 (a) Muscle cells consist of bundles of myofibrils that are segmented into thousands of tiny structures called sarcomeres that in turn contain filaments of the contractile proteins actin and myosin. As these proteins slide past each other the sarcomeres can shorten and lengthen, resulting in the muscle contracting and relaxing. (b) Sarcomeres contract at a relatively constant rate. In a hamster's diaphragm, the velocity of the sarcomere contraction is in the range of a few micrometers per second for the first 0.07 s of a muscle contraction. *(Adapted from Coirault et al., (1997), J Appl Phyiol 82:404–412 [Fig 2].)*

SET UP

The speed at any instant is the magnitude of the time derivative of the position (or displacement). The derivative is the slope of the line tangent to the position versus time curve at the time of interest, so it might be helpful to sketch out the graph of that curve. Especially because Equation 2-21 might look imposing, start by noticing that because e^{-n} is nearly zero for all but small values of n, the relationship between displacement and time is effectively linear for most of the range of time values. That's a straight line when $x - x_0$ is graphed against t, so don't let the form of Equation 2-21 concern you.

Make a graph of Equation 2-21 by using your calculator or a spreadsheet to evaluate the equation for a few values of time in the range $t \leq 0.060$ s. **Figure 2-8a** shows the curve, and **Figure 2-8b** shows an expanded view, from $t = 0.000$ s to $t = 0.006$ s. The line tangent to the curve at $t = 0.002$ s is shown in Figure 2-8b.

SOLVE

The velocity at a given time is found by taking the time derivative of $x - x_0$:

$$v = \frac{d}{dt}(x - x_0) = \frac{d}{dt}[-a(t - b(1 - e^{-ct}))]$$

Because the sarcomere is getting shorter over time, the rate of change in length— the velocity—is negative. We will therefore take the absolute value of this expression to find the speed of the end of the sarcomere.

If you aren't so comfortable with determining the derivative above, don't worry. We'll expand the expression on the right into many simple pieces, and then use the derivative rules outlined above (Equations 2-17 to 2-19) on each piece:

$$v = \frac{d}{dt}[-at + ab - abe^{-ct}]$$

Using Equations 2-17 and 2-19,

$$v = \frac{d}{dt}[-at] = -a$$

$$v = \frac{d}{dt}[ab] = 0$$

Because both a and b are constant, using Equation 2-18 gives

$$v = \frac{d}{dt}[-abe^{-ct}] = abce^{-ct}$$

Combining the equations, we find that

$$v = -a + abce^{-ct} = a(bce^{-ct} - 1)$$

We can now substitute $t = 0.002$ s and $t = 0.02$ s to get the velocities at those times:

$t = 0.002$ s: $v = (4\ \mu\text{m/s})((0.003\ \text{s})(400\ \text{s}^{-1})e^{-(400\ \text{s}^{-1})(0.002\ \text{s})} - 1) = -2\ \mu\text{m/s}$

$t = 0.02$ s: $v = (4\ \mu\text{m/s})((0.003\ \text{s})(400\ \text{s}^{-1})e^{-(400\ \text{s}^{-1})(0.02\ \text{s})} - 1) = -4\ \mu\text{m/s}$

The speeds are the absolute values of these results: 2 μm/s at $t = 0.002$ s and 4 μm/s at $t = 0.02$ s. (We have stated the results to one significant figure because the constant a in Equation 2-21 is given to only one significant digit.)

REFLECT

The speed we determined at $t = 0.002$ s should match the tangent line drawn at $t = 0.002$ s in Figure 2-8b. A quick measurement confirms that our answer is reasonable. As time changes from 0.002 s to 0.004 s, the change in length of the sarcomere goes from about -0.003 to about -0.007 μm, which gives an average speed of 2 μm/s.

During the part of the contraction represented by the linear portion of the curve in Figure 2-7b that lasts about 0.07 s, the sarcomere shortens from 2.2 to 1.9 μm. The constant velocity during that time is

$$v_{\text{constant}} = \frac{1.9\ \mu\text{m} - 2.2\ \mu\text{m}}{0.07\ \text{s}} = -4.3\ \mu\text{m/s}$$

So the speed is 4.3 μm/s. Again, this is in agreement with our calculated answer.

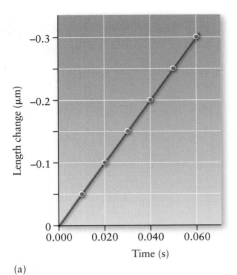

(a)

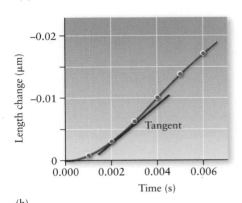

(b)

Figure 2-8 The curve in (a) shows the behavior over the time of a single contraction. The curve in (b) shows the initial, nonlinear behavior of the sarcomere.

✱ What's Important 2-1

The words *speed* and *velocity* both describe how fast an object moves, but they are not synonymous. Speed refers to how fast an object moves. An object's velocity includes both its speed as well as information about the direction it's moving.

2-2 Acceleration

Cheetahs can go from "0 to 60" in 3 s; that is, they can go from stationary to a speed of 60 mi/h in 3 s. **Acceleration** is the parameter that allows us to quantify how speed and velocity change with time; it is the time rate of change of velocity, the derivative of velocity with respect to time:

$$a = \frac{dv}{dt}$$

This equation is directly analogous to Equation 2-16, the definition of velocity. We can follow a course similar to the one we used for velocity to find a general equation for acceleration in terms of the other motion variables. For example, we can define **average acceleration** in a way similar to Equation 2-2:

$$a = \frac{v - v_0}{t - t_0} = \frac{\Delta v}{\Delta t} \tag{2-22}$$

Because the SI units of velocity are meters per second (m/s) the units of acceleration must be meters per second *per second* or m/s^2.

Got the Concept 2-3
Large and Small

Can acceleration be small if the velocities involved during a motion are large? Can acceleration be large if the velocities involved during a motion are small?

Got the Concept 2-4
Human Accelerations

A slow speed for humans is perhaps 1 m/s, which is a casual walking speed. The fastest most of us move is when we're flying on a commercial aircraft, around 250 m/s. What is a reasonable range for the accelerations that we experience?

Using a reference time $t_0 = 0$ for constant acceleration, we rearrange Equation 2-22 to get

$$v - v_0 = at$$

During constant acceleration the change in an object's velocity during any time interval is its acceleration multiplied by the elapsed time. The last equation is often written

$$v = v_0 + at \tag{2-23}$$

We will expand the description of motion to include nonconstant acceleration in Chapter 3, but for now we confine our discussion to examples of constant acceleration. Under such conditions, velocity changes at a constant rate; this is a good approximation of the way many things in your daily life move.

Physicist's Toolbox 2-2   The Velocity versus Time Plot

We have seen that a graphical representation of motion is helpful in understanding the physics. In general, when velocity is plotted against time, the slope of the line represents the acceleration. Consider Equation 2-23, which we can write as $v = at$, where v_0 is set to 0 for simplicity. When acceleration is constant, Equation 2-23 generates a straight line (**Figure 2-9a**). The value of a is the slope of that line. Even if velocity is changing, so that the graph of velocity versus time is not a straight line, the slope of the curve at any specific time point is still the acceleration at that time (**Figure 2-9b**). Just as we noted on the displacement versus time plot (Figure 2-4), the line tangent at any point on a velocity versus time curve therefore represents the acceleration of the motion at that instant.

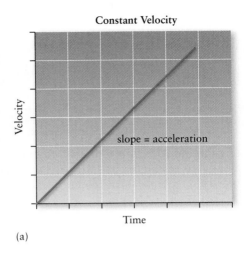

Constant Velocity

slope = acceleration

Velocity

Time

(a)

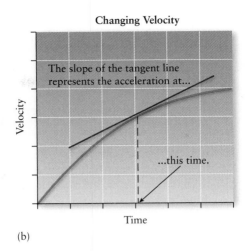

Changing Velocity

The slope of the tangent line represents the acceleration at...

...this time.

Velocity

Time

(b)

Figure 2-9 The slope of a plot of velocity versus time gives the acceleration. (a) When the curve is a straight line, acceleration is the slope of the line. (b) When the curve is not a straight line, the acceleration at any point is given by the slope of the line tangent to the curve at that time.

? Got the Concept 2-5
Acceleration Is a Slope

The velocity of a toy car as a function of time is shown in **Figure 2-10**. (a) At about what time, or in what approximate time interval, is the car accelerating at the highest rate? (b) At about what time, or in what approximate time interval, is the car accelerating at the lowest rate?

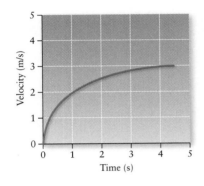

Figure 2-10 This graph shows the velocity of a toy car as a function of time.

Consider the motion of an object with a changing velocity. Recall that velocity is defined as the displacement that occurs during an infinitesimally short period of time divided by that short time interval. As we saw, that statement leads to Equation 2-16, which defines velocity as the derivative of position with respect to time. We can apply the definition by first separating the numerator from the denominator of the derivative, and then integrating. (Yes, it's okay to treat the derivative as if it were a fraction! The definition of the derivative, after all, comes from taking the limit of a fraction.) Taking

$$v = \frac{dx}{dt} \tag{2-16}$$

we get

$$dx = v\, dt$$

The expression dx represents the extremely small change in the displacement of a moving object that occurs during the extremely small time interval dt. The expression is mathematically interesting, but to understand the motion of, say, a sprinter running a 100-m race, we will want to know the total displacement of the runner after some time interval that is *not* small. Because a differential such as dx makes only a tiny contribution to the total displacement, to determine the total displacement of an object after accelerating for a time interval t we need to add together each of the small dx contributions. The mathematical symbol for summing many infinitesimally small contributions to get a total is the integral, so in this case, writing

$$\int_{x_0}^{x} dx$$

is the calculus shorthand equivalent for the process of adding up all of the infinitesimally small changes in displacement of a moving object, starting from an initial displacement equal to x_0 and ending when the object has reached the displacement x. Following an accelerating object from an initial displacement x_0 at time t_0 to a final displacement x at time t, we would then integrate as follows:

$$\int_{x_0}^{x} dx = \int_{t_0}^{t} v \, dt \qquad (2\text{-}24)$$

See the **Math Tutorial** for more information on Integrals.

Math Box 2-2 An Aside on Integration

It is not our intention to derive the calculus required to integrate the expression. We do need an operational definition, however, to enable us to move from the differential dx to an expression for the displacement x itself. First, the integral of a variable u to some power n is given by

$$\int u^n \, du = \left(\frac{1}{n+1} \right) u^{n+1}$$

Also, note that any constant term can be taken out of the integration process, for example:

$$\int c \, u^n \, du = c \int u^n \, du = c \left(\frac{1}{n+1} \right) u^{n+1}$$

where c is a constant.

When an initial value and a final value (a lower bound and an upper bound) are assigned to the variable u, the value of the integral is determined by evaluating the resulting expression for each value and subtracting:

$$\int_{u_{\text{initial}}}^{u_{\text{final}}} c \, u^n \, du = c \left(\frac{1}{n+1} \right) u^{n+1} \Big|_{u_{\text{initial}}}^{u_{\text{final}}} = c \left(\frac{1}{n+1} \right) u_{\text{final}}^{n+1} - c \left(\frac{1}{n+1} \right) u_{\text{initial}}^{n+1}$$

or

$$\int_{u_{\text{initial}}}^{u_{\text{final}}} c \, u^n \, du = c \left(\frac{1}{n+1} \right) (u_{\text{final}}^{n+1} - u_{\text{initial}}^{n+1}) \qquad (2\text{-}25)$$

For example,

$$\int_{2}^{4} x^3 \, dx = \frac{1}{4} x^4 \Big|_{2}^{4} = \frac{1}{4} (4^4 - 2^4) = 60$$

Using Equation 2-25, the integral on the left side of Equation 2-24 is then straightforward, where the variable x is substituted for u and $n = 0$:

$$\int_{x_0}^{x} dx = x \Big|_{x_0}^{x} = x - x_0$$

The integral on the right side of Equation 2-24 does not match the form of Equation 2-25, because when v is changing, the right-side integral contains two variables (v and t), and that case is not covered by Equation 2-25. However, Equation 2-23 provides a way to express velocity as a function of time, so the integral on the right side of Equation 2-24 becomes

$$\int_{t_0}^{t} v \, dt = \int_{t_0}^{t} (v_0 + at) \, dt$$

Recall that we have agreed to consider only cases in which acceleration is constant. With both v_0 and also a constant, the integral can be broken into two parts, both of which follow the form of Equation 2-25:

$$\int_{t_0}^{t} (v_0 + at) \, dt = \int_{t_0}^{t} v_0 \, dt + \int_{t_0}^{t} at \, dt = v_0 t \Big|_{t_0}^{t} + \frac{1}{2} at^2 \Big|_{t_0}^{t}$$

To make the form of the result more transparent, set $t_0 = 0$, so

$$\int_{t_0}^{t} (v_0 + at)dt = v_0 t - v_0 0 + \frac{1}{2}at^2 - \frac{1}{2}a0^2 = v_0 t + \frac{1}{2}at^2$$

Combining the two expressions gives

$$x - x_0 = v_0 t + \frac{1}{2}at^2 \tag{2-26}$$

Note again that the equation applies only when acceleration is constant. The motion of the cheetah shown on the first page of this chapter is an example of the application of Equation 2-26.

Let's return briefly to Equation 2-24, which describes the relationship between displacement, velocity, and time for a moving object:

$$\int_{x_0}^{x} dx = \int_{t_0}^{t} v\, dt \tag{2-24}$$

We just considered an accelerating object, but what if velocity is constant? In that case, both integrals can be carried out directly according to Equation 2-25:

$$\int_{x_0}^{x} dx = x \Big|_{x_0}^{x} = x - x_0$$

and

$$\int_{t_0}^{t} v\, dt = v \int_{t_0}^{t} dt = v \Big|_{t_0}^{t} = v(t - t_0)$$

Combining the equations gives

$$x - x_0 = v(t - t_0)$$

which is exactly Equation 2-2, our fundamental motion equation:

$$v = \frac{x - x_0}{t - t_0}$$

The lesson here is that the definition of velocity expressed as a derivative leads directly back to a fundamental motion equation. Said another way, derivatives and integrals are the natural "language" of motion. Once you get comfortable with the language of calculus, the physical phenomena are seen more clearly and arrived at more easily.

★ What's Important 2-2

For linear motion, acceleration quantifies how speed changes with time. Acceleration is the derivative of velocity with respect to time.

2-3 Motion under Constant Acceleration

Everything we need to know about motion under constant acceleration is contained in just two equations, Equations 2-23 and 2-26:

$$v = v_0 + at$$

$$x - x_0 = v_0 t + \frac{1}{2}at^2$$

Just as a reminder, these two equations use $t_0 = 0$ as the reference time, so the variable t_0 does not appear in either one.

The power of physics is in summarizing complex concepts and processes with one or only a few equations in terms of only a small number of variables. Here, we have our first example of that capability. Just two equations with only five variables describe all constant acceleration motion. That powerful simplicity makes solving problems involving motion straightforward.

You learned long ago that you need to have at least as many equations as unknown variables in order to completely solve a problem. If there are two unknowns, you need two equations. Three unknowns require three equations, and so on. We have seen that motion can be described using five variables ($x - x_0$, v_0, v, a, and t) and two equations. Two equations can only be solved completely if there are only two unknown variables. That statement means that we need to know three of the five variables to solve any particular motion problem.

Consider this typical problem:

A car accelerates from $v_0 = 20.0$ m/s to $v = 24.0$ m/s at a uniform acceleration $a = 2.00$ m/s^2. How far does it travel during that process?

The problem asks us to find "how far." In the language of physics, "how far" corresponds to displacement, or $x - x_0$. So, to *solve* the equations of motion, we must find a way to write a single equation having $x - x_0$ on the left side of the equal sign and only variables that have known values on the right side. Once we have such an equation, it's a simple matter to substitute the values for all variables on the right side and to compute an answer. How many variables are unknown in the problem? Three variables—v_0, v, and a—are given in the problem statement, which leaves two of the five motion variables unknown. Two equations and two unknowns mean the problem is solvable.

To keep track of which variables are known and which are not, we recommend that you use a *Know/Don't Know table*. To make the table, start with two columns. List the five variables in one column and fill in the values of each known variable in the other column as shown in **Table 2-2a**. Notice that the two unknown variables can be clearly identified.

Now we put a question mark next to the variable that we'd like to determine and an "X" in the last remaining empty space in the right column, as in **Table 2-2b**. The X is the most powerful entry in the table! It means "eliminate this variable." To eliminate a variable, rearrange one of the two motion equations so that the variable is alone on the left side of the equal sign, and then substitute the expression on the right side of the equation into the other motion equation. According to the Know/Don't Know table, we need to eliminate time in this problem. Although we're free to choose either of the motion equations to solve for t, the first equation (for speed, Equation 2-23) is easier to rearrange:

$$v = v_0 + at$$
$$at = v - v_0$$
$$t = \frac{v - v_0}{a}$$

Now substitute the relationship for t into the second motion equation (for displacement):

$$x - x_0 = v_0 t + \frac{1}{2}at^2 = v_0\left(\frac{v - v_0}{a}\right) + \frac{1}{2}a\left(\frac{v - v_0}{a}\right)^2$$

Notice that the time variable has been eliminated, which means we have accomplished the goal of expressing $x - x_0$ only in terms of the known variables v, v_0, and a. All that's left to do, then, is to simplify:

$$x - x_0 = \frac{v_0 v - v_0{}^2}{a} + \frac{1}{2}a\left(\frac{v^2 + v_0{}^2 - 2vv_0}{a^2}\right)$$

Table 2-2
Know/Don't Know Table

(a)

Variable	Know/Don't Know
$x - x_0$	
v_0	20.0 m/s
v	24.0 m/s
a	2.00 m/s^2
t	

(b)

Variable	Know/Don't Know
$x - x_0$	?
v_0	20.0 m/s
v	24.0 m/s
a	2.00 m/s^2
t	X

$$x - x_0 = \frac{v_0 v - v_0^2}{a} + \frac{\frac{1}{2}v^2 + \frac{1}{2}v_0^2 - vv_0}{a}$$

$$x - x_0 = \frac{\frac{1}{2}v^2 - \frac{1}{2}v_0^2}{a}$$

or

$$x - x_0 = \frac{1}{2a}(v^2 - v_0^2)$$

We can now get a numeric value by substituting in the given values:

$$x - x_0 = \frac{1}{2(2.00 \text{ m/s}^2)}(24.0^2 \text{ m}^2/\text{s}^2 - 20.0^2 \text{ m}^2/\text{s}^2) = 44.0 \text{ m}$$

Notice that finding the solution using a Know/Don't Know table can be summarized in only a few steps.

Physicist's Toolbox 2-3 Creating a Know/Don't Know Table

SET UP
1. Make a Know/Don't Know table.
2. Fill all known values into the table.
3. Mark the unknown variable of interest with a question mark.
4. Mark the remaining unknown variable with an X.

SOLVE
Solve one motion equation for the variable marked with the X, and eliminate the variable by substituting the resulting expression into the second equation. If necessary, rearrange the last result so that the variable of interest is alone on the left side of the equal sign. Finally, substitute the known values, and compute an answer.

REFLECT
You should make sure that your answers are dimensionally consistent and the units of the answers are correct. In addition, check to make sure the magnitudes and signs of the answers conform to your expectations.

Example 2-7 The Car, Again

A car accelerates from $v_0 = 20$ m/s to $v = 24$ m/s over a distance of $x - x_0 = 44$ m (Figure 2-11a). How long does the acceleration take?

 Go to Interative Exercise 2-2 for more practice dealing with constant acceleration.

SET UP
Follow the method of creating a Know/Don't Know table, as shown in Figure 2-11b. As you can see from the table, we need to eliminate acceleration by solving one of the motion equations for a and then substituting the resulting expression into the other motion equation. The result will be an equation that contains t, the variable in which we're interested, together with the three variables of known value. If necessary, we'll then rearrange that result to express t in terms of the other three variables.

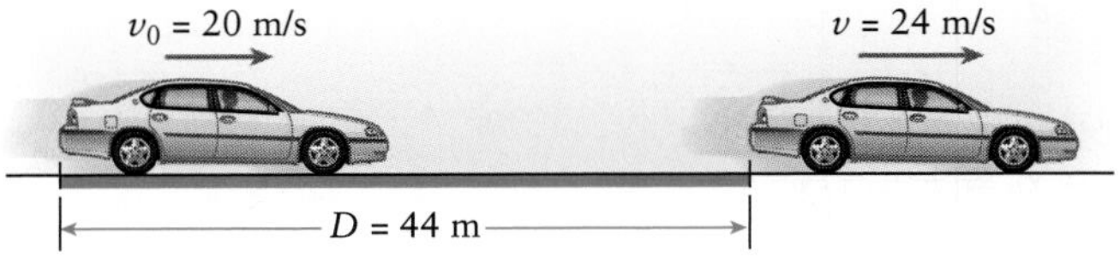

Variable	Know/Don't Know
$x - x_0$	44 m
v_0	20 m/s
v	24 m/s
a	✗
t	?

(a)   (b)

Figure 2-11 (a) A car accelerates from $v_0 = 20$ m/s to $v = 24$ m/s over a distance of $x - x_0 = 44$ m. How long does the acceleration take? (b) A Know/Don't Know table guides the solution. The red X suggests that acceleration should be eliminated to find time, represented by the question mark.

SOLVE

We are free to start with either motion equation, but in this case it's easier to use Equation 2-23 first:

$$v = v_0 + at \tag{2-23}$$

$$at = v - v_0$$

$$a = \frac{v - v_0}{t}$$

We now substitute the relationship into Equation 2-26:

$$x - x_0 = v_0 t + \frac{1}{2}\left(\frac{v - v_0}{t}\right)t^2$$

$$x - x_0 = v_0 t + \left(\frac{1}{2}v - \frac{1}{2}v_0\right)t$$

$$x - x_0 = \frac{1}{2}vt + \frac{1}{2}v_0 t$$

or

$$x - x_0 = \frac{1}{2}(v + v_0)t$$

Our goal is to find an expression in which t is alone on the left side of the equation, so we rearrange:

$$t = \frac{2(x - x_0)}{v + v_0}$$

We can now get a numeric value:

$$t = \frac{2(44 \text{ m})}{24 \text{ m/s} + 20 \text{ m/s}} = 2.0 \text{ s}$$

REFLECT

Is 2.0 s a reasonable answer for the time required for a car to accelerate from 20 m/s to 24 m/s over a distance of 44 m? One way to check the answer is to determine how long it would take the car to cover 44 m assuming no acceleration, that is, at the initial speed of 20 m/s. Equation 2-5, a rearrangement of Equation 2-2 with t_0 set to zero, gives

$$t = \frac{x - x_0}{v} = \frac{44 \text{ m}}{20 \text{ m/s}} = 2.2 \text{ s}$$

If the car maintained its initial speed, it would take 2.2 s to travel 44 m. If the car accelerates it will need less time to go 44 m, so we conclude that a time of 2.0 s is reasonable.

Practice Problem 2-7 A car accelerates from rest at a constant acceleration of 3 m/s^2. How fast is it moving after 3 s?

Example 2-8   Hit the Wall

David Purley, a British Formula One driver, was racing to qualify for the 1977 British Grand Prix when the throttle in his engine got stuck wide open. He hit a wall straight on and at full speed. His body went from an initial speed of about 48 m/s to a final speed of 0 m/s in a distance of 0.66 m. (He suffered multiple fractures but lived to tell the tale.) What was his acceleration during the crash?

SET UP

Start by making a sketch (**Figure 2-12a**). Then, follow the method of creating a Know/Don't Know table, as shown in **Figure 2-12b**. As you can see from the table, we need to eliminate time by solving one of the motion equations for t and then substituting the resulting expression into the other motion equation. The result will be an equation that includes a, the variable in which we're interested, and the three variables of known value. If necessary, we'll then rearrange that resulting expression so that a is expressed in terms of the other three variables. Also, because the motion was along a straight line and all in the same direction, we'll use the driver's speed at any time as his velocity, which allows us to use the motion equations (Equations 2-23 and 2-26) directly.

SOLVE

We can start with either motion equation, but here it's easier to use Equation 2-23 first:

$$v = v_0 + at \tag{2-23}$$

or

$$t = \frac{v - v_0}{a}$$

We now substitute the relationship into Equation 2-26:

$$x - x_0 = v_0\left(\frac{v - v_0}{a}\right) + \frac{1}{2}a\left(\frac{v - v_0}{a}\right)^2$$

Although we will want to simplify the equation, you can see that the goal has been accomplished; we have found a relationship for a in terms of the three known variables. Although it is almost always best to substitute the known values only in the last step of a problem, here if we substitute $v = 0$ m/s immediately the result becomes clear:

$$x - x_0 = v_0\left(\frac{-v_0}{a}\right) + \frac{1}{2}a\left(\frac{-v_0}{a}\right)^2 = -\frac{v_0^2}{a} + \frac{1}{2}\left(\frac{v_0^2}{a}\right) = -\frac{1}{2}\left(\frac{v_0^2}{a}\right)$$

so

$$a = \frac{-v_0^2}{2(x - x_0)}$$

Success! The expression gives a in terms only of variables of known value. We can therefore plug in those values and get a numeric solution:

$$a = \frac{-(48 \text{ m/s})^2}{2(0.66 \text{ m})} = -1.7 \times 10^3 \text{ m/s}^2$$

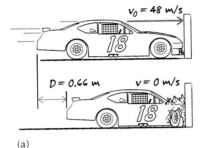

(a)

Variable	Know/Don't Know
$x - x_0$	0.66 m
v_0	48 m/s
v	0 m/s
a	?
t	✗

(b)

Figure 2-12 (a) A racecar driver decelerates from an initial velocity of 48 m/s to a final velocity of 0 m/s over a distance of only 0.66 m. (b) A Know/Don't Know table helps find the magnitude of his acceleration during the crash. The X indicates that time should be eliminated in order to find acceleration, shown by the question mark.

REFLECT

First, let's consider the sign of the answer. Notice that the velocity is positive while the acceleration is negative, that is, acceleration and speed have opposite signs. Because acceleration is the time rate of change of velocity, the difference means that speed is decreasing over time. That's certainly what happened to the racecar driver—he *slowed down*. When the speed (the magnitude of the velocity) of an object decreases, that is, when acceleration and velocity have opposite sign, the change in velocity can be referred to as **deceleration**. What about the magnitude of the deceleration? Certainly 1700 m/s² is much larger than any numbers we've encountered so far, but is there any way to tell if it's in the right range? Consider a car that is advertised to go "from zero to sixty" in 5.4 s, a typical value for commercially available sports cars. After converting 60 mi/h to 26.8 m/s, we can use Equation 2-22 to calculate the acceleration of the car to be $a = 5.0$ m/s². The speed of the racecar is about twice the maximum speed of the car in the advertisement. In addition, the time duration of the racecar driver's deceleration must be tenths or even hundredths of second, compared to 5.4 s. The time for the driver's deceleration is perhaps a hundred or more times smaller than the acceleration for the car in the ad. It shouldn't be surprising, then, to find the absolute magnitude of the racecar driver's acceleration to be a few hundred times the magnitude of the acceleration of the car in the advertisement.

Practice Problem 2-8 A commercial airplane accelerates from $v_0 = 0$ m/s to $v = 80$ m/s as it travels 1600 m down the runway before taking off. How long does it take for the plane to take off?

 Watch Out

Deceleration does not mean the acceleration is negative.

A deceleration results in an object's speed decreasing in magnitude. The speed of an object moving with a velocity that is negative, however, will *increase* as it slows down. For example, car traveling at −20 m/s slows down to go −15 m/s, but −15 is greater than −20. For that reason, a negative value of acceleration does not in and of itself indicate deceleration. An object is decelerating—slowing down—when its acceleration and velocity have opposite signs.

 What's Important 2-3

Solving problems that involve motion under constant acceleration requires only two equations that depend on only five variables: $x - x_0$ (displacement), v_0 (initial velocity), v (final velocity), a (acceleration), and t (time). Often we know the values of three of the motion variables and our goal is to find the value of the fourth by eliminating the fifth of the variables from the two equations—the Know/Don't Know table approach.

2-4 Gravity at the Surface of Earth

It takes 90 s to ride 119 m to the top of the Giant Drop tower at Dreamworld in Queensland, Australia (Figure 2-13). The return trip takes about 5 s. Dreamworld built the motors that haul the riders up, but it's gravity that pulls them back down, screaming all the way!

Perhaps the most important case of constant acceleration is the motion of objects near the surface of Earth such as the Giant Drop gondolas that fall under

Figure 2-13 At Dreamworld in Queensland, Australia, it takes riders 90 s to ascend to the top of the Giant Drop tower, but they free-fall back down the central column in only about 5 s. (*Courtesy Dreamworld Press Office.*)

the influence of Earth's gravitational force. In this chapter, we ignore air resistance, the effect that air has in slowing down moving objects. Therefore, results in this chapter are idealized and apply best to small speeds of relatively short duration. So, neglecting effects of air resistance and only considering vertical, straight-line motion, Equations 2-23 and 2-26 correctly describe motion near the surface of Earth due to gravity. Remember that the equations require acceleration to be constant.

We will eventually be able to show that the acceleration due to gravity at the surface of Earth is constant by using physics principles (Chapter 10), but for now we'll support the claim by using experimental results. **Figure 2-14** shows a strobe image of a billiard ball dropped from rest and photographed at intervals of 0.1 s. Careful measurements of the position and velocity of the ball yield the values shown in **Table 2-3**. Note that we have used the convention that the initial position of the ball is $y_0 = 0$ m, and that as the ball falls it has more and more negative y displacement. Similarly, the velocity of the ball increases in magnitude as it falls. The positions and velocities are plotted against time in **Figure 2-15**.

Figure 2-15b tells us that the acceleration due to gravity is constant at the surface of Earth, and Figure 2-15a confirms it. Because the graph of velocity versus time shown in Figure 2-15b is a straight line it has a constant slope, and therefore the acceleration of the motion must be constant. The slope of the line can be

Figure 2-14 As the ball falls its velocity increases. The distance the ball falls in each subsequent (and equal) time interval gets larger and larger. (*Richard Megna/ Fundamental Photographs.*)

Table 2-3 Position and Speed of a Falling Ball		
Time (s)	Position (m)	Velocity (m/s)
0.0	0.00	0.00
0.1	−0.05	−0.98
0.2	−0.20	−1.96
0.3	−0.44	−2.94
0.4	−0.78	−3.92
0.5	−1.23	−4.91
0.6	−1.77	−5.89

Figure 2-15 (a) The position versus time graph shows that the ball accelerates as it falls. (b) The velocity versus time graph shows that acceleration is constant. The initial position of the ball is $y_0 = 0$ m, and as the ball falls the position has more and more negative y values (Table 2-3).

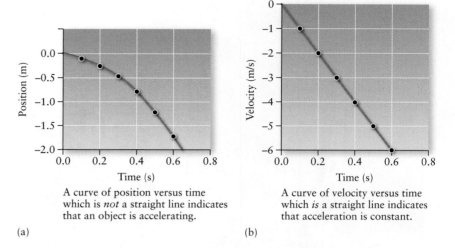

A curve of position versus time which is *not* a straight line indicates that an object is accelerating.

A curve of velocity versus time which *is* a straight line indicates that acceleration is constant.

(a)

(b)

obtained using two sets of time and velocity values (t_1, v_1) and (t_2, v_2) from Table 2-3; for example, $(t_1, v_1) = (0.2\ \text{s}, -1.96\ \text{m/s})$ and $(t_2, v_2) = (0.4\ \text{s}, -3.92\ \text{m/s})$:

$$\text{slope} = a = \frac{v_2 - v_1}{t_2 - t_1} = \frac{-3.92\ \text{m/s} - (-1.96\ \text{m/s})}{0.4\ \text{s} - 0.2\ \text{s}} = -9.8\ \text{m/s}^2$$

We can confirm the value of the acceleration by comparing the position versus time graph in Figure 2-15a with the prediction we can make by using the value of $a = -9.8\ \text{m/s}^2$ in the equation that describes displacement as a function of acceleration and time (Equation 2-26). For example, with initial speed equal to 0 m/s, when $t = 0.3$ s and $a = -9.8\ \text{m/s}^2$, Equation 2-26 predicts

$$\text{displacement} = \frac{1}{2}(-9.8\ \text{m/s}^2)(0.3\ \text{s})^2 = -0.44\ \text{m}$$

which agrees with the measured value (Table 2-3). Our experiment has therefore shown that acceleration due to gravity near the surface of Earth is constant. We have also determined the value of the acceleration.

Physicists use the variable g to describe the acceleration due to gravity near the surface of Earth. The specific value of g varies by a small amount from place to place on Earth, but we will use the accepted average value $g = 9.8\ \text{m/s}^2$. We can then rewrite Equations 2-23 and 2-26 (assuming no air resistance) to describe vertical motion near the surface of Earth:

$$v = v_0 - gt \tag{2-27}$$

$$y - y_0 = v_0 t - \frac{1}{2}gt^2 \tag{2-28}$$

Because these equations describe motion in the vertical direction, it makes sense to use the variable y rather than x to denote position, so we replace x with y in going from Equation 2-26 to Equation 2-28. We designate upward to be the positive y direction, so the acceleration due to gravity is in the negative y direction, accounting for the sign of the second term in Equation 2-28. However, although the choices of the sign of the acceleration and variables used for position cause Equations 2-27 and 2-28 to look different from Equations 2-23 and 2-26, careful inspection should convince you that the physics is *exactly* the same. You might even wonder why we don't just continue to use Equations 2-23 and 2-26 and not write Equations 2-27 and 2-28 at all. Using the former equations wouldn't be unreasonable, but in this case a separate set of equations makes working with motion in the vertical direction just a bit more straightforward.

> ! **Watch Out**
> **Objects accelerate downward under the influence of gravity, but the value of _g_ is positive.**
>
> An object dropped under the influence of Earth's gravity falls down toward the surface. Because downward is most naturally assigned to be the negative direction, students sometimes incorrectly assign a negative value to g in Equations 2-27 and 2-28. Consider Equation 2-27. The minus sign multiplied by a negative value for g would make the second term in the equation positive. What crazy results that calculation would produce—it would predict, for example, that a ball thrown straight up would go faster and faster on the way up!
>
> The minus sign that represents the direction of the acceleration under the influence of gravity can be understood by comparing Equation 2-27 with 2-23, or Equation 2-28 with 2-26. In both cases, when the motion equation is applied to the specific case of gravity, a becomes $-g$; but the value of g is 9.8 m/s^2. You know that a ball thrown up must go slower and slower as it rises, so let your experience in the world guide you if you get confused about the sign of g.

We now work some problems associated with motion under the influence of gravity at the surface of Earth. We caution you, however, that the results of these problems are not nearly as important as the process by which they are obtained.

Example 2-9 How High?

You throw a ball straight up with initial speed $v_0 = 15.0$ m/s. How high does it go? Ignore any effects due to air resistance.

SET UP
We draw a sketch (**Figure 2-16a**) and make a Know/Don't Know table to solve the problem. At first glance, however, it might seem that there isn't enough information because the problem statement appears to provide only one value while the Know/Don't Know table approach requires three. But because the motion occurs near the surface of Earth, the acceleration must have magnitude $g = 9.8$ m/s^2, pointing vertically down, so that's one more known value. And if we analyze the motion starting when the ball is released until it reaches its peak, the speed at the end of that motion must be $v = 0$ m/s. Notice that as the ball rises its velocity is positive

How high?

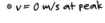
● $v = 0$ m/s at peak

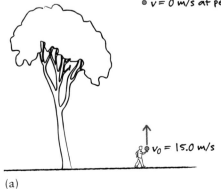

$v_0 = 15.0$ m/s

(a)

Variable	Know/Don't Know
$y - y_0$	?
v_0	15.0 m/s
v	0 m/s
a	$-g$
t	X

(b)

Figure 2-16 (a) A ball is thrown straight up. (b) A Know/Don't Know table helps find the height of the ball at the peak of its motion. The X indicates that time should be eliminated in order to find height, shown by the question mark.

and, by our definition, when the ball falls its velocity is negative. The only way for the velocity of the ball to go from being positive to being negative is to have a value of $v = 0$ m/s sometime during the flight of the ball! So we can now complete the Know/Don't Know table, shown in **Figure 2-16b**. The table identifies the time variable as the one to eliminate. By eliminating t, we can solve for $y - y_0$ as a function of the three known variables.

SOLVE
To eliminate time, first rearrange Equation 2-27 for t:

$$v = v_0 - gt \tag{2-27}$$

$$v - v_0 = -gt$$

$$t = \frac{v - v_0}{-g} = \frac{v_0 - v}{g}$$

Now substitute the relationship into Equation 2-28 to eliminate t:

$$y - y_0 = v_0\left(\frac{v_0 - v}{g}\right) - \frac{1}{2}g\left(\frac{v_0 - v}{g}\right)^2 \tag{2-29}$$

We have found a relationship for $y - y_0$ in terms of the three known variables. Although it is almost always best to substitute the known values only during the last step of a problem, the clarity of the solution will be enhanced if we substitute $v = 0$ m/s immediately:

$$y - y_0 = v_0\left(\frac{v_0}{g}\right) - \frac{1}{2}g\left(\frac{v_0}{g}\right)^2 = \frac{v_0^2}{g} - \frac{1}{2}\frac{v_0^2}{g} = \frac{1}{2}\frac{v_0^2}{g}$$

So, finally, we can obtain a numeric result:

$$y - y_0 = \frac{1}{2}\frac{(15.0 \text{ m/s})^2}{(9.8 \text{ m/s}^2)} = 11.5 \text{ m}$$

REFLECT
First, does it seem reasonable that a person could throw a ball 11.5 m into the air? That height is the height of a medium-sized tree or a three- to four-story house; so, yes, it's reasonable that a person could throw a ball that high. But is the claim that the ball is released with initial speed $v_0 = 15$ m/s reasonable? If you watch baseball, you might know that major league pitchers can throw a fastball between 90 and 100 mi/h (40 to 45 m/s). Most of us can't throw a ball quite that fast (the authors certainly can't!) but a speed of 15 m/s (or about one-third of what is probably the maximum speed with which a human can throw a ball) doesn't seem unlikely.

Practice Problem 2-9 You throw a ball straight up with initial speed $v_0 = 30.0$ m/s. How high does it go? Ignore any effects due to air resistance.

Example 2-10 How Long?

You throw a ball straight up with initial velocity $v_0 = 15.0$ m/s. How long is it in the air before it reaches its maximum height? Ignore any effects due to air resistance.

SET UP
If the problem seems a lot like Example 2-9, you're right. The only differences are apparent in the sketch (**Figure 2-17a**) and in the Know/Don't Know table (**Figure 2-17b**). We now want to eliminate $y - y_0$ in order to solve for t. In this

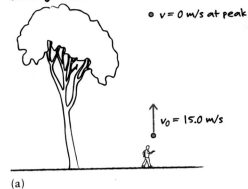

How long does the ball take to reach its peak height?

• v = 0 m/s at peak

$v_0 = 15.0$ m/s

(a)

Variable	Know/Don't Know
$y - y_0$	$\times$
v_0	15.0 m/s
v	0 m/s
a	$-g$
t	?

(b)

Figure 2-17 (a) A ball is thrown straight up. (b) A Know/Don't Know table helps find the time it takes the ball to reach the peak of its motion. The X indicates that displacement should be eliminated in order to find time, shown by the question mark.

case, we don't need to do any work to eliminate $y - y_0$, because one of the two fundamental motion equations, Equation 2-27, does not contain $y - y_0$. So start from Equation 2-27 and rearrange for t.

SOLVE

$$v = v_0 - gt \qquad (2\text{-}27)$$

$$v - v_0 = -gt$$

$$t = \frac{v - v_0}{-g} = \frac{v_0 - v}{g}$$

If we use $v = 0$ m/s at the peak, we can write this specifically as an expression for the time it takes the ball to reach the peak of its motion:

$$t_{\text{peak}} = \frac{v_0 - 0 \text{ m/s}}{g} = \frac{v_0}{g}$$

Using the given values in the problem statement, we find

$$t_{\text{peak}} = \frac{15.0 \text{ m/s}}{9.8 \text{ m/s}^2} = 1.53 \text{ s}$$

It takes 1.53 s for the ball to reach the peak of its motion.

REFLECT
To determine if our answer is valid, let's consider the motion of the ball from the moment it is released upward to the moment it returns to the height from which it was thrown. For that motion, the initial height y_0 and the final height y are the same, so $y - y_0 = 0$. With that, we can easily solve Equation 2-28 for t_{total}, the total time the ball is in the air:

$$y - y_0 = 0 = v_0 t_{\text{total}} - \frac{1}{2} g t^2_{\text{total}}$$

$$0 = v_0 - \frac{1}{2} g t_{\text{total}}$$

$$t_{\text{total}} = \frac{2v_0}{g} = \frac{2(15.0 \text{ m/s})}{9.8 \text{ m/s}^2} = 3.06 \text{ s}$$

Notice that the total time the ball is in the air is twice the time we found it would take for the ball to rise to its peak height. That observation implies that the motions of the ball as it rises to its peak and as it comes back down are symmetric. Although we haven't proven that conclusion, it certainly makes sense. Can we prove it? In the previous problem, we uncovered the height to which the ball will rise. Why not ask how long it would take for the ball to fall from that height?

In Example 2-9, we found the peak height to be $\frac{1}{2}(v_0^2/g)$, which resulted in 11.5 m for launch speed $v_0 = 15.0$ m/s. How long would it take for the ball to fall from that height when starting from 0 m/s, the speed at the peak? From Equation 2-28 the time t_{fall} it takes the ball to fall from its peak height is

$$y - y_0 = v_0 t_{\text{fall}} - \frac{1}{2} g t_{\text{fall}}^2$$

For an initial speed of zero this becomes

$$y - y_0 = -\frac{1}{2} g t_{\text{fall}}^2$$

or

$$t_{\text{fall}} = \sqrt{\frac{-2(y - y_0)}{g}}$$

Don't be concerned about the minus sign under the square root! Because we are considering the motion of the ball starting from the peak and then falling, we presume the initial height to be $y_0 = \frac{1}{2}(v_0^2/g)$, or 11.5 m, and the final height to be $y = 0$ m. These values make $y - y_0$ negative, so there's no problem with the minus sign under the radical. The time it takes for the ball to fall from the peak all the way back down to the release height is

$$t_{\text{fall}} = \sqrt{\frac{-2\left(0 - \frac{1}{2}\frac{v_0^2}{g}\right)}{g}} = \sqrt{\frac{\frac{v_0^2}{g}}{g}} = \sqrt{\frac{v_0^2}{g^2}} = \frac{v_0}{g}$$

Numerically, $t_{\text{fall}} = v_0/g = (15 \text{ m/s})/(9.8 \text{ m/s}^2) = 1.53$ s, and by comparing either the equation or the time of 1.53 s, we've shown that the time it takes for the ball to fall is the same as the time it takes for it to rise to its peak. That's convincing evidence that the motion of the ball is symmetric around the peak height—the physics of rising is the same as the physics of falling. And because we've convinced ourselves that the motion is symmetric around the peak, finding that the total time in the air is twice the time we found for the ball to rise is convincing evidence that our answer of $t_{\text{peak}} = 1.53$ s is reasonable.

Practice Problem 2-10 You throw a ball straight up with initial speed $v_0 = 30.0$ m/s. How long is it in the air before it reaches its maximum height? Ignore any effects due to air resistance.

Example 2-11   How Fast at Height?

You throw a ball straight up with initial speed $v_0 = 15.0$ m/s. What is its velocity when the ball is 5.00 m above the release point? Ignore any effects due to air resistance.

SET UP
The sketch (**Figure 2-18a**) and the Know/Don't Know table (**Figure 2-18b**) for the problem indicate that we need to eliminate the time variable in order to solve for speed. We did the same thing in Example 2-9 and obtained Equation 2-29, which can serve as the starting point for the problem:

$$y - y_0 = v_0\left(\frac{v_0 - v}{g}\right) - \frac{1}{2}g\left(\frac{v_0 - v}{g}\right)^2 \tag{2-29}$$

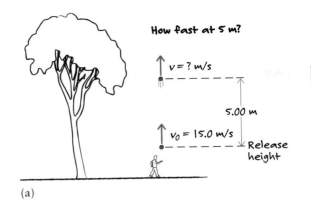

How fast at 5 m?

$v = ?$ m/s

5.00 m

$v_0 = 15.0$ m/s

Release height

(a)

Variable	Know/Don't Know
$y - y_0$	5.00 m
v_0	15.0 m/s
v	?
a	$-g$
t	X

(b)

Figure 2-18 (a) A ball is thrown straight up. (b) A Know/Don't Know table helps find the velocity when the ball is 5.00 m above the release point. The X indicates that time should be eliminated in order to find velocity, shown by the question mark.

SOLVE

The final speed is not zero here, so we need to simplify the equation in a different way than before:

$$y - y_0 = \frac{v_0{}^2 - v_0 v - \frac{1}{2}v_0{}^2 - \frac{1}{2}v^2 + v_0 v}{g}$$

or

$$y - y_0 = \frac{\frac{1}{2}v_0{}^2 - \frac{1}{2}v^2}{g} = \frac{v_0{}^2 - v^2}{2g}$$

We can rearrange to put the result in terms of v:

$$v^2 = v_0{}^2 - 2g(y - y_0)$$

To get a numeric answer, substitute the known values into the equation:

$$v^2 = (15.0 \text{ m/s})^2 - 2(9.8 \text{ m/s}^2)(5.00 \text{ m} - 0.0 \text{ m}) = 127 \text{ m}^2/\text{s}^2$$

Before we take the square root of v^2, consider that such an equation has *two* solutions—one positive value and one negative value. Why? The ball is 5.00 m above the release point at two times during its motion, once on the way up and once on the way down. By our convention, the velocity is positive when the ball is going up and negative when the ball is going down. In addition, the magnitude of the two velocities must be the same because the motion is symmetric around the peak (as we showed in Example 2-10). So, we can now take the square root with confidence:

$$v = \pm\, 11.3 \text{ m/s}$$

REFLECT

The motion of the ball starts with initial speed $v_0 = 15.0$ m/s, and the speed must decrease to $v = 0$ m/s at the peak of motion. At a height 5.00 m above the release point, we therefore expect a speed less than 15.0 m/s and greater than 0 m/s. Our answer is in that range, which means that it is physically reasonable. In addition, we find two answers for the *velocity* at a height of 5.00 m above the release point. On its way up the ball has positive velocity ($+11.3$ m/s), and on the way down it has negative velocity (-11.3 m/s).

    If you are someone who pays attention to numbers, you might recall that we found in Example 2-10 that the peak height for a ball thrown straight up with initial speed $v_0 = 15.0$ m/s to be 11.3 m. How curious that the answer to this example is also 11.3, albeit having different units. Is there a physical meaning to the similarity? Perhaps something special about the height or speeds we've chosen?

No, the similarity is just a coincidence! It's always a good idea to look for connections when studying physics, but at the same time, stay on the lookout for random curiosities that might lead you astray.

Practice Problem 2-11 You throw a ball straight up with initial speed $v_0 = 30.0$ m/s. What is its velocity when the ball is 5.00 m above the release point? Ignore any effects due to air resistance.

At the start of the section, we mentioned that riders on the Giant Drop at Dreamworld in Queensland, Australia fall from the top of a 119-m-tall tower in about 5 s. With the physics we've developed, we ought to be able to check to see if that number is reasonable. Instead of trying to get an exact answer, however, we'll estimate an answer to determine if 5 s is a reasonable time. Scientists commonly use that approach to get a sense of the answer to a question. It often takes the form of a few scribbled lines on whatever paper is handy, say, on the back of an envelope or on a napkin plucked from a dispenser in the dining hall. In this book we model the process for you in "Estimate it!" problems. You'll find these estimation problems throughout the book, and we encourage you to do similar estimations when trying to get a quick understanding of a problem.

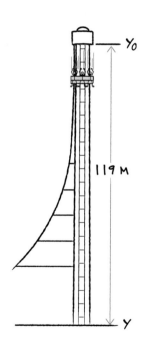

Figure 2-19 Estimate it! 2-1.

Estimate It! 2-1 Fun Fall

About how long should it take to fall 119 m from the top of the Giant Drop at Dreamworld? To make it easy, let's ignore air resistance. Let's grab a napkin and do a quick calculation (**Figure 2-19**), starting from $y - y_0 = v_0t - \frac{1}{2}gt^2$ (Equation 2-28).

SET UP

$$y - y_0 = v_0t - \frac{1}{2}gt^2$$

You would drop from rest, so $v_0 = 0$.

$$y - y_0 = -\frac{1}{2}gt^2$$

y is below y_0, so $y - y_0$ is negative. Use $-H = y - y_0$.

$$-H = \frac{1}{2}gt^2$$

$$H = \frac{1}{2}gt^2$$

$$t = \sqrt{\frac{2H}{g}}$$

SOLVE
H is about 120 m and g is about 10 m/s^2.

$$t = \sqrt{\frac{2(120 \text{ m})}{10 \text{ m/s}^2}} = \sqrt{2(12s^2)} = \sqrt{24 \ s^2} \approx 5 \text{ s}$$

REFLECT
So with a few quick strokes, we've convinced ourselves that the Giant Drop ride lasts about 5 s.

> ## ❗ Watch Out
> ### Acceleration is not zero at the peak of a projectile's motion.
>
> Objects under the influence of gravity near the surface of Earth experience an acceleration g. As long as the object stays relatively close to the surface, that acceleration is constant; we've let g equal 9.8 m/s². As an object thrown straight up rises, its speed decreases as it approaches the peak of motion, but the acceleration remains the same. Does it seem, though, that something different is occurring at the peak of the motion? Yes, it's true that at the peak, the speed is exactly 0 m/s, but students will sometimes set acceleration to 0 m/s² also. Not so! Because acceleration represents how velocity is changing and because velocity is changing from a positive value to a negative value as the object passes through the peak of motion, acceleration must have a nonzero value. On the way up, at the peak, and on the way down the acceleration that an object experiences due to Earth's gravity is always the same 9.8 m/s².

> ## ✳ What's Important 2-4
> The motion of objects near the surface of Earth is the most important case of constant acceleration we typically encounter. Physicists use the variable g to describe the acceleration due to gravity near the surface of Earth. Although the specific value of g varies by a small amount from place to place on Earth, it is common to use 9.8 m/s² as the worldwide average.

Answers to Practice Problems

2-1   40 m/s

2-2   30 m

2-4   34 m/s

2-5   No; for $V = \frac{4}{3}\pi r^3$, the rate of change $dV/dt = 4\pi r^2\, dr/dt$ changes as radius changes, so it is not constant.

2-7   9 m/s

2-8   40 s

2-9   45.9 m

2-10   3.1 s

2-11   ±28.3 m/s

Answers to Got the Concept Questions

2-1   Your average speed is the total distance divided by the total time or 6 mi/15 min, which is equal to 24 mi/h. Your instantaneous speed at any moment in time could be wildly different if you continually sped up and slowed down, or had to stop for any reason, say, for traffic lights or pedestrians. To determine instantaneous speed at any moment requires a series of distance and time measurements over short time intervals. Although you don't have that kind of information, we might guess that your instantaneous speed at any time over the last half mile is close to the average for that part of the trip, because the time interval is relatively short. For the last 30 s of the drive your average speed is 0.5 mi/30 s, or 60 mi/h. (You were lucky not to get a speeding ticket!) Although you can determine your average speed for the first half of the trip (3 mi in 5 min, or 36 mi/h) and for the last half (3 mi in 10 min, or 18 mi/h), the time intervals are too long to make any reasonable statement about instantaneous speed at specific times during those intervals.

2-2   It is not possible to determine the average velocity. We have not defined a direction to be positive; had we defined the direction to the hospital as the negative direction, the average velocity would be negative. In addition, we haven't specified that the entire trip is along the same line. If the trip to the hospital required you to turn and head back toward your starting point before reaching the hospital, your net displacement would be less than the total distance traveled, resulting in a small average velocity.

2-3   The SI units of acceleration are m/s², or m/s *per* s. Acceleration is a measure of how slowly or quickly an object's speed changes, in other words, the speed change, in m/s, every second over which that changes takes place. So, yes, acceleration can be small if the speeds involved in a motion are large. For example, a meteorite approaching Earth at 50 km/s (that's fast!) might not change speeds for a period of time as it encounters the atmosphere. During that time interval, its acceleration would be zero. But consider a baseball struck by a bat, leaving that collision at 30 m/s. The contact time between bat and ball is typically a few thousandths of a second, resulting in an acceleration of, say, (30 m/s − 0 m/s)/0.002 s. That's 15,000 m/s²! So here we have examples of an object moving at large speeds that experiences zero acceleration, and then an object moving at relatively small speeds that nonetheless experiences a rather large acceleration.

2-4   Certainly, the smallest end of human acceleration is zero or standing still. Going from standing still to a casual walking speed of 1 m/s might take 2 s. That's $a = (1$ m/s − 0 m/s)/2 s = 0.5 m/s². The commercial airliner takes about 30 s to reach the speed it needs to take off (about 80 m/s), so the acceleration is (80 m/s − 0 m/s)/ 30 s = 2.7 m/s². A champion sprinter gets up to top speed, about 10 m/s, in a little more than 1 s, which is an acceleration of (10 m/s − 0 m/s)/1 s = 10 m/s². Such acceleration is the same as you experience in free fall on an amusement park ride. We conclude that the range of accelerations typically experienced by humans is about 1 m/s² up to a few tens of m/s².

2-5   (a) The slope of the velocity versus time curve gives the rate of change in velocity, that is, the acceleration. The curve in Figure 2-10 is steepest between 0 and 1 s, so that is the time interval during which the toy car is accelerating at its highest rate. (b) The time interval during which the car is accelerating at its lowest rate corresponds to the part of the velocity versus time curve at which the slope is smallest. The slope of the curve in Figure 2-10 is zero starting at about 4 s, an acceleration of zero. Starting at about time equals 4 s, the toy car is no longer accelerating.

SUMMARY

Topic	Summary	Equation or Symbol	
acceleration	Acceleration is the rate of change of velocity versus time. In this chapter, we have considered only cases in which acceleration is constant. The SI unit of acceleration is meters per second per second, or meters per second squared (m/s²).	$a = \dfrac{dv}{dt}$	
average acceleration	The average acceleration of an object over time interval $t - t_0$ is the change in velocity $v - v_0$ divided by that time interval. Average acceleration represents the average rate of change in velocity over a time interval.	$a = \dfrac{v - v_0}{t - t_0}$	(2-22)
average speed	The average speed of an object over time interval $t - t_0$ is the net distance the object has traveled divided by that time interval.	$s_{avg} = \dfrac{\text{total distance}}{\Delta t}$	(2-14)
average velocity	The average velocity of an object over time interval $t - t_0$ is the net displacement of the object Δx divided by that time interval. Average velocity equals instantaneous velocity when the velocity of an object is constant. Average velocity represents the average rate of change in position over a time interval.	$v_{avg} = \dfrac{\Delta x}{\Delta t}$	(2-13)

constant velocity	An object travels at constant velocity when its acceleration is zero.		
deceleration	We refer to a change in velocity in which acceleration and velocity have opposite sign as deceleration. An object which is decelerating is slowing down.		
displacement	Displacement is the change in position of an object. If the initial position is x_0 and a later position is x, displacement is given by $x - x_0$. The SI unit of displacement is meters (m).		
instantaneous speed	Instantaneous speed is the speed of an object—the magnitude of its velocity—at any instant of time.		
instantaneous velocity	Although an object may be speeding up or slowing down, at any instant in time it has one specific value for velocity, its instantaneous velocity. Instantaneous velocity is the time rate of change (time derivative) of position. Instantaneous velocity can be found by taking the slope of a line tangent to a displacement versus time curve at a specific value of time.	$v = \dfrac{dx}{dt}$	(2-16)
linear motion	Linear motion is motion along a straight line.		
motion under constant acceleration	Many common physical phenomena, like free fall motion due to gravity, proceed at a constant value of acceleration. Everything we need to know about motion under constant acceleration is contained in just two equations, Equation 2-23 and Equation 2-26.	$v = v_0 + at$ $x - x_0 = v_0 t + \dfrac{1}{2}at^2$	(2-23) (2-26)
motion under constant acceleration due to gravity	Objects in free fall near Earth's surface experience constant acceleration g. The value of g varies slightly from location to location; we use $g = 9.8$ m/s^2 as a representative worldwide average. Because the speed of an object in free fall increases as it moves downward, we take acceleration as $-g$ in Equations 2-25 and 2-28 for vertical motion.	$v = v_0 - gt$ $y - y_0 = v_0 t - \dfrac{1}{2}gt^2$	(2-27) (2-28)
reference position	The description of an object's motion depends not on its position but on its change in position, or displacement. The reference position is a location we choose from which to measure displacement. We are free to choose any convenient reference position.		
speed	Speed is the magnitude of velocity, that is, the magnitude of the rate of change of position versus time. The SI units of speed are meters per second (m/s).		

time interval	Time interval is the difference between two time measurements. If the initial time measurement is t_0 and a later time measurement is t, the time interval is given by $t - t_0$. It is often convenient to set $t_0 = 0$ in order to simplify the look of an equation. In those cases, the time variable will stand alone but will nevertheless represent a time interval. The SI unit of time is the second (s).
uniform motion	Uniform motion is motion at constant speed.
velocity	The definition of velocity includes speed as well as the direction of motion. The variable v is used to represent velocity. The SI units of velocity are meters per second (m/s).

QUESTIONS AND PROBLEMS

In a few problems, you are given more data than you actually need; in a few other problems, you are required to supply data from your general knowledge, outside sources, or informed estimate.

Interpret as significant all digits in numerical values that have trailing zeros and no decimal points.

For all problems, use $g = 9.8$ m/s² for the free-fall acceleration due to gravity. Neglect friction and air resistance unless instructed to do otherwise.

• Basic, single-concept problem
•• Intermediate-level problem, may require synthesis of concepts and multiple steps
••• Challenging problem
SSM *Solution is in Student Solutions Manual*

Conceptual Questions

1. •What happens to an object's velocity when the object's acceleration is in the opposite direction to the velocity? SSM

2. •Discuss the direction and magnitude of the velocity and acceleration of a ball that is thrown straight up, from the time it leaves your hand until it returns and you catch it.

3. •Explain the difference between average *speed* and average *velocity*.

4. •Under what circumstances will the displacement and the distance traveled be the same? When will they be different?

5. •Compare the concepts of *speed* and *velocity*. Do the two quantities have the same units? When can you interchange the two with no confusion? When would it be problematic? SSM

6. •Which speed gives the largest displacement in a fixed time: 1 m/s, 1 km/h, or 1 mi/h?

7. •The manufacturer of a high-end sports car plans to present its latest model's acceleration in units of m/s² rather than the customary units of miles/hour/second. Discuss the advantages and disadvantages of such an ad campaign in the global marketplace. Would you suggest making any modifications to the plan?

8. •The upper limit of the braking acceleration for most cars is about the same magnitude as the acceleration due to gravity on Earth. Compare the braking motion of a car with a ball thrown straight upward. Both have the same initial speed. Ignore air resistance.

9. •Under what circumstance(s) will the average velocity of a moving object be the same as the instantaneous velocity? SSM

10. •Is it possible for an object to be accelerating while its speed is constant? Explain your answer.

11. •A video shows a ball being thrown up into the air and then falling. Is there any way to tell whether the video is being played backward? Explain your answer.

12. •The velocity of a ball thrown straight up decreases as it rises. Does its acceleration increase, decrease, or remain the same as the ball rises? Explain your answer.

13. •A device launches a ball straight up from the edge of a cliff so that the ball falls and hits the ground at the base of the cliff. The device is then turned so that a second, identical ball is launched straight down from the same height. Does the second ball hit the ground with a velocity that is higher than, lower than, or the same as the first ball? Explain your answer. SSM

14. •Calc Are the units for average velocity ($\Delta x/\Delta t$) the same as the units for instantaneous velocity (dx/dt)? Explain your answer.

15. •**Calc** Describe the notation that is used to indicate a second derivative. How does one read the notation unambiguously?

16. •Under what circumstances is it acceptable to omit the units during a physics calculation? What is the advantage of using SI units in *all* calculations, no matter how trivial?

17. •What are the units of the slopes of the following graphs: (a) displacement versus time, (b) velocity versus time, and (c) distance fallen by a dropped rock versus time?

18. •Is there any consistent reason why "up" can't be labeled as "negative" or "left" as "positive"? Explain why many physics professors and textbooks recommend choosing *up* and *right* as the positive directions in a description of motion.

19. •A speeding car drives by a police car (initially at rest) and continues traveling at the same constant speed. At the moment the police car catches up to the speeding car, (a) is the police car's speed less than, equal to, or greater than the speeding car; (b) is the police car's displacement less than, equal to, or greater than the displacement of the speeding car, as measured from the point at which the speeding car passed the police car; and (c) is the police car's acceleration less than, equal to, or greater than the acceleration of the speeding car? SSM

20. •Estimate the acceleration, on hitting the ground, of a painter who loses his balance and falls from a step stool. Compare this to the acceleration he would experience if he bent his knees as he hit the ground.

Multiple-Choice Questions

21. •In Figure 2-20, the velocity vector of a car is represented at successive times *t*. Which of the following best represents the acceleration vector?
  A. →
  B. ←
  C. The acceleration is first →, then ←.
  D. The acceleration is zero.
  E. The acceleration is first ←, then →.

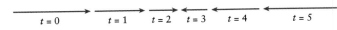

$t = 0 \qquad t = 1 \qquad t = 2 \quad t = 3 \qquad t = 4 \qquad t = 5$

Figure 2-20 Problem 21.

22. •The slope at a point on a position versus time graph of an object is
  A. the object's speed at that point.
  B. the object's instantaneous velocity at that point.
  C. the object's average velocity over a time interval centered on that point.
  D. the object's instantaneous acceleration at that point.
  E. the object's average acceleration over a time interval centered on that point.

23. •Figure 2-21 shows a position versus time graph for a moving object. At which lettered point is the object moving the fastest? SSM
  A. A
  B. B
  C. C
  D. D
  E. E

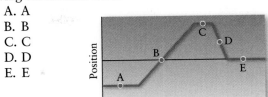

Figure 2-21 Problem 23.

24. •Figure 2-22 shows a position versus time graph for a moving object. At which lettered point is the object moving the slowest?
  A. A
  B. B
  C. C
  D. D
  E. E

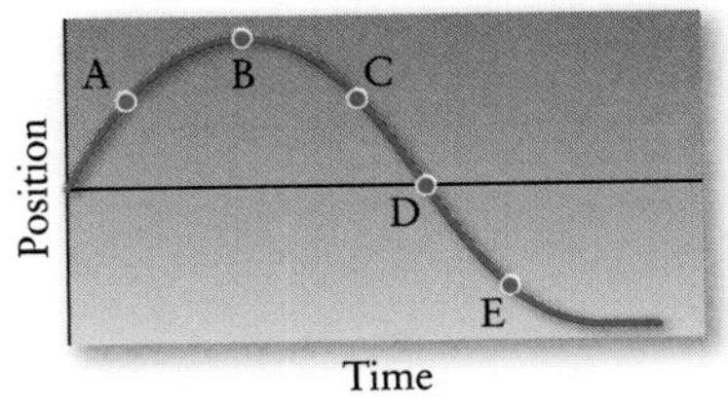

Figure 2-22 Problem 24.

25. •A person is driving a car down a straight road. The instantaneous acceleration is increasing with time, and is in the direction of the car's motion. The speed of the car is
  A. increasing.
  B. decreasing.
  C. constant.
  D. increasing but then decreasing.
  E. decreasing but then increasing.

26. •A person is driving a car down a straight road. The instantaneous acceleration is increasing with time, but is directed opposite the direction of the car's motion. The speed of the car is
  A. increasing.
  B. decreasing.
  C. constant.
  D. increasing but then decreasing.
  E. decreasing but then increasing.

27. •A person is driving a car down a straight road. The instantaneous acceleration is decreasing with time, but is in the direction of the car's motion. The speed of the car is
  A. increasing.
  B. decreasing.
  C. constant.
  D. increasing but then decreasing.
  E. decreasing but then increasing. SSM

28. •A person is driving a car down a straight road. The instantaneous acceleration is decreasing with time, but is directed opposite the direction of the car's motion. The speed of the car is

This page is intentionally left blank.

For complete end of chapter problem sets, please go to
www.whfreeman.com/kestentauck

44. •A bowling ball moves from $x_1 = 3.5$ cm to $x_2 = -4.7$ cm during the time interval from $t_1 = 3.0$ s to $t_2 = 5.5$ s. What is the ball's average velocity?

45. •What must a jogger's average speed be in order to travel 13 km in 3.25 h?

46. •**Sports** The Olympic record for the marathon is 2 h, 6 min, 32 s. The marathon distance is 26.2 mi. What is the average speed of the runner in km/h?

47. ••Kevin completes his morning workout at the pool. He swims 4000 m (80 laps in the 50-m-long pool) in 1 h. (a) What is the average velocity of Kevin during his workout? (b) What is his average speed? (c) With a burst of speed, Kevin swims one 25-m stretch in 9.27 s. What is Kevin's average speed over those 25 m? SSM

48. ••A student rides her bicycle home from physics class to get her physics book, and then heads back to class. It takes her 21 min to make the 12.2-km trip home, and 13 min to get back to class. (a) Calculate the average velocity of the student for the round-trip (from the lecture hall to home and back to the lecture hall). (b) Calculate her average velocity for the trip from the lecture hall to her home. (c) Calculate the average velocity of the woman for the trip from her home back to the lecture hall. (d) Calculate her average speed for the round-trip.

49. •A school bus takes 0.70 h to reach the school from your house. If the average velocity of the bus is 56 km/h, what is the displacement?

50. •A car traveling 80 km/h is 1500 m behind a truck traveling at 70 km/h. How long will it take the car to reach the truck?

51. •**Medical** Alcohol consumption slows people's reaction times. In a controlled government test, it takes a certain driver 0.32 s to hit the brakes in a crisis when unimpaired and 1.0 s when drunk. When the car is initially traveling at 90 km/h, how much farther does the car travel before coming to a stop when the person is drunk compared to sober? SSM

52. ••A jet takes off from SFO (San Francisco, CA) and flies to YUL (Montréal, Quebec). The distance between the airports is 4100 km. After a 1-h layover, the jet returns to San Francisco. The total time for the round-trip (including the layover) is 11 h, 52 min. If the westbound trip (from YUL to SFO) takes 48 more minutes than the eastbound portion, calculate the time for each leg of the trip. What is the average speed of the overall trip? What is the average speed *without* the layover?

53. •A trainer times his racehorse as it completes a workout on a long, straight track. The position versus time data are given below. Plot a graph and calculate the average speed of the horse between (a) 0 and 10 s, (b) 10 and 30 s, and (c) 0 and 50 s.

x (m)	t (s)
0	0
90	5
180	10
270	15
360	20
450	25
500	30
550	35
600	40
650	45
700	50

54. •**Biology** The position versus time graph for a red blood cell leaving the heart is shown in **Figure 2-24**. Determine the instantaneous velocity of the red blood cell when $t = 10$ ms. Recall, 1 ms = 0.001 s, 1 mm = 0.001 m.

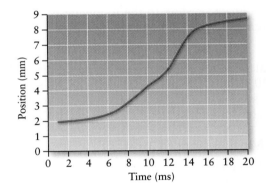

Figure 2-24 Problem 54.

55. •Use the following velocity versus time graph (**Figure 2-25**) for a kangaroo rat running in its burrow to determine its displacement for (a) 0 to 5 s, (b) 0 to 10 s, (c) 10 to 25 s, and (d) 0 to 35 s. SSM

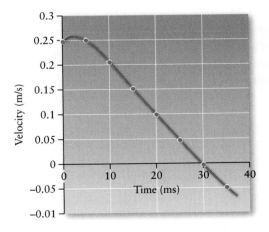

Figure 2-25 Problem 55.

56. •**Calc** Determine the following derivatives with respect to time (t):

(a) $\dfrac{d}{dt}(5t^2 + 4t + 3)$ (b) $\dfrac{d}{dt}(t^2 - 4t - 8)$

57. •**Calc** Determine the value of the following derivatives when $t = 2$:

(a) $\dfrac{d}{dt}(t^2 + t + 1)$ (b) $\dfrac{d}{dt}(2t^3 - 4t^2 - 4)$

58. •**Calc** Determine the following indefinite integrals (ignore any integration constants):

(a) $\int 6t\, dt$ (b) $\int (5t^4 + 3t^2 + 2t)\, dt$

59. •**Calc** Determine the following definite integrals:

(a) $\displaystyle\int_0^2 (12t^2 + 5t + 4)dt$

(b) $\displaystyle\int_{-2}^2 (4t^3 + 2t + 1)dt$

60. •**Calc** A rabbit starts from rest and runs in a straight line according to the following position versus time function:

$$x(t) = 50 + 2t^2 \quad \text{(SI units)}$$

(a) Calculate the displacement of the rabbit during the first 20 s. (b) Calculate the velocity of the rabbit when $t = 6$ s. (c) What is the average velocity over the time interval from 4 to 8 s?

2-3: Motion under Constant Acceleration

61. •A runner starts from rest and achieves a maximum speed of 8.97 m/s. If her acceleration is 9.77 m/s^2, how long will it take her to reach that speed? SSM

62. •A car is driving at 35 km/h and speeds up to 45 km/h in a time of 5 s. The same car later speeds up from 65 km/h to 75 km/h in a time of 5 s. Compare the constant acceleration and the displacement for each of the intervals. Give your answers for acceleration in m/s^2.

63. •A car starts from rest and reaches a maximum speed of 34 m/s in a time of 12 s. Calculate the average acceleration.

64. •The world's fastest cars are rated by the time required for them to accelerate from 0 to 60 mi/h. Convert 60 mi/h to kilometers per hour and then calculate the acceleration in m/s^2 for each of the cars on the following list:

1. Bugatti Veyron 16.4		2.4 s
2. Caparo T1		2.5 s
3. Ultima GTR		2.6 s
4. SSC Ultimate Aero TT		2.7 s
5. Saleen S7 Twin Turbo		2.8 s

65. •A Bugatti Veyron 16.4 can accelerate from 0 to 60 mi/h in 2.4 s. The Saleen S7 Twin Turbo goes "from 0 to 60" in 2.8 s. Determine the distance that the Bugatti Veyron would travel in the time it takes to reach 90 km/h, the speed limit on many Canadian roads. Compare your answer to the distance that the Saleen S7 Twin Turbo would travel while accelerating to the same final speed. SSM

66. •A horse gallops at a speed of 16 m/s. How much time is required for the horse to travel a distance of 4 m?

67. •Derive the equation that relates position to speed and acceleration but in which the time variable does not appear. Start with the basic equation for the definition of acceleration, $a = (v - v_0)/t$, solve for t, and substitute the resulting expression into the position versus time equation, $x = x_0 + v_0 t + \frac{1}{2}at^2$.

68. •A driver is moving at 30 m/s when he sees a moose standing in the road 80 m ahead. If the driver slams on the brakes, what is the minimum acceleration he must undergo to stop short of the moose and avert an accident? (In British Columbia there are over 4000 moose–car accidents each year.)

69. •**Biology** A sperm whale can accelerate at about 0.1 m/s^2 when swimming on the surface of the ocean. How far will a whale travel if it starts at a speed of 1 m/s and accelerates to a speed of 2.25 m/s? Assume the whale travels in a straight line. SSM

70. •When jumping from a standing position, Paola can flex her legs from a bent position through a distance of 20 cm. Paola leaves the ground at a speed of 4.43 m/s with her legs straight. Calculate her acceleration, assuming that it is constant.

71. ••**Calc** The position versus time function of an object is given by

$$x(t) = 12 - 6t + 3.2t^2 \quad \text{(SI units)}$$

(a) What is the displacement between $t = 4$ s and $t = 8$ s? (b) Calculate $v(t)$ of the object and evaluate the equation at $t = 3$ s. (c) At what time(s) is the velocity equal to zero? (d) Calculate $a(t)$. SSM

72. •••**Calc** The acceleration versus time function for an object that starts from rest at $t = 0$ is given by

$$a(t) = 6 \text{ m/s}^2 + (0.75 \text{ m/s}^3)t$$

(a) Determine $v(t)$. (b) Calculate the velocity of the object when $t = 5$ s. (c) Determine the displacement of the object from its starting point after 5 s has elapsed.

2-4: Gravity at the Surface of Earth

73. •A ball is dropped from rest at a height of 25 m above the ground. (a) How fast is the ball moving when it is 10 m above the ground? (b) How much time is required for it to reach the ground level? Ignore the effects of air resistance.

74. •A kiwi fruit is dropped from the roof of a high building. Complete the following table.

t (s)	y (m)	v (m/s)	a (m/s²)
0	0	0	0
1			
2			
3			
4			
5			
10			

75. •Alex climbs to the top of a tall tree while his friend Gary waits on the ground below. Alex throws down a ball at 4 m/s from 50 m above the ground at the same time Gary throws a ball up. At what speed must Gary throw a ball up in order for the two balls to cross paths 25 m above the ground? The starting height of the ball thrown upward is 1.5 m above the ground. Ignore the effects of air resistance. SSM

76. •**Biology** A fox locates its prey under the snow by slight sounds rodents make. The fox then leaps straight into the air and burrows its nose into the snow to catch its next meal. If a fox jumps up to a height of 85 cm, calculate (a) the speed at which the fox leaves the snow and (b) how long the fox is in the air. Ignore the effects of air resistance.

77. •**Medical** More people end up in U.S. emergency rooms because of fall-related injuries than from any other cause. At what speed would someone hit the ground who accidentally falls from the top step of a 6-ft-tall stepladder? (That step is usually embossed with the phrase "Warning! Do not stand on this step.") Ignore the effects of air resistance. SSM

78. •Wes and Lindsay stand on the roof of a building. Wes leans over the edge and drops an apple. Lindsay waits 1.25 s after Wes releases his fruit and throws an orange straight down at 28 m/s. Both pieces of fruit hit the ground simultaneously. Calculate the common height from which the fruits were released. Ignore the effects of air resistance.

79. •A ball is thrown straight up at 18 m/s. (a) How fast is the ball moving after 1 s? (b) After 2 s? (c) After 5 s? (d) When does the ball reach its maximum height? Ignore the effects of air resistance.

80. •A tennis ball is hit straight up at 20 m/s from the edge of a sheer cliff. Some time later, the ball passes the original height from which it was hit. (a) How fast is the ball moving at that time? (b) If the cliff is 30 m high, how long will it take the ball to reach the ground level? (c) What total distance did the ball travel? Ignore the effects of air resistance.

81. •Mary spots Bill approaching the dorm at a constant rate of 2 m/s on the walkway that passes directly beneath her window, 17 m above the ground. When Bill is 120 m away from the point below her window she decided to drop an apple down to him. (See **Figure 2-26.**) (a) How long should Mary wait to drop the apple if Bill is to catch it 1.75 m above the ground, and without either speeding up or slowing down? (b) How far from directly below the window is Bill when Mary releases the apple? (c) What is the angle between the vertical and the line of sight between Mary and Bill at the instant Mary releases the apple? Ignore the effects of air resistance. SSM

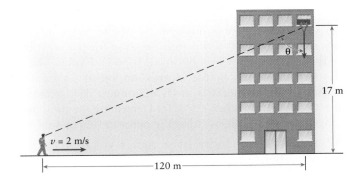

Figure 2-26 Problem 81.

General Problems

82. •A car is driving at a speed of 40 km/h toward an intersection just as the light changes from green to yellow. If the driver has a reaction time of 0.75 s and the braking acceleration of the car is -5.5 m/s², find the minimum distance x_{min} the car travels after the light changes before coming to a stop.

83. ••Two trains, traveling toward one another on a straight track, are 300 m apart when the engineers on both trains become aware of the impending collision and hit their brakes. The eastbound train, initially moving at 98 km/h, decelerates at 3.5 m/s². The westbound train, initially moving at 120 km/h, slows down at 4.2 m/s². Will the trains stop before colliding? If so, what is the distance between them once they stop? If not, what initial separation would have been needed to avert a disaster? SSM

84. ••**Biology** The cheetah is the fastest running animal in the world. Cheetahs can accelerate to a speed of 20 m/s in 2.5 s and can continue to accelerate to reach a top speed of 29 m/s. Assume the acceleration is constant until the top speed is reached and is zero thereafter. (a) Express the cheetah's top speed in miles per hour. (b) Starting from a crouched position, how long does it take a cheetah to reach its top speed and how far does it travel in that time? (c) If a cheetah sees a warthog 120 m away, how long will it take to reach lunch assuming the warthog does not move?

85. •**Medical** Very large accelerations can injure the body, especially if they last for a considerable length of time. The severity index (SI), a measure of the likelihood of injury, is defined as $SI = a^{5/2}t$, where a is the acceleration in multiples of g and t is the time the acceleration lasts (in seconds). In one set of studies of rear end collisions, a person's velocity increases by 15 km/h with an acceleration of 35 m/s². (a) What is the severity index for the collision? (b) How far does the person travel during the collision if the car was initially moving forward at 5.0 km/h?

86. ••**Medical, Calc** The velocity of the wall of a beating heart is given by the following function:

$$v(t) = 0.023\sin(7t) \quad \text{(SI units)}$$

(a) Calculate the maximum speed of the heart during one contraction. (b) Write an expression for the position of the heart as a function of time ($x(t) = ?$). Explain what the various terms in the expression mean physically. Assume that the heart is at its maximum displacement from equilibrium when $t = 0$. (c) Determine the total distance (not displacement) that the heart moves in 45 s.

87. ••**Calc** The velocity of a rocket is given by the following function:

$$v(t) = \begin{cases} 125\left(1 - e^{-0.12t}\right) & 0 < t < 22 \text{ s} \\ 116.1 & t > 22 \text{ s} \end{cases}$$

(a) What are the SI units of the numerical values 125, 0.12, and 116.1? (b) Calculate the speed of the rocket at $t = 2.2$ s and at $t = 25$ s. (c) Calculate the displacement of the rocket from 0 to 10 s. (d) Calculate the displacement of the rocket from 0 to 30 s. SSM

88. ••A man on a railroad platform attempts to measure the length of a train car by walking the length of the train and keeping the length of his stride a constant 82 cm per step. After he has paced off 12 steps from the front of the train it begins to move, in the direction opposite to his, with an acceleration of 0.4 m/s². The end of the train passes him 10 s later, after he has walked another 20 steps. Determine the length of the train car.

89. ••Blythe and Geoff compete in a 1-km race. Blythe's strategy is to run the first 600 m of the race at a constant speed of 4 m/s, and then accelerate to her maximum speed of 7.5 m/s, which takes her 1 min, and finish the race at that speed. Geoff decides to accelerate to his maximum speed of 8 m/s at the start of the race and to maintain that speed throughout the rest of the race. It takes Geoff 3 min to reach his maximum speed. Who wins the race?

90. ••**Calc** A lizard is running in a straight line according to the following:

$$x(t) = (0.20 \text{ m/s}^3)t^3 - (0.40 \text{ m/s}^2)t^2 - (0.65 \text{ m/s})t$$

(a) Determine $v(t)$. (b) Calculate the velocity when $t = 2$ s, $t = 4$ s, and $t = 10$ s. (c) When is the lizard at rest? (d) When is the lizard moving in the positive x direction? (e) When is the lizard moving in the negative x direction? (f) When does the lizard have zero acceleration? (g) What distance does the lizard travel (not its displacement) in the first 10 s?

91. ••During a test on a horizontal track, a rocket starts out from rest with an initial acceleration of 10.0 m/s². As fuel is consumed, the acceleration decreases linearly to zero in 8.00 s and remains zero after that. (a) Sketch qualitative graphs (no numbers) of the rocket's acceleration and speed as functions of time over the first 10.0 s of the motion. (b) Write the equation for the rocket's acceleration as a function of time during the first 10 s. (c) What is the speed of the rocket at (i) the instant the acceleration ceases and (ii) the end of the first 10.0 s?(d) How far does the rocket travel during the first 10.0 s of its motion?

92. •In the following graph (**Figure 2-27**) depicting a moving car, find the instantaneous acceleration at times $t = 2$ s, $t = 4.5$ s, $t = 6$ s, and $t = 8$ s.

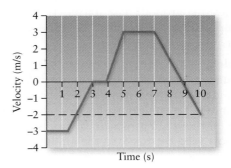

Figure 2-27 Problem 92.

93. ••A ball is dropped from an upper floor, some unknown distance above your apartment. As you look out of your window, which is 1.50 m tall, you observe that it takes the ball 0.18 s to traverse the length of the window. Determine how high above the top of your window the ball was dropped. Ignore the effects of air resistance. SSM

94. •••A lemon is thrown straight up at 15 m/s. (a) How much time does it take for the lemon to be 5 m above its release point? (b) How fast is the lemon moving when it is 7 m above its release point? (c) How much time is required for the lemon to reach a point that is 7 m above its release point? Why are there two answers to part (c)? Ignore the effects of air resistance.

95. •In the book and film *Coraline*, the title character and her new friend Wybie discover a deep well. Coraline drops a rock into the well and hears the sounds of it hitting the bottom 5.5 s later. If the speed of sound is 340 m/s, determine the depth of the well. Ignore the effects of air resistance.

96. •••Ten washers are tied to a long string at various locations, as in **Figure 2-28**. When the string is released some known height above the floor, the washers hit the floor with equal time intervals. Determine the distance between each washer ($y_n = ?$, $n = 1, 2, 3,..., 9, 10$) if the height between the lowest washer (#1) and the floor (before the string is dropped) is 10 cm.

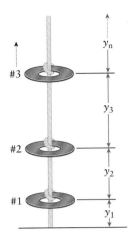

#3 y_n

y_3

#2

y_2

#1

y_1

Figure 2-28 Problem 96.

97. •••A rocket is launched straight up. It contains two stages (Stage 1 and Stage 2) of solid rocket fuel that are designed to burn for 10.0 and 5.0 s, respectively, with no time interval between them. In Stage 1, the rocket fuel provides an upward acceleration of 15 m/s². In Stage 2, the acceleration is 12 m/s². Neglecting air resistance, calculate the maximum altitude above the surface of Earth of the payload and the time required for it to return back to the surface. SSM

98. •••A lacrosse ball is hurled straight up, leaving the head of the stick at a height of 2.00 m above ground level. The ball passes by a 1.25-m-high window in a time of 0.4 s as it heads upward. Calculate the initial speed of the ball if it is 13 m from the ground to the bottom of the window. Ignore the effects of air resistance.

99. •**Biology** A black mamba snake has a length of 4.3 m and a top speed of 8.9 m/s! Suppose a mongoose and a black mamba find themselves nose to nose. In an effort to escape, the snake accelerates at 18 m/s² from rest. (a) How much time does it take for the snake to reach its top speed? (b) How far does the snake travel in that time? (c) Has the tail of the snake gone past the mongoose or does the mongoose have a chance to catch the black mamba?

100. •**Sports** Kharissia wants to complete a 1000-m race along a straight road with an average speed of 8 m/s. After 750 m, she has averaged 7.2 m/s. What average speed must she maintain for the remainder of the race in order to attain her goal?

101. ••**Sports** In April 1974, Steve Prefontaine completed a 10-km race in a time of 27 min, 43.6 s. Suppose "Pre" was at the 9-km mark at a time of 25 min even. If he accelerates for 60 s and maintains the increased speed for the duration of the race, calculate the acceleration that he had. Assume his instantaneous speed at the 9-km mark was the same as his overall average speed at that time.

102. ••An egg is thrown straight down at 1.5 m/s from a tall tower. Two seconds (2 s) later, a second egg is thrown straight up at 4.0 m/s. How far apart are the two eggs 4 s after the second egg is thrown? What is the minimum separation between the eggs and when does it occur? Ignore the effects of air resistance.

103. ••A two-stage rocket blasts off vertically from rest on a launchpad. During the first stage, which lasts for 15.0 s, the acceleration is a constant 2.00 m/s² upward. After 15.0 s, the first engine stops and the second stage engine fires, producing an upward acceleration of 3.00 m/s² that lasts for 12.0 s. At the end of the second stage, the engines no longer fire and therefore cause no acceleration, so the rocket coasts to its maximum altitude. (a) What is the maximum altitude of the rocket? (b) Over the time interval from blastoff at the launchpad to the instant that the rocket falls back to the launchpad, what are its (i) average speed and (ii) average velocity? Ignore the effects of air resistance.

104. ••**Calc** Air drag plays a significant role in the free fall of objects near Earth's surface. Suppose the acceleration of a falling object is given by the following function

$$a(v) = g - \alpha v \quad \text{(down is positive)}$$

where α is a positive constant. (a) By integrating, find the velocity of a falling object as a function of time. (b) Find the terminal velocity of an object that falls from rest starting at $t = 0$.

105. •••**Calc** Air drag is a significant problem in some situations. Suppose the acceleration of a falling object is given by the following equation

$$a(v) = g - \beta v^2 \quad \text{(down is positive)}$$

where β is a positive constant. (a) By integrating, find the velocity of a falling object as a function of time. (b) Find the terminal velocity of an object that falls from rest starting at $t = 0$.

3 Motion in Two Dimensions

(Martin Rietze/www.mrietze.com)

As clumps of molten lava shoot through the air during a volcanic eruption, they experience an acceleration due to gravity—but only in the vertical direction. The resulting parabolic path the fragments follow is a distinctive feature of all projectiles near the surface of Earth.

The glowing arcs of fireworks against the sky trace a path similar to that of streams of water shooting from a fountain (Figure 3-1a) and the graceful flight of a leaping ballet dancer (Figure 3-1b). Why does motion under the influence of gravity follow this distinctive curve? In this chapter, we'll explore motion in two dimensions and consider ways to study and characterize it.

3-1 Horizontal and Vertical Motions Are Independent

So far we have only described the motion of objects moving in a straight line in either the horizontal or vertical direction, such as a sprinter racing to the finish line or a water balloon falling from a balcony. Most of the motions we observe in our daily lives, however, such as the ones in Figure 3-1, are not limited to a single direction. To understand how to describe such motion, let's first ask what makes something move, or more precisely, what makes something change its motion.

 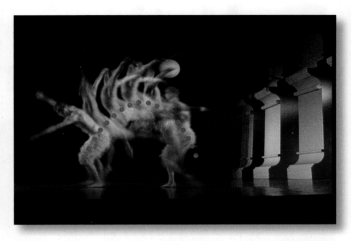

Figure 3-1 (a) Water shot into the air falls in a parabolic arc. (b) When Adam Cooper, a dancer in *Swan Lake,* leaps into the air, he traces the parabolic arc shown by the orange diamonds. *(Fountain from Toño Labra/AgeFotostock. Dancer courtesy of NVC ARTS, a Warner Music Group company.)*

Our experiences in the world suggest that the motion of an object changes when something pushes or pulls it. The something could be a bulldozer pushing against a pile of dirt or Earth's gravity pulling on a falling acorn. Such actions—pushing and pulling—only happen along a straight line because pushing and pulling can only occur in a single direction. The direction of a push or pull can change, but at any instant of time, the push or pull is like an arrow, pointing in a straight line and therefore acting only in one direction.

Consider the parabolic arc of a baseball hit over the outfield fence. Gravity only pulls in the vertical direction, so it does not affect the ball's horizontal motion. Likewise, wind blowing horizontally would affect only the horizontal motion of the ball, leaving the vertical motion unaffected. As a result, the two-dimensional motion we see can be fully analyzed as two separate, one-dimensional motions.

In which two directions should we analyze a motion? In the case of the fireworks rocket, horizontal and vertical directions are natural choices. Physics is lenient in this case, however; so any two perpendicular directions can be used to analyze a problem. Two directions that are perpendicular are said to be *orthogonal*; in a three-dimensional problem, we can choose any three mutually orthogonal directions.

Figure 3-2 The hamster traveled 100 m to return to his nest from the sunflowers, but knowing his displacement tells us nothing about the route he followed between the food source and home.

✱ **What's Important 3-1**

Pushing and pulling only happen along a straight line. The direction of a push or pull can change, but at any instant of time it can only act in one direction. The effect of applying a force—pushing or pulling—is to change an object's motion. Horizontal and vertical motions are independent and are analyzed separately to describe how something moves.

3-2 Vectors

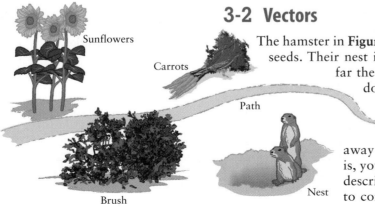

The hamster in **Figure 3-2** just returned to his mate with some sunflower seeds. Their nest is 100 m from the flower patch. Can you tell how far the hamster traveled to get back to the nest? No. You don't know whether he walked along the path, took a direct route through the brush, or maybe looped around to see if the carrots were tasty. All you know for sure is that the hamster is now 100 m away from his original position by the sunflowers; that is, you know the magnitude of his displacement. To fully describe and analyze a displacement, however, you need to consider the direction as well as the magnitude. The same is true of many physical quantities, such as velocity and acceleration. The tools that fully describe such quantities are *vectors* and *vector operations*.

A **vector** is a mathematical construct that incorporates both the magnitude and direction of a quantity. We use a half-arrow symbol written above a variable printed in boldface type to indicate a vector. In contrast, a **scalar** is a simple number that tells us the magnitude of a quantity but nothing about the quantity's direction. The magnitude of a displacement, for example, could be the number of meters an object has moved from an initial position. Therefore, although the variable x, a scalar, can represent the distance an object has been displaced, the vector $\vec{x}$ provides information about the direction as well as the magnitude of the displacement.

A vector has a specific start point and end point. Because vectors are drawn as arrows, the start point and end point of a vector are often called the tail and the tip, respectively. The length of the arrow corresponds to the magnitude of the vector. The direction of the arrow corresponds to the vector's direction. **Figure 3-3** shows a graphical representation of vector $\vec{A}$ and shows how to interpret its magnitude and direction.

In vector notation, a variable that has a circumflex above it (such as $\hat{x}$) denotes a **unit vector**; that is, a vector which has a magnitude equal to 1 (and is dimensionless). A unit vector allows us to describe the direction of a vector separately from its magnitude, although it requires a keen eye to distinguish the symbols for a vector, its magnitude, and its direction. For example, a vector $\vec{A}$ can be represented by its magnitude A and its direction $\hat{A}$ as

$$\vec{A} = A\,\hat{A}$$

In this case, A is the magnitude of the vector $\vec{A}$ and the unit vector $\hat{A}$ describes the direction in which $\vec{A}$ points. The direction of a unit vector can often be deduced from the name of the vector, for example, $\hat{x}$ points in the positive x direction.

If we multiply a vector $\vec{A}$ by a scalar c, the resulting vector points in the same direction as $\vec{A}$ but has a magnitude c times larger than the original magnitude. So if $\vec{B}$ is the result of multiplying $\vec{A}$ by c, then

$$\vec{B} = c\vec{A} = cA\,\hat{A}$$

We can construct a vector of any magnitude in a particular direction by multiplying the desired magnitude by a unit vector in that same direction. For example, a vector $\vec{A}$ of length 5 in the x direction is

$$\vec{A} = 5\hat{x}$$

Figure 3-4a shows vector $\vec{A}$.

What is the relationship between $\vec{A} = 5\hat{x}$ and $\vec{B} = -5\hat{x}$? The answer can be seen most clearly by a slight rearrangement: $\vec{B} = 5(-\hat{x})$. Both $\vec{A}$ and $\vec{B}$ have the same magnitude, 5. The directions of $\vec{A}$ and $\vec{B}$ each are given by a unit vector that multiplies the magnitude; the unit vector ($\hat{x}$) that describes the direction of vector $\vec{A}$ points in the opposite direction of the one that describes the direction of vector $\vec{B}$ ($-\hat{x}$). Two vectors that have the same magnitude but point in opposite directions are negatives of one another; $\vec{B} = -\vec{A}$. The relationship is shown graphically in **Figure 3-4b**.

Figure 3-5 shows vectors $\vec{A}$ and $\vec{B}$, both of which have the same magnitude. The tail of $\vec{B}$ is not the origin, but the angle $\vec{B}$ makes with the positive x direction is nonetheless the same as the angle $\vec{A}$ makes with positive x. $\vec{A}$ and $\vec{B}$ are parallel; they point in the same direction. We define a vector by its magnitude and direction, so $\vec{A}$ and $\vec{B}$ are equal. We can always, therefore, move a vector to a new location parallel to itself without changing the nature of the problem at hand and without changing whatever physics the vector describes.

A vector can be represented as an arrow...

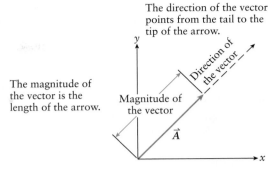

Figure 3-3 Vectors indicate both magnitude and direction.

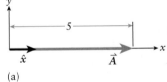

See the Math Tutorial for more information on Vectors

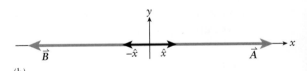

Figure 3-4 (a) Vector $\vec{A}$ is 5 units long and points in the $+x$ direction. (b) Vector $\vec{A}$ is 5 units long and points in the $+x$ direction. Vector $\vec{B}$ is 5 units long and points in the $-x$ direction. The unit vectors $\hat{x}$ and $-\hat{x}$ have a magnitude equal to 1 and are dimensionless. A unit vector allows us to represent the direction of a vector separately from its magnitude.

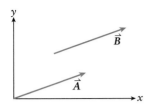

Figure 3-5 We define a vector by its magnitude and direction, so $\vec{A}$ and $\vec{B}$ are equal. We can always, therefore, move a vector to a new location without changing it, by moving its tail end to a new location and keeping its direction and magnitude the same.

✱ What's Important 3-2

A vector describes both the magnitude and the direction of a quantity. In contrast, a scalar quantity only gives magnitude. Vectors that have the same magnitude and direction are equal.

3-3 Vector Components: Adding Vectors, Analyzing by Component

Let's go back to the example of the hamster returning to his nest with sunflower seeds. Hamsters navigate by a process that relies on their knowing the distance and direction—a vector—from one location to another, so this example is not so fanciful as it might seem. Consider two possible routes that the hamster could take from the sunflowers back to the nest. One route has a stop at the carrot patch and the other leads directly back to the nest. The routes are shown in **Figure 3-6** as colored arrows. Remember that an arrow is the standard graphical representation of a vector, so as suggested in the figure, we refer to the route from the sunflowers (S) to the carrots (C) as the vector $\vec{SC}$, the route from the carrots to the nest (N) as the vector $\vec{CN}$, and the route directly from the sunflowers to the nest as the vector $\vec{SN}$.

Notice that the initial and final positions of both routes are the same, and that the magnitude of the hamster's displacement is the same regardless of which route he takes home. The combination of $\vec{SC}$ and $\vec{CN}$ is therefore the same as $\vec{SN}$. The mathematical representation of that statement is

$$\vec{SN} = \vec{SC} + \vec{CN}$$

We just added two vectors! We can define the **vector sum** of two vectors as the vector that connects the tail of the first vector to the tip of the second vector, when the two vectors are laid "tip to tail." Two vectors can be added together even if they are not tip to tail by moving one vector's tail to the other vector's tip while keeping the direction of the vectors the same. This method is demonstrated in **Figure 3-7**, where we move or *translate* the vector $\vec{A}$ to add it to $\vec{B}$. To determine the sum $\vec{C} = \vec{A} + \vec{B}$ graphically, we draw $\vec{C}$ from the tail of $\vec{B}$ to the tip of the translated copy of $\vec{A}$. In a similar way, the vector difference $\vec{C} = \vec{B} - \vec{A}$ is carried out graphically by adding the negative of $\vec{A}$ to $\vec{B}$ (**Figure 3-8**). $\vec{A}$ can also be graphically subtracted from $\vec{B}$ by connecting the tip of $\vec{A}$ to the tip of $\vec{B}$, as in the last panel of Figure 3-8.

Honeybees, such as the one shown in **Figure 3-9**, fly along a jagged path from flower to flower in search of nectar, but they fly straight back—on a beeline—to their hive. They manage the feat by keeping track of the distances and directions they've traveled in moving from place to place, and then "computing" and flying along the vector that leads them directly home. During the process, called *vector navigation*, the vector sum of the vectors that describe each segment of the outbound trip defines the route home. Other creatures, including fiddler crabs, desert ants, and hamsters, also use vector navigation.

Figure 3-6 Three vectors represent the distances and directions between the sunflowers and the carrots, between the sunflowers and the nest, and between the carrots and the nest.

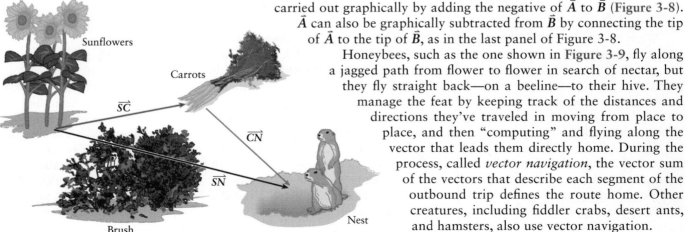

Sunflowers

Carrots

$\vec{SC}$

$\vec{CN}$

$\vec{SN}$

Nest

Brush

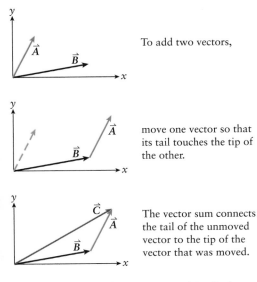

To add two vectors,

move one vector so that its tail touches the tip of the other.

The vector sum connects the tail of the unmoved vector to the tip of the vector that was moved.

Figure 3-7 We add vectors graphically by moving one so that its tail touches the tip of the other and then drawing a line from the tail of the stationary vector to the tip of one we moved.

To subtract vector $\vec{A}$ from $\vec{B}$, add the negative of $\vec{A}$ to $\vec{B}$,

Form the negative of $\vec{A}$ and use the tip-to-tail approach to add it to $\vec{B}$.

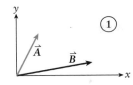

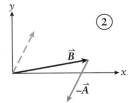

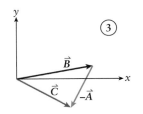

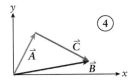

The difference between the vectors is formed by connecting the tail of $\vec{B}$ to the tip of $-\vec{A}$.

The difference between the vectors can also be formed by connecting the the tip of $\vec{A}$ to the tip of $\vec{B}$.

Figure 3-8 We subtract vectors by adding one to the negative of the other.

Figure 3-9 A honeybee collects nectar from a flower and uses vector navigation to find his way back to the hive. *(Rob Flynn/ USDA ARS.)*

Example 3-1 How Does a Hamster Get Back to Its Feeding Area?

Vector navigation in hamsters has been particularly well studied. A hamster knows both his current position relative to the nest as well as the direction and distance from the nest to nearby food sources. Using the information, the hamster can navigate directly to a food source from wherever he happens to be. Show that the direct path from a hamster's current position to a known food source can be determined by subtracting the vector describing his current position relative to the nest from the vector describing the position of the food relative to the nest.

SET UP

Start by sketching the positions of the hamster, nest, and food (Figure 3-10). The problem statement does not specify the exact locations of the nest, food, and hamster relative to one another, so we picked an arrangement that makes the problem easy to visualize. The vector $\overrightarrow{NF}$ describes the location of the food (F) relative to the nest (N); the vector $\overrightarrow{NP}$ describes the hamster's current position (P) relative to the nest; and the vector $\overrightarrow{PF}$ gives the direct path from the hamster's current position to the food.

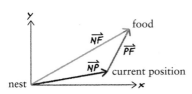

Figure 3-10 The vectors represent the current position of the hamster relative to the nest, and relative to the food as defined in the problem statement. The green vector $\overrightarrow{PF}$ represents the path the hamster needs to follow to get from his current position to the food.

SOLVE

Using the vector approach, it is clear from Figure 3-10 that

$$\overrightarrow{NF} = \overrightarrow{NP} + \overrightarrow{PF}$$

The hamster knows $\overrightarrow{NF}$ and $\overrightarrow{NP}$ but needs to determine $\overrightarrow{PF}$. Subtracting $\overrightarrow{NP}$ from both sides gives

$$\overrightarrow{PF} = \overrightarrow{NF} - \overrightarrow{NP}$$

The vector subtraction therefore defines the straight path $\overrightarrow{PF}$ from the hamster's current position directly to the food.

REFLECT

We have seen how to add vectors graphically by placing two vectors tip to tail and connecting the tail of the first vector with the tip of the second vector. Connecting the tail of $\overrightarrow{PF}$ to the tip of $\overrightarrow{NP}$ therefore gives us the vector $\overrightarrow{NF}$. In the same way, we can see from this example that when one vector ($\overrightarrow{NP}$) is subtracted from another ($\overrightarrow{NF}$), if the vectors are oriented so that their tails touch, the result is a vector which points from the tip of the vector being subtracted ($\overrightarrow{NP}$) to the tip of the other vector ($\overrightarrow{NF}$).

Practice Problem 3-1 A hamster scurries 2.0 m directly away from its nest, turns 60°, and continues in a straight line for 3.0 m. What are the magnitude and direction of the vector that describes its final position relative to the nest?

We resolve a vector into components by projecting it onto two perpendicular axes. When the vector and the axis of interest do not point in the same direction, find the component by drawing a line perpendicular to the axis up to the tip of the vector arrow. In Figure 3-11, using that approach we project $\vec{A}$ and the translated $\vec{B}$ onto the x axis and label the two projections $\vec{A}_x$ and $\vec{B}_x$. We use a unit vector that points along the positive x, y, or z axis ($\hat{x}$, $\hat{y}$, or $\hat{z}$) to describe the direction of the projections. By definition $\vec{A}_x$ and $\vec{B}_x$ point in the x direction, so

$$\vec{A}_x = A_x\hat{x}$$

and

$$\vec{B}_x = B_x\hat{x}$$

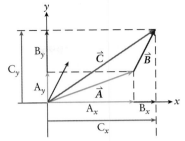

The y component of the vector C equals the sum of the y components of A and B.

The x component of the vector C equals the sum of the x components of A and B.

Figure 3-11 Vectors can be described in terms of their components along the x and y axes.

where A_x and B_x are the lengths of the projections and $\hat{x}$ is the unit vector along the x axis. Obtain the projection of $\vec{C}$ onto the x axis by using the same vertical line used to find the projection of the translated $\vec{B}$ vector. So, reading from the graph,

$$\vec{C}_x = \vec{A}_x + \vec{B}_x$$

and in particular,

$$C_x = A_x + B_x$$

The same is true if we make projections onto the y axis

$$\vec{C}_y = \vec{A}_y + \vec{B}_y$$

and

$$C_y = A_y + B_y$$

The projection of a vector along a specific axis is called the **component of the vector** in the direction of that axis. Note that we don't necessarily need to draw the component along the axis, only in the same direction as the axis. We can then summarize the method of adding vectors by component:

When $\vec{C} = \vec{A} + \vec{B}$,

$$C_x = A_x + B_x$$

and

$$C_y = A_y + B_y$$

√× *See the Math Tutorial for more information on Vectors*

For convenience, we often use the term *component* to refer to the length of the projection of a vector as well as to the vector projection itself. For example, we will call both $\vec{A}_x$ and A_x the "x component of vector $\vec{A}$," depending on the situation.

Figure 3-12 shows a vector $\vec{A}$ and its components in the x and y directions; the angle θ defines the angle that $\vec{A}$ makes with respect to the positive x axis. Notice that A_x, A_y, and a line that has a length equal to the magnitude of $\vec{A}$ together form a right triangle. The side formed by A_x is adjacent to the angle θ and the side formed by A_y is opposite to the angle θ. Most important, note that the vector $\vec{A}$ forms the hypotenuse of the triangle. In this case, we can therefore define the x and y components of vector $\vec{A}$ using A the magnitude of the vector, and the trigonometric functions sine and cosine

$$A_x = A \cos\theta \tag{3-1}$$

and

$$A_y = A \sin\theta \tag{3-2}$$

The definitions require that the vector $\vec{A}$ is the hypotenuse of the right triangle, that the adjacent side and the opposite side are A_x and A_y, respectively, and that the angle θ is defined counterclockwise from the positive x axis.

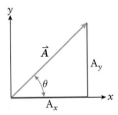

Figure 3-12 The angle θ defines the angle that $\vec{A}$ makes with respect to the positive x axis. A_x is the x component of the vector; A_y is the y component.

√× *See the Math Tutorial for more information on Trigonometry*

Watch Out

Cosine does not necessarily give the x component.

Although Equations 3-1 and 3-2 might mislead you into associating the cosine function with the x direction and the sine function with the y direction, *that would be a mistake!* The relationships between a particular trigonometric function and a particular direction only apply when the angle θ is defined counterclockwise from the positive x axis. **Figure 3-13** shows two cases in which Equations 3-1 and 3-2 do not result in the correct determination of the vector components.

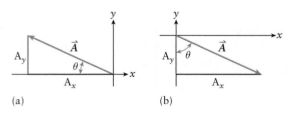

(a)   (b)

Figure 3-13 Equations 3-1 and 3-2 only apply when the angle θ is defined counterclockwise from the positive x axis. The equations do not apply in the situations represented in these diagrams.

In Figure 3-13a, a vector $\vec{A}$ of length $A = 10$ has been set at angle $\theta = 30°$ defined *clockwise* from the *negative* x axis. We are free to define the angle in whatever way is convenient. However, if we apply Equation 3-1 in this configuration, we get

$$A_x = A \cos\theta = 10 \cos 30° = 8.7$$

The result isn't correct. How do we know? Look at the figure. The x component of $\vec{A}$ is clearly negative. Equation 3-1 does not hold because the angle θ is not defined counterclockwise from the positive x axis. If we choose to use angle $\phi = 150°$, defined counterclockwise from the positive x axis, Equation 3-1 gives

$$A_x = A \cos\phi = 10 \cos 150° = -8.7$$

In Figure 3-13b, a vector $\vec{A}$ of length $A = 10$ has been set at angle $\theta = 60°$ defined counterclockwise from the *negative y* axis. We will always have a choice of how to define θ in a problem, so we could have defined θ from the positive x axis. We will encounter many situations, however, in which it is more natural to define the angle from the y axis. As you can verify from the orientation of the sides of the right triangle with respect to the angle θ, in this case

$$A_x = A \sin\theta = 10 \sin 60° = 8.7$$

The correct x component is found using sine, not cosine. So again, the x component of a vector is not always associated with cosine and the y component is not always associated with sine. This observation is just another good reason to think through problems before simply plugging numbers into an equation— you might pick the wrong equation!

In Figure 3-12, the vector $\vec{A}$ and its x and y components form a right triangle with a hypotenuse of length A, so by the Pythagorean theorem:

$$A = \sqrt{A_x^2 + A_y^2} \tag{3-3}$$

In addition,

$$\tan \theta = \frac{A_y}{A_x} \qquad (3\text{-}4)$$

Finally, we can also define any vector as the sum of its vector components:

$$\vec{A} = A_x \hat{x} + A_y \hat{y} \qquad (3\text{-}5)$$

Vectors have biomedical applications, too. An electrocardiogram, or ECG, records the electrical activity of the heart. Pacemaker cells at the top of the heart generate electrical signals that spread across the muscle, triggering a contraction. Both the directions and the magnitudes of the signals can be determined by recording them from different locations on the surface of the body. An ECG can therefore be represented by vectors, as shown in **Figure 3-14**. The vector obtained by adding the horizontal and vertical components of the ECG represents the electrical axis of the heart, which reflects the angle at which the heart is positioned in the chest cavity.

The magnitude of the electrical signal recorded between a pair of electrodes depends on the direction in which the signal spreads across the heart relative to the position of the two electrodes. For example, if the heart's signal moves exactly along the axis defined by two electrodes, the magnitude of the recording will be large. In contrast, if the signal travels in a direction approximately perpendicular to the axis of the electrodes, the recording will be small.

To find the electrical axis of the heart, one records the signal simultaneously in two directions: from right to left across the heart and from top to bottom. This is done by placing one pair of electrodes on someone's wrists and another pair on her wrist and an ankle. The recording made between the wrists measures the portion of the heart's electrical signal that spreads from the right side of the heart to the left. In other words, the wrist-to-wrist signal represents the horizontal component of the electrical signal spreading across the heart. Likewise, the recording between a wrist and an ankle closely represents the vertical component of the signal.

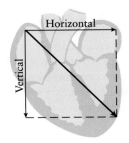

Figure 3-14 The electrical axis of the human heart is represented by the black vector superimposed over a sketch of a cross section through the organ. Like all vectors, it has both a vertical and a horizontal component.

 Go to Bio Animation 3-1 *to explore electrocardiograms and vector addition*

Example 3-2 At What Angle Is Your Heart?

Figure 3-15 shows two ECG traces, the top one recorded along the horizontal axis and the second recorded along the vertical axis. In this particular example, the amplitude of the ECG signal is 1.0 mV when recorded in the horizontal direction (between the wrists) and 1.4 mV when recorded in the vertical direction (between a wrist and an ankle). At what angle does the electrical signal propagate across the heart?

SET UP

When the electrical axis of the heart is treated as a vector, as in Figure 3-14, it can be broken down into horizontal and vertical components. In **Figure 3-16**, we plot the magnitudes of the two ECG recordings as components of a single vector that describes the heart's electrical axis. The vector drawn in red represents the horizontal component and the one drawn in blue represents the vertical component of the ECG, where down is taken as the positive y direction. The signal therefore follows the direction of the purple vector, which has the red and blue vectors as its horizontal and vertical components. We can therefore apply Equation 3-4 to determine the heart's electrical axis vector.

SOLVE

According to Equation 3-4,

$$\tan \theta = \frac{\text{vertical component}}{\text{horizontal component}}$$

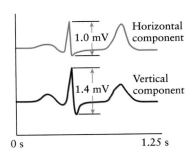

Figure 3-15 The graphs are the electrocardiogram of a single heartbeat recorded simultaneously in the horizontal and the vertical planes.

Figure 3-16 By adding the vertical and horizontal components of an electrical signal spreading across the heart, we can graphically determine the heart's electrical axis.

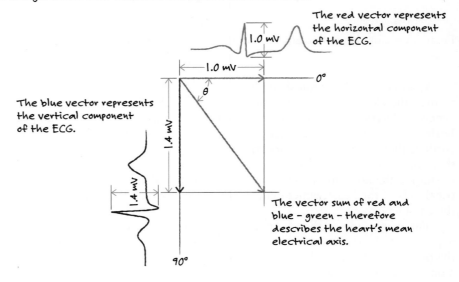

The magnitudes of the two ECG recordings are plotted as components of a single vector that describes the heart's electrical axis...

The red vector represents the horizontal component of the ECG.

The blue vector represents the vertical component of the ECG.

The vector sum of red and blue – green – therefore describes the heart's mean electrical axis.

or

$$\theta = \tan^{-1} \frac{\text{vertical component}}{\text{horizontal component}} = \tan^{-1} \frac{1.4}{1.0}$$

So

$$\theta = 0.95 \text{ rad}$$

An angle of 0.95 radians is about 54°.

REFLECT

Because the signals that trigger the heart to contract start in the top right corner and spread to the bottom of the heart, and because the signals are similar in magnitude, we should expect that the electrical axis would be tilted at an angle of about 45°. Furthermore, normal clinical values are between −30° and +90° of the horizontal axis (where +90° points straight down). Our result is therefore reasonable.

Practice Problem 3-2 The minute hand on a certain clock is twice as long as the hour hand. At exactly 3 o'clock you imagine that the two hands represent components of a vector. At what angle is the vector, measured counterclockwise from the positive horizontal direction?

? Got the Concept 3-1
Component Equal to Zero

Is it possible for a vector that has nonzero magnitude to have a component in some direction that is equal to zero? Draw an example.

Two vectors are added graphically by moving the tail of one vector to the tip of the other while keeping the directions of the vectors the same. The sum of the two is the vector drawn from the tail of the first vector to the tip of the vector that was moved. Two vectors can also be added by first resolving each vector into its x and y components and then adding each component separately. The two sums are the x and y components of the vector sum of the initial vectors.

3-4 Projectile Motion

A few professional basketball players, including the legendary Michael Jordan (Figure 3-17), can execute what some call a "flying dunk." The player leaps off the floor 4 m from the basket and glides through the air before slamming the ball through the hoop. What is it about this motion that makes the player appear to fly? In this section, we'll develop the physics necessary to analyze the flying dunk and other motions under the influence of gravity near the surface of Earth.

Equations 2-23 and 2-26 describe linear motion under constant acceleration:

$$v = v_0 + at \qquad (2\text{-}23)$$

$$x - x_0 = v_0 t + \frac{1}{2}at^2 \qquad (2\text{-}26)$$

When motion is only along a straight line, the sign ($+$ or $-$) of the displacement or velocity indicates the direction. For example, in describing motion under the influence of gravity in Chapter 2, we chose to let the positive y direction be "up" and negative y be "down." A vector is not necessary to describe motion in one dimension. To generalize Equations 2-23 and 2-26 to describe motion in more than one dimension, however, we must represent displacement, velocity, and acceleration using vectors rather than scalars because they have directions as well as magnitudes:

$$\vec{v} = \vec{v}_0 + \vec{a}t \qquad (3\text{-}6)$$

$$\vec{x} - \vec{x}_0 = \vec{v}_0 t + \frac{1}{2}\vec{a}t^2 \qquad (3\text{-}7)$$

Figure 3-17 Michael Jordan appears to fly as he executes a "flying dunk." *(Manny Millan/ Sports Illustrated/Getty Images.)*

For the most part we refer to the two aspects of a vector simply as its magnitude and direction. The magnitude of the velocity vector, however, is known as *speed*. Time is a scalar quantity.

By applying skills we developed in the previous section, we will immediately resolve the two vector equations in terms of the components of each variable in orthogonal directions. To be as general as possible, we could consider three-dimensional problems; however, in most situations relevant to us, all of the interesting physics can be studied using only two directions.

Which two directions should we use? We could choose any two directions that are perpendicular to each other, but it is often most convenient to use the horizontal (x) and vertical (y) directions to analyze motion near the surface of Earth because gravity acts vertically. We therefore rewrite Equations 3-6 and 3-7 as

$$v_x = v_{0x} + a_x t \qquad (3\text{-}8)$$

$$v_y = v_{0y} + a_y t \qquad (3\text{-}9)$$

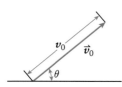

Figure 3-18 The initial velocity of an object that is launched has a magnitude v_0 at an angle θ. The magnitude and direction define the initial velocity vector $\vec{v}_0$.

$$x - x_0 = v_{0x}t + \frac{1}{2}a_x t^2 \qquad (3\text{-}10)$$

$$y - y_0 = v_{0y}t + \frac{1}{2}a_y t^2 \qquad (3\text{-}11)$$

We will primarily study motions in which an object is launched at an initial velocity of magnitude v_0 at an angle θ, as shown in **Figure 3-18**. That magnitude and direction define the initial velocity vector $\vec{v}_0$. According to Equations 3-1 and 3-2, the components of $\vec{v}_0$ can be expressed as

$$v_{0x} = v_0 \cos\theta \qquad (3\text{-}12)$$
$$v_{0y} = v_0 \sin\theta \qquad (3\text{-}13)$$

What are a_x and a_y? In general, horizontal and vertical accelerations can have a wide range of values. When an apple falls from a tree, or when a falcon dives vertically to capture prey in midair, it experiences an acceleration a_y due to gravity, but the horizontal acceleration a_x is 0. In contrast, when a jet plane takes off, a_x and a_y depend on the pilot's control of the thrust and attitude settings as well as the effect of gravity. We will limit our discussion primarily to **free-fall motion**, such as that of the apple, which is influenced only by the effect of gravity. Objects launched with no independent source of thrust exhibit that motion and follow a trajectory often referred to as **projectile motion**.

For projectile motion Equations 3-9 and 3-11 become

$$v_y = v_{0y} - gt \qquad (3\text{-}14)$$

$$y - y_0 = v_{0y}t - \frac{1}{2}gt^2 \qquad (3\text{-}15)$$

Vertical motion under the influence of gravity is discussed in more detail in Section 2-4, where we defined the acceleration due to gravity to be vertically down and to have a magnitude equal to g. We use an average value of $g = 9.8$ m/s^2 although the exact value of g varies slightly from place to place on Earth.

What about horizontal acceleration a_x? Neglecting the effects of wind and air resistance due to motion through air (which could affect both horizontal motion as well as vertical motion), there are no forces that would cause an object in free-fall motion to accelerate in the horizontal direction. We will be able to address the effects of wind and air resistance more directly in the next chapter, but for now, we state categorically that in projectile motion problems horizontal acceleration is zero:

$$a_x = 0 \text{ m/s}^2$$

For projectile motion, Equations 3-8 and 3-10 become

$$v_x = v_{0x} \qquad (3\text{-}16)$$
$$x - x_0 = v_{0x}t \qquad (3\text{-}17)$$

Interesting! In our ideal case of no wind and no air resistance, the horizontal component of velocity v_x remains constant throughout the motion. Variables that remain constant throughout a process are of great utility in solving physics problems.

 Got the Concept 3-2
Rise and Fall

A projectile rises and falls, landing a few hundred meters from its launch point. In the ideal case of no wind and no air resistance, are there any points along its trajectory at which the total acceleration is zero? Are there any points at which either the x or y component of acceleration is zero?

Equations 3-14 through 3-17 completely describe projectile motion, that is, two-dimensional motion near the surface of Earth under the influence of gravity but no other effects. Inherent in this way of describing motion is the implication that the horizontal and vertical motions can be treated separately and are independent of one another. Are they?

If the horizontal and vertical motions of an object are independent, two falling objects should exhibit the same vertical motion regardless of differences in horizontal motion. In **Figure 3-19**, the ball on the right was released from rest at the same instant that the ball on the left was thrown horizontally. Both balls fall under the influence of gravity starting with initial velocity equal to zero in the vertical direction. As you can see, the ball on the right moves horizontally while the ball on the left does not, yet both fall the same distance vertically in any selected time interval. The horizontal and vertical motions of the balls are indeed independent, so we can analyze them separately.

What is the shape of the path that a projectile follows? To understand the path of a real object launched at an initial velocity v_0 and angle θ, we need a relationship between its vertical and horizontal displacements; we need to find $y - y_0$ as a function of $x - x_0$. Equations 3-15 and 3-17 give displacement as a function of time. Using the Know/Don't Know table approach of Chapter 2, and treating $x - x_0$, v_{0x}, and v_{0y} as known values, we can find $y - y_0$ by eliminating t. First, express t in terms of $x - x_0$ and v_{0x} using Equation 3-17:

$$t = \frac{x - x_0}{v_{0x}}$$

Now substitute for t in Equation 3-15:

$$y - y_0 = v_{0y}\left(\frac{x - x_0}{v_{0x}}\right) - \frac{1}{2}g\left(\frac{x - x_0}{v_{0x}}\right)^2$$

With a slight rearrangement the shape of the path becomes clear:

$$y - y_0 = \left(\frac{v_{0y}}{v_{0x}}\right)(x - x_0) - \frac{1}{2}\left(\frac{g}{v_{0x}^2}\right)(x - x_0)^2 \qquad \textbf{(3-18)}$$

This looks a bit long, but notice that Equation 3-18 is of the form

$$y = ax + bx^2$$

The expression is the general form of the equation of a parabola, the shape shown superimposed on the dancer in Figure 3-1b. When an object is launched as in Figure 3-18, the range of values that a and b take generates a parabola that crosses $y = 0$ in two places, as shown in **Figure 3-20**. The constant b is always negative for projectiles, resulting in a path that rises and then falls. The horizontal distance that a projectile travels before returning to its launch height is called the **horizontal range**.

It is often convenient to consider $y = 0$ to be "the ground" when describing the path of a projectile, although any choice of y_0 is allowed. In addition, Equation 3-18 describes the path of the motion even if the projectile does not return to the launch height as shown in Figure 3-20. For example, Equation 3-18 correctly predicts the motion of an object launched from the top of a cliff.

We can rewrite Equation 3-18 in a more compact form using Equations 3-12 and 3-13 for the x and y components of initial velocity:

$$y - y_0 = \left(\frac{v_0 \sin\theta}{v_0 \cos\theta}\right)(x - x_0) - \frac{1}{2}\left(\frac{g}{(v_0 \cos\theta)^2}\right)(x - x_0)^2$$

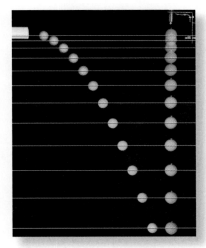

Figure 3-19 The ball on the right only moves in the vertical direction, under the force of gravity. The ball on the left has the same motion in the vertical direction but also has motion in the horizontal direction. (*Joe Henderson/Visuals Unlimited.*)

An object launched only under the influence of Earth's gravity follows a parabolic path.

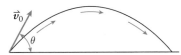

Both the velocity and angle of the path with respect to the horizontal are always changing during projectile motion.

Figure 3-20 Projectiles launched from the surface of Earth follow a parabolic path.

or

$$y - y_0 = \tan\theta(x - x_0) - \frac{g(x - x_0)^2}{2v_0^2\cos^2\theta} \tag{3-19}$$

Equation 3-19 is useful enough that we set it aside for direct use when applicable.

Example 3-3 Hitting a Ball

You hit a ball off a cliff at an initial speed $v_0 = 25$ m/s and at an initial angle $\theta = 42°$ from the horizontal (Figure 3-21). How far from the base of the cliff does the ball travel before hitting the ground $H = 30$ m below? Neglect any effects due to air resistance.

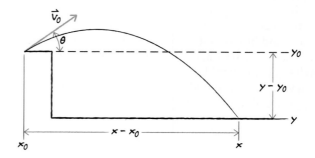

Figure 3-21 A projectile is launched from the edge of a cliff at an angle θ and velocity of magnitude v_0. After following the parabolic arc, it lands at point (x,y).

SET UP

The horizontal displacement of the ball $x - x_0$ is related to the vertical displacement $y - y_0$; for example, if the cliff were shorter, $x - x_0$ would be smaller. This result suggests that we employ Equation 3-19 to find the horizontal displacement. First and most important, notice that because the final height of the ball is lower than the initial height, $y - y_0$ must be negative. We referred to the cliff height as H in the problem statement; therefore,

$$y - y_0 = -H$$

The choice of initial position is always ours to make. For convenience, we choose to set x_0 and y_0 equal to zero in order to rewrite Equation 3-19 in a simpler form:

$$-H = (\tan\theta)\,x - \frac{g\,x^2}{2v_0^2\cos^2\theta}$$

or

$$\frac{g}{2v_0^2\cos^2\theta}x^2 - (\tan\theta)\,x - H = 0 \tag{3-20}$$

Regardless of your choice of x_0 and y_0, get in the habit of labeling those values in any picture you draw. You'll find it easier to apply the correct signs to the values you use in calculations when you label your sketches.

SOLVE

Equation 3-20 is a quadratic equation,

$$ax^2 + bx + c = 0 \tag{3-21}$$

The general solution to a quadratic equation is

$$x = \frac{-b \pm \sqrt{b^2 - 4ac}}{2a} \tag{3-22}$$

So in this case, solutions are

$$x = \frac{\tan\theta \pm \sqrt{(-\tan\theta)^2 - 4\left(\dfrac{g}{2v_0^2\cos^2\theta}\right)(-H)}}{2\left(\dfrac{g}{2v_0^2\cos^2\theta}\right)}$$

Substituting $g = 9.8$ m/s², $v_0 = 25$ m/s, $\theta = 42°$, and $H = 30$ m gives two results, $x = -24$ m and $x = +88$ m.

REFLECT

Why are there two numeric answers to the problem? What does each one tell us? Figure 3-21 suggests that $x - x_0$, which we simplified to x, should be a positive number. It is tempting, then, to take the positive root of the quadratic, $x = +88$ m, as the answer. But what is the source of the negative root? Given our choice of $x_0 = 0$ at the launch position, where is $x = -24$ m? Figure 3-22 reveals the answer. Equation 3-20 does not explicitly contain information that specifies the direction in which the ball left the cliff, and physics allows for a projectile motion path that is downward as well as upward. If the ball had been hit from the top of a tower it could go in the opposite direction and still fall the same vertical distance. In that case, if the ball had been hit at the same initial speed *downward* instead of *upward*, it would have reached the ground a horizontal distance 24 m in the opposite direction.

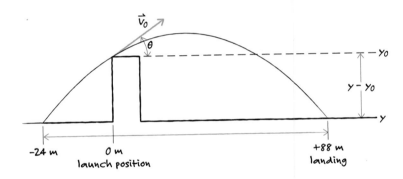

Figure 3-22 Our answer has a positive and a negative solution. Here we see the meaning of the negative answer. If the ball had been hit from the top of a tower instead of the edge of a cliff, it could have moved in the opposite direction along the x axis.

Practice Problem 3-3 A golfer, standing on a knoll 8 m above the level fairway, drives a ball with initial speed $v_0 = 32$ m/s and at an initial angle $\theta = 28°$ from the horizontal. How far does the ball travel horizontally before landing on the fairway? Neglect any effects due to air resistance.

Example 3-4 How High?

You toss a ball into the air with initial speed $v_0 = 15$ m/s and at an initial angle $\theta = 65°$ from the horizontal (Figure 3-23). Neglecting the effects of air resistance, how high does the ball go before coming back down?

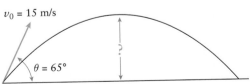

SET UP

We considered a similar problem in Example 2-9. We used a Know/ Don't Know table to determine that we should eliminate the variable t from the motion equations to find the maximum height. In creating the table, we recognized that at the peak of the motion, the velocity of the ball had to be zero. Our approach in solving this two-dimensional problem will be the same as our approach in solving the one-dimensional problem in Example 2-9. Something is different, however, between the one-dimensional problem and this problem. Unless

Figure 3-23 A ball is thrown at an angle of 65° and speed of 15 m/s. How high does it go? The black parabolic arc shows its path.

an object is thrown straight up, velocity cannot be zero at the peak of motion, because as Equation 3-16 shows, the x component of a projectile's velocity remains constant during flight. As the ball in this problem passes through its peak height, however, it reverses vertical direction and starts going back down; thus, the y component of velocity is zero at the peak. We therefore extend the method used in Example 2-9 by requiring that the y component of velocity at the peak be zero.

SOLVE

The Know/Don't Know table, shown in **Figure 3-24**, requires that we eliminate time, and using $v_{y,\,\text{peak}} = 0$ makes the elimination easier. First, by using Equation 3-14,

$$v_{y,\,\text{peak}} = v_{0y} - gt_{\text{peak}} = 0$$

so

$$t_{\text{peak}} = \frac{v_{0y}}{g} = \frac{v_0 \sin\theta}{g} \tag{3-23}$$

We can now substitute Equation 3-23 into Equation 3-15 to find the peak height of the ball, that is, $(y - y_0)_{\text{peak}}$:

$$(y - y_0)_{\text{peak}} = v_{0y}t_{\text{peak}} - \frac{1}{2}gt_{\text{peak}}^2$$

$$= v_{0y}\frac{v_{0y}}{g} - \frac{1}{2}g\left(\frac{v_{0y}}{g}\right)^2$$

$$= \frac{v_{0y}^2}{g} - \frac{1}{2}\frac{v_{0y}^2}{g}$$

$$= \frac{1}{2}\frac{v_{0y}^2}{g}$$

So, finally,

$$(y - y_0)_{\text{peak}} = \frac{1}{2}\frac{v_0^2 \sin^2\theta}{g} \tag{3-24}$$

Using the values given, the ball reaches a peak height of

$$(y - y_0)_{\text{peak}} = \frac{1}{2}\frac{(15 \text{ m/s})^2 \sin^2 65°}{9.8 \text{ m/s}^2} = 9.4 \text{ m}$$

REFLECT

In Example 2-9, we found that a ball thrown straight up at initial speed $v_0 = 15.0$ m/s reaches a maximum height of 11.5 m. Should we expect a ball thrown at an angle other than 90° to reach a lower peak height? Although we haven't developed the physics yet to answer in detail, you would probably guess that as we reduce the launch angle from 90°, a projectile travels farther horizontally but with decreasing maximum height. This observation is confirmed in **Figure 3-25**, in which paths for angles ranging from 85° (nearly straight up; the light blue line) to 45° (the purple line) are shown. The green line corresponds to a launch angle of 65°. So our answer to the problem is reasonable compared to the result in Example 2-9.

Practice Problem 3-4 You toss a ball into the air with initial speed v_0 equal to 12 m/s and at an initial angle θ equal to 85° from the horizontal. Neglecting the effects of air resistance, how high does the ball go before coming back down?

Variable	Know/Don't Know
$y - y_0$	?
v_{0y}	$v_0 \sin\theta$
$v_{y,\,\text{peak}}$	0 m/s
a_y	$-g$
t	✗
θ	65°
v_0	15 m/s

Figure 3-24 By constructing a table like this, you can immediately see which variables you already have and which one you need to eliminate in order to solve a problem.

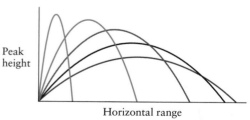

Both the horizontal range and the peak height depend on the launch angle.

Peak height

Horizontal range

Figure 3-25 The parabolas show that both the horizontal range of a projectile as well as its peak height depend on the launch angle.

Example 3-5 How Far?

You toss a ball into the air with initial speed v_0 equal to 15 m/s and at an initial angle θ equal to 65° from the horizontal (**Figure 3-26**). Neglecting any effects due to air resistance, how far has the ball traveled horizontally when it returns to the initial launch height?

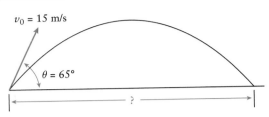

SET UP

Let's use a Know/Don't Know table in this example. To incorporate everything we know, the table shown in **Figure 3-27** is even longer than the one we used in the previous example. Let's look carefully at some of the values that we inserted into the Know/Don't Know table:

$y - y_0$: Because we are interested in the distance the ball travels before returning to its launch height, the final height y and the initial height y_0 are the same, that is, $y - y_0 = 0$.

v_x: No acceleration exists in the x direction, which means that the x component of velocity does not change with time. So the x component of the velocity of the ball is the same when it returns to its launch height as it was the instant the ball left your hand. The x component of the projectile's velocity v_{0x} equals $v_0 \cos\theta$.

v_y: At any two points at the same height along a projectile's path, the magnitude of the y component of velocity is the same; so, the magnitude of v_y, which represents the velocity when the projectile returns to its launch height, equals v_{0y} and $v_y = -v_{0y} = -v_0 \sin\theta$.

We have placed a red "X" for time in the Know/Don't Know table, which guides us to eliminate t from the motion equations in order to solve for the range ($x - x_0$).

Figure 3-26 A ball is thrown up at an angle of 65° and velocity of 15 m/s. How far does it go? The parabolic arc shows its path.

Variable	Know/Don't Know
$x - x_0$ (at range)	?
$y - y_0$ (at range)	0
v_{0x}	$v_0 \cos\theta$
v_{0y}	$v_0 \sin\theta$
v_x	$v_0 \cos\theta$
v_y (when $x - x_0$ = range)	$-v_0 \sin\theta$
a_x	0
a_y	$-g$
t	✗
θ	65°
v_0	15 m/s

Figure 3-27 The table organizes the information given in the problem statement and shows you which variable you need to eliminate in order to solve the problem.

SOLVE

For easy reference, the motion equations, broken into x and y components, are

$$v_x = v_{0x} \tag{3-16}$$

$$x - x_0 = v_{0x}t \tag{3-17}$$

$$v_y = v_{0y} - gt \tag{3-14}$$

$$y - y_0 = v_{0y}t - \frac{1}{2}gt^2 \tag{3-15}$$

To eliminate a variable, we solve one of the equations for that variable and substitute the result into one of the other equations. There are a number of ways to eliminate a variable; for example, we could solve Equation 3-15 for t and substitute into Equation 3-17. (Alternatively, we could consider substituting the equation for t into Equation 3-14, but this would be unproductive because Equation 3-14 does not involve $x - x_0$, which is our goal.) Using the knowledge that $y - y_0 = 0$ when $x - x_0$ is maximum makes eliminating a variable easier. So, the equation

$$y - y_0 = v_{0y}t - \frac{1}{2}gt^2 = 0$$

becomes

$$\frac{1}{2}gt_{\text{total}}^2 = v_{0y}t_{\text{total}}$$

where t_{total} denotes the time for the ball to complete the parabolic motion shown in Figure 3-26 or the total time of flight. So,

$$t_{\text{total}} = \frac{2v_{0y}}{g} \tag{3-25}$$

When this specific time is substituted into Equation 3-17 to eliminate t, the result is the horizontal range, $(x - x_0)_{\text{range}}$ or

$$(x - x_0)_{\text{range}} = v_{0x} t_{\text{total}} = v_{0x} \frac{2 v_{0y}}{g}$$

So,

$$(x - x_0)_{\text{range}} = \frac{2 v_{0y} v_{0x}}{g} = \frac{2 v_0^2 \sin \theta \cos \theta}{g} \tag{3-26}$$

The equation applies specifically to the case in which a projectile such as the ball is launched with initial speed v_0 and at angle θ and returns to the initial launch height. We obtain a numeric result by using the given values or

$$(x - x_0)_{\text{range}} = \frac{2 \, (15 \text{ m/s})^2 \sin 65° \cos 65°}{9.8 \text{ m/s}^2} = 17.6 \text{ m}$$

The horizontal range to two significant digits is 18 m.

REFLECT

The distance we obtained, 18 m, doesn't seem particularly short or long. Also, the length of the range as shown in Figure 3-26 is about twice the height. So, given that the peak height is about 9 m for a launch angle of 65° (see Example 3-4), if the paths shown in Figure 3-25 are properly scaled (they are), then 18 m for the range is about right.

Notice that the Know/Don't Know table we constructed for this problem (Figure 3-27) is more complicated than the tables we've used in previous problems. The added complexity arises because we must now resolve all of the motion variables (excluding time) into x and y components. Going forward, then, the Know/Don't Know table may not be an efficient way to set up problems. We will, however, use the Know/Don't Know table *approach* to identify and eliminate a variable not required in the solution of a problem.

Practice Problem 3-5 A soccer player kicks a ball at initial speed v_0 equal to 24 m/s and at an initial angle θ equal to 25° from the horizontal. Neglecting any effects due to air resistance, how far does the ball travel before hitting the ground?

Example 3-6 How Long?

You toss a ball into the air, with initial speed v_0 equal to 15 m/s and at an initial angle θ equal to 65° from the horizontal. Neglecting any effects due to air resistance, how long does it take for the ball to return to the initial launch height?

SET UP

In the previous example, we found an expression for the total time of flight for a ball to return to the launch height to be

$$t_{\text{total}} = \frac{2 v_{0y}}{g} \tag{3-25}$$

That expression is the starting point for this problem.

SOLVE

The time for the ball to return to its launch height is then

$$t_{\text{total}} = \frac{2 v_0 \sin \theta}{g} = \frac{2 (15 \text{ m/s}) \sin 65°}{9.8 \text{ m/s}^2} = 2.77 \text{ s}$$

The time of the flight, to two significant digits, is 2.8 s.

REFLECT

About 3 s is certainly physically reasonable, but a comparison of the method we used in this example with other possible approaches is of more interest to us.

During the process of eliminating t from the motion equations in Example 3-4, we found an expression for the time it takes a projectile to reach its peak to be

$$t_{peak} = \frac{v_{0y}}{g} = \frac{v_0 \sin \theta}{g} \tag{3-23}$$

Without substituting any values and by comparing Equation 3-23 with Equation 3-25, we immediately see that t_{peak} is half of t_{total}. That result gives us a glimpse into a fundamental aspect of projectile motion; it is *symmetric* around the peak of motion. The first half of the flight is a mirror image of the second half. If you could make a video of the flight of the ball in this problem, it would look the same run backward as forward.

As we noted in the previous example, the magnitude of the y component of velocity at any two points along a projectile's path is the same when the points are at the same height. This observation is also a manifestation of the symmetry of projectile motion. We can use the result to extract the time of flight in yet a different way, through Equation 3-14

$$v_y = v_{0y} - gt \tag{3-14}$$

First, solve for t

$$t = \frac{v_{0y} - v_y}{g}$$

But because of the symmetry of the projectile motion path, we also know that the magnitude of v_y (the speed when the projectile returns to its launch height) equals v_{0y} and that because the ball is going up at the start and coming down at the end, $v_y = -v_{0y}$, so

$$t_{total} = \frac{v_{0y} - (-v_{0y})}{g} = \frac{2v_{0y}}{g}$$

The expression matches our previous result. Obtaining the same answer to a problem using two different techniques gives us confidence that the answer is correct.

Practice Problem 3-6 A soccer player kicks a ball at initial speed v_0 equal to 24 m/s and at an initial angle θ equal to 25° from the horizontal. Neglecting any effects due to air resistance, how long is the ball in the air?

Example 3-7 How Does He Do That?

Figure 3-17 shows former professional basketball player Michael Jordan executing a flying dunk. Jordan, who is 6′ 6″ (about 2 m), leaps off the floor 4 m from the 10-ft-high (about 3-m-high) basket and appears to fly through the air before slamming the ball through the hoop. To analyze his motion, determine the fraction of total jump time that Jordan is higher than half the peak height of the jump. Assume that he takes off at an angle of 45°.

SET UP

Notice that, in the photograph, Jordan's head is about the same height as the basketball rim; we can therefore estimate the peak height of his jump to be 1 m, which is the difference between his height and the height of the rim. The peak height and Equation 3-24 enable us to determine the speed with which he leaves

Figure 3-28 In a flying dunk, Michael Jordan leaves the ground at an initial velocity of magnitude v_0 and at an angle θ, reaching a height H before coming back down. The dashed line marks half the maximum height.

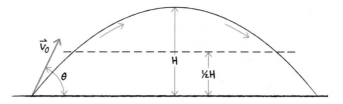

the ground. The speed and Equation 3-15 allow us to determine the time it takes for Jordan to reach the point which is half as high as the peak height. Using that information, we will be able to find the fraction of time he spends higher than half the peak height (Figure 3-28).

SOLVE

In Example 3-4, we learned that the peak height can be determined by

$$(y - y_0)_{\text{peak}} = \frac{1}{2}\frac{v_0^2 \sin^2\theta}{g} \tag{3-24}$$

Rearranging for v_0 gives

$$v_0 = \sqrt{\frac{2g(y - y_0)_{\text{peak}}}{\sin^2\theta}}$$

So having a takeoff angle of 45°, the initial speed required to reach $(y - y_0)_{\text{peak}} = 1$ m is

$$v_0 = \sqrt{\frac{2\ (9.8\ \text{m/s}^2)(1\ \text{m})}{\sin^2 45°}} = 6.3\ \text{m/s}$$

or about 6 m/s. (We are being a bit loose with the numbers but will explain more on dealing with approximate numbers in the Reflect section below.)

We can now use Equation 3-15 to determine how long it takes Jordan to reach half his peak height. Using $v_0 = 6$ m/s and $\theta = 45°$,

$$\frac{1\ \text{m}}{2} = (6\ \text{m/s})\ (\sin 45°)t - \frac{1}{2}(9.8\ \text{m/s}^2)t^2$$

or

$$4.9t^2 - 4.2t + 0.5 = 0$$

The quadratic equation in which time is the variable has two solutions (see Equations 3-21 and 3-22):

$$t = \frac{4.2 \pm \sqrt{4.2^2 - 4(4.9)(0.5)}}{2(4.9)}$$

or

$$t = 0.1\ \text{s}, 0.7\ \text{s}$$

There are two solutions because Michael Jordan reaches a height of half a meter twice, first on the way up and then later on the way down, as shown in Figure 3-29. The time that Jordan is higher than half the peak height is then

$$t_{\text{above}} = 0.7\ \text{s} - 0.1\ \text{s} = 0.6\ \text{s}$$

The total duration of his jump must be 0.8 s (0.1 s on the way up, 0.6 s above half his peak height, and 0.1 s on the way down), so he spends 75% (0.6/0.8) of the time of his jump at a height greater than half the maximum height.

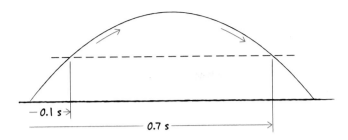

Figure 3-29 There are two solutions to our equation because Michael Jordan reaches half his maximum height twice, first on the way up and then later on the way down.

REFLECT

Although Michael Jordan's acrobatics—bending his knees and swooping the ball above his head—contribute to the effect, the principal reason Jordan appears to soar as he executes a flying dunk is that he's higher than half the peak height most of the time he's in the air. Of course, it also helps that he's in the air a relatively long time! Neither of the authors could jump off the ground for anywhere near 0.7 s, even when we were teenagers.

During this problem, we rounded off all of the computations to one digit. The value we used for Michael Jordan's peak height was estimated somewhat crudely from a photograph, and the takeoff angle we used was a pure guess. That's fine, because we were trying to get a general understanding of the problem. However, using such approximations, we can't justify calculations carried out to more than rough accuracy. In general, the number of significant digits with which we express an answer should be no more than the number of digits in the least accurate value used to solve the problem.

Got the Concept 3-3
Velocity of a Projectile

A projectile follows a parabolic arc. At what point during the trajectory is its velocity the smallest? At what point during the trajectory is the *magnitude* of velocity the smallest? At what point, if at all, will the velocity of the object be zero? Include in the trajectory only the motion of the projectile up until the instant right before it lands.

Watch Out
When a projectile is going up, the *y* component of velocity is positive. When it's going down, the *y* component of velocity is negative.

Velocity is a vector. By convention we set the direction up—away from Earth's surface—as positive, and down as negative. A projectile launched upward that eventually comes back down therefore initially has a positive *y* component of velocity, and then has a negative *y* component when returning to the ground. In between, specifically right at the peak of the motion, the *y* component of the object's velocity must transition from positive to negative. That means that at the peak of the motion the *y* component of the object's velocity is zero.

$\sqrt{x}$ **Go to Interactive Exercise 3-1**
for more practice dealing with
horizontal range

Estimate It! 3-1 Home Run

During a major league baseball game, a ball just barely clears the fence for a home run. Estimate the speed of the ball right after it was hit. Ignore the effects of air resistance.

SET UP

Use Equation 3-26 to estimate the speed of the ball

$$(x - x_0)_{\text{range}} = \frac{2v_0^2 \sin\theta \cos\theta}{g}$$

so

$$v_0 = \sqrt{\frac{g(x - x_0)_{\text{range}}}{2\sin\theta \cos\theta}}$$

SOLVE

Choose reasonable, round numbers for the distance to the wall and the angle at which the ball left the bat. Estimate $(x - x_0)_{\text{range}} = 100$ m and $\theta = 30°$.
Then,

$$v_0 = \sqrt{\frac{9.8 \text{ m/s}^2 \ (100 \text{ m})}{2 \sin 30° \cos 30°}} = 33.6 \text{ m/s} = 30 \text{ m/s}$$

REFLECT

The speed of a baseball hit by a professional ball player is typically in the range of 30 to 40 m/s.

✳ What's Important 3-4

   Projectile motion describes the trajectory of objects launched into the air with no independent source of thrust. To analyze projectile motion, we consider vertical and horizontal motions separately. When the initial and final heights of an object launched into projectile motion are the same, the first half of the flight is a mirror image of the second half.

3-5 Uniform Circular Motion

If you have ever been in a car speeding up to merge onto a highway or in a plane going faster and faster down the runway before takeoff, you have felt the effects of acceleration. In such situations, people often describe a feeling of being pushed back against their seats. The sensation arises from the *vestibular system*. Part of the inner ear, the vestibular system consists of a series of tiny, interconnected canals and chambers in the skull. Two structures sense the orientation of the head with respect to gravity. Specialized cells convert mechanical energy caused by movements of the head into electrical signals that are processed by the brain.

   Even if your speed is not changing, you might have had a similar sensation when moving in a circle, say, on an amusement park ride or in a car turning too fast or even when exiting the highway on a circular off ramp. Other parts of the vestibular system detect acceleration due to rotational motion (**Figure 3-30**). The three semicircular canals are positioned orthogonally to one another, so each detects a different spatial direction. Whenever the head rotates around the axis of one of the canals, the fluid inside it moves, bending sensory cells. The response of the brain to

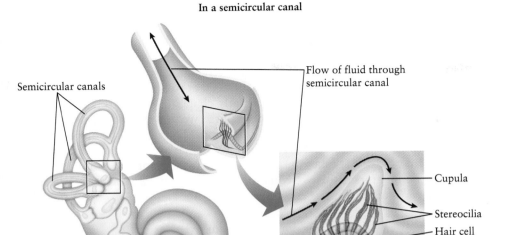

In a semicircular canal

Semicircular canals

Flow of fluid through semicircular canal

Cupula

Stereocilia

Hair cell

Support cell

Axon

Direction of body movement

Figure 3-30 The three semicircular canals in the inner ear enable one to detect rotational motion. They are positioned orthogonally in order to detect motion in each of three orthogonal rotational directions.

the resulting electrical signals it receives is the feeling of angular acceleration. Rotations of the head as small as 0.005° can be sensed in this way.

Is it possible for an object to accelerate even if its speed is constant? Acceleration is the rate of change of velocity, that is,

$$\vec{a} = \frac{d\vec{v}}{dt}$$

where we have been careful to note that both acceleration and velocity are vectors. Any change in velocity results in a nonzero acceleration. Because velocity comprises both a magnitude and a direction, there is a nonzero acceleration whenever the direction of velocity changes—even if the speed of the object remains constant.

Velocity can be written as a magnitude v multiplied by a direction $\hat{v}$

$$\vec{v} = v\hat{v}$$

We can take the derivative by applying the product rule, that is,

$$\frac{d}{dt}(xy) = \frac{dx}{dt}y + \frac{dy}{dt}x \qquad (3\text{-}27)$$

So the derivative with respect to time of the velocity vector is

$$\vec{a} = \frac{d}{dt}(v\hat{v}) = \frac{dv}{dt}\hat{v} + \frac{d\hat{v}}{dt}v \qquad (3\text{-}28)$$

The expression clearly shows that acceleration is nonzero when either speed (dv/dt), direction ($d\hat{v}/dt$), or both are changing.

We could also study the acceleration resulting from a change in direction by considering changes in the x and y components of the velocity vector. For two-dimensional motion, we have

$$a_x = \frac{dv_x}{dt} \qquad (3\text{-}29)$$

and

$$a_y = \frac{dv_y}{dt} \qquad (3\text{-}30)$$

$\sqrt{x}$ *See the Math Tutorial for more information on Differential Calculus*

Even if the speed

$$v = \sqrt{v_x^2 + v_y^2}$$

remains constant while the direction changes, the acceleration is not zero.

Consider a small bead sliding around a circular wire at constant speed v_0. **Figure 3-31a** shows the velocity vector of the bead at two positions along the wire. Because the direction of the bead is constantly changing, it must be accelerating. We determine the acceleration in three steps. First, we find the change in the x and y components of velocity between the two positions. Those form the numerators in Equations 3-29 and 3-30, respectively. Then, we divide each velocity difference by the elapsed time Δt that it takes the bead to move from point 1 to point 2. The Δt forms the denominators in the two equations. Finally, to make the fractions become derivatives, we take the limit as Δt gets small

$$a_x = \frac{dv_x}{dt} = \lim_{\Delta t \to 0} \frac{\Delta v_x}{\Delta t} = \lim_{\Delta t \to 0} \frac{v_{2,x} - v_{1,x}}{\Delta t}$$

and

$$a_y = \frac{dv_y}{dt} = \lim_{\Delta t \to 0} \frac{\Delta v_y}{\Delta t} = \lim_{\Delta t \to 0} \frac{v_{2,y} - v_{1,y}}{\Delta t}$$

As the elapsed time Δt gets shorter and shorter, the two points move closer and closer together, each approaching the point at the top of the circle. So in the limit, we find the acceleration at that specific point on the circle where $x = 0$ and $y = r$.

The x and y components of $\vec{v}_1$ and $\vec{v}_2$ are shown in **Figure 3-31b**. In that figure, we labeled the angles between the vectors and the horizontal θ, the same θ shown

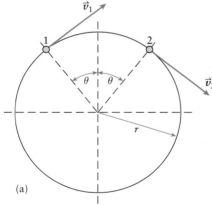

(a)

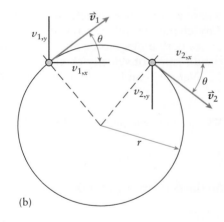

(b)

Figure 3-31 (a) A small yellow bead is sliding around a circular wire at constant speed v_0. Here we see the velocity vector of the bead at two positions along the wire. Because the direction of the bead's velocity vector is constantly changing, the bead must be accelerating.
(b) The magnitudes of the x and y components of $\vec{v}_1$ and $\vec{v}_2$ are shown by straight lines. The angles θ between the vectors and the horizontal are the same θ as shown in part (a).
(c) The proof that the angles are indeed equal is shown here, where the angle α between the vector and horizontal is seen to be equal to the central angle θ.

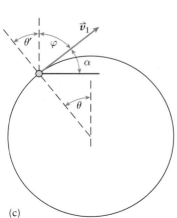

(c)

Because the velocity vector is tangent to the circle, it is perpendicular to any radius. So $\theta' + \varphi = 90°$. Because the angles φ and α together are bounded by a horizontal line and a vertical line, they also add to 90°: $\alpha + \varphi = 90°$.

Therefore $\theta' = \alpha$

Finally, angles θ and θ' are alternate interior angles formed by a traversal crossing two parallel lines and are therefore equivalent.

So $\theta = \alpha$

in Figure 3-31a. The proof that the angles are indeed equal is shown in **Figure 3-31c**, where the angle α between the vector and horizontal is shown to be equal to the central angle θ. So a_x and a_y are

$$a_x = \lim_{\Delta t \to 0} \frac{v_2 \cos\theta - v_1 \cos\theta}{\Delta t}$$

and

$$a_y = \lim_{\Delta t \to 0} \frac{v_2 \sin\theta - v_1 \sin\theta}{\Delta t}$$

The magnitude of the bead's velocity does not change, so $v_1 = v_2 = v_0$ and therefore $a_x = 0$. However, because $\vec{v}_1$ points up while $\vec{v}_2$ points down, $v_1 = v_0$ but $v_2 = -v_0$, so

$$a_y = \lim_{\Delta t \to 0} \frac{-v_0 \sin\theta - v_0 \sin\theta}{\Delta t} = \lim_{\Delta t \to 0} \frac{-2v_0 \sin\theta}{\Delta t} \qquad (3\text{-}31)$$

To carry out the limit, we need to write Δt in terms of the given variables v_0 and θ. First, because the magnitude of velocity is not changing, v_0 multiplied by Δt gives the separation distance s between the two points

$$s = v_0 \Delta t$$

In addition, the length of an arc of a circle is given by the radius multiplied by the angle subtended (in radians), as shown in **Figure 3-32**. Because the angle that subtends the distance, from Figure 3-31a, is 2θ, the arc length s is therefore equal to $r(2\theta)$. The time interval Δt can therefore be rewritten as

$$s = r(2\theta) = v_0 \Delta t$$

or

$$\Delta t = \frac{r(2\theta)}{v_0}$$

So Equation 3-31 becomes

$$a_y = \lim_{2\theta \to 0} \frac{-2v_0 \sin\theta}{(r(2\theta)/v_0)} = \lim_{\theta \to 0} \frac{-v_0^2}{r} \frac{\sin\theta}{\theta} \qquad (3\text{-}32)$$

Notice that because we expressed Δt in terms of θ, we changed the limit as well.

At first glance it may appear that the limit in Equation 3-32 approaches infinity because the denominator approaches zero. However, $\sin\theta$ in the numerator also approaches 0, so the limit should be determined using *L'Hôspital's rule*, which relates the limit of a fraction to the limit of the derivatives of the numerator and denominator.

 See the Math Tutorial for more information on Differential Calculus

The length of an arc of a circle is equal to the radius multiplied by the angle (measured in radians)

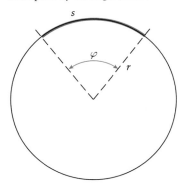

The ratio of the arc length s to the total length of the circle is the same as the ratio of the angle subtended φ to the total angle of the circle:

$$\frac{s}{2\pi r} = \frac{\varphi}{2\pi}$$

So,

$$s = r\varphi$$

Figure 3-32 The length of the arc s of a circle, shown in red, is equal to the radius r multiplied by the angle φ measured in radians.

If $f(x)$ and $g(x)$ approach 0 as x approaches c, then

$$\lim_{x \to c} \frac{f(x)}{g(x)} = \lim_{x \to c} \frac{f'(x)}{g'(x)}$$

where $f'(x) = df/dx$ and $g'(x) = dg/dx$.

In Equation 3-32, the numerator is $f(\theta) = -v_0^2 \sin \theta$, so $f'(\theta) = -v_0^2 \cos \theta$. The denominator is $g(\theta) = r\theta$, so $g'(\theta) = r$. Thus

$$a_y = \lim_{\theta \to 0} \frac{-v_0^2 \sin \theta}{r\theta} = \lim_{\theta \to 0} \frac{-v_0^2 \cos \theta}{r}$$

As θ approaches 0, the limit of $\cos \theta$ is 1, so

$$a_y = \lim_{\theta \to 0} \frac{-v_0^2 (1)}{r} = \frac{-v_0^2}{r}$$

The minus sign indicates that the vector $\vec{a}_y$ points in the negative y direction; in other words, it points toward the center of the circle.

The magnitude of the acceleration is

$$a = \sqrt{a_x^2 + a_y^2} = \sqrt{0^2 + \left(\frac{-v_0^2}{r} \right)^2}$$

$$= \frac{v_0^2}{r} \tag{3-33}$$

By taking the limit as the separation between the two points approaches zero, we have determined a_x and a_y at the top of the circular path. However, although the specific relationships apply only to one point on the circle, the conclusion that the acceleration points toward the center of the circle applies at any point, as does Equation 3-33 for the magnitude of the acceleration. In general, the **acceleration of an object moving in uniform circular motion** is

$$\vec{a}_r = \frac{v_0^2}{r}(-\hat{r}) \tag{3-34}$$

It is convention to associate the positive radial direction as away from the center, so $-\hat{r}$ indicates that the acceleration vector points toward the center of the circular path.

Example 3-8 Artificial Gravity

During the 1960s, NASA scientists began to explore mechanisms to simulate the effect of Earth's gravity in space. In movies, ring-shaped space stations such as the one that appeared in *2001: A Space Odyssey* simulated gravity by rotating. The fictional space station is described as having a diameter of 300 yards or about 275 m. If the acceleration at the edge of the ring is g (9.8 m/s²), how long would it take the station to make one complete revolution?

SET UP
Any object on the outer edge of the ring—the floor of the space station—is moving in uniform circular motion and therefore experiences acceleration according to Equation 3-34. Because the direction is in the negative $\hat{r}$ direction, this radial acceleration points toward the center of the ring. How would an inward-pointing acceleration result in a feeling similar to gravity? The floor pushing against the astronaut causes the acceleration. This pushing is toward the center of the ring, but the sensation of the floor pushing up is equivalent to feeling pushed against the floor—simulated gravity! We will be more precise about the direction of radial acceleration in Chapter 4.

Our goal is to determine the rotation rate of the space station so that the magnitude of the radial acceleration a equals the magnitude of the acceleration due to gravity g at the surface of Earth. The magnitude of the acceleration,

$$a = \frac{v_0^2}{r} \qquad (3\text{-}35)$$

does not contain a reference to rotation rate; however, the rate must be related to the rotation velocity v_0. The time T it takes for a point on the edge to rotate around one full circumference of the space station is

$$T = \frac{\pi D}{v_0} \qquad (3\text{-}36)$$

where D is the diameter of the ring. In general, the time it takes for a motion to repeat, such as a point on the edge of the space station to rotate back around to its initial position, is called the **period** of the motion. We can now combine Equations 3-35 and 3-36 to solve for the period in terms of a.

SOLVE
If we solve Equation 3-35 for velocity,

$$v_0 = \sqrt{ar} = \sqrt{a\frac{D}{2}}$$

then the time for one rotation becomes

$$T = \frac{\pi D}{v_0} = \frac{\pi D}{\sqrt{a\dfrac{D}{2}}}$$

To simulate $a = g = 9.8$ m/s² for a space station of diameter D equal to 275 m, the time for one rotation is then

$$T = \frac{\pi\,275 \text{ m}}{\sqrt{9.8\dfrac{\text{m}}{\text{s}^2}\dfrac{275 \text{ m}}{2}}} = 23.5 \text{ s}$$

For an object at the edge of the ring to experience an acceleration of 9.8 m/s², the station must rotate once every 23.5 s.

REFLECT
It may be hard to judge whether one rotation every 23.5 s is fast or slow. To get a better sense of the answer, convert the time for one rotation to velocity using Equation 3-36

$$v = \frac{\pi D}{T} = \frac{\pi\,275 \text{ m}}{23.5 \text{ s}} = 36.8 \text{ m/s}$$

This velocity is higher than the recommended top speed on the German autobahn—it's fast.

For science fiction purists, the fictional space station is described as rotating more slowly than the speed we calculated, because the rotation is intended to simulate gravity on the Moon, which is about one-sixth that on Earth.

Practice Problem 3-8 Riders on a roller coaster at Oktoberfest in Munich (**Figure 3-33**) move in several tight vertical circles. The cars on the ride travel at about 15 m/s around each loop; the loops are approximately circular with a diameter of about 15 m. Estimate the radial acceleration, in terms of g, that riders feel as they pass through the point midway up from the bottom of one of the loops.

Figure 3-33 Riders on a roller coaster at Oktoberfest in Munich experience circular motion as they traverse vertical loops. *(Amazing Images/Alamy.)*

> ✱ **What's Important 3-5**
>
> Acceleration and velocity are both vectors. Even if the speed of an object remains constant, it experiences an acceleration if its direction changes. The acceleration of an object moving in uniform circular motion is radially in, toward the center of the circle.

Answers to Practice Problems

3-1	Magnitude 4.8 m, direction 72°	**3-5**	45 m
3-2	63.4°	**3-6**	2.1 s
3-3	100 m	**3-8**	3.1 g
3-4	7.3 m		

Answers to Got the Concept Questions

3-1 Any vector that has a nonzero magnitude will always have a component of length zero in the direction perpendicular to the vector. So yes, it's always possible to find a direction in which a vector has a magnitude of zero. An example is shown in Figure 3-34.

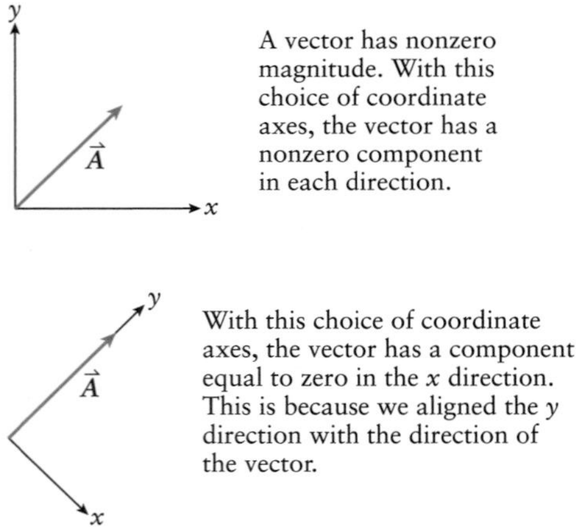

Figure 3-34

A vector has nonzero magnitude. With this choice of coordinate axes, the vector has a nonzero component in each direction.

With this choice of coordinate axes, the vector has a component equal to zero in the x direction. This is because we aligned the y direction with the direction of the vector.

3-2 The x component of acceleration is zero everywhere, because no force acts on the projectile horizontally. (To be clear, this is only in the ideal case of no wind and no air resistance.) The y component of acceleration, however, is *never* zero. Don't fall into a trap: As discussed in the Watch Out at the end of Chapter 2, the vertical acceleration is not zero at the peak of the motion. Indeed, the y component of acceleration is the constant g and never changes. (It's constant!) Finally, because the y component of acceleration is never zero, the total acceleration can't be zero, either.

3-3 The sign of velocity gives its direction. Our convention is to label upward velocity as positive. An object going up has positive velocity, and as the object falls back down from its peak its velocity is an increasingly negative number. Its final velocity, right before striking the ground, is the most negative and therefore the smallest velocity attained (a negative number is smaller than a positive one). However, the *magnitude* of velocity, which is not a signed value, is smallest at the peak of motion. At its peak, just for an instant, the object is neither going up nor down so the y component of its velocity is zero. The magnitude of the velocity is, from Equation 3-3, $v = \sqrt{(v_x^2 + v_y^2)}$. Because there is no acceleration in the x direction, v_x remains the same throughout the motion, so v is smallest when v_y is zero. For the magnitude of velocity to be zero, both v_x and v_y must be zero. The y component of velocity is zero at the peak, but because the x component of velocity does not change, v_x can only be zero if it was zero initially, or if the object was launched straight up.

SUMMARY

Topic	Summary	Equation or Symbol
component of a vector	The projection of a vector along a specific axis or in a specific direction is that vector's component along that axis. We denote a vector component with a subscript to indicate the direction. For example, A_x is the projection onto the x axis of a vector with magnitude A.	$A_{\text{dir}} = A \cos \theta$ for a vector $\vec{A}$ that makes an angle θ with respect to a particular direction.
free-fall motion	The motion of objects in free fall is influenced only by the effect of gravity.	
horizontal range	Horizontal range is the horizontal distance that a projectile travels before returning to its launch height.	
period	The period of a motion is the time it takes for a motion to repeat, for example, for a rotating object to complete one full rotation.	T
projectile motion	We refer to the trajectory followed by objects launched with no independent source of thrust as projectile motion.	$v_x = v_{0x}$ (3-16) $\\ x - x_0 = v_{0x}t$ (3-17) $\\ v_y = v_{0y} - gt$ (3-14) $\\ y - y_0 = v_{0y}t - \frac{1}{2}gt^2$ (3-15)
scalar	A scalar is a simple number that tells us the magnitude of a quantity but nothing about the quantity's direction.	
uniform circular motion (acceleration)	The direction of an object moving in a circle is constantly changing; the velocity vector is therefore also constantly changing. This is true even when the magnitude of velocity is constant. An object moving in uniform circular motion is therefore accelerating. The magnitude of the acceleration is proportional to the square of the speed v_0 and inversely proportional to the radius r of the motion. The direction of the acceleration vector is inward, toward the center of the motion.	$\vec{a}_r = \dfrac{v_0^2}{r}(-\hat{r})$ (3-34)
unit vector	The purpose of a unit vector is to define a direction in space. A unit vector has a direction and a magnitude equal to 1. A unit vector is dimensionless. Any vector can therefore be described as the product of that vector's magnitude and a unit vector that defines the vector's direction. We denote a unit vector by adding a circumflex above a variable. For example, $\hat{x}$ represents a unit vector in the (positive) x direction.	

vector	A vector is used to describe a quantity that has both magnitude and direction. We denote a vector by adding a half arrow above a variable name written in bold. For example, $\vec{A}$ represents the vector quantity A.	If vector $\vec{A}$ has magnitude A and direction defined by the unit vector $\hat{A}$ then $\vec{A} = A\,\hat{A}$
vector sum	Two vectors are added by first resolving each vector into its x and y components and then adding each component separately.	For $\vec{C} = \vec{A} + \vec{B}$ $C_x = A_x + B_x$ and $C_y = A_y + B_y$

QUESTIONS AND PROBLEMS

In a few problems, you are given more data than you actually need; in a few other problems, you are required to supply data from your general knowledge, outside sources, or informed estimate.

Interpret as significant all digits in numerical values that have trailing zeros and no decimal points.

For all problems, use $g = 9.8$ m/s^2 for the free-fall acceleration due to gravity.

- • Basic, single-concept problem
- •• Intermediate-level problem, may require synthesis of concepts and multiple steps
- ••• Challenging problem
SSM *Solution is in Student Solutions Manual*

Conceptual Questions

1. •(a) Can the sum of two vectors that have different magnitudes ever be equal to zero? If so, give an example. If not, explain why the sum of two vectors cannot be equal to zero. (b) Can the sum of three vectors that have different magnitudes ever be equal to zero? SSM

2. •What is the difference between a scalar and a vector? Give an example of a scalar and an example of a vector.

3. •Describe a situation in which the average velocity and the instantaneous velocity vectors are identical. Describe a situation in which these two velocity vectors are different. SSM

4. •(a) Explain the difference between an object undergoing uniform circular motion and an object experiencing projectile motion. (b) In what ways are these kinds of motion similar?

5. •Consider the effects of air resistance on a projectile. Describe qualitatively how the projectile's velocities and accelerations in the vertical and horizontal directions differ when the effects of air resistance are ignored and when the effects are considered.

6. •**Astro** If you were playing tennis on the Moon, what adjustments would you need to make in order for your shots to stay within the boundaries of the court? Would the trajectories of the balls look different on the Moon compared to on Earth?

7. •Explain what is meant by the magnitude of a vector.

8. •During the motion of a projectile, which of the following quantities are constant during the flight: x, y, v_x, v_y, a_x, a_y? (Neglect any effects due to air resistance.)

9. •For a given, fixed launch speed, at what angle should you launch a projectile to achieve (a) the longest range, (b) the longest time of flight, and (c) the greatest height? (Neglect any effects due to air resistance.) SSM

10. •A rock is thrown from a bridge at an angle 20° below horizontal. At the instant of impact, is the rock's speed greater than, less than, or equal to the speed with which it was thrown? Explain your answer. (Neglect any effects due to air resistance.)

11. •**Sports** A soccer player kicks a ball at an angle 60° from the ground. The soccer ball hits the ground some distance away. Is there any point at which the velocity and acceleration vectors are perpendicular to each other? Explain your answer. (Neglect any effects due to air resistance.)

12. •**Sports** Suppose you are the coach of a champion long jumper. Would you suggest that she take off at an angle less than 45°? Why or why not?

13. •An ape swings through the jungle by hanging from a vine. At the lowest point of its motion, is the ape accelerating? If so, what is the direction of its acceleration? SSM

14. •A cyclist rides around a flat, circular track at constant speed. Is his acceleration vector zero? Explain your answer.

15. •You are driving your car in a circular path on flat ground with a constant speed. At the instant you are driving north and turning right, are you accelerating? If so, what is the direction of your acceleration at that moment? If not, why not?

Multiple-Choice Questions

16. •Which of the following is not a vector?
   A. average velocity
   B. instantaneous velocity
   C. distance
   D. displacement
   E. acceleration

17. •Vector $\vec{A}$ has an x component and a y component that are equal in magnitude. Which of the following is the angle for vector $\vec{A}$ in the same x–y coordinate system? SSM
   A. 0°
   B. 45°
   C. 60°
   D. 90°
   E. 120°

18. •The vector in **Figure 3-35** has a length of 4.0 units and makes a 30° angle with respect to the y axis. What are the x component and y component of the vector?
   A. 3.5, 2.0
   B. −2.0, 3.5
   C. −3.5, 2.0
   D. 2.0, −3.5
   E. −3.5, −2.0

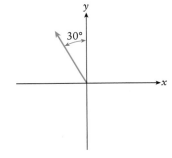

Figure 3-35 Problem 18

19. •The acceleration of a particle in projectile motion
   A. points along the parabolic path of the particle.
   B. is directed horizontally.
   C. vanishes at the particle's highest point.
   D. is vertically downward.
   E. is zero.

20. •Adam drops a ball from rest from the top floor of a building at the same time Bob throws a ball horizontally from the same location. Which ball hits the ground first? (Neglect any effects due to air resistance.)
   A. Adam's ball
   B. Bob's ball
   C. They both hit the ground at the same time.
   D. It depends on how fast Bob throws the ball.
   E. It depends on how fast the ball falls when Adam drops it.

21. •**Sports** Two golf balls are hit from the same point on a flat field. Both are hit at an angle of 30° above the horizontal. Ball 2 has twice the initial speed of ball 1. If ball 1 lands a distance d_1 from the initial point, at what distance d_2 does ball 2 land from the initial point? (Neglect any effects due to air resistance.) SSM

A. $d_2 = 0.5d_1$
B. $d_2 = d_1$
C. $d_2 = 2d_1$
D. $d_2 = 4d_1$
E. $d_2 = 8d_1$

22. •A zookeeper is trying to shoot a monkey sitting at the top of a tree with a tranquilizer gun. If the monkey drops from the tree at the same instant that the zookeeper fires, where should the zookeeper aim if he wants to hit the monkey? (Neglect any effects due to air resistance.)
   A. Aim straight at the monkey.
   B. Aim lower than the monkey.
   C. Aim higher than the monkey.
   D. Aim to the right of the monkey.
   E. It's impossible to determine.

23. •The acceleration vector of a particle in uniform circular motion
   A. points along the circular path of the particle and in the direction of motion.
   B. points along the circular path of the particle and opposite the direction of motion.
   C. is zero.
   D. points toward the center of the circle.
   E. points outward from the center of the circle.

24. •If the speed of an object in uniform circular motion is constant and the radial distance is doubled, the magnitude of the radial acceleration decreases by what factor?
   A. 2
   B. 3
   C. 4
   D. 6
   E. 1

25. •You toss a ball into the air at an initial angle 40° from the horizontal. At what point in the ball's trajectory does the ball have the smallest speed? (Neglect any effects due to air resistance.)
   A. just after it is tossed
   B. at the highest point in its flight
   C. just before it hits the ground
   D. halfway between the ground and the highest point on the rise portion of the trajectory
   E. halfway between the ground and the highest point on the fall portion of the trajectory

Estimation/Numerical Analysis

26. •If $\vec{r}$ has a magnitude of 24 and points in a direction 36° south of west, find the vector components of $\vec{r}$. Use a protractor and some graph paper to verify your answer by drawing $\vec{r}$ and measuring the length of the lines representing its components.

27. •A vector $\vec{r}$ points in the northwesterly direction and makes a 30° angle with respect to the x axis. Find

This page is intentionally left blank.

For complete end of chapter problem sets, please go to
www.whfreeman.com/kestentauck

41. ••The two vectors shown in **Figure 3-40** represent the initial and final velocities of an object during a trip that took 5 s. Find the average acceleration. Is it possible to determine whether the acceleration was uniform from the information given in the problem? SSM

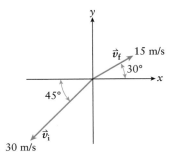

Figure 3-40 Problem 41

42. ••An object travels with a constant acceleration for 10 s. The vectors in **Figure 3-41** represent the final and initial velocities. Make a graph of the x component of the velocity versus time, the y component of the velocity versus time, and the y component of the acceleration versus time. Make certain that your graph is accurate.

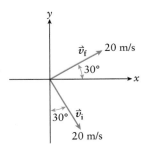

Figure 3-41 Problem 42

43. ••An object experiences a constant acceleration of 2.0 m/s² along the $-x$ axis for 2.7 s, attaining a velocity of 16 m/s in a direction 45° from the $+x$ axis. Find the initial velocity vector of the object.

44. ••Cody starts at a point 6 km to the east and 4 km to the south of a location that represents the origin of a coordinate system for a map. He ends up at a point 10 km to the west and 6 km to the north of the map origin. (a) Find his average velocity if the trip took him 4 h to complete. (b) Cody walks to his destination at a constant rate. His friend Marcus covers the distance with a combination of jogging, walking, running, and resting so that the total trip time is also 4 h. How do their average velocities compare?

3-4: Projectile Motion

45. •An object is undergoing parabolic motion as shown from the side in **Figure 3-42**. Assume the object starts its motion on ground level. For the five positions shown, draw to scale vectors representing the magnitudes of (a) the x components of the velocity, (b) the y components of the velocity, and (c) the accelerations. (Neglect any effects due to air resistance.)

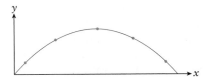

Figure 3-42 Problem 45

46. •An object undergoing parabolic motion travels 100 m in the horizontal direction before returning to its initial height. If the object is thrown initially at a 30° angle, determine the x component and the y component of the initial velocity. (Neglect any effects due to air resistance.)

47. •Five balls are thrown off a cliff at the angles shown in **Figure 3-43**. Each has the same initial velocity. Rank (a) the horizontal distance traveled, (b) the time required for each to hit the ground, and (c) the magnitude of the velocity when each hits the ground. (Neglect any effects due to air resistance.) SSM

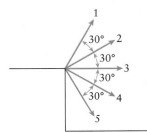

Figure 3-43 Problem 47

48. •A Chinook salmon can jump out of water with a speed of 6.3 m/s. How far horizontally can a Chinook salmon travel through the air if it leaves the water with an initial angle 40°? (Neglect any effects due to air resistance.)

49. •A tiger leaps horizontally out of a tree that is 4.00 m high. If he lands 5.00 m from the base of the tree, calculate his initial speed. (Neglect any effects due to air resistance.) SSM

50. •A football is punted at 25.0 m/s and at an angle of 30.0° above the horizon. What is the velocity vector of the ball when it is 5.00 m above ground level? Assume it starts 1.00 m above ground level. (Neglect any effects due to air resistance.)

51. ••A dart is thrown at a dartboard 2.37 m away. When the dart is released at the same height as the center of the dartboard, it hits the center in 0.447 s. At what angle relative to the floor was the dart thrown? (Neglect any effects due to air resistance.)

3-5: Uniform Circular Motion

52. •A ball attached to a string is twirled in a circle of radius 1.25 m. If the constant speed of the ball is 2.25 m/s, what is the period of the circular motion?

53. •A ball spins on a 0.870-m-long string with a speed of 3.36 m/s. Calculate the acceleration of the ball. Be sure to specify the direction of the acceleration. SSM

54. •A car travels around a circular track at a constant speed v (**Figure 3-44**). When there is (a) a change in the magnitude of the average velocity, (b) a change in the direction of the average velocity, (c) a change in the magnitude of the average acceleration, or (d) a change in the direction of the average acceleration, which of the following would occur?

    A. The car maintains a constant speed $2v$ while traveling from the starting position to point 1.

    B. The car gradually speeds up from rest to a speed v while traveling from the starting position to point 1.

    C. The car maintains a constant speed $\frac{1}{2}v$ while traveling from the starting position to point 3.

    D. The car gradually slows down to a speed of $\frac{1}{2}v$ while traveling one revolution from the starting position.

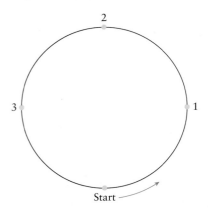

Figure 3-44 Problem 54

55. •A washing machine drum 80.0 cm in diameter starts from rest and achieves 1200 rev/min in 22 s. Assuming the acceleration of the drum is constant, calculate the net acceleration (magnitude and direction) of a point on the drum after 1.00 s has elapsed.

56. •A $\frac{1}{4}$-in-diameter drill bit accelerates from rest up to 800 rev/min in 4.33 s. Calculate the acceleration of a point on the edge of the bit once it has achieved its operating speed.

57. •In 1892 George W. G. Ferris designed a carnival ride in the shape of a large wheel. This Ferris wheel had a diameter of 76 m and rotated one revolution every 20 min. What was the magnitude of the acceleration that riders experienced? SSM

58. •Riders on a Ferris wheel of diameter 16 m move in a circle with a radial acceleration of 2.0 m/s². What is the speed of the Ferris wheel?

59. •A car races at a constant speed of 330 km/h around a flat, circular track 1.00 km in diameter. What is the car's radial acceleration in m/s²?

60. •Mary and Kelly decide they want to run side by side around a track. Mary runs in the inside lane of the track while Kelly runs in one of the outer lanes. What is the ratio of their accelerations?

61. •Astro The space shuttle is in orbit about 300 km above the surface of Earth. The period of the orbit is about 5.43×10^3 s. What is the acceleration of the shuttle? (The radius of Earth is 6.38×10^6 m.) SSM

62. •Astro We know that the Moon revolves around Earth during a period of 27.3 days. The average distance from center of Earth to the center of the Moon is 3.84×10^8 m. What is the acceleration of the Moon due to its motion around Earth?

63. ••Calculate the accelerations of (a) Earth as it orbits the Sun, (b) an artificial satellite as it orbits 320 km above Earth's surface, (c) the Moon as it orbits Earth, and (d) a car traveling along a circular path that has a radius of 50 m at a speed of 20 m/s. Identify the smallest acceleration and calculate the factor by which the other three accelerations are larger.

64. •Commercial ultracentrifuges can rotate at rates of 100,000 rpm (revolutions per minute). As a consequence, they can create accelerations on the order of 800,000g. (A "g" represents an acceleration of 9.8 m/s².) Find the distance from the rotation axis of the sample chamber in such a device. Calculate the speed of an object traveling under the given conditions.

65. •Biology In a vertical dive a peregrine falcon can accelerate at 0.6 times the free-fall acceleration (that is, at 0.6 g) in reaching a speed of about 100 m/s. If a falcon pulls out of a dive into a circular arc at this speed, and can sustain a radial acceleration of 0.6 g, what is the radius of the turn? SSM

General Problems

66. ••You observe two cars traveling in the same direction on a long, straight section of Highway 5. The red car is moving at a constant v_R equal to 34 m/s and the blue car is moving at constant v_B equal to 28 m/s. At the moment you first see them, the blue car is 24 m ahead of the red car. (a) How long after you first see the cars does the red car catch up to the blue car? (b) How far did the red car travel between when you first saw it and when it caught up to the blue car? (c) Suppose the red car started to accelerate at a rate of a equal to $\frac{4}{3}$ m/s² just at the moment you saw the cars. How long after that would the red car catch up to the blue car?

67. ••An experiment to measure the value of g is constructed using a tall tower outfitted with two sensing devices, one a distance H above the other. A small ball is fired straight up in the tower so that it rises to near the top and then falls back down; each sensing device reads out the time that elapses between the ball going up past the sensor and back down past the sensor. (a) It takes a time $2t_1$ for the ball to rise past and then come back down past the lower sensor, and a time $2t_2$ for the ball to rise past and then come back down past the upper sensor. Find an expression for g using these times and the height H. (b) Determine the value of g if H equals 25 m, t_1 equals 3 s, and t_2 equals 2 s.

68. •You drop a rock from rest from the top of a tall building. (a) How far has the rock fallen in 2.50 s? (b) What is the velocity of the rock after it has fallen 11.0 m? (c) It takes 0.117 s for the rock to pass a 2.00-m high window. How far from the top of the building is the top of the window?

69. ••Sports Steve Young stands on the 20-yard line, poised to throw long. He throws the ball at initial velocity v_0 equal to 15.0 m/s and releases it at an angle θ equal to 45.0°. (a) Having faked an end around, Jerry Rice comes racing past Steve at a constant velocity V_J equal to 8.00 m/s, heading straight down the field. Assuming that Jerry catches the ball at the same height above the ground that Steve throws it, how long must Steve wait to throw, after Jerry goes past, so that the ball falls directly into Jerry's hands? (b) As in part (a), Jerry is coming straight past Steve at V_J equal to 8.00 m/s. But just as Jerry goes past, Steve starts to run in the same direction as Jerry with V_S equal to 1.50 m/s. How long must Steve wait to release the ball so that it falls directly into Jerry's hands? SSM

70. ••You throw a ball from the balcony onto the court in the basketball arena. You release the ball at a height of 7 m above the court, with an initial velocity equal to 9 m/s at 33° above the horizontal. A friend of yours, standing on the court 11 m from the point directly beneth you, waits for a period of time after you release the ball and then begins to move directly away from you at an accelaration of 1.8 m/s². (She can only do this for a short period of time!) If you throw the ball in a line with her, how long after you release the ball should she wait to start running directly away from you so that she'll catch the ball exactly 1 m above the floor of the court?

71. ••Marcus and Cody want to hike to a destination 12 km north of their starting point. Before heading directly to the destination, Marcus walks 10 km in a direction that is 30° north of east and Cody walks 15 km in a direction that is 45° north of west. How much farther must each hike on the second part of the trip?

72. •Nathan walks due east a certain distance and then walks due south twice that distance. He finds himself 15 km from his starting position. How far east and how far south does Nathan walk?

73. •••A group of campers must decide the quickest way to reach their next campsite. (Figure 3-45) is a map of the area. One option is to walk directly to the site along a straight path 10.6 mi in length. Another option is to take a canoe down a river and then walk uphill 6.6 mi from the beach to the campsite. The campers estimate a hiking pace of mi/h on the straight path and 0.5 mi/h walking up the hill. How fast would the canoe need to

travel (assume a constant speed) in order for the second route to take less time than the first?

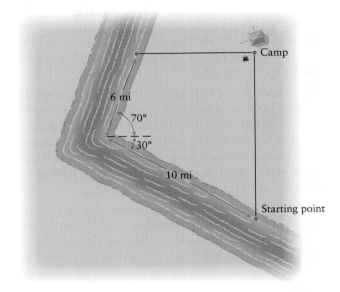

Figure 3-45 Problem 73

74. •A rock is thrown from the upper edge of a 75.0-m vertical dam with a speed of 25.0 m/s at 65.0° above the horizon. How long after throwing the rock will you (a) see it and (b) hear it hit the water flowing out at the base of the dam? The speed of sound in the air is 344 m/s. (Neglect any effects due to air resistance.)

75. ••A water balloon is thrown horizontally at a speed of 2.00 m/s from the roof of a building that is 6.00 m above the ground. At the same instant the balloon is released, a second balloon is thrown straight down at 2.00 m/s from the same height. Determine which balloon hits the ground first and how much sooner it hits the ground than the other balloon. Which balloon is moving with the fastest speed at impact? (Neglect any effects due to air resistance.) SSM

76. •An airplane releases a ball as it flies parallel to the ground at a height of 255 m (Figure 3-46). If the ball lands on the ground exactly 255 m from the release point, calculate the airspeed of the plane. (Neglect any effects due to air resistance.)

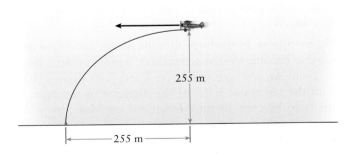

Figure 3-46 Problem 76

77. ••An airplane flying upward at 35.3 m/s and an angle of 30.0° relative to the horizontal releases a ball when it is 255 m above the ground. Calculate (a) the time of it takes the ball to hit the ground, (b) the maximum height of the ball, and (c) the horizontal distance the ball travels from the release point to the ground. (Neglect any effects due to air resistance.)

78. ••Sports In 1993, Javier Sotomayor set a world record of 2.45 m in the men's outdoor high jump. He is 193 cm (6 ft 4 in.) tall. By treating his body as a point located at half his height, and given that he left the ground a horizontal distance from the bar of 1.5 m at a takeoff angle of 65°, determine Javier Sotomayor's takeoff speed. (Neglect any effects due to air resistance.)

79. •Sports Gabriele Reinsch threw a discus 76.80 m on July 9, 1988, to set the women's world record. Assume that she launched the discus with an elevation angle of 45°, and that her hand was 2.0 m over the ground at the instant of launch. What is the initial speed of the discus required to achieve that range? (Neglect any effects due to air resistance.) SSM

80. ••Sports A boy runs straight off the end of a diving platform at a speed of 5.00 m/s. The platform is 10.0 m above the surface of the water. (a) Calculate the boy's speed when he hits the water. (b) How much time is required for the boy to reach the water? (c) How far horizontally will the boy travel before he hits the water? (Neglect any effects due to air resistance.)

81. •Astro The froghopper, a tiny insect, is a remarkable jumper. Suppose you raised a colony of the little critters on the Moon, where the acceleration due to gravity is only 1.62 m/s². If on Earth a froghopper's maximum height is h and maximum horizontal range is R, what would its maximum height and range be on the Moon in terms of h and R? Assume a froghopper's takeoff speed is the same on the Moon and on Earth.

82. ••Sports In 1998, Jason Elam kicked a record field goal. The football started on the ground 63.0 yards from the base of the goal posts and just barely cleared the 10-ft-high bar. If the initial trajectory of the football was 40.0° above the horizontal, (a) what was its initial speed and (b) how long after the ball was struck did it pass through the goal posts? (Neglect any effects due to air resistance.)

83. •••Sports In the hope that the Moon and Mars will one day become tourist attractions, a golf course is built on each. An average golfer on Earth can drive a ball from the tee about 63% of the distance to the hole. If this is to be true on the Moon and on Mars, by what factor should the dimensions of the golf courses on the Moon and Mars need to be changed relative to a course on Earth? The Moon has no atmosphere, and the effects of the thin atmosphere on Mars can be neglected for this problem.

84. •Medical In a laboratory test of tolerance for high angular acceleration, pilots were swung in a circle 13.4 m in diameter. It was found that they blacked out when they were spun at 30.6 rpm (rev/min). (a) At what acceleration (in SI units and in multiples of g) did the pilots black out? (b) If you want to decrease the acceleration by 25.0% without changing the diameter of the circle, by what percent must you change the time for the pilot to make one spin?

85. ••Biology Anne is working on a research project that involves the use of a centrifuge. Her samples must first experience an acceleration of $100g$ but then the acceleration must increase by a factor of eight. By how much will the rotation speed have to increase? Express your answer as a fraction of the initial rotation rate. SSM

86. •Medical Modern pilots can survive radial accelerations up to $9g$ (88 m/s²). Can a fighter pilot flying at a constant speed of 500 m/s and in a circle that has a diameter of 8800 m survive to tell about his experience?

87. •Sports A girl's fast pitch softball player does a windmill pitch, moving her hand through a circular arc with her arm straight. She releases the ball at a speed of 24.6 m/s. Just before the ball leaves her hand, the ball's radial acceleration is 1960 m/s². What is the length of her arm from the pivot point at her shoulder?

88. •••Calc A ball is thrown above an inclined plane with an initial speed of v_0 at an angle of θ as shown in Figure 3-47. The angle of the incline is α.

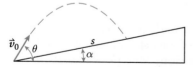

Figure 3-47 Problem 88

(a) Show that the ball travels a distance s up the incline, where

$$s = \frac{2\,v_0^2\cos\theta\sin(\theta - \alpha)}{g\cos^2\alpha}$$

(b) For what value of θ is s a maximum? What is the maximum s value? (You are required to maximize s with respect to θ for this part). Hint: $\cos(A \pm B) = \cos A\cos B \pm \sin A\sin B$ and $\sin(A \pm B) = \sin A\cos B \pm \cos A\sin B$; for example, $\cos(90° - x) = \sin x$.

4 Newton's Laws of Motion

(Mar Photographics/Alamy.)

In a tug-of-war contest, two teams pull the ends of a rope in opposite directions. When one team pulls harder than the other, the flag tied to the middle of the rope begins to move; that is, it accelerates. But if the two teams pull equally hard, the flag won't move. The physical quantity that determines how the flag moves from its initial stationary position is called *force* and has a magnitude and a direction. Equal forces acting in opposite directions can therefore cancel each other. Force is a vector!

In the previous chapters, we described some characteristics of motion without mentioning what causes it. For example, we explained that a fireworks display traces a parabolic arc across the sky because of the horizontal and vertical components of its acceleration—but what causes acceleration? What is required for a helicopter to rise into the air or for a gymnast to leap from one end of a balance beam to the other (**Figure 4-1**)? Such changes in velocity are a result of **force**.

Simply put, a force is a push or a pull. When more than one force acts on an object their directions are important; a force in one direction can cancel a force in the opposite direction. If you push on a ball, say to throw it in the air, very little force opposes your push and the ball accelerates. In contrast, if you push on a large boulder there could be enough force opposing you that no motion occurs at all. Most of the forces we will encounter are **contact forces** that involve one object touching another. Some **long-range forces**, such as gravity or the magnetic force, appear to act without any direct contact, although our understanding of forces at the fundamental level suggests that these forces are exerted when objects exchange small units of energy or matter with one another.

Figure 4-1 What forces propel the gymnast as she leaps from one end of the balance beam to the other? *(Masterfile.)*

105

In this chapter, we will examine force and how it affects acceleration.

4-1 Newton's First Law

In the late 17th century, English physicist and mathematician Sir Isaac Newton published a treatise which in part illuminated three fundamental relationships between force and motion. These ideas laid the groundwork for the physics of motion known as Newtonian mechanics; they explain nearly all physical phenomena in our everyday experience.

A central concept of Newtonian mechanics is the quantity **mass**. Mass is a measure of how much matter is present in an object—how much "stuff" it has. Mass is an intrinsic property of an object; the mass of any object is the same under all conditions. The mass of a chunk of copper is the same when it's at the bottom of the ocean as when it's on the surface of the Moon. The mass would even remain the same if the copper were placed in orbit around Earth.

For Newton, the concept of mass was intertwined with his observation that objects resist changes in their state of motion. He described this relationship as an intrinsic, unchanging property of mass and of objects. This property is defined in **Newton's first law**:

> An object at rest tends to stay at rest and an object in uniform motion tends to stay in motion with the same speed and in the same direction unless acted on by a nonzero net force.

Inertia is the tendency of an object to resist a change in motion.

Newton's first law contradicts the theories of motion developed by the ancient Greeks. Aristotle, for example, believed that every object had a natural place in the world. He believed that heavy objects, such as rocks, are naturally at rest on Earth and light objects, such as clouds, are naturally at rest in the sky. In Aristotle's view, a moving object tended to stop when it found its natural rest position. Newton, however, connected the concept of force with acceleration rather than with motion in general. He concluded that acceleration requires a nonzero net force, and motion remains constant unless a nonzero net force is applied.

? Got the Concept 4-1
On the Floor

When you suddenly apply the brakes in your car, the books that had been stacked on the back seat slide forward, hit the back of the front seat, and fall to the floor. Do the books experience a force that caused them to slide, and if so, what caused that force?

✳ What's Important 4-2

Mass is an intrinsic property of an object. A moving object will continue to move unless a net force is applied to change the motion. An object at rest will not move unless it experiences a net force.

4-2 Newton's Second Law

The relationship commonly known as **Newton's second law** is perhaps the most important for our discussion of forces. We state this law in words and then as an equation:

> The net force is the vector sum of all forces acting on a single object in a particular direction and equals the mass of the object multiplied by the acceleration of the object in that direction

$$\sum \vec{F}_{dir} = m\vec{a}_{dir} \qquad (4\text{-}1)$$

The capital Greek letter Σ (sigma) is mathematical notation for adding together terms. (See Math Box 4-1.) In this statement of Newton's second law we have mixed mathematical notation just a bit, by using the vector symbol but also adding a subscript *dir* to emphasize that we will always apply Newton's second law in only one direction at a time. In addition, although Newton's second law is sometimes presented as $F = ma$, we will almost always start our consideration of a problem from Equation 4-1 and will use $F = ma$ cautiously, because we want to ensure that we never overlook the vector nature of forces.

Math Box 4-1 Sigma

Physicists often use mathematical notation in order to represent operations that are either long or of a length not well defined. The capital Greek letter Σ (sigma), for example, is mathematical shorthand for adding together a series of terms. If N is to be found by adding four different values of n or

$$N = n_1 + n_2 + n_3 + n_4$$

the addition can be equivalently written using the summation sign

$$N = \sum_{i=1}^{4} n_i$$

Notice the role of the notation above and below the summation sign; we show the starting value of the variable that counts the terms below it and the final value above.

We might not know how many terms will be in the sum. In that case, we write

$$N = \sum n_i = n_1 + n_2 + n_3 + \cdots$$

Each force in the sum on the left side of Equation 4-1 arises as a result of the action of some agent on the object under consideration. You should be able to identify the origin of each term in the sum. The origin of a force on a block might be gravity or another block pushing on the first one. You might be the agent of a force acting on an object, if you are directly touching and pushing or pulling on the object. In general, the origin or agent of a force is that which causes the force to act on an object.

The SI unit of force is the newton (N). The unit of force in the English system is the pound (lb).

The units newtons, kilograms, meters, and seconds are related by Newton's second law (Equation 4-1). Using square brackets to indicate units,

$$\sum \vec{F}_{dir} = m\vec{a}_{dir} \rightarrow [F] = [m][a]$$

or

$$1\ N = 1\ \frac{kg \cdot m}{s^2}$$

Figure 4-2 As this fertilized sea urchin egg divides, microtubules, stained green, pull the daughter cells apart. *(Courtesy of Dr. James Grainger.)*

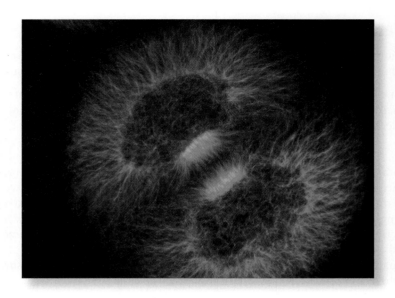

Example 4-1 Small but Powerful

Microtubules are intracellular protein molecules that help cells hold their shape (**Figure 4-2**). In addition, microtubules are responsible for various kinds of movements within cells, such as pulling apart chromosomes during cell division. Microtubules are assembled from small, identical subunits and can exert forces measured to be from a few piconewtons (pico = p = 10^{-12}) up to hundreds of nanonewtons (nano = n = 10^{-9}). The bacterium *Escherichia coli* contains a chromosome that has a mass equal to about 10^{-17} kg. If a microtubule applies a force of 1 pN to the chromosome, what would the magnitude of its acceleration be in the absence of resistive forces?

SET UP

Newton's second law states that the sum of all forces on an object in a given direction equals the mass of the object multiplied by the acceleration. Because we have chosen to neglect forces acting on the chromosome other than the one exerted by the microtubule, Equation 4-1 becomes (in the direction of the force)

$$F_{\text{microtubule}} = m_{\text{chromosome}}\, a \qquad (4\text{-}2)$$

We can rearrange this equation to find a.

SOLVE

Equation 4-2 becomes

$$a = \frac{F_{\text{microtubule}}}{m_{\text{chromosome}}}$$

Using the information in the problem statement, we find a to be

$$a = \frac{10^{-12}\,\text{N}}{10^{-17}\,\text{kg}} = 10^5\,\text{m/s}^2$$

REFLECT

This acceleration is about 10,000 times the acceleration due to gravity at the surface of Earth. Microtubules are quite strong! It is important to note, however, that the force exerted by the microtubule is opposed by resistive forces that we did not take into account, meaning that in reality the acceleration would not be this big. In addition, the accelerations generated by microtubule expansion or retraction only take place for short periods of time and over very small distances.

Practice Problem 4-1 A net force of 20 N is exerted on a rock that has a mass equal to 5.0 kg. Find the magnitude of the rock's acceleration in the absence of resistive forces.

To get a sense of Newton's second law, consider a ring of mass M that rests on a horizontal table and is pulled by three forces $\vec{F}_1$, $\vec{F}_2$, and $\vec{F}_3$. The three forces are shown in Figure 4-3 as vector arrows. Each arrow points in the direction of the force it represents, and the length of each arrow indicates the magnitude of the force. Newton's second law states that we should consider how the forces act on the ring in only one direction at a time, so let's consider Equation 4-1 in the x and y directions separately.

In the x direction, Equation 4-1 becomes

$$\sum \vec{F}_x = m\vec{a}_x$$

or

$$\vec{F}_{1x} + \vec{F}_{2x} + \vec{F}_{3x} = m\vec{a}_x$$

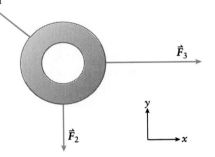

However, note that the component of $\vec{F}_1$ in the x direction is negative, and the component of $\vec{F}_2$ in the x direction is zero because $\vec{F}_2$ points only in the (negative) y direction. In terms of magnitudes of forces in the x direction,

$$-F_{1x} + 0 + F_{3x} = ma_x$$

When F_{3x} is greater than F_{1x} (as suggested in the figure), the ring will accelerate in the positive x direction and the acceleration has a magnitude of

$$a_x = \frac{F_{3x} - F_{1x}}{m}$$

The same approach works to find the magnitude of the net force in the y direction. First, Equation 4-1 becomes

$$\sum \vec{F}_y = m\vec{a}_y$$

Figure 4-3 The three forces pulling on a ring are shown here as vector arrows. The length of each arrow indicates the magnitude of the force and each arrow points in the direction of the force it represents.

or

$$\vec{F}_{1y} + \vec{F}_{2y} + \vec{F}_{3y} = m\vec{a}_y$$

Force $\vec{F}_2$ lies completely in the y direction, and $\vec{F}_{2y}$ is determined to be negative using Figure 4-3. Also, because $\vec{F}_3$ lies completely in the x direction its y component is zero. In terms of magnitudes of forces in the y direction, then,

$$F_{1y} - F_{2y} + 0 = ma_y$$

The ring will acquire an acceleration in the y direction of

$$a_y = \frac{F_{1y} - F_{2y}}{m}$$

We can now find the net magnitude of the acceleration by using Equation 3-3 or

$$a = \sqrt{a_x^2 + a_y^2}$$

We can find that the acceleration vector will point at an angle

$$\phi = \tan^{-1}\left(\frac{a_y}{a_x}\right)$$

counterclockwise from the positive x axis by using Equation 3-4.

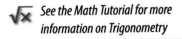

See the Math Tutorial for more information on Trigonometry

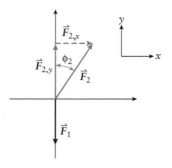

Figure 4-4 Here we see vectors representing the forces generated by two of three teams playing tug-of-war. If the net force applied to the central knot has a magnitude of 1000 N, what are the magnitude and direction of the force generated by the third team?

Example 4-2 Three-Team Tug-of-War

In a three-way game of tug-of-war, teams pull on ropes tied together at one central point. The magnitudes and directions of the forces exerted by two of the teams are shown in **Figure 4-4**; team 1 pulls in the negative y direction with a force of magnitude $F_1 = 800$ N, and team 2 pulls with a force of magnitude $F_2 = 900$ N at an angle $\phi_2 = 35°$ from the y axis. If the net force on the knot due to the three teams is $F_{net} = 1000$ N in the positive y direction, what are the magnitude and direction of the force exerted by the third team?

SET UP

A simple picture showing the forces $\vec{F}_1$ and $\vec{F}_2$ and their components (Figure 4-4) helps us apply Equation 4-1 in each direction separately. To avoid confusing the forces themselves with their components, we have been careful to distinguish between them in the figure.

The sketch contains all of the physics we need—all that remains is for us to apply Equation 4-1, making sure that we first account for all forces and that we then apply the equation in only one direction at a time.

SOLVE

The problem statement indicates that the net force due to the three teams is in the y direction; the magnitude of the x component of the net force is zero. How can there be no net force in the x direction? This statement can only be true if the direction and magnitude of $\vec{F}_3$, which is not shown, exactly cancels the sum of the x components of $\vec{F}_1$ and $\vec{F}_2$. This cancellation can be seen in the application of Equation 4-1 in the x direction:

$$\vec{F}_{1x} + \vec{F}_{2x} + \vec{F}_{3x} = \vec{F}_{net,\,x} = 0$$

so

$$\vec{F}_{3x} = -(\vec{F}_{1x} + \vec{F}_{2x})$$

Because $\vec{F}_1$ lies completely in the negative y direction, it has no x component: that is, $F_{1x} = 0$. So here the x components of $\vec{F}_3$ and $\vec{F}_2$ are equal and opposite, which is also true of their magnitudes

$$F_{3x} = -F_{2x}$$

Therefore, $F_{3x} = -F_2 \sin(\phi_2)$. Using the values given in the problem statement,

$$F_{3x} = -(900 \text{ N}) \sin 35° = -516 \text{ N}$$

We now apply Equation 4-1 in the y direction

$$\vec{F}_{1y} + \vec{F}_{2y} + \vec{F}_{3y} = \vec{F}_{net,\,y}$$

The net force in the y direction is not zero, so solving for F_{3y} we find that

$$\vec{F}_{3y} = \vec{F}_{net,y} - (\vec{F}_{1y} + \vec{F}_{2y})$$

By inspecting Figure 4-4, we see that $F_{2y} = F_2 \cos\phi_2$. Also, because the force of team 1 points in the negative y direction, $\vec{F}_{1y}$ is equal to $-F_1$. Using the given values (including $F_{net} = 1000$ N), we see that

$$F_{3y} = 1000 \text{ N} - [-800 \text{ N} + (900 \text{ N}) \cos 35°] = 1063 \text{ N}$$

Notice that the y component of the vector sum of $\vec{F}_1$ and $\vec{F}_2$ points in the negative y direction, so the force $\vec{F}_3$ due to team 3 must be greater than the 1000 N net force.

We now have the x and y components of $\vec{F}_3$. The magnitude and direction of this force can be determined using Equations 3-3 and 3-4:

$$F_3 = \sqrt{F_{3x}^2 + F_{3y}^2} = \sqrt{(-516 \text{ N})^2 + (1063 \text{ N})^2} = 1182 \text{ N}$$

$$\phi_3 = \tan^{-1}\left(\frac{F_{3y}}{F_{3x}}\right) = \tan^{-1}\left(\frac{1063 \text{ N}}{-516 \text{ N}}\right) = -64.1°$$

REFLECT

We can now redraw Figure 4-4 in which $\vec{F}_3$ is added (Figure 4-5). Notice that by using the strict definition of the angle that defines $\vec{F}_3$, ϕ_3 is negative. A negative value for the angle simply means that the angle is clockwise, rather than counterclockwise, from the x axis. A glance at the diagram suggests that our solution is right. By looking at Figure 4-5, the x components of $\vec{F}_3$ and $\vec{F}_2$ point in opposite directions and appear to have the same magnitude. Because $\vec{F}_1$ makes no contribution to the force in the x direction, and because the x components of $\vec{F}_3$ and $\vec{F}_2$ appear to cancel, we can feel comfortable that our result for $\vec{F}_3$ gives a net force of zero in the x direction. Likewise, it is reasonable to conclude that the sum of the y components of $\vec{F}_3$ and $\vec{F}_2$ is somewhat larger in magnitude and clearly opposite in direction to $\vec{F}_1$. So, we can feel comfortable that our result for $\vec{F}_3$ gives the desired force in the positive y direction.

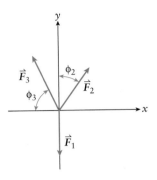

Figure 4-5 The vector arrow labeled $\vec{F}_3$ represents the force applied by the third team.

Practice Problem 4-2 The ring in Figure 4-3 is pulled by three forces $\vec{F}_1$, $\vec{F}_2$, and $\vec{F}_3$. Let $\vec{F}_1$ point in the direction 135° from the positive x direction and have a magnitude of 20.0 N, let $\vec{F}_2$ point in the negative y direction and have a magnitude of 10.0 N, and let $\vec{F}_3$ point in the positive x direction and have a magnitude of 25.0 N. Find the magnitude and direction of the net force on the ring.

In which directions should we apply Newton's second law? Equation 4-1 does not specify a particular direction, which means we are free to choose any direction in which to demand that the sum of the forces equals mass multiplied by acceleration in that direction. To fully analyze a problem requires as many directions as there are dimensions to the problem. So, for example, we should choose two orthogonal (perpendicular) directions in which to study a two-dimensional problem such as the one presented in Example 4-2. In general, it's wise to select axes that make a problem most convenient to solve, which often means aligning at least one of the axes with one of the forces.

Figure 4-6a shows three forces acting on an object. The orientation of the drawing suggests the choice of x–y coordinate axes shown in Figure 4-6b. Using that choice, $\vec{F}_1$ lies completely along the x axis, while both $\vec{F}_2$ and $\vec{F}_3$ have components along both x and y. Equation 4-1, applied in the x and y directions separately, becomes

$$x: \quad -F_1 + F_2 \cos 30° + F_3 \cos 60° = ma_x$$
$$y: \quad F_2 \sin 30° - F_3 \sin 60° = ma_y$$

But consider the choice of axes shown in Figure 4-6c. The application of Newton's second law is more straightforward by selecting axes aligned with both $\vec{F}_2$ and $\vec{F}_3$, because only $\vec{F}_1$ has components that will appear in both the x and y applications of Equation 4-1

$$x: \quad -F_1 \sin 60° + F_2 = ma_x$$
$$y: \quad F_1 \cos 60° - F_3 = ma_y$$

In either case, the physical solution must be the same, but you would probably find it easier to set up the equations when two vectors each align with an axis.

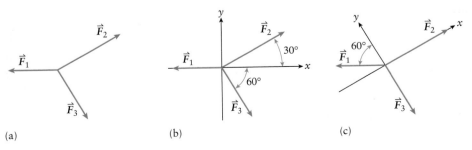

(a) (b) (c)

Figure 4-6 (a) Problems will usually be easier to solve if one of the axes we select aligns with at least one of the forces. In part (b) one of the three forces is aligned with an axis, but in part (c) two of the forces are aligned. In either case, the physical solution must be the same, but you would probably find it easier to set up the equations using the choice of axes in part (c).

> **? Got the Concept 4-2**
> **Weightless in Space**
>
> Imagine you are floating weightless on the International Space Station. Floating in front of you are two spheres, identical in size and appearance. One is made of lead and the other is made of plastic; the lead sphere is considerably more massive than the plastic one. Can you devise a simple experiment to determine which sphere is lead and which is plastic?

> **★ What's Important 4-2**
>
> Applying a force to an object causes it to accelerate. The magnitude of the acceleration depends on the mass of the object as well as the magnitude of the force applied.

4-3 Mass and Weight

How much do you weigh? If you grew up in the United States, chances are your answer will be in pounds. If not, you'll probably use kilograms. We've already declared that the SI units of mass are kilograms. Does that mean that pounds and kilograms are both units of the same quantity? Are weight and mass the same? No! Although people often use the terms *mass* and *weight* interchangeably in everyday conversation, they have different meanings in science. We'll see that although the terms are directly related, it is important to distinguish between them for the purpose of describing and studying physical systems.

In Section 4-1, we saw that mass is a property of a particular object that doesn't change even if the environment changes. **Weight**, on the other hand, is a measure of the force due to gravity that acts on an object. Your weight on the Moon, where the acceleration due to (the Moon's) gravity is less than the acceleration due to gravity on Earth's surface, is correspondingly less than your weight on Earth. Your weight is also slightly different at different locations on our planet's surface. So weight is not an intrinsic property of an object. An object's weight can vary as the object's environment changes.

Consider an object of mass m that experiences a force due only to gravity. Newton's second law equation (Equation 4-1) applied in the vertical direction becomes

$$\sum \vec{F}_y = m\vec{a}_y$$

so in terms of the magnitudes

$$F_y = ma_y = W$$

The y component of the force is due to gravity and therefore equal to weight W by the definition just given. The acceleration a_y is also due to gravity, and $a_y = g$ on Earth. Therefore on Earth,

$$W = mg \qquad\qquad (4\text{-}3)$$

In the previous chapter, we agreed to use an average value of $g = 9.8 \text{ m/s}^2$ but recognized that the exact value of g varies slightly from place to place on Earth. If the value of g varies, then the weight of any particular object must change as it is moved from place to place. *Weight is therefore not an intrinsic property of an object.* Not only does the value of acceleration due to gravity change depending on where you are on Earth, but as we will see in Chapter 10, it can be significantly different on celestial objects other than Earth. Because weight is a force, the units of weight must be newtons, the units of force. Weight is not measured in kilograms because weight is not equivalent to mass.

 ## Watch Out!
Objects in orbit are not actually weightless

You've probably seen video of astronauts in orbit around Earth in a state often referred to as "weightless." And the astronauts certainly appear to have no weight as they float effortlessly around inside their spacecraft. In fact, astronauts do have weight, in that they do experience the force due to gravity, and are accelerated by it. Their state, however, is more correctly described as free fall—the astronauts and everything in their environment is falling under the influence of gravity. Imagine that you were to leap out of a plane while standing on a scale. (You are equipped with a parachute.) Ignoring the influence of air resistance, you and the scale would both accelerate downward at g, and as a result, you would not exert a force on the scale nor would it exert one on you. For that reason you would appear weightless, even though you would be experiencing a net force and as a result, you would also be accelerating.

? Got the Concept 4-3
Mass and Weight on Earth, Mars, and Beyond

An astronaut has a mass of 60 kg when measured on Earth. What is her weight on Earth? What would her mass and weight be on Mars, where the acceleration due to gravity is about 3.7 m/s^2? Astronaut and former school teacher Barbara Morgan is shown giving a lecture on Earth in **Figure 4-7a** and flying in a space shuttle in **Figure 4-7b**. She has the same mass in both photographs. Would a standard bathroom scale show the same reading of weight for her in both?

(a)

(b)

Figure 4-7 The astronaut has the same mass on Earth as she does flying in the space shuttle. Is her weight the same in both cases? *(NASA.)*

Math Box 4-2 Converting from Newtons to Pounds

Can you lift an object that weighs 10 N? How about a 1000-N object? You may find it handy to convert between newtons and pounds using the following conversions

$$1 \text{ N} = 0.225 \text{ lb}$$
$$1 \text{ lb} = 4.448 \text{ N}$$

According to Equation 4-3, the mass of an object that weighs 10 N is

$$m = \frac{W}{g} = \frac{10 \text{ N}}{9.8 \text{ m/s}^2} \approx 1 \text{ kg}$$

A 1-L bottle of water has a mass of 1 kg; that is, it weighs about 10 N. In pounds, that's 10 multiplied by 0.225 lb or about 2.2 lb. You'd have no problem lifting that amount. This equivalence (1 L ≡ 1 kg ≡ 10 N ≅ 2.2 lb) may help you get an intuitive feel for the answer when you encounter a new problem that involves forces.

Can you lift an object that weighs 1000 N? Based on the conversion above, a 1000-N object weighs 1000 multiplied by 0.225 lb or 225 lb. That's about what, say, a textbook author might weigh. You might be able to lift that much, but we can't.

! Watch Out!
Heavier objects do not fall faster

The more massive an object is, the heavier it is and the larger the force it feels due to gravity. Yet we know that falling objects accelerate at the same rate regardless of mass if you neglect air resistance. Should an object experiencing a bigger force also experience a bigger acceleration? No! It is certainly true that a more massive object weighs more on Earth than a less massive one. Weight is a force, so the force due to gravity is larger on a more massive object. But consider inertia: A more massive object also offers a larger resistance to changes in its motion. These effects of inertia and weight cancel each other and falling objects accelerate at the same rate regardless of mass.

✱ What's Important 4-3

Mass is an intrinsic, unchanging property of an object. In contrast, weight is a force, so the weight of an object of a given mass changes as the force due to gravity on it changes. The SI unit of mass is the kilogram (kg). The SI unit of force is the newton (N).

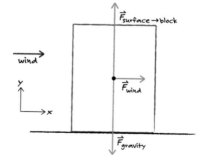

Figure 4-8 A free-body diagram uses vectors to show all of the forces acting on an object. This free-body diagram shows all of the forces acting on a block resting on a horizontal frictionless surface with the wind blowing from left to right. A properly drawn free-body diagram will guide you through any problem that involves forces.

4-4 Free-Body Diagrams

A locked door is most easily opened when you have the right key. "Keys" come in many forms, such as the answers to an exam or the map you use to navigate an unfamiliar city. The key that unlocks physics problems involving force, the map that allows you to navigate force problems, is the free-body diagram.

A **free-body diagram**, such as the one shown in Figure 4-8, is a graphical representation of all of the forces acting on an object. Figure 4-8 shows a block resting on a horizontal frictionless surface on which wind blows from left to right. A properly drawn free-body diagram serves as a map that will guide you through any problem that involves forces. The steps to creating a free-body diagram are listed in Physicist's Toolbox 4-1 and shown in Figure 4-9.

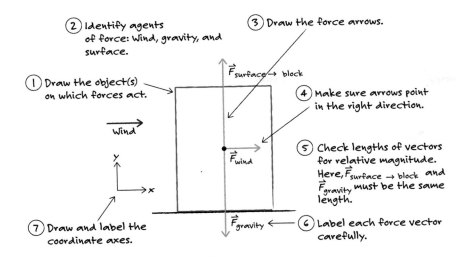

Figure 4-9 A more detailed explanation can be found in Physicist's Toolbox 4-1.

Physicist's Toolbox 4-1 Free-Body Diagrams

SET UP

1. Start a free-body diagram by drawing the object or objects on which forces act. Objects can be represented by simple shapes (for example, rectangles or circles) but it is important to get the orientations of the objects approximately correct (Figure 4-9).

2. Identify the origin or agent of any force before adding it to a free-body diagram. In most of the problems that we will encounter, two objects must be touching for one to exert a force on the other. The one exception for now is the force due to gravity. We'll come across some other exceptions later, such as the magnetic force and the force between electric charges.

SOLVE

3. Draw each force acting on an object as an arrow, where the tail end of the arrow starts at the center of the object *on which the force acts.*

4. Point the arrowhead in the direction in which the force acts.

5. Draw the arrow longer or shorter in order to represent the magnitude of a force relative to the other forces.

6. Label each force.

7. Draw and label the coordinate axes.

REFLECT

8. If there is more than one object to be studied in a system, draw all objects and forces in the same free-body diagram. Although some physicists draw a separate diagram for each object in a system, we will not do so in order to make the relationship between forces clearer. In some cases you may find it helpful to show the components of forces in a free-body diagram. If you do, use dashed lines for arrows of force components so as not to confuse the components of forces with the forces themselves.

9. One way to check your free-body diagram is to ask whether you expect a particular object to experience no acceleration in a particular direction. If so, make sure that all of the forces acting in that direction add to zero. In other words, make sure that for every force or force component acting in one direction, there is at least one force or force component acting in the opposite direction, and that the magnitudes of the sum of the forces in each of these two directions are equal.

Figure 4-10 A stationary box rests on a table. A simple free-body diagram shows the forces acting on the box. Notice that although the box exerts a force on the surface of the table, we did not include that force in the free-body diagram because in this particular case we are not interested in the motion or response of the table.

The stationary box exerts a force on the surface of the table,

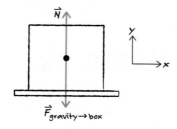

but we won't include that force in the free-body diagram unless we are interested in the motion or response of the table.

We will adopt a notation in which each force is labeled according to what object is acting on another object. For example, if a girl pushes on a box, we will label that force $\vec{F}_{\text{girl}\rightarrow\text{box}}$. However, we'll label two special forces differently to be in accordance with the convention that most physicists use. Whenever an object is supported by a surface, the force must be exerted in the direction perpendicular to, or normal to, the surface. We will label such a force $\vec{N}$ for **normal force**. Whenever a **tension force** is exerted on a string or wire attached to an object we'll label such a force $\vec{T}$.

Let's look at some examples of free-body diagrams. In **Figure 4-10**, a stationary box rests on a table. The box exerts a force on the surface of the table, but we won't include that force in the free-body diagram unless we are interested in the motion or response of the table.

In the next example, we see a box on a ramp (an inclined surface), attached by a thread to a post (**Figure 4-11**). The thread exerts a force on the post as well as on the box, but we would only include that force in the free-body diagram if we were interested in the motion or response of the post. Also, note that because the block does not accelerate in the direction along the ramp (labeled the x direction in the figure) the tension force $\vec{T}$ must cancel the x component of $\vec{F}_{\text{gravity}\rightarrow\text{box}}$. We have therefore drawn the length of the arrow for $\vec{T}$ to match the length of the x component of $\vec{F}_{\text{gravity}\rightarrow\text{box}}$. The same condition is true for the normal force $\vec{N}$ and the component of $\vec{F}_{\text{gravity}\rightarrow\text{box}}$ perpendicular to the surface. Because the box is neither leaping off the ramp nor pushing down into it, there is no acceleration in the

The thread exerts a force on the post as well as the box, but we would only include that force in the free-body diagram if we were interested in the motion or response of the post.

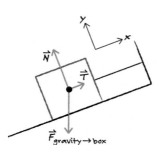

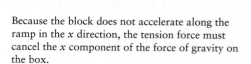

Because the block does not accelerate along the ramp in the x direction, the tension force must cancel the x component of the force of gravity on the box.

Figure 4-11 A box hangs by a thread on a ramp. The sketch on the left is a free-body diagram showing the forces acting on the box.

Because the box is neither leaping off the ramp nor pushing down into it, there is no acceleration in the direction perpendicular to the surface. Therefore, the normal force and the y component of the force of gravity on the box must be equal in length and in opposite directions.

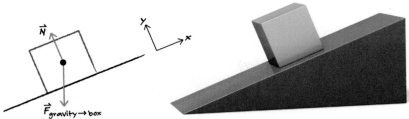

Figure 4-12 This box is accelerating down the ramp.

Note that only $\vec{F}_{\text{gravity}\rightarrow\text{box}}$ has a component in the direction along the surface of the ramp, labeled the x direction in the figure. Because there is only one force acting parallel to the ramp, there is no way for the vector sum of the forces along the ramp to be zero. According to Equation 4-1, the acceleration in this direction must be nonzero. The box must be accelerating!

direction perpendicular to the surface. So, the arrow for $\vec{N}$ and the arrow for the y component of $\vec{F}_{\text{gravity}\rightarrow\text{box}}$ must be equal in length and in opposite directions.

Let's see what happens if the box is not held back by the string. In the free-body diagram in Figure 4-12, the box is sliding down the ramp unopposed by any resistive forces. Note that only $\vec{F}_{\text{gravity}\rightarrow\text{box}}$ has a component in the direction along the surface of the ramp, labeled the x direction in the figure. Because there is only one force acting parallel to the ramp, there is no way for the vector sum of the forces along the ramp to be zero. According to Equation 4-1, a nonzero net force in that direction results in a nonzero acceleration. The box must be accelerating!

❗ Watch Out!

Velocity is not force

On first encountering forces and free-body diagrams, students are sometimes tempted to add a force in a free-body diagram that is somehow associated with the motion. But first, remember that neither velocity nor acceleration is force. The acceleration of the box in Figure 4-12 arises from the nonzero force in the x direction. Also, before adding a force on the box pointing, say, in the negative x direction, ask yourself what would cause such a force. We've already accounted for gravity and we've already accounted for the one and only thing which is touching the box (the surface of the ramp). So, there can be no force other than the two we've already drawn in Figure 4-12.

In the next example, one box rests on a horizontal table connected by a thread to a second box hanging off the table over a pulley (Figure 4-13). Two different forces acting on opposite ends of the thread are labeled $\vec{T}$ in Figure 4-13. Although these forces act in opposite directions, they must have the same magnitude because the thread is neither stretching nor becoming slack. In general, we will not consider situations in which this is allowed to happen.

Notice that when box 1 moves in the positive x direction, box 2 moves down in Figure 4-13. For that reason, "down" for box 2 is the same direction as positive x for box 1; we have added an arrow to indicate this direction on the right side of the figure.

Finally, note that we have drawn $\vec{F}_{\text{gravity}\rightarrow 2}$ somewhat longer than the tension force $\vec{T}$ which pulls in the opposite direction on box 2. This suggests that there is a net force downward on box 2; that is, box 2 will accelerate downward.

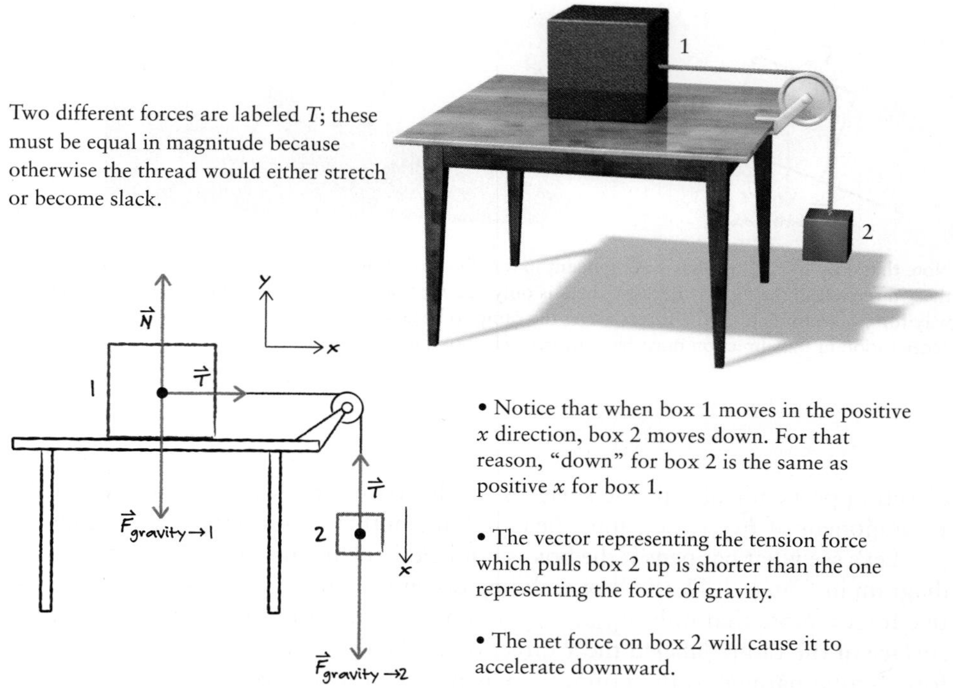

Figure 4-13 Box 2, the green one, hangs off the edge of the table by a thread connected to box 1, the purple box. The sketch at the bottom shows the forces acting on each box. Notice that when box 1 moves in the positive x direction, box 2 moves down. For that reason, "down" for box 2 is the same direction as positive x for box 1; we have added an arrow to indicate this direction on the right side of the free-body diagram for box 2.

Two different forces are labeled T; these must be equal in magnitude because otherwise the thread would either stretch or become slack.

• Notice that when box 1 moves in the positive x direction, box 2 moves down. For that reason, "down" for box 2 is the same as positive x for box 1.

• The vector representing the tension force which pulls box 2 up is shorter than the one representing the force of gravity.

• The net force on box 2 will cause it to accelerate downward.

★ What's Important 4-4

Free-body diagrams show all of the forces acting on an object. The normal force acts in a direction perpendicular to whatever surface supports an object. A force exerted on a string or wire attached to an object is called a tension force.

4-5 Newton's Third Law

Hold an egg a meter above the floor and release it. It falls to the floor as a result of the force due to gravity. But carefully release that egg on a horizontal table and it stays in place. Why? Before you answer that the table just gets in the way, recall that our focus in this chapter is force. The egg certainly exerts a force on the table as it rests on the table's surface. But because the egg is not accelerating, some *other* force must be acting on it to counter the force due to gravity. This is the force of the table acting on the egg. Forces come in pairs. In this case, the two forces in the pair are the force of the egg acting on the table and the force of the table acting on the egg.

The same principle of force pairs applies to fish. A fish swims by swishing its tail fin and the rear part of its body from side to side as shown in **Figure 4-14**. In doing so, the fish exerts a force on the surrounding water, but how does that propel the fish? As we will explore in this section, forces come in pairs, so that as the fish exerts a force on the water, the water exerts an equal but opposite force on the fish. As with any force exerted by a surface, the force of the fish on the water must be perpendicular to the surface of the fish's body, as shown in the figure. The reaction force, labeled $\vec{F}_{\text{water}\to\text{fish}}$, is in the opposite direction, clearly not in the direction that the fish is traveling. This force does, however, have a component in that direction; it is this component of $\vec{F}_{\text{water}\to\text{fish}}$, labeled $\vec{F}_{\text{thrust}}$ in Figure 4-14, that propels the fish forward.

Newton addressed this aspect of force in the third of his fundamental principles of motion. Two common restatements of **Newton's third law** are

All forces occur in pairs, and these two forces are equal in magnitude and opposite in direction.

For every force, there is an equal but opposite reaction force.

In declaring the third law, Newton added, "If you press a stone with your finger, the finger is also pressed by the stone. If a horse draws a stone tied to a rope, the horse . . . will be equally drawn back towards the stone." This quote reveals exactly how we will apply Newton's third law. Furthermore, we will use a simplified form of Newton's third law

If A pushes on B, then B pushes on A with equal force but in the opposite direction.

(You can substitute "pull" for "push" as appropriate.) The two forces involved with such an interaction are referred to as a **force pair**.

Let's return to the example of the egg resting on the table. The egg pushes on the table, so the table must push on the egg according to our restatement of Newton's third law. This second force in the pair is what we have referred to as the normal force, the force by which a surface supports an object. We could also refer to this force as a **contact force**, which is any force which arises due to one object touching another.

You might be wondering in what way the weight of the egg enters into the consideration of the forces in this case. After all, we know from Newton's second law that the force of the egg on the table is directly related to its weight. The weight of the egg is a result of Earth's gravity, so according to our simplified version of Newton's third law: "If the Earth pulls on the egg, then the egg pulls on Earth." The force pair that includes the egg's weight is therefore between Earth and the egg.

Although forces come in pairs, we are not always interested in both forces in a pair. For example, if we want to study the motion of a ball that you are about to toss into the air, we need to know the force your hand applies to the ball, but not the force that the ball exerts on your hand (**Figure 4-15**). We won't, in such cases,

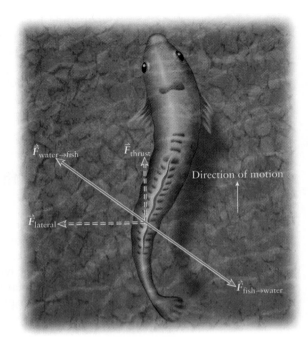

Figure 4-14 Forces come in pairs, so that as the fish exerts a force on the water, the water exerts an equal but opposite force on the fish. As with any force exerted by a surface, the force of the fish on the water must be perpendicular to the surface of the fish's body. Notice that a component of the force of the water on the fish is in the direction of motion. The thrust component is the force that propels the fish through the water.

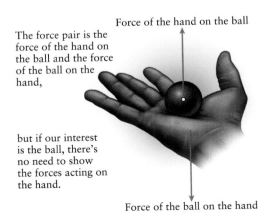

The force pair is the force of the hand on the ball and the force of the ball on the hand,

Force of the hand on the ball

but if our interest is the ball, there's no need to show the forces acting on the hand.

Force of the ball on the hand

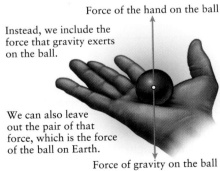

Instead, we include the force that gravity exerts on the ball.

Force of the hand on the ball

We can also leave out the pair of that force, which is the force of the ball on Earth.

Force of gravity on the ball

Figure 4-15 Although forces come in pairs, we are not always interested in both forces in a pair. For example, if we want to study the motion of a ball that you are about to toss into the air, we need to know the force your hand applies to the ball, but not the force that the ball exerts on your hand.

include both forces in a pair in the free-body diagram. In a free-body diagram representing the ball and the hand, the force of the hand on the ball is drawn, but not its pair, the force of the ball on the hand. Because we always include all of the forces acting on an object in a free-body diagram, the force of Earth's gravity on the ball is shown, but not its pair, the force of the ball on Earth. Our interest is what happens to the ball, so the forces of the ball on Earth and on the hand don't need to appear.

 Got the Concept 4-4
To Push or Not to Push, That Is the Question

You have been asked to push a large crate across a smooth floor (Figure 4-16). Having passed a physics course in high school, your friend declares that this feat isn't possible, because whatever force you apply to the box will be met with an equal but opposite reaction force. The vector sum of the forces, she argues, will be zero, resulting in zero acceleration. What's wrong with her argument?

Push

Figure 4-16 If forces come in equal and opposite pairs, how is it possible to push a large crate across the smooth floor?

 Got the Concept 4-5
Forces on Crates

Two crates are at rest, one touching the other, on a smooth floor as shown in Figure 4-17. You push on one of the crates. What are the forces on the crates?

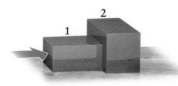

2
1

Figure 4-17 Box 1 and box 2 are next to one another on the floor. If you apply a force to box 1, represented by the red arrow, what forces act on each of the boxes?

Estimate It! 4-1 Earth Pulls on You and You Pull Back!

Earth pulls on you with a force proportional to your mass, which is the reason you have weight. If you step off the edge of a swimming pool, that force causes you to accelerate into the water at $g = 9.8$ m/s^2. According to Newton's third law, if Earth pulls on you, then you pull on Earth. According to Newton's second law, Earth accelerates as a result. Let's grab a pencil and a scrap of paper and do a quick calculation of the acceleration imparted to Earth due to your presence.

SET UP

According to Newton's third law, for every force there is an equal but opposite reaction force. The force with which Earth pulls on you is therefore equal in magnitude to the force with which you pull on Earth, or

$$F_{\text{Earth}\to\text{you}} = F_{\text{you}\to\text{Earth}}$$

Using Newton's second law, that the net force equals the product of mass and acceleration, we can find the magnitude of the resulting acceleration of Earth. Then

$$M_{\text{Earth}} a_{\text{Earth}} = m_{\text{you}} a_{\text{you}}$$

or

$$a_{\text{Earth}} = \frac{m_{\text{you}} a_{\text{you}}}{M_{\text{Earth}}}$$

SOLVE

For the purposes of an estimate we'll use big, round numbers. If you're of typical size, your mass is likely between 60 and 100 kg, so let's use 80 kg. Your acceleration is about 10 m/s^2 due to Earth's gravity. It would be difficult to estimate the mass of Earth, so we'll look that up; it's about 6×10^{24} kg. So

$$a_{\text{Earth}} = \frac{(80 \text{ kg})(10 \text{ m/s}^2)}{6 \times 10^{24} \text{ kg}} = 1.3 \times 10^{-22} \text{ m/s}^2$$

To one significant figure, the acceleration of Earth due to your gravitational pull is $1 \times 10^{-22} \text{ m/s}^2$.

REFLECT

Wow! What a small number! At that acceleration, it would take hundreds of years for Earth to move even a centimeter. So it doesn't mean much to say that Earth accelerates due to the force you exert on it.

The force with which you pull on Earth is equal in magnitude to the force with which Earth pulls on you. Because your mass is not equal to the mass of Earth, however, the resulting acceleration of Earth is not at all equal to your acceleration!

? Got the Concept 4-6
Weightless in Space, Again

In Got the Concept 4-2, we imagined floating weightless on the International Space Station and pushing on freely floating lead and plastic spheres. Considering Newton's third law, what difference would you expect as a result of pushing on the lead sphere compared to the plastic one? Assume that the mass of the lead sphere is larger than your mass, that the mass of the plastic sphere is much smaller than your mass, and also that the force you apply is the same on both spheres.

✱ What's Important 4-5

All forces created in interactions between two objects occur in pairs. The two forces in a pair are equal in magnitude and opposite in direction.

4.6 Force, Acceleration, Motion

Your friend has a pair of fuzzy dice hanging from the rearview mirror in her car. You notice that when she accelerates, the dice don't hang straight down. How can we use Newton's laws to determine for example, the angle at which they hang while the car is accelerating?

The free-body diagram is the key which unlocks any problem that involves forces. The first step in solving a force problem is to make a free-body diagram that includes all of the forces as well as the coordinate axes where the positive directions are labeled. Next, set up a mathematical description of the problem by writing Newton's second law (Equation 4-1) separately for each object in both the x and y directions. Use the free-body diagram to make sure that every force that acts on some part of the system is included in the sum of forces that forms the left side of the equation. The diagram and the equations encompass all of the physics, so all that remains is to rearrange!

Our strategy for solving force problems is summarized in Physicist's Toolbox 4-2.

Physicist's Toolbox 4-2   Solving Force Problems

SET UP

Identify all of the forces acting on the object.

SOLVE

1. Make a free-body diagram for the object. Be sure to include the forces acting on the object.

2. Choose the coordinate axes and label the positive directions.

3. Write out Newton's second law for each object and in each direction separately, using the free-body diagram as a guide.

REFLECT

Is the net force in each direction what you would expect? For example, if you expect that the object will not accelerate in a particular direction, is the vector sum of all of the forces on it in that direction equal to zero?

 Go to Interactive Exercise 4-1 for more practice dealing force problems

Example 4-3   Fuzzy Dice

A pair of fuzzy dice hang from the rearview mirror of a sports car on a thread that has negligible mass (**Figure 4-18**). At what angle to the vertical do the dice hang when the car accelerates at a constant 6.6 m/s^2?

SET UP

We apply the strategy in the Physicist's Toolbox 4-2 for force problems. First, identify all of the forces acting on one of the dice, in order to draw a complete free-body diagram. Each die feels a force due to gravity; that is, it has weight. The thread pulls on the die; the pulling force manifests itself as tension in the thread. In our free-body diagram, shown in **Figure 4-19**, we therefore draw a force due to gravity on the die pointing in the negative y direction and a tension force directed along the thread. Nothing else acts on the die so there are no other forces to draw in the free-body diagram.

The next step in our force-problem strategy is to choose the coordinate axis. Here we pick coordinates so that one of the two force vectors as well as the direction of the acceleration align with an axis, in order to make the mathematical setup more straightforward. With the choice shown in Figure 4-19, the force due to gravity is aligned with the negative y direction and the acceleration vector $\vec{a}$ points along the positive x direction. We might have considered choosing axes such that one axis points up along the thread and the other is perpendicular to it. Any choice of axes is allowed; so this choice would have worked, but the mathematical formulation would be a bit more complicated because acceleration would have nonzero components in both chosen directions.

Figure 4-18 A pair of fuzzy dice hang by a thread from the rearview mirror of a sports car. When the car accelerates the dice do not hang vertically.

The next step in our force-problem strategy is to write out Newton's second law in each direction on each object separately. Because we are treating the mass of the thread as negligible, we need to consider only the die. Equation 4-1 gives

$$x: \quad \vec{T}_x + \vec{F}_{\text{gravity}\rightarrow\text{die},x} = m_{\text{die}}\vec{a}_x$$
$$y: \quad \vec{T}_y + \vec{F}_{\text{gravity}\rightarrow\text{die},y} = m_{\text{die}}\vec{a}_y$$

To write these equations in terms of magnitudes, note that $T_x = T\sin(\theta)$ and $T_y = T\cos(\theta)$. Also, the acceleration in the x direction is $a_x = a$, while the acceleration in the y direction is zero ($a_y = 0$) because the die will remain stationary at angle θ while the car accelerates at constant acceleration a. Finally, note that the x component of the force due to gravity is zero and the y component of the force due to gravity is $F_{\text{gravity}\rightarrow\text{die},y} = -m_{\text{die}}g$. Newton's second law is then

$$x: \quad T\sin(\theta) + 0 = m_{\text{die}}a$$
$$y: \quad T\cos(\theta) - m_{\text{die}}g = m_{\text{die}}0$$

SOLVE
What's the same in these two equations? T appears in both, so one approach to finding the angle would be to solve each equation for T and set the results equal

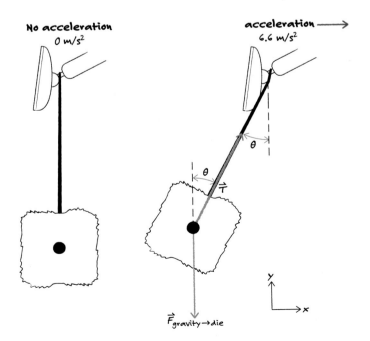

Figure 4-19 The sketch on the left shows one die hanging vertically while the car is not accelerating. The sketch on the right shows the die during an acceleration. The vector arrow along the thread represents the force exerted by the thread on the die. The downward-pointing vector arrow represents the force of gravity on the die.

to each other. Another approach would be to solve one equation for T and substitute that expression into the other equation. For example, solve the equation that represents the forces in the y direction for T and substitute this expression into the equation that represents the forces in the x direction:

$$T \cos \theta = m_{\text{die}} g$$

or

$$T = \frac{m_{\text{die}} g}{\cos \theta}$$

Then,

$$\frac{m_{\text{die}} g}{\cos \theta} \sin \theta + 0 = m_{\text{die}} a$$

or

$$g \tan \theta = a$$

So

$$\tan \theta = \frac{a}{g}$$

or

$$\theta = \tan^{-1}\left(\frac{a}{g}\right) \tag{4-4}$$

When $a = 6.6 \text{ m/s}^2$,

$$\theta = \tan^{-1}\left(\frac{6.6 \text{ m/s}^2}{9.8 \text{ m/s}^2}\right) = 34°$$

REFLECT

Let's test Equation 4-4 using a range of values of acceleration. When the car is not accelerating—which means its velocity is constant but not necessarily zero—Equation 4-4 gives $\theta = 0°$. The dice hang straight down, which is expected. What if the acceleration of the car were 9.8 m/s^2, the same as g? (Racing cars can accelerate at nearly $2g$.) If the acceleration is the same horizontally as vertically, we would expect the dice to hang at a 45° angle, which is also predicted by Equation 4-4. We conclude that our answer is reasonable.

Practice Problem 4-3 As you board an airplane you notice a thin cord with a ball at its end, hanging straight down from the ceiling. As the plane speeds down the runway before takeoff, you notice the cord hanging at an angle 11° from the vertical. What is the magnitude of the plane's acceleration at that moment?

Weight is often measured by resting an object on a scale. The surface of the scale supports the object by exerting a force on it normal to the surface. Because the scale exerts a force on the object, the object exerts an equal but opposite force on the scale according to Newton's third law. We will not always be interested in the force that an object exerts on the supporting surface. For example, we did not include this force in the free-body diagram we created for a stationary box on a horizontal surface in Figure 4-10. But in the case of an object on a scale, it is the force of the object on the scale which results in the measurement of the object's weight, often by causing a spring inside the scale to compress. The force of the object on the scale is therefore equal to its weight, and it is also equal to the normal force exerted by the surface of the scale according to Newton's third law. We put this equivalence to the test in the next example.

Example 4-4 Feel Lighter the Easy Way

One of the authors, whose mass is 95.4 kg, stands on a scale (Figure 4-20). What weight does the scale read? Hoping to feel lighter and armed with knowledge of Newton's laws, the author puts the scale on an elevator and takes a measurement while the elevator is accelerating downward at 1 m/s^2. What weight will the scale read then?

SET UP

Whether or not the scale is being used on an elevator, only two forces are exerted on the person standing on it—the force due to gravity, labeled $\vec{F}_{\text{gravity}\rightarrow\text{author}}$ in Figure 4-21, and the normal force $\vec{N}$ exerted by the supporting surface.

We selected the x and y axes for the coordinate axes, as shown in Figure 4-21, because all of the action occurs in the vertical direction.

The next step in our force-problem strategy is to write out Newton's second law equation in each direction and on each object separately. Because no forces act in the x direction, only the equation for the forces acting in the y direction is necessary:

$$y: \quad \vec{N} + \vec{F}_{\text{gravity}\rightarrow\text{author}} = m\vec{a}_y \tag{4-5}$$

Once we determine the appropriate value of $\vec{a}_y$ we can solve for $\vec{N}$, which will be the same as the reading on the scale.

SOLVE

The force due to gravity is determined by mass and the acceleration due to gravity:

$$\vec{F}_{\text{gravity}\rightarrow\text{author}} = mg$$

When the scale is not on the elevator, the net acceleration of an object resting on it is zero, so in terms of magnitudes of forces Equation 4-5 becomes

$$N - mg = 0$$

or

$$N = mg = (95.4 \text{ kg})(9.8 \text{ m/s}^2) = 935 \text{ N}$$

What changes when the author weighs himself on an elevator accelerating downward? The free-body diagram remains the same, because the forces acting on the author arise in the same way. Equation 4-5 is therefore the same as well. However, the net acceleration is no longer zero, but instead $a = 1 \text{ m/s}^2$ in the negative y direction due to the motion of the elevator. Equation 4-5 then becomes

$$N - mg = m(-a)$$

or

$$N = mg - ma = m(g - a) \tag{4-6}$$

Even before substituting the known values, we know that the normal force, which equals the reading on the scale, is smaller than the first measurement. The weight of the author riding an elevator that is accelerating downward is

$$N = (95.4 \text{ kg})(9.8 \text{ m/s}^2 - 1 \text{ m/s}^2) = 840 \text{ N}$$

Figure 4-20 One of the authors could stand to lose a few pounds.

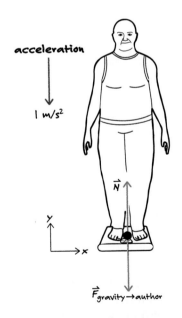

Figure 4-21 In this diagram we see that only two forces are exerted on the person standing on the scale—the force due to gravity, labeled $\vec{F}_{\text{gravity}\rightarrow\text{author}}$, and the normal force $\vec{N}$ exerted by the supporting surface.

The author weighs about 100 N less when riding on the downward accelerating elevator. The author's mass, however, does not change.

REFLECT

You may have experienced the sensation of decreased or increased weight while flying in an airplane or going on an amusement park ride. Acceleration upward increases the force that a supporting surface, such as the seat of a plane, exerts on us. That increase in the normal force is equivalent to an increase in our apparent weight. A similar phenomenon holds for acceleration downward, as in this problem.

What if the elevator were in free fall, that is, if the acceleration of the elevator is $a = 9.8$ m/s^2 downward? Substituting this value into Equation 4-6 gives

$$N = (95.4 \text{ kg})(9.8 \text{ m/s}^2 - 9.8 \text{ m/s}^2) = 0 \text{ N}$$

An object in free fall experiences *weightlessness*. As we will see later on, astronauts in orbit around Earth are in free fall, although they always "miss" hitting Earth due to their circular motion. Because they are in free fall, orbiting astronauts are weightless.

Practice Problem 4-4 After performing the experiment above, the author remains on the scale as the elevator accelerates *upward* at 1 m/s^2. What weight will the scale read?

√x̄ **Go to Picture It 4-1** *for more practice dealing with apparent weight*

Some might look at **Figure 4-22** and say that the water-skier is being pulled by the boat. As a student of physics, however, you know that can't be correct because the boat does not touch her. Instead, the boat is exerting a force on the towrope, and the towrope pulls on the skier. The force that transfers the pulling is tension. Under ideal circumstances, the rope doesn't stretch, and whatever force is applied to one end is transmitted without loss to whatever is attached to the other end.

Because the rope pulls on the skier, the skier pulls on the rope with an equal but opposite force according to Newton's third law. The boat must also pull on the rope with a force of the same magnitude, otherwise, the rope would either break or become slack. We examine this balance of tension forces in the next example.

Figure 4-22 The boat can't pull on the water-skier because it doesn't touch her. Instead, the boat exerts a force on the towrope, which in turn pulls on the water-skier. Tension, the force of the rope on her, transfers the boat's pulling force. *(Thomas Krämer/age fotostock.)*

Example 4-5   Tension in a String

A horizontal force of 7.5 N is used to pull two blocks connected by a string across a smooth horizontal surface, as shown in **Figure 4-23**. Block 1 has a mass equal to 0.5 kg, block 2 has a mass equal to 1.0 kg, and the mass of the string is so small that we can neglect it. What is the acceleration of the system?

Figure 4-23 A horizontal force pulls two blocks connected by a string across a smooth horizontal surface.

SET UP

Notice that the agent of the force pulling to the right on block 2 does not touch block 1. The tension force on block 2 is exerted by the tension in the string which arises because block 1 pulls on it. The force that block 1 exerts on the string is transmitted undiminished from the right side of the string to the left. The first step in the Solve section of our strategy in Physicist's Toolbox 4-2 is to draw these forces on a free-body diagram, as shown in **Figure 4-24**, where we have also indicated our choice of coordinate axes, selected so that the forces and acceleration are aligned with one of the axes.

 Go to Interactive Exercise 4-2 for more practice dealing with tension

   The last step in our force-problem strategy is to write out Newton's second law in each direction on each object separately:

$$\text{block 1,} \quad x: \quad \vec{T} = m_1 \vec{a}_{1,x}$$
$$y: \quad \vec{N}_1 + \vec{W}_1 = m_1 \vec{a}_{1,y}$$
$$\text{block 2,} \quad x: \quad \vec{F}_{\text{pull}} + \vec{T} = m_2 \vec{a}_{2,x}$$
$$y: \quad \vec{N}_2 + \vec{W}_2 = m_2 \vec{a}_{2,y}$$

SOLVE

Neither block experiences a net acceleration in the y direction, so $a_{1,y} = 0$ and $a_{2,y} = 0$. For this reason, the two equations describing the forces in the y direction, while true, don't elucidate this problem; the equations tell us only that the normal force equals the weight for each block. For example, on block 1

$$N_1 - W_1 = 0 \quad \rightarrow \quad N_1 = W_1$$

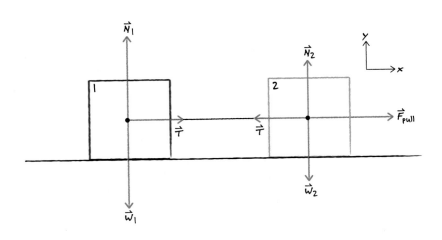

Figure 4-24 The diagram summarizes the forces acting on each block. When block 1 pulls on the string, tension arises that exerts a force on block 2.

To solve the equations that describe the forces in the x direction, note that $\vec{T}$ acts on block 2 in the negative x direction. So

$$\text{block 1,} \quad x: \quad T = m_1 a_{1,x}$$
$$\text{block 2,} \quad x: \quad F_{\text{pull}} - T = m_2 a_{2,x}$$

Because the two blocks must move together, $a_{1,y} = a_{2,y} = a$. Add these two equations so that $+T$ in the first and $-T$ in the second cancel. Then, substitute a for both accelerations in the resulting equation:

$$F_{\text{pull}} = m_1 a + m_2 a = (m_1 + m_2)a$$

so

$$a = \frac{F_{\text{pull}}}{m_1 + m_2} \tag{4-7}$$

The acceleration of the system of two blocks is

$$a = \frac{7.5 \text{ N}}{0.5 \text{ kg} + 1.0 \text{ kg}} = 5.0 \text{ m/s}^2$$

REFLECT

Considering dimensions, we should expect the result for acceleration to have a form in which force is divided by mass. In addition, we would expect both masses to play a role in the answer here. Equation 4-7 is in accordance with the two conditions.

In this example, because the two blocks move together, we could have treated the system as a single block that has a mass equal to $m_1 + m_2$. Had we done that, the Newton's second law equation representing forces acting in the x direction would read

$$F_{\text{pull}} + T - T = (m_1 + m_2)a$$

which immediately leads to Equation 4-7. We cannot, however, always assume that multiple objects in a system can be treated as one, so this approach is best used only as a way to check an answer.

Practice Problem 4-5 Suppose block 3, of mass 3.5 kg, is placed on block 1 in Figure 4-23. If the system is pulled by a horizontal force of 7.5 N, what is the acceleration of the system?

Example 4-6 Atwood's Machine

One object that has a mass equal to m_1 is connected to another object that has a mass equal to m_2 by a string of negligible mass over a pulley, as shown in Figure 4-25. Find the acceleration of the system. Assume $m_1 > m_2$. Also assume that the pulley has a mass so small that we can neglect it. This device is known as Atwood's machine, after its inventor, 18th century English mathematician George Atwood.

SET UP

The first step in the Solve section of our strategy in Physicist's Toolbox 4-2 is to make a free-body diagram, which is shown in Figure 4-26. We are interested in the forces on and the accelerations of the two objects, so we have drawn the force due to gravity on each object and the force due to the tension in the string on each object. The tension force must be the same at both ends of the string to ensure that the string neither breaks nor becomes slack.

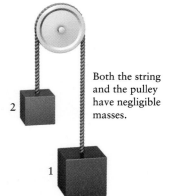

Both the string and the pulley have negligible masses.

Figure 4-25 Two boxes are connected to each other by a string that hangs over a pulley in this device called Atwood's machine.

Critical to the construction of the free-body diagram in Figure 4-26 is the choice of the coordinate axes, the next step in our force-problem strategy. First, note that there are no forces in the direction perpendicular to the string, so only one coordinate axis is necessary. We have selected the y axis to represent this axis. But also notice that when object 2 moves up object 1 moves down. If we label the upward direction for object 2 as the positive y axis, we must also associate the positive y axis with the downward direction for object 1. This result is shown by the two arrows labeled "y" in the figure.

The final step in our strategy is to write out Newton's second law equations which represent the forces acting in each direction and on each object. This problem requires two equations in the y direction, one for each of the two objects:

$$1, \quad y: \quad \vec{T} + \vec{F}_{\text{gravity}\rightarrow1} = m_1\vec{a}_{1,y} \qquad (4\text{-}8)$$

$$2, \quad y: \quad \vec{T} + \vec{F}_{\text{gravity}\rightarrow2} = m_2\vec{a}_{2,y} \qquad (4\text{-}9)$$

SOLVE

Because the two objects move together and $m_1 > m_2$, both $\vec{a}_{1,y}$ and $\vec{a}_{2,y}$ point in the same direction along the positive y axis. In addition $a_{1,y} = a_{2,y} = a$. Also observe that the tension $\vec{T}$ acts in the positive y direction on object 2 but in the negative y direction on object 1. Similarly, the force due to gravity is in the negative y direction on object 2 and the positive y direction on object 1. There is no need to guess these directions—simply compare the directions of the forces and the directions of the coordinate axis in Figure 4-26.

Rewriting in terms of magnitudes and using acceleration a for both objects, Equations 4-8 and 4-9 become

$$1, \quad y: \quad -T + m_1g = m_1a \qquad (4\text{-}10)$$

$$2, \quad y: \quad T - m_2g = m_2a \qquad (4\text{-}11)$$

where we have also replaced the forces of gravity acting on the objects with the objects' weights $W_1 = m_1g$ and $W_2 = m_2g$. Adding the two equations eliminates T:

$$m_1g - m_2g = (m_1 + m_2)a$$

or

$$a = \frac{m_1 - m_2}{m_1 + m_2}g \qquad (4\text{-}12)$$

REFLECT

We declared $m_1 > m_2$, so the numerator in the fraction in Equation 4-12 is a positive number and is smaller than m_1. The denominator is a positive number greater than m_1, so that $0 \leq a \leq g$. Because a is positive, the system accelerates and object 1 moves down while object 2 moves up.

Let's consider the result using masses for which we can guess the answer.

What if m_1 were smaller than m_2, which is contrary to the assumption we made? We should expect that the system would move in the opposite direction, that is, toward the negative y direction. For $m_1 < m_2$ the numerator in Equation 4-12 would be negative, as would the acceleration. So as expected, the system would accelerate toward the negative y direction.

Suppose $m_1 = m_2$. We would intuitively expect that the system would remain in its initial position, because the two objects would be balanced. This result is confirmed by setting the two masses equal in Equation 4-12 so that the numerator becomes zero.

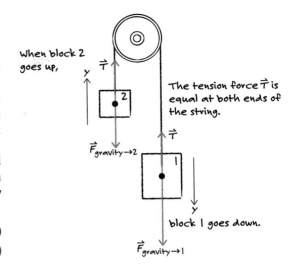

The tension force $\vec{T}$ is equal at both ends of the string.

Figure 4-26 We are interested in the forces on the boxes and in the acceleration the boxes experience, so we have drawn the force due to gravity and the force due to the tension in the string on each box. The tension force must be the same at both ends of the string to ensure that the string neither stretches nor becomes slack.

Finally, consider the case in which m_1 is much larger than m_2, that is, $m_1 \gg m_2$. This condition is nearly the same as having no object on the left side of the string, in which case object 1 would accelerate in free fall. Both $m_1 - m_2$ and $m_1 + m_2$ are hardly different from m_1 alone, so Equation 4-12 can be approximated

$$a = \frac{m_1 - m_2}{m_1 + m_2}g \approx \frac{m_1}{m_1}g$$

so

$$a \approx g$$

Again, we have used a special limiting case to confirm that our result is reasonable.

Example 4-7 Atwood's Machine, Continued

Compare the tension in the string in Example 4-6 with the forces exerted on the objects by gravity.

SET UP

Either Equation 4-10 or Equation 4-11 can be applied; for example, on block 2,

$$T - m_2g = m_2a \tag{4-11}$$

or

$$T = m_2(a + g)$$

We can now apply Equation 4-12 to Equation 4-11:

$$a = \frac{m_1 - m_2}{m_1 + m_2}g \tag{4-12}$$

SOLVE

So

$$T = m_2\left(\frac{m_1 - m_2}{m_1 + m_2}g + g\right)$$

$$T = m_2\left(\frac{m_1 - m_2}{m_1 + m_2} + 1\right)g$$

$$T = m_2\left(\frac{m_1 - m_2}{m_1 + m_2} + \frac{m_1 + m_2}{m_1 + m_2}\right)g$$

$$T = m_2\left(\frac{m_1 - m_2 + m_1 + m_2}{m_1 + m_2}\right)g$$

$$T = m_2\left(\frac{2m_1}{m_1 + m_2}\right)g$$

To compare this tension with the weights of the two objects, we rewrite T in terms of the weights $W_1 = m_1g$ and $W_2 = m_2g$:

$$T = \left(\frac{2m_1}{m_1 + m_2}\right)m_2g = \left(\frac{m_1 + m_1}{m_1 + m_2}\right)W_2$$

and

$$T = \left(\frac{2m_2}{m_1 + m_2}\right)m_1g = \left(\frac{m_2 + m_2}{m_1 + m_2}\right)W_1$$

We wrote $2m_1 = m_1 + m_1$ and $2m_2 = m_2 + m_2$ to make it more obvious that, because we declared that $m_1 > m_2$, the fraction in the first equation is always greater than 1 and the fraction in the second equation is always less than 1. Thus, we conclude that

$$T > W_2$$

and

$$T < W_1$$

REFLECT

Two forces, T and W_2, act on object 2. We now know that the tension is greater than W_2, which means that the net force points in the direction of T. The result? Object 2 is pulled up. Similarly, W_1 is greater than T, so the net force on object 1 is in the direction of the weight force. Object 1 therefore moves downward. These results underlie why the system moves as it does. The tension in the string is not large enough to prevent gravity from pulling object 1 down, but it *is* large enough to overcome the weight of object 2.

The normal force by which a surface supports an object is always perpendicular to the surface. Figure 4-27 shows the free-body diagram of a box on a ramp of angle θ; the normal force is perpendicular to the surface of the ramp. Two choices for the coordinate axes are presented, one in the horizontal and vertical directions and the other in the directions parallel and perpendicular to the ramp. Regardless of the choice we make, it will be necessary to find components of one or both of the force vectors, which demands that we can draw a right triangle with hypotenuse equal to a force vector. To do that, we need to find the relationships between the angles that will appear in such triangles and the ramp angle.

The key to analyzing problems that involve a ramp is that the ramp angle θ is the same as the angle between the vertical direction and the direction perpendicular to the ramp surface. To see that, we have redrawn the ramp in Figure 4-28, adding dashed lines to indicate the horizontal and vertical directions as well as the direction perpendicular to the ramp. Three angles labeled α, β, and γ have also been added. We can take advantage of the properties of right angles to find a relationship between them. First, note that α and β together form a right angle, so

$$\alpha + \beta = 90°$$

The angles β and γ also form a right angle, so

$$\gamma + \beta = 90°$$

Thus $\alpha = \gamma$. Also, because the dashed line labeled "x" is parallel to the base of the ramp, angles α and θ are corresponding angles and therefore congruent. The angle γ is therefore equal to the ramp angle θ. So if we need to find, say, the components of the weight (Figure 4-27) in the directions parallel and perpendicular to the ramp, we form a triangle as shown in Figure 4-29. The weight vector is the hypotenuse of the right triangle shown, yielding

$$W_\parallel = W \sin\theta$$

and

$$W_\perp = W \cos\theta$$

An easy way to remember which angle in Figure 4-28 is equal to the ramp angle is to draw this figure by making θ extremely small. A glance at Figure 4-30 leaves no doubt which of the angles β and γ is equal to θ.

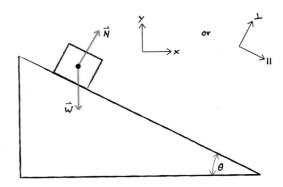

Figure 4-27 The normal force acting on an object is always perpendicular to the surface that supports it. In a free-body diagram of a box on a ramp of angle θ, the normal force is perpendicular to the surface of the ramp. Two choices for the coordinate axes are presented, one in the horizontal and vertical directions and the other in the directions parallel and perpendicular to the ramp.

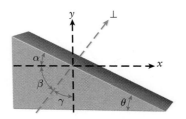

Figure 4-28 The black dashed lines indicate the horizontal and vertical directions; the red one shows the direction perpendicular to the ramp. We use the angles shown to determine the components of the forces acting on the box in directions of interest.

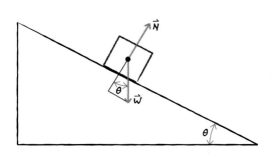

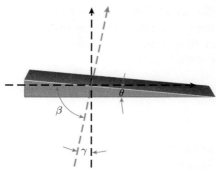

Figure 4-29 The components of the force of gravity on the box (its weight) in the directions parallel and perpendicular to the ramp can be found using the triangle drawn in red. The weight is the hypotenuse of the triangle.

Figure 4-30 An easy way to remember which angle in Figure 4-28 is equal to the ramp angle is to let θ be extremely small. This makes it obvious that γ, the angle between the vertical and the normal to the ramp, is equal to θ.

Example 4-8 Down the Slopes

A skier moves from rest at the top of a hill and reaches the bottom in 3.5 s (**Figure 4-31**). The hill makes a constant angle of 30° to the horizontal. Assuming that no forces resist her motion, how long is the hill?

Figure 4-31 The skier descends the 30° slope in 3.5 s.

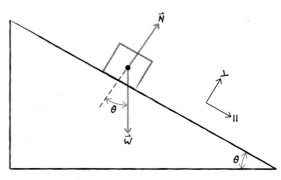

Figure 4-32 Based on the analysis in Figure 4-29 we have labeled the angle between the weight vector and the direction perpendicular to the surface of the ramp as θ, the same as the ramp angle.

SET UP
This example is a linear motion problem, in which displacement is related to initial velocity, acceleration, and time according to

$$x - x_0 = v_0 t + \frac{1}{2}at^2 \tag{2-26}$$

Because the skier starts from rest, $v_0 = 0$; so Equation 2-26 becomes

$$x - x_0 = \frac{1}{2}at^2 \tag{4-13}$$

We have been given the total time of the run down the hill. We can find the acceleration by considering the forces on the skier.

Figure 4-32 shows a free-body diagram for the skier. Note that we have drawn the skier at a generic position down the hill, and that we have represented the skier as a block. (We enjoy having a little fun when drawing sketches, but it's better to use simple shapes for objects, so as not to obscure the relationship between the forces.) Based on the analysis in Figure 4-29 we have labeled the angle between the weight vector and the direction perpendicular to the surface of the slope as θ, the same as the slope angle.

For the coordinate axes, we selected the directions parallel and perpendicular to the surface. One of the two forces, $\vec{N}$, is aligned with one of these directions and the skier's acceleration vector is aligned with the other. Next, we write separate Newton's second law equations for the skier for each direction:

$$\| : \quad \vec{N}_{\|} + \vec{W}_{\|} = m_{\text{skier}}\vec{a}_{\|} \qquad\qquad (4\text{-}14)$$

$$\perp : \quad \vec{N}_{\perp} + \vec{W}_{\perp} = m_{\text{skier}}\vec{a}_{\perp} \qquad\qquad (4\text{-}15)$$

SOLVE

We can simplify Equations 4-14 and 4-15 to find a. Because $\vec{N}$ is perpendicular to the surface of the slope, $\vec{N}_{\|} = 0$ and $\vec{N}_{\perp} = N$. Because the skier is neither leaping up off the hill nor pushing down into it, the acceleration perpendicular to the slope, $\vec{a}_{\perp}$, equals zero. We can therefore also set the parallel component of acceleration equal to the magnitude of the acceleration, that is, the magnitude of $\vec{a}_{\|}$ equals a. Finally, $W_{\|} = W \sin\theta$ and $W_{\perp} = W \cos\theta$, as the analysis in Figure 4-29 shows. So Equations 4-14 and 4-15 become

$$\| : \quad 0 + W \sin\theta = m_{\text{skier}}a$$

$$\perp : \quad N + W \cos\theta = 0$$

Only the sum of the forces along the slope is necessary to find a:

$$a = \frac{W \sin\theta}{m_{\text{skier}}}$$

The skier's weight can be expressed as $W = m_{\text{skier}}g$, so

$$a = \frac{m_{\text{skier}}g \sin\theta}{m_{\text{skier}}} = g \sin\theta \qquad\qquad (4\text{-}16)$$

Notice that the answer does not depend on the skier's mass, which is expected because the acceleration due to gravity never depends on mass.

We can now substitute this acceleration into Equation 4-13, which yields

$$x - x_0 = \frac{1}{2}g \sin\theta t^2$$

The length of the hill is

$$x - x_0 = \frac{1}{2}(9.8 \text{ m/s}^2)\sin 30°(3.5 \text{ s})^2 = 3.0 \times 10^1 \text{ m}$$

REFLECT

Let's test the result by considering the value of acceleration (Equation 4-16) in some special limiting cases for which we can guess this result. First, consider the skier on a horizontal surface, that is, a "hill" at an angle of 0°. We would expect there to be no acceleration, and Equation 4-16 gives exactly this for $\theta = 0°$. What if the hill were actually the face of a vertical cliff? In that case, $\theta = 90°$ so $a = g \sin 90° = g$. This result suggests that the skier would be in free fall, which would certainly be the case if she skied down a sheer cliff. (Don't try that at home.)

Practice Problem 4-8 A 300-m-long ski run follows a hill that makes a constant angle of 25.0° to the horizontal. Assuming that there are no forces which resist his motion, how long would it take a skier to reach the bottom of the slope, starting from the top at rest?

★ **What's Important 4-6**
The best way to understand a problem involving forces is to draw a free-body diagram.

Answers to Practice Problems

4-1	4.0 m/s^2	4-4	1100 N
4-2	Magnitude equals 11.6 N, direction is 20.9° from the positive x direction	4-5	1.5 m/s^2
		4-8	12.0 s
4-3	1.9 m/s^2		

Answers to Got the Concept Problems

4-1 The books do not slide as the result of an applied force. Because they are inside your car, the books are in motion before you hit your brakes. The books have inertia, so according to Newton's first law they tend to stay in motion—until they experience a force that causes the motion to change. *That* force is the result of the books striking the back of the front seat, which has (or is about to) come to rest.

4-2 Newton's second law holds even though the two spheres are weightless. In addition, each sphere has its same mass regardless of the force exerted by gravity. That means that you would need to exert a larger force on the lead sphere compared to the plastic sphere in order to impart the same acceleration to both. Both spheres resist the change in motion, but because the amount of resistance depends on the sphere's mass, it is harder to move the more massive one.

4-3 The astronaut's mass is the same in both cases, because mass is an intrinsic property of the astronaut and therefore does not change. The astronaut's weight does change, however, according to the magnitude of the acceleration due to gravity she experiences. On Earth, where $a_g = g = 9.8 \text{ m/s}^2$, her weight is $W_{\text{Earth}} = 588$ N. On Mars, where $a_g = 3.7 \text{ m/s}^2$, her weight is $W_{\text{Mars}} = 222$ N. And no, a standard bathroom scale would not show the same reading for Barbara Morgan's weight on Earth as it would in the space shuttle. In space she would not exert a force on the scale nor would the scale exert a force on her, so because she and the scale are both in free fall, her weight would be 0 N.

4-4 It is true that when you push on the crate, the crate pushes on you according to our simplified version of Newton's third law. However, these two forces clearly act on two different objects. Remember that one of the keys to applying Newton's third law is considering all forces only on one object at a time. There is only one horizontal force on the crate, and therefore its acceleration is not zero.

4-5 The most important aspect of this problem is that you are *not* applying a force to crate 2. How is that possible, if crate 2 accelerates, as it certainly must? Recall a fundamental rule about forces: Other than gravity and a few special exceptions that we haven't yet encountered, an object must touch another to exert a force on it. In this problem, you are not touching crate 2, so you cannot exert a force on it. But crate 1 is touching crate 2, so the contact force between the two crates causes crate 2 to move. In addition, if crate 1 pushes on crate 2, then crate 2 must push back on crate 1 according to Newton's third law. This force pair, as well as other forces on the crates, is shown in **Figure 4-33**. Notice that the force which is equal but opposite to the force you apply to crate 1 ($\vec{F}_{\text{you}\rightarrow 1}$) is not shown in Figure 4-33. This is the force that crate 1 exerts on you, and we have been asked to consider only the forces on the crates.

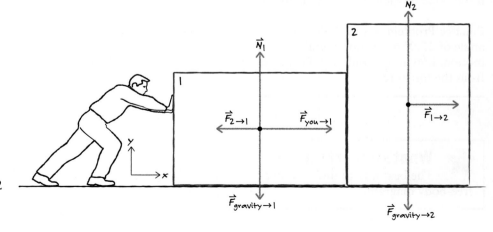

Figure 4-33 A free-body diagram shows all of the forces acting on the two crates, including the force of crate 1 pushing on crate 2 and its pair, the force of crate 2 pushing back on crate 1.

You might be tempted to draw a force on crate 2 pointing from right to left, so that the pattern of forces on both crates is the same. What would the origin or agent of such a force be? There is no object to the right of crate 2 that could exert a right-to-left force on it, and gravity acts vertically. Also, Newton's third law demands that for every force crate 2 exerts, there must be an equal but opposite force on it. Because crate 2 exerts only one horizontal force—to the left on crate 1—there is only one horizontal force acting back on it. We have already accounted for that force

($\vec{F}_{1\rightarrow2}$) so there can be no force acting to the left on crate 2.

4-6 When you push on either sphere, a reaction force pushes back on you according to Newton's third law. We realized in Got the Concept 4-2 that to achieve the same acceleration you would need to exert a stronger force on the lead sphere than on the plastic sphere. As a result, the reaction force exerted by the lead sphere is larger ("equal but opposite") than that exerted by the plastic sphere, so you will be pushed farther backward when you push on the lead sphere.

SUMMARY

Topic	Summary	Equation or Symbol
contact forces	Contact forces are the forces between two objects that are touching each other.	
force	A force is a push or a pull. When an object experiences a net force its motion changes, according to Newton's second law.	$\vec{F}$
force pair	According to Newton's third law, if A pushes on B, then B pushes on A with equal force but in the opposite direction. The two forces involved in such an interaction are a force pair.	
free-body diagram	A free-body diagram is a graphical representation of all of the forces acting on an object or system of objects.	
inertia	Inertia is the tendency of an object to resist a change in motion. Inertia is associated with mass.	
long-range forces	Some long-range forces, such as gravity or the magnetic force, appear to act without any direct contact, although the fundamental theory of physics suggests that these forces are exerted as the result of objects exchanging small units of energy or matter with one another.	
mass	The quantity mass is a measure of how much matter—material or "stuff"—an object has.	m
Newton's first law	An object at rest tends to stay at rest and an object in uniform motion tends to stay in motion with the same speed and in the same direction unless acted on by a net force.	
Newton's second law	The vector sum of all forces acting on a single object in a particular direction equals the mass of the object multiplied by the acceleration of the object in that direction.	$\sum \vec{F}_{dir} = m\vec{a}_{dir}$ (4-1)
Newton's third law	All forces occur in pairs, and these two forces are equal in magnitude and opposite in direction. If A pushes on B, then B pushes on A with equal force but in the opposite direction. For every force there is an equal but opposite reaction force.	

normal force	A normal force is the force a surface exerts to support an object. A normal force is exerted in the direction perpendicular, or normal, to the surface.	$\vec{N}$
tension force	A tension force, or tension, is a pulling force exerted on an object along a string, wire, or similar linear object attached to it.	$\vec{T}$
weight	Weight is the force that acts on an object due to gravity. The magnitude of an object's weight is the product of its mass and the acceleration due to gravity.	$W = mg$ (on Earth's surface) (4-3)

QUESTIONS AND PROBLEMS

In a few problems, you are given more data than you actually need; in a few other problems, you are required to supply data from your general knowledge, outside sources, or informed estimate.

Interpret as significant all digits in numerical values that have trailing zeros and no decimal points.

For all problems, use $g = 9.8$ m/s^2 for the free-fall acceleration due to gravity. Neglect friction and air resistance unless instructed to do otherwise.

• Basic, single-concept problem
•• Intermediate-level problem, may require synthesis of concepts and multiple steps
••• Challenging problem
SSM *Solution is in Student Solutions Manual*

Conceptual Questions

1. •According to Newton's second law, does the direction of the net force always equal the direction of the acceleration? SSM

2. •If the sum of the forces acting on an object equals zero, does this imply that the object is at rest?

3. •**Calc** What are the basic SI units (kg, m, s) for force according to Newton's second law?

$$\sum F = ma$$

How do the units compare with the differential form of this law?

$$\sum F = \frac{d(mv)}{dt}$$

4. •**Medical, Sports** Gymnastics routines are done over a padded floor to protect athletes who fall. Why is falling on padding safer than falling on concrete?

5. •A definition of the inertia of an object is that it is a measure of the quantity of matter. How does this definition compare with the definition discussed in the chapter?

6. •When constructing a free-body diagram, why is it a good idea to choose your coordinate system so that the motion of an object is along one of the axes?

7. •A certain rope will break under any tension greater than 800 N. How can it be used to lower an object weighing 850 N over the edge of a cliff without it breaking? SSM

8. •**Sports** A boxer claims that Newton's third law helps him while boxing. He says that during a boxing match, the force that his jaw feels is the same as the force that his opponent's fist feels (when the opponent is doing the punching). Therefore, his opponent will feel the same force as he feels and he will be able to fight on without any problems, no matter how many punches he receives or gives. It will always be an "even fight." Discuss any flaws in his reasoning.

9. •**Astro** We know that the Sun pulls on Earth. Does Earth also pull on the Sun? Why or why not? SSM

10. •**Sports** How can a fisherman land a 5-lb fish using fishing line that is rated at 4 lb?

11. •Tension is a very common force in day-to-day life. Identify five ordinary situations that involve the force of tension.

12. •Explain why the force that a surface exerts on an object that rests on it is called the normal force.

13. •A chair is mounted on a scale inside an elevator in the physics building at the University of California, Davis. Describe the variation in the scale reading as the elevator begins to ascend, goes up at constant speed, stops, begins to descend, goes down at a constant speed, and stops again.

14. •List all the forces acting on a soft-drink can if it were sitting on your desk.

15. •What is the net force on a bathroom scale when a 75-kg person stands on it? SSM

16. •An object of mass m is being weighed in an elevator which is moving upward with an acceleration a. What is the result if the weighing is done using (a) a spring balance and (b) a pan balance?

17. •Two forces of 30 N and 70 N act on an object. What are the minimum and maximum values for the sum of these two forces?

18. •You apply a 60-N force to push a box across the floor at constant speed. If you increase the applied force to 80 N, will the box speed up to some new constant speed or will it continue to speed up indefinitely? Assume that the floor is horizontal and the surface is uniform. Explain your answer.

19. •**Astro** Why would it be easier to lift a truck on the Moon's surface than it is on Earth? SSM

20. •**Medical** Why should the driver and passengers in a car wear seatbelts? Explain your answer.

21. •**Medical** Why does the American Academy of Pediatrics recommend that all infants sit in rear-facing car seats starting with their first ride home from the hospital? Explain your answer.

22. •**Medical** Use Newton's third law to explain the forces involved in walking.

23. •**Biology** Use Newton's third law to explain how birds are able to fly forward. SSM

Multiple-Choice Questions

24. •**Medical** A car stops suddenly during a head-on collision, causing the driver's brain to slam into the skull. The resulting injury would most likely be to which part of the brain?
    A. Frontal portion of the brain
    B. Rear portion of the brain
    C. Middle portion of the brain
    D. Left side of the brain
    E. Right side of the brain

25. •**Medical** During the sudden impact of a car accident, a person's neck can experience abnormal forces, resulting in an injury commonly known as *whiplash*. If a victim's head and neck move in the manner shown in **Figure 4-34**, his car was hit from the
    A. front.
    B. rear.
    C. right side.
    D. left side.
    E. top during a rollover. SSM

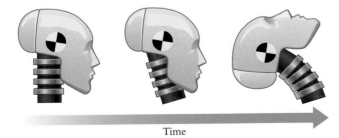

Time

Figure 4-34 Problem 25

26. •The net force on a moving object suddenly becomes zero and remains zero. The object will
    A. stop abruptly.
    B. reduce speed gradually.
    C. continue at constant velocity.
    D. increase speed gradually.
    E. reduce speed abruptly.

27. •Which has greater monetary value, a newton of gold on Earth or a newton of gold on the Moon?
    A. The newton of gold on Earth
    B. The newton of gold on the Moon
    C. The value is the same, regardless of location
    D. One cannot say without checking the weight on the Moon.
    E. The newton of gold on the Moon but only when inside a spaceship.

28. •According to Newton's second law of motion, when a net force acts on an object, the acceleration is
    A. zero.
    B. inversely proportional to the object's mass.
    C. independent of mass.
    D. inversely proportional to the net force.
    E. Both options (A) and (B) are correct.

29. •When a net force acts on an object, it
    A. is at rest.
    B. is in motion with a constant velocity.
    C. has zero speed.
    D. is accelerating.
    E. is decelerating.

30. •In the absence of a net force, an object cannot be
    A. at rest.
    B. in motion with a constant velocity.
    C. accelerating.
    D. moving with an acceleration of zero.
    E. experiencing opposite but equal forces.

31. •In the absence of a net force, an object can be
    A. at rest.
    B. in motion with a constant velocity.
    C. accelerating.
    D. at rest or in motion with a constant velocity.
    E. It's not possible to know without more information. SSM

32. •Elevators decelerate to stop at a floor. During the deceleration of an elevator traveling upward, the normal force on the feet of a passenger is _____ her weight. During the deceleration of an elevator traveling downward, the normal force on the feet of a passenger is _____ his weight.
    A. larger than; smaller than
    B. larger than; larger than
    C. smaller than; smaller than
    D. smaller than; larger than
    E. equal to; equal to

This page is intentionally left blank.

For complete end of chapter problem sets, please go to
www.whfreeman.com/kestentauck

46. •The graph in Figure 4-37 shows the force applied to a 2.5-kg object as a function of time. Calculate the velocity of the object at $t = 1$ s and $t = 4$ s. Assume the object starts from rest.

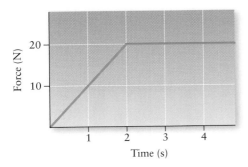

Figure 4-37 Problem 46

Problems

4-2: Newton's Second Law

47. •What is the acceleration of a 2000-kg car if the net force on the car is 4000 N?

48. •What net force is needed to accelerate a 2000-kg car at 2 m/s²?

49. •Suppose that the engine of a car delivers a maximum force of 15,000 N. In the absence of any other forces, what is the maximum acceleration this force can produce in a 1250-kg car? SSM

50. ••Applying a constant force to an object causes it to accelerate at 10 m/s². What will the acceleration of the object be if (a) the force is doubled, (b) the mass is halved, (c) the force is doubled and the mass is doubled, (d) the force is doubled and the mass is halved, (e) the force is halved, (f) the mass is doubled, (g) the force is halved and the mass is halved, and (h) the force is halved and the mass is doubled?

4-3: Mass and Weight

51. •What is the weight on Earth of a wrestler who has a mass of 120 kg?

52. •A wrestler has a mass of 120 kg. What are (a) his mass and (b) weight on the Moon, where g is 1.62 m/s²?

53. •A bluefin tuna has a mass of 250 kg. What is its weight? SSM

54. •(a) How does your mass differ as measured in Mexico City versus Los Angeles? (b) In which city is your weight greater?

55. ••An astronaut has a mass of 80 kg. How much would the astronaut weigh on Mars where surface gravity is 38% of that on Earth?

4-4: Free-Body Diagrams

56. •Draw a free-body diagram for a heavy crate being lowered by a steel cable straight down at a constant speed.

57. •Draw a free-body diagram for a person pushing a box across a smooth floor at a steadily increasing speed. Ignore the friction between the box and the smooth floor. SSM

58. •Draw a free-body diagram for a bicycle rolling down a hill. Ignore the friction between the bicycle wheels and the hill, but consider any air resistance.

59. ••A tugboat uses its winch to pull up a sinking sailboat with an upward force of 4500 N. The mass of the boat is 200 kg and the water acts on the sailboat with a drag force of 2000 N. Draw a free-body diagram for the sailboat and describe the motion of the sailboat.

60. ••Two forces are applied to a block of mass M equal to 14 kg that sits on a horizontal, frictionless table. One force $\vec{F}_1$ has a magnitude of 10 N and is directed at an angle θ_1 equal to 37° from the x direction as labeled in Figure 4-38. As a result of the forces, the block experiences an acceleration a equal to 0.5 m/s² in the x direction only. (a) Find the x and y components of the second force. (b) Draw the second force in its correct position and to scale in the picture.

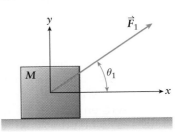

Figure 4-38 Problem 60

61. ••Three rugby players are pulling horizontally on ropes attached to a box, which remains stationary. Player 1 exerts a force F_1 equal to 100 N at an angle θ_1 equal to 60° with respect to the +x direction (Figure 4-39). Player 2 exerts a force F_2 equal to 200 N at an angle θ_2 equal to 37° with respect to the +x direction. The view in the figure is from above. Ignore friction and note that gravity plays no role in this problem! SSM

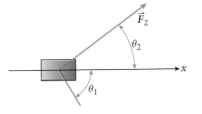

Figure 4-39 Problem 61

(a) Find the force F_3 exerted by Player 3. State your answer by giving the components of F_3 in the directions perpendicular to and parallel to the positive x direction. (b) Redraw the diagram and add the force F_3 as carefully as you can. (c) Player 3's rope breaks, and Player 2 adjusts by pulling with F_2 equal to 150 N at the same angle as before. In which direction is the acceleration of the box, relative to the direction shown? (d) In part (c), the acceleration is measured to be 10 m/s². What is the mass of the box?

62. •••Three forces act on a 2-kg object as shown in Figure 4-40. Find the magnitude of each force if the acceleration of the object is 1.5 m/s² toward the +x direction.

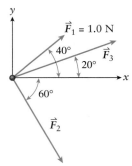

Figure 4-40 Problem 62

63. ••Box A weighs 80 N and rests on a table (Figure 4-41). A rope that connects boxes A and B drapes over a pulley so that box B hangs above the table, as shown in the figure. The pulley and rope are massless and the pulley is frictionless. What force does the table exert on box A if box B weighs (a) 35 N; (b) 70 N; (c) 90 N?

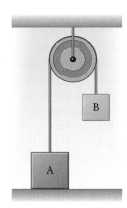

Figure 4-41 Problem 63

64. ••Students pull on two ropes attached to an object of mass M; they apply horizontal forces F_1 and F_2, respectively. Because of these forces, the object experiences an acceleration of a equal to 7 m/s² in the x direction, as shown in Figure 4-42. (The figure shows the object and force F_1 when viewed from above.) (a) Find the magnitude of F_2 and (b) make a careful drawing to show its direction, given $F_1 = 20$ N, $\theta = 30°$, and $M = 3$ kg.

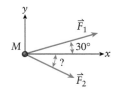

Figure 4-42 Problem 64

65. ••A 100-kg streetlight is supported equally by two ropes as shown in Figure 4-43. One rope pulls up and to the right, 40° above the horizontal; the other rope pulls up and to the left, 40° above the horizontal. Find the tension in each rope. SSM

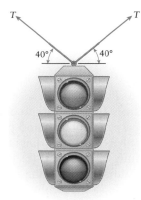

Figure 4-43 Problem 65

66. •••A 200-N sign is supported by two ropes; one rope pulls up and to the right, 30° above the horizontal, and the other pulls up and to the left 45° above the horizontal (Figure 4-44). Find the tension in each rope.

Figure 4-44 Problem 66

67. ••The distance between two telephone poles is 50 m. When a 0.50-kg bird lands on the telephone wire midway between the poles, the wire sags 0.15 m. How much tension does the bird produce in the wire? Ignore the weight of the wire.

4-5: Newton's Third Law

68. •A ball that has a weight of 40 N is falling freely toward the floor. What force does the ball exert on the floor when it hits? Neglect effects of air resistance.

69. •A locomotive pulls 10 identical freight cars. The force between the locomotive and the first car is 100,000 N and the acceleration of the train is 2 m/s². There is no friction to consider. Find the force between the ninth and tenth cars. SSM

70. •A locomotive pulls 10 identical freight cars with an acceleration of 2 m/s². How does the force between the third and fourth cars compare to the force between the seventh and eighth cars?

4-6: Force, Acceleration, Motion

71. •What is the net force on an apple that weighs 3.5 N when you hold it at rest in your hand?

72. •A block that has a mass of 0.01 kg and a block that has a mass of 2 kg are attached to the ends of a rope. A student holds the 2-kg block and lets the 0.01-kg block hang below it, then he lets go. What is the tension in the rope while the blocks are falling, before either hits the ground? Air resistance can be neglected.

73. ••A 2-kg object experiences the three forces shown in Figure 4-45. What is the acceleration of the object? Note that the forces are measured in newtons. SSM

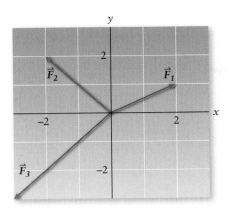

Figure 4-45 Problem 73

74. •While parachuting, a 66-kg person experiences a downward acceleration of 2.5 m/s². What is the downward force on the parachute from the person?

75. •A bicycle and 50-kg rider accelerate at 1.0 m/s² up an incline of 10° above the horizontal. What is the magnitude of the force that the bicycle exerts on the rider?

76. •A car accelerates from 0 to 100 km/h in 4.5 s. What force does a 65.0-kg passenger experience during this acceleration?

77. •A car uniformly accelerates from 0 to 28 m/s. A 60.0-kg passenger experiences a horizontal force of 400 N. How much time does it take for the car to reach 28 m/s? SSM

78. •Adam and Ben pull hand over hand on opposite ends of a rope while standing on a frictionless frozen pond. Adam's mass is 75 kg and Ben's mass is 50 kg. If Adam's acceleration is 1 m/s² to the east, what are the magnitude and direction of Ben's acceleration?

79. ••Calc A 2-kg object moves in the x direction according to the following function:

$$x(t) = 2t^2 + 3t - 5 \quad \text{(SI units)}$$

What is the force on the object after 3 s?

80. ••Two blocks M_1 and M_2 are connected by a massless string that passes over a massless pulley (**Figure 4-46**). M_2, which has a mass of 20 kg, rests on a long ramp of angle θ equal to 30°. Friction can be ignored in this problem. (a) Find the value of M_1 for which the two blocks are in equilibrium (no acceleration). (b) If the actual mass of M_1 is 5 kg and the system is allowed to move, find the acceleration of the two blocks. (c) In part (b) does M_2 move up or down the ramp? (d) In part (b), how far does block M_2 move in 2 s?

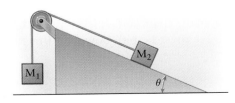

Figure 4-46
Problem 80

81. ••In **Figure 4-47**, the block on the left incline is 6 kg. Find the mass of the block on the right incline so that the system is in equilibrium (no acceleration). All surfaces are frictionless. SSM

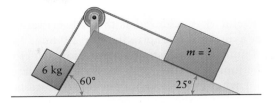

Figure 4-47 Problem 81

General Problems

82. ••A child on a sled starts from rest at the top of a 20° slope. Assuming that there are no forces resisting the sled's motion, how long will the child take to reach the bottom of the slope, 210 m from the top?

83. ••A car is proceeding at a speed of 14 m/s when it collides with a stationary car in front. During the collision, the first car moves a distance of 0.3 m as it comes to a stop. The driver is wearing her seat belt, so she remains in her seat during the collision. If the driver's mass is 52 kg, how much force does the belt exert on her during the collision?

84. ••Your friend's car runs out of fuel and you volunteer to push it to the nearest gas station. You carefully drive your car so that the bumpers of the two cars are in contact and then slowly accelerate to a speed of 2.0 m/s over the course of 1 min. If the mass of your friend's car is 1200 kg, what is the contact force between the two bumpers?

85. ••A 30-kg golden retriever stands on a digital scale. If the dog and scale are positioned in a elevator, calculate the weight of the dog when the elevator (a) accelerates at 3.5 m/s² downward, (b) when the elevator cruises down at a steady speed, and (c) when the elevator accelerates at 4.0 m/s² upward. SSM

86. ••What is the weight of a rider on the Inclinator (the lift at the Luxor Hotel in Las Vegas)? Assume the weight of the person is 588 N when at rest and that the Inclinator moves at a constant acceleration of 1.25 m/s², in a direction 39.0° above the horizontal. Assume that the rider stands vertically in the elevator car.

87. ••Describe the variation in the weight of a rider on an elevator that initially starts from rest, accelerates upward at 3 m/s², cruises upward at 4 m/s, slows to a stop at 2 m/s², then free-falls all the way to the bottom of the elevator shaft before striking springs that bring the car to a safe stop. Assume the rider has a stationary weight of 700 N.

88. ••Calc The speed of an object that has mass m moving along the x axis is given by the following function:

$$v(x) = bx^2$$

where $b = 8/(s \cdot m)$. (a) Derive an equation for the force as a function of x. (*Hint:* Use the chain rule.) (b) Calculate the force when $m = 2$ kg and $x = 10$ m.

89. ••**Biology** On average, froghopper insects have a mass of 12.3 mg and jump to a height of 428 mm. The takeoff velocity is achieved as the little critter flexes its leg over a distance of approximately 2.0 mm. Assume a vertical jump with constant acceleration. (a) How long does the jump last (the jump itself, not the time in the air), and what is the froghopper's acceleration during that time? (b) Make a free-body diagram of the froghopper during its leap (but before it leaves the ground). (c) What force did the ground exert on the froghopper during the jump? Express your answer in millinewtons and as a multiple of the insect's weight. SSM

90. ••**Sports** The tension at which a fishing line breaks is commonly called the line's "strength." A minimum strength of 300 N is needed to keep a line from breaking when the hook gets snagged on a floating log, which is initially drifting at 2.8 m/s. What is the mass of the log? Assume a constant deceleration.

91. ••**Sports** A 22-g arrow is positioned in a bow as shown in **Figure 4-48**. If the tension in the bowstring is 180 N, calculate the acceleration of the arrow when it is released. Assume the angle that is shown is 25° and the tension in the string is 100% transmitted to the arrow.

Figure 4-48 Problem 91

92. ••A fuzzy die that has a weight of 1.8 N hangs from the ceiling of a car by a massless string. The car travels on a horizontal road and has an acceleration 2.7 m/s² to the left. The string makes an angle θ with respect to the vertical, shown in **Figure 4-49**. Find the angle θ.

Figure 4-49 Problem 92

93. ••Two mountain climbers are working their way up a glacier when one falls into a crevasse (**Figure 4-50**). The icy slope can be considered frictionless. Sue's weight is pulling Paul up the 45° slope. If Sue's mass is 66 kg and if she falls 2 m in 10 s (starting from rest), find (a) the tension in the rope joining them and (b) Paul's mass. SSM

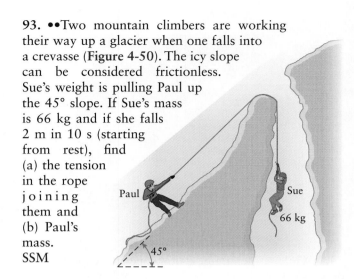

Figure 4-50 Problem 93

94. ••**Medical** A car traveling at 28 m/s hits a bridge abutment. A passenger in the car, who has a mass of 45 kg, moves forward a distance of 55 cm while being brought to rest by an inflated air bag. Assuming that the force that stops the passenger is constant, what is the magnitude of force acting on the passenger?

95. ••**Medical** During a front-end car collision from 48 km/h and with air bags inflated, the acceleration limit for the chest is $60g$ as the car comes to rest. The trunk of the body comprises about 43% of body weight. (a) For how much time does this acceleration last, assuming it is constant? (b) Draw a free-body diagram of a person during the crash. (c) What force in newtons does the air bag exert on the chest of a 72-kg person? (d) Why doesn't the force injure the person?

96. ••A 200-kg block is hoisted up by two massless pulleys as shown in **Figure 4-51**. If a force of 1500 N is applied to the massless rope, what is the acceleration of the suspended mass?

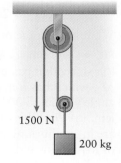

Figure 4-51 Problem 96

97. ••A window washer sits in a *bosun's chair* that dangles from a massless rope that runs over a massless, frictionless pulley and back down to the man's hand. The combined mass of man, chair, and bucket is 95.0 kg. With how much force must he pull downward to raise

himself (a) slowly at constant speed and (b) with an upward acceleration 1.5 m/s²? SSM

98. ••A 9800-kg alpine cable car hanging on a support cable is accelerated up a 30° incline at 0.78 m/s² by a second cable attached to a support tower. Assuming the cables are straight, what is the difference in tension between the section of cable above the car and the section below the car?

99. ••Two blocks are in contact on a horizontal, frictionless surface. Block 1 has a mass of 1.0 kg and block 2 has a mass of 0.5 kg. If a horizontal force with a magnitude of 7.5 N is applied to block 1, what is the magnitude of the force that acts on block 2?

100. •••Three boxes are lined up so that they are touching each other as shown in **Figure 4-52**. Box A has a mass of 20 kg, box B has a mass of 30 kg, and box C has a mass of 50 kg. If an external force (F) pushes on box A toward the right, and the force that box B exerts on box C is 200 N, find the acceleration of the boxes and the value of the external force F.

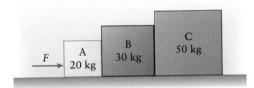

Figure 4-52 Problem 100

101. •••Two blocks connected by a light string are being pulled across a frictionless horizontal tabletop by a hanging 10-N weight (block C) (**Figure 4-53**). Block A has a mass of 2 kg. The mass of block B is only 1 kg. The blocks gain speed as they move toward the right, and the strings remain taut at all times. What are the values of the tensions T_1 and T_2? SSM

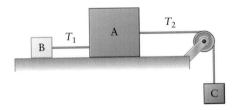

Figure 4-53 Problem 101

102. ••A 2-kg object is connected with a massless string across a massless, frictionless pulley to a 3-kg object (**Figure 4-54**). The smaller object rests on an inclined plane which is tilted at 40° as shown. Find the acceleration of the system and the tension in the string.

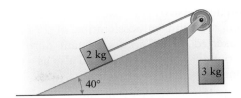

Figure 4-54 Problem 102

103. •••A 1-kg object is connected with a string to a 2-kg object, which is connected with a second string over a massless, frictionless pulley to a 4-kg object (**Figure 4-55**). The 2-kg object is connected with a second string over a massless, frictionless pulley to a 4-kg object as shown. Find the acceleration of the system and the tension in both strings. Assume the strings have negligible mass and do not stretch, and assume the level tabletop is frictionless.

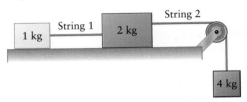

Figure 4-55 Problem 103

104. •••A 1-kg object is connected with a string to a 2-kg object, which is connected with a second string over a massless, frictionless pulley to a 4-kg object (**Figure 4-56**). The first two objects are placed onto a frictionless inclined plane. Find the acceleration of the masses and the tensions in both strings.

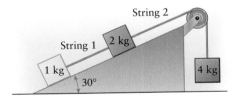

Figure 4-56 Problem 104

105. •••A compound Atwood's machine is constructed as shown in **Figure 4-57**. Calculate the acceleration of the 2-kg block when the system is released from rest.

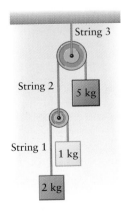

Figure 4-57 Problem 105

106. •**Medical, Astro** The space shuttle takes off vertically from rest with an acceleration of 29 m/s². What force is exerted on a 75-kg astronaut during takeoff? Express your answer in newtons and also as a multiple of the astronaut's weight on Earth.

107. ••A bulldozer pushes a mound of dirt with a constant force of 3500 N and at a constant speed of 4 m/s. At what rate does the dirt "build up" in front of the blade of the bulldozer?

108. •••**Sports** An athlete drops from rest from a platform 10 m above the surface of a 5-m-deep pool. Assuming that the athlete enters the water vertically and moves through the water with constant acceleration, what is the minimum average force the water must exert on a 62-kg diver to prevent her from hitting the bottom of the pool? Express your answer in newtons and also as a multiple of the diver's weight.

109. •••A person pulls three crates over a smooth horizontal floor as shown in **Figure 4-58**. The crates are connected to each other by identical horizontal strings A and B, each of which can support a maximum tension of 45.0 N before breaking. (a) What is the largest force that can be exerted without breaking either of the strings? (b) What are the tensions in A and B just before when string breaks?

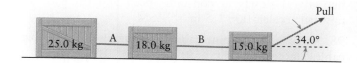

Figure 4-58 Problem 109

110. •••A 275-g hard rubber ball falls from a height of 2.20 m. After each bounce, it rebounds to 75.0% of its previous height. High-speed photos reveal that the ball is in contact with the floor for 18.4 ms during each bounce. (a) Draw a free-body diagram of the ball before it hits the floor, (b) while it is in contact with the floor, and (c) after it has bounced and is rising. (d) What average force (magnitude and direction) does the floor exert on the ball during the first bounce?

111. •••The mass of an object changes with time according to $m(t) = m_0 e^{-bt}$, where m_0 is the initial mass and b is a proportionality constant with units of s^{-1}. The velocity of the object also changes with time, according to $v(t) = at + v_0$, where v_0 is the initial velocity and a is the object's constant acceleration. (a) Determine an expression for the force on the object at any time t. (b) Determine the force when $m_0 = 2$ kg, $b = 0.16$ s^{-1}, $v_0 = 1$ m/s, $a = 6$ m/s², and $t = 3$ s.

(Ocean/Corbis.)

5 Applications of Newton's Laws

5-1 Static Friction

5-2 Kinetic Friction

5-3 Working with Friction

5-4 Drag Force

5-5 Forces and Uniform Circular Motion

An acrobat supports herself on a hanging rope. Gravity exerts a force on her but she is not accelerating downward, which means that the sum of the vertical forces exerted on her must be zero. A second force, between the acrobat and the rope, resists her motion. Notice that she has twisted the rope around her leg. The *frictional* force which arises from the large area of contact between the rope and her body is strong enough to support her entire weight against the pull of gravity.

Imagine a group of people pushing a car that has run out of gas. Even though they exert a considerable force on it, the car might not move. Before you attribute the lack of motion simply to the fact that the car is massive, consider that Newton's second law requires that unless the sum of the forces on it equals zero the car must accelerate from its initial rest position. Some force must therefore be opposing the force exerted by the people. In **Figure 5-1**, a competitor in the Vancouver 2010 Winter Olympic Games watches as a curling stone gently comes to rest after sliding more than 45 m on an icy surface. The speed of the stone has changed; according to Newton's second law, the change in speed requires the presence of a net force. Both the car and the curling stone experience a force that opposes the motion. We refer to this type of resistive force as **friction**. A force that resists the motion of an object through air or a liquid is commonly referred to as a **drag force**. Friction and drag forces always oppose motions of the objects on which they act.

5-1 Static Friction

Geckos have the remarkable ability to climb vertical, smooth surfaces as shown in **Figure 5-2**. Some insects, such as ants, bees, and flies, have sticky pads on their feet that help them gain traction. Not the gecko, however. No sticky residue is left behind after a gecko moves across a surface. Other creatures, such as the tree frog and the flying mouse, have what amounts to suction cups on their feet. It takes a bit of time to cause a suction cup both to adhere and then to release, and geckos move far too rapidly for their climbing ability to arise from suction cups. Instead, the peculiar anatomy of a gecko's feet provides the friction that allows it to climb up vertical walls. In this section we will develop an understanding of how friction works.

Figure 5-1 A competitor in the Vancouver 2010 Winter Olympic Games watches as a curling stone comes to rest after sliding more than 45 m on an icy surface. The force of friction between the stone and the ice causes the stone to slow and stop. (*Davewebbphoto/ Dreamstime.com.*)

Figure 5-2 Resistive forces such as static friction between the gecko's feet and the glass allow the animal to climb this seemingly smooth, vertical pane of glass. *(iStockphoto / Thinkstock.)*

Figure 5-3 Although the cornea feels smooth against our eyelids, in a scanning electron micrograph we see that it is actually quite rough. *(Hossler, Ph.D./Custom Medical Stock Photo.)*

The force of friction acts to prevent one surface from sliding against another. The amount of resistance to motion that two surfaces have when in contact depends on the materials in contact, as well as the force by which the two surfaces are pushed against one another. The magnitude of the frictional force may also depend on characteristics of the surfaces, for example, their roughness. A frictional force can result from the interaction of the microscopic irregularities of two surfaces that are in contact. Even surfaces that appear or feel smooth can be rough at the microscopic level, and the irregularities on one surface can catch or be impeded by the irregularities on another surface. For example, although the cornea, a dome-shaped structure that covers part of the outer surface of your eye, looks smooth and clear, a close-up image of the outer surface layer of the cornea (Figure 5-3) shows considerable irregularities, a good reason that natural lubrication is important for your eyes to function properly. It is worth noting, however, that even between surfaces that appear to be smooth, large frictional forces can arise on the molecular level due to interactions between the atoms or molecules in two surfaces.

Like your cornea, a gecko's toes look smooth, too. At the end of each toe, however, are hundreds of thousands of tiny "hairs" or *setae*, and each seta splits into hundreds of even smaller, flat pads. Each pad, aptly named a *spatula*, is only about 2×10^{-7} m across, so small that it can nestle in between the very smallest irregularities of most surfaces, resulting in a frictional force. Even on surfaces that are free of irregularities, the molecules of the spatula can get so close to the molecules in the surface that very weak intermolecular forces can come into play. The resulting force, which resists slipping along the surface, is small for any individual spatula, but when you multiply that force by the billions of spatulae on a gecko's four feet, the net force can be as much as 20 N. An average gecko weighs less than 0.5 N, which is easily supported by the total resistive force.

The direction of the frictional force is parallel to the surfaces. The magnitude of the frictional force depends on the force with which one surface presses against the other. Specifically, the frictional force f is proportional to the contact force between the two surfaces. Even if one object is pushed at an angle toward a surface, the force pair between the two—the contact force—is in the direction perpendicular to the surfaces. We have defined the contact force to be the normal force N, so

$$f \propto N \tag{5-1}$$

We could use the term F_{fric} to label the frictional force, but the use of lowercase f is conventional.

Consider applying a horizontal force to a stationary block on a horizontal surface. If the block does not accelerate, we conclude that the net force on it is zero; that is, the force of friction exactly cancels the applied force. In this case, we refer to the frictional force as "static"; **static friction** is the force that keeps one surface from sliding with respect to the other. What would happen if even less force were applied to the block? Clearly, the block would still not accelerate. In this case, the force of static friction again exactly cancels the applied force. Therefore, the frictional force is smaller when less force is applied. When an object is not moving, then, the frictional force varies up to some maximum value in order to exactly counter the applied force. We can extend Equation 5-1 to

$$f \leq \mu_s N \tag{5-2}$$

We will usually concern ourselves with the limiting case in which the frictional force is as large as it can be to prevent motion of one surface relative to another, or $f_{\text{max}} = \mu_s N$.

 Watch Out

The relationship between static friction and the normal force is an inequality.

The static force keeps an object from sliding, even as the magnitude of the force pushing on it increases. Up until the point at which the magnitude of the pushing force exceeds the maximum force of static friction, the magnitude of the frictional force exactly counters the pushing force. For that reason we express the relationship between static friction and the normal force (Equation 5-2) as an inequality.

When two contacting surfaces could be moving with respect to each other but are not, the proportionality constant μ_s in Equation 5-2 is called the **coefficient of static friction**. The value of μ_s is a function of *both* surfaces, that is, it is not a characteristic of any one material or surface. For example, it is incorrect to ask about the coefficient of static friction of, say, wood, but it *is* correct to consider the coefficient of static friction between a particular wood surface and a plastic block which rests on it. Because the expressions on both sides of Equation 5-2 have units of force, the proportionality constant μ_s must be dimensionless.

For the materials we will consider, typical values of the coefficient of static friction range between 0.1 and 0.8. For a ski on snow, μ_s is approximately equal to 0.1. The coefficient of static friction for two pieces of wood in contact is approximately equal to 0.5, and for two pieces of metal in contact μ_s is approximately 0.7. Lubrication reduces the frictional force between surfaces. In human joints, such the elbow and knee, the presence of lubricating fluid results in a coefficient of static friction μ_s approximately equal to 0.01. This relatively low coefficient of static friction reduces the wear and tear of the cartilage which supports the joints.

A block at rest on a ramp of angle θ demonstrates the relationship between the frictional force, which resists the motion of the block, the normal force, and the

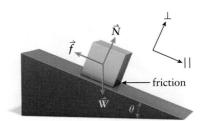

Figure 5-4 A stationary block on a ramp experiences three forces: the force due to gravity, the normal force, and a frictional force.

coefficient of static friction. The free-body diagram, shown in **Figure 5-4**, includes three forces: the force due to gravity, which points down; the normal force, which is perpendicular to the surface of the ramp; and a frictional force, which points in the direction opposite to the direction the block would move if it could. In this case we consider the block to be stationary, so the frictional force is due to static friction.

Following the method for solving problems that involve forces (see Physicist's Toolbox 4-2), set the sum of the forces on the block in each direction equal to mass multiplied by acceleration in that direction:

$$\parallel: \quad \vec{N}_{\parallel} + \vec{W}_{\parallel} + \vec{f}_{\parallel} = m_{\text{block}}\,\vec{a}_{\parallel}$$

$$\perp: \quad \vec{N}_{\perp} + \vec{W}_{\perp} + \vec{f}_{\perp} = m_{\text{block}}\,\vec{a}_{\perp}$$

 See the Math Tutorial for more information on Trigonometry

The components of the weight $\vec{W}$ in the directions parallel to and perpendicular to the surface of the ramp are $W_{\parallel} = W \sin\theta$ and $W_{\perp} = W \cos\theta$ respectively. In addition, $\vec{f}$ points in the negative parallel direction, and $\vec{N}$ points in the positive perpendicular direction, so

$$\parallel: \quad 0 + W \sin\theta - f = m_{\text{block}}(0)$$

$$\perp: \quad N - W \cos\theta + 0 = m_{\text{block}}(0)$$

where we have set $\vec{a}_{\parallel} = 0$ and $\vec{a}_{\perp} = 0$ because the block is at rest and experiencing no acceleration. We can solve the second equation for N in order to express Equation 5-2 in terms of the weight of the block:

$$N = W \cos\theta \qquad (5\text{-}3)$$

so

$$f \leq \mu_s W \cos\theta$$

We can solve the first of the Newton's second law equations for f to get

$$f = W \sin\theta$$

Combining the two equations gives

$$W \sin\theta \leq \mu_s W \cos\theta$$

If we adjust the angle of the ramp so that the block just barely stays in place without sliding down the ramp, the inequality becomes

$$W \sin\theta = \mu_s W \cos\theta$$

or

Figure 5-5 A small wooden block rests without slipping on a ramp covered with sandpaper. *(Courtesy David Tauck.)*

$$\mu_s = \frac{W \sin\theta}{W \cos\theta} = \tan\theta \qquad (5\text{-}4)$$

Watch Out

μ_s can be greater than 1.

Because typical values of the coefficient of static friction are less than 1, it is tempting to assume that μ_s is always less than 1. Equation 5-4 provides direct evidence that this assumption cannot be correct. The tangent of an angle between 45° and 90° is greater than 1, so if a ramp can be set an angle greater than 45° without an object on its surface slipping, μ_s is also greater than 1. In

Figure 5-5, a wooden block rests without slipping on a ramp covered with sandpaper. The ramp is set at angle $\theta = 50°$; so from Equation 5-4, μ_s between block and ramp is at least $\tan 50° = 1.2$.

Example 5-1 Friction in Joints

The wrist is made up of eight small bones called *carpals* that glide back and forth as you wave your hand from side to side. A thin layer of cartilage covers the surfaces of the carpals, making them smooth and slippery. In addition, the spaces between the bones contain a special fluid which provides lubrication. A bioengineer measures the contact force between two nearly planar bone surfaces in the joint as well as the force required to make them move. The contact region between the two carpal bone surfaces is indicated in the x-ray image of a human hand shown in **Figure 5-6**. The minimum force required to move the joint is 0.135 N and the contact force is 11.2 N. What is the coefficient of static friction in the joint?

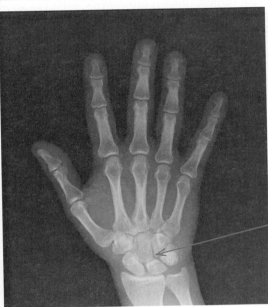

Figure 5-6 The arrow on this x-ray image of a human hand points to the contact region between two carpal bone surfaces. *(Courtesy Ted Szczepanski.)*

Some of the interfaces between the carpal bones in the wrist are nearly planar, making the friction between them analagous to the friction between a block and a flat surface.

SET UP
The free-body diagram for one of the carpal bones (bone A) is shown in **Figure 5-7**. The supporting structure of the hand pushes bone A toward bone B (and B toward A) so that each bone feels both a compression force pushing it toward the adjacent bone and a contact force due to the other bone. The contact force shown is the force of bone B acting on the adjacent bone A. The compression force shown is that of the surrounding tissue of the hand pushing on bone A. According to Equation 5-2, when $F_{applied}$ has the largest possible magnitude without causing the joint to slide, $F_{applied}$ and $F_{contact}$ are related to the coefficient of static friction by

$$F_{applied} = \mu_s F_{contact}$$

SOLVE
So, to compute an answer we substitute in known values from the problem statement

$$\mu_s = \frac{F_{applied}}{F_{contact}} = \frac{0.135 \text{ N}}{11.2 \text{ N}} = 0.012$$

REFLECT
The presence of the lubricating fluid in the space between the smooth and slippery bones is the key to minimizing the friction between the joint surfaces. A coefficient

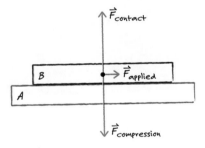

Figure 5-7 A free-body diagram shows the forces acting on two contacting carpal bones in a human wrist. The tissue of the hand pushes both bones together, so that each one feels both a compression force and a contact force due to the other bone. The contact force shown is the force of bone A acting on bone B. The compression force shown is that of bone B acting on bone A, due to the surrounding tissue.

of friction approximately equal to 0.01 is quite small. The fluid also helps reduce wear on the joints. If the bones were in direct contact, or even if the cartilage surfaces were in direct contact, our joints would wear down relatively quickly.

Practice Problem 5-1 A block that has a mass equal to 0.63 kg rests on a horizontal table. The minimum horizontal force required to move the block is 3.1 N. Find the coefficient of static friction between the block and the table.

> ## ✱ What's Important 5-1
> Frictional forces always oppose slipping, or the tendency of one object to slip along another. Drag forces oppose the motion of an object through a fluid such as air or water. We usually refer to friction as a resistive force between two surfaces that can slide with respect to each other while a drag force is one that opposes the motion of an object through a medium like air or liquid. Static friction is a frictional force between two surfaces that could slide against each other but are not moving.

5-2 Kinetic Friction

If the angle that a ramp makes to the horizontal is steep enough, the force of static friction will just barely hold a block in place on the ramp. Imagine that we now give the block a gentle push down the ramp. You might guess that it will start and continue to slide. Once the block starts to move, the frictional force is no longer enough to prevent it from moving. The block is still experiencing a resistive force, but the frictional force between two surfaces that are moving one relative to the other, **kinetic friction**, is typically less than the maximum force of static friction.

Like static friction, the force of kinetic friction between two objects is proportional to the normal force. Also like static friction, we model the force of kinetic friction as independent of the size of the contact area. The force of kinetic friction is then

$$f = \mu_k N \tag{5-5}$$

As we've seen for the coefficient of static friction, the **coefficient of kinetic friction**, μ_k, is a function of both surfaces. Typically, μ_k is close to but less than μ_s for any given pair of surfaces. For the materials we will consider, typical values of the coefficient of kinetic friction range between 0.1 and 0.8.

Consider a block moving at constant velocity on a horizontal surface, where the coefficient of kinetic friction between the two surfaces is μ_k (**Figure 5-8a**). Because velocity is constant, acceleration is zero. So, according to Newton's second law, the net force exerted on the block must be zero. The presence of the force of kinetic friction opposing the motion must be canceled by a force in the direction of motion, for example, someone pushing on the block. **Figure 5-8b** shows the free-body diagram. Note that the force of friction is in the direction opposite to the direction the block would move in the absence of friction.

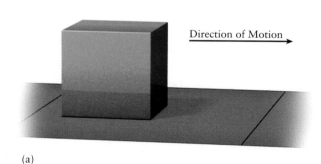

(a)

Direction of Motion

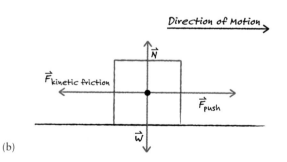

Direction of Motion

$\vec{N}$

$\vec{F}_{\text{kinetic friction}}$

$\vec{F}_{\text{push}}$

$\vec{w}$

(b)

Figure 5-8 (a) A block moves from left to right at constant velocity on a horizontal surface. (b) A free-body diagram shows the forces acting on the block. Notice that the force of friction opposes the motion.

Example 5-2 Down the Slopes—More Realistically

In Example 4-8, we considered a skier accelerating under the influence of gravity down a slope of constant angle θ. The free-body diagram in Figure 4-32 shows two forces, the skier's weight and the normal force. To make the problem more realistic, we should add the force of kinetic friction. If the coefficient of kinetic friction between her skis and the snow on the slope is μ_k, what is the skier's acceleration?

SET UP

The first step of the force problem strategy outlined in Physicist's Toolbox 4-2 is to make a free-body diagram. As shown in **Figure 5-9**, we modify the free-body diagram shown in Figure 4-32 to include the force of kinetic friction, which points up the slope because the frictional force always opposes the motion. We have also selected and labeled coordinate axes.

The final step in our force problem strategy is to write the Newton's second law equations for the skier separately for each direction:

$$\parallel: \quad \vec{N}_{\parallel} + \vec{W}_{\parallel} + \vec{f}_{\text{kinetic}\parallel} = m_{\text{skier}}\vec{a}_{\parallel} \tag{5-6}$$

$$\perp: \quad \vec{N}_{\perp} + \vec{W}_{\perp} + \vec{f}_{\text{kinetic}\perp} = m_{\text{skier}}\vec{a}_{\perp} \tag{5-7}$$

SOLVE

Using both the free-body diagram and the idea that the force of friction opposes the motion of the skis, we call the component of $\vec{f}_{\text{kinetic}}$ in the direction parallel to the slope f and set the component perpendicular to the slope equal to zero. By definition, $\vec{N}$ is perpendicular to the surface of the slope, $W_{\parallel} = W \sin\theta$, and $W_{\perp} = W \cos\theta$. Finally, the acceleration in the perpendicular direction $a_{\perp}$ equals 0 because the skier is neither jumping off the slope nor pushing down into it. This result also means that we can let the acceleration in the direction parallel to the slope $a_{\parallel} = a$. Using the simplifications and following the directions specified by the coordinate axes in Figure 5-9, we change Equations 5-6 and 5-7 into equations relating the magnitudes of the vector components:

$$\parallel: \quad 0 + W \sin\theta - f = m_{\text{skier}}a \tag{5-8}$$

$$\perp: \quad N - W \cos\theta + 0 = m_{\text{skier}}(0) \tag{5-9}$$

Equation 5-8 can be solved for the acceleration,

$$a = \frac{W \sin\theta - f}{m_{\text{skier}}}$$

However, we need to determine the magnitude of friction f to solve the problem completely. Using Equation 5-5 and the relationship for N from Equation 5-9, we can determine f to be

$$f = \mu_k N = \mu_k W \cos\theta$$

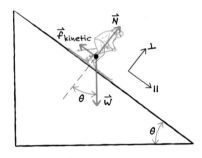

Figure 5-9 A free-body diagram shows the forces acting on a skier sliding down a hill (the ramp): the skier's weight, the normal force, and the force of kinetic friction. The force of kinetic friction (always) opposes motion.

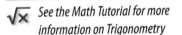

See the Math Tutorial for more information on Trigonometry

So,

$$a = \frac{W \sin\theta - \mu_k W \cos\theta}{m_{skier}}$$

To make it more obvious that the result has dimensions of acceleration, note that $W = m_{skier}\,g$, so

$$a = \frac{m_{skier}\,g \sin\theta - \mu_k\, m_{skier}\,g \cos\theta}{m_{skier}} = g(\sin\theta - \mu_k \cos\theta) \tag{5-10}$$

REFLECT

This result stands up to a number of tests. First, were μ_k to be reduced to zero, Equation 5-10 predicts $a = g \sin\theta$, which is identical to the result we obtained in Example 4-8 in which we did not consider friction. We can also consider a limiting case for the angle of the slope for which we can guess the acceleration. If the angle of the slope were $\theta = 90°$, the skier wouldn't be resting on the slope and she would be experiencing free-fall acceleration g. (She would effectively be falling over the edge of a cliff!) Equation 5-10 predicts an acceleration of g for $\theta = 90°$. This special case convinces us that the answer is reasonable.

What would the skier's acceleration be on a slope of $\theta = 0°$? Experience suggests that a skier on a horizontal surface does not accelerate, but substituting $\theta = 0°$ into Equation 5-10 results in a nonzero acceleration. The curious negative value of acceleration is a warning sign—don't ignore it. For a particular value of μ_k, there is a specific value of slope angle for which the skier will not slide and at *that* angle, acceleration is zero. For smaller angles acceleration is also zero, but Equation 5-10 can only generate a smooth continuum of answers. So for angles smaller than the minimum angle for which sliding occurs, Equation 5-10 gives an answer which is not physically reasonable! Reflecting on the meaning of the warning sign led to an interesting realization, which is why it is so important to reflect on your answer to confirm that it makes sense.

Practice Problem 5-2 A block sliding down a ramp of angle 24° accelerates at 3.0 m/s². What is the coefficient of kinetic friction between the ramp and the bottom surface of the block?

Example 5-3 Up a Ramp

The block shown in **Figure 5-10** is given a push up a ramp of angle 32° and released with an initial speed of 3.1 m/s. The coefficient of kinetic friction between block and ramp is 0.42. How far up the ramp does the block travel before coming to a stop?

SET UP

This example is a linear motion problem. Displacement along the ramp is related to initial speed, acceleration, and time according to

$$v = v_0 + at \tag{2-23}$$

and

$$x - x_0 = v_0 t + \frac{1}{2}at^2 \tag{2-26}$$

The time for the block to stop is the time for its speed to decrease to zero, so from Equation 2-23

$$0 = v_0 + at_{stop} \quad \rightarrow \quad t_{stop} = \frac{-v_0}{a}$$

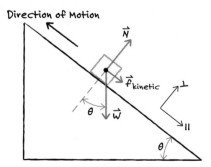

Figure 5-10 A force due to gravity (weight), a normal force, and a force due to friction act on a block being pushed up a ramp. Because the block is moving up the ramp, the force of kinetic friction points down the ramp.

Substituting this expression into Equation 2-26 yields a relationship for the displacement of the block as it comes to a stop:

$$x - x_0 = v_0 t_{\text{stop}} + \frac{1}{2} a t_{\text{stop}}^2 = v_0 \frac{-v_0}{a} + \frac{1}{2} a \left(\frac{-v_0}{a} \right)^2$$

$$= -\frac{v_0^2}{2a}$$

We are interested in the distance it takes to stop, so we take the absolute value, or

$$D_{\text{stop}} = \left| -\frac{v_0^2}{2a} \right| \tag{5-11}$$

We have been given the initial speed but need to find the acceleration. The value of a can be determined by considering the forces on the block.

Notice that the free-body diagram for the block, shown in Figure 5-10, differs from the free-body diagram for the skier in the previous example (Figure 5-9) in a significant way. Here, because the block is moving up the ramp, $\vec{f}_{\text{kinetic}}$ points down the ramp. The force of friction always opposes the motion. In choosing the coordinate axes, we have elected to make the positive $\parallel$ direction point down the ramp.

The last step in our force problem strategy is to write the Newton's second law equations for the block separately for each direction:

$$\parallel : \quad \vec{N}_\parallel + \vec{W}_\parallel + \vec{f}_{\text{kinetic}, \parallel} = m_{\text{block}} \vec{a}_\parallel \tag{5-12}$$

$$\perp : \quad \vec{N}_\perp + \vec{W}_\perp + \vec{f}_{\text{kinetic}, \perp} = m_{\text{block}} \vec{a}_\perp \tag{5-13}$$

To solve for acceleration, we need to find a way to connect the two equations. The connection will become clearer by taking advantage of the choice of coordinate axes; two of the three force vectors as well as the acceleration of the skier are aligned with one of the two axes, making the components along the other axis zero.

The component of $\vec{f}_{\text{kinetic}}$ perpendicular to the ramp is zero, so the component of friction parallel to the ramp is equal to the total force of friction. We will refer to the magnitude of the kinetic friction as f. By definition, $\vec{N}$ is perpendicular to the surface of the ramp, and $W_\parallel = W \sin \theta$ and $W_\perp = W \cos \theta$. Finally, acceleration perpendicular to the ramp is zero ($a_\perp = 0$) because the block is neither jumping off the ramp nor pushing down into it, which also means that we can let the acceleration in the direction parallel to the ramp ($a_\parallel$) equal a. Using the simplifications and following the directions specified by the coordinate axes in Figure 5-10, Equations 5-12 and 5-13 become

$$\parallel : \quad 0 + W \sin \theta + f = m_{\text{block}} a \tag{5-14}$$

$$\perp : \quad N - W \cos \theta + 0 = m_{\text{block}} (0) \tag{5-15}$$

The normal force N, which appears in the definition of f, connects the two equations.

SOLVE
Equation 5-14 can be solved for the acceleration

$$a = \frac{W \sin \theta + f}{m_{\text{block}}}$$

but we need f to solve the problem completely. Using Equation 5-5 and the relationship for N from Equation 5-15, we can determine f to be

$$f = \mu_{\text{k}} N = \mu_{\text{k}} W \cos \theta$$

So,

$$a = \frac{W \sin \theta + \mu_{\text{k}} W \cos \theta}{m_{\text{block}}}$$

See the Math Tutorial for more information on Trigonometry

Finally, use $W = m_{block}g$, so

$$a = \frac{m_{block}g \sin\theta + \mu_k m_{block}g \cos\theta}{m_{block}}$$

or

$$a = g(\sin\theta + \mu_k \cos\theta) \qquad (5\text{-}16)$$

Combining Equation 5-16 with Equation 5-11 gives the distance the block travels before stopping:

$$D_{stop} = \frac{v_0^2}{2g(\sin\theta + \mu_k \cos\theta)} \qquad (5\text{-}17)$$

so

$$D_{stop} = \frac{(3.1 \text{ m/s})^2}{2(9.81 \text{ m/s}^2)[\sin 32° + (0.42)\cos 32°]} = 0.55 \text{ m}$$

The block travels 0.55 m up the ramp before coming to a stop.

REFLECT

It does not seem unreasonable that the block would travel about half a meter up the ramp, given the initial speed, the ramp angle, and a coefficient of kinetic friction which is neither extremely small nor large. Perhaps more revealing, however, is a comparison of this answer with the distance to stop if the surface were horizontal. When the angle of the ramp is set to 0° in Equation 5-17, the stopping distance is

$$D_{stop} = \frac{(3.1 \text{ m/s})^2}{2(9.81 \text{ m/s}^2)[\sin 0° + (0.42)\cos 0°]} = 1.2 \text{ m}$$

This value is somewhat more than twice the distance we found for the block on the ramp and is also physically reasonable, giving us confidence that the answer is correct.

Practice Problem 5-3 Consider the block and ramp in Example 5-3 and shown in Figure 5-10. Find the distance the block travels up the ramp when released with an initial speed of 6.2 m/s, twice the speed used in Example 5-3, and compare it to the distance found in the example. Use the same values as in Example 5-3 for the ramp angle (32°) and the coefficient of kinetic friction between block and ramp (0.42).

? Got the Concept 5-2
   Will it Slide?

A block is launched up a ramp. Friction between the bottom of the block and the surface of the ramp causes the block to come to a stop. Will the block remain at rest after it stops or will it slide back down? If it's not possible to answer without more information, under what circumstances will it remain at rest?

✱ What's Important 5-2
   Kinetic friction is the frictional force between two surfaces that are moving relative to one another. Kinetic friction is typically less than the maximum force of static friction.

5-3 Working with Friction

In this section, we will apply our strategy for force problems (see Physicist's Tool-box 4-2) to a number of situations. The first step of that strategy, the creation of a free-body diagram, is central to solving problems in which force plays a direct role. To emphasize this, for the first three examples we will sketch the elements of the problem and make a free-body diagram, but we won't fully solve the problem.

A blue block (block 1) rests on a green block (block 2), which lies on a friction-less table as shown in Figure 5-11a. We want to study the motion of block 1 when a force is applied to block 2 from left to right. The physics that underlies the motion is contained in the free-body diagram shown in Figure 5-11b.

Excluding gravity, in order for one object to exert a force on another, the objects must be in direct contact. Because whatever is pushing on block 2 (someone's hand, perhaps) is not in contact with block 1, the force exerted on block 2 does not push on block 1. In addition, because the only object touching block 1 is block 2, any force on block 1 other than that due to gravity can only be exerted by block 2.

Block 2 supports block 1 through a normal force. That force is labeled $\vec{N}_1$ in the diagram, although we could have chosen to call it $\vec{F}_{2\to1}$ to recognize that it arises because block 2 pushes on block 1. In addition, because friction exists between the two blocks, block 2 also exerts a frictional force on block 1, which is labeled $\vec{f}_{2\to1}$ in the figure. (We hold to our convention that frictional forces are labeled with a lowercase "f.") Finally, because block 2 exerts forces on block 1, according to Newton's third law, block 1 must exert forces on block 2. The coun-terpart of $\vec{f}_{2\to1}$ is labeled $\vec{f}_{1\to2}$ and the counterpart of $\vec{N}_1$ is labeled $\vec{F}_{1\to2}$ in the free-body diagram.

Why does the frictional force $\vec{f}_{2\to1}$ point to the right in Figure 5-11b? The force of friction always opposes motion. If no friction existed, block 1 would have no horizontal force exerted on it and therefore would remain at rest in its initial position. For block 1 to move to the right along with block 2, the force of friction must point to the right.

In Figure 5-12a, a block rests on a ramp. Friction exists between the block (labeled 1) and the ramp but does not exist between the ramp and the surface under it. The ramp is given a push to the right. The force, labeled $\vec{F}_{\text{push}}$, as well as all of the other forces acting on the ramp and the block are shown in the free-body diagram in Figure 5-12b. We placed the ramp on a frictionless surface to focus attention on the friction between the ramp and the block. This simplification does not affect any other forces in the problem.

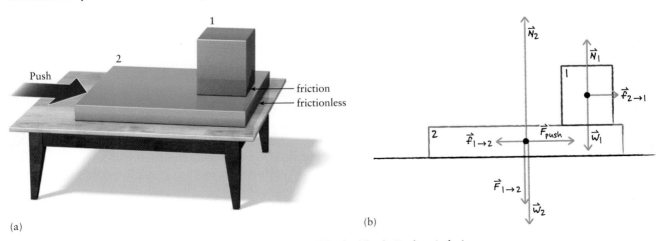

(a)   (b)

Figure 5-11 (a) A blue (block 1) rests on top of a green block (block 2) that is being pushed to the right along a frictionless tabletop. (b) A free-body diagram shows the forces acting on each block.

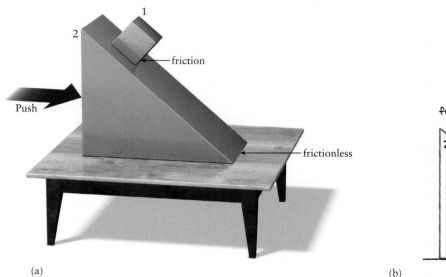

(a)

(b)

Figure 5-12 (a) A block rests on a ramp which is being pushed to the right along a frictionless surface. Friction between the block and the ramp prevents the block from sliding down. (b) A free-body diagram shows the forces acting on the block and on the ramp.

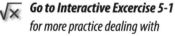

Go to Interactive Excercise 5-1
for more practice dealing with friction

The force $\vec{F}_{push}$ acts on the ramp, but not on the block. Only the ramp touches the block, so except for gravity, the forces on the block can only be exerted by the ramp. As we have seen in previous problems involving ramps, the normal force which supports the block is perpendicular to the surface of the ramp. That force is labeled $\vec{N}_1$ in the diagram. Also, because $\vec{N}_1$ is the force exerted on the block by the ramp, the counterpart of $\vec{N}_1$ in the pair of forces related by Newton's third law is the force that the block exerts on the ramp. This force is therefore labeled $\vec{F}_{1\rightarrow2}$ in the free-body diagram. In addition, because friction exists between the block and the surface of the ramp, the ramp also exerts a frictional force on the block, which we have labeled $\vec{f}_{2\rightarrow1}$ in the figure. Finally, because the ramp exerts forces on the block, according to Newton's third law, the block must exert forces on the ramp. The counterpart of $\vec{f}_{2\rightarrow1}$ in this force pair is therefore $\vec{f}_{1\rightarrow2}$.

It is instructive to consider the direction of some of the forces in Figure 5-12b. First, the frictional force $\vec{f}_{2\rightarrow1}$ points down the ramp because the force $\vec{F}_{push}$ would push the block up the ramp if the surface were frictionless. To oppose the motion, friction must point down the ramp. Once we establish the direction of $\vec{f}_{2\rightarrow1}$, we also know the direction of $\vec{f}_{1\rightarrow2}$. As the force pair counterpart of $\vec{f}_{2\rightarrow1}$, $\vec{f}_{1\rightarrow2}$ must be equal in magnitude and opposite in direction to $\vec{f}_{2\rightarrow1}$. That's why $\vec{f}_{1\rightarrow2}$ points in a direction parallel to the surface of the ramp and up rather than down. A similar argument can be made for the direction of $\vec{F}_{1\rightarrow2}$; it must point opposite to the direction of its force pair counterpart $\vec{N}_1$.

In **Figure 5-13a**, the small blue block (labeled 2) is placed on the front, vertical surface of a larger purple block (labeled 1) which rests on a frictionless, horizontal surface. Friction exists between the two blocks. A force $\vec{F}_{push}$ is applied to block 1, and we want to consider the motion of block 2.

The free-body diagram (**Figure 5-13b**) includes a number of forces with which we are now familiar, for example, the weight of each block and the normal force $\vec{N}_1$ by which the horizontal surface supports the block 1. A frictional force also acts on block 2; this force points up and opposes the motion block 2 would exhibit were there no friction.

Does block 2 experience a normal force? Yes, any time an object is supported by a surface, it experiences a normal force, so there must be a normal force here. We labeled this force $\vec{F}_{1\rightarrow2}$. Why not? After all, it is the surface of block 1 which supports block 2. Although it might surprise you, a normal force can point in any direction, including horizontally. And finally, because block 1 exerts this normal force on block 2, there must be an equal, opposite force on block 1. This force is labeled $\vec{F}_{2\rightarrow1}$ in Figure 5-13b.

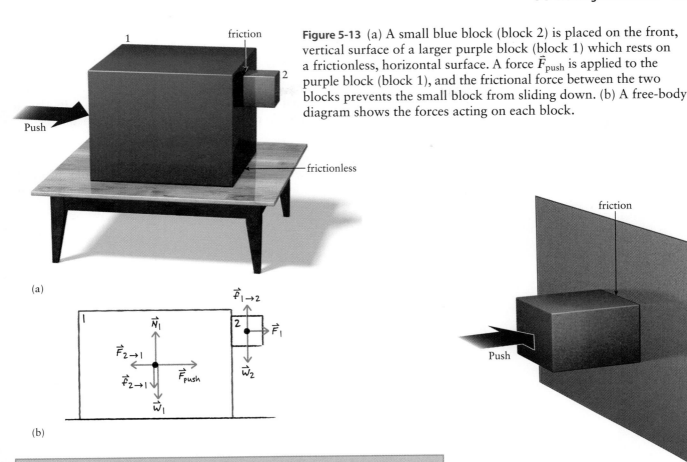

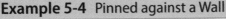

Figure 5-13 (a) A small blue block (block 2) is placed on the front, vertical surface of a larger purple block (block 1) which rests on a frictionless, horizontal surface. A force $\vec{F}_{push}$ is applied to the purple block (block 1), and the frictional force between the two blocks prevents the small block from sliding down. (b) A free-body diagram shows the forces acting on each block.

(a)

(b)

Example 5-4 Pinned against a Wall

You push a block of plastic that weighs 32 N against a wall with a force of 55 N as shown in **Figure 5-14a**. The coefficients of static friction and kinetic friction between the plastic and the wall are $\mu_s = 0.41$ and $\mu_k = 0.40$, respectively. Does the block remain motionless? If not, what is its acceleration as it slips down the wall?

SET UP
To determine whether or not the block slips, we need to establish the magnitude and direction of any forces it experiences that counter the force due to gravity. If there were no friction between the block and the wall, the block would fall straight down. The frictional force opposes the falling motion and therefore acts upward on the block. This force, labeled $\vec{f}$ in **Figure 5-14b**, is one of two forces with which the wall acts on the block. The other is the force which prevents the block from pushing into the wall; it is normal to the surface so we have labeled it $\vec{N}$. As in the case of the two blocks discussed above and shown in Figure 5-13a, the normal force is horizontal here because the supporting surface is vertical. Also, the reason we generically labeled the frictional force as $\vec{f}$ rather than as either kinetic friction or static friction is that we don't yet know whether the block will slip; that is, we don't know whether or not kinetic friction will exist between the block and the wall.

To complete the setup of the problem, write the Newton's second law equations for the block, separately in each direction. Because every force is aligned with one of the coordinate axes, we dispense with the vector notation and write the magnitudes directly:

$$x: \quad F_{push} - N = m_{block}a_x \tag{5-18}$$

$$y: \quad f - W = m_{block}a_y \tag{5-19}$$

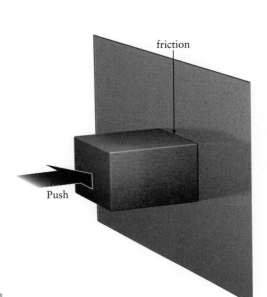

(a)

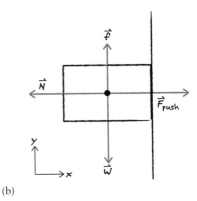

(b)

Figure 5-14 (a) A block of plastic weighing 32 N is pushed against a wall with a force of 55 N. (b) A free-body diagram shows the forces acting on the block. The frictional force, labeled $\vec{f}$, is one of two forces with which the wall acts on the block. The other is the force which prevents the block from pushing into the wall; it is normal to the surface so we have labeled it $\vec{N}$. The normal force is horizontal here because the supporting surface is vertical.

If a_y in Equation 5-19 equals zero, the block remains in its initial, at-rest position. We arrive at an answer to our problem by solving Equation 5-19.

SOLVE
Using Equation 5-19, the frictional force must equal the weight, if $a_y = 0$. Also, if $a_y = 0$, the frictional force is static. So the maximum force of static friction is

$$f = \mu_s N = W$$

Is the normal force large enough to provide the friction required to hold the block in place? Because the block is neither leaping off the wall nor pushing into it, $a_x = 0$, so

$$N = F_{push} \qquad (5\text{-}20)$$

Combining the two equations yields

$$\mu_s F_{push} = W$$

So F_{push} must be W/μ_s to keep the block in place. By substituting the values $\mu_s = 0.41$ and $W = 32$ N into $F_{push} = W/\mu_s$, the force you need to apply is

$$F_{push} = \frac{W}{\mu_s} = \frac{32\ \text{N}}{0.41} = 78\ \text{N}$$

According to the problem statement, you apply $F_{push} = 55$ N, which is not enough to keep the block of plastic in place. The block will slide down the wall.

Solve Equation 5-19 to determine the vertical acceleration:

$$a_y = \frac{f - W}{m_{block}}$$

Because the block slides, the friction between block and wall is kinetic friction or

$$f = \mu_k N \qquad (5\text{-}5)$$

We can again use Equation 5-20 to determine the force applied to the block:

$$a_y = \frac{\mu_k F_{push} - W}{m_{block}}$$

All of these variables are given ($m_{block} = W/g$), so we can determine a_y:

$$a_y = \frac{(0.40)(55\ \text{N}) - 32\ \text{N}}{(32\ \text{N}/9.81\ \text{m/s}^2)} = -3.1\ \text{m/s}^2$$

REFLECT
We have defined the positive y direction as up, so $a_y < 0$ means the block accelerates down. In addition, the magnitude of the acceleration is somewhat less than g. Both results are reasonable—we expect the block to accelerate down the wall, but the resistive nature of kinetic friction prevents it from attaining free fall.

Practice Problem 5-4 You push a plastic block that weighs 0.25 N against a wall with a force of 0.68 N as shown in Figure 5-14a. The coefficient of kinetic friction between the block and the wall equals 0.30. What is the acceleration of the block as it slips down the wall?

Example 5-5 Two Possible Directions for Friction

Two blocks, connected by a thin thread that passes over a small pulley, are released from rest. Block 1, which has a weight of 7.0 N, hangs freely while block 2 rests on a surface at an angle $\theta = 24°$ from the horizontal, as shown in **Figure 5-15a**. The coefficient of static friction between block 2 and the surface is 0.40. What weight or range of weights for block 2 will prevent the blocks from moving? Neglect the mass of the thread and the pulley.

SET UP

Our force-problem strategy says that we first create a free-body diagram, then chose the coordinate axes, label the positive direction, and finally, use the free-body diagram to write the sum of the forces on each object in each direction. **Figure 5-15b** shows a partial free-body diagram. The figure includes the weight of each block, the tension in the thread (which must be equal at both ends), and the normal force by which the angled surface (the ramp) supports block 2. The force of static friction, however, is missing.

The force of friction is parallel to the surface of the ramp. But does it point up or down? Friction always opposes motion, so the question becomes, does block 2 slide down or up?

If you answered "down," you could be correct. If you answered "up," you also could be correct. If block 2 is heavy enough, it will slide down the ramp and pull block 1 up (case 1). If block 2 is too light to keep the system in equilibrium, block 1 will fall and pull block 2 up the ramp (case 2). Friction always opposes the motion, so in case 1, $\vec{f}$ points up the ramp and in case 2, $\vec{f}$ points down. The free-body diagrams for the two cases are shown in **Figure 5-16**.

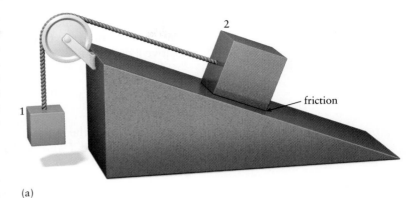

(a)

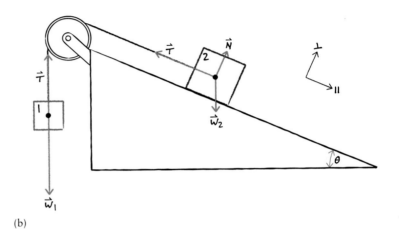

(b)

Figure 5-15 (a) Two blocks, connected by a thin thread that passes over a small pulley, are released from rest. Block 1 hangs freely and block 2 rests on the surface of a ramp. (b) A partial free-body diagram includes all of the forces that act on each block except the force of static friction.

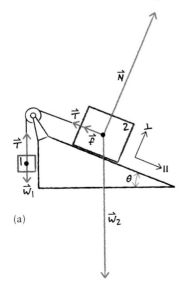

(a)

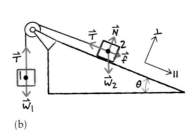

(b)

Figure 5-16 (a) Block 2 will slide down the ramp if it is heavy enough to pull up block 1. In this case the force of friction points up the ramp. (b) If block 1 is heavy enough to fall and pull block 2 up the ramp, the force of friction points down the ramp.

We now apply Newton's second law to each object in each direction separately. Note that when block 2 moves down the ramp, block 1 moves up. This requires that we treat "up" on the left side of the ramp as the same direction as "down" along the surface of the ramp. This association of directions affects the signs of the terms in the sum of the forces, for example, $\vec{T}$ on block 1 is considered to be in the positive $\parallel$ direction:

$$\text{block 1, } \parallel: \quad \vec{T} + \vec{W}_1 = m_1 \vec{a}_{1,\parallel} \tag{5-21a}$$

$$\text{block 2, } \parallel: \quad \vec{T} + \vec{W}_{2,\parallel} + \vec{f} = m_2 \vec{a}_{2,\parallel} \tag{5-21b}$$

$$\text{block 2, } \perp: \quad \vec{N} + \vec{W}_{2,\perp} = m_2 \vec{a}_{2,\perp} \tag{5-21c}$$

When generating the equations, we have recognized that $\vec{N}$ has no component in the $\parallel$ direction, and $\vec{T}$ and $\vec{W}_1$ have no components in the $\perp$ direction.

SOLVE

We will refer to the free-body diagrams to determine the signs of each term on the left side of each equation when changing vectors to scalar magnitudes. We will also require that the system remain at rest, so $a_{1,\parallel} = a_{2,\parallel} = 0$ and $a_{2,\perp} = 0$. In addition, we use the maximum static friction value, which is determined by Equation 5-2:

$$f = \mu_s N$$

In case 1, in which $\vec{f}$ points up the ramp, Equations 5-21a–c become

$$\text{block 1, } \parallel: \quad T - W_1 = 0$$

$$\text{block 2, } \parallel: \quad -T + W_2 \cos\theta - \mu_s N = 0$$

$$\text{block 2, } \perp: \quad N - W_2 \sin\theta = 0$$

The three equations include three unknown variables, W_2, N, and T, which means we can solve for all three variables completely. To obtain an expression for W_2, we want to eliminate N and T from an equation; for example, by solving the first equation for T and the third equation for N

$$\text{block 1, } \parallel: \quad T = W_1$$

$$\text{block 2, } \perp: \quad N = W_2 \sin\theta$$

and then substituting these expressions into the second equation, the sum of the forces on block 2 in the parallel direction becomes

$$-W_1 + W_2 \cos\theta - \mu_s W_2 \sin\theta = 0$$

$\sqrt{x}$ *See the Math Tutorial for more information on Trigonometry*

or

$$W_2 = \frac{W_1}{\cos\theta - \mu_s \sin\theta}$$

For $W_1 = 7.0$ N, $\theta = 24°$, and $\mu_s = 0.40$:

$$W_2 = \frac{7.0 \text{ N}}{\cos 24° - (0.40)\sin 24°} = 9.3 \text{ N}$$

How does the problem change for case 2, where block 1 falls, pulling block 2 up the ramp? Equations 5-21a and 5-21c remain the same. Only Equation 5-21b is different from case 1 to case 2, because $\vec{f}$ points up the ramp in case 1 but down the ramp in case 2. The sign of the frictional force is therefore negative in Equation 5-21b when we change vectors to scalar magnitudes. This result has the effect of changing the sign of the $\mu_s W_2 \sin\theta$ term in $-W_1 + W_2 \cos\theta - \mu_s W_2 \sin\theta$, so for case 2

$$W_2 = \frac{W_1}{\cos\theta + \mu_s \sin\theta}$$

For $W_1 = 7.0$ N, $\theta = 24°$, and $\mu_s = 0.40$:

$$W_2 = \frac{7.0 \text{ N}}{\cos 24° + (0.40)\sin 24°} = 6.5 \text{ N}$$

REFLECT

The two answers suggest that to keep the system in equilibrium, block 2 can have any weight between 6.5 and 9.3 N. If W_2 is less than 6.5 N, block 1 will fall and pull block 2 up the ramp. If W_2 is greater than 9.3 N, block 2 will slide down the ramp and pull block 1 up. In addition, the specific values allowed for the weight of block 2 are reasonable. We might suspect an error had been made if the results were much smaller or much larger than the weight of block 1.

Practice Problem 5-5 Two blocks are connected by a thin thread. Block 1 hangs freely, block 2 rests on the surface of a ramp of angle 32°, and the thread passes over a small pulley as in Figure 5-18a. The system is not moving, but both blocks begin to accelerate when a small object is placed on top of block 2. Block 1 weighs 1.2 N and block 2 weighs 1.9 N. What is the coefficient of static friction between the surface of the ramp and the bottom of block 2?

✳ What's Important 5-3

With the exception of gravity, two objects must be in direct contact in order for one to exert a force on the other. To solve force problems, first draw a free-body diagram. Use the free-body diagram to identify the forces on each object in each direction, and then apply Newton's second law to each object in each direction separately.

5-4 Drag Force

In Figure 5-17, two skydivers are falling but not accelerating. Newton's second law tells us that the net force must be zero. Obviously, the skydivers are not resting on a surface, so the force that opposes their motion is not friction as we have defined it in the previous sections. The force which resists the motion of an object through fluid is a *drag force*.

Unlike kinetic friction, which arises when two surfaces move relative to each other, drag forces depend on the speed of the moving object. In general, the magnitude of the drag force is proportional to some power n of the speed v of the object; using the variable b as the proportionality constant:

$$F_{drag} = bv^n \tag{5-22}$$

The value of b depends on the size, shape, and surface characteristics of a particular object, and may also depend on the material through which the object is moving. You may sometimes see Equation 5-22 written with a minus sign to emphasize the fact that the drag force always acts in a direction opposite to the motion of an object.

An object moving relatively slowly is often well described by setting $n = 1$ in Equation 5-22. When an object moves at high speeds through air (and other gases), the magnitude of the drag force is proportional to the square of the speed. The faster an object moves, the larger the drag force it experiences.

Consider a ball falling from rest through air. The ball always experiences a force due to gravity, so at the instant it is released it begins to accelerate at g. As a result of this acceleration, the ball's speed begins to increase. So according to Equation 5-22, the drag force, **air resistance** in this case, also increases. The larger the speed v of the ball, the larger the drag force it experiences. If the ball falls for a long enough period of time, the speed eventually gets large enough

Figure 5-17 The drag force on the skydivers due to air resistance equals the force on them due to gravity. Because the net force on them is zero, the skydivers are not accelerating. *(Masterfile.)*

Table 5-1 Terminal Velocity	
Object	m/s
Shot put	150
Skydiver	60
Hailstone	15
Ping-Pong ball	9
Raindrop	7
Parachutist	5

that the corresponding drag force equals the force of gravity and therefore the net force on the ball equals zero. The ball has ceased to accelerate; it has reached **terminal speed**. Some typical terminal speeds are shown in Table 5-1.

We can find an expression for terminal speed by setting the drag force equal to the gravitational force. For example, the terminal speed of an object that is experiencing air drag proportional to the square of its speed is

$$bv_{\text{terminal}}^2 = mg$$

or

$$v_{\text{term}} = \sqrt{\frac{mg}{b}} \tag{5-23}$$

The value of b is partially determined by how streamlined an object is. For example, the b value for a dolphin in water is about 100 times smaller than the b value of a sphere of the same cross-sectional area. Likewise, a typical car will experience 7 to 10 times less air resistance than an 18-wheeler.

Because drag force, such as air resistance, changes as the speed of an object changes, its effects are not instantaneous. For example, the time it takes to reach terminal speed depends on the object and the medium. As an example, a skydiver who has not yet opened his parachute reaches a speed which is half of terminal speed in only 3 to 5 s and hits terminal speed after 15 to 20 s.

✳ What's Important 5-4

Drag forces, such as air resistance, depend on the speed of the moving object. When the magnitude of the drag force on an object falling through air equals that of the gravitational force, the object ceases to accelerate and has reached its terminal speed.

5-5   Forces and Uniform Circular Motion

Our brains require a constant supply of oxygenated blood. To ensure that blood doesn't simply pool in our lower extremities, humans and other animals that spend most of their time upright have adapted mechanisms to compensate for the effects of gravity. Your heart supplies the force necessary to overcome the force of gravity and pump blood up to your head. If for some reason blood flow to your brain becomes inadequate, fainting is nature's way of getting you to lower your head to the level of your heart so that gravity no longer impedes blood flow to the brain.

Most of us only rarely experience vertical accelerations other than the force of gravity g. Pilots of high-speed jets, however, routinely experience large accelerations. For example, fighter pilots executing tight turns at high speeds (**Figure 5-18**) sometimes experience accelerations as high as $9g$. From Newton's second law, we know that a large force is exerted on a pilot who experiences a large acceleration. This force can cause dramatic shifts in the distribution of blood in the body, depriving the brain of oxygen and resulting first in loss of vision and then unconsciousness.

A **centripetal**, or center-seeking, **force** is required to make an object travel in a circle.

Figure 5-18 Fighter pilots executing a tight turn sometimes experience accelerations up to 9 times that of gravity. *(Stocktrek Images/Getty Images.)*

You can't throw a rock into the air and have it follow a circular path, but you can tie it to a string and swing it in a circle. In the case of rock and string, the tension in the string provides the centripetal force.

Consider the free-body diagram for a rock swinging at the end of a string shown in **Figure 5-19**. The rock moves in a circular, horizontal path of radius r.

(The string makes an angle θ to the vertical, so the radius r is not equal to L but rather $L\sin\theta$.) The acceleration of an object moving in uniform circular motion is

$$\vec{a}_r = \frac{v_0^2}{r}(-\hat{r}) \tag{3-34}$$

and points toward the center of the circle. (Recall from Section 3-5 that an object in uniform circular motion moves in a circle at a constant speed but its acceleration is nonzero because its direction is constantly changing.) The x component of $\vec{T}$ also points in toward the center of the circle; $\vec{T}_x$ is the centripetal force in this case. Noting that the negative radial direction is defined to be toward the center, we write a Newton's second law equation for the rock as

$$\vec{T}_x = m\vec{a}_x = m\frac{v_0^2}{r}(-\hat{r})$$

or

$$-T\sin\theta = -m\frac{v_0^2}{r} \tag{5-24}$$

The term on the right side of Equation 5-24 is not a force, but rather mass multiplied by acceleration as required by Newton's second law. In particular, there is no outward force on the rock; the only force on the rock other than gravity is the tension force, which points radially inward.

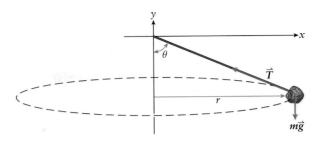

Figure 5-19 A rock swung at the end of a string moves in a circular, horizontal path of radius R. The two forces acting on the rock are the tension in the string T and the force due to gravity $m\vec{g}$.

Example 5-6 Rock on a String

A rock is attached to a string of length 0.76 m and swung in a circle at constant speed. The string makes an angle of 68° to the vertical. What is the speed of the rock? The weight of the string is small enough that it can be neglected.

SET UP
The speed of the rock and the angle that the string makes to the vertical are related through the tension in the string, as suggested by Equation 5-24. We don't, however, have a value for T, so we need to find at least one more equation containing the tension term in order to eliminate it. Equation 5-24 is the application of Newton's second law in the x direction; applying Newton's second law in the y direction gives

$$\vec{T}_y + m\vec{g} = m\vec{a}_y$$

or

$$T\cos\theta - mg = 0 \tag{5-25}$$

Because the acceleration in the y direction is 0, Equations 5-24 and 5-25 can be combined to eliminate T.

SOLVE
Equation 5-25 becomes

$$T\cos\theta = mg$$

Dividing Equation 5-24 by Equation 5-25 eliminates T:

$$\frac{-T\sin\theta}{T\cos\theta} = \frac{-m\dfrac{v_0^2}{r}}{mg}$$

 Go to Interactive Exercise 5-2
for more practice dealing with uniform circular motion

 See the Math Tutorial for more information on Trigonometry

or

$$\tan \theta = \frac{v_0^2}{gr} \tag{5-26}$$

Thus,

$$v_0 = \sqrt{gr \tan \theta} = \sqrt{gL \sin \theta \tan \theta}$$

So

$$v_0 = \sqrt{(9.81 \text{ m/s}^2)(0.76 \text{ m}) \sin(68)\tan(68)} = 4.1 \text{ m/s}$$

REFLECT

Is this a reasonable answer? The circumference of the circle traced out by the rock is $2\pi L \sin \theta = 2\pi(0.76 \text{ m}) \sin(68) = 4.4$ m. At a speed of 4.1 m/s, the rock goes around about once per second. No need to tie a rock to a string—you can probably imagine doing it and have a sense that the speed is about right.

What if the rock is swung around faster? As v_0 in Equation 5-26 increases, the angle θ also increases. In other words, the faster the rock moves, the farther from the vertical it swings, which is also reasonable. However, you can easily verify that the speed must be infinite in order to make the string perfectly horizontal ($\theta = 90°$). This result is not surprising, because when the rock is rotating in a horizontal plane, it experiences no force in the positive vertical direction. Without that force, there is no force to counteract the weight of the rock.

Practice Problem 5-6 A rock is attached to a string of length 0.76 m and swung in a circle. How fast must the rock be swung so that it moves in a nearly horizontal circle at 89.5° from the vertical? Compare the speed to the speed limit on most Canadian multi-lane highways, 100 km/h or about 62 mi/h. Assume that the weight of the string is small enough that it can be neglected.

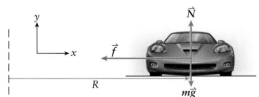

Figure 5-20 The only force acting in the horizontal direction on a car traveling at constant speed around a flat, circular curve is friction.

Friction can provide centripetal force. For example, friction between a car's tires and the road provides the centripetal force necessary for a car to maneuver around a circular curve. Consider a car traveling at constant speed around a flat, circular curve. In the free-body diagram (**Figure 5-20**), the car is moving toward us along a circular path of radius R.

The only force acting on the car in the x direction is friction. The frictional force provides the centripetal acceleration. So in the radial direction, which happens to coincide with the x direction as defined in Figure 5-20, we would write Newton's second law equation for the car as

$$\vec{f} = m\vec{a} \tag{5-27}$$

In the vertical direction, in which there is no acceleration,

$$\vec{N} + m\vec{g} = m\vec{a}_y = 0 \tag{5-28}$$

The magnitude of $\vec{f}$ on the left-hand side of Equation 5-27 equals $\mu_s N$ according to Equation 5-2, the definition of static friction. The magnitude of the right-hand side of Equation 5-27 equals $m\frac{v_0^2}{r}$ using Equation 3-33, the definition of the magnitude of acceleration for an object in uniform circular motion. So Equation 5-27 becomes

$$\mu_s N = m\frac{v_0^2}{r}$$

Equation 5-28 can be rearranged to give $N = mg$, so

$$\mu_s mg = m\frac{v_0^2}{r} \tag{5-29}$$

We have elected to consider the case when static friction is maximum. We now see the relationship between the speed of the car, the radius of the curve, and the strength of the frictional force required to keep the car on the road:

$$\mu_s = \frac{v_0^2}{gr} \qquad (5\text{-}30)$$

Notice that static friction provides the traction. Although the tires are rotating, the surfaces of the tires do not slide on the surface of the road (unless the car is skidding), so the friction is static.

? Got the Concept 5-3
Don't lock the wheels!

Most new cars come equipped with an antilock braking system (ABS). When the ABS is activated during a heavy braking situation, a car's brakes are automatically and rapidly pulsed on and off so that the wheels are less likely to "lock," that is, not turn. Why does an ABS help a driver keep control of a car in an emergency?

Banking the curve, tilting the road surface by raising the outer edge of the curve of the road relative to the inner edge, makes it easier to navigate a curve while driving. You may have experienced such banking while driving on a highway ramp. A glance at a free-body diagram for a car on a banked curve, such as the one in **Figure 5-21**, makes it clear why the bank helps. The frictional force is not shown because the direction of friction depends on whether the car is tending to slide up or down the angled surface. But even if there were no friction, a component of the normal force points toward the center of the circle. That component, called centripetal force, enables the car to follow the circular path.

We can apply Newton's second law to the car on the banked curve. In the x or radial direction,

$$\vec{N}_x = m\vec{a}$$

or

$$-N\sin\theta = m\left(-\frac{v_0^2}{r}\right) \qquad (5\text{-}31)$$

In the vertical direction, in which there is no acceleration:

$$\vec{N}_y + m\vec{g} = m\vec{a}_y = 0$$

or

$$N\cos\theta = mg \qquad (5\text{-}32)$$

Dividing Equation 5-31 by Equation 5-32 gives

$$\tan\theta = \frac{v_0^2}{gr} \qquad (5\text{-}33)$$

A comparison of Equation 5-33 with the result we obtained for a car on a horizontal curve (Equation 5-30) emphasizes the role that banking the curve plays in road design. The tilt of the banked curve is equivalent to the frictional force on a curve that is not banked.

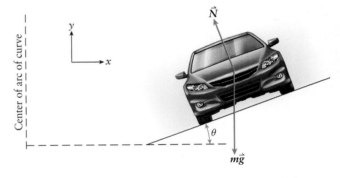

Figure 5-21 A car moves along a banked curve at a constant speed.

Example 5-7 Driving on a Curve

A car follows an unbanked curve at a constant speed of 18 m/s. If the coefficient of static friction between the tires and the road is 0.55 (a typical value when the road surface is wet), what radius of curvature is required so that the car does not skid?

SET UP

Here, we will simply apply Equation 5-30, which relates the coefficient of static friction with the speed of the car and the radius of the curve.

SOLVE

Equation 5-30 can be rearranged so that the term for the radius of the curve is the only term on the left side of the equation:

$$r = \frac{v_0^2}{g\mu_s}$$

Using the given values:

$$r = \frac{(18 \text{ m/s})^2}{(9.81 \text{ m/s}^2)(0.55)} = 60 \text{ m}$$

REFLECT

A circle that has a radius equal to 60 m is 120 m in diameter—not so different from the length of a football field, soccer pitch, or a baseball field. In other words, the curve would be relatively big. No wonder highway engineers bank the off-ramps—unbanked ones take up too much space.

Practice Problem 5-7 An unbanked highway exit follows the arc of circle of radius 23 m. The coefficient of static friction between the road and the tires of a car is 0.75 on a dry day. What is the maximum speed at which a driver could keep her car from sliding off the exit?

Figure 5-22 A chef gives the pizza dough a sizable rotation when he throws it in the air. *(Beyond Fotomedia GmbH/Alamy.)*

Example 5-8 Spinning a Pizza

A gourmet pizza chef flattens a ball of dough into a disk that has a radius of 0.15 m. He then throws the dough in the air, rotating it as he releases it (**Figure 5-22**). One region of the doughy disk has a lump, which extends from 0.05 m out to 0.15 m. Find the ratio of the force experienced by the part of the lump closer to the edge to the force experienced by the part of the lump closer to the center.

SET UP

Consider two small bits of dough in the lump, each of mass m but at different distances from the center of the spinning disk. According to Equation 3-34 each experiences a different acceleration and therefore a different force. For a tiny bit of dough a distance r from the center, the magnitude of the force it experiences is

$$F(r) = ma_r = m\frac{v_0^2}{r}$$

To compare the forces on the two bits of dough, we need to recognize that although any part of the pizza makes the same number of rotations per second, the speed of each bit varies according to its distance from the center.

SOLVE

The circumference of the path traced by any point a distance r from the center of a rotating disk is $2\pi r$, so if it takes a time T for that point to go around once, its speed is

$$v(r) = \frac{2\pi r}{T}$$

So then the force experienced by a bit of dough at radius r_1 is

$$F(r_1) = m\frac{1}{r_1}\left(\frac{2\pi r_1}{T}\right)^2 = \frac{4\pi^2 m r_1}{T^2}$$

and the force experienced by a bit of dough at radius r_2 is

$$F(r_2) = m\frac{1}{r_2}\left(\frac{2\pi r_2}{T}\right)^2 = \frac{4\pi^2 m r_2}{T^2}$$

In the ratio $F(r_2)/F(r_1)$ all of the constants cancel, leaving only

$$\frac{F(r_2)}{F(r_1)} = \frac{r_2}{r_1} = \frac{0.15 \text{ m}}{0.05 \text{ m}} = 3.0$$

REFLECT

The force on the outer part of the lump is three times the force on the inner part. This difference has the tendency to spread out the dough, making a flat, uniform crust. Many people imagine that pizza chefs throw and spin pizza dough to spread it out into a wider circle. Making a large circle of dough is certainly part of the reason, but notice that the physics of rotating objects also causes the crust to become uniform and flat.

Practice Problem 5-8 Two friends of identical mass drive around an unbanked, circular turn in a sports car. They sit 1.00 m apart. The passenger, farther from the center of curvature of the circular path, feels a radially outward force that is 20% higher than the radially outward force felt by the driver. How far from the center of curvature is the driver of the car?

Imagine you are seated in a car that accelerates rapidly on a highway. The acceleration and therefore the force exerted on you points in the direction of motion. Yet you feel pushed *back* against the seat in the direction opposite the motion of the car. According to Newton's third law, as the seat exerts a forward force on you, your body exerts an equal but opposite reaction force on the seat. It is this force that results in the sensation of being pushed backward. In the same way, the inward force associated with circular motion when you sharply turn your car, the centripetal force, results in the sensation of being pushed outward.

Watch Out

An object in circular motion experiences a force directed inward, toward the center of the circular motion.

You have probably experienced the sensation of feeling "thrown outward" while taking a tight turn in a car or riding on an amusement part ride that follows a circular path. The inward-pointing centripetal force holds you to the circular path. The car door, for example, might push against you to counter your tendency, due to your inertia, to follow a straight path. Because the door exerts a force on you, you must exert a force on the door. This force gives you the *sensation* of an outward force. However, were you to draw a free-body diagram of yourself driving around that tight curve, you would not include the outward sensation because it is not a force!

Figure 5-23 To execute a tight turn, a pilot banks the plane steeply in the direction of the turn.

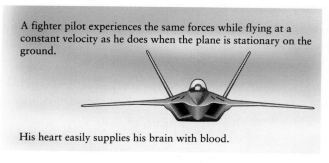

A fighter pilot experiences the same forces while flying at a constant velocity as he does when the plane is stationary on the ground.

His heart easily supplies his brain with blood.

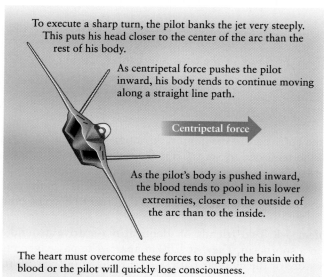

To execute a sharp turn, the pilot banks the jet very steeply. This puts his head closer to the center of the arc than the rest of his body.

As centripetal force pushes the pilot inward, his body tends to continue moving along a straight line path.

Centripetal force

As the pilot's body is pushed inward, the blood tends to pool in his lower extremities, closer to the outside of the arc than to the inside.

The heart must overcome these forces to supply the brain with blood or the pilot will quickly lose consciousness.

Let's return now to the example of the fighter pilot making a tight, fast turn. The acceleration experienced by a pilot traveling on a circular arc points toward the center of the circle, as indicated by $-\hat{r}$ in Equation 3-34. In which direction is the associated force? If the force is inward, why does blood rush to the pilot's feet when his body is aligned with his head toward the center of the arc?

To execute a tight turn, the pilot banks the plane very steeply in the direction of the turn as shown in Figure 5-23. This action causes the pilot's head to be closer to the center of the arc than the rest of his body. Now, as the centripetal force pushes the pilot inward, his body—and blood—have a tendency to continue moving along a straight line path, according to Newton's first law. The pilot's body is pushed inward by the force exerted by his seat. Because his blood is a fluid, however, the force exerted on the pilot's blood by his body (or more correctly, the walls of his blood vessels), is much smaller. For this reason the blood tends to stay behind, ending up closer to the outside rather than the center of the circular arc; in other words, the blood ends up being closer to his feet than to his head. As a result, loss of consciousness could occur due to inadequate blood flow to the brain.

To counter the effects of large accelerations fighter pilots wear gravity suits (G suits). These garments apply pressure to the legs and lower abdomen, thus preventing blood from pooling in the lower body even during extreme aerial maneuvers.

Estimate It! 5-1 Pinned to the Wall

Riders on the barrel-shaped Rotor amusement park ride stand against the inside wall of the barrel as its rotation rate increases. When the Rotor reaches full speed, the floor of the barrel drops away, leaving the riders pinned to the wall.

Estimate the minimum coefficient of static friction required to keep the riders from sliding down. The radius of the Rotor is about 4.5 m and the rotation rate is 30 rpm at full speed.

SET UP
Following our force-problem strategy, we start by making a free-body diagram. The forces on a rider on the Rotor are shown in **Figure 5-24**. For the rider to stay in place, the upward force of friction must balance the downward force due to gravity.

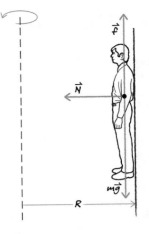

SOLVE
The force due to static friction is, by Equation 5-2,

$$f = \mu_s N$$

where the normal force is

$$N = ma_r = mv^2/R$$

Then

$$f = \mu m v^2/R$$

The velocity of the rider is equal to the distance traveled in one revolution divided the time T it takes for one revolution, or

$$v = 2\pi R/T$$

So,

$$f = \mu_s m (2\pi R/T)^2/R = \mu_s m (2\pi)^2 R/T^2$$

The frictional force must balance the force due to gravity in order for the rider to stay pinned in place, or

$$\mu_s m (2\pi)^2 R/T^2 = mg$$

To determine the coefficient of static friction, rearrange this for μ_s

$$\mu_s = gT^2/(2\pi)^2 R$$

Before we substitute values, note that the rotation rate of 30 rpm is 1 rev every 2 s, that is, T equals 2.0 s. Then

$$\mu_s = (9.8 \text{ m/s}^2)(2.0 \text{ s})^2/[(2\pi)^2(4.5 \text{ m})]$$
$$\mu_s = 0.22$$

REFLECT
This coefficient of static friction $\mu_s = 0.22$ is well within the range that we expect for ordinary materials.

Figure 5-24

⭐ What's Important 5-5
A centripetal force pushes an object in motion off its straight course, causing it to travel in a circle. The acceleration of an object moving in uniform circular motion, and therefore the centripetal force, points toward the center of the circle.

Answers to Practice Problems

Answers to Got the Concept Questions

5-1 Because the force of gravity on an object is proportional to the object's mass, mass appears on both the right- and left-hand sides of the statement of Newton's second law. These mass terms cancel, so that the acceleration of an object due to gravity is therefore independent of mass. The force of kinetic friction on an object as it moves across a surface is proportional to the normal force that the surface exerts on the object. Here, the frictional force is proportional to the normal force exerted by the ski slope on the football player. The normal force must balance the component of the football player's weight perpendicular to the slope, similar to the result we found for the block on the ramp (Equation 5-3). The force of kinetic friction is therefore also proportional to the football player's weight and therefore his mass. Because both forces acting on the football player are proportional to mass, his acceleration—essentially force divided by mass—is independent of mass. The rate at which the football player accelerates down the ski slope does not depend on his weight!

5-2 It is not possible to know whether or not the block slides back down without knowing the coefficient of static friction and the angle of the ramp. We can expect that the magnitude of the force of static friction will be larger than the magnitude of the force of kinetic friction. So although the block can continue to move against kinetic friction once set in motion, it may not be able to overcome static friction once it has stopped. Also, for any given value of the coefficient of static friction, you can imagine that setting the ramp at a steep enough angle would guarantee that the block would come to rest only momentarily and then slide back down, or that setting the ramp at some shallow angle would guarantee that the block would come to rest.

5-3 As a car moves, even though the tires are turning the surfaces of the tires are not sliding against the surface of the road. For that reason the resistive force between the tires and the road is static friction. If the tires lock while a car is in motion, however, as might happen if the brakes are applied for an extended period, the surfaces of the tires do slide against the surface of the road. The resistive force then takes the form of kinetic friction. Because the coefficient of kinetic friction is lower than the coefficient of static friction, the car is more likely to skid when the wheels are locked. By making it less likely for the wheels to lock, an ABS helps a driver maintain control of the car in an emergency braking situation.

SUMMARY

Topic	Summary	Equation or Symbol	
air resistance	Air resistance is a drag force that resists the movement of an object through air.		
centripetal force	A centripetal (center-seeking) force pushes an object in motion toward the center of a circular path.		
coefficient of friction	The coefficient of friction is the proportionality between the frictional force and the normal force with which one surface acts on another. The coefficient of friction is a property of *both* surfaces, not a characteristic of a single surface or material.	μ_s (coefficient of static friction) μ_k (coefficient of kinetic friction)	
drag force	The resistance to motion of an object through a fluid such as air or water is commonly referred to as a drag force.	$F_{drag} = bv^n$	(5-22)
friction	Friction is a force that opposes motion between two objects that are in direct contact. The frictional force is proportional to the normal force with which one surface acts on another.	$f \propto N$	(5-1)
kinetic friction	Kinetic friction is the resistive force between two surfaces that are moving relative to one another.	$f = \mu_k N$	(5-5)

| static friction | Static friction is the force that prevents one surface from sliding against another. It exactly cancels the applied force, which means that it varies as the applied force increases, up to a maximum value. | $f \leq \mu_s N$ | (5-2) |
| terminal speed | When the drag force on an object falling through air equals the gravitational force in magnitude, the net force on the object equals zero. In this case there is no acceleration; the constant speed achieved is the terminal speed. The terminal speed of an object falling near the surface of Earth is a function of the mass of the object, a proportionality constant b, and g. | $v_{\text{terminal}} = \sqrt{\dfrac{mg}{b}}$ | (5-23) |

QUESTIONS AND PROBLEMS

In a few problems, you are given more data than you actually need; in a few other problems, you are required to supply data from your general knowledge, outside sources, or informed estimate.

Interpret as significant all digits in numerical values that have trailing zeros and no decimal points.

For all problems, use $g = 9.8 \text{ m/s}^2$ for the free-fall acceleration due to gravity. Neglect friction and air resistance unless instructed to do otherwise.

• Basic, single-concept problem

•• Intermediate-level problem, may require synthesis of concepts and multiple steps

••• Challenging problem

SSM *Solution is in Student Solutions Manual*

Conceptual Questions

1. •Complete the sentence: The static frictional force between two surfaces is (never/sometimes/always) less than the normal force. Explain your answer.

2. •You want to push a heavy box of books across a rough floor. You know that the maximum value of the coefficient of static friction (μ_s) is larger than the maximum value of the coefficient of kinetic friction (μ_k). Should you push the box for a short distance, rest, push the box another short distance, and then repeat the process until the box is where you want it, or will it be easier to keep pushing the box across the floor once you get it moving?

3. •If the force of friction always opposes the motion of an object, how then can a frictional force cause an object to increase in speed? SSM

4. •A solid rectangular block has sides of three different areas. You can choose to rest any of the sides on the floor as you apply a horizontal force to the block. Does the choice of side on the floor affect how hard it is to push the block? Explain your answer.

5. •You're trying to press a book against a spot on the wall with your hand. As you get tired, you exert less force, but the book remains in the same spot on the wall. Do each of the following forces increase, decrease, or not change in magnitude when you reduce the force you are applying to the book: (a) weight, (b) normal force, (c) frictional force, and (d) maximum static frictional force?

6. •For an object moving in a circle, which of the following quantities are zero over one revolution: (a) displacement, (b) average velocity, (c) average acceleration, (d) instantaneous velocity, and (e) instantaneous centripetal acceleration?

7. •Why does water stay in a bucket that is whirled around in a vertical circle? Contrast the forces acting on the water when the bucket is at the lowest point on the circle to when the bucket is at the highest point on the circle. SSM

8. •An antilock braking system (ABS) prevents wheels from skidding while drivers stomp on the brakes in emergencies. What effect would this have on how far a car with an ABS will move before finally stopping?

9. •Give two examples of the normal force not equal to an object's weight.

10. •**Medical** A fluidlike substance called *synovial fluid* lubricates the surfaces where bones meet in joints, making the coefficient of friction between bones very small. Why is the minimum force required for moving bones in a typical knee joint different for different people, in spite of the fact that their joints have the same coefficient of friction? For simplicity, assume the surfaces in the knee are flat and horizontal.

11. •As a skydiver falls faster and faster through the air, does his acceleration increase, decrease, or remain the same? Explain your answer. SSM

12. •Why do raindrops fall from the sky at different speeds? Explain your answer.

13. •Why might your car start to skid if you drive too fast around a curve?

14. •What distinguishes the forces that act on a car driving over the top of a hill from those acting on a car driving through a dip in the road? Explain how the forces relate to the sensations passengers in the car experience during each situation.

15. •Explain how you might measure the centripetal acceleration of a car rounding a curve. SSM

16. •An object of mass M_1 rests on a horizontal table; friction exists between the object and the table (Figure 5-25). A ring of mass M_R is tied by massless strings both to the wall and to the object as shown. A second object of mass M_2 hangs from the ring. Draw a free-body diagram for the situation.

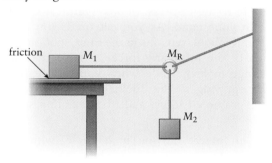

Figure 5-25 Problem 16

17. •Explain why curves in roads and running tracks are banked. SSM

18. •At low speeds, the drag force on an object moving through a fluid is proportional to its velocity. According to Newton's second law, force is proportional to acceleration. As acceleration and velocity aren't the same quantity, is there a contradiction here?

19. •James Bond leaps without a parachute from a burning airplane flying at 15,000 ft. Ten seconds later his assistant, who was following behind in another plane dives after him, wearing her parachute and clinging to one for her hero. Is it possible for her to catch up with Bond and save him?

Multiple-Choice Questions

20. •If a sport utility vehicle (SUV) drives up a slope of 45°, what must be the minimum coefficient of static friction between the SUV's tires and the road?
   A. 1.0
   B. 0.5

C. 0.7
D. 0.9
E. 0.05

21. •A block of mass m slides down a rough incline with constant speed. If a similar block that has a mass of $4m$ were placed on the same incline, it would
   A. slide down at constant speed.
   B. accelerate down the incline.
   C. slowly slide down the incline and then stop.
   D. accelerate down the incline with an acceleration four times greater than that of the smaller block.
   E. not move. SSM

22. •A 10-kg crate is placed on a horizontal conveyor belt moving with a constant speed. The crate does not slip. If the coefficients of friction between the crate and the belt are μ_s equal to 0.50 and μ_k equal to 0.30, what is the frictional force exerted on the crate?
   A. 98 N
   B. 49 N
   C. 29 N
   D. 9.8 N
   E. 0

23. •Biology The *Escherichia coli* (*E. coli*) bacterium propels itself through water by means of long, thin structures called flagella. If the force exerted by the flagella doubles, the velocity of the bacterium
   A. doubles.
   B. decreases by half.
   C. does not change.
   D. increases by a factor of four.
   E. cannot be determined without more information.

24. •A 1-kg wood ball and a 5-kg lead ball have identical sizes, shapes, and surface characteristics. They are dropped simultaneously from a tall tower. Air resistance is present. How do their accelerations compare?
   A. The 1-kg wood ball has the larger acceleration.
   B. The 5-kg lead ball has the larger acceleration.
   C. The accelerations are the same.
   D. The 5-kg ball accelerates at five times the acceleration of the 1-kg ball.
   E. The 1-kg ball accelerates at five times the acceleration of the 5-kg ball.

25. •A skydiver is falling at his terminal speed. Immediately after he opens his parachute
   A. his speed will be larger than his terminal speed.
   B. the drag force on the skydiver will decrease.
   C. the net force on the skydiver is in the downward direction.
   D. the drag force is larger than the skydiver's weight.
   E. the net force on the skydiver is zero. SSM

26. •Two rocks that are of equal mass are tied to massless strings and whirled in nearly horizontal circles at the

This page is intentionally left blank.

For complete end of chapter problem sets, please go to
www.whfreeman.com/kestentauck

43. •Draw a free-body diagram for the situation shown in **Figure 5-26**. An object of mass M rests on a ramp; there is friction between the object and the ramp. The system is in equilibrium.

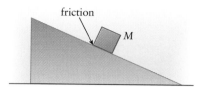

Figure 5-26 Problem 43

44. •Draw a free-body diagram for the situation shown in **Figure 5-27**. An object of mass M_2 rests on a frictionless table, and an object of mass M_1 sits on it; there is friction between the objects. A horizontal force F is applied to the object as shown.

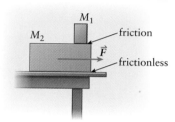

Figure 5-27 Problem 44

45. •A horizontal force of 10 N is applied to a stationary 2-kg block as shown in **Figure 5-28**. The coefficient of static friction between the block and the floor is 0.75; the coefficient of kinetic friction is 0.45. Find the acceleration of the box. SSM

Figure 5-28 Problem 45

46. •A block of mass is placed on a horizontal surface and attached to the free end of a spring. When the spring is neither stretched nor compressed, it exerts no force on the block. When the block is displaced from this position the force on it due to the spring is equal to 250 N/m times the distance the block is displaced. If the coefficient of static friction between the block and the surface is 0.24, how far from this position can be block be pulled and then released from rest so that it remains stationary?

47. •A coin that has a mass of 25 g rests on a phonograph turntable that rotates at 78 rev/min. The center of the coin is 13 cm from the turntable axis. If the coin does not slip, what is the minimum value of the coefficient of static friction between the coin and the turntable surface?

5-2: Kinetic Friction

48. •An object on a level surface experiences a horizontal force of 12.7 N due to kinetic friction. If the coefficient of kinetic friction is 0.37, what is the mass of the object?

49. •A book is pushed across a horizontal table at a constant speed. If the horizontal force applied to the book is equal to one-half of the book's weight, calculate the coefficient of kinetic friction between the book and the tabletop. SSM

50. •A 25-kg crate rests on a level floor. A horizontal force of 50 N accelerates the crate at 1 m/s². Calculate (a) the normal force on the crate, (b) the frictional force on the crate, and (c) the coefficient of kinetic friction between the crate and the floor.

51. •A block of mass M rests on a 5.00-kg block which is on a tabletop (**Figure 5-29**). A light string passes over a frictionless peg and connects the blocks. The coefficient of kinetic friction at both surfaces μ_k equals 0.330. A force of 60.0 N pulls the upper block to the left and the lower block to the right. The blocks are moving at a constant speed. Determine the mass of the upper block.

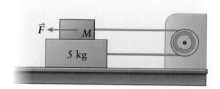

Figure 5-29 Problem 51

5-3: Working with Friction

52. •A mop is pushed across the floor with a force of 50 N at an angle of 50° (**Figure 5-30**). The mass of the mop head is 3.75 kg. Calculate the acceleration of the mop head if the coefficient of kinetic friction between the head and the floor is 0.400.

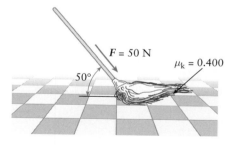

Figure 5-30 Problem 52

53. ••A taut string connects a 5-kg crate to a 12-kg crate (**Figure 5-31**). The coefficient of static friction between the smaller crate and the floor is 0.573; the coefficient of static friction between the larger crate and the floor is 0.443. Find the minimum horizontal force required to start the crates in motion. SSM

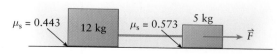

Figure 5-31 Problem 53

54. ••A 2-kg box rests on top of a 5-kg crate (Figure 5-32). The coefficient of static friction between the box and the crate is 0.667. The coefficient of static friction between the crate and the floor is 0.400. Calculate the minimum force (F) that is required to move the crate to the right and the corresponding tension (T) in the rope that connects the box to the wall when the rope is taut.

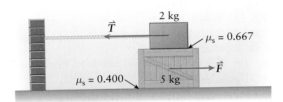

Figure 5-32 Problem 54

55. •If the coefficient of kinetic friction between a 3-kg object and a flat surface is 0.400, what force will cause the object to accelerate at 2.5 m/s²? The force is applied at an angle of 30° (Figure 5-33).

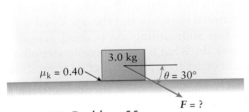

Figure 5-33 Problem 55

56. ••A 2.85-kg object is held in place on an inclined plane that makes an angle of 40° with the horizontal (Figure 5-34). The coefficient of static friction between the plane and the object is 0.552. A second object that has a mass of 4.75 kg is connected to the first object with a massless string over a massless, frictionless pulley. Calculate the initial acceleration of the system and the tension in the string once the objects are released.

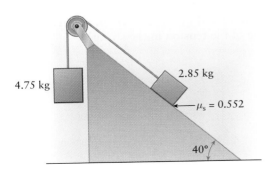

Figure 5-34 Problem 56

57. •Two blocks are connected over a massless, frictionless pulley (Figure 5-35). The mass of block 2 is 8 kg, and the coefficient of kinetic friction between block 2 and the incline is 0.22. The angle θ of the incline is 28°. Block 2 slides down the incline at constant speed. What is the mass of block 1? SSM

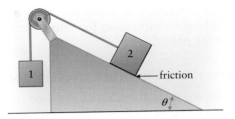

Figure 5-35 Problems 57 and 58

58. •Two blocks are connected over a massless, frictionless pulley (Figure 5-35). The mass of block 2 is 10 kg, and the coefficient of kinetic friction between block 2 and the incline is 0.20. The angle θ of the incline is 30°. If block 2 moves up the incline at constant speed, what is the mass of block 1?

5-4: Drag Force

59. •**Biology** A single-celled animal called a *paramecium* propels itself quite rapidly through water by using its hairlike *cilia*. A certain paramecium experiences a drag force $F_{drag} = -bv^2$ in water, where the drag coefficient b is approximately 0.31. What propulsion force does this paramecium generate when moving at a constant (terminal) speed v of 0.15×10^3 m/s?

60. •**Biology** The bacterium *Escherichia coli* propels itself with long, thin structures called flagella. When its flagella exert a force of 1.5×10^{-13} N, the bacterium swims through water at a speed of 20 μm/s. Find the speed of the bacterium in water when the force exerted by its flagella is 3.0×10^{-13} N.

61. •A 5-kg object in free-fall reaches a terminal velocity of 50 m/s. The drag force is proportional to speed raised to some power n; that is, $F_{drag} = -bv^n$, where b is the drag coefficient. If n equals 2.5, determine the value of the drag coefficient. SSM

62. ••If we model the drag force of the atmosphere as proportional to the square of the speed of a falling object, $F_{drag} = -bv^2$, where the value of b for a 70-kg person with a parachute is 18 kg/m. (a) What is the person's terminal velocity? (b) Without a parachute, the same person's terminal velocity would be about 50 m/s. What would be the value of the proportionality constant b in that case?

63. ••A girl rides her scooter on a hill that is inclined at 10° with the horizontal. The combined mass of the girl and scooter is 50 kg. On the way down, she coasts at a constant speed of 12 m/s, while experiencing a drag force that is proportional to the square of her velocity. What force, parallel to the surface of the hill, is required to increase her speed to 20 m/s? Neglect any other resistive forces.

64. •••**Calc** Suppose the drag force acting on a free-falling object is proportional to the velocity. The net force acting on the object would be $F = mg - bv$. (a) Using dimensional analysis, determine the units of the constant b. (b) Find an expression for velocity as a function of time for the object.

5-5: Forces and Uniform Circular Motion

65. •A 1500-kg truck rounds an unbanked curve on the highway at a speed of 20 m/s. If the maximum frictional force between the surface of the road and all four of the tires is 8000 N, calculate the minimum radius of curvature for the curve to prevent the truck from skidding off the road. SSM

66. •A hockey puck that has a mass of 170 g is tied to a light string and spun in a circle of radius 1.25 m (on frictionless ice). If the string breaks under a tension that exceeds 5.00 N, at what angular speed (revolutions per minute) will the string break?

67. •A 25-g metal washer is tied to a 60-cm-long string and whirled around in a vertical circle at a constant speed of 6 m/s. Calculate the tension in the string (a) when the washer is at the bottom of the circular path and (b) when it is at the top of the path.

68. •**Astro** What centripetal force is exerted on the Moon as it orbits about Earth at a center-to-center distance of 3.84×10^8 m with a period of 27.4 days? What is the source of the force? The mass of the Moon is equal to 7.35×10^{22} kg.

69. •A centrifuge spins small tubes in a circle of radius 10 cm at a rate of 1200 rev/min. What is the centripetal force on a sample that has a mass of 1 g? SSM

70. •**Biology** Very high-speed ultracentrifuges are useful devices to sediment materials quickly or to separate materials. An ultracentrifuge spins a small tube in a circle of radius 10 cm at 60,000 rev/min. What is the centripetal force experienced by a sample that has a mass of 0.0030 kg?

71. •At the Fermi National Accelerator Laboratory (Fermilab), a large particle accelerator, protons are made to travel in a circular orbit 6.3 km in circumference at a speed of nearly 3.0×10^8 m/s. What is the centripetal acceleration on one of the protons?

72. •What is the force that a jet pilot feels against his seat as he completes a vertical loop that is 500 m in radius at a speed of 200 m/s? Assume his mass is 70 kg and that he is located at the bottom of the loop.

73. •In the game of tetherball, a 1.25-m rope connects a 0.75-kg ball to the top of a vertical pole so that the ball can spin around the pole as shown in **Figure 5-36**. Find the tension in the rope when the ball completes 0.5 rev/s and the angle of the ball is 35° with the vertical. SSM

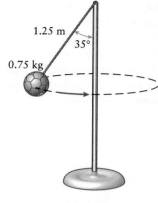

Figure 5-36 Problem 73

74. •The radius of Earth is 6.38×10^6 m and it completes one revolution in 1 day. (a) What is the centripetal acceleration of an object located on the equator? (b) What is the centripetal acceleration of an object located at latitude 40° north?

75. •**Sports** In executing a windmill pitch, a fast-pitch softball player moves her hand through a circular arc of radius 0.31 m. The 0.19-kg ball leaves her hand at 24 m/s. What is the magnitude of the force exerted on the ball by her hand immediately before she releases it?

General Problems

76. •A 150-kg crate rests in the bed of a truck that slows from 50 km/h to a stop in 12 s. The coefficient of static friction between the crate and the truck bed is 0.655. (a) Will the crate slide during the braking period? Explain your answer. (b) What is the minimum stopping time for the truck in order to prevent the crate from sliding?

77. ••The coefficient of static friction between a rubber tire and dry pavement is about 0.80. Assume that a car's engine only turns the two rear wheels and that the weight of the car is uniformly distributed over all four wheels. (a) What limit does the coefficient of static friction place on the time required for a car to accelerate from rest to 60 mph (26.8 m/s)? (b) How can friction accelerate a car *forward* when friction *opposes* motion? SSM

78. ••Two blocks are connected over a massless, frictionless pulley (**Figure 5-37**). Block m_1 has a mass of 1.0 kg and block m_2 has a mass of 0.4 kg. The angle θ of the incline is 30°. The coefficients of static friction and kinetic friction between block m_1 and the incline are μ_s equal to 0.50 and μ_k equal to 0.40, respectively. What is the value of the tension in the string?

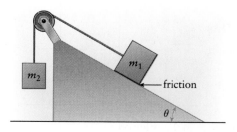

Figure 5-37 Problems 78 and 79

79. ••Two blocks are connected over a massless, frictionless pulley (Figure 5-37). Block m_1 has a mass of 1.0 kg and block m_2 has a mass of 2.0 kg. The angle θ of the incline is 30°. The coefficients of static friction and kinetic friction between block m_1 and the incline are μ_s equal to 0.50 and μ_k equal to 0.40. What is the acceleration of block m_1?

80. •A runaway ski slides down a 250-m-long slope inclined at 37° with the horizontal. If the initial speed is 10 m/s, how long does it take the ski to reach the bottom of the incline if the coefficient of kinetic friction between the ski and snow is (a) 0.10 and (b) 0.15?

81. •In a mail-sorting facility, a 2.5-kg package slides down an inclined plane that makes an angle of 20° with the horizontal. The package has an initial speed of 2 m/s at the top of the incline and it slides a distance of 12.0 m. What must the coefficient of kinetic friction between the package and the inclined plane be so that the package reaches the bottom with no speed? SSM

82. ••In Figure 5-38, two blocks are connected to each other by a massless string over a frictionless pulley. The mass of the block on the left incline is 6.00 kg. Assuming the coefficient of static friction μ_s equals 0.542 for all surfaces, find the range of values of the mass of the block on the right incline so that the system is in equilibrium.

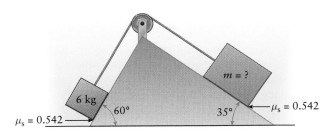

Figure 5-38 Problem 82

83. •••Calc With its sails fully deployed, a 100-kg sailboat (including the passenger) is moving at 10 m/s when the mast suddenly snaps and the sail collapses. The boat immediately starts to slow down due to the resistive drag force of the water on the boat. After 5 s, the boat's speed is only 6 m/s. If the drag force of the water is proportional to the speed of the boat, calculate how long it will take before the boat has a speed of 0.5 m/s.

84. ••The terminal velocity of a raindrop that is 4.0 mm in diameter is approximately 8.5 m/s under controlled, windless conditions. The density of water is 1000 kg/m³. Recall that the density of an object is its mass divided by its volume. (a) If we model the air drag as being proportional to the square of the speed, $F_{drag} = -bv^2$, what is the value of b? (b) Under the same conditions as above, what would be the terminal velocity of a raindrop that is 8.0 mm in diameter? Try to use

your answer from part (a) to solve the problem by proportional reasoning instead of just doing the same calculation over again.

85. ••Calc In a weird parallel universe, an object moving through a liquid experiences a drag force proportional to the inverse of the object's speed, that is, $F_{drag} = -bv^{-1}$. (a) An object of mass m has an initial speed v_0 in the liquid. Derive an equation which gives the speed as a function of time. (b) An object enters the liquid with an initial speed 10 m/s, and after 9 s its speed is 8 m/s. Determine the time it takes for the object to come to a complete stop. In both cases the drag force is the only force on the object. SSM

86. •Biomedical laboratories routinely use ultracentrifuges, some of which are able to spin at 100,000 rev/min about the central axis. The turning rotor in certain models is about 20 cm in diameter. At its top spin speed, what force does the bottom of the rotor exert on a 2.0-g sample that is in the rotor at the greatest distance from the axis of spin? Would the force be appreciably different if the sample were spun in a vertical or a horizontal circle? Why?

87. ••An amusement park ride called the Rotor debuted in 1955 in Germany. Passengers stand in the cylindrical drum of the Rotor as it rotates around its axis. Once the Rotor reaches its operating speed, the floor drops but the riders remain pinned against the wall of the cylinder. Suppose the cylinder makes 25 rev/min and has a radius of 3.5 m. Find the coefficient of static friction between the wall of the cylinder and the backs of the riders.

88. •An object of mass m_1 undergoes constant circular motion and is connected by a massless string through a hole in a frictionless table to a larger object of mass m_2 (Figure 5-39). If the larger object is stationary, calculate the tension in the string and the period of the circular motion of the smaller object. Assume that the objects have masses of 0.225 and 0.125 kg and the radius R of the circular path of the smaller object is equal to 1.00 m.

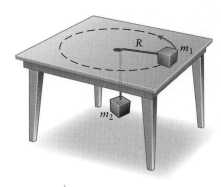

Figure 5-39 Problem 88

89. •An object that has a mass M hangs from a support by a massless string of length L (Figure 5-40). The support is rotated so that the object follows a circular path at an angle θ from the vertical as shown. The object makes N revolutions per second. Find an expression for the angle θ in terms of M, L, N, and any necessary physical constants. SSM

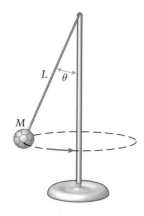

Figure 5-40 Problem 89

90. •**Medical** Occupants of cars hit from behind, even at low speed, often suffer serious neck injury from whiplash. During a low speed rear-end collision, a person's head suddenly pivots about the base of his neck through a 60° angle, a motion that lasts 250 ms. The distance from the base of the neck to the center of the head is typically about 20 cm, and the head normally comprises about 6% of body weight. We can model the motion of the head as having uniform speed over the course of its pivot. Compute your answers to the following questions to two significant figures. (a) What is the acceleration of the head during the collision? (b) What force (in newtons and in pounds) does the neck exert on the head of a 75-kg person in the collision? (As a first approximation, neglect the force of gravity on the head.) (c) Would headrests mounted to the backs of the car seats help protect against whiplash? Why?

91. •The wings of an airplane flying in a horizontal circle at a speed of 680 km/h are tilted 60° to the horizontal (**Figure 5-41**). What is the radius of the circle? Assume that the required force is provided entirely by the wings' "lift," a force perpendicular to the surface of the wings.

Figure 5-41 Problem 91

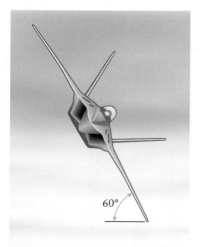

60°

92. •A curve that has a radius of 100 m is banked at an angle of 10° (**Figure 5-42**). If a 1000-kg car navigates the curve at 65 km/h without skidding, what is the minimum coefficient of static friction between the pavement and the tires?

10°

Figure 5-42 Problem 92

93. ••(a) What is the terminal velocity of a 75-kg skydiver? Assume the drag force is proportional to her speed, $F_{drag} = bv$, and b has a value of 115 kg/s. (b) When she lands, the parachutist comes to rest over a distance that is equal to the length of her legs, about 70 cm. Make a free-body diagram of her as she is slowing down once she has come into contact with the ground. (c) What is the average force that the ground exerts on her body during her landing, assuming that she has reached terminal velocity during her fall? Is she likely to get hurt?

94. ••**Calc** A small object moves through a large vat of oil. The drag force opposing the motion is given by $F_{drag} = -bv^{1/2}$. Find the velocity of the object as a function of time, in terms of its velocity v_0 when it enters the oil. Neglect the force of gravity. SSM

6 Work and Energy

(Jean-Paul Ferrero/Auscape/The Image Works.)

Animals use a significant amount of the energy gained from eating to move. Large kangaroos are among the most efficient mammals in their use of metabolic energy to generate motion. As a kangaroo comes down to the ground on each hop, energy associated with the motion is stored as elastic energy in the tendons of its hind legs. In the same way that the compressed spring in a pogo stick pushes up against a child's feet, the elastic energy stored in a kangaroo's tendons is transferred back into energy of motion. This return of energy makes hopping an efficient way for kangaroos to move.

L ife depends on energy. Nearly every living process requires energy—moving, breathing, circulating blood, digesting food, absorbing nutrients, even thinking. Indeed, we can loosely define energy as the ability to move something or to exert a force. The man in **Figure 6-1**, for example, is using energy to generate the force required to move a crate. The magnitude of the force is not, however, the full measure of the amount of effort required. If the truck bed were higher, the task would be more difficult. In a similar way, cross-country skiing for 1 km requires less energy and is less tiring than skiing for 20 km. These examples suggest that the distance over which a force is exerted plays a direct role in quantifying the effort required. We will incorporate this distance in the physical quantity called *work*.

6-1 Work

The effort required to exert a force depends on the distance over which the force is exerted. We define the quantity **work** as the product of the magnitude of force and the distance over which the force is applied. In the simplest case, in which the applied force F is constant and the directions of the force vector and the displacement are the same

$$W = Fd \qquad (6\text{-}1)$$

Figure 6-1 A man uses energy to generate the force required to move a crate. The magnitude of the force is not, however, the full measure of the amount of effort required. If the truck bed were higher, the task would be more difficult. *(Ocean/Corbis.)*

179

An object accelerates when it experiences a nonzero, net force. In the language of physics, we can now also say that an object accelerates when nonzero work is done on it. The second statement is more precise, however, because it includes a measure of the distance over which the force is exerted.

Energy and work are often defined in terms of one another. Generally **energy** is the capacity of an object or system to do work, while work can be defined as that which results in a change or transfer of energy. We will discover that it is not always possible to convert energy to work, but this somewhat loose definition will serve for now. We need to have or supply energy to do work, for example, to move a car or to walk up a hill. We will encounter a variety of forms of energy, including energy associated with motion and energy stored in the chemical bonds of molecules that make up food.

Newton's third law states that if one object exerts a force on another object, the second object exerts a force back on the first. For that reason, it is helpful to name both the agent of a force and the object on which it does work; for example, the man does work *on the crate* in Figure 6-1.

Equation 6-1 is a simplified version of the definition of work that applies to cases in which the applied force is constant and its direction is the same as that of motion. We will shortly generalize Equation 6-1 for other situations.

The unit of work, the joule (J), is named after the 19th century English physicist James Joule. His study of the relationship between motion and work led to the development of the scientific understanding of work. Using Equation 6-1,

$$1\,J = 1\,N \times 1\,m$$

or

$$1\,J = 1\,N \cdot m$$

It takes 1 J of work to lift an object that weighs 1 N through 1 m.

Example 6-1 Lifting a Water Bottle

A 1-L bottle of water has a mass of 1.0 kg. To understand the magnitude of a joule, determine the distance you must lift a 1-L bottle of water, at constant speed, in order to do 1.0 J of work.

SET UP
At constant acceleration the applied force is also constant. We can therefore apply Equation 6-1 directly. Also, the force necessary must just balance the force exerted by gravity, so

$$F = mg$$

Then, substituting mg for F and h for d into Equation 6-1 gives

$$W = mgh$$

where h is the distance through which the bottle is moved.

SOLVE
To calculate an answer we first rearrange the equation

$$h = \frac{W}{mg}$$

and then substitute our known values of variables

$$h = \frac{1.0\,J}{(1.0\,kg)(9.8\,m/s^2)} = 0.10\,m$$

One joule of work is required to raise a 1-L bottle of water by 10 cm when the bottle is moved at constant speed.

REFLECT

On a human scale, 1 J is a relatively small amount of work.

Practice Problem 6-1 How many joules of work must be expended in order to lift a 7.2-kg rock 0.95 m straight up?

Example 6-2 Pushing a Car

Three people each apply a constant force of 200 N horizontally to a car that has run out of gas. To understand the magnitude of a joule, determine the work done on the car by the three people to move it 5 m.

SET UP

The applied force is constant and in the direction in which the car moves, as shown in Figure 6-2, so Equation 6-1 applies. Note that this figure is not a free-body diagram, but instead shows the relevant force and the direction of motion.

SOLVE

We substitute the given values of force and distance into Equation 6-1 to determine the work

$$W = (3)(200\,\text{N})(5\,\text{m}) = 3000\,\text{J}$$

Together the three people do 3000 J of work on the car.

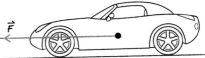

Figure 6-2 Equation 6-1 applies because the applied force is constant and in the direction in which the car moves.

REFLECT

Even if you have never had to push a car that has run out of gas, try to imagine the amount of effort required. Each of the three people in this problem did 1000 J of work on the car.

Practice Problem 6-2 Each of three people apply a constant force of 200 N horizontally to a car. How far will they move it if each does 1 J of work on the car?

How can work be determined when the applied force is not in the direction of motion? For example, a screen is pulled across a baseball diamond to smooth out the dirt in Figure 6-3a. The tension force $\vec{F}$ that the rope exerts on the screen is clearly at an angle with respect to the direction in which the screen moves. In Figure 6-3a, the red vector arrow shows the force vector and the blue arrow shows its component along the direction of motion. Two special cases support the conclusion that it is only this component of the force that contributes to the motion. First, if $\vec{F}$ were pointing straight up, the component of $\vec{F}$ in the direction of motion would be zero. Assuming the man can't lift the screen off the ground, this would result in no motion and therefore no work done on the screen. Conversely, if $\vec{F}$ were completely aligned with the direction of motion, all of the force would be applied to accelerating the screen along the ground. In general, only the component of $\vec{F}$ along the direction of motion contributes to the work done on the screen over a distance d. The component of the force along the direction of motion is $F\cos\theta$.

Including only the component of $\vec{F}$ along the direction of motion, Equation 6-1 becomes

$$W = F\cos\theta\, d \tag{6-2}$$

√x̄ *See the Math Tutorial for more information on Trigonometry*

Figure 6-3 (a) A screen is pulled across a baseball diamond to smooth out the dirt. The rope exerts a tension force $\vec{F}$ on the screen at an angle with respect to the direction in which the screen moves. The blue arrow shows the component of $\vec{F}$ along the direction of motion. (b) The work done on the screen is the product of the component of the force $\vec{F}$ along the direction of motion, shown in blue, and the distance the screen moves. That is, the work equals the product of $F \cos\theta$ and d.

The blue arrow shows the component of $\vec{F}$ in the direction of motion.

(a)

(b)

where θ is the angle between $\vec{F}$ and its component in the direction of motion. If we treat the distance d over which the screen moves as a vector, Equation 6-2 can be written more simply. We form the distance vector by multiplying the distance d that the screen moves with a unit vector $\hat{d}$ in the direction of motion

$$\vec{d} = d\,\hat{d}$$

Figure 6-3b shows the $\vec{d}$ vector. Notice that this figure is not a free-body diagram, because it shows more than just the forces. Mixing force and direction vectors in a sketch makes the interplay between the forces and the distances more clear.

If distance is a vector, Equation 6-2 can then be written as

$$W = F\,d\cos\theta = \vec{F} \cdot \vec{d} \tag{6-3}$$

We refer to the notation on the right side of Equation 6-3 as the **dot product**. There is no difference between writing the expression for work as $F\,d\cos\theta$ or $\vec{F} \cdot \vec{d}$ — both convey the same information. However, in addition to being more compact, the dot product is a visual reminder that one must consider the direction of the force and the direction of motion to determine work.

Note that when the force on an object is applied in the same direction as the object's motion, so that the angle between the force and direction vectors equals zero, Equation 6-3 reduces to $W = F\,d$ (Equation 6-1), the first expression we wrote for work.

The dot product of two vectors equals the product of the magnitudes of the vectors multiplied by the cosine of the angle between them. The dot product of two vectors is a scalar, not a vector. Work is a scalar quantity.

Mathbox 6-1 Dot Product

For two vectors $\vec{A}$ and $\vec{B}$ separated by angle θ, the dot product C is defined as

$$C = \vec{A} \cdot \vec{B} = A\,B\cos\theta \qquad\qquad \textbf{(6-4)}$$

The dot product is similar to multiplying scalar quantities. The difference between the two operations results from the directional component of vectors. Vectors can only be multiplied together when they are aligned. Built into the dot product, then, is the projection of one of the vectors in the direction of the other. You can read the mathematical operation (Equation 6-4) as "C equals A dot B."

The dot product is commutative; in other words, the order of the vectors in a dot product does not affect the result

$$\vec{A} \cdot \vec{B} = \vec{B} \cdot \vec{A}$$

The dot product is also distributive

$$\vec{A} \cdot (\vec{B} + \vec{C}) = \vec{A} \cdot \vec{B} + \vec{A} \cdot \vec{C}$$

Note three special cases of the dot product. First, the dot product of two perpendicular vectors is zero:

$$\vec{A} \cdot \vec{B} = A\,B\cos 90 = 0 \qquad\qquad \textbf{(6-5)}$$

Also, the dot product of a vector with itself is the square of the magnitude of the vector:

$$\vec{A} \cdot \vec{A} = A\,A\cos 0 = A^2 \qquad\qquad \textbf{(6-6)}$$

Finally, the dot product of two vectors that are equal in magnitude but point in opposite directions is the negative of the square of the magnitude of the vectors:

$$\vec{A} \cdot (-\vec{A}) = A\,A\cos 180 = -A^2 \qquad\qquad \textbf{(6-7)}$$

This result follows because the cosine of 180° is -1.

Equations 6-6 and 6-7 define the square of a vector.

Any force acting on an object in a direction other than perpendicular to the direction of motion does work on the object. For example, four forces act on the screen in Figure 6-3: the tension in the rope, gravity, the normal force, and friction. The free-body diagram of **Figure 6-4** shows these forces. We did not include the $\vec{d}$ vector in the diagram to avoid confusion with the forces, but instead note that the direction of motion is in the positive x direction as indicated by the dashed arrow.

The work each force does on the screen can be determined by applying Equation 6-3. These forces and the work associated with each are

Tension: $\vec{T}$ acts at angle θ with respect to the direction of motion, so
  $W_{\text{tension}} = T\,d\cos\theta$.
Gravity: The acceleration due to gravity, and therefore the weight $\vec{W}$, is
  perpendicular to the direction of motion, so according to Equation 6-5,
  the work done by gravity $W_{\text{gravity}} = 0$.
Normal: $\vec{N}$ is perpendicular to the direction of motion, so according to
  Equation 6-5, $W_{\text{normal}} = 0$.
Friction: Note that because friction is resistive, the frictional force vector
  points in the direction opposite to the motion. The angle to be applied
  to Equation 6-3 is therefore 180°, so $W_{\text{friction}} = f\,d\cos 180 = -f\,d$. The
  work done by friction on the screen has a negative value. *When a force
  opposes the motion of an object, the value of the work is negative.*

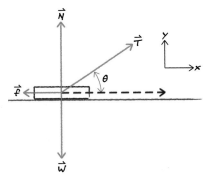

Figure 6-4 The four forces acting on the screen in Figure 6-3 are shown in a free-body diagram. The forces are the tension in the rope $\vec{T}$, the force of gravity $\vec{W}$, the normal force $\vec{N}$, and friction $\vec{f}$. The dashed arrow shows the direction of motion.

Example 6-3 Doing Work on the Moon

Earth exerts a gravitational force on the Moon of about 2×10^{20} N. Treat the orbit as a circle of radius 3.8×10^8 m to determine the work that Earth does on the Moon. (The Moon's orbit is only slightly elliptical, so it's not unreasonable to treat the orbital path as a circle.)

SET UP
According to Equation 6-3,

$$W = F\,d\cos\theta \tag{6-3}$$

Take care to use this form of the definition of work because the cosine term plays an important role. The gravitational force acts in the vertical direction, in other words, toward Earth's center. To determine the angle θ between the gravitational force and the direction of the Moon's motion, note that *at every instant* the Moon is moving tangentially to its orbital path (Figure 6-5). Therefore, the angle between the gravitational force and the Moon's direction of motion is 90°.

SOLVE
Applying Equation 6-3 with $\theta = 90°$ gives

$$W = F\,d\cos 90° = 0$$

Earth's gravity does no work on the Moon!

REFLECT
Although the force of Earth's gravity on the Moon is enormous, it does no work on the Moon because the force and the Moon's direction of motion are perpendicular to each other. This example underscores the importance of considering the angle between a force and the displacement vector of an object on which the force acts.

★ What's Important 6-1

The effort required to exert a force depends on the distance over which the force is exerted. Work is the product of force and the distance over which the force is applied. Physicists often define energy and work in terms of one another. Generally, energy is the capacity of an object or system to do work.

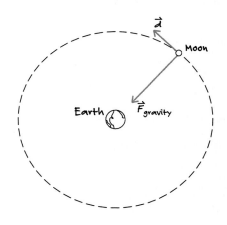

Figure 6-5 As the Moon orbits Earth, its direction of motion at every instant is tangential to its orbital path, that is, the motion is perpendicular to the direction of the gravitational force due to Earth.

6-2 The Work–Kinetic Energy Theorem

In Example 6-2, we considered three people pushing a car (Figure 6-2). The work they do on the car is the product of the force they exert and the distance over which they apply the force. Applying a net force to the car also changes its speed. Let's consider how work can be expressed in terms of an object's speed. To do so, we need to combine the definition of work (Equation 6-3) with our description of one-dimensional motion, Equations 2-23 and 2-26.

According to Newton's second law, the sum of the forces on an object of mass m results in acceleration

$$\sum \vec{F}_{\text{dir}} = m\vec{a}_{\text{dir}} \qquad (4\text{-}1)$$

In the direction of motion, Equation 4-1 becomes

$$F_{\text{net}} = m\,a$$

where a is the magnitude of the acceleration and F_{net} is the magnitude of the net force. By using Equation 6-1, the net work done on the object as it moves through a distance d can then be written

$$W_{\text{net}} = F_{\text{net}}\,d = m\,a\,d \qquad (6\text{-}8)$$

To relate this equation to the change in speed, we apply Equations 2-23 and 2-26 to describe the motion of the object as it accelerates from v_0 to v_1

$$v_1 = v_0 + at$$

$$d = v_0 t + \frac{1}{2}at^2$$

These two equations can be combined to express a in terms of v_0 and v_1 by solving the first for t and then substituting into the second:

$$d = v_0\left(\frac{v_1 - v_0}{a}\right) + \frac{1}{2}a\left(\frac{v_1 - v_0}{a}\right)^2$$

or

$$d = \frac{v_0 v_1 - v_0^2}{a} + \frac{v_1^2 + v_0^2 - 2v_1 v_0}{2a} = \frac{2v_0 v_1 - 2v_0^2 + v_1^2 + v_0^2 - 2v_1 v_0}{2a}$$

The $v_0 v_1$ terms cancel, leaving

$$d = \frac{v_1^2 - v_0^2}{2a}$$

So

$$a = \frac{v_1^2 - v_0^2}{2d}$$

When we substitute this expression into Equation 6-8, the distance term cancels and the equation becomes

$$W_{\text{net}} = m\,a\,d = m\frac{v_1^2 - v_0^2}{2}$$

or

$$W_{\text{net}} = \frac{mv_1^2}{2} - \frac{mv_0^2}{2} \qquad (6\text{-}9)$$

Evidently, work is equivalent to the difference between the final value and the initial value of the quantity formed by one-half the mass multiplied by the square of the speed. The term $\frac{1}{2}mv^2$ that appears on the right-hand side of Equation 6-9 is the **kinetic energy K**

$$K = \frac{1}{2}mv^2 \tag{6-10}$$

We also refer to it as **translational kinetic energy** in order to distinguish the energy of an object moving from one position to another along a straight line from the energy of a rotating object. Kinetic energy is of fundamental importance to the study of physical systems.

Note that while we allow velocity to be negative as well as positive (or zero) in order to show direction, kinetic energy is proportional to the square of velocity. The sign of the velocity does not, therefore, play a role. For this reason we will, in general, use the term speed when describing phenomena involving the physics of kinetic energy.

Watch Out

Kinetic energy is a scalar quantity.

We have become comfortable breaking an object's motion into components, so you might be tempted to resolve kinetic energy into components as well. However, kinetic energy depends on the square of velocity; because the square of velocity is not a vector, neither is kinetic energy. It is meaningless to set up components of kinetic energy in different directions.

Energy that can be used to do work is often referred to as **mechanical energy**. We note, however, that there are cases in which an object has or delivers energy that is not available to do work. For example, an object sliding across a surface will likely slow down due to friction. As a result, the atoms and molecules in the surfaces bounce around just a bit faster and the surfaces become slightly warmer. It is generally not possible to use this **thermal energy** transferred to the surface as heat to do work.

We can combine Equations 6-9 and 6-10 to define work as a change in energy:

$$W_{\text{net}} = K_1 - K_0 = \Delta K \tag{6-11}$$

The net work done on an object equals the change in its kinetic energy; this statement is the **work–kinetic energy theorem**. (Some texts refer to this as the "work-energy" theorem. We prefer to include the term "kinetic" to make it clear that only the energy of motion is included in the statement of Equation 6-11.) It applies only to the *net* work on an object; that is, it applies to the work done by the vector sum of all of the forces on the object.

Got the Concept 6-1

What a Hockey Puck!

A hockey player does work on a hockey puck to propel it from rest across the ice. When he applies a constant force over a certain distance, the puck leaves his stick at speed v_0. How much must he increase the distance over which he applies the same force so that the speed of the puck is $2v_0$?

The relationship between work and kinetic energy requires that the units of energy and the units of work are the same. The SI unit of kinetic energy—and energy in general—is therefore the joule. We have seen that $1\,\text{J} = 1\,\text{N} \cdot \text{m}$, and we showed that $1\,\text{N} = 1\,(\text{kg} \cdot \text{m})/\text{s}^2$ in Chapter 4. Combining these relations gives

$$1\,\text{J} = 1\left(\text{kg}\frac{\text{m}}{\text{s}^2}\right)\text{m} = 1\,\frac{\text{kg} \cdot \text{m}^2}{\text{s}^2}$$

which is consistent with the equivalence of work and energy and the form of kinetic energy in Equation 6-10.

Example 6-4 Champion Diver

In executing a dive from the 3-m springboard, a 60-kg diver passes the board on the way down at a speed of about 8 m/s and enters the water at about 11 m/s. How much work does gravity do on the diver from the moment she passes the board on the way down until she hits the water? Neglect air resistance.

SET UP
The diver's speed increases during the dive because the force of gravity does work on her. This change in speed leads directly to a change in her kinetic energy, and the work–kinetic energy theorem connects the work done to ΔK

$$W_{net} = \Delta K \qquad (6\text{-}11)$$

SOLVE
Neglecting the force due to air resistance, the net work is due to gravity. Using the definition of kinetic energy (Equation 6-10), then

$$W_{gravity} = \frac{1}{2}mv_f^2 - \frac{1}{2}mv_i^2 = \frac{1}{2}m(v_f^2 - v_i^2)$$

where v_i is the speed of the diver as she passes the board and v_f is the speed of the diver as she enters the water. So, the value of work done by gravity on the diver is:

$$W_{gravity} = \frac{1}{2}(60 \text{ kg})\left[(11 \text{ m/s})^2 - (8 \text{ m/s})^2\right] = 1.7 \times 10^3 \text{ J}$$

REFLECT
Force acting on an object does work on it, resulting in a change in speed and therefore in kinetic energy.

Practice Problem 6-4 Gravity does 4.7 J of work on a pebble of mass 1.6×10^{-1} kg dropped from a 3-m springboard with an initial speed of 0 m/s. What is the speed of the pebble as it enters the water? Neglect air resistance.

❓ Got the Concept 6-2
Pushing Boxes

You push two boxes from one side of a room to the other. The contact surfaces and areas between box and floor is the same for both, but one box is heavy while the other is light. How does the work you do on each box compare? How does the net work done on each box compare? Don't neglect friction!

You might have asked yourself, if work done depends on the distance over which a force is applied (Equation 6-3), or if work done equals the change in kinetic energy (Equation 6-11), why do you get tired when you push against a wall without moving it? Work is energy, after all, and you must expend energy to push an immovable object or to hold barbells stationary at arms' length (**Figure 6-6**). At issue here is that energy described in Equations 6-3 and 6-11 is mechanical energy, while the energy expended holding the barbells is energy stored in the chemical bonds of the ATP molecules in your muscles.

Figure 6-6 Energy must be expended, that is, work must be done, in order to hold barbells stationary at arms' length.

An interesting mechanism helps the body minimize the amount of energy expended to hold muscles stationary. The interaction between two contractile proteins, actin and myosin, underlies muscle contraction. When myosin binds to actin, it snaps into a bent shape which pulls the actin filament about 5 to 10 nm relative to the myosin. The myosin next binds to ATP, harvests energy stored in one of its chemical bonds, and then uses the energy to extend itself again to bind to actin at a point further along the filament. At each point the myosin and actin are temporarily locked together, so the muscle can remain in that position without expending more energy. Without a source of energy, however, the bond between actin and myosin does not break. This condition occurs after a living thing dies—within a few hours of death, all of the ATP in the body is consumed and the actin and myosin filaments remain in that locked position. The muscles become rigid leading to a condition known as *rigor mortis*, and also the slang term for a dead body (a "stiff").

✳ What's Important 6-2

Kinetic energy is a scalar quantity proportional to an object's mass and the square of its speed ($K = \frac{1}{2}mv^2$). The work–kinetic energy theorem says that the net work done on an object equals the change in its kinetic energy.

6-3 Applications of the Work–Kinetic Energy Theorem

Trillions of contractile proteins in the muscles of members of a bobsled team (Figure 6-7) enable them to push hard on their sled at the beginning of a race, resulting in a large starting speed. In this section, we will explore the relationship between work, force, and speed by applying the definitions of work and kinetic energy (Equations 6-3 and 6-10) and the work–kinetic energy theorem (Equation 6-11) to a variety of physical situations.

Figure 6-7 The interaction of contractile proteins with ATP in the muscles of these athletes enable them to push on their sled at the beginning of a race. *(Sampics/Corbis.)*

Example 6-5 Stopping a Car

The U.S. National Highway Transportation and Safety Administration lists the minimum braking distance for a car traveling at 20 mph (about 8.9 m/s) to be 25 ft (about 7.6 m). What is the deceleration required to stop the car in that distance, assuming the braking force is constant? For the minimum distance to stop, assume that the wheels of the car have stopped rotating, so that kinetic (sliding) friction between the tires and road result in the work necessary to reduce the car's speed.

SET UP

In many problems, especially real-world problems that have not been idealized (in contrast to the problems typically found in introductory physics books), you don't have enough information to determine the value of every variable. However, you can often find what you need by clustering some of the variables together. Here, for example, our goal is to find acceleration, but we don't need to know F and m separately as long as we can find the ratio of the two. You can see that by recognizing that the relationship between force, mass, and acceleration described by Newton's second law can be written $a = F/m$. Force appears in the definition of work and the definition of kinetic energy includes mass. Can we use the two definitions together? We can, by using the work–kinetic energy theorem.

The braking force and the distance over which the brakes are applied determine the work done on the road by the car as it comes to a stop. This work is described by Equation 6-1, but speed is not part of the definition. The work–kinetic energy theorem (Equation 6-11), however, enables us to make the connection; the change in kinetic energy depends on the initial and final speeds. In addition, because the kinetic energy depends on mass as well as speed ($K = \frac{1}{2}mv^2$) applying the work–kinetic energy theorem also enables us to consider the mass of the car in our solution of the problem.

SOLVE

According to Equation 6-3,

$$W = F\, d \cos\theta \tag{6-3}$$

Because the braking force is resistive, it points in the opposite direction of the car's motion, resulting in $\theta = 180°$. Using Equation 6-3 and $\cos 180 = -1$,

$$W = -F\, d \tag{6-12}$$

Using the work–kinetic energy theorem, we can set the work equal to the change in kinetic energy. The change in kinetic energy is

$$\Delta K = \frac{1}{2}mv_f^2 - \frac{1}{2}mv_i^2$$

and because the final velocity is zero

$$\Delta K = 0 - \frac{1}{2}mv_i^2 = -\frac{1}{2}mv_i^2$$

By the work–kinetic energy theorem, then

$$W = \Delta K = -\frac{1}{2}mv_i^2 \tag{6-13}$$

Combining Equations 6-12 and 6-13 yields

$$-Fd = -\frac{1}{2}mv_i^2 \tag{6-14}$$

The magnitude of the acceleration is related to the net force according to Newton's second law, which is $a = F/m$. The car's deceleration is then

$$\frac{F}{m} = a = \frac{v_i^2}{2d}$$

When a car traveling at initial speed $v_i = 8.9$ m/s is brought to a stop over a distance $d = 7.6$ m, the deceleration is then

$$a = \frac{(8.9 \text{ m/s})^2}{2(7.6 \text{ m})} = 5.2 \frac{\text{m}}{\text{s}^2}$$

REFLECT

You can probably imagine that the motion of a car stopping in the minimum distance would not be gentle, even in a car traveling at 20 mph. The deceleration we obtained—a magnitude more than $0.5g$—therefore seems reasonable.

Also notice that although both mass and force are a part of this problem, we were able to determine acceleration using a ratio of the two. We will often do problems in which it initially appears not enough information is provided, but as in this problem, it isn't always necessary to know the value of each variable separately to find a solution.

Practice Problem 6-5 The U.S. National Highway Transportation and Safety Administration lists the minimum braking distance for a car traveling at 20 mph (about 8.9 m/s) to be 25 ft (about 7.6 m). A car is traveling at 8.9 m/s. What is its stopping distance when it decelerates at a constant 4.4 m/s²? Assume that the wheels of the car have stopped rotating, so that kinetic (sliding) friction between the tires and road result in the work necessary to reduce the car's speed.

Example 6-6 How Far to Stop?

The National Highway Transportation and Safety Administration lists the minimum braking distance for a car traveling at 40 mph to be 101 ft. Is this reasonable given that the minimum distance to stop is 25 ft when the car's speed is 20 mph? Assume that the braking force is constant.

SET UP

In Example 6-5, we used the definition of work and the work–kinetic energy theorem to find a relationship between stopping distance and speed:

$$-Fd = -\frac{1}{2}mv_i^2 \tag{6-14}$$

We can express the minimum braking distance in terms of the speed:

$$d = \frac{m}{2F}v_i^2$$

SOLVE

The minimum braking distance is therefore proportional to the square of the speed, $d \propto v^2$. For two velocities v_1 and v_2,

$$d_1 \propto v_1^2 \quad \text{and} \quad d_2 \propto v_2^2$$

When $v_2 = 2v_1$, the second expression becomes

$$d_2 \propto (2v_1)^2 = 4v_1^2$$

Because d_1 is proportional to v_1^2, d_2 is therefore proportional to 4 multiplied by d_1. We would expect the minimum braking distance starting at 40 mph to be 4 times the minimum braking distance starting at 20 mph.

REFLECT
The reported minimum braking distances of 100 ft and 25 ft, respectively, are consistent.

Figure 6-8 In Example 6-7, we consider the work done by coiled actin in the sperm of a horseshoe crab. (*J. Hindman/ Shutterstock.*)

Example 6-7 Actin Packs a Punch

To fertilize eggs, horseshoe crabs' (Figure 6-8) sperm must penetrate two protective layers about 40 μm thick. To achieve this, a 60-μm-long bundle of coiled actin straightens out and pushes with a force of 1.9×10^{-9} N. If the actin bundle has mass on the order of 10^{-16} kg, what is the speed of the actin bundle at the end of this process?

SET UP
The nature of the force exerted at the end of the actin bundles is such that it is both constant and in the direction of the motion of the end of the bundle. For this reason we can use the definition of work offered in Equation 6-1,

$$W = F\,d$$

Using the work–kinetic energy theorem, we set the work equal to the change in kinetic energy, with initial speed equal to zero:

$$F\,d = \frac{1}{2}mv_f^2$$

SOLVE
The final speed of the actin bundle is then

$$v_f = \sqrt{\frac{2F\,d}{m}}$$

or

$$v_f = \sqrt{\frac{2\,(1.9 \times 10^{-9}\ \text{N})\,(60 \times 10^{-6}\ \text{m})}{10^{-16}\ \text{kg}}} = 48\ \text{m/s}$$

REFLECT
This speed is high, relative to, say, the speed you might travel in a car on a highway. But the process takes place over a short distance for a short period of time. The actin bundles, however, do a reasonable amount of work for something so tiny. The actin bundle does 10^{-13} J of work on the egg. To get a sense of magnitude of the work, compare the ratio of work to mass for the actin bundle penetrating the protective coating of the egg to the ratio for the people pushing the car in Example 6-2. In that problem, each person did about 1000 J of work. For a man of approximately 100 kg, the ratio of work to mass is about 1000:100 or 10 to 1. The ratio for the actin bundle in this example is about 10^{-13}:10^{-16} or 1000 to 1. The actin bundle is small, but it packs a big punch.

Practice Problem 6-7 What is the speed of a baseball if it has a mass equal to 0.145 kg and a pitcher applies a constant force of 115 N to it over a distance of 1.00 m?

Example 6-8 Bobsled Start

At the start of a race, a four-man bobsled crew pushes their 210-kg sled as fast as they can down the 40-m straight, relatively flat starting stretch (Figure 6-7). They apply a net force of 200 N aligned with the direction of motion. What is the speed of the sled right before the crew jumps in at the end of the starting stretch? Neglect effects due to friction.

SET UP

According to the work–kinetic energy theorem, the work done on the sled equals the change in its kinetic energy. The definitions of work and kinetic energy in terms of force and speed, respectively, allow us to relate the information given in the problem statement in a form that will lead to a solution.

SOLVE

We start with the work–kinetic energy theorem

$$W_{net} = \Delta K \qquad (6\text{-}11)$$

and then use the definitions of work and kinetic energy

$$F_{net}d = \frac{1}{2}mv_f^2 - \frac{1}{2}mv_i^2$$

Because the force and direction of motion are aligned, as shown in Figure 6-9, the work done is simply the product of force and distance. (Note that Figure 6-9 is not a free-body diagram, but instead shows the relevant force and the direction of motion.) The bobsled starts from rest, so $v_i = 0$ and

$$F_{net}d = \frac{1}{2}mv_f^2$$

Figure 6-9 The force acting on the bobsled is aligned with the direction of motion.

The speed at the end of the 40-m stretch is then

$$v_f = \sqrt{\frac{2F_{net}d}{m}} = \sqrt{\frac{2(200 \text{ N})(40 \text{ m})}{210 \text{ kg}}} = 8.7 \text{ m/s}$$

REFLECT

This speed is relatively close to the speed of world-class sprinters, so it seems reasonable. The typical speed of an Olympic four-man bobsled as it exits the starting stretch is actually 11 or 12 m/s; in reality the starting stretch is slightly downhill.

Practice Problem 6-8 At the start of a race, a four-man bobsled crew pushes their 210-kg sled as fast as they can down the 40-m straight, relatively flat starting stretch. Each man applies to the sled a force of 200 N, aligned with the direction of motion. What is the speed of the sled right before the crew jumps in at the end of the starting stretch? Neglect effects due to friction.

★ What's Important 6-3

The work–kinetic energy theorem allows us to explore the relationship between work, force, and speed in a variety of physical situations. Although you don't have enough information in many real-world problems to determine the value of every variable, often you *can* find what you need by clustering some of the variables together, for example, by solving for the ratio of mass and force.

6-4 Work Done by a Variable Force

An athlete trains by stretching the springlike cords of an exercise apparatus (Figure 6-10) at a constant rate. How much effort must he exert to overcome the work done on him by the cords?

Work is defined as the dot product of a force and the distance over which it is applied

$$W = F\,d\cos\theta = \vec{F}\cdot\vec{d} \qquad (6\text{-}3)$$

The distance that the athlete stretches the elastic cords is easy to measure. But the force exerted by the cords, and thus the force required to stretch them, increases the farther the free ends of the cords are moved from the unstretched position. The force therefore changes as the cords stretch. What value of force should be used to compute work?

The magnitude of the force exerted by a spring is proportional to the distance it is stretched:

$$F = -k\,x \qquad (6\text{-}15)$$

The **spring constant** k, sometimes called the *stiffness of the spring*, quantifies how easily a spring stretches. The units of k are N/m. The minus sign indicates that the force exerted by the spring always pulls or pushes the free end of the spring back toward the unstretched position.

We define the direction of the vector which describes the distance the spring is stretched as positive, where $x = 0$ marks the unstretched position of the end of the spring. Because the spring pulls back and against any attempt to stretch it, the force vector is in the opposite or negative direction, as shown in **Figure 6-11**. To make this opposition clear, we write

$$\vec{F} = -k\,\vec{x} \qquad (6\text{-}16)$$

Notice that if the spring is compressed rather than stretched, the spring force pushes the free end back toward $x = 0$. In the case of a compressed spring, the vector describing the displacement of the free end points in the negative direction and the spring force is a vector in the positive direction. The minus sign in Equation 6-16 therefore holds whether the spring is stretched or compressed. Finally, the spring is in equilibrium when the free end is at $x = 0$ because the net force is zero at that point.

Hooke's law, summarized by Equation 6-16, describes a force that increases in proportion to the displacement of a system from its equilibrium position.

Because the cords in the exercise apparatus in Figure 6-10 obey Hooke's law, the athlete must apply an increasingly greater force to stretch the cords farther and farther from equilibrium. His applied force constantly changes, so returning to the question above: What value of force should be used in Equation 6-3?

There is no correct answer to the question just posed. Equation 6-3 only applies to cases in which the force is constant, and in this situation the force is clearly changing. But what if the athlete stretches the apparatus a distance so small that we can consider the force to be constant over that distance? In that case, the application of Equation 6-3 would be justified.

When the athlete has stretched the cords a distance Δx from position x_0, the cords pull on his hand with a force of absolute magnitude $F_1 = kx_1$. If Δx is very small, the force necessary to overcome the cords' pull on his hand can be taken as the constant value F_1. Using Equation 6-3, the work done is then

$$W_1 = F_1\Delta x\cos\theta = F_1\Delta x$$

Because the force vector and the displacement vector are aligned, the angle θ between them is 0, so the cosine term equals 1. To make our conclusions more generally applicable, we will include the cosine term in the equation later.

Figure 6-10 A professional boxer trains by stretching the springlike cords of an exercise apparatus.

Figure 6-11 Regardless of whether the spring is stretched or compressed, the force it exerts on an object attached to its free end is always in the opposite direction of the displacement.

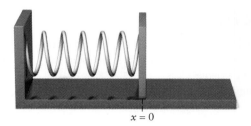

When the spring is neither stretched nor compressed, there is no net force on the free end. The location of the free end at equilibrium is labeled $x = 0$.

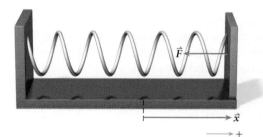

When the spring is stretched, the spring force pulls the free end back to equilibrium.

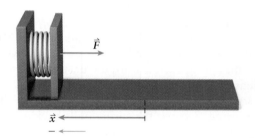

When the spring is compressed, the spring force pushes the free end back to equilibrium.

The minus sign in Hooke's law ($\vec{F} = -k\vec{x}$) reminds us that the spring force vector and the displacement from equilibrium are in the opposite direction. The spring force always acts to move the free end of the spring back toward equilibrium.

The free ends of the cords have been stretched from x_0 to x_1; the springlike cords exert a force F_1 and do work W_1 over that distance. The force required to pull the handle another Δx, from x_1 to x_2, is then $F_2 = k\,x_2$ and the work done on the cords is

$$W_2 = F_2 \Delta x$$

Again, we have assumed the spring force to be constant because the distance the cords are stretched is so small. The process is sketched in **Figure 6-12**. In general, the work required to stretch the free ends of the cords an additional distance Δx from a starting position x_i is

$$W_i = F_i \Delta x$$

Each W_i represents a small bit of the total work done. Because work is a scalar, the total work is simply the sum of all of these bits, or

$$W = \sum W_i = \sum F_i \Delta x$$

The smaller the Δx interval, the more accurate our claim becomes that each force F_i is approximately constant over that distance. We obtain the best approximation when Δx is infinitesimally small. We have previously defined the integral as the sum of an infinite number of infinitesimally small contributions, so

$$W = \int F\, dx$$

In this particular problem the force vector and the displacement vector are always aligned, so the cosine term in Equation 6-3 equals 1 and the dot product becomes hidden. In general, however, the force and displacement can be at any angle with

The spring force depends on how far the spring has been stretched. Our initial definition of work requires a constant force. How can it be applied?

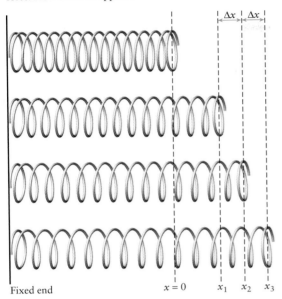

When the spring is stretched a very small distance Δx from x_1 to x_2 or from x_2 to x_3, equals 0, the force exerted by the spring is approximately constant.

The work done by the spring over each successive stretch of Δx can be determined using $W_i = F_i \Delta x$.

Δx | Δx

Fixed end $x = 0$ x_1 x_2 x_3

Figure 6-12 The work required to stretch the free end of a spring can be found by stretching it in small increments of Δx and then adding up the contributions to the total work from each successive stretch.

respect to each other. In order to make the new definition of work complete, then, we add the cosine term, or dot product, back. We also add bounds to the integral to indicate the initial and final displacements from equilibrium of the system

$$W = \int_{x_i}^{x_f} \vec{F} \cdot d\vec{x} \tag{6-17}$$

See the Math Tutorial for more information on Integrals

We can now determine the work done on the athlete by the elastic cords in the exercise apparatus shown in Figure 6-10. The result applies to any spring system.

We consider the free ends of the cords stretched from an initial position x_i to a final position x_f. Both x_i and x_f are relative to $x = 0$, the equilibrium position of the free ends of the cords. According to Equation 6-17, the work done by the cords as the device is stretched from x_i to x_f is then

$$W = \int_{x_i}^{x_f} \vec{F}_{\text{spring}} \cdot d\vec{x} = \int_{x_i}^{x_f} F_{\text{spring}} \, dx \cos\theta \tag{6-18}$$

where $\vec{F}_{\text{spring}}$ is the changing force exerted by the cords. According to Hooke's law, $\vec{F}_{\text{spring}}$ is proportional to the displacement from equilibrium, and points in the direction opposite to the displacement of the end of the spring from equilibrium. The *magnitude* of the spring force is $F_{\text{spring}} = k\,x$, and the angle θ between $\vec{F}_{\text{spring}}$ and $d\vec{x}$ is 180°. Equation 6-18 becomes

*Go to **Picture It 6-1** for more practice dealing with work by a spring*

$$W = \int_{x_i}^{x_f} kx \, dx \cos 180 = -\int_{x_i}^{x_f} kx \, dx$$

Integrating this equation (see Equation 2-25) gives

$$W = -\left(\frac{kx_f^2}{2} - \frac{kx_i^2}{2} \right)$$

$$= \frac{kx_i^2}{2} - \frac{kx_f^2}{2} \tag{6-19}$$

This expression is the work that the cords do on the athlete or the work any normal spring that has a spring constant k does on the agent which stretches it a distance x_i to x_f from its equilibrium position.

Example 6-9 Work Those Muscles

The athlete in Figure 6-10 stretches the exercise apparatus 47 cm from its unstretched length. The spring constant of the cords is 860 N/m when you assume the cords act like one spring. How much work does the device do on the athlete?

SET UP

The work done by a spring depends on its stiffness and the distance it is stretched. In this case, the initial position of the free end of the cords is the unstretched length, so $x_i = 0$.

SOLVE

Applying $x_i = 0$ makes Equation 6-19 simpler

$$W = \frac{kx_i^2}{2} - \frac{kx_f^2}{2} = -\frac{kx_f^2}{2}$$

We substitute the given values to get a numeric result

$$W = -\frac{(860 \text{ N/m})(0.47 \text{ m})^2}{2} = -95 \text{ N} \cdot \text{m} = -95 \text{ J}$$

REFLECT

First, don't ignore the minus sign. The result is less than zero because the elastic cords do work on the athlete. When a force that does work on an object has a negative value, it acts to resist motion (for example, to prevent an increase in the speed of the object). That's certainly the case here, and the reason why using a device like the one shown in Figure 6-10 is good exercise. The work that the athlete does on the elastic cords has a positive value.

The work the athlete does must be at least equal to the magnitude of the work done by the cords, so at least 95 J. In Example 6-2, we saw that in pushing a car 5 m, a person does about 1000 J of work on the car, which we imagined to be at least a modest amount of work for a human to do. The athlete in this problem does about a tenth as much work with each stretch of the cords, but he probably repeats the motion tens of times.

Practice Problem 6-9 In the previous example, we found that the exercise device does −95 J of work on the athlete as he stretches the cords 47 cm. The spring constant of the cords is 860 N/m when you assume they act like one spring. How much work does the device do on the athlete if he stretches the cords twice as far?

✳ What's Important 6-4

To compute work done by a variable force, we break the distance over which the force is applied into infinitesimally small increments, and then add up the work done over each small increment. This process leads to the integral form of the definition of work.

6-5 Potential Energy

When the weight lifter in **Figure 6-13** raises the barbell from the floor and holds it over his head, how much *net* work does he do in the process? The work–kinetic energy theorem provides a direct answer by considering the change in kinetic energy. Because both the initial and final speeds of the barbell are zero, the initial and final kinetic energies of the system are also zero. Therefore, the change in

Figure 6-13 How much *net* work does a weight lifter do as he raises a barbell over his head? *(Lee Jae-Won/Reuters/ Corbis.)*

kinetic energy is zero as well, so by the work–kinetic energy theorem, the net work done is also zero. Although physically correct, you may find that "zero" is not a particularly satisfying way to summarize the process. The weight lifter has done positive work on the barbell after all, and the system has changed in a significant way from the beginning to the end of the lift. It is this change that provides a more satisfying way to describe the process.

If the weight lifter holds and then releases the barbell while it is still on the ground, nothing happens. But if he releases the barbell while it is up above his head, a dramatic change occurs as the barbell slams to the floor. Even if releasing the barbell requires no force, once released it clearly acquires kinetic energy. So although the barbell is not moving and therefore has no kinetic energy while being held up, it has the *potential* to move.

The physical state of the barbell changes as a result of the work that the weight lifter does on it; we characterize this change as an increase in the barbell's **potential energy**. As the name implies, potential energy characterizes the potential for a system to do work or, equivalently, to move or to exert a force. Although potential energy is stored in a system, it can only be observed when converted to kinetic energy. Energy that characterizes a system's potential to acquire kinetic energy due to the action of gravity is **gravitational potential energy**.

From Equation 6-17, the work done by the athlete as he raises the barbell from initial height y_i to final height y_f is

$$W_{\text{weight lifter}} = \int_{y_i}^{y_f} \vec{F}_{\text{weight lifter}} \cdot d\vec{y}$$

Both $\vec{F}_{\text{weight lifter}}$ and $d\vec{y}$ point in the positive y direction, so $\theta = 0°$ and $\cos \theta = 1$, therefore, the work done by the weight lifter is

$$W_{\text{weight lifter}} = \int_{y_i}^{y_f} F_{\text{weight lifter}} \, dy \cos 0 = \int_{y_i}^{y_f} F_{\text{weight lifter}} \, dy$$

Imagine that the athlete in Figure 6-13 lifts the barbell up above his head at constant speed. This assumption is not strictly required, but it makes the physics a bit more clear without loss of generality. Zero acceleration means that the net force on the barbell is zero, so the force exerted by the weight lifter must be equal in magnitude to the force that gravity exerts, as suggested by **Figure 6-14**

$$|F_{\text{weight lifter}}| = |F_{\text{gravity}}| = m g$$

where m is the mass of the barbell. Thus,

$$W_{\text{weight lifter}} = \int_{y_i}^{y_f} mg \, dy = mgy \Big|_{y_i}^{y_f} = mgy_f - mgy_i = mg(y_f - y_i) \quad \text{(6-20)}$$

Work and energy are closely related; the work done by the weight lifter results in an increase in the gravitational potential energy of the barbell. The variable U is used to represent potential energy (in this case gravitational potential energy), so

$$W_{\text{weight lifter}} = \Delta U_{\text{gravity}}$$

or

$$\Delta U_{\text{gravity}} = mg(y_f - y_i) \quad \text{(6-21)}$$

If the weight lifter were to release the barbell while holding it above his head, the barbell would acquire kinetic energy due to the work done on it by gravity. To complete the connection between the work and the change in potential energy, we therefore consider the work that gravity does on the barbell as the athlete lifts it from the floor by using Equation 6-17,

$$W_{\text{gravity}} = \int_{y_i}^{y_f} \vec{F}_{\text{gravity}} \cdot d\vec{y}$$

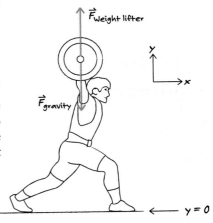

Figure 6-14 A free-body diagram shows the forces acting on the barbell in Figure 6-13.

The process is similar to that which led to Equation 6-20, except that here, $\vec{F}_{\text{gravity}}$ points in the negative y direction. The angle θ between $\vec{F}_{\text{gravity}}$ and $d\vec{y}$ is then $180°$ and $\cos\theta = -1$. This reverses the sign of the cosine term in the dot product and gives

$$W_{\text{gravity}} = -W_{\text{weight lifter}}$$

The change in gravitational potential energy equals the work done on the barbell by the weight lifter, so ΔU can also be written

$$\Delta U_{\text{gravity}} = -W_{\text{gravity}} \tag{6-22}$$

or

$$\Delta U_{\text{gravity}} = -\int_{y_i}^{y_f} \vec{F}_{\text{gravity}} \cdot d\vec{y}$$

We can generalize this expression as a definition of the change in potential energy due to the action of any force from initial position x_i to final position x_f

$$\Delta U = -\int_{x_i}^{x_f} \vec{F} \cdot d\vec{x} \tag{6-23}$$

This definition leads directly to the expression we obtained above (Equation 6-21) for the change in potential energy when an object near Earth's surface is moved from height y_i to height y_f

$$\Delta U_{\text{gravity}} = -\int_{y_i}^{y_f} \vec{F}_{\text{gravity}} \cdot d\vec{y}$$

$$= -\int_{y_i}^{y_f} mg\, dy \cos 180° = \int_{y_i}^{y_f} mg\, dy = mg(y_f - y_i)$$

We can also apply the definition of the change in potential energy to understand a spring by inserting Hooke's law for the spring force (Equation 6-16) into Equation 6-23. When a spring is stretched so that its free end is moved from x_i to x_f relative to the equilibrium position, $\vec{F}_{\text{spring}}$ points opposite to the displacement, so the angle between $\vec{F}_{\text{spring}}$ and $d\vec{x}$ is $180°$. Therefore,

$$\Delta U_{\text{spring}} = -\int_{x_i}^{x_f} \vec{F}_{\text{spring}} \cdot d\vec{x}$$

$$= -\int_{x_i}^{x_f} kx\, dx \cos 180° = \int_{x_i}^{x_f} kx\, dx = \frac{1}{2}kx_f^2 - \frac{1}{2}kx_i^2 \tag{6-24}$$

Again, we note that the minus sign in Hooke's law is the same sign that results from the cosine term in the dot product, which is why the kx term in Equation 6-24 is not negative.

You may wonder why we didn't simplify the discussion of the weight lifter and the barbell by setting $y_i = 0$ and y_f to some height, say $y_f = h$, in obtaining Equation 6-20. It is natural to think of the ground as height equal to 0. If we were to set $y_i = 0$, both the expression for the resulting work as well as that for the change in potential energy would take the simpler form mgh. But what if we designated y_i to be, say, 12 m?

The final height of the barbell would then be $y_f = 12 + h$, and according to Equation 6-20, the work done by the weight lifter would be

$$W_{\text{weight lifter}} = mg(y_f - y_i) = mg[(12 + h) - 12] = mgh \tag{6-25}$$

The result is the same as when we chose $y_i = 0$ because the work depends on the change in height, not the specific y values.

The change in gravitational potential energy is then also independent of the choice of the value of y_i. We are therefore free to set height equal to zero at any convenient reference position, or reference point, when considering both work and potential energy. As is done in Figure 6-14, label your choice of reference point in any sketch you draw to aid in the solution of an energy problem.

> ## ❗ Watch Out
> ### The magnitude of potential energy depends on the choice of reference point.
>
> The magnitude of potential energy depends on our choice of reference point. Because we are free to select any convenient reference point, the value of potential energy isn't physically meaningful. Instead, it is the *change* or *difference* in potential energy that contains the information needed to understand the evolution of a system.

Although the difference in potential energy is always physically meaningful, it is common to define the reference point so that the value of potential energy at that point is zero. For example, in the weight lifter problem,

$$\Delta U_{\text{gravity}} = U_f - U_i$$

Using $\Delta U_{\text{gravity}} = mg(y_f - y_i)$ (Equation 6-21), if we select $y_i = 0$, then

$$\Delta U_{\text{gravity}} = mgy_f$$

As a shorthand, we will write this equation as

$$U_{\text{gravity}} = mgy \qquad (6\text{-}26)$$

Note, however, that the specific value of y is determined by our choice of reference position for height, which we are free to choose, and that a specific value of U_{gravity} is only meaningful when compared to U_{gravity} at some other position. This result is examined in **Figure 6-15**, in which we compare two choices of reference position for the weight lifter problem.

In Figure 6-15a, we let $y = 0$ at the level of the floor, so that $y_i = 0$ and $y_f = 2$ m. For a barbell that has a mass $m = 100$ kg,

$$U_{\text{gravity},i} = mgy_i = 0$$

and

$$U_{\text{gravity},f} = mgy_f = (100 \text{ kg})(9.8 \text{ m/s}^2)(2 \text{ m}) = 1.96 \times 10^3 \text{ J}$$

The change in gravitational potential energy is

$$\Delta U_{\text{gravity}} = U_{\text{gravity},f} - U_{\text{gravity},i} = 1.96 \times 10^3 \text{ J} - 0 \text{ J} = 1.96 \times 10^3 \text{ J}$$

In Figure 6-15b, we let $y = 0$ at the base of the 1-m-high stage on which the weight lifter stands, so that $y_i = 1$ m and $y_f = 3$ m. In this case,

$$U_{\text{gravity},i} = mgy_i = (100 \text{ kg})(9.8 \text{ m/s}^2)(1 \text{ m}) = 0.98 \times 10^3 \text{ J}$$

and

$$U_{\text{gravity},f} = mgy_f = (100 \text{ kg})(9.8 \text{ m/s}^2)(3 \text{ m}) = 2.94 \times 10^3 \text{ J}$$

In this second case, the change in gravitational potential energy is

$$\Delta U_{\text{gravity}} = U_{\text{gravity},f} - U_{\text{gravity},i} = 2.94 \times 10^3 \text{ J} - 0.98 \times 10^3 \text{ J} = 1.96 \times 10^3 \text{ J}$$

Clearly, the particular values of $U_{\text{gravity},i}$ and $U_{\text{gravity},f}$ are different in the two cases, but the *change* in gravitational potential energy is the same. It therefore has no meaning to give a specific value of potential energy U for an object. The quantity that determines the physics is the *difference* in potential energy for two different locations of an object.

 ## Got the Concept 6-3
Choosing a Reference

A ball is dropped from the top of an 11-story building, 45 m above the ground, to a balcony on the ninth floor, 36 m above the ground. Compare the change in the ball's gravitational potential energy when the ground is considered to be at a height equal to 0 m and when the balcony is considered to be at a height equal to 0 m.

Figure 6-15 Two choices of reference point for the weight lifter problem are considered. The determination of the work required to lift the barbell is the same in both cases. (a) The position $y = 0$ is set to be the level of the floor. (b) The position $y = 0$ is set to be the base of the 1-m-high stage on which the weight lifter stands.

Same motion...

$m = 100$ kg

$y_f = 2$ m

$U_f = mgy_f = 100(9.8)(2)$
$= 1.96 \times 10^3$ J

$U_i = mgy_i = 0$

$\Delta U = U_f - U_i = 1.96 \times 10^3$ J

$y_i = 0$ m

(a)

...different reference point for height.

$y_f = 3$ m

$U_f = mgy_f = 100(9.8)(3)$
$= 2.94 \times 10^3$ J

$U_i = mgy_i = 100(9.8)(1)$
$= 0.98 \times 10^3$ J

$\Delta U = U_f - U_i = 1.96 \times 10^3$ J

$y_i = 1$ m

1 m

$y_0 = 0$ m

(b)

Same potential energy difference!

In the same way that the gravitational potential energy can be written as $U_{\text{gravity}} = mgh$ (Equation 6-26), it is common to write

$$U_{\text{spring}} = \frac{1}{2}kx^2 \qquad (6\text{-}27)$$

for the potential energy stored in a spring stretched a distance x from equilibrium. Here we have set $x_i = 0$ in Equation 6-24. When using either Equation 6-26 or

Equation 6-27, remember that you are really using ΔU with the reference value of potential energy set to zero. However, note that because of the way in which the spring force is defined by Hooke's law, we are not free to set the equilibrium point, and therefore the reference point for potential energy, to any value other than $x_i = 0$.

 Watch Out

The reference point for the spring potential energy must be the equilibrium point.

Because the potential energy for a spring is proportional to the square of the distance that the spring is stretched from equilibrium, only the equilibrium position can be used as the reference point. Said another way, when it is not stretched, the end of the spring marks the only allowable reference point. Remember that although the difference in potential energy is physically meaningful, the magnitude of potential energy is not.

Estimate It! 6-1 Energy of Gas

The energy stored in 1 gal of gasoline is about 1.3×10^8 J. The authors carried out an experiment to verify this by measuring the work our car does (on the road) while burning 1 gal of gas. Traveling at a constant 40 mph on a straight, level road, our 1600-kg car gets 28 mi/gal. The work done is related to the energy released by the gasoline. We measured this work by measuring the work done by the resistive forces, such as friction and air resistance, on our car. When the car moves at constant speed, the work done by resistive forces exactly balances the work done by the car on the road. (Constant speed means no acceleration, which means no net force and therefore no net work.) To measure the resistive forces, we observed the rate at which our car decelerates after placing the transmission in neutral while the car is moving. We drove on a straight, level road at 50 mph, shifted into neutral, and recorded (relatively crudely) the time intervals as we decelerated to 40 mph and then 30 mph. Here is a summary of our data

Speed interval (mph)	Δt (s)
50 to 40	30
40 to 30	30

Estimate the energy content of 1 gal of gasoline using these data, and the fact that about 80% of the energy released in a car's engine is lost to heat, friction, and exhaust.

SET UP

Work depends on force, and force depends on acceleration. The car's change in speed therefore allows a determination of the work done. That is, the magnitude of the work done is

$$W = F_{car}d$$

or

$$W = m_{car}ad$$

SOLVE

In the first time interval, the speed of our car changed from 50 to 40 mph in 30 s, so

$$a_1 = \frac{40\ \text{mph} - 50\ \text{mph}}{30\ \text{s}} \times \frac{1609\ \text{m}}{\text{mi}} \times \frac{1\ \text{h}}{3600\ \text{s}} = -0.149\ \frac{\text{m}}{\text{s}^2}$$

In the second time interval, the speed of our car changed from 40 to 30 mph in 30 s, so

$$a_2 = \frac{30 \text{ mph} - 40 \text{ mph}}{30 \text{ s}} \times \frac{1609 \text{ m}}{\text{mi}} \times \frac{1 \text{ h}}{3600 \text{ s}} = -0.149 \frac{\text{m}}{\text{s}^2}$$

We've kept more significant digits in these intermediate results than we are justified in using for the final answer, in order to avoid introducing errors resulting from rounding.

The acceleration of the car was the same in both time intervals, so we can use a equal to -0.15 m/s^2. The mass of the car is equal to 1600 kg, and we will take d equal to 28 mi, the distance over which we expend 1 gal of gasoline. Then, converting 28 mi to meters and taking the absolute value in order to determine the magnitude, we find that

$$W = \left| (1600 \text{ kg})(-0.149 \text{ m/s}^2)\left(28 \text{ mi} \times 1609 \frac{\text{m}}{\text{mi}}\right)\right| = 1.07 \times 10^7 \text{ J}$$

A 20% efficiency means that 20% or 1/5 of the energy in the gas is used to move the car. The total energy is then 5 times the energy used.

$$E = 5W = 5.37 \times 10^7 \text{ J}$$

We therefore estimate that 1 gal of gas provides about 5×10^7 J of energy.

REFLECT

The accepted value of the energy content of one gal of gasoline is about 1.3×10^8 J (or about 3.5×10^8 J per liter). Our estimate is reasonable. Also, notice that we assumed our engine wastes far more of the energy content of the gasoline it burns than it uses to move the car forward. This is typical of today's automobile engines; we'll revisit this topic in Chapter 15.

★ What's Important 6-5

Potential energy characterizes the potential for a system to do work or, equivalently, to move or to exert a force. Although potential energy is stored in a system, it can only be observed when converted to kinetic energy. The magnitude of potential energy depends on our choice of reference point, so the physically meaningful quantity is the *difference* in potential energy between two locations.

6-6 Conservation of Energy

When the heart contracts it pushes blood into the arteries, stretching their walls to accommodate the increased volume. In between contractions, the arterial walls return to their equilibrium position, much the same way that a stretched spring does after the free end has been released. This process pushes blood through the arteries while the heart refills. Our arteries have converted the potential energy stored in the stretched "spring" of the arterial walls into the kinetic energy of moving blood. As the arteries return to their initial diameter, they do work on their environment. Energy is the ability of a system to do work.

Work is required to store energy in any spring, including arterial walls, which can be approximated as springs. By comparing the definition of potential energy (Equation 6-23) or by following the development of the relationship between $\Delta U_{\text{gravity}}$ and W_{gravity} (Equation 6-22), we see that as the spring is stretched, the work it does is equal to the negative of the spring potential energy:

$$W = -\Delta U$$

Let's attach an object of mass m to the free end of the spring. We also know the relationship between the work done by the spring on the object and the change in the kinetic energy of the object by using the work–kinetic energy theorem:

$$W = \Delta K$$

Combining these last two equations gives

$$\Delta K = -\Delta U$$

so

$$\Delta K + \Delta U = 0 \tag{6-28}$$

The sum of the kinetic energy and the potential energy is the **total mechanical energy E**. Therefore

$$\Delta K + \Delta U = \Delta(K + U) = \Delta E \tag{6-29}$$

and then

$$\Delta E = 0$$

This system obeys the law of **conservation of energy**.

The total mechanical energy of this spring does not change. Equation 6-29, although simple in form, summarizes a most important physical law, that the total mechanical energy of any system that experiences no force from something outside of the system remains constant. Energy is **conserved** when the total mechanical energy does not change. The amount of mechanical energy in an isolated system is constant, although the form that the energy takes can change. For example, energy can be transferred between potential energy and kinetic energy as in the case of the arteries. The potential energy stored in the stretched arterial walls gets transferred into kinetic energy as the walls return to their unstretched state. The work done in the process ultimately results in blood flow, which is another manifestation of the energy of the system.

 Got the Concept 6-4
How High?

A block is released from rest at vertical height H on one side of an asymmetric skateboard ramp, as shown in **Figure 6-16**. The angle from the horizontal of the left ramp is twice the angle of the right ramp. If the ramp is frictionless, how high (vertically) does the block rise on the right side before stopping and sliding back down?

Figure 6-16 A block is released from rest at vertical height H on one side of an asymmetric skateboard ramp.

Organisms convert energy from one form to another in a variety of ways. Perhaps the most fundamental energy conversion process is *photosynthesis*, the biochemical pathway by which plants store energy from the Sun. In cells that make up the leaves of most plants, organelles called *chloroplasts* contain the molecular machinery necessary to do photosynthesis. Using the Sun's energy, chloroplasts carry out a two-part process that converts atmospheric CO_2 into sugar. The energy stored in sugar's chemical bonds is a form of potential energy. As animals digest and process food, they break these chemical bonds and transform

(a) Potential Energy

(b) Kinetic Energy

Figure 6-17 (a) When a sprinter is crouched at the starting line, his leg muscles store potential energy in the chemical bonds of phosphocreatine, ATP, carbohydrates, and fat. (b) Once the race starts, the stored energy is used to drive muscle contraction, which in turn moves his legs and also generates heat.

the energy that held the carbon atoms together into a form that can be used to do work.

During the first step of photosynthesis, the energy of light hitting the leaf is used to break water molecules into hydrogen ions and oxygen. As the hydrogen ions are concentrated in an inner compartment of the chloroplast, they experience an increase in potential energy, much like lifting a ball higher and higher above the ground. The ions "fall back down" by diffusing into the outer part of the chloroplast; as they do, the ions pass through ATP synthase, a curious enzyme that has a structure not unlike a water wheel. The hydrogen ions cause the stalk of the ATP synthase to spin, which in turn pushes phosphate ions and adenosine diphosphate (ADP) molecules together to form adenosine triphosphate (ATP). The potential energy of the ion concentration gradient is first converted to the kinetic energy of moving ions which is then converted to rotational kinetic energy of ATP synthase and finally to potential energy stored in the chemical bonds of ATP. Ultimately, the energy stored in ATP is used to convert CO_2 into sugar, and sugar into the many, often large and complex molecules that make up plants.

The chemical reactions that make or break chemical bonds are not perfectly efficient. Consider the sprinter in **Figure 6-17a**, crouched at the starting line. His leg muscles store potential energy in the chemical bonds of phosphocreatine, ATP, carbohydrates, and fat. Although a lot of that energy is used to drive muscle contraction, which in turn moves his legs after the starting gun fires (**Figure 6-17b**), the rest generates heat instead of motion. That's why the sprinter gets hot when he runs.

> ## ✴ What's Important 6-6
> The total mechanical energy of a system is the sum of its kinetic energy and its potential energy. The amount of mechanical energy in an isolated system is constant, although the form that the energy takes can change. In other words, energy is conserved.

6-7 Nonconservative Forces

We have made the claim that the energy of a system is conserved. Perhaps you wonder if this statement is always true—imagine, for example, a block sliding on a horizontal surface and then coming to a stop. The kinetic energy has clearly decreased, while the gravitational potential energy remained unchanged. Has the total energy of the system changed? The possibility for confusion arises out of the definition of the system under consideration. Yes, the total (mechanical) energy of the block has changed, but not of the block and surface taken together. The frictional force which slows the block causes both the block and the surface it slides across to become slightly warmer. Although the energy associated with that change in temperature is not as obvious to us as the energy of motion or the stored energy of an object raised above the ground, it remains in the system.

What exactly is the "system" in this block example? Certainly the surface must be taken into account. Should we also include the surrounding air, which offers resistance to the motion of the block and therefore contributes to its slowing down? Are there other forces that do work on the block, thereby transferring energy into forms that are not readily observed? To avoid these complications, it is often more straightforward to consider the system to be only the block, and to classify forces that act on the block, such as friction and air resistance, as **external forces**. Because the work done by external forces appears to cause energy not to be conserved, they are called **nonconservative forces**.

We emphasize that nonconservative forces only appear to cause the total energy to decrease. We can resolve the ambiguity by considering only the total mechanical energy. A **conservative force** acting on an object leaves the sum of the kinetic energy and potential energy unchanged, while a nonconservative force does not. Gravitational forces and forces due to springs are examples of conservative forces. Friction and air resistance are nonconservative forces.

The work done on an object by a conservative force does not depend on the path the object takes from one point to another. For example, the work that gravity does on a mountain climber as she ascends from base camp to the summit is the same whether she follows a winding trail to the peak or climbs up a sheer cliff. The only factor on which the work depends is her change in (vertical) height.

When the application of a conservative force results in an object following a closed path, that is, a path that leads from some initial position back to that same position, the net work is zero. This is true for *any* closed path because the net displacement is zero. Such is certainly the case for a simple harmonic oscillator in which we ignore dissipative (nonconservative!) forces. For the ideal pendulum, for example, the work done by gravity on the pendulum bob is zero over one full period, which means that it rises to the same height on every cycle.

The work done by a nonconservative force on an object does depend on the path. In **Figure 6-18**, identical blocks slide down two ramps of the same height but of different angle. The potential energy at the top of the ramps is the same for both blocks, and were the ramps frictionless, the kinetic energy of both blocks would be the same at the bottom of the ramps as well. If there is friction between the blocks and the ramps, however, that nonconservative force acts over a longer distance on the block on the left. Therefore, it will have less kinetic energy at the bottom of the ramp than the block on the right. The work done by friction *is* path-dependent.

The energy transferred by nonconservative forces can take a number of forms, for example, thermal energy (which we will see later is associated with the motion of atoms or molecules in a substance) or chemical energy in the bonds between atoms that comprise the molecules of a substance. We can lump all other possible forms of energy together to extend Equation 6-28

$$\Delta K + \Delta U + \Delta E_{\text{other}} = 0$$

Figure 6-18 Identical blocks slide down two ramps. When there is friction between the blocks and the ramps, the block on the left experiences the nonconservative force over a longer distance and therefore has less kinetic energy at the bottom of the ramp than the block on the right.

Or, because the energy transfer ΔE_{other} is due to the work W_{nc} done on the system by external, nonconservative forces, we find that

$$\Delta K + \Delta U + W_{\text{nc}} = 0$$

If we consider the energies at two times, generically "initial" (i) and "final" (f), then

$$K_{\text{i}} - K_{\text{f}} + U_{\text{i}} - U_{\text{f}} + W_{\text{nc}} = 0$$

or

$$K_{\text{i}} + U_{\text{i}} = K_{\text{f}} + U_{\text{f}} - W_{\text{nc}} \qquad (6\text{-}30)$$

Note that it is not necessary for "initial" to be at the start of a process, nor "final" to be at the end. The energies at *any* two times will be equal.

The work W_{nc} in Equation 6-30 can be either negative or positive. When W_{nc} is positive, so that the last term in Equation 6-30 is negative, the initial mechanical energy of the system $K_{\text{i}} + U_{\text{i}}$ is greater than the final mechanical energy $K_{\text{f}} + U_{\text{f}}$. This might lead you to conclude that energy has been lost due to the nonconservative forces, but that is not the case; *the total energy of the entire system is always conserved.*

✱ What's Important 6-7

A conservative force, such as a gravitational force and the force of a spring, leaves the sum of the kinetic and potential energies unchanged. Forces such as friction and air resistance are termed nonconservative because they result in mechanical energy being converted to forms of energy that cannot be converted back into either kinetic energy or potential energy.

6-8 Using Energy Conservation

Quantities that are conserved are powerful tools for doing physics. In Chapter 2, we discovered that asking, "What's the same?" often helps us gain insight into physical processes and to solve physics problems. A quantity that is conserved is always the same! As long as we know that total energy is conserved in a process, we are free to select any two points in the process and be guaranteed that, although the amount of energy in any particular form might change from one to the other, the total energy will be the same. For example, the total mechanical energy—kinetic plus potential—of the barbell in Figure 6-13 is the same when held above the weight lifter's head, when it has fallen halfway to the floor, and right before it hits the ground. The amounts of energy in the forms of kinetic energy and potential energy are changing, but recognizing that the total energy is the same all the way down easily enables us to determine the speed of the barbell at any height.

The energy of a system is unchanged as any particular process evolves, that is, the energy of the system at any two times must be equal. We will refer to the two positions or times generically as initial and final, so

$$E_{\text{i}} = E_{\text{f}}$$

In this section, we will deal only with nonconservative forces such as friction and air resistance which are also dissipative, and therefore decrease the mechanical energy of a system. In these cases the nonconservative forces oppose the tendency of an object to move, and the work done on the object is negative. This can be seen directly from equation 6-18 because the angle between the force and direction vectors is 180°

$$W_{nc} = \int_{x_i}^{x_f} F_{nc}\, dx \cos\theta = \int_{x_i}^{x_f} F_{nc}\, dx \cos 180 = -\int_{x_i}^{x_f} F_{nc}\, dx$$

A negative value of W_{nc} has the effect of making the final term in Equation 6-30 positive, which as we saw above makes it appear that the system has lost energy. To remove confusion from the sign of the last term in Equation 6-30, when the nonconservative forces are dissipative we find it convenient to rewrite the equation as

$$K_i + U_i = K_f + U_f + |W_{nc}| \tag{6-31}$$

Equation 6-31 reminds us not to let an extra minus sign creep into our consideration of conservation of energy.

To get an intuitive understanding of why we added absolute value signs to the last term in Equation 6-31, consider the example of the block moving at a given initial speed on a horizontal surface until it eventually slides to a stop. The block has non-zero kinetic energy initially but zero kinetic energy at the end of the slide. Also, because the surface is horizontal the initial and final potential energy terms cancel. Equation 6-31 can then be simplified as $K_i = |W_{nc}|$. Kinetic energy is always greater than or equal to zero but W_{nc} is negative in this case. Taking the absolute value of W_{nc} is necessary to avoid the physically incorrect conclusion that kinetic energy is negative. In general, rather than memorizing a rule for the sign of the term associated with the nonconservative forces, just check to make sure that the initial and final energies balance.

Watch Out

Work due to nonconservative forces always reduces the sum of kinetic energy and potential energy.

The work done by nonconservative forces reduces the sum of a system's kinetic energy and potential energy. In the absence of nonconservative forces, the total mechanical energy of the system remains the same even as kinetic energy is transformed into potential energy or vice versa. For example, in the absence of nonconservative forces, an egg thrown straight up returns to its initial height with the same kinetic energy it had initially. However, when a nonconservative force such as air resistance acts on the egg, its peak height and its final velocity are both lower as a result.

As always, begin the solution to any conservation of energy problem by drawing a simple sketch. Be sure to label all reference values. If an object experiences a gravitational force, it is usually convenient to set height equal to zero at the same point at which you choose to set the reference point for potential energy. If an object feels a force due to a spring, label the equilibrium position of the spring in order to define the displacement of the spring from equilibrium. Figure 6-19a shows the setup for a generic case of a block of mass m that experiences both a spring force and the force due to Earth's gravity.

Write out each term in Equation 6-31 and include a potential energy term for each force that an object in the system experiences. If an object experiences both a gravitational force and a spring force, Equation 6-31 becomes

$$\frac{1}{2}mv_i^2 + mgy_i + \frac{1}{2}kx_{c,i}^2 = \frac{1}{2}mv_f^2 + mgy_f + \frac{1}{2}kx_{c,f}^2 + |W_{nc}|$$

$$\tag{6-32}$$

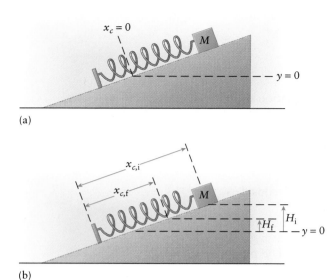

(a)

(b)

Figure 6-19 (a) A block on a ramp experiences both a spring force and the force due to Earth's gravity, but no frictional or other nonconservative forces. (b) We have labeled all of the variables needed to describe the physics.

Equation 6-32 can serve as the starting point for solving motion problems by applying the principle of energy conservation.

If nonconservative forces act on an object, write out the W_{nc} term in Equation 6-32 according to the definition of work (Equation 6-17). For example, if friction $\vec{f}$ acts on the object,

$$\frac{1}{2}mv_i^2 + mgy_i + \frac{1}{2}kx_{c,i}^2 = \frac{1}{2}mv_f^2 + mgy_f + \frac{1}{2}kx_{c,f}^2 + \left| \int_{x_i}^{x_f} \vec{f} \cdot d\vec{x} \right|$$

where we used x_c to indicate the displacement of a spring from equilibrium in order to distinguish it from x_i and x_f, the initial and final positions of the object over which friction acts. The subscript "c" stands for "compression," although x_c can also represent the amount that a spring is stretched from its equilibrium position.

Label as many of the variables as possible in your sketch. In **Figure 6-19b**, the spring has been stretched and we have picked an arbitrary position to study the system after the block has been released. If values have been given for any variables (for example, the initial height H_i and final height H_f shown in Figure 6-19), you may find it helpful to list these values underneath the variables in Equation 6-32 and to identify the variable of interest in some way, for example, with a question mark. This is similar to the method of using a Know/Don't Know table to solve motion problems (Chapter 2). Suppose we want to know the speed of a block when it reaches a height $H_f = 1.5$ m. If the block starts from rest at height $H_i = 3.0$ m and we neglect friction, we would write

$$\frac{1}{2}mv_i^2 + mgy_i + \frac{1}{2}kx_{c,i}^2 = \frac{1}{2}mv_f^2 + mgy_f + \frac{1}{2}kx_{c,f}^2 + \left| \int_{x_i}^{x_f} \vec{f} \cdot d\vec{x} \right|$$

$$v_i = 0 \quad y_i = H_i \qquad\qquad v_f = ? \quad y_f = H_f \qquad\qquad f = 0$$
$$\qquad\qquad = 3.0 \qquad\qquad\qquad\qquad = 1.5$$

Values for $x_{c,i}$ and $x_{c,f}$ can be determined using trigonometry.

For a given problem, some of the terms in Equation 6-32 may be zero and some may cancel from the right side to the left side. As you get more comfortable with the conservation of energy approach to problem solving, you may find you can identify such situations quickly and simply not include those terms at the start of the problem. We recommend, however, that at least at first you follow this standard approach to avoid, for example, setting terms equal to zero improperly.

Finally, substitute any variables which represent values specific to the problem into Equation 6-32. We urge you not to substitute numeric values until the last step of the problem to make your work as transparent as possible. For the object and spring shown in Figure 6-19, Equation 6-32 reduces to

$$\frac{1}{2}mv_i^2 + mgH_i + \frac{1}{2}kx_{c,i}^2 = \frac{1}{2}mv_f^2 + mgH_f + \frac{1}{2}kx_{c,f}^2$$

This expression can now be rearranged algebraically to find any unknown variable of interest. Finally, substitute in the known values you wrote underneath the terms when writing out Equation 6-32. Yes, in this particular problem you could have, for example, substituted zero for v_i and not written the first term in the previous equation. We recommend that you resist this temptation, at least at first, because it may lead you to eliminate terms when they are *not* zero.

Applying the principle of energy conservation to motion problems is a powerful technique and often more straightforward than using the constant acceleration motion equations.

Example 6-10 How High, Again

You throw a ball straight up with initial speed $v_0 = 15.0$ m/s. How high does it go above the point where you release it? Ignore any effects due to air resistance.

SET UP

This problem is represented by the sketch shown in Figure 6-20. Equation 6-32 is a statement of the equivalence of total energy at any two points in the evolution of a physical system. Because we have included a term for every form of energy with which we are familiar, Equation 6-32 is the generic starting point for any problem that we will approach by applying energy conservation. Because there are no springs and no resistive forces in this problem, we can immediately drop the spring potential energy terms and the work done by nonconservative forces

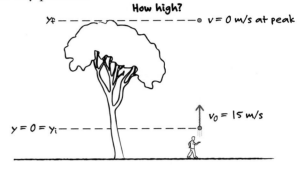

$$\frac{1}{2}mv_i^2 + mgy_i + \frac{1}{2}kx_{c,i}^2 = \frac{1}{2}mv_f^2 + mgy_f + \frac{1}{2}kx_{c,f}^2 + |W_{nc}| \quad (6\text{-}32)$$

$$\frac{1}{2}mv_i^2 + mgy_i = \frac{1}{2}mv_f^2 + mgy_f \quad (6\text{-}33)$$

Based on the problem statement and our sketch, we list the values of the variables as

$$\frac{1}{2}mv_i^2 + mgy_i = \frac{1}{2}mv_f^2 + mgy_f$$

$$v_i = 15 \quad y_i = 0 \quad v_f = 0 \quad y_f = ?$$

Figure 6-20 How high does a ball go when thrown straight up?

SOLVE

Simplify Equation 6-33 by eliminating the terms that make no contribution

$$\frac{1}{2}mv_i^2 = mgy_f$$

and solve for the unknown variable of interest y_f

$$y_f = \frac{v_i^2}{2g}$$

We can now substitute the given values to get a numeric result

$$y_f = \frac{(15.0 \text{ m/s})^2}{2(9.8 \text{ m/s}^2)} = 11.5 \text{ m}$$

REFLECT

If you had drawn the sketch for this problem, would you have set $y = 0$ at the height of the ground? There's nothing wrong with that choice, because any reference point is allowed. But because we are free to choose, why not select $y = 0$ as the starting height of the ball, which enables us to find the final height of the ball above the release point more directly? That is the power of the freedom to choose any reference point!

   Also, note that this problem is identical to Example 2-9. In Chapter 2, we solved the problem using the equations that describe motion. The answer is the same both ways—as it should be. But if you compare the two solutions (do it!), you'll certainly conclude that applying energy conservation is easier and more straightforward. That is the power of applying a conservation principle!

 Go to Picture It 6-2 *for more practice dealing with energy conversation*

Practice Problem 6-10 In the preceding example, a ball is thrown straight up with initial speed 15.0 m/s and rises 11.5 m above the release point before returning to Earth. What initial velocity is required to make the ball rise twice as high? Ignore any effects due to air resistance.

It is rarely necessary to do problems involving conservation of energy in more than a single step. Because total energy remains constant, we can select any two time points to sum all of the energy in a system and set the results equal. You can therefore analyze any number of intermediate instants during a physical process, setting the total energy at each successive point equal. The power of the conservation of energy approach, however, is that the technique works just as well without considering any intermediate points. In the next example, we compare a solution which employs an intermediate step with one that does not.

Example 6-11 A Ramp and Then Free Fall

A block is released from rest at the top of a frictionless ramp, at height H_1 above the base of the ramp. As shown in **Figure 6-21a**, the ramp ends at height H_2 above the base, so that the block flies off and follows a two-dimensional free-fall path until it hits the ground. Find the speed of the block in terms of the given variables when it strikes the ground.

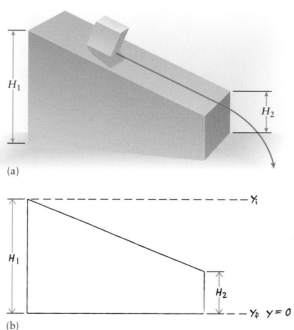

(a)

(b)

Figure 6-21 (a) A block is released from rest at the top of a frictionless ramp, at height H_1 above the base of the ramp. The ramp ends at height H_2 above the base, so that the block flies off and follows a two-dimensional free-fall path until it hits the ground. (b) The variables necessary to describe the physics are shown in a sketch.

SET UP

You might be tempted to do this problem in two steps: first finding the speed of the block when it leaves the ramp and, second, finding the speed of the block after falling from height H_2. There is no need to complete this problem in two steps, however. Because energy is conserved, we must find the same value of total energy at any point in the motion, for example, at the top of the ramp and right before the block strikes the ground. We therefore label the initial and final heights, and the reference value of height, as shown in **Figure 6-21b**. It is convenient but not required to set either y_i or y_f equal to zero.

Equation 6-32 is the generic starting point for any problem that we will approach by applying energy conservation. Because the block does not interact with a spring, and because the ramp is frictionless, we can immediately drop the spring potential energy terms and the work done by nonconservative forces

$$\frac{1}{2}mv_i^2 + mgy_i + \frac{1}{2}kx_{c,i}^2 = \frac{1}{2}mv_f^2 + mgy_f + \frac{1}{2}kx_{c,f}^2 + |W_{nc}|$$

(6-32)

$$\frac{1}{2}mv_i^2 + mgy_i = \frac{1}{2}mv_f^2 + mgy_f$$

$$v_i = 0 \quad y_i = H_1 \quad v_f = ? \quad y_f = 0$$

SOLVE

Use the values that are given in the problem statement or determined by the choice of reference height and reduce the statement of conservation of energy to

$$mgH_1 = \frac{1}{2}mv_f^2$$

The speed of the block as it strikes ground is therefore

$$v_f = \sqrt{2gH_1}$$

REFLECT

Would doing the problem in two steps give the same answer? Let's quickly redo the problem. First, the block slides from its initial position to the end of the ramp:

$$\frac{1}{2}mv_i^2 + mgy_i = \frac{1}{2}mv_f^2 + mgy_f$$

$$v_i = 0 \quad y_i = H_1 \quad v_f = v_2 = ? \quad y_f = H_2$$

So,

$$mgH_1 = \frac{1}{2}mv_2^2 + mgH_2$$

or

$$v_2 = \sqrt{2g(H_1 - H_2)}$$

Now the block falls from height H_2 to $y = 0$ starting from speed

$$v_2 = \sqrt{2g(H_1 - H_2)}$$

$$\frac{1}{2}mv_i^2 + mgy_i = \frac{1}{2}mv_f^2 + mgy_f$$

$$v_i = \sqrt{2g(H_1 - H_2)} \quad y_i = H_2 \quad v_f = ? \quad y_f = 0$$

So,

$$\frac{1}{2}m[\sqrt{2g(H_1 - H_2)}]^2 + mgH_2 = \frac{1}{2}mv_f^2$$

Using the values given in the problem statement, the car's speed when it reaches the lowest point on the track is

$$gH_1 - gH_2 + gH_2 = \frac{1}{2}v_f^2$$

$$v_f = \sqrt{2gH_1}$$

As we expected, this answer is identical to the one we obtained doing the problem in one step. We just needed more lines of work to find it.

Practice Problem 6-11 A block is released with speed v_i (not from rest) from the top of a frictionless ramp, a height H_1 above the base of the ramp. As shown in Figure 6-21a, the ramp ends at height H_2 above the base, so that the block flies off and follows a two-dimensional free-fall path until it hits the ground. Find the speed of the block in terms of the given variables when it strikes the ground.

Gravitational potential energy can have negative values. We are free to assign $y = 0$ to any height, and our choice could result in $y < 0$ at some point for an object under consideration. Because gravitational potential energy depends on height, $U_{gravity}$ will be less than zero when $y < 0$. Ultimately, however, it is the change in potential energy rather than specific values of gravitational potential energy that is physically significant. Negative values for the change in $U_{gravity}$ are also allowed—anytime the final height is lower than the initial height, $\Delta U_{gravity}$ is less than zero.

Example 6-12 Roller Coaster

Cars on the roller coaster at LunarLand experience no air resistance and no friction. A roller coaster car starts from rest at the top of the first big incline, a height 115 m above the surface of the Moon. What is the car's speed when it reaches the lowest point on the track, which rests on the surface? The acceleration due to gravity on the surface of the Moon is $g_{Moon} = 1.62 \text{ m/s}^2$.

SET UP

Figure 6-22 shows a sketch of the problem. You might be tempted to set $y = 0$ at the base of the roller coaster, but to emphasize the point that we are free to set the reference value of y at any height, we selected the top of the incline as the reference point.

Equation 6-32 is the generic starting point for any problem that we will approach by applying energy conservation. Because there are no springs in the system and because the problem statement stipulates that there are no resistive forces, we can immediately drop the spring potential energy terms and the work done by nonconservative forces

$$\frac{1}{2}mv_i^2 + mg_{Moon}y_i + \frac{1}{2}kx_{c,i}^2 = \frac{1}{2}mv_f^2 + mg_{Moon}y_f + \frac{1}{2}kx_{c,f}^2 + |W_{nc}| \quad (6\text{-}32)$$

$$\frac{1}{2}mv_i^2 + mg_{Moon}y_i = \frac{1}{2}mv_f^2 + mg_{Moon}y_f$$

SOLVE

By our choice of reference and from the sketch, $y_i = 0$ and $y_f = -H$. The value of y_f is negative because it is below the height we chose to label $y = 0$

$$\frac{1}{2}mv_i^2 + mg_{Moon}y_i = \frac{1}{2}mv_f^2 + mg_{Moon}y_f$$

$$v_i = 0 \qquad y_i = 0 \qquad v_f = ? \qquad y_f = -H$$

So,

$$0 = \frac{1}{2}mv_f^2 + mg_{Moon}(-H)$$

Using the values given in the problem statement, the car's speed when it reaches the lowest point on the track is

$$v_f = \sqrt{2g_{Moon}H} = \sqrt{2(1.62 \text{ m/s}^2)(115 \text{ m})} = 19.3 \text{ m/s}$$

REFLECT

Because energy is conserved, the speed of the car at the bottom of the roller coaster does not depend on the path down. We expect, then, that the form of the solution to this problem should be identical to that of Example 6-11, which after all is just an object taking another path to get from one height to another. The results are the same—final speed $v_f = \sqrt{2gh}$, where h is the vertical drop. It made no difference that in Example 6-11 we set $y = 0$ at the lowest point but for the roller coaster we set the reference at the highest point.

Figure 6-22 What is the speed of a roller coaster car when it reaches the lowest point of the track?

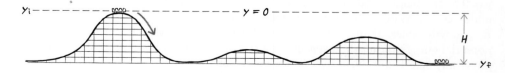

Practice Problem 6-12 Cars on the roller coaster at LunarLand experience no air resistance and no friction. A roller coaster car is moving at 15.0 m/s when it starts to roll up an incline and is barely moving when it reaches the top. How high (vertically) is the peak of the incline above its lowest point? The acceleration due to gravity on the surface of the Moon is $g_{Moon} = 1.62$ m/s^2.

The principle of conservation of energy applies equally well to an object on the end of a spring as it does to an object experiencing only a force due to gravity. As suggested by the complete statement of energy conservation (Equation 6-32), the presence of a spring requires the inclusion of spring potential energy terms. We examine this situation in the next example.

Example 6-13 Marble on a Spring

A child's toy shoots a marble into a maze with a horizontal spring that has a spring constant equal to 11 N/m. The marble has a mass of 16 g. If the spring is compressed 3.4 cm and then released, what is the speed of the marble when it leaves the spring? Assume that the marble experiences no resistive forces.

SET UP

Equation 6-32 is the generic starting point for any problem that we will approach by applying energy conservation. To help identify the reference points for the potential energy terms, draw a sketch like the one shown in **Figure 6-23**. First, because the height of the marble remains unchanged during the process we are investigating, the initial and final gravitational potential energy terms are equal, so we can drop them. The equilibrium position of the spring has been labeled $x = 0$. Also, no resistive forces act in this problem, so the work done by nonconservative forces is zero. So Equation 6-32,

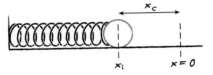
A marble is placed against the free end of a spring which has been compressed.

Figure 6-23 What is the speed of a marble placed at the end of a compressed spring and then released?

$$\frac{1}{2}mv_i^2 + mgy_i + \frac{1}{2}kx_{c,i}^2 = \frac{1}{2}mv_f^2 + mgy_f + \frac{1}{2}kx_{c,f}^2 + |W_{nc}|$$

becomes

$$\frac{1}{2}mv_i^2 + \frac{1}{2}kx_{c,i}^2 = \frac{1}{2}mv_f^2 + \frac{1}{2}kx_{c,f}^2$$

SOLVE

Before the marble is released its velocity v_i is equal to zero. The marble leaves the spring when the spring is no longer compressed, that is, $x_{c,f} = 0$. We can then list the values of the variables as

$$\frac{1}{2}mv_i^2 + \frac{1}{2}kx_{c,i}^2 = \frac{1}{2}mv_f^2 + \frac{1}{2}kx_{c,f}^2$$

$$v_i = 0 \quad x_{c,i} = 0.034 \quad v_f = ? \quad x_{c,f} = 0$$

So,

$$\frac{1}{2}kx_{c,i}^2 = \frac{1}{2}mv_f^2$$

Using the values given in the problem statement, the speed of the marble when it leaves the spring is

$$v_f = \sqrt{\frac{kx_{c,i}^2}{m}} = \sqrt{\frac{(11 \text{ N/m})(-0.034 \text{ m})^2}{0.016 \text{ kg}}} = 0.89 \text{ m/s}$$

REFLECT

Two aspects of this example are worth noting. First, although you might guess that the value of x_c should be taken as negative because the marble starts out to the left of the position marked as $x = 0$, notice that x is squared in the definition of spring potential energy. So a minus sign ultimately plays no role; the amount that a spring is either compressed or stretched can always be taken as positive.

Also, consider that the final speed depends on the mass of the marble in this problem. In contrast, in problems that only involve gravitational potential energy (for example, Example 6-12), the answer does not depend on the mass of the object under consideration. Why is this problem different? The answer lies in a difference between the gravitational force and the spring force. The force that Earth's gravity exerts on an object, $F_{gravity} = mg$, depends on mass and therefore so does the gravitational potential energy. In contrast, neither the force that a spring exerts, $F_{spring} = -kx$, nor the associated spring potential energy depends on mass. So, in the balancing of kinetic energy and potential energy (for example, $\Delta K = -\Delta U$), mass is eliminated from the equality in the gravitational case but not when a spring is present.

Practice Problem 6-13 How far should the spring in the child's toy described in Example 6-13 be compressed so that the marble leaves the spring at a speed of 1.8 m/s, about twice the speed we found in Example 6-13? Assume that the marble experiences no resistive forces.

We have identified three general kinds of forces: gravitational forces, spring forces, and resistive forces. With these forces in mind, we wrote a statement of energy conservation in the most general way by including terms due to each of these in Equation 6-32. When more than one force acts on an object, each may contribute to the potential energy of the system, as we see in the following example.

Example 6-14 A Ramp and a Spring

A child's toy uses a spring that has a spring constant k of 36 N/m to shoot a block up a 30° ramp (**Figure 6-24**). The block has a mass M of 8.0 g. How far d along the ramp does the block slide after the spring is compressed 4.2 cm and released? Neglect friction between the block and the surface of the ramp.

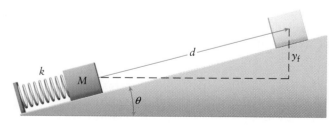

Figure 6-24 A spring is used to shoot a block up a ramp. How far along the ramp does the block slide after the spring is compressed 4.2 cm and released?

SET UP

Equation 6-32 is the generic starting point for any problem that we will approach by applying energy conservation. Because friction between the block and the surface can be neglected, we can immediately drop the term which represents work done by nonconservative forces. However, because the block will change height and experience a spring force, we will leave the gravitational potential energy and spring potential energy terms:

$$\frac{1}{2}mv_i^2 + mgy_i + \frac{1}{2}kx_{c,i}^2 = \frac{1}{2}mv_f^2 + mgy_f + \frac{1}{2}kx_{c,f}^2 + |W_{nc}| \qquad (6\text{-}32)$$

$$\frac{1}{2}mv_i^2 + mgy_i + \frac{1}{2}kx_{c,i}^2 = \frac{1}{2}mv_f^2 + mgy_f + \frac{1}{2}kx_{c,f}^2$$

SOLVE

Note that we have employed the convention that the spring compression is always taken as positive. Also note that the variable x represents the displacement of the end of the spring from equilibrium, not the direction perpendicular to the y direction. In addition, because the block starts to move only when the spring is released, the initial speed v_i of the block is zero. Because we are interested in finding y_f when the block comes to rest up the ramp, the final speed v_f is also zero. Finally, because the block is no longer touching the spring when it comes to rest up the ramp, $x_{c,f} = 0$. These values simplify the statement of energy conservation

$$\frac{1}{2}mv_i^2 + mgy_i + \frac{1}{2}kx_{c,i}^2 = \frac{1}{2}mv_f^2 + mgy_f + \frac{1}{2}kx_{c,f}^2$$

$$v_i = 0 \quad y_i = 0 \quad x_{c,i} = 0.042 \quad v_f = 0 \quad y_f = ? \quad x_{c,f} = 0$$

So,

$$\frac{1}{2}kx_{c,i}^2 = mgy_f$$

Using the values given in the problem statement, the block rises to a height of

$$y_f = \frac{k}{2mg}x_{c,i}^2 = \frac{36 \text{ N/m}}{2(0.0080 \text{ kg})(9.8 \text{ m/s}^2)}(0.042 \text{ m})^2 = 0.405 \text{ m}$$

The block rises 40.5 cm vertically. To find the distance the block slides along the ramp, we make use of the right triangle formed by the base of the ramp, the vertical height, and the distance along the ramp. Using a ramp angle $\theta = 30°$, the distance d along the ramp is

$$d = \frac{y_f}{\sin\theta} = \frac{0.405}{\sin 30°} = 0.81$$

The block slides 81 cm along the ramp.

Go to Interactive Exercise 6-1 for more practice dealing with energy conversation

REFLECT

In writing out a statement of energy conservation in this example, we included potential energy terms arising from two different forces. Each form of potential energy requires us to select a reference point. We are always free to choose any convenient point to set as the reference; however, the reference point for height (gravitational potential energy) does not need to be the same as the reference point for the spring compression (spring potential energy).

Practice Problem 6-14 A child's toy uses a spring that has a spring constant k equal to 36 N/m to shoot a block of mass 8.0 g up a ramp of angle θ equal to 30° (Figure 6-24). The spring is compressed 4.2 cm and released. What is the speed of the block at the instant it loses contact with the spring? Neglect friction between the block and the surface of the ramp.

Energy is not a vector quantity. This important aspect of energy is explored in the next example.

Example 6-15 A Pogo Stick

United States Patent 6390956, "Adjustable spring rate pogo stick," describes a pogo stick that has a spring mechanism which can be adjusted to accommodate riders of a range of weights. When the spring constant is adjusted to its maximum value of 20,300 N/m, a 100-kg kangaroo standing on the pogo stick partially compresses the spring; if the kangaroo jumps on the pogo stick, it compresses the spring an additional 0.25 m. If the kangaroo and pogo stick are at a forward angle of 60° from the horizontal when the spring returns to the equilibrium length, how fast are they moving at that moment?

SET UP

Experience with the problems of two-dimensional motion may tempt you to set up a coordinate axis and work with separate components in conserving energy. Don't! Energy is not a vector. Instead, apply the requirement that the total mechanical energy is the same when the pogo stick spring is compressed as when it has sprung back to the equilibrium length. We neglect dissipative (nonconservative) forces, so we can immediately drop the last term in Equation 6-32:

$$\frac{1}{2}mv_i^2 + mgy_i + \frac{1}{2}kx_{c,i}^2 = \frac{1}{2}mv_f^2 + mgy_f + \frac{1}{2}kx_{c,f}^2 + |W_{nc}| \qquad (6\text{-}32)$$

$$\frac{1}{2}mv_i^2 + mgy_i + \frac{1}{2}kx_{c,i}^2 = \frac{1}{2}mv_f^2 + mgy_f + \frac{1}{2}kx_{c,f}^2$$

SOLVE

Figure 6-25 shows our convenient choice of reference points. For the height to be used in the gravitational potential terms, $y_i = 0$ is set at the point of maximum compression of the spring. The final height corresponds to the instant when the spring is neither compressed nor stretched. However, note that the y_f term is not the spring compression of $x_{c,i} = 0.25$ m, but rather

$$y_f = x_{c,i}\sin\theta$$

because the pogo stick is tilted. We write out the values of each variable that will simplify the statement of energy conservation

$$\frac{1}{2}mv_i^2 + mgy_i + \frac{1}{2}kx_{c,i}^2 = \frac{1}{2}mv_f^2 + mgy_f + \frac{1}{2}kx_{c,f}^2$$

$$v_i = 0 \quad y_i = 0 \quad x_{c,i} = 0.25 \text{ m} \quad v_f = ? \quad y_f = x_{c,i}\sin\theta \quad x_{c,f} = 0$$

So,

$$\frac{1}{2}kx_{c,i}^2 = \frac{1}{2}mv_f^2 + mgx_{c,i}\sin\theta$$

or, after rearranging,

$$v_f = \sqrt{\frac{k}{m}x_{c,i}^2 - 2gx_{c,i}\sin\theta}$$

Using the numbers given in the problem statement,

$$v_f = \sqrt{\frac{20300 \text{ N/m}}{100 \text{ kg}}(0.25 \text{ m})^2 - 2(9.8 \text{ m/s}^2)(0.25 \text{ m})\sin 60°} = 2.9 \text{ m/s}$$

REFLECT

If you've ever watched someone riding a pogo stick, you know that the stick leaves the ground at about the same speed as someone jumping. In Example 3-7, we saw that a professional athlete jumps up with a speed of about 6 m/s, so our answer, about half of that, must be in the right range. While reflecting on this problem,

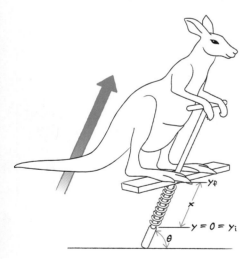

Figure 6-25 A kangaroo jumps on a pogo stick, compressing its spring. This sketch shows a choice of reference points convenient for determining the speed of the kangaroo at the instant that the spring returns to its equilibrium length.

however, you should know that more important than the final value is the recognition that kinetic energy does not have direction nor a component in a particular direction.

Our convention of always taking as positive the distance a spring is compressed or stretched comes into play in this problem. This term is squared when used to determine spring potential energy, so the choice of sign cannot affect the result. However in this problem we found that the final height y_f is equal to $x_{c,i} \sin \theta$; the sign of y_f does depend on the sign of $x_{c,i}$. Figure 6-25 resolves the question of the sign—clearly y_f must be positive, so our choice to make $x_{c,i}$ positive is the correct one.

The action of friction and other nonconservative forces may appear to drain energy from a system. This drain can be understood either by expanding the definition of the system under consideration, or by treating the energy transferred through the work done by the nonconservative force as a separate term in the statement of energy conservation. In the first view, for example, we would include the ice surface as part of the system for a hockey puck sliding to a stop on a rink. Energy is conserved because although mechanical energy leaves the puck, it appears as thermal energy in the puck as well as the ice. In the second view, the surface would not be considered part of the system, but the initial and final energies of the puck are balanced by the inclusion of the work due to nonconservative forces (W_{nc}) in Equation 6-32.

Example 6-16 A Nonconservative Force

A block is given an initial speed of 3.8 m/s on a horizontal surface. The coefficient of kinetic friction μ_k between the surface and the block is 0.35. How far does the block move before coming to a stop? See Figure 6-26.

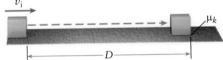

SET UP
The block comes to a stop because the frictional force does work on it, transferring kinetic energy from the block to thermal energy of the two surfaces. We will account for the energy using the W_{nc} term in Equation 6-32. Because the block does not change height and because it encounters no springs, however, we can drop the gravitational potential energy and spring potential energy terms

Figure 6-26 A block moves on a horizontal surface that offers resistance to motion. How far does the block move before coming to a stop?

$$\frac{1}{2}mv_i^2 + mgy_i + \frac{1}{2}kx_{c,i}^2 = \frac{1}{2}mv_f^2 + mgy_f + \frac{1}{2}kx_{c,f}^2 + |W_{nc}| \qquad (6\text{-}32)$$

$$\frac{1}{2}mv_i^2 = \frac{1}{2}mv_f^2 + |W_{nc}|$$

SOLVE
The block comes to rest, that is, $v_f = 0$, so,

$$\frac{1}{2}mv_i^2 = |W_{nc}| \qquad (6\text{-}34)$$

The distance the block moves is connected to this equation through the definition of work. From Equation 6-17,

$$|W_{nc}| = \left| \int_{x_i}^{x_f} \vec{f} \cdot d\vec{x} \right| = \left| \int_{x_i}^{x_f} f \, dx \cos 180° \right|$$

The angle between the force due to kinetic friction $\vec{f}$ and the displacement vector $d\vec{x}$ is 180° because $\vec{f}$ points in the direction opposite to the motion. The magnitude of the kinetic friction f is $\mu_k N$ (Equation 5-5), or because the block is on a horizontal surface

$$f = \mu_k mg$$

So

$$|W_{nc}| = \int_{x_i}^{x_f} \mu_k mg \, dx = \mu_k mg (x_f - x_i) = \mu_k mgD \qquad (6\text{-}35)$$

where D is the distance the block moves before coming to a stop. Equation 6-34 is then

$$\frac{1}{2} mv_i^2 = \mu_k mgD \qquad (6\text{-}36)$$

Using the values given in the problem statement, the distance the block moves before coming to a stop is

$$D = \frac{v_i^2}{2\mu_k g} = \frac{(3.8 \text{ m/s})^2}{2(0.35)(9.8 \text{ m/s}^2)} = 2.1 \text{ m}$$

REFLECT

In this problem, the mechanical energy is not conserved. There is kinetic energy initially but not at the end, and the potential energy of the block neither increases nor decreases. Such an energy change could be taken to mean that energy has been lost from the system, if the system is taken to be only the block. In that case, we would say that the work done by friction has transferred energy from the block. But the total energy must be conserved; when the system is defined as the block and surface together, the initial (kinetic) energy equals the thermal energy equivalent to the work done by friction.

Practice Problem 6-16 A hockey puck is given an initial speed of 3.8 m/s on a horizontal ice surface. The coefficient of kinetic friction μ_k between the puck and the ice is 0.08. How far does the puck move before coming to a stop?

When a frictional force acts on an object, information about the change in the speed of an object leads directly to an understanding of the coefficient of kinetic friction, as seen in the next example.

Example 6-17 Speed Up, Slow Down

A block is released from rest at vertical height H above the base of a frictionless ramp. After sliding off the ramp the block encounters a rough, horizontal surface and comes to a stop after moving a distance H, as shown in **Figure 6-27**. What is the coefficient of kinetic friction between the block and the horizontal surface?

SET UP

The frictional force does work on the block, transferring its kinetic energy to thermal energy and eventually causing the block to stop. We will account for this energy using the W_{nc} term in Equation 6-32, which is simplified by dropping the spring potential terms

$$\frac{1}{2} mv_i^2 + mgy_i + \frac{1}{2} kx_{c,i}^2 = \frac{1}{2} mv_f^2 + mgy_f + \frac{1}{2} kx_{c,f}^2 + |W_{nc}| \qquad (6\text{-}32)$$

$$\frac{1}{2} mv_i^2 + mgy_i = \frac{1}{2} mv_f^2 + mgy_f + |W_{nc}|$$

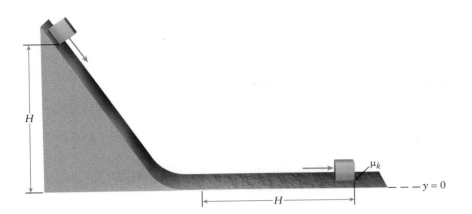

Figure 6-27 A block is released from rest above the base of a frictionless ramp, slides down, encounters a rough, horizontal surface, and comes to a stop.

SOLVE

As we have seen (Example 6-11), it isn't necessary to break a problem such as this one into separate steps, for example, by finding the speed of the block at the bottom of the ramp and then creating a separate step to examine the motion of the block from the base of the ramp to the position where it comes to rest. Instead, we simplify the conservation of energy statement by setting the initial and final values of speed to zero. Using $v_i = 0$ and $v_f = 0$ and the relationship for $|W_{nc}|$ (Equation 6-35) from Example 6-16, we note the simplifying substitutions:

$$\frac{1}{2}mv_i^2 + mgy_i = \frac{1}{2}mv_f^2 + mgy_f + |W_{nc}|$$

$$v_i = 0 \quad y_i = H \quad v_f = 0 \quad y_f = 0 \quad |W_{nc}| = \int_{x_i}^{x_f} \mu_k mg \, dx$$

So

$$mgH = \int_{x_i}^{x_f} \mu_k mg \, dx$$

or

$$mgH = \mu_k mg\,(x_f - x_i) = \mu_k mgH$$

The m, g, and H terms cancel, leaving

$$\mu_k = 1$$

REFLECT

In Chapter 5 we noted that for materials we typically encounter, the range of values of the coefficient of kinetic friction is about 0.1 to 0.8. The value $\mu_k = 1$ here suggests, as you might have expected, that the coefficient of friction must be relatively large in order to bring the block to a stop over the same distance through which it dropped vertically.

Practice Problem 6-17 A block is released from rest at vertical height 2.4 m above the base of a frictionless ramp. After sliding off the ramp the block encounters a rough, horizontal surface and eventually comes to a stop. The coefficient of kinetic friction between the block and the horizontal surface is 0.80. How far does the block travel along the horizontal surface before coming to a stop?

★ What's Important 6-8

Although the amount of energy in any particular form might change during some process, the total energy remains constant. The action of friction and other nonconservative forces may appear to drain energy from a system, but this can be understood by expanding the boundaries of the system under consideration.

Answers to Practice Problems

6-1	67 J		6-10	21.2 m/s
6-2	0.5 cm		6-11	$v_f = \sqrt{v_i^2 + 2gH_1}$
6-4	7.7 m/s		6-12	69.4 m
6-5	9.0 m		6-13	6.9 cm
6-7	39.8 m/s		6-14	2.7 m/s
6-8	17 m/s		6-16	9.2 m
6-9	−380 J		6-17	3.0 m

Answers to Got the Concept Questions

6-1   The work–kinetic energy theorem relates the work done on a system to the change in its kinetic energy. Work is determined by the force and the distance over which it is applied, so the distance is directly related to the change in kinetic energy. When an object starts from rest, the change in kinetic energy (from Equation 6-10) is proportional to the square of the final velocity, so here distance is proportional to the square of velocity. To double the velocity therefore requires that the force be applied over four times the distance.

6-2   The work you do on each box is given by Equation 6-3; the work you do on each box depends on the applied force and the distance the box moves. The weight of the box does not determine the force required—the same force applied to a heavier box would simply result in a lower acceleration, but this does not affect the work done.

The boxes are initially not moving, so initial kinetic energies are zero. After you push a box across the room, the box is again not moving, so its final kinetic energy is zero. The change in kinetic energy is therefore zero. So, the net work done on the box is also zero (regardless of the weight of the box) according to the work–kinetic energy theorem (Equation 6-11). Although you do work to move the box, the force of kinetic friction between the floor and the box also does work on the box. Because the frictional force points in the direction opposite to the motion, you must use the angle of 180° in the cosine term in Equation 6-3 to determine the work done by the frictional force, and cos 180° = −1. The work done by friction is therefore negative, so the sum of the work done by you and the work done by friction is zero. Also, because the normal force must equal the weight of the box and because the box does not move in the vertical direction, the work done on the box by each of those forces is zero. So the net work done on each box is zero, and this result is independent of the weight of each box.

6-3   The change in gravitational potential energy is the same in both cases: ΔU does not depend on the choice of reference point. To convince yourself, consider $\Delta U_{gravity} = mg(y_f - y_i)$ (Equation 6-21) in the context of Figure 6-28. When the ground is considered to be at a height equal to zero, $y_i = 45$ m and $y_f = 36$ m, so $y_f - y_i = 36$ m − 45 m = −9 m. When the height of the balcony is considered to be equal to zero, $y_i = 9$ m and $y_f = 0$ m, so $y_f - y_i$ is again −9 m.

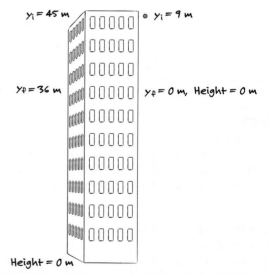

Figure 6-28 The change in gravitational potential energy does not depend on the choice of reference point.

6-4   Mechanical energy is conserved, which requires that the change in kinetic energy plus the change in potential energy equals zero (Equation 6-28). Here we take the initial position of the block as its starting point on the left, and the final position as the point on the right where it momentarily comes to a stop. The initial and final kinetic energies of the block are therefore both zero, so the change in kinetic energy is zero. This means that the difference between the initial and final gravitational potential energies must also be zero; in other words, the initial gravitational potential energy equals the final gravitational potential energy. The block rises up to vertical height H on the right side of the ramp.

SUMMARY

Topic	Summary	Equation or Symbol
conservation of energy	The mechanical energy in an isolated system is conserved, that is, the sum of the kinetic energy K and potential energy U does not change over time. Work done by nonconservative forces W_{nc} reduces the final mechanical energy of a system.	$K_i + U_i = K_f + U_f + \lvert W_{nc} \rvert$ (6-31)
conservative force	A conservative force acting on an object leaves the sum of the kinetic and potential energy unchanged.	
dot product	The dot product of two vectors equals the product of the magnitudes of the vectors multiplied by the cosine of the angle between them. The dot product of two vectors is a scalar.	$C = \vec{A} \cdot \vec{B} = A\,B \cos\theta$ (6-4)
energy	Energy is the capacity of an object or system to do work.	
external forces	A force that acts on a system from outside is external.	
gravitational potential energy	Energy that characterizes a system's potential to acquire kinetic energy due to the action of gravity is gravitational potential energy. The change in gravitational potential energy depends on the change in (vertical) height.	$\Delta U_{gravity} = mg(y_f - y_i)$ (6-21)
Hooke's law	Hooke's law describes a force that increases in proportion to the displacement of a system from its equilibrium position.	$\vec{F} = -k\vec{x}$ (6-16)
kinetic energy	Kinetic energy is the energy embodied in the motion of an object.	$K = \dfrac{1}{2} mv^2$ (6-10)
mechanical energy	Mechanical energy is the sum of the kinetic energy and potential energy in a system.	
nonconservative forces	Nonconservative forces convert mechanical energy in a system to other forms of energy that cannot be converted back to either potential or kinetic energy.	

potential energy	Potential energy characterizes the potential for a system to do work or, equivalently, to move or to exert a force. It can only be observed when converted to kinetic energy.	$$\Delta U = -\int_{x_i}^{x_f} \vec{F} \cdot d\vec{x}$$	(6-23)
spring constant	The spring constant, sometimes called the stiffness of the spring, quantifies how easily a spring stretches.	k	
thermal energy	Thermal energy is not available to do work.		
translational kinetic energy	Translational kinetic energy is the kinetic energy of an object moving from one position to another along a straight line.		
work	Work is the result of a force applied to an object over a distance. Work results in the transfer of energy.	$$W = \int_{x_i}^{x_f} \vec{F} \cdot d\vec{x}$$	(6-17)
work–kinetic energy theorem	The net work done on an object equals the change in its kinetic energy.	$W_{net} = \Delta K$	(6-11)

QUESTIONS AND PROBLEMS

In a few problems, you are given more data than you actually need; in a few other problems, you are required to supply data from your general knowledge, outside sources, or informed estimate.

Interpret as significant all digits in numerical values that have trailing zeros and no decimal points.

For all problems, use $g = 9.8$ m/s^2 for the free-fall acceleration due to gravity. Neglect friction and air resistance unless instructed to do otherwise.

• Basic, single-concept problem

•• Intermediate-level problem, may require synthesis of concepts and multiple steps

••• Challenging problem

SSM *Solution is in Student Solutions Manual*

Conceptual Questions

1. •Using Equation 6-2, how can the work done on an object be equal to zero?

2. •The energy that is provided by your electric company is *not* sold by the joule. Instead, you are charged by the kilowatt-hour (typically 1 kWh $\cong$ $0.25 including taxes). Explain why this makes sense for the average consumer. It might be helpful to read the electrical specifications of an appliance (such as a hair dryer or a blender).

3. •Why do seasoned hikers step *over* logs that have fallen in their path rather than stepping *onto* them? SSM

4. •A common classroom demonstration involves holding a bowling ball attached by a rope to the ceiling close to your face and releasing it from rest. In theory, you should not have to worry about being hit by the ball as it returns to its starting point. However, some of the most "exciting" demonstrations *have* involved the professor being hit by the ball! (a) Why should you expect not to be hit by the ball? (b) Why *might* you be hit if you actually perform the demonstration?

5. •One of your classmates in physics reasons, "If there is no displacement, then a force will perform no work." Suppose the student pushes with all of her might against a massive boulder, but the boulder doesn't move. (a) Does she expend energy pushing against the boulder? (b) Does she do work on the boulder? Explain your answer.

6. •Your roommate lifts a cement block and carries it across the room. Is the net work done by her on the block positive, negative, or zero? Explain your answer.

7. •Can the normal force ever do work on an object? Explain your answer. SSM

8. •Can kinetic energy ever have a negative value? Explain your answer.

9. •Can a change in kinetic energy ever have a negative value? Explain your answer.

10. •Define the concept of "nonconservative force" in your own words. Give three examples.

11. •Analyze the types of energy that are associated with the "circuit" that a snowboarder follows from the bottom of a mountain, to its peak, and back down again. Be sure to explain all changes in energy. SSM

12. •The work–kinetic energy theorem relates the work done to the change in kinetic energy. Does it require more work to swing a large sledgehammer with a slower speed or a smaller claw hammer with a faster speed (which factor is more significant, the speed or the mass)?

13. •A satellite orbits around Earth in a circular path at a high altitude. Explain why the gravitational force does zero work on the satellite.

14. •Model rockets are propelled by an engine containing a combustible solid material. Two rockets, one twice as heavy as the other, are launched using engines that contain the same amount of propellant. (a) Is the kinetic energy of the heavier rocket less than, equal to, or more than the kinetic energy of the lighter one? (b) Is the launch speed of the heavier rocket less than, equal to, or more than the launch speed of the lighter one?

15. •When does the kinetic energy of a rock that is dropped from the edge of a high cliff reach its maximum value? Answer the question (a) when the air resistance is negligible and (b) when there is significant air resistance. SSM

16. •Why are the ramps for disabled people quite long instead of short and steep?

17. •Can gravitational potential energy have a negative value? Explain your answer.

18. •Bicycling to the top of a hill is much harder than coasting back down to the bottom. Do you have more gravitational potential energy at the top of the hill or at the bottom? Explain your answer.

19. •Calc When is it necessary to use an integral to calculate the work done by a force acting over a distance? Assume the force acts parallel to the displacement for this question. SSM

20. •A rubber dart can be launched over and over again by the spring in a toy gun. Where is the energy generated to launch the dart?

Multiple-Choice Questions

21. •Box 2 is pulled up a rough incline by box 1 in an arrangement as shown in Figure 6-29. How many forces are doing work on box 2?
   A. 1
   B. 2
   C. 3
   D. 4
   E. 5

Figure 6-29 Problem 21

22. •Figure 6-30 shows four situations in which a box slides to the right a distance d across a frictionless floor as a result of applied forces, *one* of which is shown. The magnitudes of the forces shown are identical. Rank the four cases in order of increasing work done on the box during the displacement by the force shown.

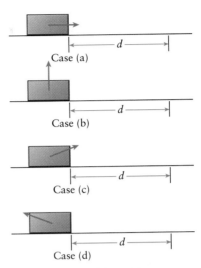

Case (a)

Case (b)

Case (c)

Case (d)

Figure 6-30 Problem 22

23. •A box is dragged across a floor by a force $\vec{F}$ which makes an angle θ with the horizontal (Figure 6-31). If the magnitude of the force is held constant but the angle θ is increased up to 90°, the work done by the force
   A. remains the same.
   B. increases.
   C. decreases.
   D. first increases, then decreases.
   E. first decreases, then increases. SSM

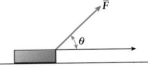

Figure 6-31 Problem 23

24. •A car moves along a level road. If the car's velocity changes from 60 mph due east to 60 mph due west, the kinetic energy of the car
   A. remains the same.
   B. increases.
   C. decreases.
   D. first increases, then decreases.
   E. first decreases, then increases.

25. •A boy swings a ball on a string at constant speed in a circle that has a circumference equal to 6 m. What is the work done on the ball by the 10-N tension force in the string during one revolution of the ball?
   A. 190 J
   B. 60 J
   C. 30 J
   D. 15 J
   E. 0

This page is intentionally left blank.

For complete end of chapter problem sets, please go to
www.whfreeman.com/kestentauck

the field. (a) If the speed of the cooler is constant throughout the trip, calculate the work done by the assistant. (b) How much work is done by the force of gravity on the cooler of water?

43. ••The statue of Isaac Newton is crated and moved from the Oxford Museum of Natural History and Science for cleaning. The mass of the statue and the crate is 150 kg. As the statue slides down a ramp inclined at 40°, the curator pushes up, parallel to the ramp's surface, so that the crate does not accelerate (Figure 6-32). If the statue slides 3 m down the ramp, and the coefficient of kinetic friction between the crate and the ramp is 0.54, calculate the work done on the crate by each of the following: (a) gravity, (b) the curator, (c) friction, and (d) the normal force between the ramp and the crate. SSM

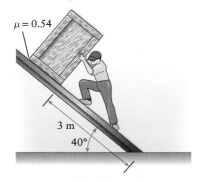

Figure 6-32 Problem 43

44. •(a) Calculate the magnitude of each of the following vectors. (b) Find the angle between the +x axis and each of the four vectors.

$$\vec{A} = 3\hat{x} + 6\hat{y}$$
$$\vec{B} = -6\hat{x} + 6\hat{y}$$
$$\vec{C} = 2\hat{x} - 3\hat{y}$$
$$\vec{D} = -3\hat{x} - 4\hat{y}$$

45. •Find the angle between vectors $\vec{A}$ and $\vec{B}$

$$\vec{A} = 6\hat{x} + 6\hat{y}$$
$$\vec{B} = -6\hat{x} + 6\hat{y}$$

Make a sketch to confirm your calculation.

46. •Prove that the angle between two parallel vectors is zero. Two parallel vectors would obey the relationship

$$\vec{A} = x\hat{x} + y\hat{y}$$
$$\vec{B} = a(x\hat{x} + y\hat{y})$$

where a is a constant.

6-2: The Work–Kinetic Energy Theorem

47. •A bumblebee has a mass of about 0.25 g. If its speed is 10 m/s, calculate its kinetic energy. SSM

48. •A 1250-kg car moves at 20 m/s. How much work is required from the engine to increase the car's speed to 30 m/s?

49. •A small truck has a mass of 2100 kg. How much work is required to decrease the speed of the vehicle from 22 m/s to 12 m/s on a level road?

50. •A 10-kg block rests on a level floor; the coefficient of kinetic friction between the block and the floor is 0.44. A force of 200 N acts for 4 m. How fast is the block moving after being pushed 4 m? Assume it starts from rest.

51. ••Calculate the final speed of the 2-kg object that is pushed for 22 m by the 40-N force on a level, frictionless floor (Figure 6-33). Assume the object starts from rest. SSM

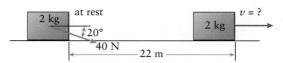

Figure 6-33 Problem 51

6-3: Applications of the Work–Kinetic Energy Theorem

52. •A force of 1200 N pushes a man on a bicycle forward. Air resistance pushes against him with a force of 800 N. If he starts from rest and is on a level road, how fast will he be moving after 20 m? The mass of the bicyclist and his bicycle is 90 kg.

53. •Sports A 0.15-kg baseball rebounds off of a wall. The rebound speed is one-third of the original speed. By what percent does the kinetic energy of the baseball change in the collision with the wall? Where does the energy go?

54. •A book slides across a level, carpeted floor at an initial speed of 4 m/s and comes to rest after 3.25 m. Calculate the coefficient of kinetic friction between the book and the carpet. Assume the only forces acting on the book are friction, weight, and the normal force.

55. •Sports A catcher in a baseball game stops a pitched ball that was originally moving at 44 m/s over a distance of 12.5 cm. The mass of the ball is 0.15 kg. What is the average force that the glove imparts to the ball during the catch? Comment on the force that the catcher's hand experiences during the catch. SSM

56. •A 325-g model boat floats on a pond. The wind on its sail provides a force of 1.85 N that points 25° north of east. The drag force of the water on the boat is 0.75 N, toward the west. If the boat starts from rest and heads east for a distance of 3.55 m, how fast is it moving?

6-4: Work Done by a Variable Force

57. •Calculate the work done from $x = 0$ to $x = 10$ m by the one-dimensional force depicted in the following graph (**Figure 6-34**).

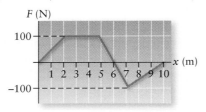

Figure 6-34 Problem 57

58. •Calculate the work done in moving an object from $x = 0$ to $x = 2$ m, when the force acting on it is a function of position given by $F(x) = 2x^2 + 3x$ (**Figure 6-35**). (The values 2 and 3 carry SI units.)

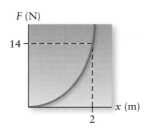

Figure 6-35 Problem 58

59. •Calc An object attached to the free end of a horizontal spring of constant 450 N/m is pulled from a position 12 cm beyond equilibrium to a position 18 cm beyond equilibrium. Calculate the work the spring does on the object. SSM

60. •••Calc Calculate the kinetic energy of an object of mass m on the end of a spring of mass m_s. The speed of any part of the spring depends on the distance from the fixed end (in other words, part of the spring is at rest at the fixed end and the velocities of the parts of the spring increase in a linear fashion up to v at the other end of the spring).

61. •A 5.0-kg object is attached to one end of a horizontal spring that has a negligible mass and a spring constant of 250 N/m. The other end of the spring is fixed to a wall. The spring is compressed by 10 cm from its equilibrium position and released from rest. (a) What is the speed of the object when it is 8.0 cm from equilibrium? (b) What is the speed when the object is 5.0 cm from equilibrium? (c) What is the speed when the object is at the equilibrium position?

62. •Calc The force acting on an object is given by $F(x) = 12x + 2x^2 - 0.25x^3$. (The multiplicative constants carry SI units.) Calculate the work done on the object if it starts from rest and moves from $x = 0$ to $x = 1$ m.

63. •Calc The force that acts on an object is given by $\vec{F} = 3x\hat{i} + 4y\hat{j}$. (The multiplicative constants carry SI units.) Calculate the work done on the object by the force when the object moves from the origin (0,0) to the point (3,4). SSM

64. •Calc The force that acts on an object is given by $\vec{F} = 2x\hat{i} + 3y\hat{j} + 0.2z^2\hat{k}$. (The multiplicative constants carry SI units.) Calculate the work done on the object by the force when the object moves from the origin (0,0,0) to the point (2,8,3).

6-5: Potential Energy

65. •What is the gravitational potential energy relative to the ground of a 1-N Gravenstein apple that is hanging from a limb 2.5 m above the ground?

66. •Pilings are driven into the ground at a building site by dropping a 2000-kg object onto them. What change in gravitational potential energy does the object undergo if it is released from rest 18 m above the ground and ends up 2 m above the ground?

67. •A 40.0-kg boy steps on a skateboard and pushes off from the top of a hill. What change in gravitational potential energy takes place as the boy glides down to the bottom of the hill, 4.35 m below the starting level? SSM

68. •How much additional potential energy is stored in a spring that has a spring constant of 15.5 N/m if the spring starts 10 cm from the equilibrium position and ends up 15 cm from the equilibrium position?

69. •A spring that has a spring constant of 200 N/m is oriented vertically with one end on the ground. (a) What distance does the spring compress when a 2-kg object is placed on its upper end? (b) By how much does the potential energy of the spring increase during the compression?

70. •An external force moves a 3.5-kg box at a constant speed up a frictionless ramp. The force acts in a direction parallel to the ramp. (a) Calculate the work done on the box as it is pushed up the 5 m ramp to a height of 3 m. (b) Compare the value with the change in gravitational potential energy that the box undergoes as it rises to its final height.

71. •Over 630 m in height, the Burj Dubai is the world's tallest skyscraper. What is the change in gravitational potential energy of a $20 gold coin (33.5 g) when it is carried from ground level up to the top of the Burj Dubai? Neglect any slight variations in the acceleration due to gravity. SSM

72. •For a great view and a thrill, check out EdgeWalk at the CN Tower in Toronto, where you can walk on the roof of the tower's main pod, 356 m above the ground. What is the gravitational potential energy relative to the surface of Earth of a 65-kg sightseer on EdgeWalk? Neglect any slight variations in the acceleration due to gravity.

73. •A spring that is compressed 12.5 cm from its equilibrium position stores 3.33 J of potential energy. Determine the spring constant.

6-6: Conservation of Energy

74. •A ball is thrown straight up at 15 m/s. At what height will the ball have one-half of the initial speed?

75. •A water balloon is thrown straight down at 12 m/s from a second floor window, 5.0 m above ground level.

How fast is the balloon moving when it hits the ground? SSM

76. •A gold coin (33.5 g) is dropped from the top of the Burj Dubai building, 630 m above ground level. In the absence of air resistance, how fast would it be moving when it hit the ground?

77. •A 30-kg child rides a 9-kg sled down a frictionless ski slope. At the bottom of the hill, her speed is 7 m/s. If the slope makes an angle of 15° with the horizontal, how far did she slide on the sled?

78. •During a long jump, four-time Olympic champion Carl Lewis' center of mass rose about 1.2 m from the launch point to the top of the arc. What minimum speed did he need at launch if he was traveling at 6.6 m/s at the top of the arc?

79. •A pendulum is constructed by attaching a small metal ball to one end of a 1.25-m-long string that hangs from the ceiling (**Figure 6-36**). The ball is released when it is raised high enough for the string to make an angle of 30.0° with the vertical. How fast is it moving at the bottom of its swing? Does the mass of the ball affect the answer? SSM

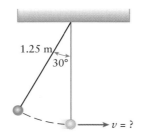

Figure 6-36 Problem 79

80. •An ice cube slides without friction down a track as shown in **Figure 6-37**. Calculate the speed of the cube at points B, C, D, and E. The ice cube has no speed at point A.

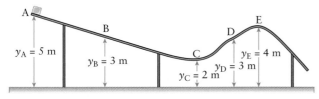

Figure 6-37 Problem 80

6-7: Nonconservative Forces

81. •A driver loses control of her car when she slams on the brakes, leaving 88-m-long skid marks on the level road. The coefficient of kinetic friction is estimated to be 0.48. How fast was the car moving when the driver hit the brakes?

82. ••A 65-kg woman steps off a 10-m diving platform and drops straight down into the water. If she reaches a depth of 4.5 m, what is the average resistance force exerted on her by the water? Ignore any effects due to air resistance.

83. •A skier leaves the starting gate at the top of a ski jump with an initial speed of 4 m/s (**Figure 6-38**). The starting position is 120 m higher than the end of the ramp, which is 3 m above the snow. Find the final speed

of the skier if he lands 145 m down the 20° slope. Assume there is no friction on the ramp, but air resistance causes a 50% loss in the final kinetic energy. The GPS reading of the elevation of the skier is 4212 m at the top of the jump and 4017 m at the landing point. SSM

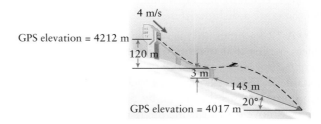

Figure 6-38 Problem 83

6-8: Using Energy Conservation

84. •An 18-kg suitcase falls from a hot-air balloon that is floating at a height of 385 m above the surface of Earth. The suitcase has an initial speed of 0 m/s but it reaches a speed of 30 m/s just before it hits the ground. Calculate the percentage of the initial energy that is "lost" to air resistance.

85. •A bicyclist maintains a constant speed of 4 m/s up a hill that is inclined at 10° with the horizontal. Calculate the work done by the person and the work done by gravity if the bicycle moves a distance of 20 m up the hill. The combined mass of the rider and the bike is 90 kg.

86. •A child slides down a snow-covered slope on a sled. At the top of the hill, her mother gives her a push at a speed of 1 m/s to get her started. The frictional force acting on the sled is one-fifth of the combined weight of the child and the sled. If she travels for a distance of 25 m and her speed at the bottom is 4 m/s, calculate the angle that the hill makes with the horizontal.

87. ••The coefficient of restitution, e, of a ball hitting the floor is defined as the ratio of the speed of the ball right after it rebounds from the impact to the speed of the ball right before it hits the floor, that is, $e = v_{\text{after}}/v_{\text{before}}$. (a) Derive a formula for the coefficient of restitution when a ball is released from an initial height H and rebounds to a final height h. (b) Calculate the coefficient of restitution for a golf ball that bounces to a height of 60 cm after having been released from a height of 80 cm. SSM

General Problems

88. •The force that acts on an object is given by $\vec{F} = 2.2\hat{x} + 4.5\hat{y}$. Calculate the work done on the object when the force results in a displacement given by $\vec{d} = 12\hat{x} + 20\hat{y}$. (The multiplicative constants in both expressions carry SI units.)

89. ••A force of magnitude 50 N does 24 J of work on an object as it undergoes a displacement given by the vector $\vec{d} = 2\hat{x} + 2\hat{y}$. (The multiplicative constants carry SI units.) Find direction of the force, using $\hat{x}, \hat{y}$ notation.

90. •You push a 20-kg crate at constant velocity up a ramp inclined at an angle of 33° to the horizontal. The coefficient of kinetic friction between the ramp and the crate, μ_k is equal to 0.2. How much work must you do to push the crate a distance d of 2 m?

91. ••A 12-kg block (M) is released from rest on a frictionless incline that makes an angle of 28°, as shown in **Figure 6-39**. Below the block is a spring that has a spring constant of 13500 N/m. The block momentarily stops when it compresses the spring by 5.5 cm. How far does the block move down the incline from its rest position to the stopping point? SSM

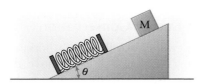

Figure 6-39 Problem 91

92. •**Sports** A man on his luge (total mass of 88 kg) emerges onto the horizontal straight track at the bottom of the hill with an initial speed of 28 m/s. If the luge and rider slow at a constant rate of 2.8 m/s², what work is done on them by the force that slows them?

93. •**Biology** An adult dolphin is about 5 m long and weighs about 1600 N. How fast must he be moving as he leaves the water in order to jump to a height of 2.5 m? Ignore any effects due to air resistance.

94. •A 3-kg block is placed at the top of a track consisting of two frictionless quarter circles of radius R equal to 2 m connected by a 7-m-long, straight, horizontal surface (**Figure 6-40**). The coefficient of kinetic friction between the block and the horizontal surface is μ_k equal to 0.1. The block is released from rest. Find an expression that describes where the block stops, as measured from the left end of the horizontal surface, marked $x = 0$ on the figure.

Figure 6-40 Problem 94

95. •An object is released from rest on a frictionless ramp of angle θ_1 equal to 60°, at a (vertical) height H_1 equal to 12 m above the base of the ramp (**Figure 6-41**). The bottom end of the ramp *merges smoothly* with a second frictionless ramp that rises at angle θ_2 equal to

37°. (a) How far up the second ramp, as measured along the ramp, does the object slide before coming to a momentary stop? (b) When the object is on its way back down the second ramp, what is its speed at the moment that it is a (vertical) height H_2 equal to 7 m above the base of the ramp? SSM

Figure 6-41 Problem 95

96. •A child's game involves shooting a block of mass M toward a target by having it slide down a frictionless, straight ramp, and then up and off the end of a second ramp. The target is a horizontal distance d equal to 3.6 m from the end of the second ramp. When the block is released a height h_1 equal to 2.6 m above the base of the ramps, it falls short of the target by 0.8 m. How high h_2 above the base of the ramps should the block be released so that it just hits the target? Neglect any effects of air resistance.

97. •**Biology** An average froghopper insect has a mass of 12.3 mg and reaches a maximum height of 290 mm when its takeoff angle is 58° above the horizontal. (a) Find the takeoff speed of the froghopper. (b) How much of the energy stored in its legs just before the leap is used for the jump? Express your answer in microjoules (μJ) and in joules per kilogram of body mass (J/kg).

98. •A 20-g object is placed against the free end of a spring (k equal to 25 N/m) that is compressed 10 cm (**Figure 6-42**). Once released, the object slides 1.25 m across the tabletop and eventually lands 1.60 m from the edge of the table on the floor, as shown. Is there friction between the object and the tabletop? If there is, what is the coefficient of kinetic friction? The sliding distance on the tabletop includes the 10-cm compression of the spring and the tabletop is 1.00 m above the floor level.

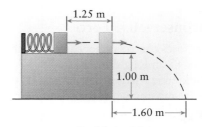

Figure 6-42 Problem 98

99. ••A 1.0-kg object is attached by a thread of negligible mass, which passes over a pulley of negligible mass, to a 2.0-kg object. The objects are positioned so that they are the same height from the floor and then released from rest. Find the speed of the objects when they are separated vertically by 1.0 m.

100. •You may have noticed runaway truck lanes while driving in the mountains. These gravel-filled lanes are designed to stop trucks that have lost their brakes on mountain grades. Typically such a lane is horizontal (if possible) and about 35 m long. We can think of the ground as exerting a frictional drag force on the truck. If the truck enters the gravel lane with a speed of 55 mph (24.6 m/s), use the work–kinetic energy theorem to find the minimum coefficient of kinetic friction between the truck and the lane to be able to stop the truck.

101. •In 2006, the United States produced 282×10^9 kilowatt-hours (kWh) of electrical energy from 4138 hydroelectric plants ($1.00 \text{ kWh} = 3.60 \times 10^6 \text{ J}$). On average, each plant is 90% efficient at converting mechanical energy to electrical energy, and the average dam height is 50.0 m. (a) At 282×10^9 kWh of electrical energy produced in one year, what is the average power output per hydroelectric plant? (b) What total mass of water flowed over the dams during 2006? (c) What was the average mass of water per dam and the average volume of water per dam that provided the mechanical energy to generate the electricity? (The density of water is 1000 kg/m^3.) (d) A gallon of gasoline contains $45.0 \times 10^6 \text{ J}$ of energy. How many gallons of gasoline did the 4138 dams save?

102. •**Astro** In 2006, NASA's Mars Odyssey orbiter detected violent gas eruptions on Mars, where the acceleration due to gravity is 3.7 m/s^2. The jets throw sand and dust about 75 m above the surface. (a) What is the speed of the material just as it leaves the surface? (b) Scientists estimate that the jets originate as high-pressure gas speeds through vents just underground at about 160 km/h. How much energy per kilogram of material is lost due to nonconservative forces as the high-speed matter forces its way to the surface and into the air?

103. •A 3.0-kg block is released up a ramp of angle θ equal to 37° at initial velocity v_0 equal to 20 m/s. Between the block and the ramp, the coefficient of kinetic friction is μ_k equal to 0.50 and the coefficient of static friction is μ_s equal to 0.80. How far up the ramp (in the direction along the ramp) does the block go before it comes to a stop?

104. •A small block of mass M is placed halfway up on the inside of a frictionless, circular loop of radius R (Figure 6-43). The size of the block is very small compared to the radius of the loop. Determine an expression for the minimum downward speed with which the block must be released in order to guarantee that it will make a full circle.

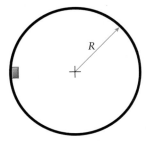

Figure 6-43 Problem 104

Challenge Problems

105. ••An object that weighs 25 N is placed on a horizontal, frictionless table in front of a spring that has a constant k (Figure 6-44). When the spring is compressed and released, the object shoots off the table and is caught on the end of a 3-m-long massless string. The object then moves in a circular path. The string doesn't break unless the spring is compressed more than 1 m. The maximum tension the string can sustain is 200 N. What is the value of the spring constant? (Ignore the fact that it would be odd for a string to be standing straight up as shown.)

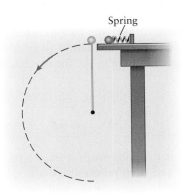

Figure 6-44 Problem 105

106. •••The potential energy of an electron in an atom is given approximately by $U(r) = a/r^6 - b/r^{12}$, where a and b are positive constants. Find the force $F(r)$ exerted on the electron. *Hint:* The relationship between force and potential, Equation 6-23, can be written $F(r) = -dU/dr$. (b) Find the units of the two constants, a and b.

107. •••A 1-kg object has a potential energy function given by $U(x) = 3(x - 1) - (x - 3)^3$. The graph in Figure 6-45 represents the object's potential energy versus position. The energies indicated on the vertical axis are evenly spaced. (a) Determine the numerical values of x_1 and x_3. (b) Describe the motion of the particle if the total energy is E_2. (c) What is the particle's speed at $x = x_1$ if its total energy, E, equals 58 J? (d) Sketch the graph of the particle's force as a function of x. Indicate x_1 and x_3 on your graph. (e) The particle is released from rest at $x = \frac{1}{2}x_1$. Find its speed as it passes through $x = x_1$.

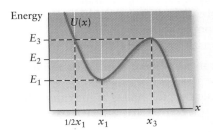

Figure 6-45 Problem 107

MCAT® Section

The section that follows includes material from previously administered MCAT® items and is reprinted with permission of the Association of American Medical Colleges (AAMC).

Passage 13 (81–85)

Tennis balls must pass a rebound test before they can be certified for tournament play. To qualify, balls dropped from a given height must rebound within a specified range of heights. Measuring rebound height can be difficult because the ball is at its maximum height for only a brief time. It is possible to perform a simpler indirect measurement to calculate the height of rebound by measuring how long it takes the ball to rebound and hit the floor again. The diagram below illustrates the experimental setup used to make the measurement.

The ball is dropped from a height of 2.0 m, and it hits the floor and then rebounds to a height h. A microphone detects the sound of the ball each time it hits the floor, and a timer connected to the microphone measures the time (t) between the two impacts. The height of the rebound is $h = gt^2/8$ where $g = 9.8$ m/s^2 is the acceleration due to gravity. Care must be taken so that the measured times do not contain systematic error. Both the speed of sound and the time of impact of the ball with the floor must be considered.

Four balls were tested using the method. The results are listed in the table below.

Ball	Time (s)
A	1.01
B	1.05
C	0.97
D	1.09

81. If NO air resistance is present, which of the following quantities remains constant while the ball is in the air between the first and second impacts?
   A. Kinetic energy of the ball
   B. Potential energy of the ball
   C. Momentum of the ball
   D. Horizontal speed of the ball

82. What measurement is made by the timer?
   A. Duration of the impacts
   B. Time that the ball is at maximum height

C. Time between the first and second bounces
D. Decrease in time of successive bounces

83. Balls C and D failed the test, while balls A and B passed. A ball that had which of the following measured times would definitely pass the test?
   A. 0.95 s
   B. 0.99 s
   C. 1.03 s
   D. 1.07 s

84. What percentage of its original potential energy did ball C lose between the start of the experiment and its rebound to maximum height?
   A. 24%
   B. 42%
   C. 51%
   D. 58%

85. With what approximate vertical speed does a ball in the experiment strike the floor?
   A. 4 m/s
   B. 6 m/s
   C. 10 m/s
   D. 20 m/s

Passage X (Questions 134–138)

The study of the flight of projectiles has many practical applications. The main forces acting on a projectile are air resistance and gravity. The path of a projectile is often approximated by ignoring the effects of air resistance. Gravity is then the only force acting on the projectile. When air resistance is included in the analysis, another force F_R is introduced. F_R is proportional to the square of the velocity, v. The direction of the air resistance is exactly opposite the direction of motion. The equation for air resistance is $F_R = pv^2$, where p is a proportionality constant that depends on such factors as the density of the air and the shape of the projectile.

Air resistance was studied by launching a 0.5-kg projectile from a level surface. The projectile was launched with a speed of 30 m/s at a 40° angle to the surface. (Note: Assume air resistance is present unless otherwise specified. Acceleration due to gravity is $g = 9.8$ m/s^2; sin 40° = 0.64; cos 40° = 0.77.)

134. If a 0.5-kg projectile is launched straight up and is given an initial kinetic energy of 259 J, to what maximum height will the projectile rise? (Note: Assume that the effects of air resistance are negligible.)
   A. 26 m
   B. 51 m
   C. 102 m
   D. 204 m

135. Which of the following statements is (are) true about the kinetic energy of the projectile when it returns to the elevation from which it was launched?

    I. It is the same as when it was launched.

    II. It is dependent on the initial velocity.

    III.It is dependent on the value of p.

    A. II only

    B. I and II only

    C. I and III only

    D. II and III only

136. What is the approximate total kinetic energy of the projectile at the highest point in its path? (Note: Assume that the effects of air resistance are negligible.)

    A. 92 J

    B. 133 J

    C. 184 J

    D. 225 J

137. Which of the following graphs best illustrates the relationship between the total speed of the projectile (v) and its horizontal distance from the launch point (x)? (Note: Assume that the effects of air resistance are negligible and that the left axis represents the location of the launch point.)

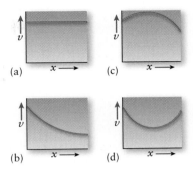

138. What is the magnitude of the horizontal component of air resistance on the projectile at any point during the flight? (Note: v_x = horizontal speed.)

    A. $(pv^2) \cos 40°$

    B. $pv^2/2$

    C. $(pv_x^2) \sin 40°$

    D. pv_x^2

7 Linear Momentum

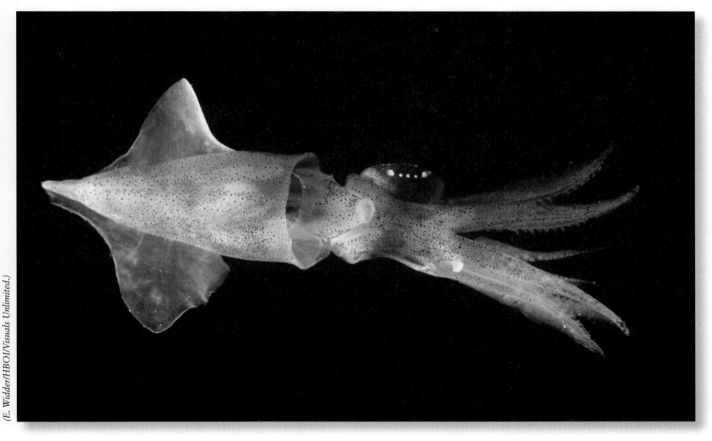

(E. Widder/HBOI/Visuals Unlimited.)

Cephalopods such as this squid propel themselves by filling a large muscular cavity with water and then forcefully shooting it out through the nozzle-shaped opening near the head. When the squid is not moving its *momentum*, a quantity which depends on mass and velocity, is zero. For a moving squid, we will find that the total momentum of the squid and the expelled water remains unchanged, so that when the water comes out in one direction, the squid moves in the opposite direction. Velocity and therefore momentum are vector quantities, so the momentum acquired by the squid must cancel the momentum of the water in order for the total momentum to remain unchanged.

You've probably seen someone experiencing what's happening to the person in the photo (**Figure 7-1**). As the skateboarder loses his balance and falls in one direction, the skateboard flies off in the opposite direction. This motion can be understood using the quantity *momentum*, which depends on the masses and velocities of the objects in a system. We have seen before that answering the question "What's the same?" often leads directly to the solution to a problem. Total momentum remains unchanged as many of the processes we study evolve. In this chapter we will see the powerful role that momentum plays in understanding the physics of moving objects.

Figure 7-1 As the skateboarder loses his balance and falls in one direction, the skateboard flies off in the opposite direction. The skateboarder should be wearing a helmet! *(Superstock.)*

233

7-1 Linear Momentum

The skateboarder in Figure 7-1 is probably most interested in the vertical motion of his fall, which, of course, is under the influence of gravity. In this section, however, we will focus on the horizontal motion. What are the forces acting in the horizontal plane? As the skateboarder begins to fall, his feet push against the skateboard and the skateboard pushes back against his feet according to Newton's third law. **Figure 7-2** shows the horizontal components of these two forces. Because the force of his feet on the skateboard and the force of the skateboard on his feet are a force pair, they are of equal magnitude. The net horizontal force within the system is therefore zero, so we need something more than forces to understand the horizontal motion. (There are some frictional forces involved, for example, between the wheels and the ground, but the net horizontal force is still zero.)

Can we use the concept of energy to understand the motion? The skateboard and the skateboarder both acquire kinetic energy. Based on our discussion of the conservation of energy in the previous chapter, you might wonder if the kinetic energy arises from a transfer of gravitational potential energy, say, from the skateboarder falling. This transfer can't be the whole explanation, however, because the same kind of horizontal motion—the skateboard going in one direction and the skateboarder in the other—would result if the skateboarder had slipped essentially sideways on a horizontal surface, in which case gravitational potential would not play a role. The motion is therefore not the result of a transformation of potential energy to kinetic energy. We need something more than energy to understand the horizontal motion.

The quantity momentum will allow us to look more deeply at motions such as those of the skateboard and skateboarder. Recall that in the previous chapter we defined work as the product of a force and the distance over which the force is applied. This definition led to the formulation of kinetic energy and ultimately to the powerful realization that the total energy of a system is conserved. In a similar way, we will find another useful quantity, momentum, by considering the product of a force and the time interval over which the force is applied. The application of a force imparts momentum to an object, but the final momentum also depends on the duration of the force. For this reason, we define **momentum** as force multiplied by the time over which it is applied.

When the applied force is constant, the momentum is simply the applied force multiplied by the total time interval. When the force varies as a function of time, however, we follow an approach similar to the one we took in studying work.

Figure 7-2 A free-body diagram shows the forces on the skateboarder and the skateboard. As the skateboarder falls but before he loses contact with the skateboard, his feet push against the skateboard and the skateboard pushes back against his feet. According to Newton's third law, the forces are a pair and therefore are equal in magnitude.

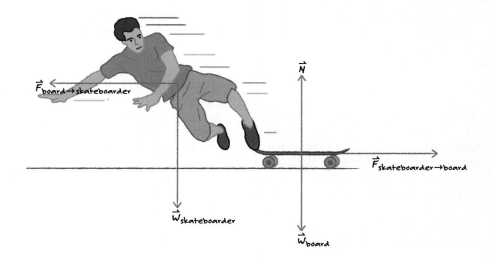

For a force $\vec{F}$ applied from an initial position x_i to a final position x_f, we defined the quantity work as

$$W = \int_{x_i}^{x_f} \vec{F} \cdot d\vec{x} \qquad (6\text{-}17)$$

See the Math Tutorial for more information on Integrals

in which an infinite number of small contributions of the product of (a varying) force and distance are added together. In an analogous way, we will define the change in momentum $\Delta\vec{p}$ to be

$$\Delta\vec{p} = \int_{t_i}^{t_f} \vec{F}\, dt \qquad (7\text{-}1)$$

Note that because $\vec{F}$ is a vector but dt is not, there is no dot product in $\int_{t_i}^{t_f} \vec{F}\, dt$, and the result of this integral is therefore a vector quantity. The integral can be interpreted as the sum of an infinite number of contributions to the change in momentum of an object, where each contribution is the force applied to the object multiplied by the infinitesimally small time interval over which the force is applied. The bounds on the integral indicate that the force $\vec{F}$ is applied from an initial time t_i to a final time t_f.

By using Newton's second law, the net force on an object in a given direction results in an acceleration in that direction:

$$\vec{F} = m\vec{a} = m\frac{d\vec{v}}{dt}$$

So the integral becomes

$$\int_{t_i}^{t_f} \vec{F}\, dt = \int_{t_i}^{t_f} m\frac{d\vec{v}}{dt}\, dt \qquad (7\text{-}2)$$

Any derivative can be treated as a fraction, so the two dt terms in the integral on the right-hand side of Equation 7-2 cancel. The result is

$$\int_{t_i}^{t_f} \vec{F}\, dt = \int_{v_i}^{v_f} m\, d\vec{v}$$

Note that because we changed the differential on the right-hand side of the equation from dt to $d\vec{v}$ we had to change the bounds on the integral to match. So

$$\int_{t_i}^{t_f} \vec{F}\, dt = m \int_{v_i}^{v_f} d\vec{v} = m\vec{v}\Big|_{\vec{v}_i}^{\vec{v}_f} = m\vec{v}_f - m\vec{v}_i \qquad (7\text{-}3)$$

The integral $\int_{t_i}^{t_f} \vec{F} dt$ is evidently equivalent to the difference between the final and initial values of the quantity formed by the product of mass and velocity. We have already uncovered two fundamental quantities that involve mass and motion—force, which depends on mass and acceleration, and kinetic energy, which depends on mass and the square of speed. The product of mass and velocity, momentum, is also of fundamental importance to the study of physical systems. The variable p denotes momentum:

$$\vec{p} = m\vec{v} \qquad (7\text{-}4)$$

The quantity described in Equation 7-4 is referred to as **linear momentum** to distinguish it from *angular momentum*, the momentum of an object rotating around a fixed point. It is common, however, for physicists to refer to linear momentum as simply "momentum," and we will follow that convention. In addition, as we already noted, momentum is a vector quantity. From Equation 7-4 you can see that the momenta of two objects traveling in opposite directions will have opposite signs.

The units of momentum are the units of mass multiplied by the units of velocity:

$$[p] = [m][v] = \text{kg}\frac{\text{m}}{\text{s}}$$

There is no SI unit that is equivalent to 1 $(\text{kg} \cdot \text{m})/\text{s}$.

Example 7-1 Momentum of Two Objects

Compare the magnitude of the momentum of a 0.15-kg baseball thrown at 42 m/s with the magnitude of the momentum of a 0.0042-kg bullet fired at a speed of 900 m/s.

SET UP

We evaluate the magnitude of the momentum of an object using Equation 7-4, that is, $p = mv$.

SOLVE

For the baseball:

$$p_{\text{ball}} = (0.15 \text{ kg})(42 \text{ m/s}) = 6.3 \text{ (kg} \cdot \text{m)/s}$$

For the bullet:

$$p_{\text{bullet}} = (4.2 \times 10^{-3} \text{ kg})(900 \text{ m/s}) = 3.8 \text{ (kg} \cdot \text{m)/s}$$

REFLECT

Both mass and speed contribute to the magnitude of the momentum of an object. Here, the baseball, which we've given a speed typical of a major league fastball, has a greater momentum than the bullet. Even though the bullet's speed is much greater than that of the baseball, the ball's larger mass more than compensates.

Practice Problem 7-1 An unmanned X-43 aircraft has a mass of about 1300 kg and has been flown at a speed of nearly 4000 m/s. Compare the magnitude of the momentum of the X-43 during this flight with that of a peregrine falcon (the fastest bird) which has a mass of 0.75 kg and flies at 80 m/s.

Example 7-2 California on the Move

Earth's crust is broken into a number of tectonic plates. Those of us who live in California know that a region of the west coast of our state is part of the Pacific Plate. On average, the Pacific Plate moves about 2 cm per year relative to the North American Plate. The mass of the Pacific Plate has been estimated to be about 5×10^{21} kg. How fast would *Genesis*, the world's most massive cruise ship at 2×10^8 kg, need to travel relative to the North American Plate to have the same momentum as the Pacific Plate?

SET UP

We set the magnitude of the momentum of the Pacific Plate equal to that of the ship:

$$p_{\text{ship}} = p_{\text{plate}}$$

or

$$m_{\text{ship}}v_{\text{ship}} = m_{\text{plate}}v_{\text{plate}}$$

SOLVE

The speed of the ship is then

$$v_{\text{ship}} = \frac{m_{\text{plate}}v_{\text{plate}}}{m_{\text{ship}}}$$

Before we substitute values into the equation above, we need to convert the speed of the plate into SI units:

$$v_{\text{plate}} = \left(\frac{2\,\text{cm}}{1\,\text{y}}\right)\left(\frac{1\,\text{m}}{100\,\text{cm}}\right)\left(\frac{1\,\text{y}}{365\,\text{days}}\right)\left(\frac{1\,\text{day}}{24\,\text{h}}\right)\left(\frac{1\,\text{h}}{3600\,\text{s}}\right) = 6 \times 10^{-10}\,\text{m/s}$$

Using the values given in the problem statement, the speed of the ship is

$$v_{\text{ship}} = \frac{(5 \times 10^{21}\,\text{kg})\,(6 \times 10^{-10}\,\text{m/s})}{2 \times 10^{8}\,\text{kg}} = 1.5 \times 10^{4}\,\text{m/s}$$

REFLECT

The greatest speed ever achieved by humans is about $11\,\text{km/s}$ ($1.1 \times 10^{4}\,\text{m/s}$), the speed necessary to send a rocket to the Moon or Mars. So the cruise ship would have to travel, well, really fast to carry the same momentum as the Pacific Plate.

Practice Problem 7-2 A modest-sized cruise ship weighs about $1.1 \times 10^{8}\,\text{N}$ fully loaded and travels at speeds of more than $10\,\text{m/s}$. How fast must a 747 aircraft, weighing about $4.4 \times 10^{6}\,\text{N}$ fully loaded, travel to have the same momentum as this cruise ship moving at $10\,\text{m/s}$?

✳ What's Important 7-1

Momentum is defined as force multiplied by the time over which it is applied. The application of a force imparts momentum to an object, but the final momentum also depends on the duration of the force.

7-2 Conservation of Momentum

"What's the same?" We've already seen that answering that question can be a powerful tool for understanding physical processes, for example, in analyzing two simultaneous motions (in Chapter 2) and in considering the conservation of energy (in Chapter 6). Momentum is also conserved in many commonly occurring circumstances, giving us yet another tool to help us study and understand physical systems.

Equation 7-3 leads to the definition of momentum as mass multiplied by velocity, so we can now write it more directly in terms of the change in momentum:

$$\int_{t_i}^{t_f} \vec{F}\, dt = m\vec{v}_f - m\vec{v}_i = \vec{p}_f - \vec{p}_i \tag{7-5}$$

where the initial and final times are identified by the subscripts i and f, respectively. The net force $\vec{F}$ is zero in many natural phenomena; in such cases, the integral $\int_{t_i}^{t_f} \vec{F}\, dt$ must also be zero, which means that

$$\vec{p}_f - \vec{p}_i = 0$$

or

$$\vec{p}_i = \vec{p}_f \tag{7-6}$$

The total momentum of an object does not change when the net force acting on it is zero. In the language of physics, *when the net force on an object is zero, total linear momentum is conserved*. This is a statement of the **conservation of momentum**.

How can we apply the concept of momentum to a system that consists of more than one object? Certainly the net force is zero when a system experiences no net external force because all internal forces must come in pairs and therefore cancel. When two billiard balls collide, for example, it is likely that the normal force and the force due to gravity on each cancel. The only other forces are internal forces, those that each ball exerts on the other. These two internal forces form a force pair, so according to Newton's third law they must be equal in magnitude and opposite in direction. The internal forces therefore cancel, so the net force on the system is zero. The total momentum of the system is conserved. For the rest of this chapter we will therefore usually confine ourselves to situations in which the net force is zero. We will therefore apply the statement of momentum conservation without repeating the assertion that net force is zero.

Equation 7-6 is a vector statement, so the direction of motion of each object in a system must be considered when determining the total momentum. Unlike the kinetic energy of an object, which does not depend on the direction of the object's velocity, the momentum of an object moving in the positive x direction and the momentum of an object moving in the negative x direction have opposite signs. The total kinetic energy of the system of the two objects moving in opposite directions would be positive, but it is possible that the two momenta of the objects could even cancel completely when added together.

Example 7-3 Pushing Off

Figure skaters Jim and Sarah face each other while standing stationary on the ice and push against each other's hands (Figure 7-3). Jim finds himself moving backward at 2.0 m/s. Jim's mass is 50% more than Sarah's. How fast is Sarah moving, and in which direction? Frictional forces can be ignored at the moment immediately after Jim and Sarah are no longer touching.

SET UP

The momentum of the system comprising Jim and Sarah is conserved because the net force on the two together is zero. Jim pushes on Sarah and Sarah pushes on Jim; according to Newton's third law, these two internal forces are equal and opposite, and therefore cancel. The forces acting vertically—the normal force and the force due to gravity—also cancel. In addition, we can consider the system of the two skaters to be isolated from other external forces. (The system would not be isolated if, say, a hockey player approached the two skaters and pushed on them.) We can therefore set the initial momentum of the system equal to the momentum of the system at any later time.

SOLVE

The initial momentum of the system is the sum of Jim's initial momentum and Sarah's initial momentum:

$$\vec{p}_i = \vec{p}_{i,J} + \vec{p}_{i,S}$$

Momentum depends on velocity (Equation 7-4), so because both initial velocities are zero, the initial momentum is also zero. Equation 7-6 requires that the final momentum also be zero:

$$\vec{p}_f = m_J \vec{v}_{f,J} + m_S \vec{v}_{f,S} = 0$$

or

$$m_S \vec{v}_{f,S} = -m_J \vec{v}_{f,J}$$

The magnitude of Sarah's velocity after the two push off is then

$$v_{f,S} = -\frac{m_J}{m_S} v_{f,J}$$

Figure 7-3 Ice skaters Jim and Sarah, initially stationary, push against each other's hands.

Using $m_J = 1.5m_S$ and $v_{f,J} = 2.0$ m/s,

$$v_{f,S} = -\frac{1.5m_S}{m_S}(2.0 \text{ m/s}) = -3.0 \text{ m/s}$$

REFLECT
The sign of Sarah's final velocity is negative; that is, the two skaters' velocities are in opposite directions. Also, because Sarah is less massive than Jim, her velocity is larger than his velocity after they push against each other. Neither of these conclusions is surprising.

Practice Problem 7-3 Reconsider the preceding problem, this time with Sarah (who is very strong) holding a rock such that the mass of Sarah and the rock together is four times Jim's mass. If Jim again finds himself moving backward at 2.0 m/s, how fast is Sarah moving? Frictional forces can be ignored at the moment immediately after Jim and Sarah are no longer touching.

Equation 7-6, a general statement of momentum conservation, tells us that the momentum of a system at any two arbitrary times is the same. A system, of course, can contain any number of separate objects. Say a system consists of three particles that have masses m_1, m_2, and m_3. Regardless of the way the particles interact, the total momentum of the system will remain the same even if the velocity of any individual particle does not. For these three particles, Equation 7-6 becomes

$$m_1\vec{v}_{1,i} + m_2\vec{v}_{2,i} + m_3\vec{v}_{3,i} = m_1\vec{v}_{1,f} + m_2\vec{v}_{2,f} + m_3\vec{v}_{3,f} \qquad (7\text{-}7)$$

The same approach applies when the number of objects in a system changes, for example, when an object breaks apart or when two separate objects end up sticking together. If an object of mass M breaks up into two smaller objects of masses m_1 and m_2, Equation 7-6 becomes

$$M\vec{v}_i = m_1\vec{v}_{1,f} + m_2\vec{v}_{2,f} \qquad (7\text{-}8)$$

where $\vec{v}_i$, $\vec{v}_{1,f}$, and $\vec{v}_{2,f}$ are the velocity of the original object and the velocities of the two fragments, respectively. The total mass remains the same.

As in both of these last examples, the sum of the momenta of *all* objects in a system remains unchanged. *Conversely, the momentum of any particular object within a larger system will not, in general, be conserved.* When two billiard balls collide, the sum of their momenta remains unchanged, but the momentum of either ball is not likely to be the same before and after the collision. What if one of the balls breaks apart during the collision? The total momentum of all of the fragments will also not likely equal the momentum of that ball before it breaks apart. Only the total momentum is a conserved quantity, so the vector sum of the momenta of the ball that didn't break and all of the fragments of one that did will equal the sum of momenta of the two balls before the collision.

Momentum is a vector quantity, which requires that momentum be conserved in any direction along which a system is analyzed. For example, if we establish an x–y coordinate axis for a system of three objects, then the total of both the x and the y components of momentum must each be conserved separately:

$$m_1\vec{v}_{1x,i} + m_2\vec{v}_{2x,i} + m_3\vec{v}_{3x,i} = m_1\vec{v}_{1x,f} + m_2\vec{v}_{2x,f} + m_3\vec{v}_{3x,f}$$

$$m_1\vec{v}_{1y,i} + m_2\vec{v}_{2y,i} + m_3\vec{v}_{3y,i} = m_1\vec{v}_{1y,f} + m_2\vec{v}_{2y,f} + m_3\vec{v}_{3y,f}$$

When writing a statement of momentum conservation *in terms of the magnitudes of momenta*, the sign of each term reflects the direction an object moves. For example, if object 3 is moving in the negative x direction at the beginning of the

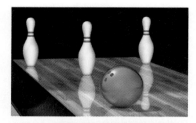

Figure 7-4 In this sequence, a bowling ball approaches a pin and then strikes it. In the third image, the base of the pin is moving rapidly to the right while the ball has been slightly deflected to the left.

problem and object 2 is moving in the negative x direction at the end, we would write

$$m_1 \vec{v}_{1x,i} + m_2 \vec{v}_{2x,i} - m_3 \vec{v}_{3x,i} = m_1 \vec{v}_{1x,f} - m_2 \vec{v}_{2x,f} + m_3 \vec{v}_{3x,f}$$

as the statement of conservation of momentum in the x direction.

Example 7-4 Bowling

The sequence of images in **Figure 7-4** shows a bowling ball that has a mass M_{ball} and an initial velocity $\vec{v}_{ball,i}$ striking a stationary pin that has a mass M_{pin}. The collision occurs in the second image, and you can see that the ball strikes a bit to the left of the horizontal center of the pin. The third image shows what happens after the collision, when the ball is moving on a path to the left of its original direction and the pin is shooting off to the right. This motion is sketched in **Figure 7-5a**, and the velocity vectors of the ball and pin are shown in Figure 7-5b. (a) Write the equations which describe conservation of momentum in the collision. (b) The ball moves away from the collision at an angle θ equal to 25°, and the pin at angle ϕ equal to 65°.

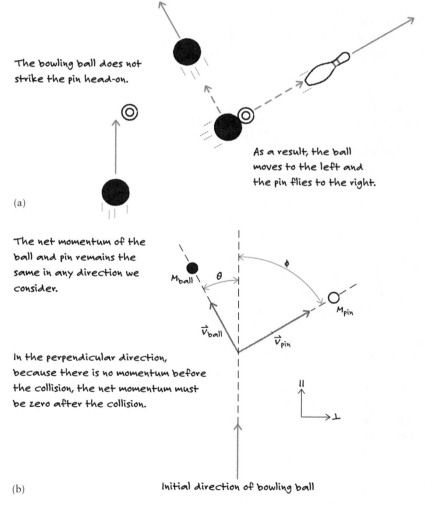

Figure 7-5 (a) This sketch shows the motion of the ball and pin as a result of the collision. (b) Here we plot the velocity vectors and angles associated with the collision of the bowling ball and pin.

The ball is 3.5 times more massive than the pin. How much faster than the ball is the pin moving after the two collide? Consider frictional forces between the pins and the floor, and between pins, to be negligible.

SET UP

The momentum of the system is conserved because the system experiences no net force, so

$$M_{ball}\vec{v}_{ball,i} + M_{pin}\vec{v}_{pin,i} = M_{ball}\vec{v}_{ball,f} + M_{pin}\vec{v}_{pin,f}$$

(The internal, pin-on-pin forces all cancel, the normal force and the force due to gravity on each pin cancel, and we have reasonably decided to neglect frictional forces.) The initial velocity of the pin is zero, so we can drop the second term

$$M_{ball}\vec{v}_{ball,i} = M_{ball}\vec{v}_{ball,f} + M_{pin}\vec{v}_{pin,f} \tag{7-9}$$

In addition, because momentum is a vector, the component of momentum in any direction we choose must also be conserved. The ball and pin move in two dimensions after the collision, so we need to write statements of conservation of momentum for each of the two directions. We are free to choose any convenient coordinate axes, and we have selected the axes parallel and perpendicular to the bowling ball's original direction as shown in Figure 7-5b. This choice is good because the net momentum in the direction perpendicular to the ball's original direction must be zero before the collision: The ball only has velocity in the direction it's moving (of course!) and the pin is stationary. It therefore makes the problem easier to describe mathematically if we choose coordinate axes parallel and perpendicular to the ball's initial motion.

SOLVE

(a) We want to write Equation 7-9 separately in the perpendicular and parallel directions. In the parallel direction, the speed of the ball before the collision is v_0. So,

$$M_{ball}v_0 = M_{ball}v_{ball}\cos\theta + M_{pin}v_{pin}\cos\phi$$

where we have used the notation shown in the figure that the final velocities of the ball and pin are v_{ball} and v_{pin}, respectively. In the perpendicular direction, the speed of the ball before the collision is zero, so

$$0 = -M_{ball}v_{ball}\sin\theta + M_{pin}v_{pin}\sin\phi \tag{7-10}$$

The first term on the right is negative because the ball is moving in the negative direction as defined by the coordinate axes we chose.

(b) Our goal is to find the speed of the pin after the collision relative to the speed of the ball. We therefore rearrange Equation 7-10 for v_{pin} in terms of v_{ball}:

$$v_{pin} = \frac{M_{ball}}{M_{pin}}\frac{\sin\theta}{\sin\phi}v_{ball} \tag{7-11}$$

The ratio M_{ball}/M_{pin} is given in the problem statement as 3.5 and the angles θ and ϕ are also known, so we can obtain a numeric solution:

$$v_{pin} = 3.5\frac{\sin 25°}{\sin 65°}v_{ball} = 1.6v_{ball}$$

After the collision, the speed of the pin is 1.6 times that of the ball.

REFLECT

Momentum is a vector, so the net momentum of the ball and pin together must not only remain unchanged, but the sum of the components of total momentum must be conserved in any direction we choose. Because the ball only has velocity in the direction it's moving and the pin is stationary, the problem becomes easier to

 See the Math Tutorial for more information on Trigonometry

describe mathematically when we choose coordinate axes parallel and perpendicular to the ball's initial motion. This choice leads directly to the statement of conservation of momentum in the direction perpendicular to the initial motion (Equation 7-10) which in turn enables us to find the pin's velocity in terms of the velocity of the ball.

Practice Problem 7-4 A billiard ball traveling at 1.7 m/s strikes two others that are initially stationary. The first one stops and the other two leave the collision each at 45° from the direction that the first ball was moving. The three balls are identical. What is the speed of the two moving billiard balls immediately after the collision?

Consider a stationary object that breaks up into two smaller objects that have masses m_1 and m_2. Because the initial momentum is zero, conservation of momentum requires that the final momentum is also zero:

$$0 = m_1\vec{v}_1 + m_2\vec{v}_2 \tag{7-12}$$

where $\vec{v}_1$ and $\vec{v}_2$ are the velocities of the two fragments, or

$$m_2\vec{v}_2 = -m_1\vec{v}_1$$

In terms of the speeds:

$$m_2 v_2 = -m_1 v_1$$

This result makes it clear that the two fragments must move away in opposite directions in order for momentum to be conserved.

? Got the Concept 7-1
Breaking Up Is Hard to Do

An atomic nucleus disintegrates into a number of fragments as it moves through an experimental apparatus. You observe a large fragment continuing to move in the original direction and a small fragment moving in a perpendicular direction (Figure 7-6). Is it possible that the nucleus only disintegrated into those two fragments? If not, is it ever possible for the nucleus to disintegrate into two fragments, with one moving in the direction perpendicular to the initial one?

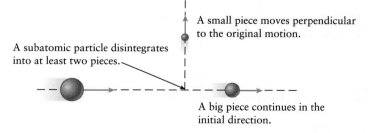

A small piece moves perpendicular to the original motion.

A subatomic particle disintegrates into at least two pieces.

A big piece continues in the initial direction.

Figure 7-6 An atomic nucleus disintegrates into a number of fragments as it moves through an experimental apparatus.

Example 7-5 Superheroes Conserve Momentum

A movie superhero stops an enormous floating platform from crashing into Earth. He comes to rest supporting the platform above his head with his feet on the roof of a house and with a mighty push, he flings the platform back up at a speed of about 1 km/s. If the mass of the platform is about 10^{10} kg, at what speed will the superhero and Earth recoil (move backward, away from the push against the floating platform)? Ignore the likelihood that in pushing against the platform the superhero will crush the house. The mass of Earth is 5.98×10^{24} kg.

SET UP
At the moment that the superhero holds the platform above his head, both are at rest with respect to Earth. The net linear momentum of all three (Earth, the superhero, and the platform) is therefore zero, and must remain zero in order for momentum to be conserved. This situation is equivalent to the breakup of a stationary object into two pieces discussed above, so the mathematical statement of momentum conservation is similar to Equation 7-12:

$$0 = m_{(E+SH)}\vec{v}_{(E+SH)} + m_P\vec{v}_P \qquad (7\text{-}13)$$

where E, SH, and P represent Earth, the superhero, and the platform, respectively.

SOLVE
We know that the two velocities must be in opposite directions. By using Equation 7-13, $\vec{v}_{(E+SH)}$ and $\vec{v}_P$ have opposite signs:

$$\vec{v}_{(E+SH)} = -\frac{m_P}{m_{(E+SH)}}\vec{v}_P$$

Because the mass of the superhero must be miniscule compared to the mass of Earth, the mass of the two together is approximately that of Earth alone ($m_{(E+SH)} \approx m_E$). Thus,

$$\vec{v}_{(E+SH)} \approx -\frac{m_P}{m_E}\vec{v}_P$$

The magnitude of the velocity of Earth and superhero together is then

$$v_{(E+SH)} = \frac{1 \times 10^{10}\,\text{kg}}{5.98 \times 10^{24}\,\text{kg}} 1 \times 10^3\,\text{m/s} = 2 \times 10^{-12}\,\text{m/s}$$

REFLECT
We used a positive value for the velocity of the platform; the resulting negative value for the vector $\vec{v}_{(E+SH)}$ indicates that Earth and the superhero move in the opposite direction of the platform. Also, while the speed we found is quite small, it's not zero.

Practice Problem 7-5 A 75-kg astronaut floating next to a 350-kg satellite pushes against it so that the satellite acquires a speed of 1.2 m/s. At what speed does the astronaut move as a result?

✱ What's Important 7-2
When the net force on an object is zero, total linear momentum is conserved; in other words, the sum of the momenta of all objects in a system remains unchanged. However, the momentum of any particular object within a larger system may not be conserved.

Figure 7-7 Linear momentum is conserved in the collision of two racing cars if the cars don't spin after the collision. *(AP Photo/Dale Davis.)*

Figure 7-8 Momentum is conserved at the instant that the boy jumps on his skim board. *(Bill Stevenson/Photolibrary.)*

7-3 Inelastic Collisions

Momentum is conserved when two or more objects collide. For example, the collision of two racing cars, as in **Figure 7-7**, conserves momentum. The physics of collisions and conservation of momentum also underlies the motions of bowling pins (Example 7-4) and a boy about to jump on a skim board (**Figure 7-8**). In each example, momentum is conserved at the instant of the collision. We begin by considering the special case in which two objects moving along a common line collide and either briefly or permanently stick together. Such collisions are termed **inelastic**; in a *completely inelastic collision* the objects don't separate afterward. A collision between two cars, in which the cars deform and then bounce apart, is inelastic. The collision would be completely inelastic should the bumpers lock together so that the cars don't separate.

Imagine that an object with a mass m_1 and an initial velocity $\vec{v}_1$ strikes and sticks to an object with a mass m_2, as shown in **Figure 7-9**. For simplicity, we let the second object be at rest before the collision occurs. How fast does the combined object move after the collision?

Conservation of momentum requires that the net momentum of the system remain unchanged from any time before the collision to any time after the collision. We therefore require that

$$m_1\vec{v}_1 + m_2\vec{v}_2 = (m_1 + m_2)\vec{V}$$

where we indicate that the two objects stick together by writing the mass of the final state as $m_1 + m_2$. The combined object leaves the collision with velocity $\vec{V}$. The net linear momentum of the system before the collision is in the same direction as $\vec{v}_1$, so for momentum to be conserved, the net momentum after the collision must also be aligned with $\vec{v}_1$. Because m_2 is initially at rest, we substitute $v_2 = 0$ into the equation:

$$m_1\vec{v}_1 = (m_1 + m_2)\vec{V}$$

or

$$\vec{V} = \frac{m_1}{m_1 + m_2}\vec{v}_1 \tag{7-14}$$

The final velocity is the initial velocity multiplied by a fraction which depends on the two masses. That fraction is never greater than one—of course—and approaches zero when the mass of the moving object (m_1) is negligibly small compared to the stationary one (m_2). In general, Equation 7-14 enables us to determine the magnitude of the velocity of the combined object in all cases and also confirms that the direction of the combined object is the same as the direction of the original motion.

Is mechanical energy conserved when objects collide and stick together? And in particular, because we can examine collisions on a horizontal surface so that no change in gravitational potential energy is possible, we can also ask whether kinetic energy is conserved. Before the collision, the moving object (m_1) carries the kinetic energy of the system:

$$K_i = \frac{1}{2}m_1 v_1^2$$

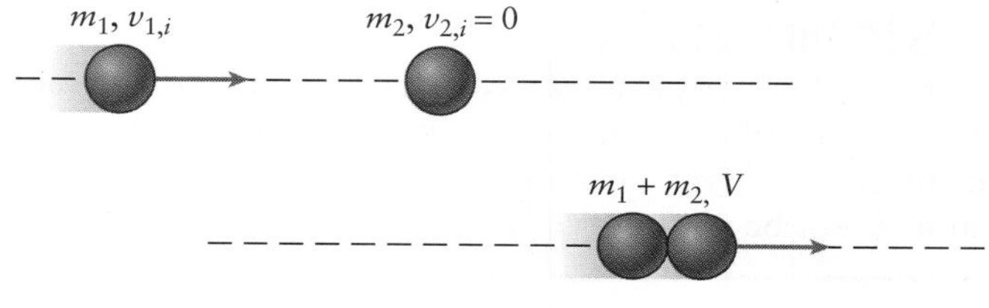

Figure 7-9 In a completely inelastic collision an object strikes and sticks to a second object. In this particular case we have let the second object be initially stationary.

$m_1, v_{1,i}$ $m_2, v_{2,i} = 0$

$m_1 + m_2, V$

After the collision, the kinetic energy of the system is determined by the combined mass of both objects and their final speed:

$$K_f = \frac{1}{2}(m_1 + m_2)V^2$$

Using Equation 7-14, we know the final velocity $\vec{V}$ in terms of the initial velocity of the moving object and the masses of both objects. The final speed is the magnitude of the final velocity, so

$$K_f = \frac{1}{2}(m_1 + m_2)\left(\frac{m_1}{m_1 + m_2}v_1\right)^2$$

or

$$K_f = \frac{1}{2}\frac{m_1^2}{m_1 + m_2}v_1^2 \tag{7-15}$$

To see the relationship between the final kinetic energy and the initial kinetic energy more clearly, we can rewrite Equation 7-15 as

$$K_f = \left(\frac{m_1}{m_1 + m_2}\right)\frac{1}{2}m_1 v_1^2$$

So,

$$K_f = \left(\frac{m_1}{m_1 + m_2}\right)K_i \tag{7-16}$$

The fraction in parentheses in Equation 7-16 must be smaller than 1, because even the very smallest value for m_2 will cause the denominator to be larger than the numerator. The final kinetic energy is therefore less than the initial kinetic energy. The difference between the two energies is not lost; it is transformed into a nonmechanical form such as thermal energy. We now see that a change in kinetic energy characterizes inelastic collisions; the final kinetic energy is less than the initial kinetic energy.

When the mass of the moving object (m_1) is negligibly small compared to that of the stationary one (m_2), the final kinetic energy of the system predicted by Equation 7-16 is essentially zero. In this extreme case of an inelastic collision, effectively all of the initial kinetic energy is transformed into nonmechanical forms. For example, a lump of clay dropped to the ground "sticks" to the surface, but Earth's recoil is negligibly small so that we can justifiably say that there is zero kinetic energy after the collision.

? Got the Concept 7-2
Splitting Logs

As a young man, one of the authors used a sledge hammer and a wedge to split logs for the fireplace, similar to the process shown in **Figure 7-10**. How would you explain why the metal wedge got hot as it was repeatedly struck by the sledge hammer?

Example 7-6 Car Crash

A car that has a mass equal to 1500 kg and is traveling at 4.5 m/s strikes a stationary car that has a mass equal to 1000 kg. During the collision, the bumpers of the two cars lock together (**Figure 7-11**). What is the speed of the two connected cars

Figure 7-10 Why does the metal wedge get hot as it is repeatedly struck by the sledge hammer in order to split the log? *(Jim Holden/Alamy.)*

Figure 7-11 The bumpers of two cars lock together during a collision.

$v_i = 4.5$ m/s

V

√x **Go to Interactive Exercise 7-1**
*for more practice dealing with
inelastic collisions*

as they leave the collision? What fraction of the initial kinetic energy is converted to nonmechanical forms of energy?

SET UP

Conservation of total linear momentum allows us a direct path to finding the final velocity of the two cars. The total momentum of the system before the collision is due only to the first car, so we write

$$m_1 \vec{v}_1 = (m_1 + m_2)\vec{V} \tag{7-17}$$

where m_1 and m_2 are the masses of the two cars and $\vec{V}$ is the velocity of the two cars stuck together.

Although momentum is conserved, this collision is clearly inelastic so kinetic energy is not conserved. To determine how much energy is converted to heat and other nonmechanical forms, we need to compare the kinetic energies of the system before and after the collision.

SOLVE

Solving Equation 7-17 for the magnitude of the final velocity gives

$$V = \frac{m_1}{m_1 + m_2} v_1 \tag{7-18}$$

Using the values given in the problem statement, we calculate the final speed to be

$$V = \frac{1500 \text{ kg}}{1500 \text{ kg} + 1000 \text{ kg}} 4.5 \text{ m/s} = 2.7 \text{ m/s}$$

The difference between the initial and final kinetic energies $(K_i - K_f)$ is the amount of kinetic energy converted to nonmechanical forms. This difference can be expressed as a fraction of the initial kinetic energy by dividing by K_i:

$$\text{fractional } \Delta K = \frac{K_i - K_f}{K_i} = 1 - \frac{K_f}{K_i}$$

In terms of the masses and speeds, the fractional difference is then

$$\text{fractional } \Delta K = 1 - \frac{\frac{1}{2}(m_1 + m_2)V^2}{\frac{1}{2}m_1 v_1^2} \tag{7-19}$$

The fraction of the initial kinetic energy converted to nonmechanical forms is

$$\text{fractional } \Delta K = 1 - \frac{\frac{1}{2}(1500 \text{ kg} + 1000 \text{ kg})(2.7 \text{ m/s})^2}{\frac{1}{2}(1500 \text{ kg})(4.5 \text{ m/s})^2} = 0.40$$

That is, 40% of the initial kinetic energy is transformed into nonmechanical forms.

REFLECT

In this particular example, 40% of the initial energy is no longer in the form of kinetic energy after the collision. Some of this energy deforms the cars and the rest becomes heat.

A physicist would probably not have substituted values into Equation 7-19 but instead would have first simplified the equation algebraically. Waiting to substitute values often leads to a deeper understanding, but it also can help prevent computation errors by involving fewer mathematical steps. Using V from Equation 7-18, Equation 7-19 becomes

$$\text{fractional } \Delta K = 1 - \frac{\frac{1}{2}(m_1 + m_2)\left(\frac{m_1}{m_1 + m_2}v_1\right)^2}{\frac{1}{2}m_1 v_1^2}$$

This equation might look complicated, but multiple cancellations immediately lead to an equation that is much simpler:

$$\text{fractional } \Delta K = 1 - \frac{m_1}{m_1 + m_2} = \frac{m_2}{m_1 + m_2} \tag{7-20}$$

Finding a numerical result using Equation 7-20 involves fewer numbers than using Equation 7-19, so we're less likely to make a mistake. Equation 7-20 also shows us something fundamental about the physics: the loss of kinetic energy does not depend on the initial speed but only on the masses of the objects involved in the collision. We see, for example, that the initial and final kinetic energies are essentially the same when the moving object is far more massive than the stationary one, and that the final kinetic energy of the system approaches zero as the mass of the moving object gets increasingly small compared to the mass of the object initially at rest.

Practice Problem 7-6 A lump of red clay collides with and sticks to a lump of brown clay that is initially stationary. After the collision, the kinetic energy of the combined lump is half of the kinetic energy of the initially moving one. What is the mass of the lump of brown clay in terms of the mass of the lump of red clay?

To gain insight into the physics of inelastic collisions, let's consider the completely inelastic collision shown in Figure 7-9 for three special cases: (1) the two objects have the same mass, (2) the moving object is far more massive than the stationary one, and (3) the moving object is of negligible mass compared to the stationary one.

Special Case: The Objects Have the Same Mass

When the masses of the objects are the same (that is, when $m_1 = m_2 = m$), the magnitude of the final velocity (from Equation 7-14) becomes

$$V = \frac{m_1}{m_1 + m_2}v_1 = \frac{m}{2m}v_1 = \frac{1}{2}v_1$$

When a moving object collides with and sticks to a stationary one of the same mass, the combined system moves away from the collision at half the initial speed.

Special Case: The Moving Object Is More Massive than the One at Rest

Whenever one quantity in an expression is significantly larger than another, it is acceptable to ignore the smaller quantity if the two quantities are added or subtracted.

So when the mass of the moving object is much larger than the one at rest (that is, when $m_1 \gg m_2$), we can neglect m_2 in the denominator where it is added to m_1:

$$V = \frac{m_1}{m_1 + m_2} v_1 \approx \frac{m_1}{m_1} v_1 \approx v_1$$

In this case, the combined object moves away from the collision at the same speed with which the first object approached. This conclusion makes sense—the stationary object is so small that the moving one is essentially unaffected by the collision. You can imagine that when an elephant runs into a hovering fly, the elephant would slow down only imperceptibly. (If you're worried about the fact that the elephant is pushing itself forward, imagine it sliding on ice skates!)

Special Case: The Moving Object Is Less Massive than the One at Rest

Finally, consider the collision when the mass of the moving object is much smaller than the one at rest (that is, when $m_1 \ll m_2$). Once again, we can neglect the smaller mass when the two masses are added, so

$$V = \frac{m_1}{m_1 + m_2} v_1 \approx \frac{m_1}{m_2} v_1 \approx (0) v_1 \approx 0 \qquad (7\text{-}21)$$

Because the denominator m_2 is so large compared to the numerator m_1, the fraction is nearly zero. So in this case, the final speed is negligibly small. Think of the collision of a small blob of clay and something large, say, Earth. You can create such a collision by dropping a blob of clay to the ground. The clay sticks to the surface, but because the mass of the clay is so small compared to the mass of Earth, we observe no net motion of the final state of the system.

 Watch Out

Momentum is always conserved during inelastic collisions, even when the collision is completely inelastic.

We just considered the special case of a completely inelastic collision between a blob of clay and Earth. We observe no net motion after such a collision, suggesting that the final momentum of the system is zero. Yet the clay certainly has nonzero momentum as it falls to the ground. Does that imply that momentum was not conserved? No! As indicated in Equation 7-21, the final speed of the two objects stuck together is approximately zero, but that doesn't mean *exactly* zero. The final speed is so very small that it is beyond our ability to detect it. Because the mass of Earth is so large, the product of Earth's mass and the final speed will be equal to the initial momentum. Momentum is conserved, even in this special case.

✱ What's Important 7-3

An inelastic collision is one in which two objects collide and stick together either briefly or permanently. Objects don't separate after a completely inelastic collision. Momentum is conserved in an inelastic collision, but kinetic energy is not.

7-4 Contact Time

During a car collision, airbags inflate rapidly and prevent serious injuries by cushioning the impact between passengers and hard surfaces in the vehicle. We intuitively sense that airbags are softer than any surface in the car. From a physics perspective, the protection airbags offer comes from extending both the distance over which the stopping force is applied as well as the time of contact between the accident victims and the inside surfaces of the car. What happens during a collision depends not only on the masses and velocities of the objects involved, but also on the **contact time**, the time during which the colliding objects touch.

Earlier in this chapter, we arrived at the definition of momentum by considering the application of a force over a time interval (Equation 7-5):

$$\int_{t_i}^{t_f} \vec{F}\, dt = \vec{p}_f - \vec{p}_i$$

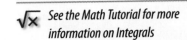

See the Math Tutorial for more information on Integrals

The difference between the final and initial values of momentum, or **impulse** $\vec{J}$, is mathematically equal to the sum of all the infinitesimally small changes in momentum between the initial and final times of interest:

$$\vec{J} = \vec{p}_f - \vec{p}_i = \int_{p_i}^{p_f} d\vec{p} \tag{7-22}$$

so

$$\int_{t_i}^{t_f} \vec{F}\, dt = \int_{p_i}^{p_f} d\vec{p}$$

The integrand on the left, $\vec{F}\, dt$ is equivalent to the integrand on the right, $d\vec{p}$. We can therefore define force in terms of momentum and time

$$\vec{F} = \frac{d\vec{p}}{dt} \tag{7-23}$$

You may notice that this expression leads directly back to Newton's second law:

$$\vec{F} = \frac{d(m\vec{v})}{dt} = m\frac{d\vec{v}}{dt} = m\vec{a}$$

In the same way that we related average velocity $v_{avg} = \Delta x / \Delta t$ with instantaneous velocity $v = dx/dt$ in Chapter 2, we can find the average force that one of the colliding objects exerts on the other from Equation 7-23. The magnitude of average force is then

$$F_{avg} = \frac{\Delta p}{\Delta t} \tag{7-24}$$

The average force is inversely proportional to the contact time Δt. So, for example, by increasing the time that a person's body is in contact with a car's surfaces, an inflated air bag reduces the average force exerted on a person during a crash.

> **? Got the Concept 7-3**
> **Bend Your Knees!**
> If you jump down from a table, why is it less painful if you bend your knees when landing?

Example 7-7 Follow Through

Tennis players are taught to continue to swing the racket after striking the ball (an action called *follow-through*). The rationale is that by increasing the time that the ball and racket are in contact, the ball might have a larger speed as it leaves the

Figure 7-12 Tennis players continue to swing the racket after striking the ball in order to increase the time that the ball and racket are in contact. This imparts a higher speed to the ball. *(AP Photo/Kathy Willens.)*

racket. Using her racket, a player can exert a constant average force of 560 N on a 57-g tennis ball during an overhand serve (**Figure 7-12**). How fast does the ball leave her racket if the contact time between them is 0.0050 s? What would the speed of the ball be if she improved her follow-through and increased the contact time to 0.0065 s? During an overhand serve, the tennis racket strikes the ball when it is essentially motionless.

SET UP

The relationship between the speed of the ball after impact with the racket and the time of contact can be defined through the change in momentum. Using Equation 7-24,

$$F_{avg} = \frac{\Delta p}{\Delta t} = \frac{mv_f - mv_i}{\Delta t} = \frac{mv_f}{\Delta t}$$

We use $v_i = 0$ because the ball is essentially at rest before the collision.

SOLVE

So,

$$v_f = \frac{F_{avg}\Delta t}{m}$$

When the average force is constant, the ball's velocity is directly proportional to the contact time. In particular, when $\Delta t = 0.0050$ s,

$$v_f = \frac{(560\ \text{N})(0.0050\ \text{s})}{0.057\ \text{kg}} = 49\ \text{m/s}$$

When $\Delta t = 0.0065$ s,

$$v_f = \frac{(560\ \text{N})(0.0065\ \text{s})}{0.057\ \text{kg}} = 64\ \text{m/s}$$

REFLECT

When the average force is constant, the change in momentum and therefore the change in velocity during a collision depend directly on the contact time. Here, increasing the contact time from $\Delta t = 0.0050$ s to $\Delta t = 0.0065$ s results in an increase of 15 m/s in the speed of the tennis ball.

Practice Problem 7-7 A soccer player's foot contacts the 0.43-kg ball for 8.6 ms during a kick. What is the speed of the ball immediately after being kicked if the player exerts a force of 1700 N on it?

Estimate It! 7-1 Enormous Force!

When major league baseball player Derek Jeter hits a home run, the baseball is in contact with his bat for about 0.0007 s. Assuming that the ball leaves Jeter's bat at about the same speed as the pitch (about 95 mph) estimate the average force Jeter's bat exerts on the ball. Major league baseballs weigh between 5.00 and 5.25 oz.

SET UP

We use Equations 7-23 and 7-24 to write the average force as a vector

$$\vec{F}_{avg} = \frac{\Delta \vec{p}}{\Delta t} = \frac{\vec{p}_f - \vec{p}_i}{\Delta t}$$

So

$$\vec{F}_{avg} = \frac{m\vec{v}_f - m\vec{v}_i}{\Delta t}$$

The initial and final velocity vectors are in opposite directions, so the velocities have opposite signs. We choose to let v_f be positive so v_i is negative, and then

$$F_{avg} = \frac{mv_f + mv_i}{\Delta t}$$

SOLVE
We have assumed that the ball strikes and leaves the bat at the same speed, so $v_f = v_i \approx 95$ mph or 42 m/s. Let m equal 5.125 oz or 0.15 kg and $\Delta t = 0.0007$ s.
So

$$F_{avg} = \frac{(0.15 \text{ kg})(42 \text{ m/s}) + (0.15 \text{ kg})(42 \text{ m/s})}{0.0007 \text{ s}}$$

$$= 1.8 \times 10^4 \text{ N}$$

REFLECT
The average force exerted on the ball is 1.8×10^4 N—over 2 tons. That's an enormous force to be generated by a human (even a professional baseball player!), but it only lasts for a short time.

✴ What's Important 7-4

The average force that one object exerts on another during a collision is proportional to the change in momentum and inversely proportional to the contact time, which is the length of time objects touch each other during a collision. The difference between the final and initial values of momentum is called impulse.

7-5 Elastic Collisions

Drop a soft rubber ball, and even on the first bounce it won't go nearly as high as the height from which you dropped it. At the peak of the bounce, the kinetic energy of the ball is zero and the gravitational potential energy is less than it was when you first dropped it. Clearly some of the mechanical energy has changed into a form not immediately visible, so this is certainly an example of an inelastic collision. If you dropped a steel ball bearing or a Super Ball® above a smooth concrete floor, however, you might reasonably expect it to bounce up nearly as high as the original drop point. In this case, the amount of energy dissipated in the ball–floor collision is relatively small. There are a number of situations in physics in which energy remains nearly, and sometimes completely, unchanged in a collision. A collision in which energy remains constant is referred to as an **elastic collision**.

? Got the Concept 7-4
Elastic and Inelastic

Which collisions are elastic and which are inelastic? (a) A ball drops from rest and does not bounce up to its original height. (b) After a ball that has kinetic energy equal to 200 J hits a stationary ball on a horizontal surface, the first ball then has 125 J of kinetic energy and the second has 75 J of kinetic energy. (c) A dog leaps in the air and catches a ball.

Figure 7-13 An object collides elastically with a second object that is initially at rest.

To analyze elastic collisions, let's consider the impact of a moving object on one at rest. If necessary, the analysis can be easily extended to the case when both objects are initially moving. Momentum is always conserved during a collision, and because the collision is elastic, energy is also conserved. So to determine, for example, how the final velocities depend on the masses of the objects and the initial velocity, we set both the initial and final momenta of the system equal and the initial and final energies of the system equal.

For the purpose of this study, we will consider the case of a one-dimensional collision, that is, one in which the final motions take place along the line of the original motion.

Figure 7-13 shows an object that has a mass m_1 and an initial speed $v_{1,i}$ striking an object that has a mass m_2 and is at rest. Conservation of momentum requires that the net momentum of the system remain unchanged from any time before the collision to any time after the collision. We therefore require that

$$m_1 \vec{v}_{1,i} + m_2 \vec{v}_{2,i} = m_1 \vec{v}_{1,f} + m_2 \vec{v}_{2,f}$$

or, because we have allowed m_2 to be initially at rest:

$$m_1 \vec{v}_{1,i} = m_1 \vec{v}_{1,f} + m_2 \vec{v}_{2,f} \tag{7-25}$$

Energy is also conserved because the collision is elastic. Because we can choose to study a system immediately before and immediately after a collision, so that no change in gravitational potential energy can occur, we can consider only the kinetic energy of the objects. Thus,

$$\frac{1}{2} m_1 v_{1,i}^2 = \frac{1}{2} m_1 v_{1,f}^2 + \frac{1}{2} m_2 v_{2,f}^2 \tag{7-26}$$

Treating the variables m_1, m_2, and $v_{1,i}$ in Equations 7-25 and 7-26 as known quantities, our goal is to find the velocities of the objects after the collision, $\vec{v}_{1,f}$ and $\vec{v}_{2,f}$.

Start by solving Equations 7-25 and 7-26 for one of the two unknowns, for example, $v_{2,f}$. Equation 7-26 becomes

$$\frac{1}{2} m_2 v_{2,f}^2 = \frac{1}{2} m_1 v_{1,i}^2 - \frac{1}{2} m_1 v_{1,f}^2$$

or

$$m_2 v_{2,f}^2 = m_1 v_{1,i}^2 - m_1 v_{1,f}^2 = m_1 \left(v_{1,i}^2 - v_{1,f}^2 \right) \tag{7-27}$$

After you've seen expressions such as Equation 7-27 a few times, you'll come to recognize that $v_{1,i}^2 - v_{1,f}^2$ can be factored into $(v_{1,i} + v_{1,f})(v_{1,i} - v_{1,f})$, a "trick" that is often useful in simplifying equations. Can a term of the form $(v_{1,i} + v_{1,f})$ or $(v_{1,i} - v_{1,f})$ be found in our momentum relationship? Yes! Because we are considering a one-dimensional collision, Equation 7-25 can also be written in terms of the speeds:

$$m_1 v_{1,i} = m_1 v_{1,f} + m_2 v_{2,f}$$

or

$$m_2 v_{2,f} = m_1 v_{1,i} - m_1 v_{1,f} = m_1 \left(v_{1,i} - v_{1,f} \right) \tag{7-28}$$

We therefore divide Equation 7-27 by Equation 7-28, so that the common $(v_{1,i} - v_{1,f})$ term can be divided out:

$$\frac{m_2 v_{2,f}^2}{m_2 v_{2,f}} = \frac{m_1 \left(v_{1,i}^2 - v_{1,f}^2 \right)}{m_1 \left(v_{1,i} - v_{1,f} \right)} = \frac{\left(v_{1,i} + v_{1,f} \right) \left(v_{1,i} - v_{1,f} \right)}{v_{1,i} - v_{1,f}}$$

So,

$$v_{2,f} = v_{1,i} + v_{1,f} \tag{7-29}$$

Equation 7-29 does not provide a relationship for the unknown variable $v_{2,f}$ in terms only of known variables, but we can substitute the result back into Equation 7-28 in order to eliminate one unknown:

$$m_2(v_{1,i} + v_{1,f}) = m_1(v_{1,i} - v_{1,f})$$

or

$$v_{1,f} = v_{1,i}\left(\frac{m_1 - m_2}{m_1 + m_2}\right) \tag{7-30}$$

This equation is an expression for $v_{1,f}$ in terms only of known variables, which is halfway to our goal. To find a relationship for the other unknown, we substitute the expression for $v_{1,f}$ into Equation 7-29:

$$v_{2,f} = v_{1,i} + \frac{m_1 - m_2}{m_1 + m_2}v_{1,i}$$

$$= v_{1,i}\left(1 + \frac{m_1 - m_2}{m_1 + m_2}\right)$$

$$= v_{1,i}\left(\frac{m_1 + m_2}{m_1 + m_2} + \frac{m_1 - m_2}{m_1 + m_2}\right)$$

So,

$$v_{2,f} = v_{1,i}\left(\frac{2m_1}{m_1 + m_2}\right) \tag{7-31}$$

Equations 7-30 and 7-31 form a compact, elegant summary of the elastic collision of two objects. The way we have written these relationships emphasizes that the final speeds of both objects is proportional to the original speed, and that the multiplicative factor depends on the masses of the two objects.

In which directions do the two objects move after the elastic collision? Because the term in parentheses in Equation 7-31 is positive regardless of the values of m_1 and m_2, the sign of $v_{2,f}$ must be the same as the sign of $v_{1,i}$. Object 2 therefore always moves in the same direction as the initial motion of object 1. Object 1 does not necessarily continue in its initial direction; however, notice that the term in parentheses in Equation 7-30 can be equal to zero, less than zero, or greater than zero depending on the values of m_1 and m_2. These values would result in object 1, in turn, becoming stationary after the collision, leaving the collision in a direction opposite to its initial motion, or leaving the collision in the same direction as its initial motion. The three possibilities are shown in **Figure 7-14**.

In an elastic collision of moving object 1 with stationary object 2,

the final velocities of the objects are $v_{1,f} = v_{1,i}\left(\frac{m_1 - m_2}{m_1 + m_2}\right)$ and $v_{2,f} = v_{1,i}\left(\frac{2m_1}{m_1 + m_2}\right)$.

When $m_1 < m_2$, the term in parentheses is negative, so object 1 bounces back.

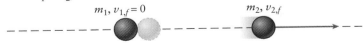

When $m_1 = m_2$, the term in parentheses is zero, so object 1 ends up at rest.

When $m_1 > m_2$, the term in parentheses is positive, so object 1 continues to move forward.

Figure 7-14 Three examples of an elastic collision of one object with a second, stationary object.

Physicist's Toolbox 7-1   Elastic Collisions of Two Objects

A study of the elastic collision between two objects, like many physical phenomena, encompasses a number of special cases. Physicists, however, like simplicity! Equations 7-30 and 7-31 form an elegant and compact summary of the elastic collision of two objects.

An object of mass m_1 moving at velocity $\vec{v}_{1,i}$ collides elastically with a second object of mass m_2 so that all motion remains linear. The second object is initially at rest. By conserving both kinetic energy and momentum, the speeds of the two objects after the collision are found to be

$$v_{1,f} = v_{1,i}\left(\frac{m_1 - m_2}{m_1 + m_2}\right) \tag{7-30}$$

$$v_{2,f} = v_{1,i}\left(\frac{2m_1}{m_1 + m_2}\right) \tag{7-31}$$

As we have seen, the final speeds of both objects are proportional to the original speed, and the multiplicative factor depends on the masses of the two objects. In addition, the direction of each object is indicated by the sign of its final velocity. If the sign of an object's final velocity is the same as the sign of $v_{1,i}$, the object leaves the collision in the same direction as the first object was moving initially. If the final velocity has the opposite sign of the initial velocity, the object is moving in the direction opposite to the initial motion. For example, if $v_{1,f}$ is negative while $v_{1,i}$ is positive, the moving object has bounced backward after colliding with the initially stationary one.

Like we did in our exploration of inelastic collisions (Section 7-3), we can gain insight into the physics of elastic collisions by considering three special cases involving the relative masses of the two objects: when the moving object is far more massive than the stationary one, when the two objects have the same mass, and finally, when the moving object is of negligible mass compared to the stationary one.

Special Case: The Moving Object Is More Massive than the One at Rest

 Go to Picture It 7-1 for more practice dealing with elastic collisions

In this case, the mass of object 1, which was initially at rest, is negligibly small compared to the mass of object 2 which was initially moving. Imagine a boulder colliding with a stationary pebble. Although the mass of object 2 is not zero, we have seen that whenever one quantity in an expression is significantly larger than another, it is acceptable to ignore the smaller one if the two quantities are added or subtracted. We will therefore treat m_2 as negligible in Equations 7-30 and 7-31:

$$v_{1,f} = v_{1,i}\left(\frac{m_1 - m_2}{m_1 + m_2}\right) \approx v_{1,i}\left(\frac{m_1 - 0}{m_1 + 0}\right)$$

$$v_{1,f} = v_{1,i}$$

and

$$v_{2,f} = v_{1,i}\left(\frac{2m_1}{m_1 + m_2}\right) \approx v_{1,i}\left(\frac{2m_1}{m_1 + 0}\right)$$

$$v_{2,f} = 2v_{1,i}$$

In this extreme case, the motion of the massive object 1 is unaffected by the collision, but the second, much smaller object 2 flies off at high speed relative to the initial velocity of object 1. The motion of the boulder is unchanged as a result of the collision, but the pebble is affected in a dramatic way.

Special Case: The Objects Have the Same Mass

When the masses of the objects are the same (that is, when $m_1 = m_2$), we can set both masses equal to m in Equations 7-30 and 7-31:

$$v_{1,f} = v_{1,i}\left(\frac{m_1 - m_2}{m_1 + m_2}\right) = v_{1,i}\left(\frac{m - m}{m + m}\right)$$

$$v_{1,f} = 0$$

and

$$v_{2,f} = v_{1,i}\left(\frac{2m_1}{m_1 + m_2}\right) = v_{1,i}\left(\frac{2m}{m + m}\right)$$

$$v_{2,f} = v_{1,i}$$

In this case, the object that is initially moving comes to a stop during the collision, and the second object leaves the collision at speed equal to the initial speed of the first.

Special Case: The Moving Object Is Less Massive than the One at Rest

In this case, the mass of the object initially in motion is negligibly small compared to the mass of the object initially at rest. Imagine the pebble colliding with a stationary boulder. Again, we will neglect the smaller mass, in this case m_1, in Equations 7-30 and 7-31:

$$v_{1,f} = v_{1,i}\left(\frac{m_1 - m_2}{m_1 + m_2}\right) \approx v_{1,i}\left(\frac{0 - m_2}{0 + m_2}\right)$$

$$v_{1,f} = -v_{1,i}$$

and

$$v_{2,f} = v_{1,i}\left(\frac{2m_1}{m_1 + m_2}\right) \approx v_{1,i}\left(\frac{2(0)}{0 + m_2}\right)$$

$$v_{2,f} = 0$$

In this extreme case, the pebble is reflected back from the collision at the same speed as it had initially, but the boulder is unaffected by the impact.

Example 7-8 A Dragonfly and an Elephant

The Australian dragonfly may be the world's fastest insect. A dragonfly that has a mass equal to 1.1 g and is traveling at 16 m/s collides with a stationary elephant that has a mass equal to 9900 kg (Figure 7-15). Assume that the collision is completely elastic, and that the elephant is on a frictionless ice rink. Find the velocity (magnitude and direction) of both the dragonfly and the elephant immediately after the impact.

SET UP
Momentum is always conserved in a collision, and kinetic energy is conserved in this example because we declared the collision to be completely elastic. Equations 7-30 and 7-31 give the final velocities of two objects involved in a one-dimensional elastic collision. Using the subscripts "F" for the fly and "E" for the elephant, these expressions become

$$v_{F,f} = v_{F,i}\left(\frac{m_F - m_E}{m_F + m_E}\right) \tag{7-30}$$

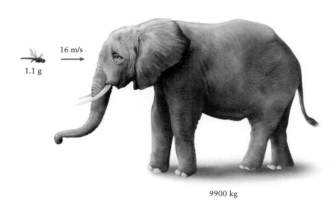

16 m/s

1.1 g

9900 kg

Figure 7-15 A dragonfly collides elastically with an elephant—what is the velocity of each after the collision?

and

$$v_{E,f} = v_{F,i}\left(\frac{2m_F}{m_F + m_E}\right) \qquad (7\text{-}31)$$

SOLVE

Using the values given in the problem statement,

$$v_{F,f} = 16 \text{ m/s}\left(\frac{0.0011 \text{ kg} - 9900 \text{ kg}}{0.0011 \text{ kg} + 9900 \text{ kg}}\right) = -16 \text{ m/s}$$

and

$$v_{E,f} = 16 \text{ m/s}\left(\frac{2(0.0011 \text{ kg})}{0.0011 \text{ kg} + 9900 \text{ kg}}\right) = 3.6 \times 10^{-6} \text{ m/s}$$

REFLECT

There are three significant aspects of these results that demand our attention. First, notice that the magnitude of the final speed of the dragonfly, 16 m/s, is the same as dragonfly's original speed. However, because $v_{F,f}$ is negative, the direction of the dragonfly's final motion is opposite to its original direction. The dragonfly bounces directly back as a result of the collision.

The magnitude of the final velocity of the elephant is extremely small. The velocity $v_{E,f} = 3.6 \times 10^{-6}$ m/s is equivalent to 1 m every $1/(3.6 \times 10^{-6})$ s, or 1 m per 2.8×10^5 s. At this rate, it would take about 3 days for the elephant to move 1 m. In other words, the elephant moves imperceptibly slowly as a result of the impact.

Finally, note that this collision meets the conditions of the special case we discussed in which the mass of the incoming object m_1 is negligibly small compared to the mass of the stationary object m_2. Based on the approximations that arise from $m_1 \ll m_2$, we concluded that the moving object leaves the collision with its speed unchanged but in the opposite direction, and that the stationary object appears unaffected by the impact. Both conclusions are borne out in the numerical results of the problem.

Practice Problem 7-8 A 9000-kg elephant runs at 5.0 m/s directly into a berry of the *Synsepalum dulcificum* plant, sometimes called the West African miracle fruit tree. The berry has a mass of 0.0040 kg. Find the velocity (magnitude and direction) of both the elephant and berry immediately after impact. Assume that the collision is completely elastic, and that air resistance can be neglected.

Example 7-9 Two Steel Balls

A 14-g steel ball traveling horizontally at 25 m/s strikes a second, stationary steel ball that has a mass equal to 210 g. The stationary ball is suspended from a long thread of negligible mass as shown in **Figure 7-16**. After the balls collide elastically, how high will the second ball rise?

SET UP

We have to solve this problem in two parts. First, we need to determine how much energy the initially stationary ball acquires during the collision by calculating its speed immediately after the collision. Then we need to calculate how high it swings as its kinetic energy is converted to potential energy. From Equation 7-31, we know the velocity of the ball immediately after the collision. We also know that energy is conserved both during the collision and during the upward swing of the ball and thread, although it will be shared between kinetic energy and gravitational potential energy. Finally, we know that at the top of the swing the speed of the ball is zero and all of the energy is gravitational potential energy.

SOLVE
According to Equation 7-31, the speed of the second (initially stationary) ball immediately after the collision is

$$v_2 = v_1 \left(\frac{2m_1}{m_1 + m_2} \right) \tag{7-32}$$

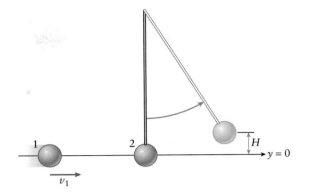

Figure 7-16 A steel ball collides elastically with a second ball suspended from a long thread of negligible mass. How high does the second ball rise?

Because this quantity is not the final speed of the initially stationary ball, we labeled it v_2 rather than $v_{2,f}$ to avoid confusion. The sum of the potential energy and the kinetic energy the ball acquires as a result of this speed remains unchanged from the instant after the collision to the point at which the ball has swung up to its highest point. For the second part of the motion, we use the subscripts "A" and "T" to indicate the instants right after the collision and at the top of the swing, respectively. Then,

$$K_A + U_A = K_T + U_T$$

or

$$\frac{1}{2}m_2 v_{2,A}^2 + m_2 g y_A = \frac{1}{2}m_2 v_{2,T}^2 + m_2 g y_T$$

Notice that the speed of the second ball immediately after the collision, $v_{2,A}$, is represented by v_2 in Equation 7-32. As shown in Figure 7-16, because we are free to define the reference height $y = 0$ at any place that is convenient, we set the height of the second ball before the collision to be 0, so that $y_A = 0$. The reference height also defines y_T as H, the solution to the problem. Finally, note that at the top of the swing, the second ball is momentarily at rest, so $v_{2,T} = 0$. Thus,

$$\frac{1}{2}m_2 v_1^2 \left(\frac{2m_1}{m_1 + m_2} \right)^2 + 0 = 0 + m_2 g H$$

or

$$H = \frac{2m_1^2 v_1^2}{(m_1 + m_2)^2 g} \tag{7-33}$$

So,

$$H = \frac{2(0.014 \text{ kg})^2 (25 \text{ m/s})^2}{(0.014 \text{ kg} + 0.210 \text{ kg})^2 (9.8 \text{ m/s}^2)} = 0.50 \text{ m}$$

REFLECT
Equation 7-33 enables us to consider extreme values of the masses. Were the initially moving ball less massive, for example, the resulting height to which second ball rises would be smaller. This result makes sense both intuitively and also from our study of the special cases of elastic collisions; when the initially moving object is small relative to the initially stationary one, very little of the kinetic energy gets transferred to the stationary object. Conversely, Equation 7-33 suggests that when the mass of the initially stationary ball is negligible compared to the moving one, nearly all of the kinetic energy gets transferred. This results in a final speed of the initially stationary ball equal to twice the initial speed of the moving one, and also results in the maximum possible height. By using Equation 7-33, the maximum height is

$$H \approx \frac{2m_1^2 v_1^2}{(m_1 + 0)^2 g} = \frac{2v_1^2}{g}$$

This expression might look familiar—it is the height to which a vertical projectile of initial speed $2v_1$ rises. (See, for example, Example 2-9, but make sure you take the initial speed in the example as twice the speed of the initially moving ball here.)

Practice Problem 7-9 A 14-g steel ball traveling horizontally at 25 m/s strikes a second, stationary steel ball that is 4 times as massive. The stationary ball is suspended from a long thread of negligible mass as shown in Figure 7-16. After an elastic collision, how high will the second ball rise?

> ✱ **What's Important 7-5**
> A collision is defined to be elastic when the total mechanical energy of the colliding objects is the same after the collision as before. Momentum is conserved in both elastic and inelastic collisions.

7-6 Center of Mass

How often do you recharge your cell phone? What is your heart rate when you exercise? How long does it take you to walk to class? How tall are your friends in physics class? Although there might be specific numeric answers to each of these questions, a more powerful way to address them is by way of an *average*. The average is a mathematical tool that provides a way to summarize a collection of potentially complicated information, perhaps even with a single number. In this section, we apply the concept of an average to study the motion of a system of objects.

To find the average of a set of N numbers, add them together and divide by N. The average of 4, 8, and 18, for example, is 10. The average of the set of values (4, 4, 8, 8, 8, 10) is

$$\overline{n_i} = \frac{1}{6}\sum_{i=1}^{6} n_i = \frac{4 + 4 + 8 + 8 + 8 + 10}{6} = 7$$

where n_i represents the ith value in the set. Putting a bar over a variable is one conventional way to indicate the average value of that variable. Notice that because some of the values in the sets occur more than once, we could write the average as

$$\overline{n_i} = \frac{2(4) + 3(8) + 1(10)}{6} = 7$$

or

$$\overline{n_i} = \frac{2}{6}(4) + \frac{3}{6}(8) + \frac{1}{6}(10) = 7 \tag{7-34}$$

Each term in the sum includes the fraction of the entire set comprised by the value that follows it—two-sixths of the set has value 4, three-sixths of the set has value 8, and one-sixth of the set has value 10. In writing the sum this way, we have defined a **weighted average** in which each value affects the result to a greater or lesser extent depending on how often it appears in the set. Here, *weight* does not refer to a force due to gravity, but rather the amount that each value contributes to the result.

We can apply the method of the weighted average to find a single position that represents a group of objects. Imagine a waiter supporting on one hand a tray that has a bowl of soup on one side and a small mint on the other (Figure 7-17). To balance the tray, the waiter supports it at a position much closer to the soup than the mint. This single position, which is the average of the positions of the objects weighted by how massive (or heavy) each one is, represents the position of both the soup and the mint. The average position, as opposed to the *weighted* average

Figure 7-17 A waiter balances a serving tray on one hand by placing more massive objects closer to his hand and less massive ones farther away.

position, of the soup and mint would be halfway between the two objects, but that position is not physically useful. In contrast, the weighted average position, or **center of mass**, represents the position of a large, complex object or system of objects as a particular point.

Let's be more precise. Consider two small objects with masses of m_1 and m_2, located a distance from the origin of x_1 and x_2, respectively, as shown in **Figure 7-18**. We know intuitively that when the masses are not equal, as in the case of the soup and mint, the representative position of the system will be closer to the more massive object. In determining this position, the distance to each object should be weighted so that the more massive object has a larger effect on the result. In direct analogy to the weighted average of numbers shown by example in Equation 7-34, we find the location of the center of mass of the two objects by weighting the position of each object by the fraction of the mass of the total system that each carries:

$$\overline{x} = \frac{m_1}{m_1 + m_2}x_1 + \frac{m_2}{m_1 + m_2}x_2$$

For a system of N objects, we generalize the weighted average as

$$\overline{x} = \frac{m_1}{M_{tot}}x_1 + \frac{m_2}{M_{tot}}x_2 + \cdots + \frac{m_N}{M_{tot}}x_N = \frac{1}{M_{tot}}\sum_{i=1}^{N} m_i x_i$$

where M_{tot} is the mass of all of the objects added together. The weighted average position has a number of applications; perhaps most important, it provides a way to represent the position of a large object by a single point in space, often simplifying otherwise complex problems. For example, the momentum of a system of objects can be determined by taking the product of the total mass of the system and the velocity of the center of mass.

We will refer to the position of the center of mass as x_{CM} or

$$x_{CM} = \frac{1}{M_{tot}}\sum_{i=1}^{N} m_i x_i \qquad (7\text{-}35)$$

Just as the center of mass is a single point that represents the entire system, the motion of the center of mass represents the motion of the system as a whole. For example, look at Figure 3-1b, a stop-action photograph of a leaping ballet dancer. The path traced by any particular point on the dancer's body is almost certainly not described by a simple curve. However, the path traced out by the center of mass, shown by the orange diamonds, is exactly the parabola dictated by the physics of two-dimensional projectile motion. The equations that describe motion treat the entire system or object as if it were concentrated at a single point, the center of mass.

$\sqrt{x}$ **Go to Picture It 7-2** *for more practice dealing with center of mass*

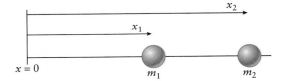

Figure 7-18 The center of mass of a system depends on the distance each object is from an arbitrarily selected reference point as well as the mass of each object.

Example 7-10 Weighted Average

To explore how the weighted average works in the center of mass relationship (Equation 7-35), determine the center of mass of the two objects shown in Figure 7-18. Let the objects be at positions $x_1 = 2.0$ m and $x_2 = 8.0$ m. Do the calculation (a) first by letting m_1 and m_2 both be 3.0 kg and (b) then by letting m_2 be much more massive than m_1, taking $m_1 = 3.0$ kg and $m_2 = 33$ kg.

SET UP

In both cases we will apply Equation 7-35. For a system that consists of two objects, the equation becomes

$$x_{CM} = \frac{1}{m_1 + m_2}(m_1 x_1 + m_2 x_2) = \frac{m_1 x_1 + m_2 x_2}{m_1 + m_2} \qquad (7\text{-}36)$$

SOLVE
For part (a),

$$x_{CM} = \frac{(3.0\,\text{kg})(2.0\,\text{m}) + (3.0\,\text{kg})(8.0\,\text{m})}{3.0\,\text{kg} + 3.0\,\text{kg}} = 5.0\,\text{m}$$

For part (b), we expect the weighted average position to be closer to object 2 because m_2 is larger than m_1. Equation 7-36, our version of Equation 7-35 simplified for a two-object system, gives

$$x_{CM} = \frac{(3.0\,\text{kg})(2.0\,\text{m}) + (33\,\text{kg})(8.0\,\text{m})}{3.0\,\text{kg} + 33\,\text{kg}} = 7.5\,\text{m}$$

REFLECT
Notice that in part (a), the position 5.0 m from the origin is exactly halfway between the objects, as expected when two objects have the same mass. In part (b), however, the larger weighting factor that multiplies the distance to object 2 does indeed cause the center of mass to be closer to the more massive object.

Practice Problem 7-10 Determine the center of mass of the two objects shown in Figure 7-18, letting the objects be at positions $x_1 = 2.0$ m and $x_2 = 8.0$ m as in Example 7-10, but letting m_1 equal 33 kg and m_2 equal 3.0 kg.

 Watch Out

No special origin is required to determine the center of mass.

Don't let the choice of origin necessary to determine the center of mass confuse you: The location you determine for the center of mass does not depend on your choice. For example, in the previous example we determined the center of mass of two objects of mass 3.0 kg, with one 2.0 m and the other 8.0 m from the origin. The result is $x_{CM} = 5.0$ m, exactly halfway between the two objects. Would we get a different result if the origin were farther down the x axis, so that $x_1 = 32$ m and $x_2 = 38$ m? According to Equation 7-35,

$$x_{CM} = \frac{3.0\,\text{kg}}{6.0\,\text{kg}}\,32\,\text{m} + \frac{3.0\,\text{kg}}{6.0\,\text{kg}}\,38\;\text{m} = 35\,\text{m}$$

This result of 35 m looks different from the original value of 5.0 m, but notice that relative to the new origin $x_{CM} = 35$ m is once again halfway between the two objects. The numeric result is different, but the position in space is not. You are free to choose any origin you find convenient to determine center of mass.

According to Newton's law of motion, the total linear momentum of an object or a system of objects is conserved when no net force acts on it. An object or a system of objects that experiences a net force of zero does not accelerate. Because the center of mass is a single point that represents an entire system, we therefore conclude that the center of mass does not accelerate if the system experiences a net force of zero. In particular, if the center of mass of such a system is at rest, it must remain at rest. If internal forces cause an element in the system to move, some other part of the system must also move in such a way that the center of mass remains stationary.

? Got the Concept 7-5
A Boy, a Ball, and a Boat

A boy stands at one end of a long, narrow boat. Relative to a point outside the boat, how far is the boy from his initial position after he throws a ball so that it lands at the other end of the boat? How do the assumptions you make about the masses of the boy, boat, and ball affect your answer?

Example 7-11 An Ant on a Twig

A narrow, uniform twig that has a length equal to 15 cm and a mass equal to 2.2 g floats in a swimming pool so that one end touches the edge of the pool (Figure 7-19a). A 0.35-g ant, initially resting at the other end of the twig, crawls directly toward the edge of the pool. Assuming that the twig experiences no resistive forces due to the water, how far is the ant from the edge of the pool when it reaches the other end of the twig?

SET UP

Because the twig is uniform, we'll assume its center of mass to be at the center of the twig. The center of mass of the ant is near its center too, and the center of mass of the system is somewhere in between. Regardless of its position, however, the center of mass of the ant and twig system is at rest at the start of the problem, and because no external forces act on the system, the center of mass does not move as the ant moves. We will therefore write an expression for the center of mass position before and after the ant has crawled across the twig, and set the two equal.

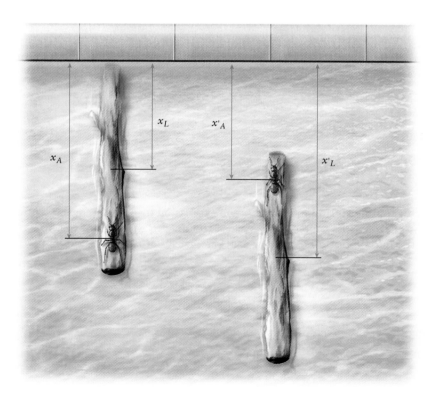

Figure 7-19 (a) An ant crawls along a narrow, uniform twig floating in a swimming pool. (b) The final position of the ant is determined by the motion of both the ant and the twig.

SOLVE

Using Figure 7-19a as a guide, before the ant begins to crawl the center of mass x_{CM} of the ant and twig is given by Equation 7-35

$$x_{CM} = \frac{m_A}{m_A + m_T} x_A + \frac{m_T}{m_A + m_T} x_T$$

where the x_A and x_T are the positions of the center of mass of the ant and twig, respectively. After the ant crawls to the other end of the twig, the center of mass is

$$x'_{CM} = \frac{m_A}{m_A + m_T} x'_A + \frac{m_T}{m_A + m_T} x'_T$$

where x'_A and x'_T are the new positions of the ant and twig. The position of the center of mass remains unchanged, that is, $x'_{CM} = x_{CM}$, so

$$\frac{m_A}{m_A + m_T} x_A + \frac{m_T}{m_A + m_T} x_T = \frac{m_A}{m_A + m_T} x'_A + \frac{m_T}{m_A + m_T} x'_T$$

or

$$m_A x_A + m_T x_T = m_A x'_A + m_T x'_T \tag{7-37}$$

Equation 7-37 is compact, but it contains two unknown variables, x'_A and x'_T. We need a way to connect x'_A and x'_T to solve for x'_A. Figure 7-19b provides the clue. The center of mass of the twig is half the length of the twig $(L/2)$ farther from the edge of the pool than the ant, or

$$x'_T = x'_A + \frac{L}{2}$$

Equation 7-37 becomes

$$m_A x_A + m_T x_T = m_A x'_A + m_T\left(x'_A + \frac{L}{2}\right)$$

We can solve for x'_A, the final position of the ant using

$$m_A x_A + m_T x_T = m_A x'_A + m_T x'_A + m_T \frac{L}{2}$$

$$m_A x_A + m_T x_T - m_T \frac{L}{2} = (m_A + m_T) x'_A$$

So

$$x'_A = \frac{m_A x_A + m_T x_T - m_T \dfrac{L}{2}}{m_A + m_T}$$

We can combine terms and simplify the expression by noting that the center of mass of the twig is at its center, so $x_T = L/2$ (relative to the edge of the pool), and also that the initial position of the ant is at the far end of the twig, that is, $x_A = L$. So

$$x'_A = \frac{m_A L + m_T \dfrac{L}{2} - m_T \dfrac{L}{2}}{m_A + m_T} = \frac{m_A}{m_A + m_T} L \tag{7-38}$$

Using the values given in the problem statement,

$$x'_A = \frac{0.00035\ \text{kg}}{0.00035\ \text{kg} + 0.0022\ \text{kg}} 0.15\ \text{m} = 0.021\ \text{m}$$

The ant ends up 2.1 cm from the edge of the pool.

REFLECT

As the ant crawls, it exerts a force in the opposite direction on the twig. The twig moves away from the edge of the pool, but because the twig is far more massive than the ant, it moves a relatively small distance compared to the ant.

We can test the result by considering extreme values of the masses. Suppose the ant were far less massive than the twig, that is, $m_A \ll m_T$. We would expect that the twig would barely move as the ant crawled, much in the same way that an ocean liner doesn't move when a passenger walks along the deck. Letting m_A be negligible compared to m_T in Equation 7-38 gives

$$x'_A \approx \frac{0}{0 + m_T} L = 0$$

A final position for the ant of zero puts the ant up against the edge of the pool, which means that the twig didn't move. Our expectation is borne out.

Suppose the ant were far more massive than the twig, that is, $m_A \gg m_T$. In such a case, we would expect the ant to remain in the same place, while the twig slides out from underneath it. Letting m_T be negligible compared to m_A in Equation 7-38 gives

$$x'_A \approx \frac{m_A}{m_A + 0} L = L$$

The ant's final position is a distance L from the edge of the pool, which is where it started. Again, our expectation is borne out. Both of the extreme cases give us confidence that our result is correct.

Practice Problem 7-11 If the mass of the twig in the preceeding problem were only 0.025 g, determine the ant's distance from the edge of the pool when it reaches the other end of the twig.

An isolated system experiences no net external forces. The absence of a net force results in two different implications that will help us solve problems. First, linear momentum is conserved in an isolated system. Second, in a system that experiences no net force, the center of mass does not accelerate and it remains at rest if it was initially at rest.

Example 7-12 Pushing Off, Again

In Example 7-3, we considered ice skaters Jim and Sarah, who face each other while stationary and then push against each other's hands. We found Sarah's velocity by demanding that linear momentum be conserved. Repeat the problem by considering the motion of their center of mass. Remember that Jim's mass is 50% more than Sarah's, and that he moves backward at 2.0 m/s after the push.

SET UP

The center of mass of the Jim and Sarah system is close to the point where their hands meet. The center of mass position is at rest at the start of the problem, and must remain at rest because no net external forces act on the Jim and Sarah system. We will therefore determine an expression for the center of mass position at any time after the push, and require that it not move because it is at rest before the push occurs.

SOLVE

By using Equation 7-35, the center of mass of the Jim (J) and Sarah (S) system is

$$x_{CM} = \frac{m_J}{m_J + m_S} x_J + \frac{m_S}{m_J + m_S} x_S$$

Our goal is to find Sarah's velocity after the push; we can introduce velocities by taking the derivative with respect to time of both sides:

$$\frac{d}{dt}x_{CM} = \frac{d}{dt}\left(\frac{m_J}{m_J + m_S}x_J + \frac{m_S}{m_J + m_S}x_S\right)$$

$$= \frac{m_J}{m_J + m_S}\frac{d}{dt}x_J + \frac{m_S}{m_J + m_S}\frac{d}{dt}x_S$$

$$= \frac{m_J}{m_J + m_S}v_J + \frac{m_S}{m_J + m_S}v_S$$

See the Math Tutorial for more information on Differential Calculus

The term $dx_{CM}/dt = 0$, thus

$$\frac{m_S}{m_J + m_S}v_S = -\frac{m_J}{m_J + m_S}v_J$$

or

$$v_S = -\frac{m_J}{m_S}v_J$$

Using the values given in the problem statement,

$$v_S = -\frac{1.5m_S}{m_S}(2.0\text{ m/s}) = -3.0\text{ m/s}$$

REFLECT

The answer is identical to the result we obtained in Example 7-3. In that problem, we applied the principle of conservation of momentum. Here, we required that the center of mass remain at rest. We obtain the same result because the two approaches are simply different manifestations of the same fundamental physics.

✱ What's Important 7-6

The weighted average position of a system is its center of mass. It is often convenient to represent the position of a large, complex object or system of objects as if the entire system were located at the center of mass.

Answers to Practice Problems

7-1   X-43: 5.2×10^6 (kg·m)/s, falcon: 60 (kg·m)/s

7-2   250 m/s

7-3   −0.5 m/s

7-4   1.2 m/s

7-5   5.6 m/s

7-6   Mass of red equals mass of brown

7-7   34 m/s

7-8   Elephant: 5.0 m/s, berry: 10 m/s

7-9   5.1 m

7-10 2.5 m

7-11 0.14 m

Answers to Got the Concept Questions

7-1 There must be at least three separate fragments after the nucleus breaks apart, as shown in **Figure 7-20a**. The nucleus initially has linear momentum only in the direction of its velocity, because the direction of the momentum vector is the same as the direction of the velocity vector. The component of the initial momentum in the direction perpendicular to the motion is therefore zero. The component of the final momentum of the system in the perpendicular direction must then also be zero, because linear momentum is conserved in any direction we consider. As shown in Figure 7-6, the large fragment does not contribute to the linear momentum in the perpendicular direction. The only way to ensure that the *sum* of the momenta in the perpendicular direction is zero is if the nucleus breaks up into at least three fragments, two of which have a component of velocity in the direction perpendicular to the original path and in opposite directions.

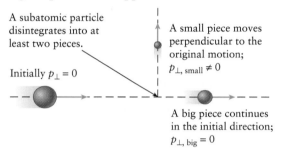

A subatomic particle disintegrates into at least two pieces.

A small piece moves perpendicular to the original motion; $p_{\perp, \text{small}} \neq 0$

Initially $p_\perp = 0$

A big piece continues in the initial direction; $p_{\perp, \text{big}} = 0$

There must be at least one more piece, so that the sum of the momenta of all pieces remains zero:

$$p_{\perp, \text{small}} + p_{\perp, \text{big}} + p_{\perp, \text{other}} = 0$$

(a)

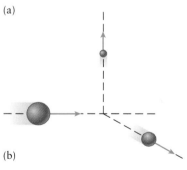

(b)

Figure 7-20

It is possible for the nucleus to break into only two fragments with one moving in a direction perpendicular to the original one, as shown in **Figure 7-20b**. The requirement is that the other fragment (the larger one in the figure) must acquire some momentum in the perpendicular direction to ensure that the net component of momentum in that direction is zero. As in the first case, the component of the initial momentum (that of the nucleus) perpendicular to the direction of motion is zero, so the final momentum in the perpendicular direction must also be zero.

Notice, too, that the final speed of the larger fragment is greater than the initial speed, because momentum is conserved in the original direction of motion.

7-2 The rapidly moving, relatively massive head of the sledge hammer carries a significant amount of kinetic energy. After striking and remaining in contact with the wedge, the sledge hammer and wedge move relatively slowly; a considerable fraction of the energy of the system is no longer kinetic. The energy has been transformed into thermal energy. The wedge (and the hammer's head) gets hot.

7-3 By bending your knees, you extend the time over which the collision between your feet and the ground takes place. The time between when your feet first touch the ground and when your body stops moving in reaction to the impact is the equivalent to contact time that appears in Equation 7-24. By increasing the contact time, you decrease the average force exerted on your body by the ground.

7-4 (a) Some of the ball's energy has been transformed to heat and other nonmechanical forms during the collision with the floor; otherwise it would have bounced back to its original height. Because energy was not conserved, the collision was inelastic. (b) The collision of the two balls is elastic. In reality, such a collision could not be completely elastic, but we have stipulated that the kinetic energy of the system is the same before and after the collision. (c) The dog undergoes an inelastic "collision" with the ball, which must be the case because the two leave the collision stuck together.

7-5 There are no external forces on the boy, boat, and ball system, so the center of mass of the three remains stationary even when the boy throws the ball. If we assume that the boat is far more massive than the boy and the ball, the center of mass of the system is very close to the center of mass of the boat alone because the weighting factors for the positions of the boy and ball in Equation 7-35 will be small. The boat hardly moves when the boy throws the ball because the center of mass computation is still overwhelmed by the mass of the boat. So, the boy's position relative to a point outside the boat hardly changes. But suppose the boat's hull is made from carbon fiber and therefore has a relatively small mass, and, just to examine an extreme case, that the ball is far more massive than the boy or boat. (We offer no explanation as to how the boy would manage to throw the ball.) In this case, the center of mass of the boy, boat, and ball system must be close to the center of mass of the ball, so when the ball is relocated to the other end of the boat, the ball will barely move *relative to a point outside the boat* because the center of mass of the system again will not move! The boy and boat must shift the length of the boat in the direction opposite to the motion of the ball in order for this to occur. **Figure 7-21** shows these two cases.

Figure 7-21

(a) **Massive boat**

The center of mass of the system is close to the center of mass of the boat, because the boat is so massive relative to the boy and the ball.

Center of mass
of system

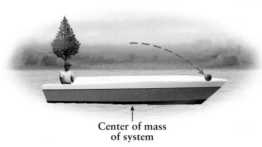

Compare the center of mass position to the fixed position of the tree. The center of mass doesn't move, even after the ball is thrown.

Center of mass
of system

(b) **Massive ball**

Notice that the center of mass is close to the larger, more massive, ball.

Center of mass
of system

Again, the center of mass doesn't move after the ball is thrown. You can see that by comparing the center of mass position with the position of the tree.

Center of mass
of system

SUMMARY

Topic	Summary	Equation or Symbol	
center of mass	The weighted average position, or center of mass, can be taken as the representative position of a large, complex object or system of objects.	$x_{CM} = \dfrac{1}{M_{tot}} \displaystyle\sum_{i=1}^{N} m_i x_i$	(7-35)
conservation of momentum	The total momentum of an object or system of objects does not change when the net force acting on it is zero.	$\vec{p}_i = \vec{p}_f$	(7-6)

contact time	Contact time is the time during which colliding objects touch.	
elastic collision	A collision is defined to be elastic when the total mechanical energy of the colliding objects is the same after the collision as before.	
impulse	Impulse is the difference between the final and initial values of momentum.	$\vec{J} = \vec{p}_f - \vec{p}_i = \int_{p_i}^{p_f} d\vec{p}$ (7-22)
inelastic collision	An inelastic collision is one in which two objects stick together after colliding, either briefly or completely. In a completely inelastic collision the objects don't separate after colliding. Momentum is conserved in an inelastic collision, but kinetic energy is not.	
linear momentum	Linear momentum is the momentum of an object moving in a straight line.	$\vec{p} = m\vec{v}$ (7-4)
momentum	The application of a force imparts a change in momentum $\Delta\vec{p}$ to an object. The magnitude of the momentum depends on the applied force and also its duration.	$\Delta\vec{p} = \int_{t_i}^{t_f} \vec{F} dt$ (7-1)
weighted average	In a weighted average each value is allowed to have a greater or lesser effect on the result according to how often it occurs in the group of values being averaged.	

QUESTIONS AND PROBLEMS

In a few problems, you are given more data than you actually need; in a few other problems, you are required to supply data from your general knowledge, outside sources, or informed estimate.

Interpret as significant all digits in numerical values that have trailing zeros and no decimal points.

For all problems, use $g = 9.8 \text{ m/s}^2$ for the free-fall acceleration due to gravity. Neglect friction and air resistance unless instructed to do otherwise.

• Basic, single-concept problem

•• Intermediate-level problem, may require synthesis of concepts and multiple steps

••• Challenging problem

SSM *Solution is in Student Solutions Manual*

Conceptual Questions

1. •Using the common definition of the word *impulse*, comment on physicists' choice to define impulse as the change in momentum.

2. •Calc Starting from Newton's second law, explain how a collision that is free from external forces conserves momentum. In other words, explain how the momentum of the system remains constant with time.

3. •If the mass of a basketball is 18 times that of a tennis ball, can they ever have the same momentum? Explain your answer. SSM

4. •Two objects have equal kinetic energies. Are the magnitudes of their momenta equal? Explain your answer.

5. •A glass will break if it falls onto a hardwood floor but not if it falls from the same height onto a padded, carpeted floor. Describe the different outcomes in the collision between a glass and the floor in terms of fundamental physical quantities.

6. •A child stands on one end of a long wooden plank that rests on a frictionless icy surface. (a) Describe the motion of the plank when she runs to the other end of the plank. (b) Describe the motion of the center-of-mass of the system. (c) How would your answers change if she had walked the plank rather than run down it?

7. •Based on what you know about center of mass, why is it potentially dangerous to step off of a small boat before it comes to a stop at the dock? SSM

8. •A man and his large dog sit at opposite ends of a rowboat, floating on a still pond. You notice from shore that the boat moves as the dog walks toward his owner. Describe the motion of the boat from your perspective.

9. •**Astronomy** An asteroid (2007 WD5) passed between Earth and Mars in 2007. Scientists initially estimated a 4% chance that the 50-m-wide asteroid would collide with Mars. If the asteroid had collided with Mars, could it have knocked the planet out of its orbit? Explain your answer. (For more information on Near Earth Objects, see neo.jpl.nasa.gov/index.html.)

10. •**Calc** An object is subjected to a force that varies with time. Describe the mathematical process that you would use to calculate the impulse that is given to the object. Does your answer depend on the shape of the force versus time graph?

11. •How would you determine if a collision is elastic or inelastic?

12. •A recent US patent application describes a "damage avoidance system" for cell phones, which, upon detecting an impending, uncontrolled impact with a surface, deploys an air-filled bag around the phone. Explain how this could protect the cell phone from damage.

13. •Cite two examples of totally inelastic collisions that occur in your daily life.

14. •Why is conservation of energy alone not sufficient to explain the motion of a Newton's cradle toy, shown in **Figure 7-22**? Consider the case when two balls are raised and released together.

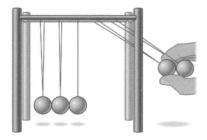

Figure 7-22 Problem 14

15. •**Calc** (a) Describe how you would determine the center of mass of a uniform sheet of metal. (b) Describe how you would determine the center of mass if a small hole had been cut through the sheet. SSM

16. •After being thrown into the air, a lit firecracker explodes at the apex of its parabolic flight. Is momentum conserved before or after the explosion? Is the mechanical energy conserved? What is the path of the center-of-mass? Neglect the effects of air resistance.

17. •An arrow shot into a straw target penetrates a distance that depends on the speed with which it strikes the target. How does the penetration distance change if the arrow's speed is doubled? Be sure to list all the assumptions you make while arriving at your answer.

Multiple-Choice Questions

18. •A large semitrailer truck and a small car have equal momentum. How do their speeds compare?
    A. The truck has a much higher speed than the car.
    B. The truck has only a slightly higher speed than the car.
    C. Both have the same speed.
    D. The truck has only a slightly lower speed than the car.
    E. The truck has a much lower speed than the car.

19. •A tennis player smashes a ball of mass m horizontally at a vertical wall. The ball rebounds at the same speed v with which it struck the wall. Has the momentum of the ball changed, and if so, by how much?
    A. mv
    B. 0
    C. $\frac{1}{2}mv$
    D. $2mv$
    E. $4mv$. SSM

20. •You throw a bouncy rubber ball and a wet lump of clay, both of mass m, at a wall. Both strike the wall at speed v, but while the ball bounces off with no loss of speed, the clay sticks. What is the change in momentum of the clay and ball, respectively?
    A. 0; mv
    B. mv; 0
    C. 0; $-2mv$
    D. $-mv$; $-mv$
    E. $-mv$; $-2mv$

21. •Consider a completely inelastic, head-on collision between two particles that have equal masses and equal speeds. Describe the velocities of the particles after the collision.
    A. The velocities of both particles are zero.
    B. Both of their velocities are reversed.
    C. One of the particles continues with the same velocity and the other comes to rest.
    D. One of the particles continues with the same velocity and the other reverses direction at twice the speed.
    E. More information is required to determine the final velocities.

22. •An object is traveling in the positive x direction with speed v. A second object that has half the mass of the first is traveling in the opposite direction with the same speed. The two experience a completely inelastic collision. The final x component of the velocity is
    A. 0
    B. $v/2$
    C. $v/3$

This page is intentionally left blank.

For complete end of chapter problem sets, please go to
www.whfreeman.com/kestentauck

t (s)	F (N)	t (s)	F (N)
12	25	19	25
13	25	20	25
14	25	21	20
15	25	22	15
16	25	23	10
17	25	24	5
18	25	25	0

Problems

7-1: Linear Momentum

38. •A 10,000-kg train car moves east at 15 m/s. Determine the momentum of the train car.

39. •**Sports** The magnitude of the instantaneous momentum of a 57-g tennis ball is 2.6 (kg·m)/s. What is its speed? SSM

40. •Determine the initial momentum, final momentum, and change in momentum of a 1250-kg car initially backing up at 5 m/s, then moving forward at 14 m/s.

41. •**Sports** What is the momentum of a 135-kg defensive lineman running at 7 m/s?

42. •One ball has four times the mass and twice the speed of another. (a) How does the momentum of the more massive ball compare to the momentum of the less massive one? (b) How does the kinetic energy of the more massive ball compare to the kinetic energy of the less massive one?

43. •A girl who has a mass of 55 kg rides her skateboard to class at a speed of 6 m/s. (a) What is her momentum? (b) If the momentum of the skateboard itself is 30 (kg·m)/s, what is its mass? SSM

7-2: Conservation of Momentum

44. •A 2-kg object is moving east at 4 m/s when it collides with a 6-kg object that is initially at rest. After the completely elastic collision, the larger object moves east at 1 m/s. Find the final velocity of the smaller object after the collision.

45. •A 3-kg object is moving toward the right at 6 m/s. A 5-kg object moves to the left at 4 m/s. After the two objects collide completely elastically, the 3-kg object moves toward the left at 2 m/s. Find the final velocity of the 5-kg object.

46. •Blythe and Geoff are ice-skating together. Blythe has a mass of 50 kg and Geoff has a mass of 80 kg. Blythe pushes Geoff in the chest when both are at rest, causing him to move away at a speed of 4 m/s. (a) Determine Blythe's speed after she pushes Geoff. (b) In what direction does she move?

47. ••An object of mass $3M$, moving in the $+x$ direction at speed v_0, breaks into two pieces of mass M and $2M$ as shown in **Figure 7-24**. Find the final velocities of the resulting pieces in terms of v_0. SSM

Figure 7-24 Problem 47

48. •In a game of pool, the cue ball is rolling at 2 m/s in a direction 30° north of east when it collides with the eight ball (initially at rest). The mass of the cue ball is 170 g but the mass of the eight ball is only 156 g. After the completely elastic collision, the cue ball heads off 10° north of east and the eight ball moves off due north. Find the final speeds of each ball after the collision.

49. •A 1000-kg car moving at 30 m/s due north collides completely elastically with a 1250-kg truck heading 45° north of east. After the collision the truck is moving due north and the car is moving due east. Find the initial speed of the truck.

50. •A 0.10-kg firecracker is hanging by a light string from a tree limb. The fuse is lit and the firecracker explodes into three pieces: a small piece (0.01 kg), a medium-sized piece (0.03 kg), and a large piece (0.06 kg). Assume the firecracker is at the origin of a coordinate system such that the $+z$ axis points straight up, the $+x$ axis points due east, and the $+y$ axis points due north. The fragments fly off according to the following momentum vectors:

$$\text{Largest fragment:} \quad \vec{p}_L = p_{Lx}\,\hat{x} - p_{Lz}\hat{z}$$
$$\text{Medium fragment:} \quad \vec{p}_M = p_{My}\,\hat{y} + p_{Mz}\,\hat{z}$$
$$\text{Smallest fragment:} \quad \vec{p}_S = -p_{Sx}\,\hat{x} - p_{Sy}\,\hat{y}$$

After the explosion, the component of velocity of the smallest piece in both the x and y directions is 4 m/s. ($v_{Sx} = v_{Sy} = 4$ m/s.) The velocity of the largest fragment is 1 m/s in the z direction ($v_{Lz} = 1$ m/s). Determine the velocity vector of each fragment immediately after the explosion.

51. •**Biology** During mating season, male bighorn sheep establish dominance with head-butting contests which can be heard up to a mile away. When two males butt heads, the "winner" is the one that knocks the other backward. In one contest a sheep of a mass 95 kg and moving at 10 m/s runs directly into a sheep of mass 80 kg moving at 12 m/s. Which ram wins the head-butting contest? SSM

52. ••One way that scientists measure the mass of an unknown particle is to bounce a familiar particle, such as a proton or electron, off the unknown particle in a bubble chamber. The initial and rebound velocities of the familiar particle is measured from photographs of the bubbles it creates as it moves; the information is used to determine the mass of the unknown particle. (a) If a known particle of mass m and initial speed v_0 collides elastically, head-on with a stationary unknown particle and then rebounds with speed v, find the mass of the unknown particle in terms of m, v, and v_0. (b) If the known particle is a proton and the unknown particle is a neutron, what will be the recoil speed of the proton and the final speed of the neutron?

7-3: Inelastic Collisions

53. •A 10,000-kg train car moving due east at 20 m/s collides with and couples to a 20,000-kg train car that is initially at rest. Find the common velocity of the two-car train after the collision.

54. •A large fish has a mass of 25 kg and swims at 1.0 m/s toward and then swallows a smaller fish that is not moving. If the smaller fish has a mass of 1.0 kg, what is the speed of the larger fish immediately after it finishes lunch?

55. •A 5.0-kg howler monkey is swinging due east on a vine. It overtakes and grabs onto a 6.0-kg monkey also moving east on a second vine. The first monkey is moving at 12 m/s at the instant it grabs the second, which is moving at 8 m/s. After they join on the same vine, what is their common speed? SSM

56. •A 1200-kg car is moving at 20 m/s due north. A 1500-kg car is moving at 18 m/s due east. The two cars simultaneously approach an icy intersection where, with no brakes or steering, they collide and stick together. Determine the speed and direction of the combined two-car wreck immediately after the collision.

57. •Sports An 85-kg linebacker is running at 8 m/s directly toward the sideline of a football field. He tackles a 75-kg running back moving at 9 m/s straight toward the goal line (perpendicular to the original direction of the linebacker). As a result of the collision both players momentarily leave the ground. Determine their common speed and direction immediately after they collide.

7-4: Contact Time

58. •A sudden gust of wind exerts a force of 20 N for 1.2 s on a bird that had been flying at 5 m/s. As a result, the bird ends up moving in the opposite direction at 7 m/s. What is the mass of the bird?

59. •Determine the force exerted on your hand as you catch a 0.200-kg ball moving at 20 m/s. Assume the time of contact is 0.025 s. SSM

60. •Determine the impulse delivered over the first 10 s to the object acted on by the force described by the graph (Figure 7-25).

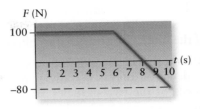

Figure 7-25 Problem 60

61. •Sports A baseball of mass 0.145 kg is thrown at a speed of 40 m/s. The batter strikes the ball with a force of 25,000 N; the bat and ball are in contact for 0.5 ms. Assuming that the force is exactly opposite to the original direction of the ball, determine the final speed of the ball.

62. •A 5-kg object is constrained to move along a straight line. Its initial speed is 12 m/s in one direction, and its final speed is 8 m/s in the opposite direction. Complete the graph of force versus time with appropriate values for both variables (Figure 7-26). Several answers are correct, just be sure that your answer is internally consistent.

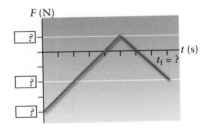

Figure 7-26 Problem 62

63. ••Calc A single force acts in the x direction on an object that is initially stationary. The force is described by $F(t) = 9t^2 + 8t^3$, where the constants carry SI units. Determine the momentum in the x direction of the object after the force has acted for 1 s. SSM

64. •Sports A baseball bat strikes a ball when both are moving at 31.3 m/s (relative to the ground) toward each other. The bat and ball are in contact for 1.20 ms, after which the ball is traveling at a speed of 42.5 m/s. The mass of the bat and the ball are 850 g and 145 g, respectively. Find the magnitude and direction of the impulse given to (a) the ball by the bat and (b) the bat by the ball. (c) What average force does the bat exert on the ball? (d) Why doesn't the force shatter the bat?

7-5: Elastic Collisions

65. •A 2-kg ball is moving at 3 m/s toward the right. It collides elastically with a 4-kg ball that is initially at rest. Determine the velocities of the balls after the collision.

66. •A 0.170-kg ball is moving at 4.00 m/s toward the right. It collides elastically with a 0.155-kg ball moving at 2.00 m/s toward the left. Determine the final velocities of the balls after the collision.

67. •A 10-kg block of ice is sliding due east at 8 m/s when it collides elastically with a 6-kg block of ice that is sliding in the same direction at 4 m/s. Determine the velocities of the blocks of ice after the collision. SSM

68. •A neutron traveling at 2×10^5 m/s collides elastically with a deuteron that is initially at rest. Determine the final speeds of the two particles after the collision. The mass of a neutron is 1.67×10^{-27} kg, and the mass of a deuteron is 3.34×10^{-27} kg.

7-6: Center of Mass

69. •Find the coordinates of the center of mass of the three objects shown in Figure 7-27. Distances are in meters.

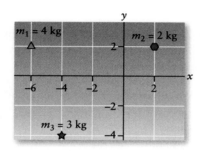

Figure 7-27 Problem 69

70. •Four beads each of mass M are attached at various locations to a hoop of mass M and radius R (Figure 7-28). Find the center of mass of the hoop and beads.

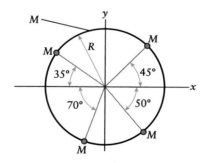

Figure 7-28 Problem 70

71. •What are the coordinates of the center of mass for the three objects shown in Figure 7-29? The uniform rod has a mass of 10 kg, has a length of 30 cm, and is located at $x = 50$ cm. The oval football has a mass of 2 kg, a semimajor axis of 15 cm, a semiminor axis of 8 cm, and is centered at $x = -50$ cm. The spherical volleyball has a mass of 1 kg, a radius of 10 cm, and is centered at $y = -30$ cm. SSM

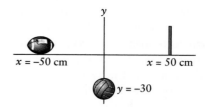

Figure 7-29 Problem 71

General Problems

72. •**Sports** A major league baseball has a mass of 0.145 kg. Neglecting the effects of air resistance, determine the momentum of the ball when it hits the ground if it falls from rest on the roof of the Metrodome in Minneapolis, Minnesota, a height of 60 m.

73. •Forensic scientists can determine the speed at which a rifle fires a bullet by shooting into a heavy block hanging by a wire. As the bullet embeds itself in the block, the block and embedded bullet swing up; the impact speed is determined from the maximum angle of the swing. (a) Which would make the block swing higher, a 0.204 Ruger bullet of mass 2.14 g and muzzle speed 1290 m/s or a 7-mm Remington Magnum bullet of mass 9.71 g and muzzle speed 948 m/s? Assume the bullets enter the block right after leaving the muzzle of the rifle. (b) Using your answer in part (a), determine the mass of the block so that when hit by the bullet it will swing through a 60.0° angle. The block hangs from wire of length 1.25 m and negligible mass. (c) What is the speed of an 8.41-g bullet that causes the block to swing upward through a 30.0° angle?

74. •Sally finds herself stranded on a frozen pond so slippery that she can't stand up or walk on it. To save herself, she throws one of her heavy boots horizontally, directly away from the closest shore. Sally's mass is 60 kg, the boot's mass is 5 kg, and Sally throws the boot with speed equal to 30 m/s. (a) What is Sally's speed immediately after throwing the boot? (b) Where is the center of mass of the Sally–boot system, relative to where she threw the boot, after 10 s? (c) How long does it take Sally to reach the shore, a distance of 30 m away from where she threw the boot? For all parts, assume the ice is frictionless.

75. •A friend suggests that if all the people in the United States dropped down from a 1-m-high table at the same time, Earth would move in a noticeable way. To test the credibility of this proposal, (a) determine the momentum imparted to Earth by 300 million people, of an average mass of 65 kg, dropping from 1 m above the surface. Assume no one bounces. (b) What change in Earth's speed would result? SSM

76. •A 65-kg person jumps off a fence that is 2 m high and bends his knees when landing to avoid breaking a leg. Estimate the average force exerted on his feet by the ground if the landing is (a) stiff-legged and (b) with bent legs. Assume his center of mass moves 1.0 cm during impact when he lands with stiff legs, and 50.0 cm when he bends his legs while landing.

77. •You have been called to testify as an expert witness in a trial involving a head-on collision. Car A weighs 1500 lb and was traveling eastward. Car B weighs 1100 lb and was traveling westward at 45 mph. The cars locked bumpers and slid eastward with their wheels locked for 19 ft before stopping. You have measured the coefficient of kinetic friction between the tires and the pavement to be 0.75. How fast (in miles per hour) was car A traveling just before the collision? (This problem uses English units because they would be used in a U.S. legal proceeding.)

78. •A 5000-kg open train car is rolling at a speed of 20.0 m/s when it begins to rain heavily and 200 kg of water collects quickly in the car. If only the total mass has changed, what is the speed of the flooded train car? For simplicity, assume that all of the water collects at one instant, and that the train tracks are frictionless.

79. •An 8000-kg open train car is rolling at a speed of 20.0 m/s when it begins to rain heavily. After water has collected in the car it slows to 19 m/s. What mass of water has collected in the car? For simplicity, assume that all of the water collects at one instant, and that the train tracks are frictionless. SSM

80. •An open rail car of initial mass 10,000 kg is moving at 5 m/s when rocks begin to fall into it from a conveyor belt. The rate at which the mass of rocks increases is 500 kg/s. Find the speed of the train car after rocks have fallen into the car for a total of 3 s.

81. •A skier and a snowboarder of equal mass collide in an area where the snow is level. Just before the collision, the skier was moving south at 8 m/s and the snowboarder was moving west at 12 m/s. After the unplanned meeting, the skier slid 45° south of west and the snowboarder slid 45° north of west. Find the speed of each athlete after the collision, assuming that the collision was completely elastic.

82. •Sports The sport of curling is quite popular in Canada. A curler slides a 19.1-kg stone so that it strikes a competitor's stationary stone at 6.4 m/s before moving at an angle of 120° from its initial direction. The competitor's stone moves off at 5.6 m/s. Find the final speed of the first stone and the final direction of the second one.

83. ••Biology Lions can run at speeds up to approximately 80 km/h. A hungry, 135-kg lion running northward at top speed attacks and holds onto a 29-kg Thomson's gazelle running eastward at 60 km/h. Find the speed and direction of travel of the lion–gazelle system just after the lion attacks. SSM

84. •Biology The mass of a pigeon hawk is twice that of the pigeons it hunts. Suppose a pigeon is flying north at 23 m/s when a hawk swoops down, grabs the pigeon and flies off (Figure 7-30). The hawk was flying north at 35 m/s, at an angle 45° below the horizontal, at the instant of the attack. Find the final velocity vector of the birds.

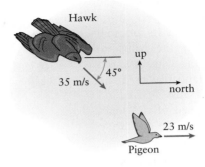

Figure 7-30 Problem 84

85. •A 12-g bullet is fired into a block of wood at 250 m/s (Figure 7-31). The block is attached to a spring that has a spring constant of 200 N/m. The block, bullet, and spring all move to the right a distance of 30 cm, as shown, before momentarily coming to a stop. Determine the mass of the wooden block. The spring compresses 30 cm when the bullet becomes embedded in the block. SSM

Figure 7-31 Problem 85

86. •In a ballistic pendulum experiment, a small marble is fired into a cup attached to the end of a pendulum. If the mass of the marble is 0.0075 kg and the mass of the pendulum is 0.250 kg, how high will the pendulum swing if the marble has an initial speed of 6 m/s? Assume that the mass of the pendulum is concentrated at its end.

87. •••Calc Suppose the force that acts on an object in the z direction is given by $F_z(t) = t^2 e^{-t^2}$. Determine the momentum of the object in the z direction after the force has been acting for a long time, assuming that the object starts from rest. Note that $\int_0^\infty x^2 e^{-ax^2} dx = \sqrt{\pi/16a^3}$.

88. ••Calc The momentum of a particle in the x direction is given by $p_x(t) = 3 \sin(5t) + 2t^2$, where the constants carry SI units. Determine the net force in the x direction acting on the particle, first at an arbitrary time t and then for the specific case when $t = 2$ s.

89. ••Calc The momentum of a particle in the y direction is given by $p_y(t) = 3e^{-t} \cos(5t) + 3t^{-1/2}$, where

the constants carry SI units. Determine the net force in the y direction acting on the particle as function of time.

90. •••A 2.5-kg object elastically collides with a 3.6-kg object that is initially at rest. The small object has a speed of 4 m/s and travels at an angle of θ_1 with its original direction after the collision; the larger object has a speed of 2.5 m/s and travels at an angle of θ_2 (**Figure 7-32**). Find the initial speed of the small object and the two scattering angles, θ_1 and θ_2.

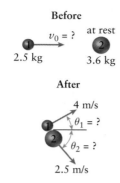

Before

$v_0 = ?$ at rest

2.5 kg   3.6 kg

After

4 m/s

$\theta_1 = ?$

$\theta_2 = ?$

2.5 m/s

Figure 7-32 Problem 90

91. ••A 0.075-kg ball is thrown at 25 m/s toward a brick wall. (a) Determine the impulse that the wall imparts to the ball when it hits and rebounds at 25 m/s in the opposite direction. (b) Determine the impulse that the wall imparts to the ball when it hits and rebounds at an angle of 45°. (c) If the ball thrown in part (b) contacts the wall for 0.01 s, determine the magnitude and direction of the force that the wall exerts on the ball.

92. ••**Calc** What are the center of mass coordinates of a uniform sheet of metal if it is formed into the triangular shape shown in **Figure 7-33**?

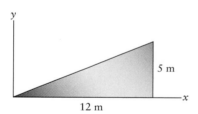

5 m

12 m

Figure 7-33 Problem 92

93. ••**Calc** Find the center of mass coordinates of a sheet of uniform metal shaped into an isosceles triangle as shown in **Figure 7-34**. SSM

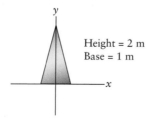

Height = 2 m
Base = 1 m

Figure 7-34 Problem 93

94. ••**Calc** A thin trapezoidal plate has a mass M, a base length L, a height H, and an upper length $L/2$. Find the position of the center of mass without using results you may know for the center of mass of other objects. Express your answer in terms of H.

95. •••**Calc** The force vector that acts on an object is given by

$$\vec{F}(t) = \left(\frac{14}{t^2 + 1}\right)\hat{x} + \left(\frac{12}{t^2 - 1}\right)\hat{y}$$

where the constants carry SI units. Determine the momentum as a function of time, assuming that the object starts from rest.

96. ••The mass of an oxygen nucleus is about 16 times that of a proton. A proton traveling at 5×10^5 m/s collides completely elastically with an oxygen nucleus and leaves the interaction at 5×10^5 m/s. (a) What fraction of the initial kinetic energy is transferred to the oxygen nucleus? (b) Determine the final directions of the proton and the oxygen nucleus.

97. ••**Calc** A circular hole of radius $R/4$ is cut from a thin, uniform disk of mass M and radius R. The center of the hole is located at $x = R/2$ and $y = 0$ as shown in **Figure 7-35**. Find the center of mass of the resulting object, expressing your answer in terms of M and R.

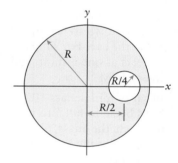

y

R

$R/4$

x

$R/2$

Figure 7-35 Problem 97

98. •••**Calc** Determine the center of mass of a solid hemisphere of mass M and radius R, relative to the center of the base of the hemisphere.

99. •••**Calc** A rocket gains forward thrust as it ejects stored gases from its tail section, as shown in **Figure 7-36**. In a time interval of Δt the rocket ejects gas of mass Δm with a speed of v_g relative to the rocket. In the sketch, $m(t)$ is the mass of the rocket plus the stored fuel at any time, and v is the speed of the rocket at that instant. Determine the net thrust acting on the rocket at that time.

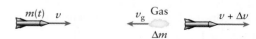

$m(t)$ v v_g Gas $v + \Delta v$

Δm

Figure 7-36 Problem 99

(AGENCE NATURE/Natural History Picture Archive.)

8 Rotational Motion

Falling cats usually land on their feet. The physics behind this graceful motion is a complex dance of rotations. By bending in the middle, the cat can rotate her front and hindquarters separately. Extending her front legs and tucking in her rear legs allows her back legs to turn more rapidly. Later in the fall, she extends her back legs, which slows down their rotation and leaves them in the landing position as her front legs come vertical. Each of these motions can be described by the rotational quantities uncovered in this chapter.

Physical quantities such as kinetic energy and linear momentum depend on speed, which we defined as the change in position divided by the change in time. Does the windmill shown in **Figure 8-1** have kinetic energy and momentum? Although its *position* hasn't changed since it was built, the rotating blades are constantly moving; there must be kinetic energy and momentum associated with this motion. In this chapter, we examine the physics associated with rotations.

8-1 Rotational Kinetic Energy

To determine the kinetic energy of a windmill, consider the rotation of one of its blades around the fixed central axis, as shown at two successive times in **Figure 8-2**. Imagine that the rigid blade is divided up into many small pieces, the first being at the rotation point (fixed axis) and each of the other pieces farther and farther out along the blade. We identify one of these pieces using the subscript i, and specify both the mass m_i and the distance r_i from the rotation point. In the figure, the blade is shown rotating through an angle $\Delta\theta$ in a time Δt.

Figure 8-1 Although each windmill remains in one place and therefore has no translational kinetic energy, its blades rotate in the wind. In this chapter we add rotation to our discussion of motion. *(ImageState)*

Figure 8-2 Each small element of a blade of a windmill is moving around the rotation axis and therefore contributes to the kinetic energy associated with the rotation of the blade.

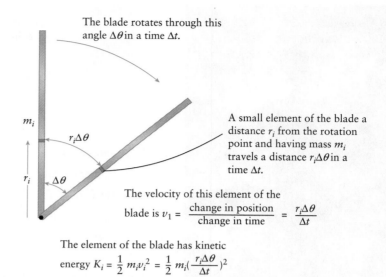

The blade rotates through this angle $\Delta\theta$ in a time Δt.

A small element of the blade a distance r_i from the rotation point and having mass m_i travels a distance $r_i\Delta\theta$ in a time Δt.

The velocity of this element of the blade is $v_1 = \dfrac{\text{change in position}}{\text{change in time}} = \dfrac{r_i\Delta\theta}{\Delta t}$

The element of the blade has kinetic energy $K_i = \frac{1}{2}m_iv_i^2 = \frac{1}{2}m_i(\dfrac{r_i\Delta\theta}{\Delta t})^2$

The position of each small element of the blade is changing versus time, so each has a defined velocity and therefore kinetic energy. The *i*th element, shown in the figure, moves a distance $r_i\Delta\theta$ in time Δt, so its speed is

$$v_i = \frac{r_i\Delta\theta}{\Delta t} \tag{8-1}$$

(The relationship between the distance along the arc of a circle its radius, and the angular extent of the arc is presented in Figure 3-32.) The kinetic energy of this element is

$$K_i = \frac{1}{2}m_iv_i^2 = \frac{1}{2}m_i\left(\frac{r_i\Delta\theta}{\Delta t}\right)^2$$
$$= \frac{1}{2}m_ir_i^2\left(\frac{\Delta\theta}{\Delta t}\right)^2 \tag{8-2}$$

Notice that the term in parentheses in the last step doesn't depend on which element of the blade we selected, because both the angle through which the blade rotates and the time it takes are the same for all parts of the blade. For a rigid object, this **angular velocity** serves as a convenient way to characterize the rotation in a way that is independent of size or shape. The Greek letter omega (ω) is the conventional symbol for angular velocity:

$$\omega = \frac{\Delta\theta}{\Delta t} \tag{8-3}$$

The units of angular velocity are radians per second (rad/s). As a reminder, radians are a unit of angle; there are 2π radians in a circle. By convention, ω is positive when $\Delta\theta$ is counterclockwise and negative when $\Delta\theta$ is clockwise.

At any instant, the rate at which the rotation angle changes is the same across all pieces of a rigid object, so angular velocity ω is also the same. This does not mean, however, that ω is constant over time, but rather, unlike (linear) velocity, angular velocity does not vary across a rotating, rigid object. If the rotation rate varies with time, then Equation 8-3 defines the **average angular velocity**. We define the angular velocity at any instant, **instantaneous angular velocity**, by letting the time interval Δt be infinitesimally small:

$\sqrt{x}$ *See the Math Tutorial for more information on Differential Calculus*

$$\omega = \lim_{\Delta t \to 0} \frac{\Delta\theta}{\Delta t} = \frac{d\theta}{dt} \tag{8-4}$$

Instantaneous angular velocity is defined as the derivative of angle with respect to time in the same way that instantaneous velocity is defined as the derivative of position with respect to time (Equation 2-16).

The relationship between speed and the radius (Equation 8-1) of an object's rotation can be rewritten as $v = r\omega$ or

$$\omega = \frac{v}{r} \tag{8-5}$$

for an object or a piece of an object a distance r from the rotation point moving at v. Because angular velocity ω is the same for any element of a rotating object, Equation 8-5 tells us that the speed of any element of a rotating object increases at the same rate that the distance r from the rotation point increases.

The expression for the kinetic energy of the ith element of the blade (Equation 8-2) can be written in terms of ω:

$$K_i = \frac{1}{2}m_i r_i^2 \omega^2$$

Kinetic energy is a scalar (not a vector), so the kinetic energy of the entire rotating windmill blade can be found by adding up the kinetic energy of each of the elements of the blade:

$$K = \sum K_i = \sum \frac{1}{2}m_i r_i^2 \omega^2$$

We need to take a bit of care in doing the sum. We must break the blade into pieces so small that there is only one value of r for each piece. If these pieces were large, the distance from the fixed rotation axis would be different for different parts of the piece, and we wouldn't know what value of r to enter into the sum.

Values that don't change with the subscript i, such as $\frac{1}{2}$ and ω^2, can be taken out of the sum:

$$K = \frac{1}{2}\left(\sum m_i r_i^2\right)\omega^2 \tag{8-6}$$

See Mathbox 8-1.

Math Box 8-1   Removing a Constant Term from a Sum

Constant terms can always be removed from a sum. Prove it to yourself by trying it with a few numbers. For example, in the sum $2(3) + 2(4) + 2(5)$, the constant value 2 can be pulled out from each term:

$$2(3) + 2(4) + 2(5) = 2(3 + 4 + 5)$$

We could have written this expression as

$$\sum_{n=3}^{5} 2n = 2\sum_{n=3}^{5} n = 2(3 + 4 + 5)$$

So for $K = \sum \frac{1}{2}m_i r_i^2 \omega^2$, because $\frac{1}{2}$ and ω are constant, both terms can be taken out of the sum:

$$K = \frac{1}{2}\omega^2 \sum m_i r_i^2$$

or

$$K = \frac{1}{2}\left(\sum m_i r_i^2\right)\omega^2$$

We choose to write the ω^2 term after the sum only to make this expression have the same form as the expression for linear kinetic energy

$$K = \frac{1}{2} \times \text{something} \times (\text{rate})^2.$$

The term in parentheses in Equation 8-6, the **rotational inertia**, or **moment of inertia**, characterizes the motion of a rigid object that is allowed to rotate. As we'll see more clearly later in this chapter, the moment of inertia plays a role in rotational motion similar to the role that mass plays in linear motion. Moment of inertia is commonly represented by the variable I:

$$I = \sum m_i r_i^2 \tag{8-7}$$

Equation 8-7 requires that the separate elements of mass m_i and distance r_i from the rotation axis be small. Moment of inertia is the topic of the next section. The SI units of I are kilogram-square meters (kg $\cdot$ m^2).

We will call the kinetic energy of a rigid object rotating around a fixed axis **rotational kinetic energy** to differentiate it from the *translational* (or linear) kinetic energy $K_{\text{translational}} = \frac{1}{2}mv^2$. By substituting our definition of moment of inertia (Equation 8-7) into Equation 8-6, we define rotational kinetic energy as

$$K_{\text{rotational}} = \frac{1}{2}I\omega^2 \tag{8-8}$$

By comparing the definitions of translational and rotational kinetic energies, notice that mass (in the translational kinetic energy equation) corresponds to the moment of inertia (in the rotational kinetic energy equation). Whereas we interpreted mass as a property of matter that represents the resistance of an object to a change in translational velocity, the moment of inertia represents the resistance of an object to a change in rotational or angular velocity. In the same way that we defined inertia as the tendency of an object to resist a change in translational motion, we can define rotational inertia as the tendency of an object to resist a change in rotational motion.

Both translational and rotational kinetic energies depend on the mass of the moving object. However, the moment of inertia (Equation 8-7) and therefore the rotational kinetic energy also depends on how the mass is distributed with respect to the axis of rotation. A bit of mass far from the rotation axis has a larger effect on the value of the moment of inertia than the same amount of mass close to the axis.

Two objects that have the same mass and move at the same translational speeds have the same kinetic energies even if their shapes are different. However, even at the same angular velocity, these two objects will likely have different rotational kinetic energies when rotated because the different shapes will result in different values for the moment of inertia. For example, when both are rotated at the same angular velocity, a bicycle wheel, essentially a ring with negligibly small mass near the center, has more rotational kinetic energy than a flat, uniform disk of the same size and mass because more of the mass of the wheel is farther from the rotation axis (Figure 8-3). Indeed, two *identical* objects rotated around different *axes* will likely have different rotational kinetic energies, because again, the distribution of mass relative to the rotation axis will be different for the two. The rotational kinetic energy of a rod rotating around an axis perpendicular to the rod and through its end is four times larger than the rotational kinetic energy of the rod when it rotates at the same angular velocity around an axis perpendicular to the rod and through its center (Figure 8-4).

These two wheels have the same mass and size, but the first is a ring with no mass at the center while the other is a uniform disk.

Sliding at the same linear velocity, both have the same linear kinetic energy.

When both rotate at the same angular velocity, the ring has greater rotational kinetic energy. More of its mass is farther from the rotation axis, resulting in a larger moment of inertia around that axis.

Figure 8-3 The two wheels have the same mass and size, but one is a uniform disk and the other is a ring with none of its mass in the center. When the wheels slide with the same velocity, without rotating, they have the same kinetic energy. When they rotate at the same angular velocity, the uniform disk has less rotational kinetic energy and a smaller moment of inertia because more of its mass is closer to the rotation axis.

The kinetic energy of a rod rotating around an axis perpendicular to the rod and through its end is four times larger...

...than the kinetic energy of the rod when it rotates at the same angular velocity around an axis perpendicular to the rod and through its center.

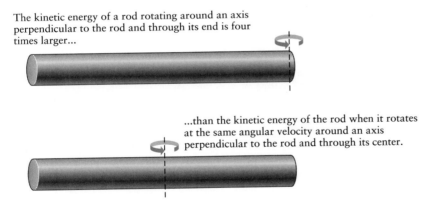

The more mass there is farther from the rotation axis, the larger the moment of inertia, and the larger the rotational kinetic energy for any given angular velocity.

Figure 8-4 The magnitude of the kinetic energy of a rotating rod depends on the location of the rotation axis relative to the distribution of the mass of the rod. The moment of inertia is larger when more of the mass is farther from the rotation axis.

 **Got the Concept 8-1**
Swinging a Phone

Tie a cell phone to the end of a very light thread. Holding on to the other end of the thread, would you be able to swing the cell phone around in a horizontal circle? (Careful, the answer might not be what you expect!)

? Got the Concept 8-2
Flight

The physics which underlies the flight characteristics of birds and flying insects is complicated. Their maneuverability has as much to do with the contributions their wings make to the moments of inertia around their roll, pitch, and yaw axes (Figure 8-5) as with the aerodynamic characteristics of the wings. A bird's wings can be as much as 15% of the total body mass, while an insect's wings are typically considerably less. In general, would you expect a flying insect or a bird to be able to maneuver more quickly in flight?

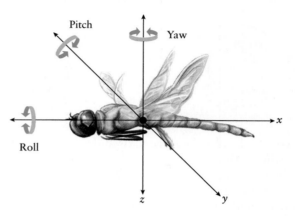

Figure 8-5 Flying objects such as insects, birds, and planes can rotate around the x, y, and z axes. A rotation around the x axis is called a roll, a rotation around the y axis is called a pitch, and a rotation around the z axis is called a yaw. (Dudley, R. (2002). Mechanisms and Implications of Animal Flight Maneuverability. *Integrative and Comparative Biology* , 42:135–140.)

Example 8-1 Whirl an Object

When the physicist in Figure 8-6 whirls a small red object in a nearly horizontal circle at the end of a 0.30-m-long string, the object makes 5 rev/s. Treating the object as if all its mass were concentrated at a single point and neglecting the mass of the string, how much rotational kinetic energy must the physicist supply to cause this motion to occur? The object has a mass equal to 0.20 kg.

SET UP

To determine rotational kinetic energy,

$$K_{\text{rotational}} = \frac{1}{2}I\omega^2 \tag{8-8}$$

we need to know the moment of inertia I of the system around the rotation axis as well as the angular velocity ω. By neglecting the mass of the string and treating the red object as if all its mass were concentrated at a single point, we can find the moment of inertia using Equation 8-7:

$$I = \sum m_i r_i^2 = m_{\text{object}} r_{\text{object}}^2$$

SOLVE

The rotational kinetic energy of the small red object is

$$K_{\text{rotational}} = \frac{1}{2} m_{\text{object}} r_{\text{object}}^2 \omega^2$$

Figure 8-6 Physicists have more fun than physiologists. *(Courtesy David Tauck)*

To compute a numeric result, convert $\omega = 5$ rev/s to rad/s:

$$\omega = \frac{5 \text{ rev}}{\text{s}} \times \frac{2\pi \text{ rad}}{1 \text{ rev}} = \frac{10\pi \text{ rad}}{\text{s}}$$

So

$$K_{\text{rotational}} = \frac{1}{2}(0.20 \text{ kg})(0.30 \text{ m})^2\left(\frac{10\pi \text{ rad}}{\text{s}}\right)^2 = 8.9 \text{ J}$$

REFLECT

The physicist must supply nearly 9 J of energy to rotate the small red object—not a lot of energy, but not a little, either. From what height would he need to drop the object to impart the same energy? Setting gravitational potential energy $U = mgh$ equal to the result of 8.9 J gives $h = 4.5$ m. That's probably almost two and a half times the physicist's height, so it's reasonable to conclude that it takes a modest amount of effort on his part to swing the object.

Practice Problem 8-1 When the physicist in Figure 8-6 whirls a small red object of mass 0.20 kg in a nearly horizontal circle at the end of a 0.30-m-long string, he imparts 15 J of energy to the object. Treating the object as if all its mass were concentrated at a single point and neglecting the mass of the string, how many revolutions per second does the object make?

> ✳ **What's Important 8-1**
> The kinetic energy of a rigid object rotating around a fixed axis is called rotational kinetic energy. It depends not only on the mass and angular velocity of an object, but also on how the mass of the object is distributed with respect to the axis of rotation.

8-2 Moment of Inertia

The astronaut in **Figure 8-7** would find that even in a weightless environment it's harder to cause a massive object to accelerate than it is a less massive one. A woman pulling on a massive vault door has to contend with not only the mass of the door but also the distribution of the mass with respect to the hinges. Inertia, the tendency of an object to resist a change in translational motion, and the object's mass are really two aspects of the same physics. The moment of inertia plays the same role in rotational physics that mass does for linear motion.

The moment of inertia of a very small object that has a mass m and rotates around an axis a distance r away is given by $I = mr^2$. For an object to be considered small, the distance from the rotation axis to all points on the object must be the same; the object in **Figure 8-8a** is small enough, but the object in **Figure 8-8b** is not. The moment of inertia of a large object rotating around a given axis can be found by imagining the object broken into many small pieces, finding the contribution $I_i = m_i r_i^2$ that each piece makes to the moment of inertia, and then adding up the separate contributions. This method leads to the same relationship we discovered for the moment of inertia by considering the kinetic energy of rotation:

$$I = \sum m_i r_i^2 \tag{8-7}$$

The moment of inertia of an object can be determined only with respect to a specific rotation axis. Except in the special case of objects which are symmetric in some way, an object likely has a different moment of inertia around each different rotation axis.

Figure 8-7 Inertia is the tendency of an object to resist a change in translational motion. Even in orbit around Earth, it is difficult for an astronaut to cause a massive object to accelerate because of its inertia. *(NASA)*

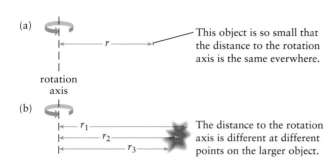

Figure 8-8 (a) The distance from the rotation axis to any part of a very small object is the same. (b) In contrast, different parts of a large object will be different distances from the rotation axis. *(Ocean/Corbis)*

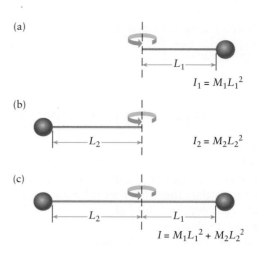

Figure 8-9 Moment of inertia is additive. The moments of inertia of objects 1 and 2 are I_1 and I_2, respectively. If we combine the two objects into one single object, its moment of inertia is the sum of I_1 and I_2.

The moment of inertia is additive. If you know the moments of inertia of two objects around some rotation axis and you attach them to form a single object, the moment of inertia of the new object—around the same axis—is the sum of the moments of inertia of the two separate objects.

In **Figure 8-9a**, a small sphere that has a mass M_1 is attached to a Styrofoam rod that has a negligible mass and a length L_1. The moment of inertia of the sphere when it rotates around the axis passing through the end of the rod is $I_1 = M_1 L_1^2$. Similarly, the moment of inertia of another small sphere that has a mass M_2 and rotates at the end of a rod that has a negligible mass and length L_2 is $I_2 = M_2 L_2^2$ (**Figure 8-9b**). When the ends of the two rods are attached and the combined object is rotated around the same axis, as in **Figure 8-9c**, the moment of inertia is

$$I = I_1 + I_2 = M_1 L_1^2 + M_2 L_2^2$$

Moment of Inertia of a Thin, Uniform Rod Rotating around One End

The moment of inertia of a large object is defined by

$$I = \sum m_i r_i^2 \qquad (8\text{-}7)$$

where we have imagined the object broken into many small pieces identified by the subscript i. Each piece must be so small that the distance from the rotation axis to all points on the piece is the same. To guarantee that this is always true, the pieces must be of infinitesimal size. To remind us that each piece is of infinitesimal mass, we write the mass term as dm. This allows us to write the sum in Equation 8-7 as the integral:

$$I = \int dm\, r^2$$

Or, to write the moment of inertia in a more standard form,

$$I = \int r^2\, dm \qquad (8\text{-}9)$$

√x *See the Math Tutorial for more information on Integrals*

Again, note that in converting the sum to an integral, the mass m_i of each piece of an object becomes dm, the mass of an infinitesimally small mass element. The bounds on the integral are set up so that every bit of mass of an object is included in the integration.

To get some hands-on experience finding the moment of inertia of an object that can rotate, let's consider a thin, uniform rod of length L and mass M that rotates around one end. Such a rod is shown in **Figure 8-10**. An arbitrarily selected, infinitesimal slice of the rod of mass dm is shown. Note that we exaggerated the size of dm in the figure; in reality, even the thinnest line we could draw would be too thick because mathematically dm must be infinitesimally small. To emphasize this we have labeled the width of dm as dr, a differential element of distance along the rod.

To carry out the integral in Equation 8-9, we need to assign an upper bound and a lower bound to sweep up every possible mass element in the rod. Warning! Don't be too quick to insert 0 to L as the bounds; the values of the bounds of an integral must match the differential variable of integration. As it stands, the differential is dm, so the differential variable is m, and 0 and L are *not* valid values of mass. We can, however, directly change the variable of integration from mass to length. Because the rod is uniform, the length of the infinitesimal slice of the rod is the same proportion to the total length of the rod as the mass of the slice is to the total mass of the rod:

$$\frac{\text{length of slice}}{\text{length of rod}} = \frac{\text{mass of slice}}{\text{mass of rod}}$$

So,

$$\frac{dr}{L} = \frac{dm}{M}$$

or

$$dm = \frac{M}{L}\,dr \tag{8-10}$$

This relationship is valid whenever the mass of a rod—or any object we can treat as one-dimensional—is uniformly distributed.

Inserting Equation 8-10 into the expression for the moment of inertia (Equation 8-9) gives

$$I = \int r^2 \frac{M}{L} dr = \frac{M}{L} \int r^2\,dr$$

The distance from the rotation axis of the mass of the slice at one end of the rod is $r = 0$, and at the other end, $r = L$, so 0 and L are the lower and upper bounds of the variable over which the integral is evaluated:

$$I = \frac{M}{L} \int_0^L r^2\,dr \tag{8-11}$$

Solving the integral leads us to an expression for the moment of inertia:

$$I = \frac{M}{L} \frac{r^3}{3}\bigg|_0^L = \frac{M}{L}\left(\frac{L^3}{3} - \frac{0^3}{3}\right) = \frac{ML^2}{3}$$

The moment of inertia of a thin rod that has a mass M and a length L and rotates around one end is $ML^2/3$.

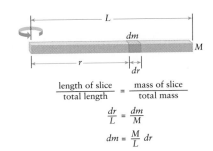

$$\frac{\text{length of slice}}{\text{total length}} = \frac{\text{mass of slice}}{\text{total mass}}$$

$$\frac{dr}{L} = \frac{dm}{M}$$

$$dm = \frac{M}{L}\,dr$$

Figure 8-10 The moment of inertia of a large object is the sum of the moments of inertia of each infinitesimally small piece of the object.

Watch Out!

An object does not have "a" moment of inertia. Rather, it has a moment of inertia defined for rotation around *each specific choice* of rotation axis.

If your friend asks, "What is the moment of inertia of that object?" it could be a trick question! To determine the different values of radius in Equation 8-7 requires that we first specify the axis around which the object rotates. Any object can be made to rotate around any number of axes, even one like a DVD that commonly rotates around a particular axis. The axis does not even have to pass through the object that rotates around it—imagine tying a string to the edge of a DVD and swinging it around in a circle above your head. The DVD would be rotating around an axis that does not pass through it. Three possible rotation axes are shown for a DVD in **Figure 8-11**. Make sure you identify the axis of rotation before determining the moment of inertia of an object.

Figure 8-11 The moment of inertia depends on the specific rotation axis of the object. For example, the moment of inertia of a DVD depends on whether the rotation axis is the usual one in the center of the disk (top left), some other point on the disk (bottom left), or even a point outside the disk (right). You must always identify the axis of rotation before determining the moment of inertia of an object.

Moment of Inertia of a Thin, Uniform Rod Rotating around Its Center

A thin, uniform rod of length L and mass M rotates around its center in **Figure 8-12**. What is the moment of inertia of the rod around this axis, and how does it compare to the moment of inertia of the rod rotating around one end?

The difference that the choice of rotation axis makes is evident in a comparison of Figures 8-12 and 8-10. The values of r which identify the slices of the rod at the two ends are different in the two cases. When the rotation axis goes through one end of the rod, r varies from 0 to L. When the rotation axis goes through the center of the rod, r must vary from $-L/2$ to $+L/2$ for the integration to cover the entire rod. In other words, the difference between the moment of inertia of the rod rotating around one end and the moment of inertia of the rod rotating around the center is in the bounds of the integral. We simply change the bounds in Equation 8-11 to r varying from $-L/2$ to $+L/2$ or

$$I = \frac{M}{L}\int_{-L/2}^{+L/2} r^2\, dr$$

$$= \frac{M}{L}\frac{r^3}{3}\bigg|_{-L/2}^{+L/2} = \frac{M}{L}\left(\frac{(+L/2)^3}{3} - \frac{(-L/2)^3}{3}\right)$$

$$= \frac{M}{L}\left(\frac{L^3}{24} - \frac{-L^3}{24}\right) = \frac{ML^2}{12}$$

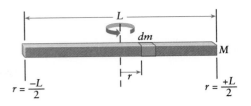

Figure 8-12 A thin, uniform rod of length L and mass M rotates around its center. The moment of inertia is found by adding the contribution that each infinitesimally small mass element dm makes to the total.

The moment of inertia of a thin rod of mass M and length L that rotates around its center is $ML^2/12$. This value is one-fourth the value of the moment of inertia of the rod when it rotates around one end. We expect the moment of inertia to be smaller in this case because moment of inertia varies as the square of r, and the mass of the rod is distributed closer to the rotation axis when the rod rotates around its center. When the rod rotates around its center no part of it is farther than $L/2$ from the axis, but when the rod rotates around one end, half of it is farther than $L/2$ from the axis.

Got the Concept 8-3
Moment of Inertia of Two Rods

Two thin, uniform rods that each have a mass M and a length L are attached at their centers to form an "X" shape. What is the moment of inertia when this configuration is rotated around the axis which passes through their centers, perpendicular to the plane of the two rods?

Got the Concept 8-4
Moment of Inertia of Four Rods

Four thin, uniform rods that each have a mass $M/2$ and a length $L/2$ are attached at their ends to form an "X" shape. What is the moment of inertia when this configuration is rotated around the axis which passes through the center of the "X" perpendicular to the plane of the rods?

Moment of Inertia of a Thin, Uniform Ring Rotating around an Axis Perpendicular to the Plane of the Ring and through Its Center

A thin, uniform ring of radius R and mass M rotates around an axis perpendicular to the plane of the ring and through its center as shown in Figure 8-13. An arbitrarily selected, infinitesimal slice of the ring of mass dm is shown. Again, for clarity we exaggerated the size of dm in the figure; dm must be infinitesimally small. To emphasize this, we have labeled the angular extent of dm as $d\theta$, a differential element of angle along the ring. Note that although the ring occupies a two-dimensional space, we can treat the mass distribution as one-dimensional because the mass is distributed along the circumference of a circle. In effect, the ring is a thin, uniform rod which has been bent into a circle.

To solve the integral in Equation 8-9, we need to assign an upper bound and a lower bound, to sum every possible mass element in the ring. By converting from

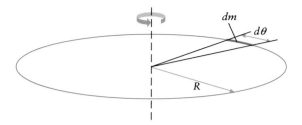

Figure 8-13 A thin, uniform ring of radius R rotates around an axis perpendicular to the plane of the ring and through its center. The moment of inertia is found by adding the contribution that each infinitesimally small mass element dm makes to the total. One arbitrarily selected mass element, which extends over an infinitesimally small angle $d\theta$, is shown.

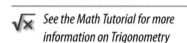 *See the Math Tutorial for more information on Trigonometry*

an integral in terms of mass to one in terms of angle, we can add up every mass element by allowing θ to vary from 0 to 2π. To make this conversion, note that because the ring is uniform, the angular extent of an infinitesimal slice of the ring is the same proportion to the total angle of the ring as the mass of the slice is to the total mass:

$$\frac{\text{angle of slice}}{\text{angle of ring}} = \frac{\text{mass of slice}}{\text{mass of ring}}$$

The total angular extent of the ring is 2π radians, so

$$\frac{d\theta}{2\pi} = \frac{dm}{M}$$

or

$$dm = \frac{M}{2\pi} d\theta \qquad (8\text{-}12)$$

This relationship is valid because the mass of the ring is uniformly distributed.

Inserting Equation 8-12 into the expression for the moment of inertia (Equation 8-9) gives

$$I = \int r^2 \frac{M}{2\pi} d\theta = \frac{M}{2\pi} \int r^2 d\theta$$

Note that the variable r is the radial distance to any particular infinitesimal piece of the object over which we integrate. Be careful to distinguish between the variable and the value that variable takes in any specific case. For this ring, the distance from the rotation axis to any piece is always R, which can be taken out of the integral because it is constant. Also, to include in the integral every infinitesimal slice of the ring, we need to integrate in a complete circle, so θ ranges from 0 to 2π:

$$I = \frac{MR^2}{2\pi} \int_0^{2\pi} d\theta \qquad (8\text{-}13)$$

So

$$I = \frac{MR^2}{2\pi} \theta \Big|_0^{2\pi} = \frac{MR^2}{2\pi} (2\pi - 0) = MR^2$$

The moment of inertia of a thin, uniform ring of mass M and radius R that rotates around an axis perpendicular to the plane of the ring and through its center is MR^2.

Moment of Inertia of a Thin, Uniform Disk Rotating around an Axis Perpendicular to the Plane of the Disk and through Its Center

A thin, uniform disk of radius R and mass M rotates around an axis perpendicular to the plane of the disk and through its center. Figure 8-14 shows an arbitrarily selected, infinitesimal piece of the disk of mass dm. Note that because the disk is thin we can treat the mass distribution as two-dimensional, which requires dm to be two-dimensional as well. To emphasize the requirement that dm be infinitesimally small, we have defined the piece of the disk as being bounded by inner radius r and outer radius $r + dr$, and having angular extent $d\theta$.

We need to assign an upper bound and a lower bound to the moment of inertia integral (Equation 8-9) so that the integration includes every possible mass element in the disk. Because the disk is uniform we can set up a proportion analogous to the one we used in the case of the rod and the ring. Note, however, that

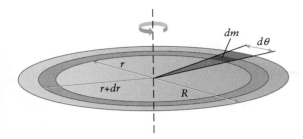

Figure 8-14 A thin, uniform disk of radius R rotates around an axis perpendicular to the plane of the disk and through its center. The moment of inertia is found by adding the contribution that each infinitesimally small mass element dm makes to the total. One arbitrarily selected mass element, a small slice of an annulus of the disk, is shown in dark gray. The mass element is defined by the region from a distance r from the rotation axis to r plus an infinitesimally small distance dr, and extends an infinitesimally small distance $d\theta$ along the annulus.

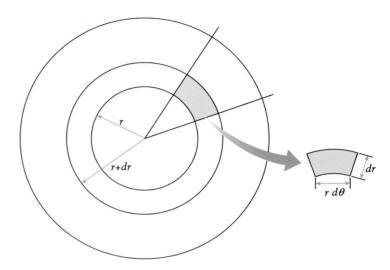

Figure 8-15 Because both the width of the annulus and the angular extent of the slice are infinitesimally small, we can treat the piece shown in dark blue as if it were a rectangle of length $r\,d\theta$ and width dr.

because the infinitesimal dm is two-dimensional, the proportionality is in terms of area, or

$$\frac{\text{area of piece}}{\text{area of disk}} = \frac{\text{mass of piece}}{\text{mass of disk}}$$

It's helpful to know the shape of the dm piece in order to determine its area. Don't be fooled by the appearance of the sample piece shown in Figure 8-14—we can treat the shape of dm as rectangular. It may appear that the outer edge of dm is much wider than the inner edge, but this is due to our exaggeration of the overall size of dm. Because the angular extent $d\theta$ of each piece of the disk is infinitesimally small, the lengths of the inner and outer sides of dm are also infinitesimally small, so we can treat them as mathematically equivalent.

Because dm is a rectangle, its area is the product of the length of the two sides. As shown in **Figure 8-15**, this area is $(r\,d\theta)(dr)$, so

$$\frac{r\,d\theta\,dr}{\pi R^2} = \frac{dm}{M}$$

or

$$dm = \frac{M}{\pi R^2} r\,dr\,d\theta \tag{8-14}$$

We have been careful to distinguish between r, the radial distance to any particular infinitesimal patch of the disk over which we will integrate, and R, the constant radius of the entire disk. The relationship for dm in Equation 8-14 is valid because the mass of the disk is uniformly distributed.

Inserting Equation 8-14 into the expression for the moment of inertia (Equation 8-9) gives

$$I = \int\int r^2 \frac{M}{\pi R^2} r\,dr\,d\theta = \frac{M}{\pi R^2} \int\int r^3\,dr\,d\theta$$

It is necessary to write two integrals because the expression for dm involves two differentials. However, the r and θ variables are independent of one another, so we can treat the expression as two separate integrals or

$$I = \frac{M}{\pi R^2} \int r^3 \, dr \int d\theta$$

Finally, we set the bounds to include every possible infinitesimal slice of the disk. The smallest and largest values of radius are $r = 0$ and $r = R$, respectively, and angle θ runs from 0 to 2π:

$$I = \frac{M}{\pi R^2} \int_0^R r^3 \, dr \int_0^{2\pi} d\theta \tag{8-15}$$

So

$$I = \frac{M}{\pi R^2} \left. \frac{r^4}{4} \right|_0^R \left. \theta \right|_0^{2\pi} = \frac{M}{\pi R^2} \left(\frac{R^4}{4} - 0 \right)(2\pi - 0)$$

$$= \frac{M}{\pi R^2} \frac{2\pi R^4}{4} = \frac{MR^2}{2}$$

The moment of inertia of a thin, uniform disk of mass M and radius R that rotates around an axis perpendicular to the plane of the disk and through its center is $MR^2/2$.

When rotating about the same axis, the moment of inertia of a thin, uniform disk is half that of a thin, uniform ring of the same mass and radius. You should expect the disk to have a smaller moment of inertia, because the moment of inertia of an object is strongly influenced by how far the mass is from the rotation axis. All of the mass of the ring is located a distance R from the axis, while only a fraction of the mass of the disk is that far from the axis. Therefore, the moment of inertia of the ring must be larger than the moment of inertia of the disk.

Math Box 8-2 Checking a Differential Area

We needed to devise an expression for the area of an infinitesimally small patch of the disk shown in Figure 8-14 in order to determine the moment of inertia. This approach is the same regardless of the shape or number of dimensions of the object in question. It is therefore useful, and often straightforward, to check whether the expression we devise is correct. For a disk rotating around its center, for example, we integrated over infinitesimally small regions dA that, as shown in Figure 8-15, we assigned area $r \, dr \, d\theta$:

$$dA = r \, dr \, d\theta$$

The integral of the area elements dA over the entire disk must give the area of the disk. To convince ourselves that the form we constructed for a tiny piece of area is correct, let's integrate dA over the entire disk; we should get $A = \pi R^2$. The bounds on radius r and angle θ are $0 < r < R$ and $0 < \theta < 2\pi$, respectively, in order to cover the entire disk, so

$$\int_{\text{Disk}} dA = \int_0^R r \, dr \int_0^{2\pi} d\theta$$

$$= \left. \frac{r^2}{2} \right|_0^R \left. \theta \right|_0^{2\pi} = \left(\frac{R^2}{2} - 0 \right)(2\pi - 0) = \pi R^2$$

Although this is not a proof, it gives us confidence that the differential element of area we constructed is correct.

The moments of inertia of a variety of objects and rotation axes are given in Table 8-1.

✳ **What's Important 8-2**

Rotational inertia is the tendency of an object to resist a change in rotational motion. It depends not only on the mass of an object but also on how the mass is distributed with respect to the axis of rotation.

Table 8-1
Moments of Inertia of Uniform Bodies of Various Shapes*

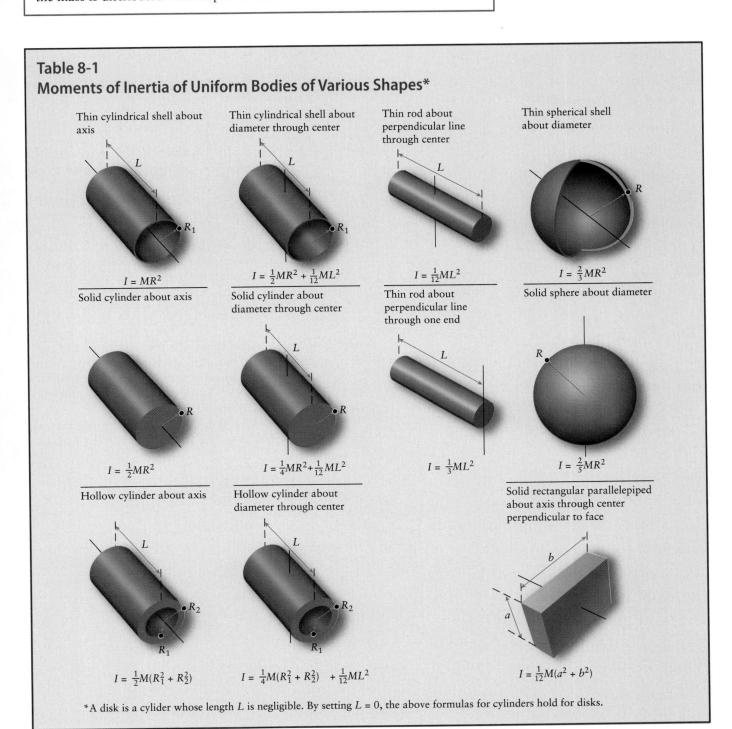

Thin cylindrical shell about axis
$$I = MR^2$$
Solid cylinder about axis

Thin cylindrical shell about diameter through center
$$I = \tfrac{1}{2}MR^2 + \tfrac{1}{12}ML^2$$
Solid cylinder about diameter through center

Thin rod about perpendicular line through center
$$I = \tfrac{1}{12}ML^2$$
Thin rod about perpendicular line through one end

Thin spherical shell about diameter
$$I = \tfrac{2}{3}MR^2$$
Solid sphere about diameter

$$I = \tfrac{1}{2}MR^2$$
Hollow cylinder about axis

$$I = \tfrac{1}{4}MR^2 + \tfrac{1}{12}ML^2$$
Hollow cylinder about diameter through center

$$I = \tfrac{1}{3}ML^2$$

$$I = \tfrac{2}{5}MR^2$$
Solid rectangular parallelepiped about axis through center perpendicular to face

$$I = \tfrac{1}{2}M(R_1^2 + R_2^2)$$

$$I = \tfrac{1}{4}M(R_1^2 + R_2^2) + \tfrac{1}{12}ML^2$$

$$I = \tfrac{1}{12}M(a^2 + b^2)$$

*A disk is a cylinder whose length L is negligible. By setting $L = 0$, the above formulas for cylinders hold for disks.

8-3 The Parallel-Axis Theorem

Finding the moment of inertia of an object, even one with a symmetrical shape, can be challenging for certain choices of rotation axis. Imagine, for example, a uniform disk rotating around an axis perpendicular to the plane of the disk but passing through a point near its edge, as in **Figure 8-16a**. In this case setting the bounds necessary to find the moment of inertia would be cumbersome. However, a curious relationship exists between the moment of inertia of an object when it rotates around its center of mass, which is often easy to find, and the moment of inertia when the object rotates around any other parallel axis.

Let's say you know the moment of inertia I_{CM} of an object when it rotates around an axis passing through its center of mass. The moment of inertia for a rotation around any other parallel axis is

$$I = I_{CM} + Mh^2 \qquad (8\text{-}16)$$

where M is the mass of the object and h is the distance between the two axes. This relationship is known as the **parallel-axis theorem**.

To see the parallel-axis theorem in action, consider a thin, uniform rod that has a mass M, a length L, and its center of mass at the center of the rod. In Section 8-2, we determined the moment of inertia of a similar rod rotating around an axis perpendicular to the rod and through its center (Figure 8-12). As summarized in Table 8-1,

$$I_{CM} = \frac{ML^2}{12}$$

In Section 8-2, we also found the moment of inertia for a thin, uniform rod of mass M and length L that rotates around an axis perpendicular to the rod and through one end (Figure 8-10) to be $I = ML^2/3$. Can the parallel-axis theorem reproduce this result?

The end of the rod is $L/2$ from the center, so the axis through the end is $h = L/2$ from the center of mass. By Equation 8-16,

$$I = \frac{ML^2}{12} + M\left(\frac{L}{2}\right)^2$$

$$= \frac{ML^2}{12} + \frac{ML^2}{4}$$

$$= \frac{ML^2}{12} + \frac{3ML^2}{12}$$

$$= \frac{4ML^2}{12}$$

$$= \frac{ML^2}{3}$$

(a)

(b)

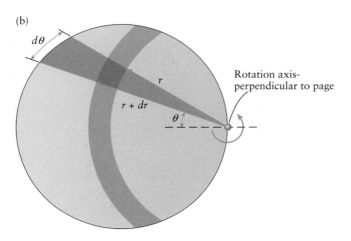

Figure 8-16 (a) A thin, uniform disk of radius R and mass M rotates around a point near the edge of the disk. (b) Finding the moment of inertia would be complex without applying the parallel-axis theorem.

The parallel-axis theorem does indeed reproduce the correct moment of inertia for the rod rotating about its end.

The parallel-axis theorem is particularly useful for determining the moment of inertia when the shape of an object or the orientation of the axis makes the integral in Equation 8-9 difficult, as in the next example.

Example 8-2 The Edge of a Disk

Find the moment of inertia of a thin, uniform disk of radius R and mass M that rotates around a pin pushed through a small hole near the edge of the disk as in Figure 8-16a.

SET UP

The center of mass of the disk is at its center. In Section 8-2, we found the moment of inertia of a thin, uniform disk rotating around an axis perpendicular to the plane of the disk and passing through its center to be $I = MR^2/2$. We can therefore use the parallel-axis theorem to find the moment of inertia around the axis specified in this problem.

SOLVE

The edge of the disk is a distance $h = R$ from the center, so by the parallel-axis theorem (Equation 8-16),

$$I = \frac{MR^2}{2} + MR^2$$
$$= \frac{3MR^2}{2}$$

REFLECT

Finding the moment of inertia of the disk rotating around a point on its edge was straightforward using the parallel-axis theorem. The integral definition of moment of inertia (Equation 8-9) would not be. Although the integral would be similar to the one we used to find the moment of inertia for the disk rotating around its center (Equation 8-15), determining the bounds would be challenging! As suggested by the differential patch of area shown in **Figure 8-16b**, the mass dm would be the same as that in Equation 8-15. But what, for example, are the upper bounds on r and θ to cover the entire disk? As you can see from the figure, the largest value of r depends on the angle θ and the largest value of θ depends on the size of the disk. As a result the integral approach is not nearly as straightforward as using the parallel-axis theorem.

Practice Problem 8-2 Find the moment of inertia of a thin, uniform disk of radius R and mass M that rotates around a pin pushed through a small hole halfway between the center and the edge of the disk.

★ What's Important 8-3

The parallel-axis theorem describes the relationship between the moment of inertia of an object when it rotates around its center of mass, which is often easy to find, and the moment of inertia when the object rotates around any other parallel axis.

Figure 8-17 Competitors in the soap box derby must use approved wheels. A wheel design that reduces the amount of the car's potential energy that goes into the rotational kinetic energy of the wheels results in the car going faster down the hill. *(George Tiedemann/NewSport/Corbis)*

8-4 Conservation of Energy Revisited

Every year in the All American Soap Box Derby (**Figure 8-17**) thousands of boys and girls race in homemade cars powered only by the force of gravity pulling them down a hill. Although the organizers encourage creativity in car design, they disqualify competitors who do not use officially approved wheels. How would different kinds of wheels affect the outcome of the race? Conservation of energy lies at the heart of the answer.

Before the start of the race, a car at the top of a hill has gravitational potential energy relative to the bottom of the hill but no kinetic energy. The car gains speed as it rolls down the race course—a transformation of potential energy into kinetic energy. As we saw in Section 6-6, energy must be conserved. We write

$$K_i + U_i = K_f + U_f + |W_{nc}| \tag{6-31}$$

as a general statement of energy conservation, noting that a separate potential energy term must be included for each force an object in the system experiences and W_{nc} is work done by nonconservative forces. Because a rotating object has rotational kinetic energy separate from any translational kinetic energy it possesses, we will now augment Equation 6-31 to include both a rotational kinetic energy as well as a translational kinetic energy term:

$$K_{\text{translational},i} + K_{\text{rotational},i} + U_i = K_{\text{translational},f} + K_{\text{rotational},f} + U_f + |W_{nc}| \tag{8-17}$$

As the car rolls down the racecourse, some of its initial gravitational potential energy is transformed to translational kinetic energy. The rate at which the translational kinetic energy increases is directly related to the car's linear acceleration. But we now see that some of the initial gravitational potential energy is transformed into the rotational kinetic energy of the wheels. The more potential energy that goes into rotational kinetic energy, the less energy is available to make the car go fast (translational kinetic energy). The opposite is also true; when less energy is required to rotate the wheels, more of the potential energy can be transformed into translational kinetic energy, resulting in a higher linear velocity. Specially designed wheels could give a competitor an unfair advantage. In all cases, we assume that the wheels roll without slipping down the hill.

? **Got the Concept 8-5**
Faster Wheels

Which type of wheels would allow a soapbox derby car to go faster, uniform disks or wheels that look like conventional bicycle tires where most of the mass is along the rim of the wheels? Assume that both types of wheels have the same mass.

Rolling, Slipping, and Sliding

A moving circular or cylindrical object can slide, roll without slipping, or slide and roll at the same time. When sliding, as in Figure 8-18a, the same point (or points) on the object remains in contact with the surface at all times. No rotational kinetic energy is present because the object doesn't rotate. Rotational kinetic energy is introduced when the object rotates, for example, when rolling. A special case of rolling motion occurs when the edge of the disk does not slip relative to the surface, as in Figure 8-18b. Notice that the colored thread that has been wrapped around the circumference of the object unwinds as the object rolls, marking the distance traveled. The distance the disk moves along the surface in one rotation exactly equals the circumference of the circle because the edge of the disk does not move relative to the surface.

The object in Figure 8-18 could also rotate and slide at the same time, in which case a fixed point on the edge of the disk would move relative to the surface as the object moved. We could in principle analyze motion like this, by including a rotational kinetic energy term as well as a dissipative term due to sliding (kinetic) friction in Equation 8-17. In practice, however, considering the phenomenon of sliding is complicated by the need to know how much sliding occurs, which in turn determines how much energy is taken up by rotation and how much is dissipated by friction. For this reason, we will only deal with cases in which objects roll without slipping.

Objects that roll without slipping nevertheless experience a retarding, frictional force. We tend to neglect this **rolling friction**, because it is usually small compared to other effects, but without it a rolling object would never come to a stop. Like static and kinetic frictional forces, the force of rolling friction, is proportional to the normal force acting on an object.

Figure 8-19a shows a completely rigid, rolling disk. Because the disk is rigid, the contact between the disk and the surface is a point or a line directly below the rotation axis. The normal force therefore points radially in toward the rotation axis and, as a result, has no component along the direction of motion. The normal force has no effect on the rotation because it is directed toward the rotation axis. Under ideal conditions, then, a completely rigid object experiences no rolling friction. No object (no material) is perfectly rigid, however, which means that all objects deform when in contact with a supporting surface. That deformation results in the contact between object and surface being spread out over an area, as in Figure 8-19b. Because the leading edge of the rolling object is coming down to

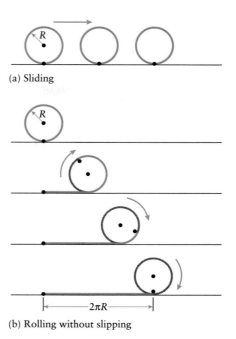

(a) Sliding

(b) Rolling without slipping

Figure 8-18 A disk can slide, roll without slipping, or slide and roll at the same time. (a) An object does not rotate when it slides, so it acquires no rotational kinetic energy. (b) When a disk rolls without slipping relative to the surface, in one rotation the disk moves a distance exactly equal to the circumference of the disk.

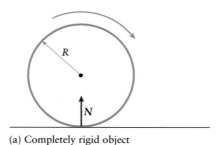

(a) Completely rigid object

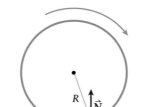

The leading edge of the rolling object is coming down to the surface, while the trailing edge is lifting off the surface. The net normal force can be taken as acting in front of the rotation axis.

The normal force $\vec{N}$ acting forward of the axis counters the rotation, and has the effect of opposing the motion.

(b) A deformed object

Figure 8-19 (a) A completely rigid, rolling disk only contacts the surface at a point directly below the rotation axis. The normal force therefore points toward the rotation axis and has no effect on the rotation. Under ideal conditions a completely rigid object experiences no rolling friction. (b) A nonrigid, rolling disk contacts the surface at more than one point below the rotation axis. The net normal force acts in front of the rotation axis and opposes the motion.

the surface while the trailing edge is lifting off the surface, the net normal force can be taken as acting in front of the rotation axis. This force counters the rotation, and has the effect of opposing the motion. Rolling friction arises from the deformation of a rolling object.

Let's compare the speed of a uniform disk to that of a hoop, when both rolling without slipping down a ramp. Both the disk and the hoop have the same radius R and mass M and both traverse a vertical distance H (**Figure 8-20**).

Total energy is conserved in both cases. The frictional forces will be negligibly small for typical materials, so we can neglect the last term in Equation 8-17 the expressions of energy conservation for the disk and the hoop have the same form:

$$K_{\text{translational, i}} + K_{\text{rotational, i}} + U_i = K_{\text{translational, f}} + K_{\text{rotational, f}} + U_f$$

Because both the disk and the hoop are initially at rest, $K_{\text{translational, i}}$ and $K_{\text{rotational, i}}$ equal zero for both the disk and the hoop. So the previous expression for either the disk or hoop becomes

$$Mgh_i = \frac{1}{2}Mv_f^2 + \frac{1}{2}I\omega_f^2 + Mgh_f$$

Here, h_i and h_f are the heights of the top and bottom of the ramp, v_f is the translational speed at the bottom of the ramp, and ω_f is the rotational velocity at the bottom of the ramp. The height of the ramp H is $h_i - h_f$, so

$$\frac{1}{2}Mv_f^2 = Mgh_i - Mgh_f - \frac{1}{2}I\omega_f^2 = MgH - \frac{1}{2}I\omega_f^2 \qquad (8\text{-}18)$$

To compare the speeds of the disk and hoop at the bottom of the ramp, we need to solve this equation for v_f in terms of the variables that define the problem: R, M, and H. This requires that both I and ω_f be expressed in terms of those variables as well.

The moment of inertia of a uniform disk that has radius R, mass M, and rotates around an axis through the center and perpendicular to the plane of the object is

$$I_{\text{disk}} = \frac{MR^2}{2}$$

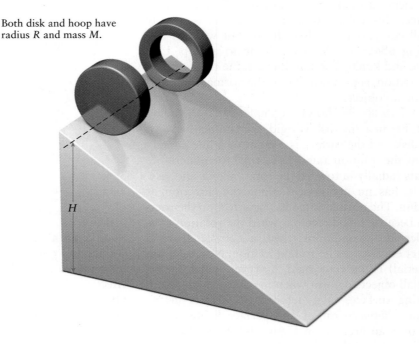

Both disk and hoop have radius R and mass M.

H

Figure 8-20 A disk and a hoop have the same radius R and mass M. Both are allowed to roll from rest without slipping down a ramp of height H.

The moment of inertia of a hoop that has a radius R and a mass M and rotates around an axis through the center and perpendicular to the plane of the object is

$$I_{hoop} = MR^2$$

The angular velocity term can be understood by using Figure 8-18b, in which a circular object of radius R rolls at a constant rate through one full rotation without slipping. As we noted above, the disk moves linearly a distance equal to the circumference when it rolls through one full rotation. The change in position of the center of the object is then equal to $2\pi R$. Let the time for one full rotation be Δt, so that the linear velocity is

$$v = \frac{\Delta x}{\Delta t} = \frac{2\pi R}{\Delta t}$$

Similarly, the angular velocity is

$$\omega = \frac{\Delta \theta}{\Delta t} = \frac{2\pi}{\Delta t}$$

so

$$v = \left(\frac{2\pi}{\Delta t}\right)R = \omega R \tag{8-19}$$

Equation 8-19 is a general relationship between the linear velocity and angular velocity of a circular object rolling without slipping.

We can now write Equation 8-18 for the disk and the hoop separately, using the appropriate moment of inertia (either I_{disk} or I_{hoop}) and the relationship between the linear velocity and angular velocity ($\omega = v/R$, from Equation 8-19). For the disk,

$$\frac{1}{2}Mv_{disk,\,f}^2 = MgH - \frac{1}{2}I_{disk}\left(\frac{v_{disk,\,f}}{R}\right)^2 = MgH - \frac{1}{2}\frac{MR^2}{2}\left(\frac{v_{disk,\,f}}{R}\right)^2$$

$$v_{disk,\,f}^2 = 2gH - \frac{1}{2}v_{disk,\,f}^2$$

$$\frac{3}{2}v_{disk,\,f}^2 = 2gH$$

$$v_{disk,\,f} = \sqrt{\frac{4}{3}gH}$$

For the hoop,

$$\frac{1}{2}Mv_{hoop,\,f}^2 = MgH - \frac{1}{2}I_{hoop}\left(\frac{v_{hoop,\,f}}{R}\right)^2 = MgH - \frac{1}{2}MR^2\left(\frac{v_{hoop,\,f}}{R}\right)^2$$

$$v_{hoop,\,f}^2 = 2gH - v_{hoop,\,f}^2$$

$$2v_{hoop,\,f}^2 = 2gH$$

$$v_{hoop,\,f} = \sqrt{gH}$$

As we should expect, the speed of the hoop as it comes off the ramp is smaller than the speed of the disk ($\sqrt{gH}$ compared to $\sqrt{4gH/3}$), because more of the hoop's initial gravitational potential energy is converted into rotational kinetic energy as the hoop rolls down the ramp. Both objects will accelerate, but at any given moment, the speed of the hoop will be smaller than that of the disk. In a race, the disk would win.

> ### ? Got the Concept 8-6
> #### Hoop and Block
>
> A hoop and a block are released from rest down adjacent ramps that have identical angles. The hoop rolls without slipping and the block slides without friction, both starting from the same height. Which one reaches the bottom of the ramp first?

Estimate It! 8-1 Rolling Spider

To escape predators, the golden wheel spider of the Namib Desert extends its legs like spokes on a wheel and rolls rapidly down sand dunes. Estimate the speed such a spider might attain after rolling 1 m down a 15° slope.

SET UP

We can treat the spider as an object that rolls without slipping down a ramp of angle 15°. As shown in the sketch in **Figure 8-21**, the spider travels a distance L while dropping a vertical distance H. Energy is conserved as the spider rolls down the sand dune, so we start from Equation 8-17. Note that the spider has no kinetic energy at the instant it starts to roll. We'll also neglect frictional forces, which we would expect to be small. If we declare the initial height of the spider to be H, then the final height is zero; the initial potential energy U_i then equals the sum of the final translational and rotational kinetic energies $K_{\text{translational, f}}$ and $K_{\text{rotational, f}}$:

$$U_i = K_{\text{translational, f}} + K_{\text{rotational, f}}$$

or

$$M_{\text{spider}}gH = \frac{1}{2}M_{\text{spider}}v^2 + \frac{1}{2}I_{\text{spider}}\omega^2$$

where M_{spider} is the mass of the spider, v is the translational speed of the spider, and ω is its angular velocity.

We can approximate the spider as a uniform disk rotating around its central axis. Even when the spider sticks out its legs as spokes, most of its mass is concentrated near the rotation axis, and the spider is also relatively flat. The effective radius R_{eff} of the disk is equal to that of the main part of the spider's body. From Table 8-1, the moment of inertia of the spider is then

$$I_{\text{spider}} = \frac{M_{\text{spider}}R_{\text{eff}}^2}{2}$$

SOLVE

Conservation of energy therefore leads to

$$M_{\text{spider}}gH = \frac{1}{2}M_{\text{spider}}v^2 + \frac{1}{2}\frac{M_{\text{spider}}R_{\text{eff}}^2}{2}\omega^2$$

Notice that the mass of the spider appears in each term, so it can be canceled. Also, from Equation 8-19, v is equal to ωR, where R is the radius of the spider *including* its spokelike legs. So

$$gH = \frac{1}{2}v^2 + \frac{1}{2}\frac{R_{\text{eff}}^2}{2}\frac{v^2}{R^2}$$

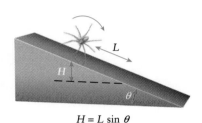

$H = L \sin\theta$

Figure 8-21

or

$$v = \sqrt{\frac{2gH}{1 + R_{eff}^2/2R^2}}$$

From Figure 8-21, H equals $L \sin 15°$, where L is given as 1 m. We want to use round, but reasonable, values in doing estimations; based on our experience with spiders let's take R_{eff} equal to 0.2 cm and R equal to 0.5 cm. To the level of significance of these values, we can also use 10 m/s^2 as an approximate value for g. Then

$$v \approx \sqrt{\frac{2(10 \text{ m/s}^2)(1 \text{ m} \sin 15°)}{1 + (0.002 \text{ m})^2/2(0.005 \text{ m})^2}} = 2.2 \text{ m/s}$$

To one significant figure, we estimate the speed of the spider to be 2 m/s.

REFLECT

Our estimate certainly seems reasonable—golden wheel spiders can attain speeds of approximately 1 m/s.

✴ What's Important 8-4

A rotating object has rotational kinetic energy, a quantity separate from any translational kinetic energy it possesses. An application of the statement of conservation of energy therefore includes terms for both rotational kinetic energy as well as translational kinetic energy.

8-5 Rotational Kinematics

Video information recorded on a DVD zips around in a circle as the disk rotates. Although the data recorded near the outer edge of the DVD travels farther in one revolution of the disk than data recorded near the disk's center, both parts of the disk make the same one revolution. In a similar way, the speed of information at the outer edge passing the reader in the DVD player is higher than the speed at which information recorded near the center passes the reader. Yet both parts of the disk make the same number of revolutions in any given time. In this section, we will address the correspondence between the quantities that describe linear motion, such as distance and speed, and the quantities that describe rotational motion, such as angle and angular velocity.

You probably noticed the similarities between the equations that define translational kinetic energy (Equation 6-10) and rotational kinetic energy (Equation 8-8):

$$K_{translational} = \frac{1}{2}mv^2 = \frac{1}{2}m\left(\frac{dx}{dt}\right)^2$$

$$K_{rotational} = \frac{1}{2}I\omega^2 = \frac{1}{2}I\left(\frac{d\theta}{dt}\right)^2$$

 See the Math Tutorial for more information on Differential Calculus

The form of these equations reveals both a similarity and a difference between mass and moment of inertia. Both are properties of an object, and moment of inertia is related to mass. But while mass quantifies the tendency of an object to resist a change in linear motion, moment of inertia quantifies the tendency of the object to resist a change in rotational motion. This is inferred from the presence of m and linear velocity dx/dt in $K_{translational}$, but I and angular velocity in $K_{rotational}$. The relationship between x and θ is also apparent: x measures linear displacement and θ measures angular displacement, the angle through which an object has rotated.

We can extend the relationship between x and θ to the derivatives of these quantities with respect to time. The rotational equivalent of linear velocity v is angular velocity ω; this result can be seen by comparing $K_{\text{translational}}$ and $K_{\text{rotational}}$, or by taking the derivatives of x and θ with respect to time:

$$v = \frac{dx}{dt}$$

$$\omega = \frac{d\theta}{dt}$$

And finally, as linear acceleration is the derivative of linear velocity with respect to time, we can define an analogous quantity **angular acceleration** α for objects that are rotating:

$$a = \frac{dv}{dt}$$

$$\alpha = \frac{d\omega}{dt} \tag{8-20}$$

Angular displacement is measured in radians. The SI units of ω are

$$[\omega] = \frac{\text{rad}}{\text{s}}$$

Therefore, the units of α are radians per square second (rad/s^2). We can also describe both angular displacement and angular velocity in terms of the number of revolutions an object makes. Another useful set of units for ω is therefore

$$[\omega] = \frac{\text{rev}}{\text{s}}$$

A full circle is 2π rad, so a rate of 1 rev/s is equal to 2π rad/s. In addition, angular velocity can be converted into an equivalent linear speed by recognizing that a point at radius r from the rotation axis travels a distance $2\pi r$ in one revolution. So an angular velocity of 1 rev/s is also equal to $2\pi r$ m/s.

The correspondence between the linear variables x, v, a, and m and the rotational variables θ, ω, α, and I enable a translation of sorts between the equations that describe linear kinematics and equations that describe rotational kinematics. In Chapter 2, we developed two fundamental equations that describe motion in one dimension under constant acceleration:

$$v = v_0 + at \tag{2-23}$$

$$x - x_0 = v_0 t + \frac{1}{2}at^2 \tag{2-26}$$

We can now use similar equations to describe rotational motion under constant angular acceleration:

$$\omega = \omega_0 + \alpha t \tag{8-21}$$

$$\theta - \theta_0 = \omega_0 t + \frac{1}{2}\alpha t^2 \tag{8-22}$$

As we did for linear kinematics, we set the initial value of angular velocity to be ω_0. We let the time variable "start" at $t_0 = 0$ in accordance with the standard convention. We also define an initial angular position θ_0, so that angular displacement is $\theta - \theta_0$.

? Got the Concept 8-7
DVD

Information recorded on a DVD is evenly spaced along a long spiral that spans most of the surface of the disk. DVD players read the information at a constant rate. Should the disk rotate faster, slower, or at the same rate as the player reads information recorded closer and closer to the center of the DVD?

Equations 2-23 and 2-26 completely describe linear motion when acceleration is constant and enable us to solve a wide range of problems. Using just the two equations we could, for example, find the distance a hockey puck travels given its initial velocity or we could find the constant acceleration required for a car to attain a certain velocity after starting from rest. In the same way, Equations 8-21 and 8-22 fully describe rotational motion when angular acceleration is constant. The variables are different, but the physics is exactly the same as that of linear motion discussed in Chapter 2.

Example 8-3 A Stopping Top

A top spinning at 25 rad/s (4.0 rev/s) comes to a complete stop in 42 s. Assuming the top decelerates at a constant rate, how many revolutions does it make before coming to a stop?

SET UP
The two fundamental linear kinematics equations (Equations 2-23 and 2-26) contain five variables; solving any problem requires identifying three known variables and then eliminating a fourth to find the value of the unknown variable. The method of creating a Know/Don't Know table, useful for determining a value for the unknown variable in linear kinematics, can also be applied here. As you can see from the Know/Don't Know table in **Table 8-2**, we need to eliminate angular acceleration by solving one of the rotational motion equations for α and then substituting the resulting expression into the other equation. The final equation will include the three variables of known value (ω_0, ω, and t) and $\theta - \theta_0$. We can solve for $\theta - \theta_0$ in terms of ω_0, ω, and t, from which we can determine the number of revolutions the top makes before stopping.

Table 8-2

Variable	Know/Don't Know
$\theta - \theta_0$	?
ω_0	25 rad/s
ω	0 rad/s
α	X
t	42 s

SOLVE
We can choose to solve either of the rotational motion equations for α to eliminate angular acceleration, but it's more straightforward to solve Equation 8-21. We can also set the final value of angular velocity (ω) to zero, as indicated in the Know/Don't Know table, leading to

$$\alpha = -\frac{\omega_0}{t}$$

Substituting this expression into Equation 8-22 gives

$$\theta - \theta_0 = \omega_0 t + \frac{1}{2}\left(-\frac{\omega_0}{t}\right)t^2 = \omega_0 t - \frac{1}{2}\omega_0 t$$
$$= \frac{1}{2}\omega_0 t$$

All of the variables on the right side of the resulting equation are known, so we can compute a numeric answer:

$$\theta - \theta_0 = \frac{1}{2}\left(25\,\frac{rad}{s}\right)(42\,s) = 5.3 \times 10^2\,rad$$

The question asks us to express our answer in terms of revolutions, so we also write

$$\theta - \theta_0 = \frac{1}{2}\left(4\,\frac{rev}{s}\right)(42\,s) = 84\,rev$$

REFLECT
The result, $\theta - \theta_0 = \frac{1}{2}\omega_0 t$, can be interpreted graphically. Because the angular deceleration is constant, the angular velocity decreased linearly, as shown in **Figure 8-22a**. In any short time interval Δt such as the one shown in the figure,

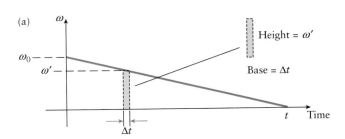
(a)

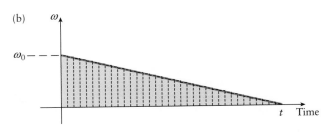
(b)

Figure 8-22

the number of revolutions the top makes is the product of Δt and the angular velocity ω during that time (or the average angular velocity, if the time interval is not infinitesimally short). As you can see, the product is the area of a narrow rectangle, and adding the areas of the rectangles over the total stopping time of the top (Figure 8-22b) gives the area under the ω versus t line. The shape of that area is a triangle, the area of which is $\frac{1}{2} \times$ base $\times$ height, or $\frac{1}{2}\omega_0 t$.

Practice Problem 8-3   A top spinning at 25 rad/s (4.0 rev/s) slows down and comes to a stop after making 36 revolutions. Assuming the top decelerates at a constant rate, how long does it take to come to a stop?

> ✱ **What's Important 8-5**
> Every quantity we use to describe linear (translational) motion has a rotational analog. For example, the angular acceleration of a rotating object is the derivative of angular velocity with respect to time, so it is analogous to the linear acceleration of an object moving along a straight line.

8-6 Torque

Where do you push on a door to open it easily? Perhaps without knowing why, you've learned that it's easiest to open a door by pushing on a spot far from the hinges. You would never push near the hinges, such as the person in Figure 8-23a, because you learned a long time ago that it's nearly impossible to open a door that way. In which direction do you push on a door? The best choice is perpendicular to the plane of the door and far from the hinges (Figure 8-23b). In this section, we will see why opening the door depends not just on the magnitude of the applied force, but also on both the direction of the force vector and the distance between the rotation axis and the position at which the force is applied.

Torque τ is the rotational analog of force and takes into account the distance r between a force F and the rotation axis as well as the angle φ between the $\vec{r}$ vector and the $\vec{F}$ vector. (The $\vec{r}$ vector points from the rotation axis to the point at which the force is applied.) The magnitude of torque is

$$\tau = rF\sin\varphi \tag{8-23}$$

The quantities on which torque depends are shown in Figure 8-24.

Notice that $F\sin\varphi$ on the right side of the relationship is the component of the applied force in the direction perpendicular to the $\vec{r}$ vector. The r and $\sin\varphi$

Figure 8-23 (a) It's hard to open a door by pushing it close to the hinges. (b) The door opens effortlessly when force is exerted far from the hinge, and perpendicular to the plane of the door.

(a)   (b)

terms taken together describe the perpendicular distance from the axis of rotation to the line along which the force acts. Applying a given force for a large value of this distance, called the **lever arm** or **moment arm**, provides a mechanical advantage that, for example, allows you to open a massive door with relatively little effort.

The SI units of torque are evident from Equation 8-23:

$$[\tau] = [r][F][\sin\varphi] = \text{m} \cdot \text{N}$$

By convention, the units are written in the opposite order, as N · m, and referred to as newton-meters.

 ## Got the Concept 8-8
Socket Wrench

A socket wrench can be used to loosen a bolt. A common trick to turn a bolt that has become frozen in place is to slide a section of pipe over the handle of the wrench and turn the bolt while gripping the end of the pipe. Why does this work? Why do many experienced mechanics tend to avoid using this trick?

Using the definition of torque (Equation 8-23), we see that the lever arm can either amplify or reduce the effect that an applied force has on rotating an object. When the lever arm is large (for example, when the application of the force is relatively far from the rotation axis), a small force generates a large torque. Humans and other animals take advantage of the power of the lever arm in the arrangement of muscles and bones as shown in Figure 8-25. The point at which the muscle is attached to the lower jawbone, for example, is far from the joint around which the jaw rotates, resulting in a torque large enough to crack a nut between your back teeth. In contrast, one end of the biceps muscle attaches to the bone of the upper arm and the other to the lower arm just below the elbow (Figure 8-26). This muscle–joint arrangement doesn't generate a large torque relative to the size of the muscle because the length of the lever arm is relatively small. The anatomy

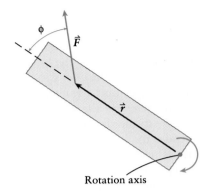

Figure 8-24 Torque τ is the rotational analog of force and takes into account the distance r between where a force F is applied and the rotation axis. Torque also takes into account the angle φ between the force vector $\vec{F}$ and the $\vec{r}$ vector that points from the rotation axis to the point at which the force is applied.

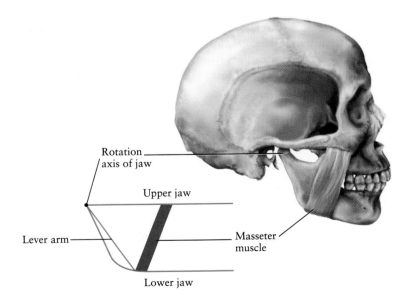

Figure 8-25 The arrangement of muscles and bones takes advantage of the power of the lever arm. The masseter muscle connects the upper jawbone to a point on the lower jawbone that is rather far from the joint around which the jaw rotates. This orientation allows you to generate a torque large enough to crack a macadamia nut between your back teeth.

Figure 8-26 One end of the biceps muscle attaches to the bone of the upper arm and the other end attaches to the lower arm just below the elbow. This muscle–joint arrangement doesn't generate a large torque relative to the size of the muscle because the length of the lever arm is relatively small. A small change in the length of the biceps, however, results in a large, fast movement of the hand.

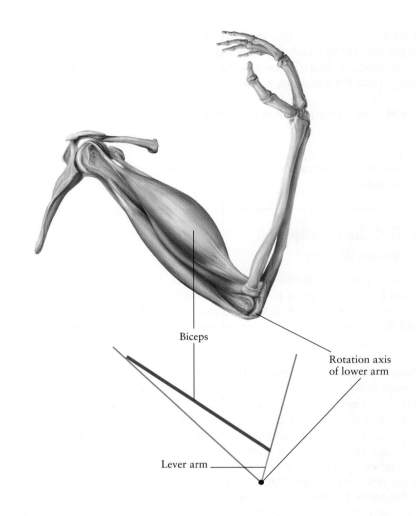

Biceps

Rotation axis
of lower arm

Lever arm

does provide an advantage, however; a small change in the length of the biceps produces a large, fast movement of the hand at the end of the arm.

In Section 8-5, we described a translation between quantities necessary to describe linear motion and those used to describe rotational motion. The quantities mass m and moment of inertia I play analogous roles, as do linear and angular displacements (x and θ, respectively), linear and angular velocities (v and ω), and linear and angular accelerations (a and α). We now see that force and torque are also analogs. This conclusion allows us to write a rotational equivalent of Newton's second law by substituting rotational analogs for the linear quantities:

$$\sum \vec{F}_{\mathrm{dir}} = m\vec{a}_{\mathrm{dir}} \tag{4-1}$$

becomes

$$\sum \vec{\tau} = I\vec{\alpha} \tag{8-24}$$

Don't treat this expression as something new; the physics supporting it is still Newton's second law. We have simply applied Newton's second law to rotational motion.

You may be surprised to find that both torque and angular acceleration are vectors. We will look more carefully at the vector nature of rotational quantities in Section 8-8; until then, we will consider only the magnitudes of the quantities.

Example 8-4 A Simple Pulley

A block has a mass M_{block} and is attached to one end of an inelastic thread that has negligible mass. The other end of the thread is wrapped around the circumference of a thin, uniform cylinder that has a radius R and a mass $M_{cylinder}$ and is allowed to rotate around its central axis as shown in **Figure 8-27**. At what rate, relative to g, does the block accelerate as the thread unwinds without slipping?

SET UP
The block exerts a force on the thread which in turn exerts a force the edge of the cylinder. This force results in a torque on the cylinder which causes it to rotate.

Because this example is fundamentally a force problem, let's use the strategy developed in Chapter 4 for problems involving forces. The first step is to make a free-body diagram, shown in **Figure 8-28**. The second step is to write an expression for the situation using Newton's second law and the free-body diagram. In terms of the magnitudes, the sum of the forces on the box in the y direction is

$$M_{box}g - T = M_{box}a \qquad (8\text{-}25)$$

If we can find the tension T, this expression will provide a solution for the acceleration of the box.

The tension T exerts a torque on the cylinder, and we know two ways to express this torque. Using Equation 8-23,

$$\tau = RT\sin\varphi \qquad (8\text{-}26)$$

and using Equation 8-24,

$$\tau = I\alpha \qquad (8\text{-}27)$$

We now have three equations, all of which depend on either tension, acceleration, or both. Yes, the last equation depends on angular acceleration rather than linear acceleration, so we will also need to find a way to relate these two quantities.

SOLVE
Because the thread unwinds without slipping, the linear acceleration of the edge of the disk is equal to the downward acceleration of the box. We can find a relationship between the linear acceleration and angular acceleration of the cylinder by using Equation 8-5, which directly relates a linear quantity to a rotational quantity:

$$\omega = \frac{v}{r}$$

Using the definition of angular acceleration (Equation 8-20),

$$\alpha = \frac{d\omega}{dt} = \frac{d}{dt}\left(\frac{v}{R}\right) = \frac{1}{R}\frac{dv}{dt} = \frac{a}{R}$$

The second of the two torque equations (Equation 8-27) can now be written in terms of linear acceleration using $\alpha = a/R$. Together with Equation 8-26 and Equation 8-25, we now have three equations and two unknowns, a solvable problem. For convenience, here are the three equations:

$$\tau = I\frac{a}{R} \qquad (8\text{-}28)$$

$$\tau = RT\sin\varphi$$

$$M_{box}g - T = M_{box}a$$

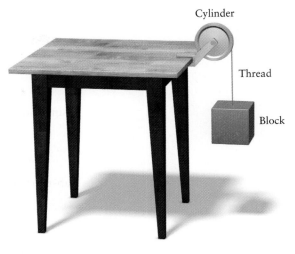

Figure 8-27 A block is attached to one end of an inelastic thread that has negligible mass. The other end of the thread is wrapped around the circumference of a thin, uniform cylinder of nonnegligible mass that can rotate around its central axis.

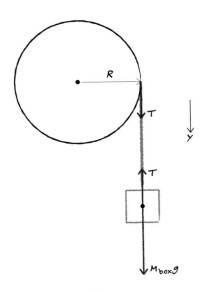

Figure 8-28 A free-body diagram shows the forces acting on a block hanging from a thread wrapped around a pulley of nonnegligible mass.

The first equation gives τ in terms of a. The second equation gives T in terms of τ (step 2 below), and the last equation gives a in terms of T (step 1 below). We're done, except for the algebra!

$$1: \quad a = \frac{M_{box}g - T}{M_{box}} = g - \frac{T}{M_{box}}$$

$$2: \quad T = \frac{\tau}{R \sin\varphi}$$

Using the expression for τ from Equation 8-28, the relationship in step 2 becomes

$$T = \frac{Ia}{R^2 \sin\varphi}$$

Substituting this expression into the step 1 equation gives

$$a = g - \frac{Ia}{R^2 M_{box} \sin\varphi}$$

All of the variables other than a are known. From Table 8-1, the moment of inertia of the rotating cylinder is $M_{cylinder}R^2/2$. The angle φ between the direction in which the thread is pulled and the vector that extends from the center of the cylinder to the point at the thread comes off the cylinder is 90°. As shown in **Figure 8-29**, whenever an object that has been wrapped around a circular surface is pulled, the tension force is always tangential to the edge, so $\sin\varphi$ equals 1 regardless of the direction in which the thread is pulled. By using all this information,

$$a = g - \left(\frac{M_{cylinder}R^2}{2}\right)\left(\frac{a}{R^2 M_{box}}\right)$$

or

$$a = g - \frac{M_{cylinder}a}{2M_{box}}$$

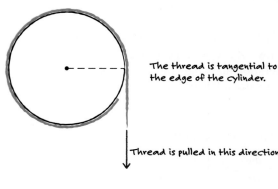

The thread is tangential to the edge of the cylinder.

Thread is pulled in this direction.

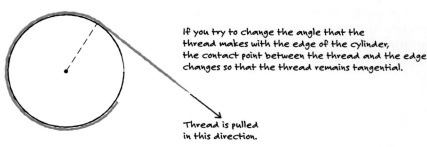

If you try to change the angle that the thread makes with the edge of the cylinder, the contact point between the thread and the edge changes so that the thread remains tangential.

Thread is pulled in this direction.

Figure 8-29 A thread is wrapped around the circumference of a cylinder. Regardless of the direction in which the thread is pulled, the tension force is always tangential to the edge of the cylinder.

Gather the terms that contain a on the left side of the equation

$$a\left(1 + \frac{M_{\text{cylinder}}}{2M_{\text{box}}}\right) = g$$

and then solve for a

$$a = \frac{2M_{\text{box}}}{2M_{\text{box}} + M_{\text{cylinder}}}g$$

REFLECT

The acceleration of the box is proportional to g. Let's test the result for limiting cases of the mass of the cylinder. When the cylinder has negligible mass, the multiplicative factor is 1 and $a \approx g$. That is surely reasonable; when the cylinder is treated as having no mass, the box undergoes free-fall motion. A consideration of energy supports this result; no kinetic energy is required to rotate a cylinder that has no mass, so all of the box's initial potential energy transforms into translational kinetic energy as the box falls. When M_{cylinder} is very large compared to M_{box}, the denominator of the multiplicative factor overwhelms the numerator and results in a fraction tending to 0, so $a \approx 0$. This result also makes sense. A large mass results in the cylinder having a large moment of inertia around the rotation axis, so the torque $(\tau = I\alpha)$ produces only a small angular acceleration.

Got the Concept 8-9
Human Jaw

The human jaw can rotate around three axes as shown in **Figure 8-30**. Chewing is accomplished primarily by rotations around the axis labeled y. The moments of inertia around each of the axes have been measured. For a lower jaw that has a mass approximately equal to 0.4 kg, typical values for the moments of inertia are $I_y = 3 \times 10^{-4}$ kg · m^2 and $I_z = 9 \times 10^{-4}$ kg · m^2. Does the anatomy of the jaw favor up–down motions or side-to-side motions? Is more or less torque required to open your jaw or to rotate it from side to side?

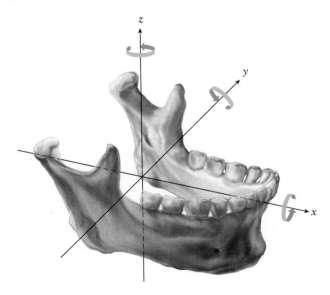

Figure 8-30 The human jaw rotates to some degree about the x, y, and z axes.

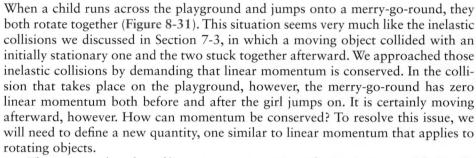

What's Important 8-6

Torque is the rotational analog of force. Torque depends on the magnitude of the applied force, and also on the lever (or moment) arm, the perpendicular distance between the axis of rotation and the line along which the force is applied. Torque is maximum when the force is applied in the direction perpendicular to the direction of the lever arm.

Figure 8-31 A girl has run across the playground and jumped on a merry-go-round, which is free to rotate around its center. (*Lane V. Erickson/Shutterstock. com*)

8-7 Angular Momentum

When a child runs across the playground and jumps onto a merry-go-round, they both rotate together (**Figure 8-31**). This situation seems very much like the inelastic collisions we discussed in Section 7-3, in which a moving object collided with an initially stationary one and the two stuck together afterward. We approached those inelastic collisions by demanding that linear momentum is conserved. In the collision that takes place on the playground, however, the merry-go-round has zero linear momentum both before and after the girl jumps on. It is certainly moving afterward, however. How can momentum be conserved? To resolve this issue, we will need to define a new quantity, one similar to linear momentum that applies to rotating objects.

The rotational analog of linear momentum is *angular momentum*, traditionally represented by either a lowercase or uppercase L. We will use L to represent angular momentum. Using the translation between linear quantities and rotational quantities described in Section 8-5, we can write a rotational equivalent of linear momentum by substituting rotational analogs for the linear quantities. Just as linear momentum is defined by

$$\vec{p} = m\vec{v}$$

we define **angular momentum** as the product of angular velocity and moment of inertia:

$$\vec{L} = I\vec{\omega} \tag{8-29}$$

As with torque and angular acceleration, angular momentum and angular velocity are vectors. We will look more carefully at the vector nature of rotational quantities in Section 8-8; until then, we will consider only the magnitudes of these quantities.

To understand angular momentum L, let's extend the analogy between linear momentum and angular momentum to other linear quantities. We determined linear momentum through its relationship to force, for example,

$$\vec{F} = \frac{d\vec{p}}{dt} \tag{7-23}$$

We have seen that torque $\vec{\tau}$ is the rotational analog of force $\vec{F}$, so torque and angular momentum are related by

$$\vec{\tau} = \frac{d\vec{L}}{dt} \tag{8-30}$$

In cases in which the net torque is zero, $d\vec{L}/dt$ is zero and angular momentum is constant. In the same way that linear momentum is conserved when the net force on a system is zero, *angular momentum is conserved when the net torque on a system is zero*. In particular, angular momentum is conserved in rotating systems that experience no external forces or torques.

Figure 8-32 In the children's game of tetherball, players hit a ball at the end of a rope attached to the top of a tall pole. Players try to wind the rope around the pole by hitting the ball in opposite directions. *(Michael Newman/PhotoEdit)*

> **? Got the Concept 8-10**
> **Tetherball**
> In the children's game of tetherball, a rope attached to the top of a tall pole is tied to a ball. Players hit the ball in opposite directions in an attempt to wind the rope around the pole. As the ball circles the pole, does the speed of the ball decrease, stay the same, or increase? Explain in terms of rotational kinematics. Treat the rope as having negligible mass and neglect any resistive forces.

Example 8-5 Figure Skater

A figure skater executing a "scratch spin" gradually pulls her arms in toward her body while spinning on one skate. During the spin her angular velocity increases from 1.5 rad/s (approximately 1 revolution every 2 s) to 19 rad/s (approximately 3 revolutions per 1 s). By what factor does her moment of inertia around her central axis change as she pulls in her arms?

SET UP

We treat the ice as frictionless so that the ice skater's angular momentum can be considered constant. Her moment of inertia is not constant, however, because the distribution of her mass around the rotation axis changes as she brings in her arms. The magnitude of angular momentum is then $L = I_{start}\omega_{start}$ (from Equation 8-29) at the start of the spin and $L = I_{end}\omega_{end}$ at the end. The two expressions are equal because angular momentum is conserved.

SOLVE

We begin by setting the expressions for angular momentum at the start and end of the spin equal:

$$I_{start}\omega_{start} = I_{end}\omega_{end}$$

The ratio of the moments of inertia is

$$\frac{I_{end}}{I_{start}} = \frac{\omega_{start}}{\omega_{end}}$$

So, the factor by which her moment of inertia around her central axis changes as she pulls in her arms is

$$\frac{I_{end}}{I_{start}} = \frac{1.5 \text{ rad s}}{19 \text{ rad/s}} \approx \frac{1}{13}$$

REFLECT

The skater's moment of inertia decreases by a factor of approximately 13. This might seem large, because her arms (and one leg) are a relatively small fraction of her total mass. Moment of inertia depends on the square of distance (Equations 8-7 and 8-9), however, so holding her arms and leg close to her body rather than extended has a significant effect on the skater's moment of inertia.

Practice Problem 8-5 A figure skater executing a scratch spin starts spinning on one skate at 1.5 rad/s (approximately 1 revolution every 2 s) and gradually pulls her arms in toward her body. If this reduces her moment of inertia by a factor of four (from I_0 to $I_0/4$), what is her angular velocity after she has pulled her arms in completely?

We can extend the description of angular momentum as a rotational analog of linear momentum. From Equation 8-30, the magnitude of torque is

$$\tau = \frac{dL}{dt}$$

Torque is generated by a force applied at a distance r from the axis around which an object can rotate, $\tau = rF \sin \varphi$ (Equation 8-23), so,

$$\frac{dL}{dt} = rF \sin \varphi$$

Finally, we can apply the definition of force as the derivative of momentum with respect to time by using Equation 7-23 to relate angular momentum to linear momentum:

$$\frac{dL}{dt} = r\frac{dp}{dt} \sin \varphi$$

or

$$dL = r\, dp \sin \varphi$$

where φ is the angle between the $\vec{r}$ and $\vec{p}$ vectors. Integrating both sides gives

$$L = rp \sin \varphi \qquad (8\text{-}31)$$

The angular momentum that an object has as it moves with respect to a rotation axis depends on the distance from the axis and direction of motion, as well as the linear momentum.

We can now address the problem posed in the opening paragraph of this section, in which a girl runs across the playground and jumps onto a merry-go-round that was initially at rest. The girl and the merry-go-round begin to rotate rapidly. We now understand that the girl and merry-go-round have angular momentum while rotating. We have also seen that angular momentum is conserved for a system which experiences no external forces. Is there angular momentum in this system *before* the collision, so that we can apply a conservation statement? Yes, objects moving linearly have angular momentum with respect to a fixed point. You might think of an object moving along a straight line as if it were rotating around a fixed point with an ever-changing radius from the axis.

In Figure 8-33, the girl moves in the direction shown at constant momentum p. We apply Equation 8-31 and $\vec{r}$ points from the center of the merry-go-round to the position of the girl at any instant in time, which defines φ as shown in the figure. Notice that the radius R of the merry-go-round and the distance r form one leg and the hypotenuse of a right triangle, so at any instant of time:

$$R = r \sin \varphi$$

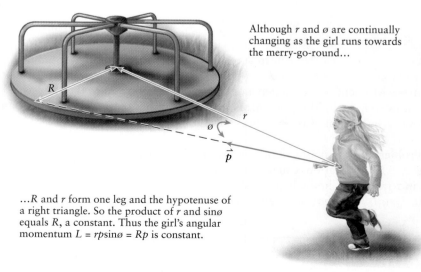

Although r and $\varnothing$ are continually changing as the girl runs towards the merry-go-round...

...R and r form one leg and the hypotenuse of a right triangle. So the product of r and $\sin\varnothing$ equals R, a constant. Thus the girl's angular momentum $L = rp\sin\varnothing = Rp$ is constant.

Figure 8-33

By using Equation 8-31, the girl's angular momentum with respect to the center of the merry-go-round is

$$L = Rp \tag{8-32}$$

Not only does the girl have angular momentum with respect to the merry-go-round, but because both R and p are constant, L is also constant. The magnitude of the angular momentum of the girl and merry-go-round is therefore constant both before and after the collision. If this were not true it would not be possible to apply a requirement that angular momentum is conserved.

Estimate It! 8-2 Girl and Merry-Go-Round

Estimate the number of revolutions per second that the girl (Figure 8-31) makes after jumping on the merry-go-round from a running start.

SET UP
We can determine rotations per second from angular velocity. In addition, angular momentum is conserved, so the angular momentum of the girl just as she jumps onto the merry-go-round equals the total angular momentum of the girl and merry-go-round after she has jumped onto it. The girl's initial angular momentum is

$$L_{\text{girl, i}} = Rm_{\text{girl}}v_i$$

from Equation 8-32, where R is the radius of the merry-go-round, m_{girl} is her mass, and v_i is her speed at the moment she jumps on. The total angular momentum is the sum of contributions from the girl and also the rotation of the merry-go-round:

$$L_{\text{total, f}} = I_{\text{girl+mgr}}\omega = \left(m_{\text{girl}}R^2 + \frac{m_{\text{mgr}}R^2}{2} \right)\omega$$

SOLVE
Setting the initial and final angular momenta equal gives

$$Rm_{\text{girl}}v_i = \left(m_{\text{girl}}R^2 + \frac{m_{\text{mgr}}R^2}{2} \right)\omega$$

or

$$\omega = \frac{m_{\text{girl}}v_i}{\left(m_{\text{girl}} + \dfrac{m_{\text{mgr}}}{2} \right)R}$$

For our estimate, we want to use big, round, but reasonable values. Let the mass and speed of the girl be 30 kg and 3 m/s, respectively, both reasonable guesses. The merry-go-round must be much more massive, say, 100 kg. We estimate the radius of the merry-go-round to be 2 m. Then

$$\omega \approx \frac{(30\ \text{kg})(3\ \text{m/s})}{(30\ \text{kg} + 50\ \text{kg})(2\ \text{m})} \approx 0.56\ \text{rad/s}$$

Notice that angular velocity carries units of radians per second, and because there are 2π radians in one revolution of the merry-go-round, this is

$$\omega \approx 0.56\ \text{rad/s}\ (1\ \text{rev}/2\pi\ \text{rad}) \approx 0.09\ \text{rev/s}$$

or about one revolution every 11 s.

REFLECT
Our answer, about 1 rev every 11 s, seems reasonable. However, the mass we used for the merry-go-round is really just a best guess. No doubt we could have chosen from a range of reasonable values for m_{mgr}, which would in turn result in a

range of final answers. There is a precise mathematical way to determine how an uncertainty in the value of a quantity affects the final answer, but we can get a feeling for the range of answers by computing values using the high end and low end of the range that we believe is reasonable.

We used $m_{mgr} = 100$ kg. Let's consider the range 50 kg $< m_{mgr} <$ 150 kg. The smallest value of the mass gives the highest estimate of ω, and the largest value gives the lowest estimate of ω. When $m_{mgr} = 50$ kg, $\omega = 0.8$ rad/s or about 1 rev every 8 s. When $m_{mgr} = 150$ kg, $\omega = 0.4$ rad/s or about 1 rev every 15 s. As estimates, all of the answers are reasonable, suggesting that our estimate of the mass of the merry-go-round is a good one.

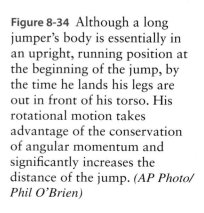

√x̄   *See the Math Tutorial for more information on Significant Figures*

> **✱ What's Important 8-7**
>
> Angular momentum is the rotational analog of linear momentum. Angular momentum is conserved in rotating systems that experience no external forces or torques. The angular momentum that an object has as it moves with respect to a rotation axis depends on the distance from the axis and direction of motion, as well as the linear momentum.

8-8 The Vector Nature of Rotational Quantities

Figure 8-34 shows long jumper Mike Powell winning the silver medal at the 1992 Barcelona Olympics. Although his body is essentially in an upright, running position at the beginning of the jump, by the time he lands his legs are out in front of his torso. This significantly increases the distance of his jump. Although his legs have rotated forward, his torso hasn't rotated with them. Yet while up in the air, Powell has nothing on which to exert a force that would change his position. What physics underlies his motion?

We have established that angular momentum is conserved. As Powell leaves the ground, the net angular momentum of his body is zero or nearly so. Once in the air, he brings his legs up and forward (counterclockwise as seen from the vantage point of the photograph), resulting in nonzero angular momentum. For angular momentum to be conserved, some other part of his body must rotate in such a way that the two contributions to his net angular momentum cancel. Long jumpers learn to rotate their arms rapidly clockwise while in the air, which helps rotate their upper torso clockwise. The angular momentum associated with rotation in one direction can cancel angular momentum associated with rotation in the opposite direction.

Figure 8-34 Although a long jumper's body is essentially in an upright, running position at the beginning of the jump, by the time he lands his legs are out in front of his torso. His rotational motion takes advantage of the conservation of angular momentum and significantly increases the distance of the jump. *(AP Photo/ Phil O'Brien)*

(a)

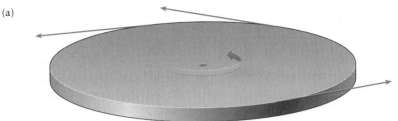

Which one vector represents the
rotational motion? (None of them)

(b) The angular momentum vector
points in a direction perpendicular
to the rotation plane.

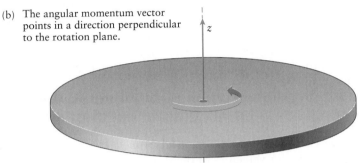

Is angular momentum up or down? The
direction is up in a right-handed sense. Curl
the fingers on your right hand in the direction
of motion and stick your thumb straight out;
your thumb points in the direction of the
angular momentum vector.

Figure 8-35 (a) No single vector
that lies in the plane of rotation
indicates the direction of rotation.
(b) The angular momentum
vector $\vec{L}$ points in a direction
perpendicular to the plane of
rotation.

The most natural and easiest way to understand this property of angular momentum is to consider angular momentum as a vector.

In which direction does the angular momentum vector $\vec{L}$ point? We need to
account for the direction of rotation, but as shown in **Figure 8-35a**, there is no way
to align a single vector in the plane of the rotation to indicate the direction. The
angular momentum points along the axis of rotation, but in which of the two directions? By convention, the specific direction is given in a right-handed sense, often
summarized by the **right-hand rule** (**Figure 8-36**). Curl the fingers of your right
hand in the direction of motion and stick your thumb straight out. By the right-
hand rule, your thumb points in the direction of the angular momentum vector.
Figure 8-35b shows the angular momentum vector for a rotating disk. How does
this apply to the long jumper? If more than one element of a system is rotating, the

Right hand rule: Curl
the fingers on your
right hand from $\vec{A}$ to $\vec{B}$
along the closest path.
Stick out your thumb;
it points in the direction
of $\vec{C}$, the result of the
cross product $\vec{A} \times \vec{B}$.

$\vec{C} = \vec{A} \times \vec{B}$

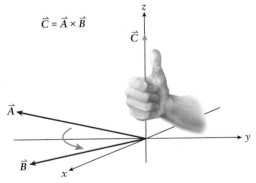

$\vec{A}$ and $\vec{B}$ lie in the xy plane in this example.
$\vec{C}$ points in the positive z direction.

Figure 8-36 The right-hand
rule allows us to determine
the direction of the angular
momentum vector of a
rotating object.

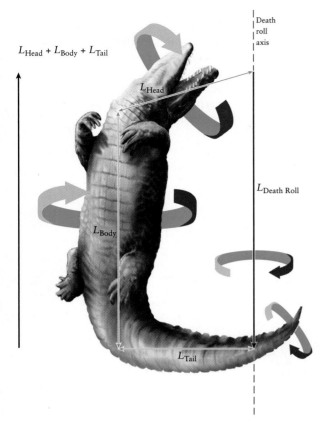

$L_{\text{Head}} + L_{\text{Body}} + L_{\text{Tail}}$

L_{Head}

Death roll axis

L_{Body}

$L_{\text{Death Roll}}$

L_{Tail}

and tail around different axes. The sum of the angular momentum around each axis results in the death roll, a rotation in the opposite direction of the separate spins. This rotation results in a torque, and therefore a force, large enough to detach pieces of meat from the alligator's prey.

net angular momentum $\vec{L}$ is the vector sum of the angular momenta of each individual element. Therefore, the angular momentum of the jumper's arms and upper torso cancels out that of his legs. This stabilizes his motion so that both arms and legs are forward at the end of the jump, resulting in a longer flight, and perhaps a gold medal.

Let's explore this physics more deeply by considering the feeding behavior of alligators, which involves rotations of their head, body, and tail. The structure of an alligator's jaws and teeth prevents these reptiles from cutting large prey into chunks small enough to swallow. The problem is exacerbated by the fact that alligators have no leverage when they clamp onto their prey while swimming—their legs are too short to hold the animal and they can't push against the river bottom. However, by tightly biting the prey and then executing a special spinning maneuver, they can exert forces large enough to remove bite-sized pieces of food. This "death roll" arises from the conservation of angular momentum.

After the alligator has securely grabbed hold of its prey, it bends its tail and head to one side, forming a C-shape as shown in Figure 8-37. The shape is crucial to the execution of the death roll because it enables the animal to spin its head, body, and tail around different axes. The angular velocities of the three rotations are shown as large blue arrows in the figure. You can use the right-hand rule to verify that the angular momentum vectors associated with the three rotations, labeled $\vec{L}_{\text{head}}$, $\vec{L}_{\text{body}}$, and $\vec{L}_{\text{tail}}$, point in the directions shown. The net angular momentum generated by the animal is the vector sum $\vec{L}_{\text{head}} + \vec{L}_{\text{body}} + \vec{L}_{\text{tail}}$, shown as a black arrow pointing directly upward in the figure.

The total angular momentum of the alligator remains constant, however, because the alligator experiences no net torque. Therefore, as a result of rotating its head, body, and tail the alligator must acquire a compensatory, net rotation *in the opposite direction* of the separate spins. This is the death roll. As the alligator rotates around the death roll axis, it acquires the angular momentum $\vec{L}_{\text{death roll}}$ shown as a black arrow pointing downward in the figure. This angular momentum has the same magnitude as $\vec{L}_{\text{head}} + \vec{L}_{\text{body}} + \vec{L}_{\text{tail}}$ but points in the opposite direction, so the two vectors cancel. During the death roll, the alligator's tail, body, and head rotate out of the page toward you in the orientation shown in the figure. This rotation results in a torque, and therefore a force, large enough to rip hunks of meat from the prey. The force generated by the rolling maneuver is estimated to be well over 130 N for an adult alligator.

 Got the Concept 8-11
Falling Cat

Everyone knows that a falling cat usually lands on his feet. This result requires the cat to rotate itself into an upright position while falling. The keys to a cat's ability to start and stop rotating in midair is bending his body in the middle at the start of the fall, and either pulling in or extending his legs. Explain how the actions help the cat right himself while falling.

? Got the Concept 8-12
Falling Cat Part 2

After a falling cat bends in the middle (see Got the Concept 8-11) he pulls in his front legs, extends his rear legs, and rotates the front and back segments of his body. After a moment, the cat extends his front legs and pulls in his rear legs while continuing to rotate the front and rear parts of his body. Explain how this helps the cat right himself while falling. Consider the moment of inertia of each part around its rotation axis.

? Got the Concept 8-13
Tides

Ocean tides are evidence that Earth and the Moon are connected by the forces each exerts on the other (Figure 8-38). The rotation of Earth, the rotation of the Moon, and the orbit of the Moon around Earth all contribute to the angular momentum of the Earth–Moon system. The rotation rate of the Moon is stable (it doesn't change) but the distance between Earth and the Moon *is* changing, and the rate at which Earth rotates is slowing down. Is the Moon getting farther from or closer to Earth? Explain your answer.

Figure 8-38 Earth and the Moon are connected by the forces each exerts on the other. (*Courtesy David Tauck*)

! Watch Out!
The Moon is getting farther, not closer, to Earth.

When confronted with the previous problem (Got the Concept 8-13) many students guess that the Moon, over time, will get closer and closer to Earth, much the way old satellites eventually spiral down and burn up in Earth's atmosphere. The orbits of the Moon and of those satellites are different, however. Satellites experience air resistance as they move through the atmosphere, which is thin but still exists at the altitude at which they orbit Earth. This drag force saps the kinetic energy of a satellite, causing the satellite's orbit to have a smaller and smaller radius. The Moon does not experience such resistive forces, but the Moon and Earth are connected by gravitational forces. The forces allow the transfer of angular momentum, or, alternatively, energy, between the two parts of the system. As the Moon gains the energy that Earth loses, its orbit is pushed farther from Earth.

Angular momentum $\vec{L}$ and torque $\vec{\tau}$ are vectors. In Section 8-7, we saw how the two are related (in Equation 8-30). We also defined torque in terms of lever arm $r \sin \varphi$ and force F

$$\tau = rF \sin \varphi \tag{8-23}$$

This statement gives the magnitude of the torque as a function of the distance r from the rotation axis at which a force F is applied. Both the distance and the force are vectors, leading to the vector equivalent of Equation 8-23:

$$\vec{\tau} = \vec{r} \times \vec{F} \tag{8-33}$$

We can also write angular momentum in vector notation by taking the vector equivalent of Equation 8-31:

$$\vec{L} = \vec{r} \times \vec{p} \tag{8-34}$$

The symbol "×" represents the mathematical operation known as a ***cross product.*** As you can infer by comparing Equations 8-33 and 8-23, the result of taking the cross product of two vectors is a vector. The magnitude of the vector is the product of the magnitudes of the two vectors and the sine of the angle in between them. The cross product of $\vec{A}$ and $\vec{B}$ is

$$\vec{C} = \vec{A} \times \vec{B} = AB \sin\varphi \qquad (8\text{-}35)$$

where φ is the angle defined according to convention to go from $\vec{A}$ to $\vec{B}$. $\vec{C}$ points in the direction perpendicular to both $\vec{A}$ and $\vec{B}$ and is defined by the right-hand rule. When you curl the fingers on your right hand along the shortest path from $\vec{A}$ to $\vec{B}$, your outstretched thumb points in the direction of $\vec{C}$. The elements of the cross product and the right-hand rule are shown in Figure 8-36.

As defined in Equation 8-33, the torque $\vec{\tau}$ is perpendicular to both $\vec{r}$ and $\vec{F}$ according to the definition of the cross product. This is an ambiguous statement, however; the torque could point in either of the two directions away from the plane formed by $\vec{r}$ and $\vec{F}$. The ambiguity is resolved according to the convention embodied by the right-hand rule—by curling your fingers from $\vec{r}$ to $\vec{F}$ along the shortest path, your outstretched thumb points along the direction of $\vec{\tau}$. This is demonstrated in **Figure 8-39** for two different forces applied to the end of a rod that can rotate around an axis through the other end.

By convention, the angle φ in the cross product $\vec{A} \times \vec{B} = AB \sin\varphi$ is defined counterclockwise from $\vec{A}$ to $\vec{B}$. This is also shown for the two cases in Figure 8-39. Notice how the angle convention affects the result. In the first case, in which φ is less than 180°, the cross product is positive because the sine of an angle less than 180° is positive. In the second case, φ is greater than 180°, and because $\sin\varphi$ is negative, the cross product is as well. The difference in sign is simply another way to tell that the cross product $\vec{r} \times \vec{F}$ is in opposite directions for the two forces shown in the figure.

It is important to recognize that no motion and no change in motion occur in the direction of the torque vector. In fact, the vector direction associated with the rotational quantities ($\vec{\tau}$, $\vec{L}$, $\vec{\omega}$, $\vec{\theta}$, and $\vec{\alpha}$) is a mathematical convention only. For example, a net torque will result in a change in motion, but that angular accelera-

Figure 8-39 The torque vector $\vec{\tau}$ is perpendicular to both $\vec{r}$ and $\vec{F}$. The perpendicular direction could be either up or down relative to the plane formed by $\vec{r}$ and $\vec{F}$; the ambiguity is resolved according to the convention embodied by the right-hand rule. In these examples, torque points upward when the object rotates counterclockwise and downward when it rotates clockwise.

To apply the right hand rule correctly, curl your fingers from the direction of $\vec{r}$ to $\vec{F}$ along the shortest path.

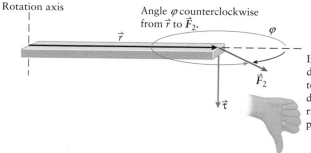

Rotation axis

$\vec{r}$　　$\vec{F}_1$　$\vec{\tau}$　φ

Angle φ is defined counterclockwise from the direction of $\vec{r}$ to $\vec{F}_1$. In this case φ corresponds to the shortest path from the direction of $\vec{r}$ to $\vec{F}_1$.

Rotation axis

Angle φ counterclockwise from $\vec{r}$ to $\vec{F}_2$.

$\vec{r}$　φ　$\vec{F}_2$　$\vec{\tau}$

In this case the counterclockwise definition of φ does not correspond to the shortest path from the direction of $\vec{r}$ to $\vec{F}_1$. To apply the right hand rule, use the shortest path from $\vec{r}$ to $\vec{F}_1$.

tion does not take place in the direction of the torque vector. The direction of the torque vector is a mathematical tool that ensures that the relationships between angular quantities remain consistent.

Example 8-6 Balancing on a Seesaw

Alan and Bob sit on a seesaw. Alan weighs 1.5 times as much as Bob. When Alan sits 0.8 m from the center of the seesaw, where should Bob sit so they just balance?

SET UP
When the system, which comprises Alan, Bob, and the seesaw, is balanced, there are no linear and no angular accelerations. Another way to say this is that there are no net forces and no net torques. No net linear acceleration means that when the system is released from rest it acquires no net linear motion. Of course, that would be true even if Alan and Bob didn't balance each other, because the system as a whole would move up or down. No angular acceleration means that when the system is released from rest, the seesaw doesn't rotate around the center point, so that's the condition we want to require.

One of our definitions of torque is

$$\sum \vec{\tau} = I\vec{\alpha} \qquad (8\text{-}24)$$

We now know that the net torque ($\sum \vec{\tau}$) can be zero if two vector contributions to the net torque cancel. Torque arises in the system due to the boys' weights, which are exerted at distances r_A and r_B, respectively, from the rotation axis of the seesaw. Can the torques $\vec{\tau}_A$ and $\vec{\tau}_B$ be made to cancel, that is, are they in opposite directions?

First, let's use the right-hand rule to determine the directions of $\vec{\tau}_A$ and $\vec{\tau}_B$. Look at Figure 8-40 and curl the fingers on your right hand from the direction of $\vec{r}_A$ to $\vec{W}_A$. Your outstretched thumb should point into the plane of the page. When you curl the fingers on your right hand from the direction of $\vec{r}_B$ to $\vec{W}_B$, your thumb points out of the plane of the page. So, yes, $\vec{\tau}_A$ and $\vec{\tau}_B$ can be made to cancel because they point in opposite directions.

We can come to the same conclusion by considering angle in the definition of torque $\tau = rF \sin \varphi$ (Equation 8-23). We have $\tau_A = r_A W_A \sin \varphi_A$ and $\tau_B = r_B W_B \sin \varphi_B$, and because the angles are defined counterclockwise from the direction of $\vec{r}$ to the direction of $\vec{W}$, φ_A is more than 180° while φ_B is less than 180°. The angles are shown in Figure 8-40. A glance at the sine curve shows that $\sin \varphi_A$ is therefore less than 1 while $\sin \varphi_B$ is more than 1. The magnitude of the torques τ_A and τ_B are therefore of opposite sign. Again, this means we can get them to cancel.

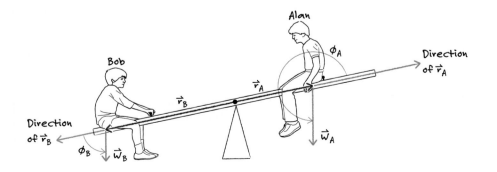

Figure 8-40 A sketch shows the relationship between the variables needed to analyze a problem in which Alan and Bob sit on opposite sides of a seesaw.

When $\phi_B = \phi_A + 180°...$

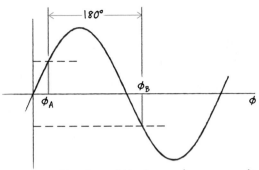

The sines of the two angles are equal in magnitude but opposite in sign:

$$\sin \phi_A = -\sin \phi_B$$

Figure 8-41 When two angles differ by 180°, the sines of the angles are equal in magnitude but opposite in sign.

In terms of the magnitudes, then, to get zero net angular acceleration we require

$$\tau_A = \tau_B$$

or

$$r_A W_A \sin \varphi_A = r_B W_B \sin \varphi_B$$

SOLVE
Solving for r_B

$$r_B = \frac{r_A W_A \sin \varphi_A}{W_B \sin \varphi_B}$$

We have been given $W_A = 1.5 W_B$ and also $r_A = 0.8$ m. It might seem, however, that we've hit a roadblock because no information was given about the angles. But let's ask: what else do we know? Inspection of Figure 8-40 reveals a relationship between φ_A and φ_B, that is, $\varphi_A = \varphi_B + 180°$. Figure 8-41 shows that when $\varphi_A = \varphi_B + 180°$, $\sin \varphi_A = -\sin \varphi_B$.

So by using the values given and our insights we get

$$r_B = \frac{(0.8 \text{ m})(1.5 W_B)(-\sin \varphi_B)}{W_B \sin \varphi_B} = -1.2 \text{ m}$$

Bob must sit 1.2 m from the center in order to balance Alan. The minus sign tells us that Bob must sit on the opposite side of the center from Alan.

REFLECT
Notice the advantage of the lever arm in this problem. Bob, who is lighter than Alan, can balance the seasaw by sitting farther from the center. A small force can give rise to a large torque when the lever arm is large.

Something happened in the process of solving this problem that is also important to note. Although our first step in solving a problem is to plan out (**Set Up**) the solution, it isn't always possible to anticipate every variable or relationship we'll need. In solving this problem, we didn't anticipate the need to know φ_A and φ_B. When we got stuck, however, we asked, "What else do we know?" which lead to the realization that the angles φ_A and φ_B are related, and that, in turn, provided the last step in solving the problem.

Practice Problem 8-6 Alan and Bob sit on a seesaw. When Alan, who weighs 1.5 times as much as Bob, sits 1.2 m from the center of the seesaw, Bob sits 1.8 m from the center. If Alan changes his position so that he is half as far from the center, where must Bob move to keep the seesaw balanced?

⭐ **What's Important 8-8**
Angular momentum and torque are vectors, so both have direction as well as magnitude. To determine the direction of the angular momentum vector, curl the fingers of your right hand in the direction of motion and stick out your thumb. According to the right-hand rule, your thumb points in the direction of the angular momentum vector. To determine the direction of the torque vector, curl the fingers of your right hand in the direction from the radial vector to the force vector along the shortest path, and stick your thumb straight out. Your outstretched thumb points along the direction of the torque vector.

Answers to Practice Problems

8-1 6.5 rev/s

8-2 $3MR^2/4$

8-3 18 s

8-5 6.0 rad/s

8-6 0.9 m from the center

Answers to Got the Concept Questions

8-1 Although you probably would be able to swing the cell phone in a circle, it's possible that you might not. The answer depends on the moment of inertia of the thread and cell phone together, which in turn depends on the length of the thread. For a relatively short thread, the amount of energy required to swing the cell phone would be small; you would easily be able to swing the phone in a circle. Although unlikely, what if the thread were, say, 10 m long? Even for a small mass and a small angular velocity, the amount of rotational kinetic energy required to rotate the system would be large. For a long enough thread, the amount of energy would be more than you could supply, and you would not be able to swing the cell phone in a circle.

8-2 Because a bird's wings are a larger fraction of its total mass than the contribution that an insect's wings make to its mass, the bird has more of its mass farther from the roll axis compared to the insect. For this reason the moment of inertia around that axis will be larger for the bird than for the insect. More energy is therefore required to cause the bird to turn (rotate) around that axis, making the bird less responsive to forces exerted by its wings. As a result, the typical bird is less maneuverable than, say, a dragonfly.

8-3 The moment of inertia of a thin, uniform rod rotating around its center is $ML^2/12$. Because the moment of inertia is additive, the moment of inertia of the two rods together is $I = ML^2/12 + ML^2/12 = ML^2/6$.

8-4 The moment of inertia of a thin, uniform rod that has a mass M and a length L rotating around its end is $ML^2/3$. In this problem, each of the four rods of mass $M/2$ and length $L/2$ has a moment of inertia of

$$I = \frac{(M/2)(L/2)^2}{3} = \frac{ML^2}{24}$$

when rotating around its end. Because the moment of inertia is additive, the moment of inertia of the four rods together is

$$I = \frac{ML^2}{24} + \frac{ML^2}{24} + \frac{ML^2}{24} + \frac{ML^2}{24} = \frac{ML^2}{6}$$

Notice that the final configuration of the four rods is identical to that of the two rods in the previous problem, so it is not surprising that the result is the same.

8-5 The car would go faster if its wheels were shaped like uniform disks. A wheel like a bicycle tire, where most of the mass is far from the rotation axis, will have a larger moment of inertia than one where the mass is uniformly distributed. A larger moment of inertia means more energy is required to rotate the wheel, so less energy can be transformed into translational kinetic energy.

8-6 Because it does not rotate, all of the block's gravitational potential energy is transformed into translational kinetic energy. Regardless of the moment of inertia of the hoop around its rotation axis, at least some of its gravitational potential energy will be transformed into rotational kinetic energy. As a result, the hoop's linear speed at any instant will be smaller than the linear speed of the block. The block reaches the bottom of the ramp first.

8-7 To read information at a constant rate, the reader must pass the same distance along the spiral in a given amount of time regardless of radius. Distance in this case is the length of a circular arc, which is given by the product of angle θ and radius r. In order for the product of angle and radius to remain constant, the angular displacement in any fixed amount of time must change. So as the information to be read gets closer to the center of the disk, the rotation rate must increase. In general, in order for the linear extent of information passing the reader in a given amount of time to remain constant, the angular velocity must be higher when the distance of the information to the center of the disk is smaller.

8-8 Extending the handle of the wrench with the pipe increases the distance r between the rotation axis and the place the force is applied. Torque (Equation 8-23) increases as r increases even when the force remains the same, making it more likely that the bolt will rotate. Of course, mechanics avoid this method because the application of even a small force may result in a torque large enough to break the bolt or the wrench!

8-9 The torque required to create a certain angular acceleration is proportional to the moment of inertia (Equation 8-24). Because I_y is typically three times smaller than I_z, it therefore requires three times less torque to cause the jaw to open and close than to rotate it from side to side. The anatomy of the jaw favors up–down motions.

8-10 The ball speeds up. Striking the ball imparts a fixed amount of angular momentum, and because we neglect resistive forces, there is no external torque on it. (Yes, there is a force on the ball due to tension in the rope. This force is nearly radial, however, so the angle between $\vec{r}$ and $\vec{F}$ in the definition of torque in Equation 8-23 is nearly zero, resulting in a torque that is nearly zero.) Angular momentum is therefore conserved, or nearly so, but because the distance of the ball from the rotation axis decreases as the rope loops around the pole, its moment of inertia is decreasing. From $\vec{L} = I\vec{\omega}$ (Equation 8-29), when I decreases the magnitude of ω must increase in order for the magnitude of L to remain constant.

8-11 By bending in the middle, the cat divides his body into two segments, each of which can rotate around a different axis. The cat uses his muscles to rotate his front section into an upright position while simultaneously rotating his hindquarters slightly. This small twist cancels the angular momentum generated by the rotation of the front part of his body. Next, the opposite situation occurs: as the rear of the animal rotates into alignment with the front, he slightly twists his forequarters. Because the net angular momentum of the cat must remain zero, the cat's body must twist to cancel the angular momentum associated with the front and rear of his body separately.

8-12 By pulling in his front legs, the cat reduces the moment of inertia around the rotation axis of the [PE] front part of his body. Similarly, extending his rear legs increases the moment of inertia around the rotation axis of the rear part of his body. During the first part of the cat's twisting motion, the front of his body rotates through a much larger angle than the rear. When the front legs are nearly in the correct position for a safe landing, the cat extends his front legs, increasing the moment of inertia and slowing the rotation of the front of his body. Tucking in his rear legs has the opposite effect, which allows the rear of the cat's body to swing farther into the right position for landing. Because there is never a net rotation of the cat's body, when the front and rear legs are vertical both rotations stop and the cat lands with his feet beneath him.

8-13 No external torques act on the Earth–Moon system, so the net angular momentum of the system must remain constant. The angular momentum associated with Earth's rotation, $L_{\text{Earth}} = I_{\text{Earth}}\omega_{\text{Earth}}$, decreases as the rate of rotation ω_{Earth} decreases. Because the rate of the Moon's rotation does not change, the angular momentum of the Moon's revolution around Earth must increase in order for the net angular momentum to remain constant. The Moon is moving farther from Earth, increasing its moment of inertia around the axis of rotation (which passes through Earth).

SUMMARY

Topic	Summary	Equation or Symbol	
angular acceleration	The angular acceleration α of a rotating object is the derivative of angular velocity with respect to time. Angular acceleration gives the rate at which rotation rate is changing.	α	
angular momentum	Angular momentum L is the rotational analog of linear momentum. Angular momentum is the product of angular velocity $\vec{\omega}$ and moment of inertia I, or equivalently, the cross product of the radial vector $\vec{r}$ of an object from a fixed point and the linear momentum $\vec{p}$ of the object.	$\vec{L} = I\vec{\omega}$ $L = rp\sin\varphi$ $\vec{L} = \vec{r} \times \vec{p}$	(8-29) (8-31) (8-34)
angular velocity	For a rigid object, angular velocity ω characterizes the speed of rotation in a way that is independent of size or shape of the rotating object. Angular velocity ω is related to the linear velocity v of an object or an element of an object and its distance r from the axis of rotation.	$\omega = \dfrac{v}{r}$	(8-5)
average angular velocity	Average angular velocity ω is the average angular velocity over a finite period of time Δt.	$\omega = \dfrac{\Delta\theta}{\Delta t}$	(8-3)

instantaneous angular velocity	Instantaneous angular velocity is the angular velocity of an object at any instant of time. Instantaneous angular velocity is defined as the limit of average angular velocity over an infinitesimally short time, or equivalently, the derivative of angular position with respect to time.	$\omega = \lim\limits_{\Delta t \to 0} \dfrac{\Delta \theta}{\Delta t} = \dfrac{d\theta}{dt}$ (8-4)
lever arm	The lever arm is the perpendicular distance between the axis of rotation and the line along which a force is applied.	
moment arm	For a force applied to an object which can rotate, the lever arm is also referred to as the moment arm.	
moment of inertia	Moment of inertia (or rotational inertia) I represents the resistance of an object to a change in angular velocity. I depends on both the mass of an object and how the mass is distributed with respect to the axis of rotation (Equation 8-9) or, for multiple objects, the mass of each m_i and the distance r_i each is from the axis of rotation.	$I = \displaystyle\int r^2 \, dm$ (8-9) $I = \displaystyle\sum m_i r_i^2$ (8-7)
parallel-axis theorem	The parallel-axis theorem describes the relationship between I_{CM}, the moment of inertia of an object rotating around its center of mass, and the moment of inertia when the object rotates around any other parallel axis running though a point a distance h from the center of mass.	$I = I_{CM} + Mh^2$ (8-16)
right-hand rule	The right-hand rule is used to determine the direction of a vector formed by the cross product of two others. For example, when you curl the fingers of your right hand in the direction of motion and stick your thumb straight out, according to the right-hand rule your thumb points in the direction of the angular momentum vector.	
rolling friction	Rolling friction is the retarding, frictional force experienced by objects that roll without slipping. The force of rolling friction, like static and kinetic frictional forces, is proportional to the normal force acting on an object.	
rotational inertia	Rotational inertia is the equivalent of moment of inertia.	I
rotational kinetic energy	The kinetic energy of a rigid object rotating around a fixed axis is called rotational kinetic energy. It depends not only on the mass and angular velocity of an object, but also on how the mass of the object is distributed with respect to the axis of rotation.	$K_{rotational} = \dfrac{1}{2} I \omega^2$ (8-8)
torque	Torque τ is the rotational analog of force. Torque depends on the magnitude F of the applied force, but also on the lever arm $r \sin \varphi$, the perpendicular distance from the axis of rotation to the line along which the force acts.	$\tau = rF \sin \varphi$ (8-23) $\vec{\tau} = \vec{r} \times \vec{F}$ (8-33)

QUESTIONS AND PROBLEMS

In a few problems, you are given more data than you actually need; in a few other problems, you are required to supply data from your general knowledge, outside sources, or informed estimate.

Interpret as significant all digits in numerical values that have trailing zeros and no decimal points.

For all problems, use $g = 9.8$ m/s^2 for the free-fall acceleration due to gravity. Neglect friction and air resistance unless instructed to do otherwise.

• Basic, single-concept problem

•• Intermediate-level problem, may require synthesis of concepts and multiple steps

••• Challenging problem

SSM *Solution is in Student Solutions Manual*

Conceptual Questions

1. •In your own words, define rotation and revolution and compare the concepts of rotational motion and orbital motion (revolution) for a planet in our solar system.

2. •What are the units of angular velocity (ω)? Why are there factors of 2π present in the equations of rotational motion?

3. •Why is it critical to define the axis of rotation when you set out to find the moment of inertia of an object? SSM

4. •Explain how an object moving in a straight line can have a nonzero angular momentum.

5. •What are the units of the following quantities: (a) rotational kinetic energy, (b) moment of inertia, and (c) angular momentum?

6. •Explain which physical quantities change when an ice skater moves her arms in and out as she rotates in a pirouette. What causes her angular velocity to change, if it changes at all?

7. •While watching two people on a seesaw, you notice that the person at the top always leans backward, while the person at the bottom always leans forward. Why do the riders do this? Assuming they are sitting equidistant from the pivot point of the seesaw, what, if anything, can you say about the relative masses of the two riders? SSM

8. •The five solids shown in **Figure 8-42** in cross section have equal heights, widths, and masses. The axes of rotation are located at the center of each object and are perpendicular to the plane of the paper. Rank the moments of inertia from greatest to least.

9. •Referring to the time-lapse photograph of a falling cat at the beginning of this chapter, do you think that a cat will fall on her feet if she does not have a tail? Explain your answer using the concepts of this chapter.

10. •What is the difference in rotational kinetic energy between two balls each tied to a light string and spinning in a circle with a radius equal to the length of the string? The first ball has a mass m, a string of length L, and rotates at a rate of ω. The second ball has a mass $2m$, a string of length $2L$, and rotates at a rate of 2ω.

11. •**Calc** In your own words, describe the quantity dm that is found in the integral equation for moment of inertia:

$$I = \int r^2 dm \quad \text{SSM}$$

12. •Describe any inconsistencies in the following statement: *The units of torque are* N · m, *but that's not the same as the units of energy.*

13. •A student cannot open a door at her school. She pushes with ever-greater force, and still the door will not budge! Knowing that the door does push open, is not locked, and a minimum torque is required to open the door, give a few reasons why this might be occurring.

14. •Analyze the following statement and determine if there are any physical inconsistencies:

While rotating a ball on the end of a string of length L, the rotational kinetic energy remains constant as long as the length and angular speed are fixed. When the ball is pulled inward and the length of the string is shortened, the rotational kinetic energy will remain constant due to conservation of energy, but the angular momentum will not, because there is an external force acting on the ball to pull it inward. The moment of inertia and angular speed will, of course, remain the same throughout the process because the ball is rotating in the same plane throughout the motion.

15. •A freely rotating turntable moves at a steady angular velocity. A glob of cookie dough falls straight down and attaches to the very edge of the turntable. Describe which quantities (angular velocity, angular acceleration, torque, rotational kinetic energy, moment of inertia, or angular momentum) are conserved during the process and describe qualitatively what happens to the motion of the turntable. SSM

16. •Describe what a "torque wrench" is and discuss any difficulties that a Canandian auto or bicycle mechanic might have working with an American mechanic's tools (and vice versa).

Hoop   Cube   Solid cylinder   Solid sphere   Hollow sphere

Figure 8-42 Problem 8

17. •In describing rotational motion, it is often useful to develop an analogy with translational motion. First, write a set of equations describing translational motion. Then write the rotational analogs (for example, $\theta = \theta_0 \dots$) of the translational equations (for example, $x = x_0 + v_0 t + \frac{1}{2} a t^2$) using the following legend:

$$x \Leftrightarrow \theta \quad v \Leftrightarrow \omega \quad a \Leftrightarrow \alpha \quad F \Leftrightarrow \tau \quad m \Leftrightarrow I$$
$$p \Leftrightarrow L \quad K \Leftrightarrow K_{\text{rotational}}$$

18. •In Chapter 4, you learned that the mass of an object determines how that object responds to an applied force. Write a rotational analog to that idea based on the concepts of this chapter.

19. •Using the rotational concepts of this chapter, explain why a uniform solid sphere beats a uniform solid cylinder which beats a ring when the three objects "race" down an inclined plane while rolling without slipping. SSM

20. •Which quantity is larger: the angular momentum of Earth rotating on its axis each day or the angular momentum of Earth revolving about the Sun each year? Try to determine the answer without using a calculator.

21. •Define the SI unit radian. The unit appears in some physical quantities (for example, the angular velocity of a turntable is 3.5 rad/s) and it is omitted in others (for example, the translational velocity at the rim of a turntable is 0.35 m/s). Because the formula relating rotational and translational quantities involves multiplying by a radian ($v = r\omega$), discuss when it is appropriate to include radians and when the unit should be dropped.

22. •Describe how many unique ways a vector cross product ($\vec{C} = \vec{A} \times \vec{B}$) can equal zero.

23. •State the steps used to apply the right-hand rule when determining the vector direction of a cross product.

24. •A hollow cylinder rolls without slipping up an incline, stops, and then rolls back down. Which of the following graphs in **Figure 8-43** shows the (a) angular acceleration and (b) angular velocity for the motion? Assume that up the ramp is the positive direction.

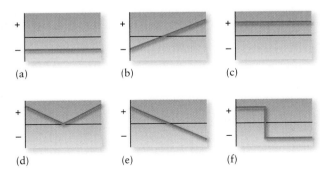

Figure 8-43 Problem 24

25. •Consider a situation where a merry-go-round, starting from rest, speeds up in the counterclockwise direction. It eventually reaches and maintains a maximum rotational velocity. After a short time the merry-go-round starts to slow down and eventually stops. Assume the accelerations experienced by the merry-go-round have constant magnitudes.

    (a) Which graph in **Figure 8-44** describes the angular velocity as the merry-go-round speeds up?

    (b) Which graph describes the angular displacement as the merry-go-round speeds up?

    (c) Which graph describes the angular velocity as the merry-go-round travels at its maximum velocity?

    (d) Which graph describes the angular displacement as the merry-go-round travels at its maximum velocity?

    (e) Which graph describes the angular velocity as the merry-go-round slows down?

    (f) Which graph describes the angular displacement as the merry-go-round slows down?

    (g) Draw a graph of the torque experienced by the merry-go-round as a function of time during the scenario described in the problem. SSM

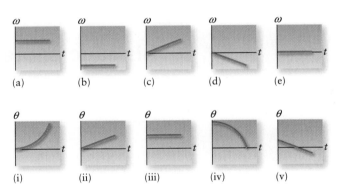

Figure 8-44 Problem 25

26. •Rank the torques exerted on the bolts in A–D (**Figure 8-45**) from least to greatest. Assume the wrenches and the force F are identical.

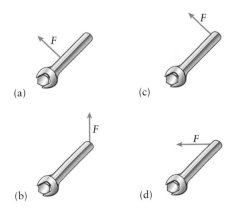

Figure 8-45 Problem 26

This page is intentionally left blank.

For complete end of chapter problem sets, please go to
www.whfreeman.com/kestentauck

ramps approximately three-quarters of a circle connecting two orthogonal freeways. SSM

38. •A fan is designed to last for a certain time before it will have to be replaced (planned obsolescence). The fan only has one speed (at a maximum of 750 rpm) and it reaches the speed in 2 s (starting from rest). It takes the fan 10 s for the blade to stop once it is turned off. If the manufacturer specifies that the fan will operate up to 1 billion rotations, estimate how many days will you be able to use the fan.

39. •Estimate the torque you apply when you open a door in your house. Be sure to specify the axis that your estimate refers to.

40. •Make a rough estimate of the moment of inertia of a pencil that is spun about its center by a nervous student during an exam.

41. •Estimate the moment of inertia of a figure skater as she rotates about the longitudinal axis that passes straight down through the center of her body into the ice. Make this estimation for the extreme parts of a pirouette (arms fully extended versus arms drawn in tightly).

42. •Estimate the angular displacement (in radians *and* degrees) of Earth in one day of its orbit around the Sun.

43. •Estimate the angular speed of the apparent passage of the Sun across the sky of Earth (from dawn until dusk).

44. •Estimate the angular acceleration of a lone sock that is inside a washing machine which starts from rest and reaches the maximum speed of its spin cycle in typical fashion.

45. •Estimate the angular momentum about the center of rotation for a "skip-it ball" that is spun around on the ankle of a small child (the child hops over the ball as it swings around and around the child's feet). SSM

46. •Estimate the moment of inertia of Earth about its central axis as it rotates once in a day. Try to recall (or guess) the mass and radius of Earth before you look the data up, just to see how close your estimate would be.

47. •Using a spreadsheet and the data below, calculate the angular speed of the rotating object over the first 10 s. Calculate the angular acceleration from 15 to 25 s. If the object has a moment of inertia of 0.25 kg · m² about the axis of rotation, calculate the torque during the following time intervals: 0 < t < 10 s, 10 s < t < 15 s, and 15 s < t < 25 s.

t (s)	θ (rad)	t (s)	θ (rad)
0	0	5	1.75
1	0.349	6	2.10
2	0.700	7	2.44
3	1.05	8	2.80
4	1.40	9	3.14

t (s)	θ (rad)	t (s)	θ (rad)
10	3.50	18	6.48
11	3.50	19	8.53
12	3.49	20	11.0
13	3.50	21	14.1
14	3.51	22	17.6
15	3.51	23	21.6
16	3.98	24	26.2
17	5.01	25	31.0

Problems

8-1: Rotational Kinetic Energy

48. •What is the angular speed of an object that completes 2 rev every 12 s? Give your answer in rad/s.

49. •A car rounds a curve with a translational speed of 12 m/s. If the radius of the curve is 7 m, calculate the angular speed in rad/s.

50. •Convert the following:
$$45 \text{ rev/min} = 10_____\text{rad/s}$$
$$33\tfrac{1}{3} \text{ rpm} = _____\text{rad/s}$$
$$2\pi \text{ rev/s} = _____\text{rad/s}$$

51. •Calculate the angular speed of the Moon as it orbits Earth (recall, the Moon completes one orbit about Earth in 27.4 days and the Earth–Moon distance is 3.84×10^8 m). SSM

52. •If a 0.25-kg point object rotates at 3 rev/s about an axis that is 0.5 m away, what is the kinetic energy of the object?

53. •What is the rotational kinetic energy of an object that has a moment of inertia of 0.28 kg · m² about the axis of rotation when its angular speed is 4 rad/s?

54. •What is the moment of inertia of an object that rotates at 13 rev/min about an axis and has a rotational kinetic energy of 18 J?

55. What is the angular speed of a rotating wheel that has a moment of inertia of 0.33 kg · m² and a rotational kinetic energy of 2.75 J? Give your answer in both rad/s and rev/min. SSM

8-2: Moment of Inertia

56. •What is the combined moment of inertia for the three point objects about the axis O in **Figure 8-47**?

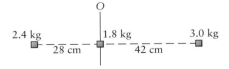

Figure 8-47 Problem 56

57. •What is the combined moment of inertia of three point objects ($m_1 = 1.0$ kg, $m_2 = 1.5$ kg, $m_3 = 2.0$ kg)

tied together with massless strings and rotating about the axis O as shown in **Figure 8-48**?

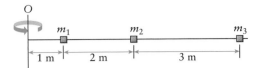

Figure 8-48 Problem 57

58. •••Calc What is the moment of inertia of a thin, uniform washer that has an inner radius of r_i and an outer radius of r_o about the axis that passes through the center of the washer (**Figure 8-49**)? The mass of the washer is M.

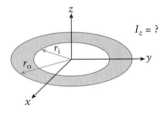

Figure 8-49 Problem 58

59. ••Calc Show that the moment of inertia about the central axis for a hollow cylinder of radius R, length L, and mass M is MR^2 (**Figure 8-50**). You may assume that the cylinder's mass density is uniform throughout its outer surface area.

Figure 8-50 Problem 59

60. ••A baton twirler in a marching band complains that her baton is defective (**Figure 8-51**). The manufacturer specifies that the baton should have an overall length of 60 cm and a total mass between 940 and 950 g (there are two 350-g objects on each end). Also according to the manufacturer, the moment of inertia about the central axis passing through the baton should fall between 0.075 and 0.080 kg · m². The twirler (who has completed a class in physics) claims this is impossible. Who's right? Explain your answer.

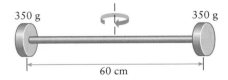

Figure 8-51 Problem 60

61. ••What is the moment of inertia of a steering wheel about the axis that passes through its center? Assume the rim of the wheel has a radius R and a mass M. Assume that there are five radial spokes that connect in the center as shown in **Figure 8-52**. The spokes are each thin rods of length R, of mass $\frac{1}{2}M$, and evenly spaced around the wheel. SSM

Figure 8-52 Problem 61

8-3: The Parallel-Axis Theorem

62. •Using the parallel-axis theorem, calculate the moment of inertia for a solid, uniform sphere about an axis that is tangent to its surface (**Figure 8-53**).

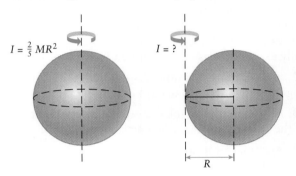

Figure 8-53 Problem 62

63. •Calculate the moment of inertia for a uniform, solid cylinder (mass M, radius R) if the axis of rotation is tangent to the sides of the cylinder as shown in **Figure 8-54**.

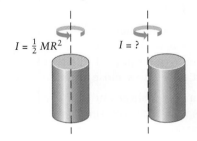

Figure 8-54 Problem 63

64. •Calculate the moment of inertia for a thin rod that is 1.25 m long and is 2.25 kg. The axis of rotation passes through the rod at a point one-third of the way from the left end (**Figure 8-55**).

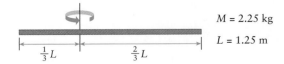

Figure 8-55 Problem 64

65. •Calculate the moment of inertia of a thin plate that is 5 cm × 7 cm in area and has a mass density of 1.5 g/cm². The axis of rotation is located at the left side, as shown in (Figure 8-56).

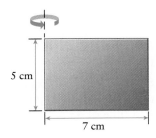

Figure 8-56 Problem 65

66. •Calculate the radius of a solid sphere of mass M that has the same moment of inertia about an axis through its center of mass as a second solid sphere of radius R and mass M which has the axis of rotation passing tangent to the surface and parallel to the CM axis (Figure 8-57).

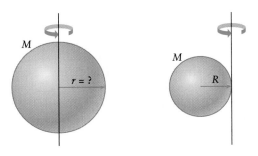

Figure 8-57 Problem 66

67. ••Two uniform, solid spheres (one has a mass M and a radius R and the other has a mass M and a radius $2R$) are connected by a thin, uniform rod of length $3R$ and mass M (Figure 8-58). Find the moment of inertia about the axis through the center of the rod. SSM

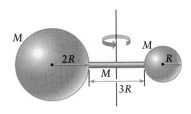

Figure 8-58 Problem 67

68. ••What is the moment of inertia of the sphere–rod system shown in Figure 8-59 when the sphere has a radius R and a mass M and the rod is thin and massless and has a length L? The sphere–rod system is spun about an axis A.

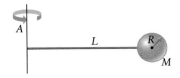

Figure 8-59 Problem 68

69. •What is the moment of inertia through the hinges for a door that is 0.81 m wide, 1.78 m long, and has a mass of 9.0 kg?

70. ••What is the moment of inertia of a hollow sphere that has a mass M, a radius R, and is attached to a solid cylinder that has a length $2R$, a radius r equal to $\frac{1}{2}R$, and a mass M (Figure 8-60)? Assume the axis of rotation passes through the center of the cylinder and the center of the sphere.

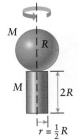

Figure 8-60 Problem 70

71. ••Use integration methods to calculate the moment of inertia of an unsharpened pencil about an axis through the end that has the eraser. The wooden part of the pencil has a length of $7L/8$ and a mass of M. The eraser has a length of $L/8$ and a mass M. SSM

72. •••What is the moment of inertia of the weighted rod about its left end (Figure 8-61)? Assume the rod has a length L. From the left end to the half-way point at $\frac{1}{2}L$, the mass of the rod is M. From the half-way point to the right end, the mass of the rod is $4M$. All segments of the rod are uniform in mass density.

Figure 8-61 Problem 72

8-4: Conservation of Energy Revisited

73. •Calculate the final speed of a uniform, solid cylinder of radius 5 cm and mass 3 kg that starts from rest at the top of an inclined plane that is 2 m long and tilted at an angle of 25° with the horizontal. Assume the cylinder rolls without slipping down the ramp.

74. •Calculate the final speed of a uniform, solid sphere of radius 5 cm and mass 3 kg that starts with a translational speed of 2 m/s at the top of an inclined plane that is 2 m long and tilted at an angle of 25° with the horizontal. Assume the sphere rolls without slipping down the ramp.

75. •A spherical marble that has a mass of 50 g and a radius of 0.5 cm rolls without slipping down a loop-the-loop track that has a radius of 20 cm. The marble starts from rest and *just barely* clears the loop to emerge on the other side of the track. What is the minimum height that the marble must start from to make it around the loop?

76. •A billiard ball of mass 160 g and radius 2.5 cm starts with a translational speed of 2 m/s at point A on the track as shown in Figure 8-62. If point B is at the top of a hill that has a radius of curvature of 60 cm, what is the normal force acting on the ball at point B? Assume the billiard ball rolls without slipping on the track.

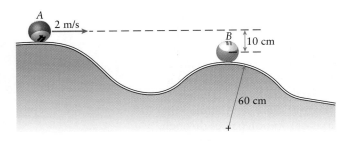

Figure 8-62 Problem 76

77. •**Sports** A bowling ball that has a radius of 11 cm and a mass of 5 kg rolls without slipping on a level lane at 2 rad/s. Calculate the ratio of the translational kinetic energy to the rotational kinetic energy of the bowling ball. SSM

78. •**Astro** Earth is approximately a solid sphere, has a mass of 5.98×10^{24} kg, a radius of 6.38×10^6 m, and completes one rotation about its central axis each day. Calculate the rotational kinetic energy of Earth as it spins on its axis.

79. •**Astro** Calculate the translational kinetic energy of Earth as it orbits the Sun once each year (the Earth–Sun distance is 1.50×10^{11} m). Compare the translational kinetic energy to the rotational kinetic energy calculated in the previous problem.

80. •A potter's flywheel is made of a 5-cm-thick, round slab of concrete that has a mass of 60 kg and a diameter of 35 cm. This disk rotates about an axis that passes through its center, perpendicular to its round area. Calculate the angular speed of the slab about its center if the rotational kinetic energy is 15 J. Express your answer in both rad/s and rev/min.

81. ••**Sports** A Frisbee (160 g, 25 cm in diameter) spins a rate of 300 rpm with its center balanced on a fingertip. What is the rotational kinetic energy of the Frisbee if the disc has 70% of its mass on the outer edge (basically a thin ring 25-cm in diameter) and the remaining 30% is a nearly flat disk 25-cm in diameter? If 60% of the energy goes into increasing the height of the disc, estimate the final value above the release point for the Frisbee. SSM

8-5: Rotational Kinematics

82. •••**Calc** A rotating disc starts from rest. How many radians will the disc rotate in 18 s if the angular acceleration is given by the following relation?

$$\alpha(t) = 0.2t^2 - 1.25t + 12 \qquad \text{(SI units)}$$

83. •A spinning top completes 6000 rotations before it starts to topple over. The average speed of the rotations is 800 rpm. Calculate how long the top spins before it begins to topple.

84. •A child pushes a merry-go-round that has a diameter of 4 m and goes from rest to an angular speed of 18 rpm in a time of 43 s. Calculate the angular displacement and the average angular acceleration of the merry-go-round. What is the maximum tangential speed of the child if she rides on the edge of the platform?

85. •Jerry twirls an umbrella around its central axis so that it completes 24 rotations in 30 s. If the umbrella starts from rest, calculate the angular acceleration of a point on the outer edge. What is the maximum tangential speed of a point on the edge if the umbrella has a radius of 55 cm?

86. •Prior to the music CD, stereo systems had a phonographic turntable on which vinyl disk recordings were played. A particular phonographic turntable starts from rest and achieves a final constant angular speed of $33\frac{1}{3}$ rpm in a time of 4.5 s. How many rotations did the turntable undergo during that time? The classic Beatles album *Abbey Road* is 47 min and 7 s in duration. If the turntable requires 8 s to come to rest once the album is over, calculate the total number of rotations for the complete start-up, playing, and slowdown of the album.

87. •A CD player varies its speed as it moves to another circular track on the CD. A CD player is rotating at 300 rpm. To read another track, the angular speed is increased to 450 rpm in a time of 0.75 s. Calculate the angular acceleration in rad/s^2 for the change to occur. SSM

88. •**Astro** A communication satellite circles Earth in a geosynchronous orbit such that the satellite remains directly above the same point on the surface of Earth. What angular displacement (in radians) does the satellite undergo in 1 h of its orbit? Calculate the angular speed of the satellite in rev/min and rad/s.

89. •Suppose a roulette wheel is spinning at 1 rev/s. How long will it take for the wheel to come to rest if it experiences an angular acceleration of -0.02 rad/s^2? How many rotations will it complete in that time?

8-6: Torque

90. •A driver applies a horizontal force of 20 N (to the right) to the top of a steering wheel, as shown in **Figure 8-63**. The steering wheel has a radius of 18 cm and a moment of inertia of 0.097 kg · m^2. Calculate the angular acceleration of the steering wheel about the central axis.

Figure 8-63 Problem 90

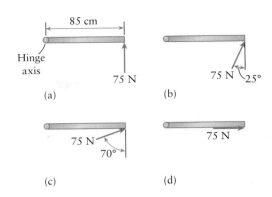

Figure 8-66 Problem 93

91. •**Medical** When the palmaris longus muscle in the forearm is flexed, the wrist moves back and forth (**Figure 8-64**). If the muscle generates a force of 45 N and it is acting with an effective lever arm of 22 cm, what is the torque that the muscle produces on the wrist? Curiously, over 15%

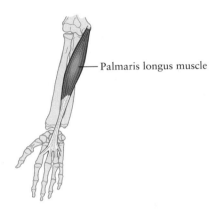

Figure 8-64 Problem 91

of all Caucasians lack this muscle; a smaller percentage of Asians (around 5%) lack it. Some studies correlate the absence of the muscle with carpal tunnel syndrome.

92. •A torque wrench is used to tighten a nut on a bolt. The wrench is 25 cm long and a force of 120 N is applied at the end of the wrench as shown in **Figure 8-65**. Calculate the torque about the axis that passes through the bolt.

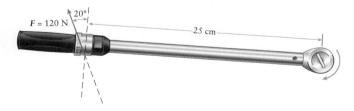

Figure 8-65 Problem 92

93. •An 85-cm-wide door is pushed open with a 75-N force. Calculate the torque about an axis that passes through the hinges in each of the cases in **Figure 8-66**. SSM

94. •••A 50-g meter stick is balanced at its midpoint (50 cm). A 200-g weight is hung with a light string from the 70-cm point and a 100-g weight is hung from the 10-cm point (**Figure 8-67**). Calculate the clockwise and counterclockwise torques for the three forces about the following axes: (a) the 0-cm point, (b) the 50-cm point, and (c) the 100-cm point.

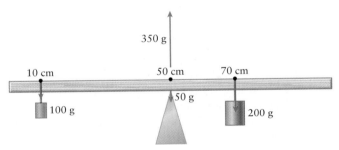

Figure 8-67 Problem 94

95. •A robotic arm lifts a barrel of radioactive waste (**Figure 8-68**). If the maximum torque delivered by the arm about the axis O is 3000 N · m and the distance r in the diagram is 3 m, what is the maximum mass of the barrel?

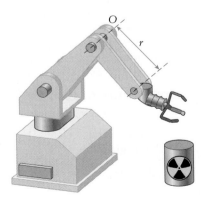

Figure 8-68 Problem 95

96. •A normal human being typical adult can deliver about 10 N · m of torque when attempting to open a twist-off cap on a bottle. What is the maximum force that the average person can exert with his fingers if most bottle caps are about 2 cm in diameter?

97. •What is the torque about your shoulder axis if you hold a 10-kg-barbell in one hand straight out and at shoulder height? Assume your hand is 75 cm from your shoulder. SSM

8-7: Angular Momentum

98. •What is the angular momentum about the central axis of a thin disk that is 18 cm in diameter, has a mass of 2.5 kg, and rotates at a constant 1.25 rad/s?

99. •What is the angular momentum of a 300-g tetherball when it whirls around the central pole at 60 rpm and at a radius of 125 cm?

100. •**Astro** Calculate the angular momentum of Earth as it orbits the Sun. Recall that the mass of Earth is 5.98×10^{24} kg, the distance between Earth and the Sun is 1.50×10^{11} m, and the time for one orbit is 365.3 days.

101. •**Astro** Calculate the angular momentum of Earth as it spins on its central axis once each day. Assume Earth is approximately a uniform, solid sphere that has a mass of 5.98×10^{24} kg and a radius of 6.38×10^6 m.

102. •What is the speed of an electron in the lowest energy orbital of hydrogen, of radius equal to 5.29×10^{-11} m. The mass of an electron is 9.11×10^{-31} kg and its angular momentum in this orbital is 1.055×10^{-34} J · s.

103. •What is the angular momentum of a 70-kg person riding on a Ferris wheel that has a diameter of 35 m and rotates once every 25 s? SSM

104. •A professor sits on a rotating stool that spins at 10 rpm while she holds a 1-kg weight in each of her hands. Her outstretched arms are 0.75 m from the axis of rotation, which passes through her head into the center of the stool. When she draws the weights in toward her body her angular speed increases to 20 rpm. Neglecting the mass of her arms, how far are the weights from the rotational axis at the increased speed?

8-8: The Vector Nature of Rotational Quantities

105. ••The position vector that locates an object relative to a given axis at the origin is given by the following:

$$\vec{r} = 3\hat{x} + 2\hat{y} + \hat{z}$$

The force vector that acts on the object is given by the following:

$$\vec{F} = 10\hat{x} - 20\hat{y} + 5\hat{z}$$

Calculate the torque vector that the force creates about the axis. The position vector is expressed in units of meters and the force vector is in newtons. Use the determinant method of calculating cross products:

$$\vec{A} \times \vec{B} = \begin{vmatrix} \hat{x} & \hat{y} & \hat{z} \\ A_x & A_y & A_z \\ B_x & B_y & B_z \end{vmatrix}$$

106. ••At one instant in time, a 2-kg object is located relative to an axis that passes through the origin by the following position vector:

$$\vec{r} = -0.5\hat{x} + 2\hat{y} + 0.75\hat{z}$$

The velocity vector that describes the object's motion at that time is given as

$$\vec{v} = \hat{x} - 3\hat{y} - 5\hat{z}$$

Calculate the angular momentum vector for the object.

107. •••Using two mathematical representations of the cross product, find the angle between the two vectors $\vec{A}$ and $\vec{B}$:

$$\vec{A} = 3\hat{x} + 4\hat{y} + 2\hat{z}$$
$$\vec{B} = 5\hat{x} - 2\hat{y} - 3\hat{z}$$

108. •••**Calc** The angular momentum vector for a rotating object is given by the following:

$$\vec{L} = 3t\hat{x} + 4t^2\hat{y} + 0.5t^3\hat{z}$$

Calculate the torque as a function of time associated with the rotational motion (about the same axis). What is the magnitude of the torque at $t = 2$ s?

109. •••**Calc** What is the angular momentum vector as a function of time associated with a rotating mass if the torque vector is given by the following?

$$\vec{\tau} = 3\cos(\pi t)\hat{x} + 4\cos(\pi t)\hat{y}$$

Assume that the angular momentum is zero at $t = 0$. What is the magnitude of the angular momentum at $t = 0.5$ s? SSM

General Problems

110. •The outside diameter of the playing area of an optical Blu-ray disc is 11.75 cm and the inside diameter is 4.5 cm. When viewing movies, the disc rotates so that a laser maintains a constant linear speed relative to the disc of 7.5 m/s as it tracks over the playing area. (a) What are the maximum and minimum angular speeds (in rad/s and rpm) of the disc? (b) At which location of the laser on the playing area do the speeds occur? (c) What is the average angular acceleration of a Blu-ray disc as it plays an 8.0-h set of movies?

111. •A table saw has a 25-cm-diameter blade that rotates at a rate of 7000 rpm. It is equipped with a safety mechanism that can stop the blade within 5 ms if something like a finger is accidentally placed in contact with the blade. (a) What angular acceleration occurs if the saw starts at 7000 rpm and comes to rest in 5 ms? (b) How many rotations does the blade complete during the stopping period?

112. •In 1932 Albert Dremel of Racine, Wisconsin, created his rotary tool that has come to be known as a dremel. (a) Suppose a dremel starts from rest and achieves an operating speed of 35,000 rev/min. If it requires 1.2 s for the tool to reach operating speed and it is held at that speed for 45 s, how many rotations has the bit made? Suppose it requires another 8.5 s for the tool to return to rest. (b) What are the angular accelerations for the start-up and the slowdown periods? (c) How many rotations does the tool complete from start to finish?

113. ••A baton is constructed by attaching two small objects that each have a mass M to the ends of a rod that has a length L and a uniform mass M. Find an expression for the moment of inertia of the baton when it is rotated around a point $3/8\ L$ from one end.

114. •••Calc Calculate the moment of inertia of a solid, uniform sphere of mass M and radius R about an axis passing through the center of the sphere. The volume element for spherical polar coordinates is $dV = (dr)(r\sin\theta\, d\varphi)(r\,d\theta)$ and the limits are $0 < r < R$; $0 < \theta < \pi$; $0 < \varphi < 2\pi$.

115. •Medical, Sports On average both arms and hands together account for 13% of a person's mass, while the head is 7.0% and the trunk and legs account for 80%. We can model a spinning skater with his arms outstretched as a vertical cylinder (head + trunk + legs) with two solid uniform rods (arms + hands) extended horizontally. Suppose a 62-kg skater is 1.8 m tall, has arms that are each 65 cm long (including the hands), and a trunk that is 35 cm in diameter. If the skater is initially spinning at 70 rpm with his arms outstretched, what will his angular velocity be (in rpm) when he pulls in his arms until they are at his sides parallel to his trunk?

116. •Find an expression for the moment of inertia of a spherical shell (for example, the peel of an orange) that has a mass M, a radius R and rotates about an axis which is tangent to the surface.

117. •Because of your success in physics class you are selected for an internship at a prestigious bicycle company in its research and development division. Your first task involves designing a tire rim that has a total mass of 1 kg, a radius of 50 cm, spokes with a mass of 10 g, and a moment of inertia of 0.255 kg · m². (a) How

many spokes are necessary to construct the wheel? (b) What is the mass of the rim? SSM

118. ••Two beads that each have a mass M are attached to a thin rod that has a length $2L$ and a mass $M/8$. The beads are initially each a distance $L/4$ from the center of the rod. The whole system is set into uniform rotation about the center of the rod, with initial angular frequency $\omega = 20\pi$ rad/s. If the beads are then allowed to slide to the ends of the rod, what will the angular frequency become?

119. ••A uniform disk that has a mass M of 0.3 kg and a radius R of 0.27 m rolls up a ramp of angle θ equal to 55° with initial velocity v of 4.8/s. If the disk rolls without slipping, how far up the ramp does it go?

120. •In a new model of a machine, a spinning solid spherical part of radius R must be replaced by a ring of the same mass which is to have the same kinetic energy. Both parts need to spin at the same rate, the sphere about an axis through its center and the ring about an axis perpendicular to its plane at its center. (a) What should the radius of the ring be in terms of R? (b) Will both parts have the same angular momentum? If not, which one will have more?

121. •Many 2.5-in-diameter (6.35 cm) disks spin at a constant 7200 rpm operating speed. The disks have a mass of about 7.5 g and are essentially uniform throughout with a very small hole at the center. If they reach their operating speed 2.5 s after being turned on, what average torque does the disk drive supply to the disk during the acceleration?

122. •Sports At the 1984 Olympics, the great diver Greg Louganis won one of his 10 gold medals for the reverse $3\frac{1}{2}$ summersault tuck dive. In the dive, Louganis began his $3\frac{1}{2}$ turns with his body tucked in at a maximum height of approximately 2.0 m above the platform, which itself was 10.0 m above the water. He spun uniformly $3\frac{1}{2}$ times and straightened out his body just as he reached the water. A reasonable approximation is to model the diver as a thin uniform rod 2.0 m long when he is stretched out and as a uniform solid cylinder of diameter 0.75 m when he is tucked in. (a) What was Louganis' average angular velocity as he fell toward the water with his body tucked in? *Hint:* How long did it take him to reach the water from his highest point? (b) What was his angular velocity just as he stretched out? (c) How much did Louganis' rotational kinetic energy change while extending his body if his mass was 75 kg?

123. ••Medical The bones of the forearm (radius and ulna) are hinged to the humerus at the elbow (Figure 8-69). The biceps muscle connects to the bones of the forearm about 2 cm beyond the joint. Assume the forearm has a mass of 2 kg and a length of 0.4 m. When the humerus and the biceps are nearly vertical and the forearm is horizontal, if a person wishes to hold an object of mass M so that her forearm remains motionless, what is the relationship between the force exerted by the biceps muscle and the mass of the object? (In other words, find a mathematical expression between force and mass.)

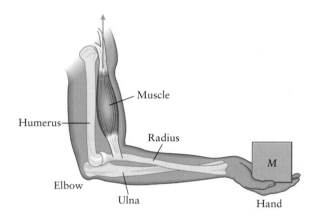

Figure 8-69 Problem 123

124. ••Medical The femur of a human leg (mass 10 kg, length 0.9 m) is in traction (**Figure 8-70**). The center of gravity of the leg is one-third of the distance from the pelvis to the bottom of the foot. Two objects, with masses m_1 and m_2, are hung using pulleys to provide upward support. A third object of 8 kg is hung to provide tension along the leg. The body provides tension as well. (a) What is the mathematical relationship between m_1 and m_2? Is this relationship unique in the sense that there is only one combination of m_1 and m_2 that maintains the leg in static equilibrium? (b) How does the relationship change if the tension force due to m_1 is applied at the leg's center of mass?

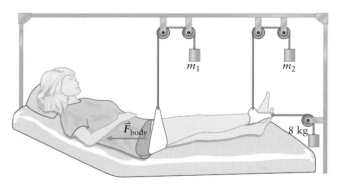

Figure 8-70 Problem 124

125. •Astro It is estimated that 60,000 tons of meteors and other space debris accumulates on Earth each year. Assume the debris is accumulated uniformly across the surface of Earth. (a) How much does Earth's rotation rate change per year? (That is, find the change in angular velocity.) (b) How long would it take the accumulation of debris to change the rotation period by 1 s? SSM

126. •Astro Suppose we decided to use the rotation of Earth as a source of energy. (a) What is the maximum amount of energy we could obtain from this source? (b) By the year 2020 the projected rate at which the world uses energy is expected to be 6.4×10^{20} J/y. If energy use continues at that rate, for how many years would the spin of Earth supply our energy needs? Does this seem long enough to justify the effort and expense involved? (c) How long would it take before our day was extended to 48 h instead of 24 h? Assume that Earth is uniform throughout.

127. •Astro In a little over 5 billion years, our Sun will collapse to a white dwarf approximately 16,000 km in diameter. Assume that, starting now, its rate of collapse will remain constant and that it will lose 20% of its mass. (a) What will our Sun's angular momentum and rotation rate be as a white dwarf? (Express your answers as multiples of its present-day values.) (b) Compared to its present value, will the Sun's rotational kinetic energy increase, decrease, or stay the same when it becomes a white dwarf? If it does change, by what factor will it change? The radius of the Sun is presently 6.96×10^8 m.

128. •Astro (a) If all the people in the world (~6.9 billion) lined up along the equator would Earth's rotation rate increase or decrease? Justify your answer. (b) How would the rotation rate change if all people were no longer on Earth? Assume the average weight of a human being is 70 kg.

129. •A 1000-kg merry-go-round (that is, a solid cylinder) has 10 children, each with a mass of 50 kg, located at the axis of rotation (thus you may assume the children have no angular momentum at that location). Describe a plan to move the children such that the angular velocity of the merry-go-round decreases to one-half its initial value.

130. •One way for pilots to train for the physical demands of flying at high speeds is with a device called the "human centrifuge." It involves having a pilot travel in circles at high speeds so that they can experience forces greater than their own weight (1 g = mass $\times$ 9.8 m/s^2). The diameter of the NASA device is 17.8 m.

(a) Suppose a pilot starts at rest and accelerates at a constant rate so that he undergoes 30 rev in 2 min. What is his angular acceleration (in rad/s^2)?

(b) What is his angular velocity (in rad/s) at the end of that time?

(c) After the 2-min period, the centrifuge moves at a constant speed. The g-force experienced is the centripetal force keeping the pilot moving along a circular path. What is the g-force experienced by the pilot?

(d) If the pilot can tolerate 12 g's in the horizontal direction, how long would it take the centrifuge to reach that state if it accelerates at the rate found in part (a)?

131. •••The moment of inertia of a rolling marble is $I = \frac{2}{5}MR^2$, where M is the mass of the marble and R is the radius. The marble is placed in front of a spring that has a constant k and has been compressed a distance x_c. The spring is released and *as the marble comes off the spring* it begins to roll without slipping. *Note:* The static friction that causes rolling without slipping does not do work. (a) Derive an expression for the time it takes for the marble to travel a distance D along the surface after it has lost contact with the spring. (b) Show that your answer for part (a) has the correct units. SSM

132. •••**Calc** A "pie slice" is cut from a thin, circular sheet of metal of radius R. The slice has mass M. Two such slices are then attached by a massless circular rim to form the object shown in **Figure 8-71a**.

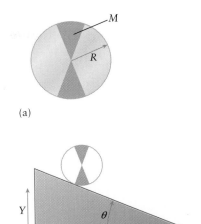

(a)

(b)

Figure 8-71 Problem 132

(a) Starting from the definition $I = \int r_\perp^2 \, dm$, find the moment of inertia of the configuration when it rotates around an axis perpendicular to the plane of the circle and through the center of the circle. Express your answer in terms of M and R.

(b) The object is allowed to roll (without slipping) from rest down a ramp of angle θ (**Figure 8-71b**). The bottom of the wheel starts a vertical distance Y above the bottom of the ramp. Find an expression for the linear velocity of the wheel as it leaves the ramp. Your answer should be in terms of the given variables and physical constants. *Note:* The static friction that causes rolling without slipping does not do work.

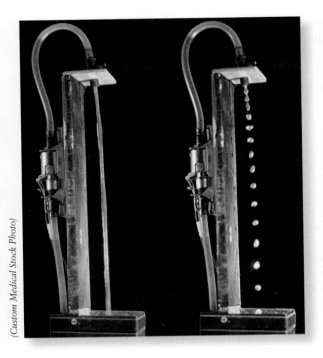

(Custom Medical Stock Photo)

9 Elasticity and Fracture

9-1   Tensile Stress and Strain   9-3 Shear Stress and Strain

9-2   Volume Stress and Strain   9-4 Elasticity and Fracture

Although most small water pumps produce what appears to be a continuous stream (left), illumination with a strobe light (right) reveals that the pump actually expels discrete pulses of liquid. The human heart works in much the same way, ejecting blood into the aorta for about 300 ms and then resting for about 500 ms. Yet blood flow to tissues is continuous and not pulsatile. As the heart forces blood into the arteries, not all of it moves forward down the vessel; some pushes against the arterial walls causing them to stretch. Even when the heart relaxes, blood keeps moving because the walls of the arteries spring back to their original diameter, pushing the blood along through the vessels. In this chapter, we explore the physics of stretching and squeezing.

In Chapters 1–8, we treated force as an agent that changes the speed and direction of an object. Applying a net force to an object can result in linear acceleration. Applying force to an object that is free to rotate can result in torque which in turn causes angular acceleration. However, it is complicated to apply these physical relationships to a lump of taffy as it is pulled and stretched (**Figure 9-1**). Parts of the taffy move, but it's unlikely that we can simply apply Newton's second law to understand the motion. Certainly the law applies to each small mass element of the taffy, but finding the net response of the entire mass would require a complex calculation. Moreover, the stretching and deformation of the taffy as it is pulled are the most interesting aspects of the motion. Some materials, such as taffy, exhibit a permanent change of shape when stretched. Other materials, such as your ear lobes or the tip of your nose, will snap back to their original shape after being stretched. Of course, even an elastic material such as rubber could remain irreversibly deformed, or even break apart, if a large enough force were applied. In this chapter, we examine the physics of stretching and materials.

9-1 Tensile Stress and Strain

Anyone who follows sports has probably heard of an athlete felled by a torn *anterior cruciate ligament* or *ACL*. The ACL, shown in **Figure 9-2**, is one of four ligaments that stabilize the knee joint by preventing one of the lower leg bones (the *tibia*) from sliding forward or twisting during jumping or rapid accelerations. A ligament is like an elastic band of tissue that connects one bone to another. The ACL connects the upper leg bone (the *femur*) to the *tibia*. As the bones rotate around the axis passing through the knee, long lever arms can exert large forces on the ACL. How does the ACL respond?

In Chapter 6, we encountered Hooke's law:

$$\vec{F} = -k\vec{x} \qquad (6\text{-}16)$$

Figure 9-1 How does taffy respond to a stretching force? Does it exhibit a permanent change of shape when stretched or does it snap back to its original shape? In this chapter we focus on the physics of stretching, squeezing, and compressing materials. *(Dennis MacDonald/ PhotoEdit.)*

333

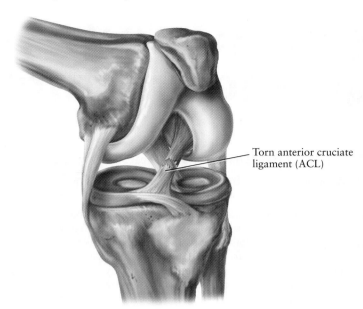

which relates the force $\vec{F}$ applied to the end of a spring and the distance $\vec{x}$ it stretches.

Hooke's law applies equally well both to springs and to materials and objects that can stretch like a spring, as long as x is small compared to the unstretched length of the spring, material, or object.

The spring constant k in Hooke's law, which has SI units of newtons per meter (N/m), is a measure of the stiffness of the spring. Particularly for solid objects that can stretch, this constant depends not only on the material from which the object is made but also on the object's length and cross-sectional area. The longer an object is, the more it stretches for a given force. So when two objects made from the same material have the same cross-sectional area, k is smaller for the longer of the two. The thicker an object (that is, the larger the cross-sectional area) the *less* it stretches in response to a given force, which results in a larger value of k (**Figure 9-3**).

Figure 9-2 The ACL is one of four ligaments connecting leg bones together at the knee. It stabilizes the joint by preventing one of the lower leg bones from sliding forward or twisting during jumping or rapid accelerations.

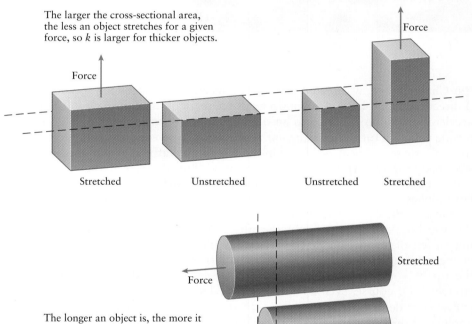

The larger the cross-sectional area, the less an object stretches for a given force, so k is larger for thicker objects.

The longer an object is, the more it stretches for a given force, so k is smaller for longer objects.

Figure 9-3 For an object that can stretch, the change in length is proportional to the magnitude of the applied force, as long as the change in length is small. The proportionality constant k depends on the material that makes up the object as well as its length and cross-sectional area.

The mathematical relationship that describes the observations drawn from Figure 9-3 is

$$k \propto \frac{A}{L_0}$$

where L_0 is the unstretched length and A is the cross-sectional area. The proportionality constant between k and A/L_0 is the **Young's modulus** (elastic modulus) Y. So

$$k = Y\frac{A}{L_0}$$

The magnitude of the Hooke's law force is then

$$F = Y\frac{A}{L_0}\Delta L \tag{9-1}$$

where ΔL is the change in length of the spring or object.

The specific value of Young's modulus depends on the material; Y for various materials is given in the second column of **Table 9-1**. The SI units of Y are newtons per square meter (N/m^2). A material that has a small Young's modulus is easily stretched; when Y is small, the force required to cause any particular change in length ΔL is less than when Y is large. Young's modulus for a typical rubber band (which is easily stretched) is around $5 \times 10^6\ N/m^2$ while the Young's modulus for steel (which is hard to stretch) is about $2 \times 10^{11}\ N/m^2$. That Y for steel is so large compared to Y for a rubber band means that much more force is needed to stretch a band of steel as far as a rubber band of the same length and cross-sectional area. Measured values of Young's modulus for a human ACL are on the order of $1 \times 10^8\ N/m^2$; so, an ACL is about 20 times less stretchable than a rubber band, but about 2000 times more stretchable than steel.

Table 9-1   Young's Modulus, Shear Modulus, and Bulk Modulus of Various Materials

Material	Young's Modulus ($10^9\ N/m^2$)	Bulk Modulus ($10^9\ N/m^2$)	Shear Modulus ($10^9\ N/m^2$)
Aluminum	70	70	25
Brass	100	80	40
Concrete	30	13	15
Iron	190	70	65
Nylon	3	—	4.1
Rubber band	0.005	—	0.003
Steel	200	140	78
Air	—	$1.01 \times 10^5\ N/m^2$	—
Ethyl alcohol	—	1.0	—
Water	—	2.0	—
Human ACL	0.1	—	—
Human lung	—	$1.5–9.8 \times 10^3\ N/m^2$	—
Pig endothelial cell	—	—	$2 \times 10^4\ N/m^2$

? Got the Concept 9-1
Two Different Rods

Two rods, which have identical cross sections, are made of different materials and have different lengths. The same stretching force is applied to both. One rod, made of material A, stretches 1 cm while the other, made of material B, stretches 2 cm. Can you identify which material has the greater Young's modulus? *Hint:* Consider the quantities on which Young's modulus depends.

Example 9-1 ACLs and Steel

A human anterior cruciate ligament that has a length equal to 2.7 cm and a steel rod that has a length equal to 54.0 m are subjected to the same stretching force. Both have the same cross-sectional area. Find the ratio of the change in length of the ACL to the change in length of the steel rod. Use $Y_S = 2 \times 10^{11} \, \text{N/m}^2$ and $Y_{ACL} = 1 \times 10^8 \, \text{N/m}^2$.

SET UP

The change in length of an object put under tension F is proportional to its initial length L_0 and inversely proportional to its cross-sectional area A. The proportionality constant is Young's modulus Y; we can rearrange Equation 9-1 to represent this change in length:

$$\Delta L = \frac{1}{Y} \frac{L_0}{A} F$$

The change in length of the ACL is then

$$\Delta L_{ACL} = \frac{1}{Y_{ACL}} \frac{L_{0,ACL}}{A_{ACL}} F_{ACL}$$

Similarly, the change in length of the steel rod is

$$\Delta L_S = \frac{1}{Y_S} \frac{L_{0,S}}{A_S} F_S$$

SOLVE

When the ACL and steel rod have the same cross-sectional area and are subjected to the same stretching force, the ratio of the change in lengths is

$$\frac{\Delta L_{ACL}}{\Delta L_S} = \frac{Y_S}{Y_{ACL}} \frac{L_{0,ACL}}{L_{0,S}}$$

Using the numbers given in the problem statement,

$$\frac{\Delta L_{ACL}}{\Delta L_S} = \left(\frac{2 \times 10^{11} \, \text{N/m}^2}{1 \times 10^8 \, \text{N/m}^2} \right) \left(\frac{0.027 \, \text{m}}{54.0 \, \text{m}} \right) = 1$$

REFLECT

Although Young's modulus for the ACL is much smaller than for steel (that is, the ACL is far more elastic than steel), both objects experience the same change in length in this example. That's because the Young's modulus for steel is 2000 times larger than the Young's modulus for the ACL, but the steel rod is 2000 times longer in the beginning.

Practice Problem 9-1 A typical cross-sectional area of a human ACL is $4.4 \times 10^{-5} \, \text{m}^2$. What applied stretching force is required to cause a 0.10% change in the length of an ACL?

Notice that the right-hand side of Equation 9-1,

$$F = Y \frac{A}{L_0} \Delta L$$

could be written in terms of a dimensionless fraction $\Delta L / L_0$. This quantity, the **tensile strain**, is a measure of how much an object stretches as a fraction of its total length. As noted above, the longer an object, the more it stretches for a given force, but the fractional change in length $\Delta L / L_0$ of any object remains the same as long as the ratio of F to A remains constant as shown in Figure 9-4. So, it makes more physical sense to describe stretching in terms of the ratio F/A and the tensile strain:

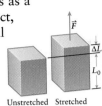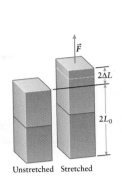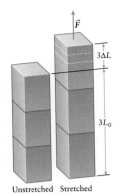

Unstretched Stretched   Unstretched Stretched   Unstretched Stretched

$$\frac{F}{A} = Y \frac{\Delta L}{L_0} \qquad (9\text{-}2)$$

The quantity F/A is the **tensile stress**. The units of tensile stress are the units of force divided by the units of area, that is,

$$[\text{tensile stress}] = \frac{N}{m^2}$$

Tensile stress, sometimes labeled σ, is a measure of both how much force is exerted on an object and also the area over which that force is applied, when caused by a small increase in force. Consider, for example, how it feels to stand on a floor compared to how it might feel to stand on a nail. In both cases, the force of the supporting surface on you is equal to your weight, so the force is the same. However, when that force is applied over the surface area of your feet you will be comfortable, while it will likely hurt to stand on the nail because the force exerted on you is applied over a much smaller area.

The units of both Young's modulus and tensile stress are newtons per square meter (N/m^2). Tensile stress is defined as the force on an object divided by the area over which the force is applied, and while the definition of Young's modulus does not explicitly depend on force or area, it must arise from similar considerations to tensile stress. Indeed, we will encounter other quantities in our study of elasticity that carry the same units and that are related to force divided by area. In general, a force divided by the area over which it is applied is *pressure*. We will look more carefully at pressure in Chapter 11.

Finally, Equation 9-2 can be written

$$\text{Tensile stress} = \text{constant} \times \text{tensile strain} \qquad (9\text{-}3)$$

This equation is referred to as Hooke's law—it is a more general form of the relationship we have used to describe the stretching of a spring. Hooke's law applies only when the tensile strain is small; that is, it only applies when the change in length is small relative to the overall initial length. **Compressive strain**, in which an object is subjected to a force that squeezes rather than stretches it along a single direction, and **compressive stress** both follow Hooke's law.

Figure 9-4 Tensile strain is a measure of how much an object stretches as a fraction of its total length. Although a longer object stretches more for a given stretching force, the fractional change in length $\Delta L / L_0$ remains the same as long as the ratio of the stretching force F to the cross-sectional area remains constant.

 Watch Out

The scientific definitions of stress and strain are slightly different than those in common usage.

Stress is force divided by the area, or cross-sectional area, over which it is applied. Strain describes the change in length (or volume or shape) of an object as a result of stress. It is therefore possible for an object to undergo stress without strain—you pull on the end of a rod without it stretching in a measurable way—but not possible for an object to experience strain without stress.

> ## ✱ What's Important 9-1
>
> Young's modulus Y is a measure of how easy it is to stretch an object. The change in length relative to the initial length, $\Delta L/L_0$, is a dimensionless measure of an object's stretchability called the tensile strain. Tensile stress is the force exerted on an object divided by the area over which that force is applied. Compressive stress and compressive strain are analogous terms that describe the response of an object to a force that squeezes rather than stretches it in a single direction.

9-2   Volume Stress and Strain

A scuba diver who ascends to the surface too quickly risks decompression sickness, a potentially fatal condition commonly known as "the bends." The physics of gases dissolved in liquids explains both the problem and its treatment. The deeper a person dives, the more gases dissolve in his body fluids. If the diver then ascends too quickly, gases pop out of solution and form bubbles, a process not too different from what happens when you open a bottle of your favorite carbonated beverage. The bubbles can be large enough to block blood vessels, which is one cause of the painful symptoms associated with the bends. Fortunately, a treatment exists. Victims of decompression sickness are placed in a sealed chamber, such as the one shown in Figure 9-5, which can be pressurized by pumping in air. The increased pressure in such a *hyperbaric chamber* squeezes on the person inside, causes the bubbles to shrink, and restores blood flow in the person's body. With the slow release of the excess air from the hyperbaric chamber, the person is able to exhale the excess gas that had been dissolved in his body before the bubbles formed. The physics behind the shrinking gas bubbles is similar to that of tensile stress and strain.

Figure 9-5 Victims of decompression sickness are placed in a sealed, hyperbaric chamber which can be pressurized by pumping in air. The increased pressure squeezes on the person and causes the gas bubbles in his body to shrink. (*U.S. Navy photo by MCSN Andrew Breese*)

In the last section, we saw that for objects that don't stretch too much when a stretching or tension force is applied, there is a linear relationship (Equation 9-3) between the tensile stress and tensile strain:

$$\text{tensile stress} = \text{constant} \times \text{tensile strain}$$

The proportionality constant is Young's modulus Y, so we also wrote this as Equation 9-2:

$$\frac{F}{A} = Y\frac{\Delta L}{L_0}$$

Both equations apply when a net force is applied to an object along one dimension, and when the object changes shape (length) in that same dimension. Neither condition is true of the gas bubbles in the diver's body. In the hyperbaric chamber, the bubbles experience a force squeezing in from every direction and change shape in three dimensions. With only a slight modification to the concepts of stress and strain, however, we can also describe the physics of the forces acting on the bubbles.

First, when an object experiences a force from all directions, and in particular, when that force always acts in a direction perpendicular to the surface, the ratio of force to area is called the **volume stress**. A material that experiences a volume stress and is compressible undergoes a change in volume (rather than in length as for a tensile stress), so we define **volume strain** as the change in volume divided by the initial volume. Hooke's law still holds, that is, there is a linear relationship between the volume stress and volume strain:

$$\text{volume stress} = \text{constant} \times \text{volume strain}$$

The proportionality constant is the **bulk modulus** B, and we can express the physics of this type of compressibility as

$$\frac{F}{A} = -B\frac{\Delta V}{V_0} \tag{9-4}$$

The minus sign indicates that as the force increases the volume decreases. Also, in the same way that the quantity F in tensile stress is the net stretching force, the force F in Equation 9-4 is the net compression force. In the case of volume stress, because any volume must necessarily have an inside and an outside, the net force is the difference between the force pushing in and the force pushing out. However, note that when a system experiences a volume change due to a net compression force, by definition the net compression force is zero once the new, equilibrium volume is reached. In Equation 9-4, the area A is the surface area over which the compression force is applied.

The units of volume stress are newtons per square meter (N/m^2), the same as those for tensile stress. Volume strain, the change in volume divided by initial volume, is clearly dimensionless, so bulk modulus also carries units of N/m^2, or pressure. Values of the bulk modulus for some materials are given in the fourth column of Table 9-1.

By rearranging Equation 9-4, we can more clearly see how the change in volume of an object is related to the bulk modulus:

$$\Delta V = -\frac{1}{B}\frac{F}{A}V_0$$

For the given applied force F, initial volume V_0, and surface area A, the larger B is, the smaller the change in volume. Thus, a material that has a relatively large bulk modulus is relatively incompressible; in other words, its volume doesn't change much even when it experiences a large volume stress. Said another way, the bulk modulus relates the fractional change in the volume of a substance as the pressure it experiences changes. Again, pressure is the application of a force over the surface area of an object or substance.

From the entries in Table 9-1 you can see that materials such as iron and concrete have relatively large bulk moduli; they are hard to compress, which makes them ideal materials with which to construct buildings. Likewise, air, which has a bulk modulus more than a million times smaller than that of steel, is relatively easy to compress. Under physiologic conditions, the bulk modulus of the human lung varies between about 1.5×10^3 and 10^4 N/m^2 in young adults and increases with age.

? Got the Concept 9-2
Two Balloons

Two balloons are filled to the same size—one with water and one with air. When the balloons are subjected to the same volume stress (the same force per unit surface area pushing in on the balloon), which of the two will exhibit the largest change in volume?

Example 9-2 Bubbles Rising

A scuba diver who is 10 m below the surface releases a spherical air bubble 2.0 cm in diameter and watches it rise to the surface. The air in the bubble initially pushes out on the surrounding water with the same force as the weight of all the water above the bubble pushing in. When the bubble has risen to the surface, however, the volume stress on the bubble decreases by 1.0×10^5 N/m^2 due to the decreasing amount of water above the bubble. This decrease causes the bubble to expand. What is the radius of the air bubble as it breaks the surface? Would the diver be able to detect the change in the bubble's size?

SET UP

The relationship between volume stress and volume strain, Equation 9-4, gives the change in volume versus initial volume. Volume stress is determined by the difference between the force pushing in on the bubble and the force pushing out. Because the bubble is sealed as it rises (air neither gets in nor escapes), the amount of trapped air doesn't change. Thus, the change in volume of the bubble is due only to the change in the volume stress, which in turn is due to the change in the inward force exerted by the weight of the water above the bubble. The volume stress F/A is therefore equal to the change in inward force of 10^5 N/m^2.

We want the change in radius, so we must rewrite Equation 9-4 in terms of radius. The volume of a sphere is $V = \frac{4}{3}\pi r^3$, so the change in volume from the initial volume V_0 to the final volume V_f is

$$\Delta V = V_f - V_0 = \frac{4}{3}\pi r_f^3 - \frac{4}{3}\pi r_0^3 = \frac{4}{3}\pi \left(r_f^3 - r_0^3\right)$$

√x *See the Math Tutorial for more information on Geometry*

The volume strain is then

$$\frac{\Delta V}{V_0} = \frac{\frac{4}{3}\pi \left(r_f^3 - r_0^3\right)}{\frac{4}{3}\pi r_0^3} = \frac{r_f^3 - r_0^3}{r_0^3}$$

From this expression, we should be able to determine the final radius of the bubble.

SOLVE

Using B_{air} as the bulk modulus of air, Equation 9-4 becomes

$$\frac{F}{A} = -B_{air}\frac{r_f^3 - r_0^3}{r_0^3}$$

or

$$\frac{F}{A} = -B_{air}\frac{r_f^{\,3}}{r_0^{\,3}} + B_{air}$$

So,

$$r_f^{\,3} = r_0^{\,3}\left(1 - \frac{F}{A}\frac{1}{B_{air}}\right)$$

or

$$r_f = r_0\left(1 - \frac{F}{A}\frac{1}{B_{air}}\right)^{1/3}$$

We can now substitute numbers into the final equation. The bulk modulus of air from Table 9-1 is $1.01 \times 10^5\,\text{N/m}^2$ and the volume stress is given as $F/A = -1.0 \times 10^5\,\text{N/m}^2$. (The volume stress is negative because the volume stress on the bubble decreases as it rises.) Thus,

$$r_f = r_0\left(1 - (-1.0 \times 10^5)\frac{1}{1.01 \times 10^5}\right)^{1/3} = (2.0\ \text{cm})(1.26) = 2.5\ \text{cm}$$

The bubble's radius grows from 2.0 cm to 2.5 cm as it rises 10 m.

The change in radius of the air bubble, half of a centimeter, is certainly easily measured, although it would probably be hard for the scuba diver to see it from 10 m away.

REFLECT

Consider the implications of these results on decompression sickness. When a diver ascends too quickly, not only do bubbles form, they get bigger as the person rises to the surface.

Practice Problem 9-2 A balloon filled with air to a volume of $0.03\ \text{m}^3$ is attached to a rock and thrown into a lake. When it sinks to the bottom the balloon experiences a volume stress of $0.33 \times 10^5\,\text{N/m}^2$. What is the volume of the balloon at the bottom of the lake?

Estimate It! 9-1 Scuba Tank

The scuba diver in **Figure 9-6** carries a cylindrical air tank filled with compressed air. The volume stress on the walls of the cylinder—the force per unit area—is about $2 \times 10^7\,\text{N/m}^2$, or about 200 times what it would be if the tank were opened to the environment above the water. Using the photo to estimate sizes, to about what volume will the air in the tank expand if the tank is opened at the surface?

SET UP

The change in the volume ΔV of the air depends on the initial volume V_0 and bulk modulus B_{air} of air, as well as the volume stress F/A, a relationship that can be obtained from Equation 9-4:

$$\Delta V = -\frac{V_0}{B}\frac{F}{A}$$

To use the relationship above we need to know the initial volume of the air, that is, the volume of the tank. Yes, we could look up the exact size of the tank. Because we are making an estimate, however, we only need to use big, round numbers, as long as they're reasonable. We can estimate the volume of the tank by comparing its dimensions to the height of the diver. His height H is about nine tank diameters,

Figure 9-6 A scuba diver carries a cylindrical air tank filled with compressed air. *(Sami Sarkis/ Getty Images)*

or alternatively, about three tank lengths; that is, the diameter D of the tank is about $H/9$ and the length L of the tank is about $H/3$.

SOLVE

The initial volume of the air is the volume of the tank:

$$V_0 = \pi \left(\frac{D}{2} \right)^2 L \approx \pi \left(\frac{H/9}{2} \right)^2 (H/3)$$

 See the Math Tutorial for more information on Geometry

An average person is about 2 m tall to one significant figure, so the volume of the tank is about

$$V_0 \approx \pi \left(\frac{2\,\text{m}}{18} \right)^2 \left(\frac{2\,\text{m}}{3} \right) = 0.026\,\text{m}^3$$

The bulk modulus of air, from Table 9-1, is about $1 \times 10^5\,\text{N/m}^2$. The volume stress on the air decreases by about a factor of 200 when the tank is opened at the surface, so it goes from $2 \times 10^7\,\text{N/m}^2$ to $1 \times 10^5\,\text{N/m}^2$ (atmospheric pressure). The change is about $-2 \times 10^7\,\text{N/m}^2$. So then

$$\Delta V = -\frac{V_0}{B}\frac{F}{A} = -\frac{0.026\,\text{m}^3}{1 \times 10^5\,\text{N/m}^2}(-2 \times 10^7\,\text{N/m}^2) = 5.2\,\text{m}^3$$

To one significant figure, the air in the tank expands to a volume of about $5\,\text{m}^3$ when the tank is opened at the surface.

REFLECT

Based on our estimate of tank height and diameter, the volume of the tank is about $3 \times 10^{-2}\,\text{m}^3$. This value is small compared to the volume of the air which has been squeezed into the tank, about $5\,\text{m}^3$ at sea level by our estimate. We should have expected this result, of course, given that the force the air exerts on the walls of the tank is so large compared to the force of the air in the atmosphere at the surface of the water.

> ### ✳ What's Important 9-2
> When an object experiences a force from all directions, the ratio of force to area (F/A) is called the volume stress. Volume strain is the change in volume of an object divided by its initial volume $(\Delta V/V_0)$ when it experiences a volume stress. Volume stress equals the product of the volume strain and a proportionality constant, the bulk modulus B.

9-3 Shear Stress and Strain

Physical activity increases blood flow to active muscles. One mechanism by which the increase occurs is triggered by the deformation of cells lining the inside surfaces of arteries. The force of blood pushing on the exposed surfaces of the cells deforms them in the direction of blood flow. The change in shape is neither a stretch nor a compression, the kind of deformations that result from the tensile, compressive, and volume stresses that we have already discussed. Instead, the cells deform in a way similar to a cube of Jell-O being pushed parallel to its top surface, as in **Figure 9-7**. In response to deformation, endothelial cells release nitric oxide gas which diffuses to muscle cells in the arterial walls, causing them to relax. This is a classic physiological control mechanism. When the muscle cells relax the diameter of the artery increases; this immediately reduces the deformation of the endothelial cells, the original stimulus to release nitric oxide, so that the release of nitric oxide subsides. At the basis of this process is *shear stress*, a phenomenon caused by a force applied parallel (tangential) to the face of an object. An object is said to experience shear when one face is made to move or slide relative to the opposite face. Shear stress can also cause an object to twist.

Figure 9-7 When pushed from one side, a cube of Jell-O deforms so that the top surface moves relative to the bottom surface. The Jello-O is experiencing shear. *(Courtesy David Tauck)*

Like the other types of stress we've encountered, **shear stress** is defined as a force, the shear force, per unit area. A shear force is parallel to the surface that moves and the area is of the surface in line with the force. In **Figure 9-8**, a force $\vec{F}$ is applied to the top surface of an object. It is the component of $\vec{F}$ parallel to the top surface, labeled $\vec{F}_{\parallel}$, that causes shear stress. Notice that unlike our treatments of tensile stress and volume stress, in which we only needed to consider the magnitude of the applied force, here we explicitly treat the force as a vector because the physics depends on a component of the force. The units of shear stress are newtons per square meter (N/m^2).

Like tensile stress and volume stress, *shear stress* obeys Hooke's law. In other words, shear stress is proportional to a strain, in this case shear strain:

$$\text{shear stress} = S \times \text{shear strain} \qquad (9\text{-}5)$$

As we have seen, strain is a measure of the extent to which an object is deformed. The change in length divided by initial length defines both tensile strain and compressive strain; the change in volume divided by initial volume defines volume strain. Similarly, the deformation of an object in response to a shear stress is quantified by the displacement of one surface relative to the other, compared to the distance between the surfaces. The proportionality constant in Equation 9-5 is the **shear modulus S**, a measure of the rigidity of a material. The more rigid a material is, the larger its shear modulus. For example, concrete, which has a shear modulus of $15 \times 10^9 \, N/m^2$ (Table 9-1), is about 5000 times more rigid than rubber, the shear modulus of which is $0.003 \times 10^9 \, N/m^2$.

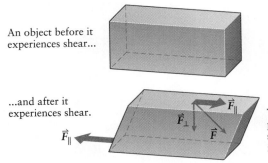

An object before it experiences shear...

...and after it experiences shear.

$\vec{F}_{\parallel}$ $\vec{F}_{\perp}$ $\vec{F}_{\parallel}$ $\vec{F}$

The component of the force parallel to the surface that slides (here the top surface) is the shear force.

The area in the definition of shear stress is the area of the top surface.

Figure 9-8 The component of force $\vec{F}$ parallel to the surface that slides results in shear stress.

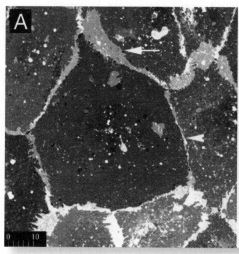

shear strain = $\frac{x}{h}$

Figure 9-9 Shear strain quantifies the deformation of an object in response to a shear stress. Shear strain is characterized by the relative displacement (x) of one surface with respect to the other and the distance (h) between them.

We define shear strain using the notation in **Figure 9-9**: x represents the relative displacement of the surfaces and h is the distance between them. The deformation is then characterized by

$$\text{shear strain} = \frac{x}{h} \quad (9\text{-}6)$$

The net forces on objects that are fixed in place, such as the cells lining walls of arteries, are zero. Notice that the object in Figure 9-8 experiences a force tangential to its lower surface, and that this force is equal in magnitude and opposite in direction to the force along the upper face. The net force on the object in the tangential direction is therefore zero.

Water and other fluids cannot sustain a force applied tangentially to their surfaces. Instead of deforming like the cube of Jell-O (Figure 9-7), a liquid experiencing a force tangential to its surface flows rather than deforms. Because liquids (and also gases) cannot sustain a shear force, the shear modulus is only defined for solids.

Example 9-3 Endothelial Cells and Shear

Endothelial cells lining the inside surface of an artery experience shear stress as a result of blood flow. In an *in vitro* study of arterial endothelial cells from pigs, the shear strain was proportional to the shear stress when the stress was $8.6 \times 10^3 \, \text{N/m}^2$. The shear modulus of a typical endothelial cell is $2.0 \times 10^4 \, \text{N/m}^2$. At what angle φ does one surface of a cell move relative to the opposite surface? A normal endothelial cell (left) and a cell experiencing shear (right) are shown in **Figure 9-10**; the angle φ is labeled.

SET UP
The shear strain is defined by Equation 9-6 as

$$\text{shear strain} = \frac{x}{h}$$

using the same notation as in Figure 9-9. Notice that the fraction x/h is also $\tan \varphi$. Because shear stress is proportional to shear strain (Equation 9-5), we can therefore write

$$\text{shear stress} = S \times \tan \varphi$$

where S is the shear modulus. This expression can be solved for φ.

$\sqrt{x}$ *See the Math Tutorial for more information on Trigonometry*

Figure 9-10 A cell experiencing no shear is shown on the left; a cell experiencing shear is shown on the right. *(American Journal of Pathology, 2004, 164: 1211–1223, Assembly and Reorientation of Stress Fibers Drivers Morphological Changes to Endothelial Cells Exposed to Shear Stress, Sabrena Noria, Feng Xu, Shannon McCue, Mara Jones, Avrum I. Gotlieb and B. Lowell)*

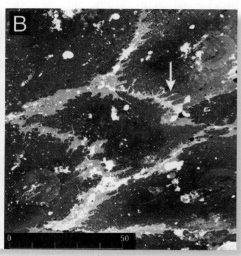

SOLVE
Solving for φ gives

$$\varphi = \tan^{-1}\left(\frac{\text{shear stress}}{S}\right)$$

Using the values given in the problem statement,

$$\varphi = \tan^{-1}\left(\frac{8.6 \times 10^3\,\text{N/m}^2}{2.0 \times 10^4\,\text{N/m}^2}\right) = \tan^{-1}(0.43)$$

or

$$\varphi = 23°$$

REFLECT
Endothelial cells in arteries are highly sensitive to shear stress due to fluid flow. The cells can undergo dramatic shape changes, as shown in Figure 9-10. Moreover, these changes in shape stimulate the cells to release nitric oxide, which causes the underlying muscle cells to relax as the gas diffuses. This classic negative feedback mechanism enlarges the diameter of the artery, reducing shear and increasing blood flow.

Example 9-4 Earthquake Damage

The dedication plaque mounted at the base of a building was originally 0.80 m high, 0.50 m long, and 0.10×10^{-2} m thick. During an earthquake the plaque is deformed so that its top surface is shifted 0.08 m relative to the bottom surface. What shearing force did the plaque experience during the earthquake? It was made from a metal alloy of shear modulus $0.40 \times 10^{11}\,\text{N/m}^2$.

SET UP
As with other forms of stress, the shear stress is defined by the shear force divided by area

$$\text{shear stress} = \frac{F}{A}$$

For shear stress A is the area of the object parallel to the shear force, as indicated in Figure 9-8. In addition, shear stress is proportional to shear strain (Equation 9-5), so

$$\frac{F}{A} = S \times \text{shear strain}$$

Shear strain can be written in terms of the height of the plaque (h) and the relative shift of the two surfaces (x) according to Equation 9-6, so

$$\frac{F}{A} = S \times \frac{x}{h}$$

We can solve the equation for the force applied to the plaque.

SOLVE
Solving for F yields

$$F = S \times A \times \frac{x}{h}$$

Using the values given in the problem statement,

$$F = (0.40 \times 10^{11}\,\text{N/m}^2)(0.50\,\text{m} \times 0.10 \times 10^{-2}\,\text{m})\left(\frac{0.08\,\text{m}}{0.80\,\text{m}}\right) = 2.0 \times 10^6\,\text{N}$$

REFLECT

The shear force required to deform the dedication plaque is 2.0×10^6 N. This force is about the same as the weight of the Statue of Liberty. That's a lot of force!

Practice Problem 9-4 A plaque mounted to the base of a building was originally half as high and half as thick, but the same length as the dedication plaque described in the previous problem. It was made from a metal alloy of shear modulus 0.40×10^{11} N/m². If the plaque experienced a shearing force of 2.0×10^6 N during an earthquake, by how far would the top surface be shifted relative to the bottom surface?

★ **What's Important 9-3**

Shear stress occurs when a force is applied parallel to the face of an object and causes that face to move relative to the opposite face. The shear modulus is a measure of the rigidity of a material and is only defined for solids. Shear strain describes the extent to which an object is deformed by shear stress.

9-4 Elasticity and Fracture

Many structures of your body, such as your arteries and skin, are stretchable. Some, such as your ears and nose, are bendable. After you stretch your earlobe or bend the end of your nose by applying a modest force, it quickly returns to its original shape. The ACL (anterior cruciate ligament) of the human knee is stretchable, too. It's not as stretchable as a rubber band, of course, but the ACL is elastic enough to accommodate the forces the knee typically experiences. For the typical forces that structures in our bodies experience, the tensile (or compressive) strain is small enough that Hooke's law (Equation 9-3) applies. The applicability of Hooke's law means that a stretched object returns to its initial shape (much like a spring) when the force is removed, so the change in shape that results from tensile stress is reversible. It is possible, however, for the force on an object such as the ACL to exceed the normal range. When the force is large enough, the object deforms in a way from which it cannot recover; this almost always results in tearing.

As long as the strain is relatively small, the relationship between stress and strain obeys Hooke's law. An object subjected to a tensile or compressive stress that results in a relatively small strain will deform, but only temporarily. After the stress is removed, the object returns to its initial shape. In a graph of stress versus strain, the corresponding portion of the curve is called the **elastic regime** and is linear. An idealized stress versus strain curve is shown in **Figure 9-11**; the elastic regime, where stress equals a constant multiplied by strain, is drawn in blue.

Stretchable biological tissue such as the connective tissue that supports and connects structures in the body does not initially respond linearly to stress. These tissues get their rigidity from the protein collagen and their elastic properties from the protein elastin (**Figure 9-12**). Elastic fibers are formed from a jumble of elastin molecules. So when a stress is applied to soft tissue, the process of the collagen fibers uncrimping and the elastic fibers aligning initially results in a nonlinear regime in the stress–strain relationship. This so-called **toe region** is evident in the idealized stress versus strain curve for biological tissue shown in **Figure 9-13**.

If the stress on an object is increased so that the change in length becomes large relative to the object's initial length, the deformation of the object becomes irreversible. This **plastic regime** is drawn in green on the stress versus strain curve in

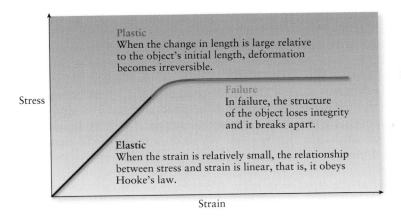

Figure 9-11 An idealized stress versus strain curve shows how a system responds to a tensile or compressive stress.

Figure 9-11. A spring stretched only a small distance from equilibrium returns to its original length, but when stretched into the plastic regime, as was the spring shown in **Figure 9-14**, it is permanently deformed and can no longer return to its original shape. Notice that once an object enters the plastic regime, a small increase in stress can result in a large increase in strain. In other words, a small increase in applied force can result in a large change in length.

When the stress on an object is large enough so that its deformation can be described by the plastic regime, the structure of the material starts to lose its integrity. This characteristic is the onset of *failure*, which eventually leads to the object breaking apart. In biological tissues, failure may be referred to as fracture, rupture, or tearing. Although some biological tissue can enter the plastic regime, structures in the body typically undergo partial or complete failure once a stress is applied that exceeds the ability of the tissue to spring back to its initial shape. In the idealized stress versus strain curve for biological tissue (Figure 9-13), once the tissue can no longer be described by the linear, elastic regime it experiences failure. This region of the stress versus strain curve is drawn in red in the figure. In the ACL, for exam-

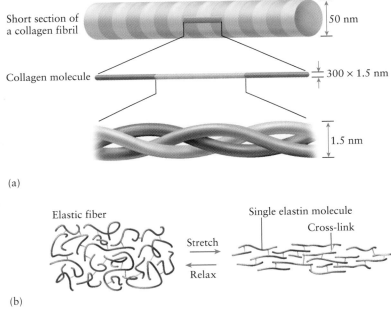

Figure 9-12 Connective tissues get their rigidity from the protein collagen and their elastic properties from the protein elastin.

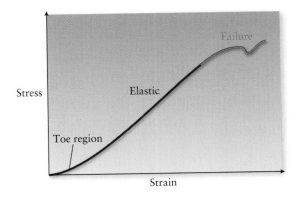

Figure 9-13 When a stress is applied to biological tissue, the collagen fibers initially uncrimp and the elastic fibers align, resulting in a nonlinear toe region in the stress versus strain curve.

Figure 9-14 When a spring is stretched too far it becomes irreversibly deformed. *(Tom Patagnes.)*

ple, fibers begin to break apart, resulting in a partial tear. A large enough stress on the ACL causes it to break into two separate pieces.

We quantify the point at which a material under tensile stress fails by its **tensile strength,** the maximum tensile stress it can withstand before it either irreversibly deforms or breaks. The strength of a material is often quantified by three similar but slightly different values. The **yield strength** is the tensile stress at which the material enters the plastic, permanent deformation regime. The **ultimate strength** is the maximum tensile stress the material can withstand without breaking apart. The **breaking strength** is the tensile stress which does result in the material breaking apart, called **failure.**

For certain materials, for example, ductile metals (metals that can be drawn out into wires), the cross-sectional area decreases as tensile force is increased. The total amount of metal must remain the same, so the decrease in cross-sectional area (referred to as "necking") results in an additional contribution to the change in length. As a result, the curve of stress versus strain plateaus and then decreases before breaking occurs. So for materials such as metals, the ultimate strength is larger than the breaking strength. This result is evident in **Figure 9-15**, an idealized stress versus strain curve for a ductile metal.

Figure 9-15 The transition from the elastic to the plastic region of the stress versus strain curve determines the yield strength of a material. Yield strength is the tensile stress for which results the material becoming permanently deformed. For ductile metals, the stress versus strain curve plateaus; the ultimate strength marks the peak of the stress–strain curve.

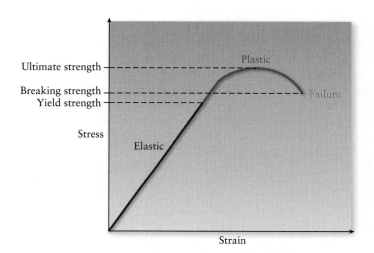

Example 9-5   Maximum Force for a Human ACL

The ultimate strength of a young person's ACL (anterior cruciate ligament) is, on average, $3.8 \times 10^7 \, \text{N/m}^2$. A typical cross-sectional area of the ACL is $4.4 \times 10^{-5} \, \text{m}^2$. What is the force exerted on the ACL when the tensile stress is maximum?

SET UP
Stress is defined as the tensile force applied to an object divided by the object's cross-sectional area. The ultimate stress is therefore the maximum force divided by the cross-sectional area:

$$\text{ultimate stress} = \frac{F_{\text{max}}}{A}$$

We can rearrange this equation to solve for the maximum force.

SOLVE
Rearranging the equation gives

$$F_{\text{max}} = A \times \text{ultimate stress}$$

Using the values given in the problem statement, the maximum force is

$$F_{max} = (4.4 \times 10^{-5}\,\text{m}^2) \times (3.8 \times 10^7\,\text{N/m}^2) = 1.7 \times 10^3\,\text{N}$$

REFLECT
On the surface of Earth, an object that has a weight of 1700 N has a mass of more than 170 kg. That's about the size of an adult female grizzly bear. By the way, the heaviest professional American football players, those who play offensive tackle, have an average mass of about 150 kg. So a knee ligament can withstand a sizeable force.

Practice Problem 9-5 What is the ultimate stress of a ligament if its cross-sectional area is $2.1 \times 10^{-5}\,\text{m}^2$ and the maximum force it can withstand is $1.3 \times 10^3\,\text{N}$?

Example 9-6 ACL Breaking Strain

At the point of failure, the tensile strain on the ACL of a young person is approximately 60% beyond its unstretched length. If the typical length of an ACL is $2.7 \times 10^{-2}\,\text{m}$, how far can it stretch before breaking?

SET UP
For an object experiencing a tensile stress, tensile strain is defined as how much an object stretches as a fraction of its total length. The strain at the point of failure is therefore the maximum change in length divided by the initial length or

$$\text{breaking strain} = \frac{\Delta L_{max}}{L_0}$$

We can rearrange this equation to solve for the maximum change in length.

SOLVE
The rearrangement gives

$$\Delta L_{max} = L_0 \times \text{breaking strain}$$

Using the values given in the problem statement,

$$\Delta L_{max} = 2.7 \times 10^{-2}\,\text{m} \times 0.60 = 1.6 \times 10^{-2}\,\text{m}$$

REFLECT
The ACL stretches 1.6 cm before it tears completely. If that seems like a large stretch, it is—1.6 cm is 60% of the original length of the ACL. Of course, this number shouldn't be a surprise as it comes directly from the measurement of maximum strain of 60% before failure.

Practice Problem 9-6 A relatively inelastic connecting rod of unstretched length 20 cm can only be stretched to a length of 21 cm before breaking. What is the breaking strain of the connecting rod?

Example 9-7 The Point of No Return

The yield strength of a young person's ACL is approximately $3.3 \times 10^7\,\text{N/m}^2$. By what percentage of its initial length can the ACL stretch before it will no longer return to its original length intact? Use $Y_{ACL} = 1.1 \times 10^8\,\text{N/m}^2$ as the value for Young's modulus for an ACL.

SET UP

The yield strength is the stress on the ligament described by the upper end of the linear part of the stress versus strain curve. We can therefore apply the linear relationship of Hooke's law (Equation 9-2)

$$\frac{F}{A} = Y\frac{\Delta L}{L_0}$$

where F/A is the tensile stress and $\Delta L/L_0$ is the tensile strain. The strain at which the ACL begins to deform irreversibly is then defined by

$$\left(\frac{F}{A}\right)_{yield} = Y\left(\frac{\Delta L}{L_0}\right)_{irreversible}$$

We can rearrange to solve for the strain beyond which the ACL cannot recover.

SOLVE

Rearranging the above equation gives

$$\left(\frac{\Delta L}{L_0}\right)_{irreversible} = \frac{1}{Y}\left(\frac{F}{A}\right)_{yield}$$

Using the values given in the problem statement,

$$\left(\frac{\Delta L}{L_0}\right)_{irreversible} = \frac{1}{1.1 \times 10^8\,\text{N/m}^2}\,(3.3 \times 10^7\,\text{N/m}^2) = 0.30 = 30\%$$

REFLECT

Yield strength is the stress below which an object obeys the "small stretch" limit of Hooke's law. When an object experiences a stress below the yield strength it should return to its initial shape undamaged. Here, we have found that the human ACL is elastic enough to withstand being stretched by a factor of 30% while still being able to return undamaged to its normal length. In the previous worked example, we discovered that a human ACL tears completely when stretched beyond about 60% of its initial length. For strains in between 30% and 60% some damage, for example, a partial tear, can occur.

✳ What's Important 9-4

When an object is subjected to a stress that results in a relatively small strain, it deforms but returns to its initial shape after the stress is removed (the elastic regime). If the stress on an object is increased so that the change in length becomes large relative to the object's initial length, the deformation of the object becomes irreversible (the plastic regime) and represents the onset of failure, which eventually leads to the object breaking apart.

Answers to Practice Problems

9-1 4.4 N

9-2 0.02 m³

9-4 0.08 m

9-5 $6.2 \times 10^7\,\text{N/m}^2$

9-6 0.05

Answers to Got the Concept Questions

9-1 No! The amount that an object stretches when put under tension depends on its original length. Although the rod made from material B stretches farther, its initial length could have been long enough to allow it to stretch more than the rod made from material A, even if Y for material A is smaller. According to Equation 9-1, when the applied force is the same,

$$Y_A \frac{A}{L_{0,A}} \Delta L_A = Y_B \frac{A}{L_{0,B}} \Delta L_B$$

so

$$\Delta L_B = \frac{Y_A}{Y_B} \frac{L_{0,B}}{L_{0,A}} \Delta L_A$$

ΔL_B will be larger than ΔL_A when $Y_A L_{0,B}$ is larger than $Y_B L_{0,A}$. This result can be true even when Y_B is larger than Y_A, depending on the relative initial lengths of the rods.

9-2 The volume of the balloon filled with air will change the most. From Table 9-1 you can see that the bulk modulus of water is about 2×10^4 times larger than that of air. Water is therefore about 2×10^4 times *less* compressible than air.

SUMMARY

Topic	Summary	Equation or Symbol
breaking strength	Breaking strength is the tensile stress which results in a material breaking apart.	
bulk modulus	The bulk modulus B is the proportionality constant between the fractional volume change $\Delta V/V_0$ an object experiences and the force F, spread out over the object's surface area A, that results in that volume change. The bulk modulus is a measure of the compressibility of a material.	$\dfrac{F}{A} = -B\dfrac{\Delta V}{V_0}$ (9-4)
compressive strain	Compressive strain is the equivalent of tensile strain for an object that is subjected to a force that squeezes it, rather than stretches it, along a single direction.	
compressive stress	Compressive stress is the equivalent of tensile stress for an object that is subjected to a force that squeezes it, rather than stretches it, along a single direction.	
elastic modulus	Elastic modulus is another name for Young's modulus.	
elastic regime	An object subjected to a tensile or compressive stress in the elastic regime deforms but then returns to its initial shape after the stress is removed. The relationship between stress and strain is linear in the elastic regime.	
failure	Failure occurs when the stress on an object is so large that the material starts to lose its structural integrity.	
plastic regime	When an object experiences stress in the plastic regime, the change in its length becomes large relative to its initial length and the deformation of the object becomes irreversible.	
shear modulus	The shear modulus S is a measure of the rigidity of a material.	

shear strain	Shear strain is measure of how much an object deforms under a shear stress, specifically how far one face moves relative to the opposite face of an object, relative to the distance h between the faces.	$\text{shear strain} = \dfrac{x}{h}$ (9-6)
shear stress	Shear stress results when a force F is applied parallel (tangential) to the face of an object, causing one face of the object to move or slide relative to the opposite face. Shear stress can also cause an object to twist.	
tensile strain	The tensile strain, a dimensionless fraction, is a measure of how much an object stretches (ΔL) relative to its initial length (L_0).	$\text{tensile strain} = \Delta L / L_0$
tensile strength	The tensile strength is the maximum tensile stress a material can withstand before it begins to irreversibly deform.	
tensile stress	Tensile stress is a measure of how much force F is exerted on an object relative to the area A over which that force is applied, when the length of the object changes as a result of the force. The units of tensile stress are N/m^2. The proportionality constant between tensile stress and tensile strain is Young's modulus Y.	$\dfrac{F}{A} = Y\dfrac{\Delta L}{L_0}$ (9-2)
toe region	The toe region is the regime in which the change in length, often of biological tissue, initially does not grow linearly as stress is increased.	
ultimate strength	The ultimate strength is the maximum tensile stress the material can withstand.	
volume strain	Volume strain, a dimensionless fraction, is a measure of how much the volume of an object changes (ΔV) relative to its initial volume (V_0).	$\text{volume strain} = \Delta V / V_0$
volume stress	Volume stress is a measure of how much force F is exerted on an object relative to the area A over which that force is applied, when the volume of the object changes as a result of the force. The units of volume stress are N/m^2. The proportionality constant between volume stress and volume strain is the bulk modulus B. The stress–strain relationship includes a minus sign because an increase in applied force results in a decrease in volume.	$\dfrac{F}{A} = -B\dfrac{\Delta V}{V_0}$ (9-4)
yield strength	Yield strength is the tensile stress at which the material becomes permanently deformed.	
Young's modulus	The distance an object stretches under an applied force is, under the right conditions, proportional to force. The proportionality constant k is itself proportional to the ratio of the cross-sectional A area of the object and its unstretched length L_0. Young's modulus Y is the proportionality constant between k and A/L_0. Young's modulus is a measure of the stretchability of a material or object.	$k = Y\dfrac{A}{L_0}$

QUESTIONS AND PROBLEMS

In a few problems, you are given more data than you actually need; in a few other problems, you are required to supply data from your general knowledge, outside sources, or informed estimate.

Interpret as significant all digits in numerical values that have trailing zeros and no decimal points.

For all problems, use $g = 9.81 \text{ m/s}^2$ for the free-fall acceleration due to gravity. Neglect friction and air resistance unless instructed to do otherwise.

- • Basic, single-concept problem
- •• Intermediate-level problem, may require synthesis of concepts and multiple steps
- ••• Challenging problem

SSM *Solution is in Student Solutions Manual*

Conceptual Questions

1. •(a) What is the difference between Young's modulus and bulk modulus? (b) What are the units of these two physical quantities?

2. •Define the term *yield stress*.

3. •Is it possible for a long cable hung vertically to break under its own weight? Explain your answer. SSM

4. •Define the following terms: tensile strength, ultimate strength, and breaking strength.

5. •Devise a simple way of determining which modulus (Young's, bulk, or shear) is appropriate for any given stress–strain problem.

6. •Give a few reasons why Hooke's law is intuitively obvious and a few reasons why it is counterintuitive.

7. •In Figure 9-15, there are two strain values that correspond to the breaking strength. Can you explain how a single metal sample can have two different strain values for one breaking strength? SSM

8. •In some recent studies, it has been shown that women are more susceptible to torn ACLs than men when competing in similar sports (most notably in soccer and basketball). What are some reasons why this disparity might exist?

9. •Describe the small stretch limit of Hooke's law for a spring.

10. •A 2″ × 4″ pine stud is securely clamped at one end to an immovable object. A heavy weight hangs from the free end of the wood causing it to bend. (a) Which part of the plank is under compression? (b) Which part of the plank is under tension? (c) Is there any part that is neither stretched nor compressed?

11. •Calc When working with the compression of a gas, it is often convenient to define a related quantity known as the compressibility, κ. The compressibility is the inverse of the bulk modulus: $\kappa = 1/B$. What is the equation for compressibility of a gas in terms of volume (V) and pressure (P)? Write your answer as a derivative and as a ratio of finite differences. SSM

12. •(a) Describe some common features of strain that were defined in this chapter. (b) We encountered three types of strain (tensile, volume, and shear). What are some distinguishing features of these quantities?

13. •With regard to metals, why is the ultimate strength larger than the breaking strength?

14. •Why are tall mountains typically shaped like cones rather than a straight vertical columnlike structure?

15. •The shear modulus (S) is sometimes known as the *rigidity*. Can you explain why rigidity is an appropriate synonym for this physical quantity? SSM

16. •Is it possible, when tightening the lug nuts on the wheel of your car, to use too much torque and break off one of the bolts? Explain your answer.

17. •What can cause nylon tennis racket strings to break when they are hit by the ball?

18. •Human skin is under tension like a rubber glove that has had air blown into it. Why does skin acquire wrinkles as people get older?

19. •Biology The leg bone of a cow has an ultimate strength of about $150 \times 10^6 \text{ N/m}^2$ and a maximum strain of about 1.5%. The antler of a deer has an ultimate strength of about $160 \times 10^6 \text{ N/m}^2$ and a maximum strain of about 12%. Explain the relationship between structure and function in these data.

Multiple-Choice Questions

20. •The units for strain are
   A. N/m.
   B. N/m².
   C. N.
   D. N·m².
   E. none of the above.

21. •The units for stress are
   A. N/m.
   B. N/m².
   C. N.
   D. N·m.
   E. N·m².

22. •When tension is applied to a metal wire of length L, it stretches by ΔL. If the same tension is applied to a metal wire of the same material with the same cross-sectional area, but of length of $2L$, by how much will it stretch?
   A. ΔL
   B. $2\,\Delta L$
   C. $0.5\,\Delta L$
   D. $3\,\Delta L$
   E. $4\,\Delta L$

23. •A steel cable lifting a heavy box stretches by ΔL. If you want the cable to stretch by only half of ΔL, by what factor must you increase its diameter?
   A. 2
   B. 4
   C. $\sqrt{2}$
   D. $1/2$
   E. $1/4$ SSM

24. •A wire is stretched just to its breaking point by a force F. A longer wire made of the same material has the same diameter. The force that will stretch it to its breaking point is
   A. larger than F.
   B. smaller than F.
   C. equal to F.
   D. much smaller than F.
   E. much larger than F.

25. •Two solid rods have the same length and are made of the same material with circular cross sections. Rod 1 has a radius r, and rod 2 has a radius $r/2$. If a compressive force F is applied to both rods, their lengths are reduced by ΔL_1 and ΔL_2, respectively. The ratio $\Delta L_1/\Delta L_2$ is
   A. $1/4$
   B. $1/2$
   C. 1
   D. 2
   E. 4

26. •A wall mount for a television consists in part of a mounting plate screwed or bolted flush to the wall. Which kinds of stresses play a role in keeping the mount securely attached to the wall?
   A. compression stress
   B. tension stress
   C. shear stress
   D. bulk stress
   E. A, B, and C

27. •When choosing building construction materials, what kinds of materials would you choose, all other things being equal?
   A. materials with a relatively large bulk modulus
   B. materials with a relatively small bulk modulus

C. either materials with a large or a small bulk modulus
D. it doesn't matter as long as the building is not too tall
E. materials with a relatively small shear modulus SSM

28. •A book is pushed sideways, deforming it as shown in Figure 9-16. To describe the relationship between stress and strain for the book in this situation, you would use
   A. Young's modulus.
   B. bulk modulus.
   C. shear modulus.
   D. both Young's modulus and bulk modulus.
   E. both shear modulus and bulk modulus.

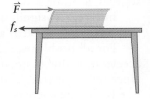

Figure 9-16 Problem 28

29. •A steel cable supports an actor as he swings onto the stage. The weight of the actor stretches the steel cable. To describe the relationship between stress and strain for the steel cable, you would use
   A. Young's modulus.
   B. bulk modulus.
   C. shear modulus.
   D. both Young's modulus and bulk modulus.
   E. both shear modulus and bulk modulus.

Estimation/Numerical Analysis

30. •Estimate the Young's modulus for (a) a rubber band and (b) a wooden pencil.

31. •Estimate the Young's modulus for a strip of paper. SSM

32. •Estimate the shear modulus for a chilled stick of butter taken from a refrigerator. Describe how this value would change as the butter warms to room temperature.

33. •Estimate the force needed to break a bone in your arm.

34. •Estimate the force needed to bend a bar of $\frac{1}{2}$-in. rebar. (Rebar is made from iron and is used to reinforce concrete.)

35. •Estimate the force needed to puncture a 0.5-cm-thick sheet of aluminum with a 1-cm-diameter rivet.

36. •Estimate the shear strain for a typical athletic shoe in a basketball game.

37. •The following data are associated with an alloy of steel. Plot a graph of stress versus strain for the alloy. What are (a) the yield strength, (b) the ultimate strength, (c) the Young's modulus, and (d) the point of rupture for the material? SSM

This page is intentionally left blank.

For complete end of chapter problem sets, please go to
www.whfreeman.com/kestentauck

9-2: Volume Stress and Strain

53. •A rigid cube (each side is 0.1 m) is filled with water and frozen solid. When water freezes it expands about 9%. How much pressure is exerted on the sides of the cube? *Hint:* Imagine trying to squeeze the block of ice back into the original cube. SSM

54. •A cube of lead (each side is 5 cm) is pressed equally on all six sides with forces of 100,000 N. What will the new dimensions of the cube be after the forces are applied ($B = 46 \times 10^9 \, \text{N/m}^2$ for lead)?

55. •A spherical air bubble has a radius of 4 cm when it is 8 m below the surface of a freshwater lake. What is the radius of the bubble immediately before it reaches the surface?

56. •A sphere of copper that has a radius of 5 cm is compressed uniformly by a force of 2×10^8 N. Calculate the change in volume of the sphere and the final radius. The bulk modulus for copper is $140 \times 10^9 \, \text{N/m}^2$.

9-3: Shear Stress and Strain

57. •A brass nameplate is 2 cm × 10 cm × 20 cm in size. If a force of 200,000 N acts on the upper left side and the bottom right side (**Figure 9-17**), find the shear strain (x/h) and the angle φ. SSM

Figure 9-17 Problem 57

58. •A steel door is slammed shut when an earthquake occurs. There are shear forces acting on the door (**Figure 9-18**). If the shear strain is 0.005, calculate the force acting on the door with dimensions 0.044 m × 0.81 m × 2.03 m.

Figure 9-18 Problem 58

59. ••A force of 5×10^6 N is applied tangentially at the center of one side of a brass cube. The angle of shear φ is measured to be 0.65°. Calculate the volume of the original cube.

60. •**Medical, Biology** In asthmatic patients, an increased thickness of the airways causes a local reduction in stress through the airway walls. The effect can be as much as a 50% reduction in the local shear modulus of the airways of an asthmatic patient as compared to those of a healthy person. Calculate the ratio of the shear strain in an asthmatic airway to that of a healthy airway.

61. •**Medical** In regions of the cardiovascular system where there is steady laminar blood flow, the shear stress on cells lining the walls of the blood vessels is about 20 dyne/cm². If the shear strain is about 0.008, estimate the shear modulus for the affected cells. Note 1 dyne = 1 g · cm/s² and 1 N = 10⁵ dyne. SSM

62. •An enormous piece of granite that is 200 m thick and has a shear modulus of $50 \times 10^9 \, \text{N/m}^2$ is sheared from its geologic formation with an earthquake force of 275×10^9 N. The area on which the force acts is a square of side x as shown in **Figure 9-19**. Find the value of x if the shear force produces a shear strain of 0.125.

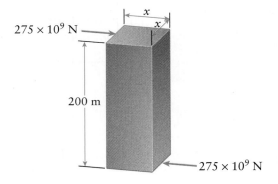

Figure 9-19 Problem 62

9-4: Elasticity and Fracture

63. •Steel will ultimately fail if the shear stress exceeds $400 \times 10^6 \, \text{N/m}^2$. Determine the force required to shear a steel bolt that is 0.50 cm in diameter.

64. •The stress–strain graph for an idealized spring with a spring constant of 100 N/m is shown in **Figure 9-20**. If the spring is composed of steel wire that is 0.1 cm² in cross-sectional area with a length of 12 cm, identify the different points on the graph. Note that point D should be a numerical value (with units of N/m²) and assume that point E corresponds to a strain of 1.00 (100%).

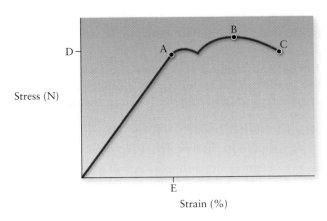

Figure 9-20 Problem 64

65. •A theater rigging company uses a safety factor of 10 for all its ropes, which means that all ultimate breaking strengths will be overestimated by a factor of 10 just to be safe. Suppose a rope with an ultimate breaking strength of 10,000 N is tied with a knot that decreases rope strength by 50%. (a) If the rope is used to support a load of 1000 N, what is the safety factor? (b) Will the rigging company be able to use the rope with the knot? SSM

General Problems

66. •A 50-kg air-conditioning unit slips from its window mount, but the end of the electrical cord gets caught in the mounting bracket. In the process the cord (which is 0.5-cm in diameter) stretches from 3.0 m to 4.5 m. What is the Young's modulus for the cord?

67. •**Medical** The largest tendon in the body, the Achilles tendon connects the calf muscle to the heel bone of the foot. This tendon is typically 25.0 cm long, 5.0 mm in diameter, and has a Young's modulus of $1.47 \times 10^9 \, \text{N/m}^2$. If an athlete has stretched the tendon to a length of 26.1 cm, what is the tension (in newtons and pounds) in the tendon?

68. ••Two rods have the same diameter and are welded together as shown in **Figure 9-21**. The shorter rod has a Young's modulus of Y_1, the longer one has a Young's modulus of Y_2. Calculate the combined Young's modulus for the system.

Figure 9-21 Problem 68

69. •**Biology** Spiders spin their webs from silk, one of the strongest naturally occurring materials known. Each thread is typically 2.0 μm in diameter and the silk has a Young's modulus of $4.0 \times 10^9 \, \text{N/m}^2$. (a) How many strands would be needed to make a rope 9.0 m long that would stretch only 1.00 cm when supporting a pair of 85-kg mountain climbers? (b) Assuming that there is no appreciable space between the parallel strands, what would be the diameter of the rope? Does the diameter seem reasonable for a rope that mountain climbers might carry? SSM

70. ••A brass sphere that has a radius of 6 cm sinks 2000 m to the bottom of the ocean. By how much does the volume shrink due to the enormous pressure at that depth and what is the new radius? Neglect any temperature changes.

71. •A 2.8-carat diamond is grown under a high pressure of $58 \times 10^9 \, \text{N/m}^2$. (a) By how much does the volume of a spherical 2.8-carat diamond expand once it is removed from the chamber and exposed to atmospheric pressure? (b) What is the increase in the diamond's radius? One carat is 200 mg and the density of diamond is 3.52 g/cm³. The bulk modulus for diamond is $200 \times 10^9 \, \text{N/m}^2$.

72. •A glass marble that has a diameter of 1 cm is dropped into a graduated cylinder that contains 20 cm of mercury. (a) By how much does the volume of the marble shrink while at the bottom of the mercury? (b) What is the corresponding change in radius associated with the compression? The bulk modulus of glass is $50 \times 10^6 \, \text{N/m}^2$.

73. •**Sports** During the 2004 Olympic clean-and-jerk weight lifting competition, Hossein Rezazadeh lifted 263.5 kg. Mr. Rezazadeh himself had a mass of 163 kg. Ultimately, the weight is all supported by the tibia (shin bone) of the lifter's legs. The average length of a tibia is 385 mm and its diameter (modeling it as having a round cross-section) is about 3.0 cm. Young's modulus for bone is typically about $2.0 \times 10^{10} \, \text{N/m}^2$. (a) By how much would the lift compress the athlete's tibia, assuming that the bone is solid? (b) Does this seem to be a significant compression? (c) Is it necessary to include the lifter's weight in your calculations? Why or why not? SSM

74. •When a house is moved, it is gradually raised and supported on wooden blocks. A typical house averages about 120,000 lb (54,000 kg). The house is supported uniformly on six stacks of blocks of Douglas fir wood (which has a Young's modulus of $13 \times 10^9 \, \text{N/m}^2$). Each block is 25 cm by 75 cm. (a) If the wood is stacked 1.5 m high, by how much will the house compress the supporting stack of blocks? (b) Can the blocks with stand the compression?

75. ••(a) What diameter is needed for a steel cable to support a large diesel engine with a mass of 4000 kg? (b) By how much will the 10-m-long cable stretch once the engine is raised up off the ground? Assume the ultimate breaking strength of steel is $400 \times 10^6 \, \text{N/m}^2$.

76. •**Sports** A runner's foot pushes on the ground as shown in **Figure 9-22**. The 25-N shearing force is distributed over an area of 15 cm² and a 1-cm-thick sole. If the shear angle θ is 5.0°, what is the shear modulus of the sole?

Figure 9-22 Problem 76

77. •Sports A runner's foot pushes off on the ground as shown in **Figure 9-23**. The 28-N shearing force is distributed over an area of 20 cm² and a 1-cm-thick sole. If the shear modulus of the sole is 1.9×10^5 N/m², what is the shear angle θ?

Figure 9-23 Problem 77

78. •••Biology A particular human hair has a Young's modulus of 4.0×10^9 N/m² and a diameter of 150 μm. (a) If a 250-g object is suspended by the single strand of hair that is originally 20.0 cm long, by how much will the hair stretch? (b) If the same object were hung by an aluminum wire of the same dimensions as the hair, by how much would the aluminum stretch? (Try to do this part without repeating the previous calculation, but use proportional reasoning instead. Also, see Table 9-1.) (c) If we think of the strand of hair as a spring, what is its spring constant? (d) How does the hair's spring constant compare with that of ordinary springs in your physics laboratory?

79. •Astro The spherical bubbles near the surface of a glass of water on Earth, where the atmosphere exerts 1.0×10^5 N/m² over the surface of each bubble, are 2.5 mm in diameter. If the glass of water is taken to Mars, where the atmosphere exerts 650 N/m² over the bubble surface, what will be the diameter of the bubbles? (See Table 9-1.) SSM

80. •Biology, Medical The femur bone in the human leg has a minimum effective cross section of 3.0 cm². How much compressive force can it withstand before breaking? Assume the ultimate strength of the bone to be 1.7×10^8 N/m².

81. •Biology, Medical At its narrowest point the femur bone in the human leg resembles a hollow cylinder that has an outer radius of roughly 1.1 cm and an inner radius of just about 0.48 cm. Assuming that the ultimate strength of the bone is 1.7×10^8 N/m², how much force will be required to rupture it?

82. •••A beam is attached to a vertical wall with a hinge. The mass of the beam is 1000 kg and it is 4 m long. A steel support wire is tied from the end of the beam to the wall, making an angle of 30° with the beam (**Figure 9-24**). (a) By summing the torque about the axis passing through the hinge, calculate the tension in the support wire. Assume the beam is uniform so that the weight acts at its exact center. (b) What is the minimum cross-sectional area of the steel wire so that it is not permanently stretched? Recall the important values for steel: the Young's modulus is 200×10^9 N/m², the shear modulus is 78×10^9 N/m², the bulk modulus is 140×10^9 N/m², the yield stress (elastic limit) is 290×10^6 N/m², and the ultimate breaking strength is 400×10^6 N/m².

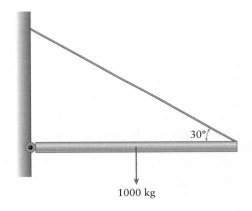

Figure 9-24 Problem 82

83. ••Biology A representative average of cortical bone properties for humans and cows are presented in **Table 9-2**.

(a) From the data, calculate which species (human or cow) will most likely sustain a transverse break in a bone. Explain your reasoning.

(b) From the data, explain how a cow's bones are much more capable of supporting their extreme weight in comparison to a human's bones.

(c) Are the data consistent with the fact that the maximum running speed of a cow is about 7.5 m/s while a human can run at about 10 m/s?

(d) A bone in a woman's leg has an effective cross-sectional area of 3 cm². If the leg is 35 cm long, how much

Table 9-2

Property	Human Value	Bovine Value
Elastic modulus transverse	17.4×10^9 N/m²	20.4×10^9 N/m²
Elastic modulus longitudinal	9.6×10^9 N/m²	11.7×10^9 N/m²
Shear modulus	3.5×10^9 N/m²	4.1×10^9 N/m²
Tensile yield stress longitudinal	115×10^6 N/m²	141×10^6 N/m²
Tensile ultimate stress longitudinal	133×10^6 N/m²	156×10^6 N/m²
Tensile ultimate stress transverse	51×10^6 N/m²	50×10^6 N/m²
Compressive yield stress longitudinal	182×10^6 N/m²	196×10^6 N/m²
Compressive yield stress transverse	121×10^6 N/m²	50×10^6 N/m²
Compressive ultimate stress longitudinal	95×10^6 N/m²	237×10^6 N/m²
Compressive ultimate stress transverse	133×10^6 N/m²	178×10^6 N/m²
Tensile ultimate strain	2.9–3.2%	0.67–0.72%
Compressive ultimate strain	2.2–4.6%	2.5–5.2%

http://www.engin.umich.edu/class/bme456/bonefunction/bonefunction.htm

compressive force can it withstand before breaking? How much will the bone compress if it is subjected to a force one tenth the magnitude of the force that breaks it?

Challenge Problems

84. •••Calc A bar with a square cross-sectional area is subject to a constant tensile stress (**Figure 9-25**). Consider a thin slab within the bar making an angle of θ with the horizontal. Find the following:

(a) the forces on the upper surface of the slab parallel and perpendicular to the plane,

(b) the tensile stress and the shear stress at that plane,

(c) the angle θ_{max} for which the tensile stress is a maximum value, and

(d) the angle θ_{max} for which the shear stress is a maximum value.

Figure 9-25 Problem 84

85. •••Poisson's ratio is defined as the induced strain divided by the primary strain:

$$\nu = -\frac{\Delta w/w}{\Delta \lambda/\lambda} = -\frac{\Delta t/t}{\Delta \lambda/\lambda}$$

The primary stress acts along the longitudinal axis (λ) of an object and the induced stress is found along the axes perpendicular to that direction. Consider a rectangular object made from a material that has a Young's modulus of Y and a bulk modulus of B (**Figure 9-26**). The original dimensions of the object are λ, w, and t. After it is stressed, these dimensions are $\lambda - \Delta\lambda$, $w + \Delta w$, $t + \Delta t$.

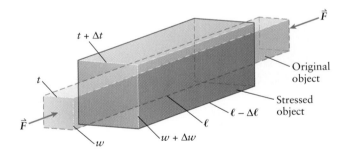

Figure 9-26 Problem 85

When the object is immersed in a fluid and symmetrically stressed (in the manner described above), prove that the following relationship is true: SSM

$$B = \frac{Y}{3(1 - 2\nu)}$$

10 Gravitation

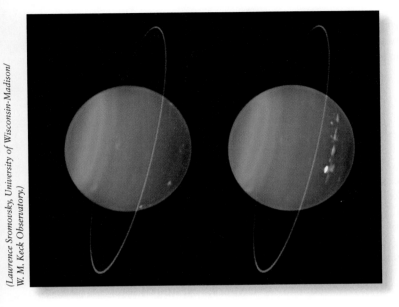

In the 17th century, Sir Isaac Newton deduced that large objects, such as stars, planets, and moons, exert forces on each other. His insight came to be known as the universal law of gravitation and supported Johannes Kepler's earlier idea that one object orbiting another traces out an elliptical path. It is difficult to capture this path in a photograph. However, the rings of Jupiter, Saturn, Neptune, and Uranus are composed of uncountable numbers of bits of ice and rock spread out around the orbital paths, so the particles' entire orbits are dramatically visible. The rings of Uranus can be seen in this false-color, composite image. Although nearly circular, the rings are actually elliptical. (The bright patches are clouds in Uranus' thick atmosphere.)

(Laurence Sromovsky, University of Wisconsin-Madison/ W. M. Keck Observatory.)

We have seen that objects fall to the ground under the influence of Earth's gravity. In the mid-17th century, Isaac Newton wondered whether the orbit of the Moon around Earth was simply a different manifestation of the same underlying physical phenomenon. Newton's reasoning is something like the following: If a cannon were fired horizontally on level ground, the cannonball would fall back to Earth following a smooth arc (ignoring the effects of air resistance). The larger the initial speed of the ball, the farther it would land from the cannon. What would happen, Newton asked himself, if the cannon were fired horizontally from the top of a tall mountain, so that as the cannonball fell toward Earth's surface, the surface curved away from its path? And what if the cannonball were fired so energetically that the rate at which it fell just matched the rate at which the surface fell away? This, he reasoned, would result in the cannonball completing a full orbit, always falling but never hitting the surface. Figure 10-1 shows Newton's sketch of his thought experiment. Ultimately, Newton successfully developed a description of the gravitational force that explained free fall near the surface of Earth, the motion of the Moon and, indeed, the motions of all celestial objects.

10-1 Newton's Universal Law of Gravitation

Newton's **universal law of gravitation** describes the gravitational force that one object exerts on another. In words, it states that

> The gravitational force that one particle exerts on another is proportional to the product of their masses and inversely proportional to the square of the distance between their centers. The force is

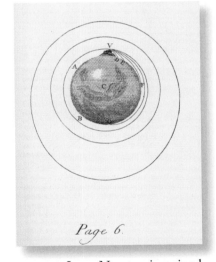

Page 6.

Figure 10-1 Isaac Newton imagined that if a cannonball fired horizontally from the top of a mountain had enough energy, the rate at which it fell could be made to match the rate at which Earth's surface fell away and that the cannonball would end up orbiting Earth. Newton created this sketch for his book *Principia* (first published July 5, 1687). *(Smithsonian Institution.)*

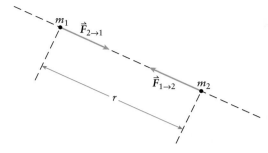

Figure 10-2 Each of two infinitesimally small objects, of mass m_1 and m_2, exert a gravitational force on the other. Object m_2 exerts force $\vec{F}_{2\rightarrow1}$ on m_1. Object m_1 exerts force $\vec{F}_{1\rightarrow2}$ on m_2. Each force is directed along the line connecting the two objects. The magnitude of the force is inversely proportional to the square of the separation distance r.

attractive, and directed along the line that connects the centers of the two particles.

The choice of the word *particle* implies an object that is infinitesimally small. The gravitational force that one particle exerts on the other is

$$\vec{F} = -\frac{Gm_1m_2}{r^2}\hat{r} \tag{10-1}$$

where m_1 and m_2 are the masses of two particles separated by distance r as shown in Figure 10-2. The proportionality constant G, which has been measured to better than one part in 10^8, is approximately

$$G = 6.67 \times 10^{-11}\,\text{N}\cdot\text{m}^2/\text{kg}^2$$

G is sometimes referred to as the *universal constant of gravitation* because it has the same value regardless of the masses of the objects and regardless of their separations. In addition, as best we know, G has the same value everywhere throughout our galaxy and beyond.

Newton's universal law of gravitation can be generalized to account for objects such as moons, planets, and stars, that is, objects that are not pointlike. Generally, as long as two relatively spherical objects are far enough apart so that their sizes are small compared to the distance between their centers, the gravitational force between them is determined by treating the mass of each as if it were concentrated at the center of the object. This means that r in Equation 10-1 is the center-to-center distance. With this modification, we can consider the force between large objects, such as moons, planets, and stars. Because G is small, however, the gravitational force between two objects is large only when one or both are relatively massive. The gravitational force between objects that have masses typical of our experiences on Earth is therefore quite small.

The minus sign in Equation 10-1 indicates that the gravitational force is attractive. This means that every object pulls other objects *toward it* by a gravitational force. We could write Equation 10-1 as

$$\vec{F} = \frac{Gm_1m_2}{r^2}(-\hat{r}) \tag{10-2}$$

to remind us of the attraction. Equation 10-2 makes it clear that the magnitude of the force vector is Gm_1m_2/r^2 and the direction of the vector is $-\hat{r}$. We have defined the positive radial direction to point away from the origin or outward (see the discussion of Equation 3-34), so the gravitational force points inward, toward the object that exerts it.

Wait! The direction of the gravitational force was just described as pointing toward the object exerting it. According to Equation 10-1, two objects generate the gravitational force. Which one is exerting the force and which is experiencing it? Newton understood that the gravitational force is symmetric. Two objects each attract the other with gravitational forces that are equal in magnitude but opposite in direction. For example, the arrow pointing down to the right, and labeled $\vec{F}_{2\rightarrow1}$ in Figure 10-2 is the force of object 2 pulling on object 1. However, object 1 also exerts a force of equal magnitude on object 2. The second force, the arrow pointing up and to the left and labeled $\vec{F}_{1\rightarrow2}$ is none other than the equal and opposite force required by Newton's third law. (If A pulls on B, then B pulls on A with equal force but in the opposite direction.) The two forces form a force pair.

Estimate It! 10-1 You and Your Physics Book

Estimate the magnitude of the attractive force due to gravity between you and the hardback edition of this physics book if it were on the desk in front of you. For simplicity, treat both you and the book as pointlike particles.

SET UP

From Equation 10-1, the magnitude of the gravitational force between two objects is

$$F = \frac{Gm_1m_2}{r^2}$$

To complete the estimation you'll need round, reasonable values for your mass, the mass of the book, and the distance in between. For the purpose of the answer below, let's say your mass is about 80 kg, which was about the mass of each of the authors when we were in college. You can estimate the mass of the book by comparing it to some easily available objects of known weight and mass. For example, the book is heavier than a 2-L bottle of water; 2 L of water has a mass of 2 kg. The book is lighter than a gallon of water; a gallon of water has a mass of about 3.78 kg. So a reasonable estimate of the mass of the book, to one significant digit, is 3 kg. As the book sits on your desk, it is somewhat less than a half a meter away, so we'll take r equal to 0.4 m as the center-to-center distance. We have treated both you and the book as particles for the purpose of estimating the distance between you.

SOLVE

The magnitude of the force is

$$F = \frac{(6.67 \times 10^{-11}\,\text{N} \cdot \text{m}^2/\text{kg}^2)\,(80\,\text{kg})\,(3\,\text{kg})}{(0.4\,\text{m})^2} = 1 \times 10^{-7}\,\text{N}$$

REFLECT

The gravitational attraction between you and a hardback version of this physics book is about 10^{-7} N. Small? Yes! The force is equivalent to the weight of perhaps a few specks of dust. It is also clearly far smaller than the force the book experiences due to Earth's gravity, as well as the force of friction between the book and the desk. So you wouldn't, for example, expect to see the motion of the book change as a result of your presence—no matter how attracted you are to the book!

Example 10-1 The Force of Earth on You

Find the magnitude of the gravitational force of Earth on you (Figure 10-3). For the purposes of the problem, take your mass to be 65.0 kg.

SET UP

The magnitude of the gravitational force that Earth exerts on is proportional to the product of the masses and inversely proportional to the square of the distance between you and Earth's center, as described in Equation 10-1. The center-to-center distance is the radius of Earth R_E. The magnitude of the force is then

$$F = \frac{GM_Em_Y}{R_E^2} \tag{10-3}$$

where M_E is the mass of Earth and m_Y is your mass. Your size is so small compared to Earth that we can neglect it.

SOLVE

To find a numerical result, we substitute into the equation the known values, which are the mass of Earth ($M_E = 5.98 \times 10^{24}$ kg), the mean radius of Earth ($R_E = 6.38 \times 10^6$ m), and your mass ($m_Y = 65.0$ kg):

$$F = \frac{(6.67 \times 10^{-11}\,\text{N} \cdot \text{m}^2/\text{kg}^2)\,(5.98 \times 10^{24}\,\text{kg})\,(65.0\,\text{kg})}{(6.38 \times 10^6\,\text{m})^2}$$

$$= 6.37 \times 10^2\,\text{N}$$

Figure 10-3 Find the magnitude of the gravitational force between Earth and you.

Note: Your size has been greatly exaggerated in this figure! The radius of Earth is much, much larger than you are.

R_E

REFLECT

Weight, as we commonly use the term, is the force of Earth's gravity on an object on or close to Earth's surface. The weight of a typical person is often in the range of 500 to 1000 N. Our calculation of your weight using Newton's universal law of gravitation is therefore reasonable. See the next worked example for more discussion about Earth's gravity.

Practice Problem 10-1 A giant ball of twine in the Ripley's Believe It or Not Museum in Branson, Missouri, is about 2.1 m in radius and has a mass of about 8100 kg. Find the magnitude of the gravitational force between the ball of twine and a ladybug of mass 2.1×10^{-5} kg walking on its surface.

Example 10-2 Big *G* and Little *g*

In the previous worked example, we concluded that Newton's universal law of gravitation gives a reasonable value for the weight of an object on or close to Earth's surface. Find the downward free-fall *acceleration* due to gravity of an object near Earth's surface.

SET UP

Newton's second law (Equation 4-1) requires that the sum of the forces in any direction on an object be equal to the object's mass multiplied by its acceleration in that direction. From Equation 10-2, the gravitational force on an object of mass m near Earth's surface is

$$\vec{F} = \frac{GM_E m}{R_E^2}(-\hat{r})$$

where R_E is the mean radius of Earth and M_E is Earth's mass. The gravitational force is the only force in the vertical direction if you neglect air resistance, so by Newton's second law,

$$\frac{GM_E m}{R_E^2}(-\hat{r}) = mg(-\hat{r})$$

The term on the right is the object's mass multiplied by its acceleration due to gravity on Earth's surface. We are explicit about the sign convention; both the force due to gravity and the acceleration vector point down toward the center of Earth.

SOLVE

Canceling m from both sides of the equation and writing the new expression in terms of magnitudes,

$$\frac{GM_E}{R_E^2} = g \qquad\qquad (10\text{-}4)$$

Does this equation give the expected result? Using $M_E = 5.98 \times 10^{24}$ kg and $R_E = 6.38 \times 10^6$ m,

$$g = \frac{GM_E}{R_E^2} = \frac{(6.67 \times 10^{-11}\,\text{N} \cdot \text{m}^2/\text{kg}^2)\,(5.98 \times 10^{24}\,\text{kg})}{(6.38 \times 10^6\,\text{m})^2} = 9.80\,\text{m/s}^2$$

Note that after canceling units in the calculation, the result is N/kg. However $1\,\text{N} = 1\,\text{kg} \cdot \text{m/s}^2$, so $1\,\text{N/kg} = 1\,\text{m/s}^2$.

REFLECT

Newton's universal law of gravitation correctly determines the accepted value of g, the acceleration due to gravity at Earth's surface. The relationship we use for weight on Earth's surface, $W = mg$, is simply an application of Newton's universal law of gravitation.

Practice Problem 10-2 Find the downward free-fall acceleration of a ladybug due to the gravitational attraction of the giant ball of twine in the Ripley's Believe It or Not Museum in Branson, Missouri. Assume that the insect is crawling on the surface of the ball of twine.

How did Newton convince himself and other scientists that the relationship expressed in Equation 10-2 was correct? As a scientist he no doubt would have preferred to use his equation to predict the outcome of an experiment and then to show that the prediction was correct. However, he faced two challenges. First, although he did not know the value of G, it was clear to him that it had to be small. As a result, the gravitational forces in any experiment he devised that did not involve enormously massive objects would be too small for him to accurately measure. His experiment would therefore have to use a massive object, such as Earth or the Moon. This led to Newton's second challenge: The masses of Earth and the Moon were not known in his time.

What *was* known in Newton's time about the two closest, massive objects? The distance from Earth to the Moon D_{EM} had been determined by astronomers by measuring the difference in the Moon's position relative to a distant star when observed from two places on Earth. This difference is known as *parallax*; a method for measuring distance using parallax is shown in **Figure 10-4**. The time T_M it takes for the Moon to complete an orbit around Earth was also known, because it is easily measured by direct observation. Although time does not appear in Newton's description of gravitational force, time plays a role in acceleration which in turn is related to force. In addition, g, the acceleration due to gravity at Earth's surface, had been well measured in Newton's time. Finally, as we saw in Examples 10-1 and 10-2, the radius of Earth R_E is required to apply Newton's universal law of gravitation on the surface of Earth. First measured by the Greek astronomer Eratosthenes in around 200 BCE, Earth's radius was relatively well known in Newton's time. The method Eratosthenes used is described in **Figure 10-5**.

The Moon's apparent position relative to a distant star changes when observed from two different places. Please note that objects and distances are not drawn to scale.

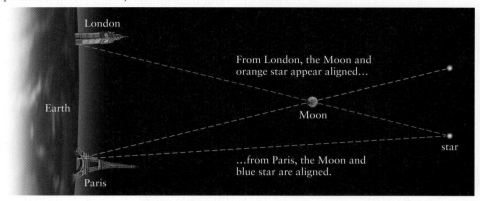

The Earth-to-Moon distance is determined from the right triangle which includes the line between the observations points and an angle related to θ.

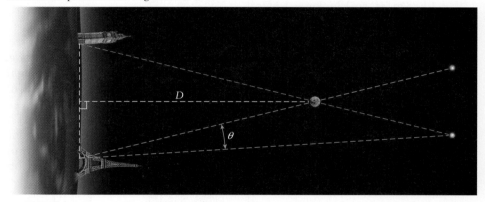

Figure 10-4 The distance from Earth to the Moon can be determined by measuring the difference in the Moon's position relative to a distant star when observed from two places on Earth.

As the story is told, on the day of the summer solstice Eratosthenes observed the shadow cast by a post in the city of Alexandria. Distances are not drawn to scale.

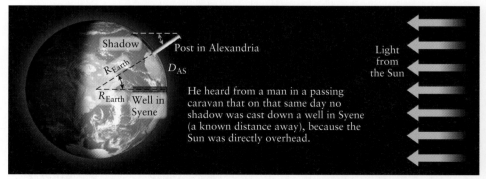

The length of the shadow and the distance from Alexandria to Syene (DAS, exaggerated in the sketch) are nearly straight lines. Eratosthenes used the two triangles to determine Earth's radius.

Figure 10-5 The Greek astronomer Eratosthenes measured the radius of Earth in around 200 BCE, by analyzing the triangles formed by the shadow cast by a vertical post in Alexandria and a water well in Syrene.

In the *Principia*, Newton gives the distance from Earth to the Moon D_{EM} as 60 Earth radii, Earth's circumference as 123,249,400 Paris feet (a unit of distance in use during Newton's time), and T_M as 27 days, 7 h, and 43 min. In SI units, $D_{EM} = 3.8 \times 10^8$ m, $R_E = 6,372,003$ m, and $T_M = 2,360,580$ s.

So Newton's test of his universal law of gravitation could have been to use it to determine the distance from Earth to the Moon, and then compare it to the known value. From Equation 10-2, the force due to Earth's gravity on the Moon is

$$\vec{F}_M = \frac{GM_E M_M}{D_{EM}^2}(-\hat{r})$$

where M_E is the mass of Earth and M_M is the mass of the Moon. Earth's gravitational force is the only force of any consequence on the Moon, so it must equal the mass of the Moon multiplied by the Moon's acceleration according to Newton's second law. (Yes, every celestial object exerts some force on the Moon, but those forces are quite small even for nearby objects such as Mars and Venus.) The Moon's orbit, while elliptical, is close to circular, so the Moon's acceleration is approximately that of uniform circular motion. From Equation 3-34,

$$\vec{a}_M = \frac{v_M^2}{D_{EM}}(-\hat{r})$$

so

$$\frac{GM_E M_M}{D_{EM}^2}(-\hat{r}) = M_M \frac{v_M^2}{D_{EM}}(-\hat{r})$$

or

$$\frac{GM_E}{D_{EM}^2} = \frac{v_M^2}{D_{EM}} \tag{10-5}$$

Newton didn't know G or M_E, but as we saw in Example 10-2 the product of G and M_E can be obtained from g and R_E. From Equation 10-4,

$$GM_E = gR_E^2$$

Also, the speed of the Moon is given by the distance the Moon travels in one orbit divided by the time required for one orbit, T_M:

$$v_M = \frac{2\pi D_{EM}}{T_M}$$

Substituting the last two relationships into Equation 10-5,

$$\frac{gR_E^2}{D_{EM}^2} = \frac{\left(\dfrac{2\pi D_{EM}}{T_M}\right)^2}{D_{EM}}$$

or

$$D_{EM} = \left(\frac{gR_E^2 T_M^2}{4\pi^2}\right)^{1/3}$$

Using Newton's values, which are converted to SI units,

$$D_{EM} = \left(\frac{(9.81 \text{ m/s}^2)(6,372,003 \text{ m})^2(2,360,580 \text{ s})^2}{4\pi^2}\right)^{1/3}$$

$$= 3.8 \times 10^8 \text{ m}$$

This prediction, based on Newton's universal law of gravitation, is in excellent agreement with the measured center-to-center distance between Earth and the Moon.

10-2 The Shell Theorem

The variable r in Newton's universal law of gravitation is the center-to-center distance between two particles. Using r in this way is equivalent to treating an extended, uniform, and spherical object as if all of its mass were concentrated at the object's center. What if a spherical object is hollow? How can we determine the gravitational force it exerts on another object? In particular, how does Newton's universal law of gravitation apply if a second spherical object is placed *inside* the hollow one? Newton asked himself these questions in order to generalize his law to include extended as well as pointlike objects.

To answer such questions, we start by recognizing that the net force on an object can be expressed as the vector sum of every separate force that acts on it. So, if we treat a solid object as an assemblage of very small or pointlike pieces, the gravitational force it exerts on another object can be found by adding the forces (vector sum) that each piece exerts on the other object separately.

According to Newton, the gravitational force between two particles points along the center-to-center line. This result is immediately evident from finding the net force by breaking a larger object into many pieces and then adding the contributions from each piece to the total force. Consider a large solid sphere of radius R; its mass is uniformly distributed and exerts a gravitational force on a second, relatively small object as shown in **Figure 10-6**. Two of the many tiny pieces of the

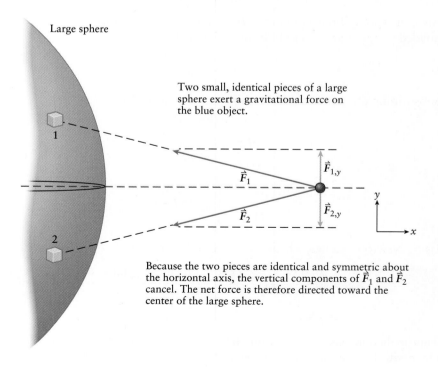

Large sphere

Two small, identical pieces of a large sphere exert a gravitational force on the blue object.

Because the two pieces are identical and symmetric about the horizontal axis, the vertical components of $\vec{F}_1$ and $\vec{F}_2$ cancel. The net force is therefore directed toward the center of the large sphere.

Figure 10-6 The gravitational force that a large sphere exerts on a small object is found by breaking the sphere into infinitesimally small pieces and adding vectorially the forces that each piece exerts on the small object.

large sphere are shown, along with the gravitational force each exerts on the small object. The two pieces were chosen carefully: Each is a mirror reflection of the other around the center-to-center line. As a result of this choice, the vertical components $\vec{F}_{1,y}$ and $\vec{F}_{2,y}$ of the two forces cancel because they are of equal magnitude but point in opposite directions. The vector sum $\vec{F}_{1,y} + \vec{F}_{2,y}$ therefore has no vertical component and points horizontally toward the center of the large sphere. We can find a mirror reflection for every small piece of the large sphere, so when the forces due to all pieces are added using vector addition, the resulting net force will also have no vertical component. The gravitational force between the two objects therefore points along the center-to-center line.

If the large uniform sphere in Figure 10-6 were replaced by a uniform spherical shell (a hollow sphere), the vertical components of properly chosen force pairs would still cancel, and the direction of the gravitational force on the small object would again be along the center-to-center line.

What if the small object is inside the hollow sphere, as in Figure 10-7? Once again, let's consider the forces that two infinitesimally small pieces of the sphere exert on the second object. (Take the shell of the sphere to be infinitesimally thin, so that the pieces are also infinitesimally small.) We are free to choose the pieces for our convenience, just like the case in which the small object is outside the hollow sphere. We again select one piece to be a reflection of the other, although here the reflection is around the vertical dashed line as well as the center-to-center line. Also, the two pieces shown are not equidistant from the small object, but we can make the magnitudes of the forces F_1 and F_2 equal, as shown, by selecting the mass of piece 1 to be greater than piece 2. (The sizes of the pieces have been exaggerated in the figure for clarity.) In particular, we need to select the pieces so that their masses are in the same ratio as the squares of their distances to the small object. The magnitude of the forces on the small object due to pieces 1 and 2 are

$$F_1 = \frac{GMm_1}{r_1^2}$$

$$F_2 = \frac{GMm_2}{r_2^2}$$

where m_1 and m_2 are the masses of the pieces, M is the mass of the object inside the hollow sphere, and r_1 and r_2 are the distances between M and the piece 1 and

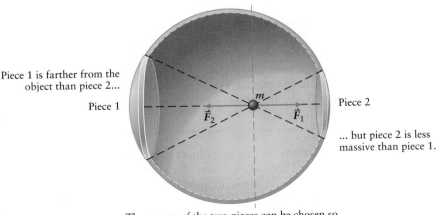

The net gravitational force on an object inside a hollow sphere is zero!

Piece 1 is farther from the object than piece 2...

Piece 1

$\vec{F}_2$ $\vec{F}_1$ m

Piece 2

... but piece 2 is less massive than piece 1.

The masses of the two pieces can be chosen so that $\vec{F}_1$ equals $\vec{F}_2$. These forces cancel.

Figure 10-7 The net gravitational force on an object inside a hollow sphere is zero.

piece 2, respectively.) We chose the mass of piece 1 to be greater than the mass of piece 2 by the ratio of the squares of their distances to the small object

$$m_1 = \frac{r_1^2}{r_2^2} m_2$$

so F_1 becomes

$$F_1 = \frac{GM\left(\dfrac{r_1^2}{r_2^2} m_2\right)}{r_1^2} = \frac{GMm_2}{r_2^2}$$

As a result of the choice of m_1 relative to m_2, F_1 equals F_2. Because the two forces point in opposite directions, they cancel.

The dashed red line in Figures 10-7 and 10-8 represents a plane which divides the hollow sphere into a left and right side. For every small piece of the hollow sphere we consider on the left, there is a corresponding piece on the right. The relative masses are chosen so that the forces balance; for every piece of the hollow sphere there is a corresponding piece that exactly cancels the gravitational force of the first. The forces due to all pieces are added. Although we will not present the proof, it can be shown that when the resulting net force is zero. No gravitational force exists on an object inside the hollow uniform sphere!

The gravitational force between a hollow uniform sphere (a shell) and a second object is summarized by the **shell theorem**:

- For an object outside a hollow uniform sphere of uniform mass distribution, the shell exerts a gravitational force on the object as if the mass of the shell were concentrated at its center.
- The shell of a hollow sphere of uniform mass distribution exerts no net gravitational force on an object inside it.

By considering a large, solid sphere, such as Earth, as an assemblage of an infinite number of thin, hollow spheres, we can see how Newton was able to apply his universal law of gravitation to objects other than pointlike particles. Each shell exerts a force on an object outside it as if its mass were concentrated at its center, so the net force of the large sphere is equivalent to adding the mass of each separate shell (which gives the total mass of the sphere) and treating *it* as if it were at the center of the sphere.

> **Watch Out**
>
> **The net force on an object inside a hollow sphere is not zero.**
>
> The shell theorem tells us that a hollow sphere exerts zero gravitational force on an object inside it. However, the object still experiences forces exerted by objects other than the sphere. For example, while a marble placed inside a hollow sphere that rests on the surface of Earth experiences no net gravitational force due to the sphere, it is nevertheless affected by the pull of Earth's gravity so, for example, it would experience a normal force when resting on the inner surface of the sphere. The sphere does not shield the object from forces exerted by objects other than the sphere.

The gravitational force exerted by any object can be found by considering it as an assemblage of pieces and then taking the vector sum of the forces that each piece exerts separately. Imagine a solid uniform sphere treated as a series of thin shells that each have a mass m and are one inside the other, much like the layers of an onion. An object placed somewhere inside the sphere will be outside some of the layers and inside the rest. In Figure 10-9, a small blue sphere of mass m has been

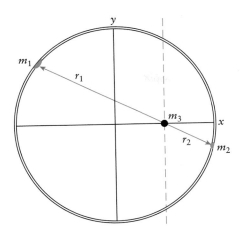

Figure 10-8 According to the shell theorem, an object inside a solid sphere experiences a gravitational force due only to the mass of the sphere closer to the center.

placed at the bottom of a hole drilled radially into a spherical planet that has a uniformly distributed mass M. According to the shell theorem, the blue sphere experiences a gravitational force due to the layers colored yellow because it lies outside of those shells. Because the blue sphere is inside the layers colored green, those shells of mass exert no gravitational force on the blue sphere.

Let's call the total mass of the yellow-colored layers in Figure 10-8 M', where clearly $M' < M$. The gravitational force of the planet on the blue sphere is then

$$F = \frac{GM'm}{r^2} \tag{10-6}$$

where r is the radius of the largest yellow layer, not R, the radius of the planet.

We can determine M' as a function of r. Because the mass is uniformly distributed through the volume, M' is a fraction of the total mass M. That fraction is the ratio of the volume out to radius r and the total volume of the sphere:

$$M' = \frac{\frac{4}{3}\pi r^3}{\frac{4}{3}\pi R^3}M = \frac{r^3}{R^3}M$$

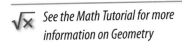 *See the Math Tutorial for more information on Geometry*

So Equation 10-6 becomes

$$F = \frac{G\left(\frac{r^3}{R^3}M\right)m}{r^2} = \frac{GMm}{R^3}r \tag{10-7}$$

 Watch Out

The net gravitational force on an object at the center of a solid sphere is zero.

The variable r in Equation 10-6 is a measure of the distance between the center of a solid sphere and an object located somewhere inside it. On first glance, it appears that when the object is placed at the center of the sphere (so that $r = 0$), the gravitational force on it due to the sphere becomes infinitely large. Note, however, that the mass closer to the center than the object or M' is also zero when the object is at the center of the sphere. The effect of this is most easily seen in Equation 10-7, in which M' has been expressed in terms of r; when r is zero, the net force is also zero.

Example 10-3 Acceleration due to Gravity at the Bottom of a Deep Hole

Determine the acceleration g' due to Earth's gravity at the bottom of a hole drilled radially to a depth $D = 12$ km into Earth. (This is about the depth of the deepest hole ever drilled.) Assume Earth's mass is uniformly distributed, and express your answer as a fraction of g, the acceleration due to gravity at the surface.

SET UP

Earth's gravitational force on an object placed at the bottom of the hole is less than the force it experiences on the surface because only the fraction of Earth's mass that is deeper than the bottom of the hole exerts a force on the object. The smaller force results in a smaller acceleration due to gravity. As in Example 10-2, the gravitational force and the acceleration due to gravity are related; according to Newton's second law, for an object of mass m, the force (given by Equation 10-7) equals m multiplied by the acceleration g' according:

$$\frac{GM_{\mathrm{E}}m}{R_{\mathrm{E}}^3}r = mg'$$

Here M_{E} and R_{E} are Earth's mass and radius, respectively, and r is the distance from Earth's center to the bottom of the hole, not the depth of the hole.

SOLVE

Canceling m from both sides of the equation and writing the new expression in terms of magnitudes gives

$$\frac{GM_{\mathrm{E}}}{R_{\mathrm{E}}^3}r = g'$$

We can express this equation as a fraction of g, which we found to be

$$g = \frac{GM_{\mathrm{E}}}{R_{\mathrm{E}}^2}$$

on Earth's surface (Equation 10-4), so

$$g' = \frac{GM_{\mathrm{E}}}{R_{\mathrm{E}}^3}r = \frac{r}{R_{\mathrm{E}}}\frac{GM_{\mathrm{E}}}{R_{\mathrm{E}}^2} = \frac{r}{R_{\mathrm{E}}}g$$

Finally, the distance from the bottom of the hole to the center of Earth is given by

$$r = R_{\mathrm{E}} - D$$

so, by substitution we find an expression for g'

$$g' = \frac{R_{\mathrm{E}} - D}{R_{\mathrm{E}}}g = \left(1 - \frac{D}{R_{\mathrm{E}}}\right)g$$

Using the numbers given in the problem statement, the acceleration due to gravity at the surface is

$$g' = \left(1 - \frac{12 \times 10^3\,\mathrm{m}}{6.38 \times 10^6\,\mathrm{m}}\right)g = 0.98\,g$$

REFLECT

The value of g' is close to g, which should not be surprising because the depth of the hole, 12×10^3 m, is only a small fraction of Earth's radius. However, assuming no other forces act on an object at the bottom of the hole (a bad assumption!), the difference between g' and g is certainly large enough to be measured.

Practice Problem 10-3 If the acceleration due to gravity at the bottom of a hole is 90% of that on the surface, how deep is the hole?

> ## ✳ What's Important 10-2
> The shell theorem summarizes the gravitational force between a uniform, hollow sphere (a spherical shell) and a second object. For an object outside a hollow sphere of uniform mass distribution, the shell exerts a gravitational force on the object as if the mass of the shell were concentrated at its center. For an object inside a hollow sphere of uniform mass distribution, the shell exerts no net gravitational force on it. An object inside a uniform, hollow sphere nevertheless experiences gravitational forces exerted by objects other than the sphere.

10-3 Gravitational Potential Energy

Energy is conserved. In Chapter 6, we uncovered this powerful law of physics: The sum of the kinetic energy, potential energy, and energy dissipated by nonconservative forces in a system remains constant. The form of potential energy depends on the force or forces at work. For example, we use $U_{gravity} = mgy$ (Equation 6-26) as shorthand for the potential energy acquired by an object of mass m raised a vertical distance y near Earth's surface. Let's find the form of potential energy associated with the force described by Newton's universal law of gravitation.

We define the change in potential energy ΔU as the negative of the work done by a force over a given distance. For an object that moves from x_i to x_f as a result of a force F,

$$\Delta U = -\int_{x_i}^{x_f} \vec{F} \cdot d\vec{x} \qquad (6\text{-}23)$$

 See the Math Tutorial for more information on Integrals

For example, at Earth's surface the magnitude of the force due to gravity on an object is equal to the object's mass multiplied by g. The force vector points vertically down, and by convention, displacement y is increasingly positive in the upward direction. The angle between $\vec{F}$ and $d\vec{y}$ is therefore 180°, so

$$\vec{F} \cdot d\vec{y} = mg\,dy\,\cos 180° = -mg\,dy$$

The change in potential energy is then

$$\Delta U = -\int_{y_i}^{y_f} -mg\,dy = mgy_f - mgy_i$$

Remember that a specific value of potential energy is only meaningful when compared to the value of potential energy at some other position. Using the conventional choice of reference point, the potential energy equals 0 at the height equal to 0 ($U = 0$ at $y_i = 0$). Also using the variable y to measure height (so $y_f = y$), we get the familiar form of gravitational potential energy $U = mgy$. When free to do so, objects move toward positions of lower potential energy, so an object lifted from the surface to height y will fall back down.

But be careful! This form of gravitational potential energy only applies on or near Earth's surface, where the acceleration due to gravity is (relatively) constant over a range of heights close to the surface and where the gravitational force on an object is therefore $F = mg$. To determine the gravitational potential energy in any other circumstance, we need to repeat the integral in Equation 6-23 and use force defined by Newton's universal law of gravitation. The force in Newton's universal law of gravitation (for example, in Equation 10-2) acts in the radial direction only, so we'll write Equation 6-23 in terms of r:

$$\Delta U = -\int_{r_i}^{r_f} \vec{F} \cdot d\vec{r}$$

We again start by considering the dot product. By convention, radial vectors point out (or away) from the center of the object where $r = 0$. The gravitational force points in toward $r = 0$, so the angle between $\vec{F}$ and $d\vec{r}$ is 180°, thus,

$$\vec{F} \cdot d\vec{r} = \frac{Gm1m2}{r^2}dr\,(\cos 180°) = -\frac{Gm1m2}{r^2}dr$$

and

$$\Delta U = -\int_{r_i}^{r_f} -\frac{Gm_1m_2}{r^2}dr = Gm_1m_2 \int_{r_i}^{r_f} \frac{dr}{r^2}$$

Here the variable r is the distance between the two objects, and in particular, the center-to-center separation when one or both objects is not pointlike, so,

$$\Delta U = Gm_1m_2\left(-\frac{1}{r}\Big|_{r_i}^{r_f}\right) = Gm_1m_2\left(\frac{1}{r_i} - \frac{1}{r_f}\right)$$

Again, it is convenient to set the reference point so that U equals zero at the starting position. By convention, we choose the initial separation r_i equal to infinity, so that $1/r_i$ equals 0. For an object of mass m_2 and a center-to-center distance r from an object of mass m_1, the potential energy equals

$$U = -\frac{Gm_1m_2}{r} \tag{10-8}$$

This represents the work done to bring the two objects initially separated by an infinite distance to a final separation of r, as shown in **Figure 10-9**. Because this form of gravitational potential energy gives zero at $r = \infty$ and is increasingly negative as r gets smaller, an object lifted to a radius r from the center of a massive object will fall back toward the center.

Don't be concerned that our convention results in negative values for gravitational potential energy. Remember that only the *differences* between potential energy values are physically meaningful. Specific values are determined by our choice of reference point, which is made as a matter of convenience. Refer back to the discussion on the choice of reference point in Section 6-5 to reacquaint yourself with this aspect of potential energy.

The gravitational potential is the work done to bring the objects from being separated by an infinite distance...

$r = \infty$

...to a separation of distance r.

r

Figure 10-9 The gravitational potential energy of two objects separated by a distance r is equivalent to the work done to bring the two objects to that separation, having initially been separated by an infinite distance.

Example 10-4 How High (Yet Again)?

In Example 6-10, we considered how high a ball would go when thrown straight up near the surface of Earth if we neglect air resistance. In that problem, the ball is given an initial speed $v_0 = 15.0$ m/s, and rises to a peak height of 11.5 m. How high above Earth's surface does an object rise when given a much larger initial vertical speed of 8.00×10^3 m/s?

SET UP

We will follow the approach used to solve Example 6-10 starting with a sketch (Figure 10-10).

Energy is conserved. The only force experienced by the projectile is due to gravity, so according to the conservation of energy,

$$K_i + U_{\text{gravity, i}} = K_f + U_{\text{gravity, f}}$$

In Example 6-10, the ball was given an initial speed small enough so that it never got far from Earth's surface. We were therefore justified in using $U_{\text{gravity}} = mgy$ for gravitational potential energy in that case. It is unlikely that the acceleration due to gravity will remain constant over the range of heights covered by the projectile in this problem, however. For that reason we need to use Equation 10-8:

$$\frac{1}{2}mv_i^2 - \frac{GM_Em}{r_i} = \frac{1}{2}mv_f^2 - \frac{GM_Em}{r_f} \tag{10-9}$$

where M_E is the mass of Earth and m is the mass of the projectile. We list the values of the variables underneath the terms, based on the problem statement and our sketch:

$$\frac{1}{2}mv_i^2 - \frac{GM_Em}{r_i} = \frac{1}{2}mv_f^2 - \frac{GM_Em}{r_f}$$

$$v_i = v_0 \quad r_i = R_E \quad v_f = 0 \quad r_f = H + R_E$$

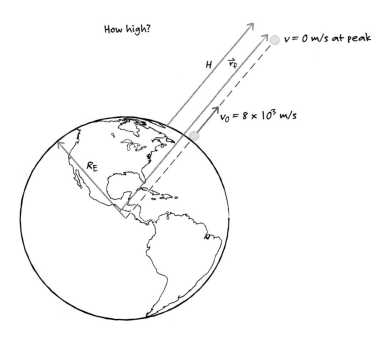

Figure 10-10 An object is launched straight up from Earth's surface. How high does it go?

Notice that, as shown in Figure 10-10, r_f, the final distance of the object from the center of Earth, is the sum of Earth's radius and H, the maximum height the object reaches above the surface.

SOLVE

We simplify Equation 10-9:

$$\frac{1}{2}v_0^2 - \frac{GM_E}{R_E} = -\frac{GM_E}{H + R_E}$$

and rearrange the expression to solve for H:

$$H = -\frac{GM_E}{\frac{1}{2}v_0^2 - \frac{GM_E}{R_E}} - R_E$$

We could algebraically rearrange the equation further to simplify it, but in this case it's easier to substitute the given values into this equation to get a numeric result:

$$H = -\frac{(6.67 \times 10^{-11}\,\text{N}\cdot\text{m}^2/\text{kg}^2)(5.98 \times 10^{24}\,\text{kg})}{\frac{1}{2}(8.00 \times 10^3\,\text{m/s})^2 - \frac{(6.67 \times 10^{-11}\,\text{N}\cdot\text{m}^2/\text{kg}^2)(5.98 \times 10^{24}\,\text{kg})}{6.38 \times 10^6\,\text{m}}}$$

$$- 6.38 \times 10^6\,\text{m}$$

or

$$H = 6.69 \times 10^6\,\text{m}$$

REFLECT

The final height is a bit more than one Earth radius, so Figure 10-10 is a reasonable representation. More important, the value of the acceleration due to Earth's gravity at that height is certainly (see the next example) much smaller than on the surface. For that reason, we had to use $U = -GM_E m/r$ and not $U = mgy$ for gravitational potential energy.

Practice Problem 10-4 An object launched vertically from Earth reaches a maximum height equal to 2 Earth radii from the surface. Find the launch speed.

Example 10-5 g above Earth's Surface

In Example 10-2, we showed that the value of g, the acceleration due to gravity on the surface of Earth, can be obtained from Newton's universal law of gravitation. The same approach should yield the correct value of acceleration due to gravity at any distance from Earth's center. Find the acceleration due to gravity at a height 3.6×10^7 m above the surface, the typical altitude of communications satellites and about six Earth radii above the surface.

SET UP

As in Example 10-2, we begin by setting the gravitational force on an object equal to its mass m multiplied by acceleration:

$$\frac{GM_E m}{r^2}(-\hat{r}) = mg'(-\hat{r})$$

Compare this equation with the second equation in the Setup section of Example 10-2. Here, we use r rather than R_E in the denominator of the gravitational force because the object is not on Earth's surface. We also use g' for the acceleration because we don't expect the value to be $g = 9.81 \text{ m/s}^2$.

SOLVE

Canceling m from both sides of the equation and writing the new expression in terms of magnitudes gives

$$\frac{GM_E}{r^2} = g'$$

Using the values given in the problem statement and $r = R_E + H$:

$$g' = \frac{GM_E}{r^2} = \frac{(6.67 \times 10^{-11}\,\text{N}\cdot\text{m}^2/\text{kg}^2)\,(5.98 \times 10^{24}\,\text{kg})}{(6.38 \times 10^6\,\text{m} + 3.6 \times 10^7\,\text{m})^2} = 0.22 \text{ m/s}^2$$

REFLECT

As we expected, the value of acceleration due to Earth's gravity is considerably less at a height of about six Earth radii compared to on Earth's surface.

Another useful way to see this is to express g' in terms of g. We have written

$$g = \frac{GM_E}{R_E^2} \qquad (10\text{-}4)$$

so

$$g' = \frac{GM_E}{r^2} = \frac{GM_E}{R_E^2}\frac{R_E^2}{r^2} = g\frac{R_E^2}{r^2}$$

Here, r is $H + R_E$. The value of H, 3.6×10^7 m, is about $6R_E$, so

$$g' \approx g\frac{R_E^2}{(R_E + 6R_E)^2} = \frac{gR_E^2}{(7R_E)^2} = \frac{g}{49}$$

The fraction $1/49$ is approximately $1/50$ or 0.02. So at a height approximately equal to six Earth radii above the surface, the acceleration due to gravity is 2% of the value on the surface.

Practice Problem 10-5 Find the acceleration due to gravity at a height 2.0×10^7 m above Earth's surface, the typical altitude of global positioning system (GPS) satellites.

Escape Speed

A ball thrown vertically reaches a peak height and eventually returns to the surface. As we saw in Example 10-4, the larger the initial speed, the farther away from the surface the projectile goes before returning to the surface. Is it possible to launch a projectile with an initial speed large enough so that it *never* returns? You can imagine that even launched with an enormously large speed and with no forces other than Earth's gravitational attraction, the object would still reach some final height and then fall back to Earth. We can only be certain that it would not fall back if its final height is an infinite distance from Earth, so that an infinite amount of time is required for the object to return. An infinite amount of time is equivalent to "never!"

Conservation of energy enables us to determine the minimum speed required. We have written a statement of energy conservation when the only force is due to gravity as described by Newton's universal law of gravitation:

$$\frac{1}{2}mv_i^2 - \frac{GM_Em}{r_i} = \frac{1}{2}mv_f^2 - \frac{GM_Em}{r_f} \qquad (10\text{-}9)$$

The projectile starts on the surface, so $r_i = R_E$. When the final distance is at an infinite distance, $r_f = \infty$. In addition, the speed is zero at infinite distance, or $v_f = 0$ so,

$$\frac{1}{2}mv_i^2 - \frac{GM_E m}{R_E} = \frac{1}{2}m(0)^2 - \frac{GM_E m}{\infty} = 0$$

or

$$v_i = \sqrt{\frac{2GM_E}{R_E}}$$

The expression can be generalized for any object of mass M and radius R:

$$v_{escape} = \sqrt{\frac{2GM}{R}} \tag{10-10}$$

Escape speed is the minimum vertical launch speed required so that a projectile never returns to the object from which it was launched.

Got the Concept 10-1
Space Shuttle

About 8 min after lifting off the launchpad, the space shuttle would be traveling at its orbital speed (its maximum speed) of about 7.8 km/s. This speed is enormous by human standards. Is it large enough to send the shuttle hurtling into space—never to return?

What's Important 10-3

Gravitational potential energy represents the work required to bring two objects together that were initially separated by an infinite distance. Escape speed is the vertical launch speed required so that a projectile never returns to the object from which it was launched.

10-4 Kepler's Laws

Some 60 years before Newton published his universal law of gravitation, German mathematician and astronomer Johannes Kepler stated three laws that describe the orbit of one object around another, for example, the Moon around Earth. Kepler developed these laws by studying observations of the positions of the planets made by the Danish astronomer Tycho Brahe.

Kepler's laws describe the shape of celestial orbits, the speed at which one object orbits another, and the time it takes an object to orbit another. We will write them in terms of a planet orbiting the Sun, but they apply equally well to any object in orbit around another.

The law of orbits	All planets trace out an elliptical path, where the Sun is at one focus.
The law of areas	A line joining a planet and the Sun sweeps out equal areas in equal intervals of time, regardless of the position of the planet in the orbit.

The law of periods   The square of the period of a planet's orbit is proportional to the cube of the **semimajor axis**. The semimajor axis is half the length of the long axis of the elliptical orbit.

The Law of Orbits

Kepler's law of orbits states that all celestial orbits trace out elliptical paths. An ellipse describes all points such that the sum of the distances to two fixed points (each called a *focus*) is constant. In **Figure 10-11a**, A and B show two points on the ellipse; the sum of the distances from each point to the two foci is the same:

$$r_{A,1} + r_{A,2} = r_{B,1} + r_{B,2} = \text{constant}$$

An ellipse can be characterized by the semimajor axis a, the **semiminor axis** b, and the eccentricity e. As shown in **Figure 10-11b**, a is half the length of the long axis of the ellipse and b is half the length of the short axis. The product of a and e gives the distance from the center of the ellipse to one of the foci. The eccentricity, a number between 0 and 1, describes how flat the ellipse is. The closer e is to 1, the more an ellipse is stretched out. At the other extreme, when $e = 0$, both foci are located at the center of the ellipse, the semimajor and semiminor axes are the same length, and the ellipse is circular. A circle is an ellipse in which the semimajor and semiminor axes are the same.

The orbits of the planets in our solar system are all nearly circular; the most elliptical are Mercury's orbit, which has $e = 0.21$, and Mars' orbit, which has $e = 0.09$. Even Mercury's orbit is still rather circular; at $e = 0.21$, the length of Mercury's semiminor axis is about 0.98 times the semimajor axis, so they are very close to being equal. Comets, however, have highly elliptical orbits. Halley's comet, which makes a full revolution around the Sun once every 76 years, has an eccentricity of 0.967. The orbital paths of the planets, Pluto, and Halley's comet are shown in **Figure 10-12**.

The description of orbits as ellipses assumes that the object being orbited (for example, the Sun in the case of the planets and comets of our solar system) is fixed in place at one focus of the ellipse. This, of course, can't be exactly correct—after all, the gravitational force is symmetric, so whatever force the Sun exerts on a planet, the planet also exerts on the Sun. Of course, the Sun is rather more massive than all

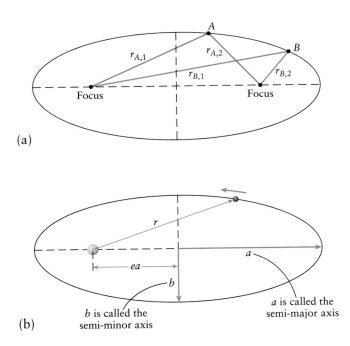

(a)

(b)

b is called the semi-minor axis

a is called the semi-major axis

Figure 10-11 (a) An ellipse describes all points such that the sum of the distances to two fixed points is constant. (b) The semimajor axis a is half the length of the long axis of the ellipse. The semiminor axis b is half the length of the short axis. The eccentricity e, a number between 0 and 1, describes how flat the ellipse is—the closer e is to 1, the more an ellipse is stretched out.

Figure 10-12 The orbits of the planets, Pluto, and Halley's comet are all elliptical. The planets and Pluto follow nearly circular paths.

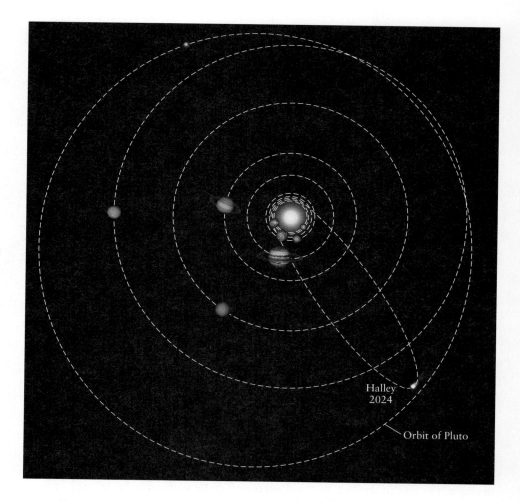

of the planets combined, so its acceleration as a result of the forces of the planets is relatively small. Careful analysis, however, reveals that the Sun does wobble a small amount due to the pull of the planets, especially Jupiter, the most massive planet. We should also note that because the solar system is moving in and with our Milky Way galaxy, the actual paths of the planets is quite complicated. Their paths are only elliptical when we treat the Sun as fixed in one position, or at least nearly so.

The Law of Areas

Figure 10-13a shows a planet at two positions in its orbit around the Sun, separated by an infinitesimally small time interval Δt. (The shape of the orbit has been exaggerated.) A line connecting the Sun and the planet sweeps out the shaded area over that time interval. For any specific value of Δt, the value of the shaded area is the same regardless of the position of the planet according to **Kepler's law of areas**. To show this statement is correct, we need a relationship between the area ΔA of the shaded region and Δt.

What is the shape of the shaded area? Because Δt is infinitesimally small, we can treat the small piece along the elliptical path as straight, which makes the shaded region a triangle. This region is shown in **Figure 10-13b**. Its area is then $\frac{1}{2}$(base × height). The height of the triangle is r. The base, as seen in the figure, is $v\,\Delta t \sin\varphi$, where φ is the angle between the radial and motion directions. The shaded area in the figure is therefore

$$\Delta A = \frac{1}{2}v\,\Delta t \sin\varphi \times r = \frac{rv\sin\varphi\,\Delta t}{2}$$

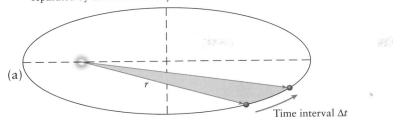

A planet is shown at two positions in its orbit around the Sun, separated by an infinitesimally small time interval Δt.

(a)

r

Time interval Δt

A line connecting the Sun and this planet sweeps out the shaded area over this time interval.

(b)

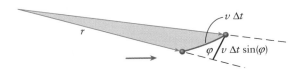

When Δt is infinitesimally small we can treat the small piece along the elliptical path as straight...

v Δt

r

φ v Δt sin(φ)

Figure 10-13 (a) The line connecting a planet to the Sun sweeps out the area shaded in green over a time interval Δt. (b) Over an infinitesimally short time period the area swept out is approximately triangular.

The numerator in the equation contains the expression for angular momentum; from Equation 8-31, $L = rp \sin \varphi$, or

$$L = r\,mv\,\sin\varphi$$

so

$$\Delta A = \frac{L\,\Delta t}{2m}$$

As we saw in Chapter 8 section 7, angular momentum is conserved in systems that experience no external forces or torques. That is certainly the case for the planet-Sun system we are considering, so L is constant. As a result, the area ΔA swept out is the same for any part of the motion that occurs over a specific value of the time interval Δt.

To explore the implications of Kepler's law of areas, consider the orbit of Halley's comet around the Sun. Two time intervals are shown in **Figure 10-14**, one when the comet is far from the Sun and one when it is close. The time intervals are the same, so according to Kepler's law of areas, the area shaded blue and the area shaded pink are also equal. Note that when close to the Sun, then, Halley's comet travels a much longer distance in the fixed time interval compared to when it is far

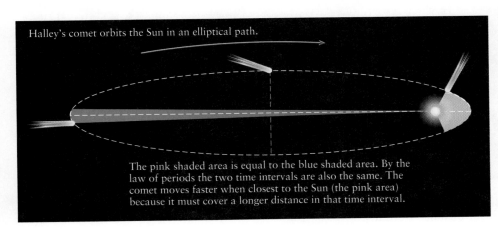

Halley's comet orbits the Sun in an elliptical path.

The pink shaded area is equal to the blue shaded area. By the law of periods the two time intervals are also the same. The comet moves faster when closest to the Sun (the pink area) because it must cover a longer distance in that time interval.

Figure 10-14 Two time intervals during the orbit of Halley's comet around the Sun are shown, one when the comet is closest to the Sun (shaded in pink) and one when the comet is farthest from the Sun (shaded in blue). According to Kepler's law of areas the area swept out in each case is the same, which requires the comet to move faster when closest to the Sun.

away. The speed of the comet is therefore larger when close to the Sun. In general, every object that orbits another speeds up as it approaches the object and slows down as it moves farther away. This result isn't surprising—the object is accelerating due to the gravitational pull of the orbited body as it gets closer and decelerating for the same reason as it recedes.

 Got the Concept 10-2
Length of a Day

The rate that Earth rotates around its axis is constant. (To be precise, Earth's rotation rate changes very slightly over long periods of time; see Example 10-8.) However, we define the length of a day as the time between when the Sun reaches its highest point in the sky on one day ("noon") and the time that the Sun reaches its highest point in the sky on the next day. Is the length of each day the same? If not, how does it change? (Note that it is common to refer to the length of a day as the length of daylight. That is not what we're thinking about in this problem!)

 Watch Out
The Earth is not closest to the Sun in summer (unless you live in the Southern Hemisphere).

Seasons on Earth are governed by the tilt of the rotational axis of our planet relative to the plane in which it moves around the Sun, not our distance from it. As you can see in **Figure 10-15**, when Earth is farthest from the Sun, which occurs during June, the Northern Hemisphere is tilted toward it. At that time the Sun's light energy falls more directly on the Northern Hemisphere. More energy from the Sun means warmer weather, so this is summertime in the Northern Hemisphere. Again, this occurs when Earth is farthest from the Sun. In a similar way, when Earth is farthest from the Sun the Southern Hemisphere is tilted away from the Sun, so sunlight falls less directly on the Southern Hemisphere. That makes it colder in the Southern Hemisphere, so June marks the height of winter there.

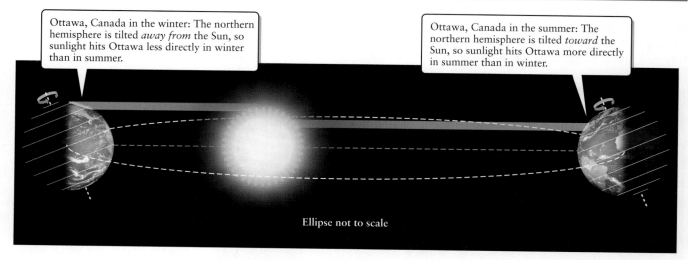

Figure 10-15 Seasons on Earth are governed by the tilt of the rotational axis of our planet relative to the plane in which it moves around the Sun, not our distance from it.

The Law of Periods

Kepler's law of periods expresses a relationship between the time it takes for an object to orbit once around another object and the length of the semimajor axis of the elliptical path. We have already seen how to connect the period of an object in a circular orbit with the constants associated with gravity. In Section 10-1, we set the force on the Moon due to Earth's gravity equal to the Moon's mass multiplied by the acceleration of uniform circular motion and obtained

$$\frac{GM_E}{D_{EM}^2} = \frac{v_M^2}{D_{EM}} \qquad (10\text{-}5)$$

We set the velocity of the Moon equal to the distance traveled in one orbit divided by the time of one orbit:

$$v_M = \frac{2\pi D_{EM}}{T_M}$$

Now we combine the two expressions again, this time using M as the mass of the orbited object, r as the radius of the orbit, and T as the period:

$$\frac{GM}{r^2} = \frac{\left(\frac{2\pi r}{T}\right)^2}{r}$$

or

$$T^2 = \frac{4\pi^2}{GM}r^3$$

This relationship also works for elliptical orbits when r is replaced by the semimajor axis a:

$$T^2 = \frac{4\pi^2}{GM}a^3 \qquad (10\text{-}11)$$

Kepler, who did not have the benefit of Newton's universal law of gravitation, stated the law of periods slightly differently than Equation 10-11. Kepler's formulation showed that the ratio of the square of the period to the cube of the mean distance to the Sun is the same for any two planets:

$$\frac{T_1^2}{a_1^3} = \frac{T_2^2}{a_2^3} \qquad (10\text{-}12)$$

? **Got the Concept 10-3**
Jupiter's Orbit

Jupiter is about 5.2 times as far from the Sun as Earth is. About how long, in years, does it take Jupiter to orbit the Sun once?

Example 10-6 How Long Is a Year?

The mean distance of Earth to the Sun is 1.496×10^{11} m. Determine the time it takes for Earth to revolve around the Sun once. To four significant figures, the mass of the Sun is M_S equals 1.989×10^{30} kg and the universal constant of gravitation G equals $6.674 \times 10^{-11}\,\text{N}\cdot\text{m}^2/\text{kg}^2$.

SET UP

The time it takes for Earth to revolve once around the Sun is the Earth's period, so it can be computed using Kepler's law of periods. Because we want to make the calculation for a single object we use the form of the law given in Equation 10-11:

$$T_E^2 = \frac{4\pi^2}{GM_S}a_E^3$$

SOLVE

$$T_E = \sqrt{\frac{4\pi^2}{GM_S}a_E^3}$$

Using the values given in the problem statement, Earth's period is

$$T_E = \sqrt{\frac{4\pi^2}{(6.674 \times 10^{-11}\,\text{N}\cdot\text{m}^2/\text{kg}^2)(1.989 \times 10^{30}\,\text{kg})}(1.496 \times 10^{11}\,\text{m})^3}$$

or

$$T_E = 3.155 \times 10^7\,\text{s}$$

REFLECT

There are 86,400 s in a day, so 3.155×10^7 s equals 365.2 days. Sounds about right.

Practice Problem 10-6 Determine the time it takes the Moon to orbit Earth once, given that the mean distance between Earth and the Moon is about 3.8×10^8 m and the mass of Earth is 5.98×10^{24} kg. Express your answer in days.

Example 10-7 Satellites

Many communications satellites orbit Earth in a circular path above the equator. By making exactly one revolution per day, such geostationary satellites remain above the same point on the surface. Other satellites, such as the International Space Station, complete one orbit about every 90 min. Determine the altitude of each type of satellite.

SET UP

The time it takes for a satellite to orbit once is governed by Kepler's law of periods. The orbited object in this case is Earth, so the mass term in Equation 10-11 is M_E, the mass of Earth:

$$T^2 = \frac{4\pi^2}{GM_E}a^3$$

Rearranging for a, the mean distance of the satellite to the center of Earth, gives

$$a = \left(\frac{GM_E}{4\pi^2}T^2\right)^{1/3}$$

The altitude H is given by

$$H = a - R_E$$

SOLVE

$$H = \left(\frac{GM_E}{4\pi^2}T^2\right)^{1/3} - R_E$$

For the geostationary satellite, for which T is 1 day or 86,400 s,

$$H = \left(\frac{(6.67 \times 10^{-11}\,\text{N}\cdot\text{m}^2/\text{kg}^2)\,(5.98 \times 10^{24}\,\text{kg})}{4\pi^2}(86{,}400\,\text{s})^2 \right)^{1/3}$$

$$- 6.38 \times 10^6\,\text{m}$$

$$= 3.59 \times 10^7\,\text{m}$$

For the International Space Station, for which T is 90 min or 5400 s,

$$H = \left(\frac{(6.67 \times 10^{-11}\,\text{N}\cdot\text{m}^2/\text{kg}^2)\,(5.98 \times 10^{24}\,\text{kg})}{4\pi^2}(5400\,\text{s})^2 \right)^{1/3}$$

$$- 6.38 \times 10^6\,\text{m}$$

$$= 0.274 \times 10^6\,\text{m}$$

REFLECT

Geostationary satellites, such as those used for satellite television and other communications, orbit far above Earth's surface. The altitude we found, almost 3.6×10^4 km, is about 6 Earth radii. Satellites such as the International Space Station, on the other hand, follow low-Earth orbits, typically only a few hundred kilometers above the surface—about 100 times closer to Earth than geostationary satellites. The low-altitude orbit enables the satellites, and the people on board in the case of the International Space Station, to pass over a wide swath of Earth's surface in a relatively short time.

Practice Problem 10-7 Jupiter's moon Ganymede, the largest moon in our solar system, orbits the planet in a nearly circular orbit at a mean radius of 1.07×10^9 m. The mass of Jupiter is 1.90×10^{27} kg. How long does it take for Ganymede to orbit Jupiter? Express your answer in Earth days.

Example 10-8 Goodbye Moon

The rotation of Earth is slowing down by about 2.3 ms every hundred years. Use that information to determine how the distance between Earth and the Moon is changing over time, by treating the total angular momentum of the Earth–Moon system as constant.

SET UP

The total angular momentum of a system that experiences no net external forces and no external torques remains constant. External forces and torques on Earth and the Moon due to other celestial objects are relatively small and tend to cancel, so the total angular momentum of the system can be treated as constant.

Three motions contribute to the angular momentum of the Earth–Moon system: the rotation of Earth around its axis, the revolution of the Moon around Earth, and the rotation of the Moon around its axis. A change in any one of the motions must be accompanied by a corresponding change in one or both of the others, so that the total remains constant. In addition, the rate of the Moon's rotation, once around its axis for every one revolution around Earth, is the most stable possible configuration. For that reason any change to its rate of rotation would make the orbit less stable, so this rate does not change. Thus, as the rate of Earth's rotation slow, the decrease in its angular momentum, is accompanied by an increase in the angular momentum of the Moon as it revolves around Earth,

$$\Delta L_{\text{Moon}} = -\Delta L_{\text{Earth}}$$

where ΔL_{Moon} is the angular momentum of the Moon as it revolves around Earth and ΔL_{Earth} is the angular momentum of Earth due to its rotation.

SOLVE

From Equation 8-29, angular momentum can be expressed in terms of moment of inertia I and angular frequency ω, so

$$\Delta \left(I_{\text{Moon}} \omega_{\text{Moon}} \right) = -\Delta \left(I_{\text{Earth}} \omega_{\text{Earth}} \right)$$

The Moon revolves around Earth with angular frequency equal to 2π divided by T_{Moon}, the period of the Moon's orbit around Earth:

$$\omega_{\text{Moon}} = \frac{2\pi}{T_{\text{Moon}}}$$

The Earth rotates at angular frequency equal to 2π divided by T_{Earth}, the time it takes Earth to rotate once:

$$\omega_{\text{Earth}} = \frac{2\pi}{T_{\text{Earth}}}$$

From Table 8-1, I_{Moon} is equal to the product of the mass of the Moon (M_{Moon}) and the Earth–Moon distance (D_{EM}) squared:

$$I_{\text{Moon}} = M_{\text{Moon}} D_{\text{EM}}^2$$

From the same table, treating Earth as a sphere rotating around a central axis,

$$I_{\text{Earth}} = \frac{2}{5} M_{\text{Earth}} R_{\text{Earth}}^2$$

where M_{Earth} and R_{Earth} are the mass and radius of Earth, respectively.

The change in angular momentum relationship therefore becomes

$$\Delta \left(M_{\text{Moon}} D_{\text{EM}}^2 \frac{2\pi}{T_{\text{Moon}}} \right) = -\Delta \left(\frac{2}{5} M_{\text{Earth}} R_{\text{Earth}}^2 \frac{2\pi}{T_{\text{Earth}}} \right)$$

We are interested in the change in D_{EM} as a function of the change in T_{Earth}, so we'll use Kepler's law of periods (Equation 10-11) to express T_{Moon} in terms of D_{EM}:

$$T_{\text{Moon}}^2 = \frac{4\pi^2}{GM_{\text{Earth}}} D_{\text{EM}}^3$$

So

$$\Delta \left(M_{\text{Moon}} D_{\text{EM}}^2 \frac{2\pi}{\sqrt{\dfrac{4\pi^2}{GM_{\text{Earth}}} D_{\text{EM}}^3}} \right) = -\Delta \left(\frac{2}{5} M_{\text{Earth}} R_{\text{Earth}}^2 \frac{2\pi}{T_{\text{Earth}}} \right)$$

or

$$\Delta \left(M_{\text{Moon}} \sqrt{GM_{\text{Earth}}} \, D_{\text{EM}}^{1/2} \right) = -\Delta \left(\frac{4\pi}{5} \frac{M_{\text{Earth}} R_{\text{Earth}}^2}{T_{\text{Earth}}} \right)$$

√x̄ *See the Math Tutorial for more information on Differential Calculus*

Yes, this relationship has become a bit complicated, but look carefully! Only D_{EM} and T_{Earth} change, so this expression will allow us to determine the change in the Earth–Moon distance (ΔD_{EM}) in terms of the change in Earth's rotation rate (ΔT_{Earth}). If the idea of finding the change in one variable with respect to the change in another sounds familiar, you're on to something: This has the feel of a derivative. Indeed, because the changes are small,

$$\frac{\Delta D_{\text{EM}}}{\Delta T_{\text{Earth}}} = \frac{dD_{\text{EM}}}{dT_{\text{Earth}}}$$

The derivative is straightforward, but requires care. It might be the most daunting derivative we need to take in this book, so for that reason, we challenge you to do it! The answer is right in front of you, so give it a try; you should get

$$\Delta D_{EM} = -\frac{8\pi}{5}\frac{M_{Earth}R_{Earth}^2}{M_{Moon}\sqrt{GM_{Earth}}}\frac{D_{EM}^{1/2}}{T_{Earth}^2}\Delta T_{Earth}$$

The period of Earth's rotation is about 1 day or 86,400 s. The change in the period is about -2.3×10^{-3} s every hundred years; the change is negative because the period is getting smaller. The mean center-to-center distance from Earth to the Moon is 3.84×10^8 m, and the values of the masses and Earth's radius are found in Appendix B:

$$\Delta D_{EM} = -\frac{8\pi}{5}\frac{(5.98 \times 10^{24}\,\text{kg})(6.38 \times 10^6\,\text{m})^2}{(7.35 \times 10^{22}\,\text{kg})\sqrt{(6.67 \times 10^{-11}\,\text{N}\cdot\text{m}^2/\text{kg}^2)(5.98 \times 10^{24}\,\text{kg})}}$$

$$\frac{(3.85 \times 10^8\,\text{m})^{1/2}}{(86,400\,\text{s})^2}(-2.3 \times 10^{-3}\,\text{s})$$

or

$$\Delta D_{EM} = 5.0\,\text{m}$$

According to our approximation, the distance between Earth and the Moon increases by approximately 5 m every hundred years.

REFLECT

Astronomical observations indicate that the Moon gets about 3.84 m farther from Earth every 100 years; our value of 5.0 m is a good approximation. It may surprise you that the Moon is getting farther away, because our experience with manufactured satellites is that their orbits eventually *decay*. But in this case, because Earth and the Moon are connected to each other by their mutual gravitational attraction, as the Earth gives up angular momentum (and energy) by rotating more slowly, the Moon gains that change in angular momentum and energy. Higher angular momentum for the Moon means an orbit with a larger radius.

✱ What's Important 10-4

Kepler's laws describe the shape of celestial orbits, the speed at which one object orbits another, and the time it takes an object to orbit another. We have written them in terms of a planet orbiting the Sun, but they apply equally well to any object in orbit around another. According to the law of orbits all planets in our solar system trace out an elliptical path with the Sun at one focus. The law of areas states that a line joining a planet and the Sun sweeps out equal areas in equal intervals of time, regardless of the position of the planet in its orbit. According to the law of periods, the square of the period of a planet's orbit is proportional to the cube of the semimajor axis or half the length of the long axis of its elliptical orbit.

Answers to Practice Problems

10-1	2.6×10^{-12} N
10-2	1.2×10^{-7} m/s^2
10-3	640 km
10-4	9.1×10^3 m/s

10-5	0.57 m/s^2
10-6	27 days
10-7	7.15 days

Answers to Got the Concept Questions

10-1 Let's hope not! We can use the shuttle's top speed in the problem statement (7.8 km/s) and Equation 10-10 to determine if the shuttle exceeds the escape speed for an object launched from Earth:

$$v_{escape} = \sqrt{\frac{2\,(6.67 \times 10^{-11}\,\text{N} \cdot \text{m}^2/\text{kg}^2)\,(5.98 \times 10^{24}\,\text{kg})}{6.38 \times 10^6\,\text{m}}}$$

$$= 11.1 \times 10^3\,\text{m/s}$$

The speed required for the shuttle (or any object) to break free of Earth's gravity, never to return, is about 11 km/s. The speed of the shuttle is large enough to launch it into orbit, but not so large that it does not return to Earth.

10-2 The Sun appears to move through the sky because of Earth's rotation around its axis. Each rotation takes 23 h 56 min 4 s. If you look straight up, observe the direction in space you are facing, and then wait 23 h, 56 min, and 4 s, you'll be pointing in the same direction again. However, if you observe the Sun at noon and wait the same amount of time, the Sun won't quite be at its highest point, because in this time Earth has moved about 1/365th of its way around the Sun. You'll have to wait an additional 3 min and 56 s—for a total of 24 h— for the Sun to be at its highest point. The exact amount of the time difference depends on Earth's speed around the Sun. When its speed is largest (that is, when Earth is closest to the Sun), it takes more time for the Sun to return to its highest point in the sky because Earth travels farther in any specific amount of time. This occurs in late December and early January. Throughout the year the length of the day varies—no two successive days are exactly the same length of time. (The tilt of Earth's rotation axis relative to the plane in which it moves around the Sun plays a role, but here we are interested in the implication of Kepler's law of areas.)

10-3 The form of Kepler's law of periods as expressed in Equation 10-12 allows us to compare the orbits of two different planets around the Sun. Comparing Earth (E) and Jupiter (J),

$$\frac{T_E^2}{a_E^3} = \frac{T_J^2}{a_J^3}$$

or

$$T_J^2 = \frac{T_E^2}{a_E^3} a_J^3$$

so,

$$T_J = T_E \sqrt{\left(\frac{a_J}{a_E}\right)^3}$$

Using 5.2 as the ratio of the distances of the planets from the Sun, the number of years Jupiter takes to orbit the Sun once is

$$T_J = (1\,\text{year})\sqrt{5.2^3} = 12\,\text{years}$$

SUMMARY

Topic	Summary	Equation or Symbol	
escape speed	Escape speed is the vertical launch speed required so that a projectile never returns to the object from which it was launched. Escape speed depends on the mass M and radius R of the object from which projectile is launched as well as the universal constant of gravitation G.	$v_{escape} = \sqrt{\dfrac{2GM}{R}}$	(10-10)
Kepler's law of areas	A line joining a planet and the Sun sweeps out equal areas in equal intervals of time, regardless of the position of the planet in the orbit.		
Kepler's law of orbits	All planets trace out an elliptical path, where the Sun is at one focus.		
Kepler's law of periods	The square of the period T of a planet's orbit is proportional to the cube of the semimajor axis a. The period also depends on the mass of the orbited object M and G, the universal constant of gravitation.	$T^2 = \dfrac{4\pi^2}{GM} a^3$	(10-11)

semimajor axis	The semimajor axis is half the length of the long axis of an ellipse.	a	
semiminor axis	The semiminor axis is half the length of the short axis of an ellipse.	b	
shell theorem	The shell theorem describes the gravitational force between a uniform, hollow sphere (a shell) and a second object. For an object outside a hollow sphere of uniform mass distribution, the shell exerts a gravitational force on the object as if the mass of the shell were concentrated at its center. For an object inside a hollow sphere of uniform mass distribution the shell exerts no net gravitational force on it.		
universal law of gravitation	The gravitational force that one particle exerts on another is proportional to the product of their masses (m_1 and m_2) and inversely proportional to the square of r, the distance between their centers. The force is attractive, and directed along the line which connects the centers of the two particles.	$\vec{F} = -\dfrac{Gm_1m_2}{r^2}\hat{r}$	(10-1)

QUESTIONS AND PROBLEMS

In a few problems, you are given more data than you actually need; in a few other problems, you are required to supply data from your general knowledge, outside sources, or informed estimate.

Interpret as significant all digits in numerical values that have trailing zeros and no decimal points.

For all problems, use $g = 9.8$ m/s^2 for the free-fall acceleration due to gravity. Neglect friction and air resistance unless instructed to do otherwise.

- • Basic, single-concept problem
- •• Intermediate-level problem, may require synthesis of concepts and multiple steps
- ••• Challenging problem
SSM *Solution is in Student Solutions Manual*

Conceptual Questions

1. •Isaac Newton probably gained no insight by being hit on the head with a falling apple, but he did believe that "the Moon is falling." Explain this statement. SSM

2. •Would the magnitude of the acceleration due to gravity near Earth's surface increase more if Earth's mass were doubled or if Earth's radius were cut in half? Justify your answer.

3. •According to Equation 10-1, what happens to the force between two objects if (a) the mass of one object is doubled and (b) if both masses are halved? SSM

4. •Every object in the universe experiences an attractive gravitational force due to every other object, but why don't you feel a pull from objects close to you?

5. •The gravitational force acts on all objects in proportion to their mass. Why don't heavy objects fall faster than light ones, if you neglect air resistance?

6. •When powering a trip from Earth to the Moon, the thrusters of a ship don't have to get the ship all the way to the Moon. Why do the thrusters only have to get the ship partway there?

7. •The Sun gravitationally attracts the Moon. Does the Moon orbit the Sun?

8. •Imagine a world where the force of gravity was proportional to the inverse *cube* of the distance between two objects ($F_{gravity} \propto r^{-3}$). How would the Kepler's law of periods be changed?

9. •Much attention has been devoted to the exact numerical power in Newton's law of universal gravitation ($F \propto r^{-2}$). Some theorists have investigated whether the dependence might be slightly larger or smaller than 2. What would be the significance (what impact would there be) if the power was not exactly 2?

10. •The shell theorem tells us that a hollow sphere exerts zero gravitational force on an object inside it. Why does a marble placed inside a hollow sphere that rests on the surface of Earth experience a nonzero net force?

11. •Why is the gravitational potential energy of two objects negative?

12. •A frog at the bottom of a well has a negative amount of gravitational potential energy with respect to the ground level where we choose to set potential energy equal to zero. Can the frog escape from the well if it jumps upward with a positive amount of kinetic energy that is larger than its negative potential energy at the bottom of the well? Explain.

13. •Earth moves faster in its orbit around the Sun during the winter in the Northern Hemisphere than it does during the summer in the Northern Hemisphere. Is Earth closer to the Sun during the Northern Hemisphere's winter or during the Northern Hemisphere's summer? Explain your answer. SSM

14. •A satellite is to be raised from one circular orbit to another, with the first closer to Earth's surface. What will happen to its period?

15. •Geostationary satellites remain stationary over one point on Earth. How is this accomplished?

16. •One enterprise that may be better performed on the International Space Station rather than on the surface of Earth is the fabrication of precision ball bearings. Explain why.

17. •Describe some ways that Newton's law of gravity may have affected human evolution. SSM

Multiple-Choice Questions

18. •According to Newton's universal law of gravitation, $\vec{F} = -\dfrac{Gm_1m_2}{r^2}\,\hat{r}$, if both masses are doubled, the force is

    A. four times as much as the original value.
    B. twice as much as the original value.
    C. the same as the original value.
    D. one-half of the original value.
    E. one-fourth of the original value.

19. •According to Newton's universal law of gravitation, $\vec{F} = -\dfrac{Gm_1m_2}{r^2}\,\hat{r}$, if the distance r is doubled, the force is

    A. four times as much as the original value.
    B. twice as much as the original value.
    C. the same as the original value.
    D. one-half of the original value.
    E. one-fourth of the original value.

20. •Which is larger, the Sun's pull on Earth or Earth's pull on Sun?
    A. The Sun's pull on Earth is larger.
    B. Earth's pull on the Sun is larger.
    C. They pull on each other equally.

    D. The Sun's pull on Earth is twice as large as Earth's pull on the much larger Sun.
    E. There is no pull or force between Earth and the Sun.

21. •Compare the weight of a mountain climber when she is at the bottom of a mountain with her weight when she is at the top of the mountain. In which case is her weight larger? SSM
    A. She weighs more at the bottom.
    B. She weighs more at the top.
    C. Both are the same.
    D. She weighs twice as much at the top.
    E. She weighs four times as much at the top.

22. •The magnitude of the gravitational force on a small object that has a mass m and is embedded at a distance r from the center of a uniform spherical cloud of mass M and radius R is F (Figure 10-16). If r is doubled, while keeping $r < R$, the magnitude of the gravitational force on the small object is now

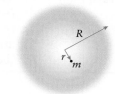

    A. F.
    B. $2F$.
    C. $F/2$.
    D. $4F$.
    E. $F/4$.

Figure 10-16 Problem 22

23. •A satellite is in a circular orbit about Earth moving at a speed of v. In order to escape the planet, the satellite must attain a speed of (SSM)
    A. v.
    B. $\sqrt{2}v$.
    C. $2v$.
    D. $v/2$.
    E. $\dfrac{v}{\sqrt{2}}$.

24. •The escape speed from planet X is v. Planet Y has the same radius as planet X but is twice as dense. The escape speed from planet Y is
    A. v.
    B. $\sqrt{2}v$.
    C. $2v$.
    D. $v/2$.
    E. $\dfrac{v}{\sqrt{2}}$.

25. •Two satellites having equal masses are in circular orbits around Earth. Satellite A has a smaller orbital radius than satellite B. Which statement is true? SSM
    A. Satellite A has more kinetic energy, less potential energy, and less mechanical energy (potential energy plus kinetic energy) than satellite B.
    B. Satellite A has less kinetic energy, less potential energy, and less mechanical energy (potential energy plus kinetic energy) than satellite B.

This page is intentionally left blank.

For complete end of chapter problem sets, please go to
www.whfreeman.com/kestentauck

40. •Determine the force of gravity between the Sun and Earth.

41. •Determine the net force of gravity acting on the Moon during an eclipse when it is directly between Earth and the Sun. SSM

42. •A star that has a mass equal to the mass of our Sun is located 50 AU from another star that has a mass which is one-half of the Sun's mass. At what point(s) will the gravitational force from the two stars on a 100,000-kg space probe be equal to zero?

43. •At what point between Earth and the Moon will a 50,000-kg space probe experience no net force? SSM

44. •Eight stars, all with the same mass as our Sun, are located on the vertices of an astronomically sized cube with sides of 100 AU. Determine the net gravitational force on one of the stars due to the other seven.

45. •Compare the weight of a 5-kg object on Earth's surface to the gravitational force between a 5-kg object that is one Earth radius from aother object of mass equal to 5.98×10^{24} kg. Use Newton's universal law of gravitation for the second part of the question.

46. •The *gravitational field* due to a planet or star is the gravitational force per unit mass in the vicinity around the planet or star; it is found by dividing the gravitational force experienced by a small object by the mass of that small object. The gravitational field is given by

$$\vec{g} = \frac{GM}{r^2}(-\hat{r})$$

where M is the mass of the object creating the gravitational field. Similar to the gravitational force, the gravitational field points radially inward, toward the center of the object creating the gravitational field. (a) Determine the gravitational field due to Earth at the location of the Moon. (b) Compare that numerical value to the centripetal acceleration of the Moon as it orbits in a circle about Earth once each 27.32 days. (c) Comment on any similarities or differences.

10-2: The Shell Theorem

47. •Determine the gravitational force between a hollow sphere that has a mass m_1 and radius r_1 and a pointlike object that has a mass m_2. The pointlike object is located a distance r_2 from the surface of the hollow sphere (Figure 10-17). SSM

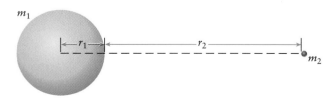

Figure 10-17 Problem 47

48. •The lowest point in the Western Hemisphere is found at Badwater in Death Valley, California where the elevation is 86 m below sea level. Using the shell theorem, determine the value of g at that location.

49. •The center of a ping-pong ball (hollow sphere that has a mass of 2.7 g and a diameter of 40 mm) is located 100 cm from the center of a basketball (hollow sphere that has a circumference of 75 cm and a mass of 600 g). Determine the gravitational force between the two balls.

50. •The highest point on Earth is Mount Everest at 8850 m above sea level. (a) Determine the acceleration due to gravity at that elevation. (b) What fractional change in the acceleration due to gravity would you find between Mount Everest and the Dead Sea (the lowest elevation on Earth at 400 m *below* sea level)?

51. •A 1-kg pointlike object is positioned at the exact center of a huge, hollow sphere that has a mass of 100,000 kg. The radius of the sphere is 10,000 m. (a) What is the mass per unit area of the uniform sphere? (b) What is the net force on the 1-kg pointlike object?

52. ••A huge, uniform sphere that has an inner radius of 9500 m and an outer radius of 10,000 m has a total mass of 100,000 kg (Figure 10-18). (a) What is the net force on a 1-kg pointlike object that is 50,000 m from the surface of the sphere? (b) What is the mass per unit volume of the sphere?

Figure 10-18 Problem 52

53. ••Spherical storage tanks are used to hold natural gas. A storage facility contains a row of five such tanks, each with a volume of 5000 m³. Determine the net gravitational force on the sphere at the end of the row of tanks due to the other tanks if they are filled with natural gas (natural gas' density is 0.8 kg/m³) and constructed from steel (steel's density is 8000 kg/m³) that is 10 cm thick. The tanks are 75 m apart (this distance is the closest one from the right edge of one tank to the left edge of the adjacent one). SSM

10-3: Gravitational Potential Energy

54. •Planet X is composed of material that has a mass density ρ. It has a radius of R. When a space probe of mass m is a distance of r from the center of planet X, it has a speed of v moving straight away from the planet. (a) What speed will the probe have (in the absence of any booster rockets) when it moves out to a distance of $2r$? (b) What is the escape velocity of the space probe from the new orbit?

55. •How much work is done by the force of gravity as a 10-kg object moves from a point that is 6000 m above sea level to a point that is 1000 m above sea level? SSM

56. •(a) What is the escape velocity of a space probe that is launched from the surface of Earth? (b) Would the answer change if the launch occurs on top of a very tall mountain? Explain your answer.

57. •The Schwarzschild radius is defined as the distance from a black hole where the escape velocity equals the speed of light. Determine the value of the Schwarzschild radius for a black hole with the mass of our Sun.

58. ••Determine the approximate size of the universe. The Schwarzschild radius is defined as the distance from a black hole where the escape velocity equals the speed of light. The radius of the visible universe can be found by calculating the Schwarzschild radius of a sphere with a uniform density equal to the critical density ρ_c, where

$$\rho_c = \frac{3H^2}{8\pi G}$$

The Hubble parameter H is currently estimated to be about 70 (km/s)/Mpc. The unit Mpc is a megaparsec. One parsec equals 30.857×10^{12} km or about 3.3 light-years.

59. ••The volume of water in the Pacific Ocean is about 7×10^8 km³. The density of seawater is about 1030 kg/m³. (a) Determine the gravitational potential energy of the Moon–Pacific Ocean system when the Pacific is facing away from the Moon. (b) Repeat the calculation when Earth has rotated so that the Pacific Ocean faces toward the Moon. (c) Estimate the maximum speed of the water in the Pacific Ocean due to the tidal influence of the Moon. For the sake of the calculations, treat the Pacific Ocean as a pointlike object (obviously a very rough approximation). SSM

60. •How much energy would be required to move the Moon from its present orbit around Earth to a location that is twice as far away? Assume the Moon's orbit around Earth is nearly circular and has a radius of 3.84×10^8 m, and that the Moon's orbital period is 27.3 days.

61. •A small asteroid that has a mass of 100 kg is moving at 200 m/s when it is 1000 km above the Moon. (a) How fast will the meteorite be traveling when it impacts the lunar surface if it is heading straight toward the center of the Moon? (b) How much work does the Moon do in stopping the asteroid if neither the Moon nor the asteroid heat up in the process? The radius of the Moon is 1.737×10^6 m. SSM

10-4: Kepler's Laws

62. •The space shuttle usually orbited Earth at altitudes of around 300 km. (a) Determine the time for one orbit of the shuttle about Earth. (b) How many sunrises per day did the astronauts witness?

63. •What is the speed of a space shuttle that orbits Earth at an altitude of 300 km?

64. •A satellite orbits Earth at an altitude of 80,000 km. Determine the time for the satellite to orbit Earth once.

65. •A satellite requires 86.5 min to orbit Earth once. Assuming a circular orbit, what is the circumference of satellite's orbit? SSM

66. •The orbital period of Saturn is 29.46 y. Determine the distance from the Sun to the planet.

67. •The orbit of Mars around the Sun has a radius that is 1.524 times greater than the radius of Earth's orbit. Determine the time required for Mars to complete one revolution.

68. ••Venus has the most circular orbit of all planets in our solar system. The eccentricity of its orbit is 0.0068. The semimajor axis of Venus' orbit is 108,210 km. Determine the apogee and perigee distances for Venus and the time required for one Venusian year.

69. •Determine the altitude of a geosynchronous satellite. Such a satellite will have precisely the same orbital period as Earth's rotational period.

70. •Given that Earth orbits the Sun with a semimajor axis of 1.000 AU and an approximate orbital period of 365.24 days, determine the mass of the Sun.

71. •The Moon orbits Earth in a nearly circular orbit that lasts 27.32 days. Determine the distance from the surface of the Moon to the surface of Earth. SSM

72. •A planet orbits a star with an orbital radius of 1 AU. If the star has a mass that is 1.75 times our own Sun's mass, determine the time for one revolution of the planet around the star.

General Problems

73. ••Calc A thin, uniform rod of length L and mass M lies along the x axis as shown in Figure 10-19. Derive an algebraic expression for the gravitational force the rod exerts on a small object of mass m located at point P, which is a distance D from the end of the rod. Hint: You will need to determine the differentially small gravitational force exerted by each differentially small mass element of the rod on the small object and then integrate to get the net force. SSM

Figure 10-19 Problem 73

74. •••Calc Suppose a thin, uniform rod with a mass M and length L is located on the y axis as shown Figure 10-20.

Derive an algebraic expression for the gravitational force exerted by the rod on a small object of mass m located at point P, which is a distance D along the x axis. *Hint*: You will need to determine the differentially small gravitational force exerted by each differentially small mass element of the rod on the small object and then integrate to get the net force. Also, the y components of the net force cancel out!

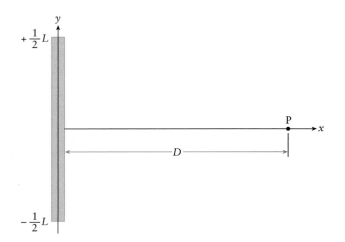

Figure 10-20 Problem 74

75. ••A satellite is launched into an elliptical orbit that has a period of 702 min. Apogee of the orbit is 39,200 km and the perigee is 560 km. (a) Determine the gravitational potential energy at the apogee and perigee of the satellite's orbit. (b) Estimate the energy required to place the satellite in orbit around the Earth.

76. •Draw an ellipse with an eccentricity of 0.3 and a semimajor axis of 5 cm. Label all the important elliptical parameters (the semiminor axis, the center, and the distance between the foci).

77. ••(a) What speed is needed to launch a rocket due east near the equator into a low-Earth orbit? Assume the rocket skims along the surface of Earth, so $r \approx R_E$. (b) Repeat the calculation for a rocket fired due west. (c) Why do you think the location for the launch of U.S. space missions is on the east coast of Florida? SSM

78. ••The former Soviet Union launched the first artificial Earth satellite, *Sputnik*, in 1957. Its mass was 84 kg and it made one orbit every 96 min. (a) Determine the altitude of *Sputnik*'s orbit above Earth's surface, assuming circular orbits. (b) What was *Sputnik's* weight in orbit and at the Earth's surface?

79. •Astro The 2004 landings of the Mars rovers *Spirit* and *Opportunity* involved many stages, resulting in each probe having zero vertical velocity about 12 m above the surface of Mars. Determine (a) the time required for the final free-fall descent of the probes and (b) the vertical

velocity at impact. The mass of Mars is 6.419×10^{23} kg and its radius is 3.397×10^6 m.

80. ••Astro The four largest of Jupiter's 63 moons are listed in the table below. (a) Complete the table and (b) determine the mass of Jupiter.

Moon	Semimajor Axis (km)	Orbital Period (days)
Io	421,700	1.769
Europa	671,034	?
Ganymede	?	7.155
Callisto	?	16.689

81. ••Astro An elliptical orbit of a planet in a distant solar system has a semimajor axis of $a = 2.25$ AU and a semiminor axis of $b = 1.75$ AU (**Figure 10-21**). Find the eccentricity, the perihelion and aphelion (closest and farthest distances from the star to the planet), and the area of the ellipse. SSM

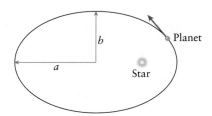

Figure 10-21 Problem 81

82. ••Biology, Astro On Earth, froghoppers can jump upward with a takeoff speed of 2.8 m/s. Suppose you took some of the insects to an asteroid. If it is small enough, they can jump free of it and escape into space. (a) What is the diameter (in kilometers) of the largest spherical asteroid from which they could jump free? Assume a typical asteroid density of 2.0 g/cm^3. (b) Suppose that one of the froghoppers jumped horizontally from a small hill on an asteroid. What would the diameter (in km) of the asteroid need to be so that the insect could go into a circular orbit just above the surface?

83. •Astro The International Space Station (ISS) orbits Earth in a nearly circular orbit that is 345 km above Earth's surface. (a) How many hours does it take for the ISS to make each orbit? (b) Some of the experiments performed by astronauts in the ISS involve the effects of weightlessness on objects. What gravitational force does Earth exert on a 10.0-kg object in the ISS? Express your answer in newtons and as a fraction of the force that Earth would exert on the object at Earth's surface. (c) Considering your answer in part (b), how can an object be considered *weightless* in the ISS?

84. ••Astro Measurements on the asteroid Apophis have shown that its aphelion (farthest distance from the

Sun) is 1.099 AU, its perihelion (closest distance from the Sun) is 0.746 AU, and its mass is 2.7×10^{10} kg. (a) Determine the semimajor axis of Apophis in astronomical units and in meters. (b) How many days does it take Apophis to orbit the Sun? (c) At what point in its orbit is Apophis traveling fastest, and at what point is it traveling slowest? (d) Determine the ratio of its maximum speed to its minimum speed.

85. ••A rack of seven spherical bowling balls (each 8 kg, radius of 11 cm) is positioned 100 cm from a point P, as shown in **Figure 10-22**. Determine the gravitational force the bowling balls exert on a ping-pong ball of mass 2.7 g centered at point P.

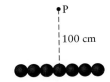

Figure 10-22 Problem 85

86. •**Astro** The Sun and solar system actually are not at rest in our Milky Way galaxy. We orbit around the center of the Milky Way galaxy once every 225,000,000 years, at a distance of 52,000 light-years. If the mass of the Milky Way were concentrated at the center of the galaxy, what would be the mass of the galaxy?

87. •••**Calc** Starting with Kepler's law of areas, prove Kepler's law of periods. *Hint*:

$$\frac{dA}{dt} = \frac{L}{2m} \qquad a = \frac{L^2}{GMm^2(1-e^2)} \qquad A = \pi a\,b$$

Begin by integrating to find the area for one period, T.

88. •••Find the net gravitational force exerted on a small, 1-kg object located at the point P by the three massive objects (**Figure 10-23**). Each of the massive objects lies on a coordinate axis as shown.

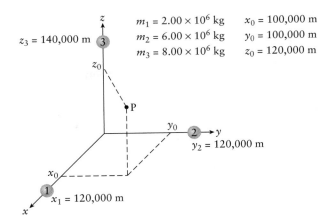

Figure 10-23 Problem 88

89. ••Locate the point(s) along the line $\overline{AB}$ where a small, 1-kg object could rest such that the net gravitational force on it due to the two objects shown is exactly zero (**Figure 10-24**). SSM

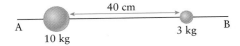

Figure 10-24 Problem 89

90. •••**Calc** One consequence of the shell theorem is that if you are inside a uniform spherical object (which is a reasonable approximation for Earth or a star), the gravitational force you experience actually increases as you move radially away from the center until you reach the outer edge of the object. The gravitational force on a small object of mass m inside a spherically symmetric object can be written as

$$\vec{F}(r) = -\Gamma m\vec{r}$$

where Γ is a constant with units of s^{-2}. Formulate an expression for the gravitational potential energy function when an object of mass m is inside such a spherical object.

11 Fluids

(Corbis.)

Objects submerged in a fluid feel a buoyant force. Because air is a fluid, balloons feel a buoyant force from the air that surrounds them. Hot air balloons float because the sum of the weight of the air inside of them, the weight of the balloon, and the weight of the basket and passengers is less than the weight of the air displaced by the balloon. The buoyant force equals the weight of the displaced air.

There's a good chance that you have a 0.5-L bottle of water nearby right now. How many water molecules are in it, do you suppose? If you guessed, "A lot," you'd be correct—there are more than 10^{25} molecules in a half liter of water. How likely is it that all 10^{25} of them are, at this instant, oriented in the same way? And how likely is it that they are all either moving in the same direction or all stationary? Not likely! To describe and understand the physics of fluids, we could try to study each fluid molecule separately and then create a bulk description of the fluid, but that process would be messy and complicated. Instead, we want to find a way to describe the properties of fluids as an average of the properties of each molecule in the fluid. As part of that representation, we will need to consider the behavior of all of those molecules together; the physics that governs fluids depends on this statistical description.

Until now we've concentrated on describing and understanding systems that can be represented as a collection of discrete, small, "point" objects. If an

397

object or a system isn't small, we found that we can imagine dividing it up into many tiny pieces and then adding up the contribution from each piece to get the desired result. During our discussion of center of mass (Section 7-6), for example, we represented a complex system as a point object by finding the right mass and right point in space at which to locate it. It would be complicated to apply that approach to a study of fluids.

Why is the center-of-mass approach difficult to apply to our study of fluids? To answer this question, let's first compare fluids to solids. The molecules in a **fluid** are not arranged in any organized structure, but that alone does not differentiate fluids from solids. One of the forms of solid ice, for example, is *amorphous*, that is, there is no organization to the frozen water molecules. The fundamental difference between a solid and a fluid is that a fluid does not have any shape of its own. A fluid flows to conform to the shape of its container.

By definition, a fluid cannot sustain a shear force, that is, a force pushing parallel to the surface. If you pour a small puddle of water on your desk, it will not be able to resist any forces pushing on it in the same direction as the surface. The water flows, spreading out until effects such as surface tension become dominant.

Both liquids and gases are fluids. Liquids, such as water, certainly match the description above, but so do gases, such as water vapor or the air that we breathe. The principal difference between liquids and gases is that although neither has a fixed shape, liquids have a fixed volume but gases do not. Take two equal amounts of air, put them into containers of different volumes and both containers will be "full."

The three parameters most useful to us in describing fluids are *density, pressure,* and (when we discuss fluids in motion) *volume flow*. Density tells us something about how tightly or loosely the molecules of a fluid are packed into a given volume. Pressure is related to the force that a fluid can exert on its surroundings or the force that the surroundings can exert on the fluid. Volume flow describes the rate at which a volume of fluid moves.

11-1 Density

Which is more massive, a piece of Styrofoam or a piece of iron? If you chose the iron, be careful! Suppose the piece of Styrofoam is as big as a house but the iron is the size of a speck of dust. The iron would be much less massive! At issue is **density**, that is, how much of the substance we can fit into a given volume. You can tell that iron is denser than Styrofoam because a given volume of iron has more mass than the same volume of Styrofoam. We define density ρ in terms of mass M and volume V:

$$\rho = \frac{M}{V} \tag{11-1}$$

From this definition, you can see that the SI units of density must be kilograms per cubic meter (kg/m^3). Table 11-1 lists the densities of various substances. One useful value is the density of water, $\rho_{water} = 1000 \ kg/m^3 = 1 \ g/cm^3$.

The densities of substances vary with pressure and temperature. For some, the variation over the range of pressures and temperatures that humans typically experience is small, but for others, such as gases, densities can vary greatly as pressure or temperature changes.

Most substances become denser as temperature decreases; the rate of that change depends on the substance. Most of the values of density in Table 11-1 are given for a standard pressure of 1 atm (Section 11-4) and temperature of 0 °C (273.15 K). The values for the gases (air and helium) are given at room temperature (20 °C or 293.15 K). Because water becomes less dense as temperature decreases from 4 °C down to 0 °C, by convention we list the densities of water, sea water, and ice at a temperature of 4 °C.

Table 11-1 Densities of Selected Substances

Substance (at 0 °C unless otherwise noted)	Density (kg/m³)
air at 20 °C	1.217
alcohol (ethanol)	0.806×10^3
aluminum	2.70×10^3
bone	1.7–2.0×10^3
copper	8.93×10^3
earth (average)	5.52×10^3
glass (common)	2.4–2.8×10^3
gold	19.3×10^3
helium	0.1786
ice	0.917×10^3
iron	7.8×10^3
lead	11.3×10^3
mercury	13.6×10^3
seawater at 4 °C	1.025×10^3
steam at 100 °C	0.6
water at 4 °C	1.000×10^3
water at 20 °C	0.998×10^3
wood (maple)	0.75×10^3
wood (pine)	0.50×10^3

Example 11-1 Bony Chickens

The bones in a typical chicken have a density ρ_{bone} equal to 2100 kg/m^3. In an unscientific survey at a local grocery store, we noticed that whole chickens weigh about 5.0 lb, or $M_{chicken} \approx 2.3$ kg. What volume would a chicken of this mass occupy if the entire chicken were composed of bone?

SET UP
Volume, mass, and density are related through the definition of density:

$$\rho = \frac{M}{V} \tag{11-1}$$

This relationship can be rearranged and applied to this problem:

$$V_{chicken} = \frac{M_{chicken}}{\rho_{chicken}}$$

SOLVE
Let the density of the chicken be the density of chicken bones, and substitute the given values into the equation to find a numeric answer:

$$V_{chicken} = \frac{M_{chicken}}{\rho_{chicken}} = \frac{M_{chicken}}{\rho_{bone}} = \frac{2.3 \text{ kg}}{2100 \text{ kg/m}^3} = 1.1 \times 10^{-3} \text{ m}^3$$

REFLECT

To put this result in perspective, we can approximate such a chicken as being spherical, and determine the chicken's diameter. Using $V = \frac{4}{3}\pi r^3$, the diameter of this spherical chicken would be about 13 cm. That diameter is clearly smaller than the diameter of a typical chicken—why the discrepancy? You've probably guessed that the original assumption (that the chicken is composed entirely of bone) is not realistic. Organisms are mostly water. Why not consider what the radius of the chicken would be if it were all water? The density of water ρ_{water} is equal to 1000 kg/m^3, so

$$V_{chicken} = \frac{M_{chicken}}{\rho_{water}} = \frac{2.3 \text{ kg}}{1000 \text{ kg/m}^3} = 2.3 \times 10^{-3} \text{ m}^3$$

√x̄ See the Math Tutorial for more
     information on Geometry

Applying $V = \frac{4}{3}\pi r^3$, a spherical chicken composed entirely of water would have a radius of about 16 cm, perhaps a bit more reasonable. And what if the chicken were composed entirely of feathers, which have a density $\rho_{feathers}$ equal to 40 kg/m^3? In this case, the volume of the chicken would be

$$V_{chicken} = \frac{M_{chicken}}{\rho_{feathers}} = \frac{2.3 \text{ kg}}{40 \text{ kg/m}^3} = 58 \times 10^{-3} \text{ m}^3$$

The spherical all-feather chicken would have a diameter of 48 cm!

Practice Problem 11-1 What is the volume of a person who weighs 670 N? Assume the person has an average density equal to that of water.

Sometimes the density of a substance will be given as a **specific gravity**. By definition, the specific gravity of a substance is the ratio of the density of the substance to the density of water at the standard temperature of 4 °C. For example, because the density of iron is 7.8×10^3 kg/m^3 (Table 11-1), the specific gravity of iron SG$_{iron}$ is

$$SG_{iron} = \frac{7.8 \times 10^3 \text{ kg/m}^3}{1000 \text{ kg/m}^3} = 7.8$$

Note that there are no units associated with specific gravity; it is a dimensionless quantity.

★ What's Important 11-1

Density describes how much of a substance fits into a given volume. The average density of an object is defined as its mass divided by its volume.

11-2 Pressure

We stated that a fluid does not have a fixed shape. You'd be surprised, for example, if you poured out a bottle of water and it retained its cylindrical shape! Fluids flow to conform to the shape of a container, which means that fluids exert a force in all directions. So, imagine air flowing into the lungs, or any gas filling a chamber. Whatever force the gas exerts on the chamber walls is distributed over the entire surface area of those walls. If the size of the chamber were to change but the amount of gas stayed the same, the force exerted by the gas would be distributed over a different area. In addition, gas exerts a force which is always perpendicular to the wall's surface. We can characterize the force exerted by a fluid using the quantity **pressure**; pressure P is the proportionality constant between the force and the area of the walls of the container. To embody the fact the force is perpendicular to the surface, we write

$$\vec{F} = P\vec{A} \tag{11-2}$$

where we treat the area as a vector. The magnitude of the area vector is the area of the surface over which the force acts, and its direction is perpendicular to the surface. The direction of the area vector, as suggested by Equation 11-2, is the same as the direction of the force vector.

The definition of pressure in Equation 11-2 assumes that the area is small enough that the force, and therefore pressure, is constant. In that case we can also define pressure as the magnitude of a force divided by the area over which it is applied:

$$P = \frac{F}{A} \qquad (11\text{-}3)$$

The SI unit of pressure, the pascal (Pa), is defined as 1 newton per square meter (N/m²):

$$1 \, \text{Pa} = 1 \frac{\text{N}}{\text{m}^2}$$

We discuss other common units of pressure in Section 11-4.

Consider this scenario: You take off your shoes and stand up, distributing your weight evenly over the surface area of your feet. The pressure on the soles of your feet is your mass divided by the surface area of the bottom of your feet and is probably not uncomfortable. But if you stood on a bed of nails it would hurt! Your weight hasn't changed, so the force you exert on the surface supporting you hasn't changed. However, because the supporting surface—the tips of a few nails—has a much smaller surface area, the pressure on your feet is much greater in this case. People can, of course, comfortably lie on a suitably prepared bed of nails. By lying flat, a person's weight is distributed over the area of tips of many nails, resulting in a tolerable pressure.

 What's Important 11-2
Pressure is the magnitude of a force divided by the area over which it is applied.

11-3 Pressure versus Depth in a Fluid

The mother of one of the authors sketched a hummingbird on a Styrofoam coffee cup which we then immersed in a pressure cooker for a few minutes. As you can see in **Figure 11-1**, the force on the cup due to the high pressure shrank the cup to about a third of its original size! Another way to achieve the same effect would be to place a Styrofoam cup outside the pressure-controlled compartment of a submarine, so that when the sub dives deep below the surface of the ocean the cup experiences an increase in pressure. At any depth below the surface of a fluid the pressure is greater than at the surface; this observation is a direct consequence of the weight of the fluid directly above an object. In addition, because there is necessarily more and more fluid above the object as depth below the surface increases, pressure increases with depth.

Let's consider the relationship between pressure and depth in a fluid by examining a volume of water within a large tank. Imagine a box-shaped volume of water within a tank filled with water that is not moving, as shown in **Figure 11-2**. The

Figure 11-1 A styrofoam cup shrunk when it experienced high pressure inside a pressure cooker. *(Courtesy David Tauck.)*

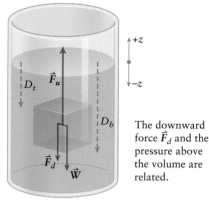

The upward force $\vec{F}_u$ and the pressure below the volume are related.

The downward force $\vec{F}_d$ and the pressure above the volume are related.

The imaginary volume is stationary because the net upward and downward forces cancel.

Figure 11-2 An imaginary box-shaped volume of water floats in a tank of water at rest. The imaginary volume remains at rest because the net upward force and the net downward force (the sum of F_d and W) are equal in magnitude.

top of the box is at a depth D_t from the surface of the water and the bottom of the box is at a depth D_b. The area of the top and bottom of the box is A. We won't put an actual box in the fluid, but instead will simply imagine a boundary that separates the box-shaped volume from the rest of the fluid.

The weight of the water above the volume exerts a force of magnitude F_d downward. Because fluids exert a force in all directions, the fluid below the box pushes upward with force of magnitude F_u. Finally, although the box-shaped volume is imaginary, we must consider the force of gravity on the water in it. The water has weight. This weight appears as a downward force of magnitude W on the water volume. Figure 11-2 shows the three forces.

We started with the assumption that the water in the tank is at rest. This assumption requires that there be no net force on the water volume, so the vector sum of the three forces acting on the volume must be zero:

$$\vec{F}_u + \vec{F}_d + \vec{W} = 0$$

We can express the upward force $\vec{F}_u$ as the pressure P_b at the bottom of the volume multiplied by the area of the bottom of the volume. Similarly, the downward force $\vec{F}_d$ equals the pressure P_t at the top of the volume multiplied by the area of the top of the volume. We can therefore rewrite the forces in terms of pressure:

$$P_b\vec{A}_b + P_t\vec{A}_t + \vec{W} = 0$$

Substituting $\vec{W} = m\vec{g}$ for the weight gives

$$P_b\vec{A}_b + P_t\vec{A}_t + m\vec{g} = 0 \tag{11-4}$$

Although the areas of the top and bottom of the volume have the same magnitude, we write them as $\vec{A}_t$ and $\vec{A}_b$ here to emphasize the different directions of the two area vectors: Because the pressure above the volume results in a downward force and the pressure below results in an upward force, we let $\vec{A}_t$ point down and $\vec{A}_b$ point up.

Let's adopt the coordinate axis z to measure displacement in the vertical direction, setting $z = 0$ at the surface of the fluid and positive z above the surface. Using this designation, then, $\vec{A}_b$ points in the positive direction and both $\vec{A}_t$ and $\vec{g}$ point in the negative direction, so Equation 11-4 can be written in terms of magnitudes of these vectors as

$$P_bA_b - P_tA_t - mg = 0 \tag{11-5}$$

We can simplify this, first, by noting the magnitude of the areas of the top and bottom of the volume are the same, $A_b = A_t = A$. Also, we can rewrite the mass m of the volume of water in terms of density and volume:

$$m = \rho V = \rho A\,\Delta z$$

where ρ is the density of water. The difference in depth between the top and bottom of the volume, $\Delta z = D_t - D_b$, is the height of the box-shaped volume. Substituting these values into Equation 11-5 results in

$$(P_b - P_t)A - \rho A\,\Delta zg = 0$$

or, by dividing each term by A,

$$P_b - P_t - \rho\,\Delta zg = 0$$

We can put this equation into a more standard form, first, by moving the negative terms to the right side of the equation:

$$P_b = P_t + \rho \Delta z g$$

and then by setting the top of the box-shaped volume to coincide with the surface of the water in the tank. Let the pressure at the surface be P_0, and note that Δz now represents the depth d at the bottom of the volume. The pressure P at depth d is then

$$P = P_0 + \rho d g$$

It is common to write the last term in a different order, so we'll stick to convention:

$$P = P_0 + \rho g d \qquad (11\text{-}6)$$

Equation 11-6 is a general one for the **pressure versus depth** below the surface of any fluid of density ρ. When g is set to 9.8 m/s², this equation applies to fluids near the surface of Earth. Be careful with the sign of the $\rho g d$ term; the standard convention is to measure d from the surface of the fluid as a positive number increasing with depth.

Note that pressure P in Equation 11-6 shows no dependence on the width of the tank of water. Pressure in a fluid depends only on depth. We can make this clearer by examining two tubes each filled with water; as shown in **Figure 11-3**, tube A is straight whereas tube B has two right angle bends, but both are filled to a height H above the bottom of the tube. In tube B, point B_1 is located directly below the opening of the tube, at depth H_1; point B_2 is located directly above the bottom of the tube, also at distance H_1 below the level of the top of the tube. The bottom of tube B is a depth H_2 below the level of points B_1 and B_2.

According to Equation 11-6, the pressure at the bottom of tube A is

$$P = P_0 + \rho g d = P_0 + \rho g H$$

We can find the pressure P_1 at point B_1 in tube B the same way:

$$P_1 = P_0 + \rho g H_1$$

Because point B_2 is at the top of lower part of tube B, we can find the pressure at the bottom of tube B relative to the pressure P_2 at B_2:

$$P = P_2 + \rho g H_2$$

But because B_1 and B_2 are at the same depth, $P_1 = P_2$, and we can combine these last two equations:

$$P = (P_0 + \rho g H_1) + \rho g H_2 = P_0 + \rho g (H_1 + H_2)$$

Moreover, because $H = H_1 + H_2$, the pressure at the bottom of tube B is the same as it is at the bottom of tube A:

$$P = P_0 + \rho g H$$

Clearly, the pressure in a tube of fluid does not depend on width, or in fact on any aspect of the shape of the tube. Pressure depends only on the depth.

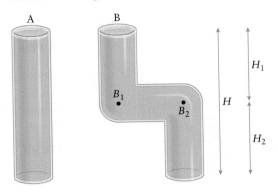

Tubes A and B are filled with the same liquid, and both have height H.

Because B_1 and B_2 are at the the same depth, there is no pressure difference between them. The pressure at the bottom of B is due to a depth change H_1 plus a depth change H_2.

Figure 11-3 Because pressure in a fluid depends only on depth below the surface, the pressures at the bottoms of tube A and tube B are the same.

Estimate It! 11-1 Pressure in the Cord

Cerebrospinal fluid fills cavities in your skull and spinal cord. Estimate the difference in pressure in the cerebrospinal fluid between the top and base of your spinal cord.

SET UP

$$P_{base} = P_{top} + \rho_{fluid}gH$$

where P_{base} and P_{top} are the pressures at the base and top of your spinal cord, respectively, ρ_{fluid} is the density of spinal fluid, and H is the height of your spinal cord.

$$P_{difference} = P_{base} - P_{top} = \rho_{fluid}gH$$

Use big, round (but reasonable!) numbers for an estimate. Let the density of spinal fluid be about that of water and assume that your spinal cord extends on the order of 1 m. To the precision of this estimate, let g be about 10 m/s^2.

SOLVE

The density of water is 1000 kg/m^3, so

$$P_{difference} \approx (1000 \text{ kg/m}^3)(10 \text{ m/s}^2)(1 \text{ m}) = 10^4 \text{ Pa}$$

REFLECT

A pressure difference of about 10^4 Pa is also what you'd experience submerged 1 m below the surface of a swimming pool filled with water.

> ✳ **What's Important 11-3**
> Pressure in a fluid increases as depth in the fluid increases. Pressure does not depend on the width of the body of fluid.

11-4 Atmospheric Pressure and Common Pressure Units

You might not be aware of it, but all of the air in the atmosphere directly above your head has weight, and that weight is pushing down on you right now! To be more precise, it's pushing in on you from every direction. Although air is relatively light, there's a lot of it in the atmosphere. Pressure on the surface of Earth due to the weight of air is considerable.

In the SI system, in which the unit of pressure is the pascal, 1 atmosphere of pressure is taken to be

$$P_{atm} = 1 \text{ atmosphere} = 1 \text{ atm} = 1.01 \times 10^5 \text{ Pa}$$

In the English system of units, weight is measured in pounds and area can be measured in square inches, so the pressure of the atmosphere is

$$P_{atm} = 1 \text{ atm} = 14.7 \frac{\text{lb}}{\text{in.}^2}$$

A pressure of 1 atm is not particularly high or particularly low, compared to the pressures we might find at the bottom of the ocean, where water pressure can exceed 10^8 Pa, or 1000 atmospheres. We can find both very high and very low pressures in the manufacturing world. High-pressure gas systems can operate at pressures of hundreds or even thousands of atmospheres of pressure, and systems that require high vacuums operate at pressures of 10^{-5} Pa or lower. A perfect vacuum would have a pressure of 0 Pa.

Example 11-2 Diver's Rule of Thumb

How much does the pressure on a diver change as he goes down to a depth 10 m below the surface of a freshwater lake?

SET UP

We know that pressure increases with depth in a fluid:

$$P = P_0 + \rho g d \tag{11-6}$$

Because P_0 is the pressure at the surface, the pressure increase at any depth is the difference between P_0 and P, the pressure at depth d, or $P - P_0$.

To get pressure difference, rearrange the pressure versus depth equation:

$$P - P_0 = \rho g d$$

SOLVE

Now substitute the known values into the equation:

$$P - P_0 = 1000 \text{ kg/m}^3 (9.8 \text{ m/s}^2) 10 \text{ m}$$

or

$$P - P_0 = 0.98 \times 10^5 \text{ Pa}$$

Going from the surface to a depth of 10 m, the diver experiences a change in pressure of 0.98×10^5 Pa.

REFLECT

Notice that this result is very close to 1.01×10^5 Pa, which is 1 atm. It is therefore a good approximation to say that our diver experiences a pressure change of 1 atm going from the surface to a depth of 10 m. Divers use this approximation as a "rule of thumb": Pressure increases approximately 1 atm for every 10 m change in depth.

Practice Problem 11-2 The Pacific Canada Pavilion exhibit at the Vancouver Aquarium Marine Science Centre in Vancouver includes a saltwater tank about 3.6 m deep. What is the pressure at the bottom of the tank? At the temperature of the tank, salt water has a density of 1023 kg/m^3. Express your answer in atm.

You've almost certainly encountered pressure measurements before, for example, when you go to the doctor or listen to a weather report. However, the value of pressure was probably not presented with units of pounds per square inch or pascals, but rather in either millimeters or inches of mercury. To understand these units of pressure, we'll imagine constructing a simple version of a barometer, a device used to measure air pressure.

Suppose you fill a tube closed at one end with mercury. While keeping the open end sealed, you turn the tube upside down into a pan partially filled with mercury. (Mercury is poisonous, so don't try this at home.) Now you release the seal on the end of the tube, which is beneath the surface of the mercury in the pan. No air can get into the tube, so an equal volume of vacuum must replace whatever volume of fluid drains from the tube. **Figure 11-4** shows this process.

A curious thing happens as the mercury starts to run out of the tube into the pan. Let's consider the system after some mercury has run out of the tube, but when a height H of mercury remains, where H is measured from the surface of the mercury in the pan. Because the pressure in a vacuum is zero, according to Equation 11-6, the pressure at depth H below the top of the mercury in the tube must be

$$P(H) = 0 + \rho_{\text{Hg}} g H$$

Although the pressure right at the surface of the mercury in the pan is atmospheric or air pressure P_{air}, its magnitude is not necessarily 1 atm because atmospheric pressure rises and falls with weather conditions. But here's the curious part: Because

Figure 11-4 A simple barometer, which measures atmospheric pressure, is created when an inverted tube of fluid like mercury is placed in a pan. Some liquid remains in the tube, supported by air pressure.

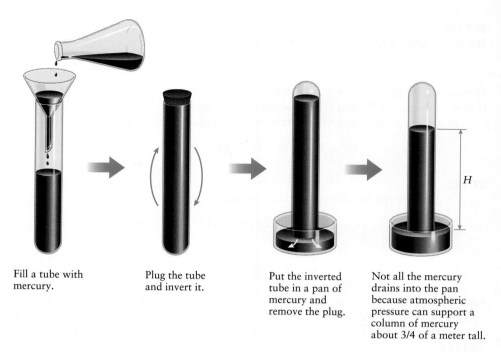

Fill a tube with mercury.

Plug the tube and invert it.

Put the inverted tube in a pan of mercury and remove the plug.

Not all the mercury drains into the pan because atmospheric pressure can support a column of mercury about 3/4 of a meter tall.

the point which is a depth H from the surface of the mercury in the tube is also at the level of the surface of mercury in the pan, the pressure $P(H)$ must also be equal to P_{air}, so

$$P(H) = \rho_{Hg}gH = P_{air}$$

This pressure exerts a force upward on the mercury in the tube; as a result of this upward force not all of the mercury can run out of the tube. When the weight of the mercury in the tube equals this upward force, the net force on the mercury is zero and the column of mercury will be in equilibrium, that is, its height H will be stable. Air pressure determines the height of a column of mercury that can be supported in the tube of our simple barometer. To find the height of a column of mercury supported by air pressure, we'll set the weight of the mercury equal to the force due to air pressure. The weight of the mercury is the density of mercury multiplied by the volume of the mercury in the column multiplied by g:

$$W_{Hg} = \rho_{Hg}Vg = \rho_{Hg}HAg$$

where A is the cross-sectional area of the tube. The upward force due to atmospheric pressure is

$$F = P_{air}A$$

Setting these equal yields

$$\rho_{Hg}HAg = P_{air}A$$

or

$$H = \frac{P_{air}}{\rho_{Hg}g}$$

As the air pressure P_{air} varies, the height of the column H goes up and down. In particular, when the air pressure is 1 atm, or 1.01×10^5 Pa, the height of the column can be determined from the density of mercury given in Table 11-1:

$$H_{1\,atm} = \frac{1.01 \times 10^5\,\text{Pa}}{13.6 \times 10^3\,\text{kg/m}^3\,(9.8\,\text{m/s}^2)} = 0.76\,\text{m}$$

A pressure of 1 atm can therefore support a column of mercury 0.76 m high.

Because the precise value of the density of mercury depends on temperature, and the precise value of the acceleration due to gravity depends on location, the scientific community has established the standard values of the density of mercury and the acceleration due to gravity; using these values, the height of a column of mercury supported by 1 atm of pressure is 760 mm. For that reason, 760 mm of mercury, or 760 mmHg, is equivalent to a pressure of 1 atm. Another unit of pressure is therefore mmHg. The torr, named after the Italian scientist and mathematician Evangelista Torricelli, is equivalent to mmHg. Torricelli invented the barometer around the middle of the 17th century.

Gauge Pressure

What is the air pressure inside a flat automobile tire? It's not zero, but rather 1 atmosphere or 14.7 lb/in.^{2} The same is true for a balloon that hasn't been inflated, or an empty, open bottle. We tend not to notice it, but the air pressure all around us is 1 atm. But if you use a tire pressure gauge to measure the air pressure in the deflated tire, it will read 0 lb/in.^{2} not 14.7 lb/in.^{2} This **gauge pressure** is the pressure above 1 atmosphere; for many applications this is a more intuitive way to think about pressure. The **absolute pressure** in a deflated tire is 1 atm.

Blood Pressure

As is often the case, calculations get a little more complicated in the body, but the relationship between pressure and depth applies directly to blood and blood pressure. The blood vessels in the body can be modeled as one of the tubes in Figure 11-3; for a standing person we would expect the pressure in the feet to be higher than in the head. (For a person who is lying flat, the effect of gravity on his blood pressure is slight because most people aren't very thick!) Because there is no surface at which to apply Equation 11-4 we can, for simplicity, assign the top of the heart in an upright person as the reference point for pressure. Compared to the pressure at the reference point, blood pressure will be higher in the feet and lower in the head. Blood pressure is usually given as two numbers, for example, "120 over 80." These values are measured in units of mmHg, and represent the high and low values of the gauge pressure in the arteries at the level of the heart. (Because the heart is a pump, the pressure in the system changes over the period of each heartbeat.) The mean blood pressure in an upright person's feet is about 100 mmHg higher than it is when it leaves the heart—about twice as high as the blood pressure measured in a person's arm. In contrast, the pressure in the head is about 50 mmHg less than at the heart. Just for fun, imagine what happens in a giraffe. The head of an adult giraffe is about 2 m higher than its heart. Several adaptations allow the giraffe to get blood to its brain; most importantly, the giraffe heart generates much higher pressure than a human heart does—about 260 mmHg!

More interesting, perhaps, is how differences in pressure in different regions of the body affect the flow of blood in the veins returning to the heart from the head and neck. Blood pressure in veins is always lower than in arteries; in fact, the pressure of blood entering the heart is very close to 0 mmHg. Because the pressure above our reference point will be even lower, the veins of the head and neck tend to collapse! So how then does blood get back to the heart from the head and neck? As blood continues to enter the veins, pressure builds up and the vessels reopen. But as the blood flows through them, the pressure will again drop and the veins will collapse again. The result is intermittent venous blood flow in the head and neck!

11-5 Pressure Difference and Net Force

During a hurricane, the pressure difference between the inside and outside of a house can lift off the roof! We have seen that pressure and force are related by the area over which a force is exerted:

$$\vec{F} = P\vec{A}$$

How can we apply this relationship to an object that has an inside and an outside?

Whatever pressure P_{out} exists outside the object causes an inward force, and whatever pressure P_{in} exists inside the object causes an outward force. Our experience with the vector addition of forces suggests that the net force on the object is

$$\vec{F}_{net} = P_{out}\vec{A}_{out} + P_{in}\vec{A}_{in}$$

where $\vec{A}_{in}$ and $\vec{A}_{out}$ are the area vectors of the boundary or walls of the object, with equal magnitude but opposite direction; $\vec{A}_{in}$ is directed outward from the object, and $\vec{A}_{out}$ is directed inward. If we establish a convention where the outward direction is positive, then

$$F_{net} = P_{out}A_{out} - P_{in}A_{in}$$

or since the magnitudes $A_{in} = A_{out} = A$,

$$\vec{F}_{net} = \Delta P\vec{A} \tag{11-7}$$

The net force on the object is proportional to the difference in pressure between outside and inside. So, for example, it is the **pressure difference** between the inside and outside of a house that can lift off the roof when a hurricane passes. Similarly, the pressure difference between the atmosphere and inside the lungs determines whether air enters or leaves the body.

The numeric value of 1 atm of pressure, taken together with the definition of pressure as the magnitude of a force divided by the area over which it is applied, leads to a puzzle. If you're an average-sized person, the surface area of your body is about 1.7 m². Using $F = PA$, the force pushing inward on your body due to air pressure is about $F = 10^5 \, \text{N/m}^2 \times 1.7 \, \text{m}^2 = 1.7 \times 10^5 \, \text{N}$, or nearly 20 tons. The air around you is pushing in on your body with a force of nearly 20 tons! Why don't you get crushed?

Air pressure doesn't crush you because the force pushing out from inside you is equal to the force pushing in on you by the atmosphere. Although the 1 atm of pressure outside of your body exerts an enormous force inward on you, the pressure inside of your body is also 1 atm, which exerts the same enormous force outward. According to Equation 11-7, the net force is therefore zero. What a relief!

Example 11-3 Submarine Hatch

The compartment of a research submarine is maintained at a pressure of P_{inside} equal to 1 atm. When the submarine has descended to depth $d = 100$ m below the surface of a freshwater lake, what force would be required to push open a hatch of area $A = 2$ m²?

SET UP

The pressure of the water pushing against the hatch is equivalent to the pressure outside the submarine; in other words, $P_{outside}$ equals the pressure at depth d. This pressure results in a force on the hatch that points inward. Air pressure inside the submarine P_{inside} results in a force on the hatch pointing outward. The pressure difference between these two therefore results in a net force on the hatch. This net force must be countered in order to push open the hatch.

According to Equation 11-7, the magnitude of the net force on the hatch is proportional to the pressure difference:

$$F_{net} = \Delta P A$$

ΔP is the difference between $P_{outside}$ and P_{inside}, so

$$F_{net} = (P_{outside} - P_{inside})A$$

SOLVE

Opening the hatch requires an applied force greater than F_{net}. Also, because $P_{outside}$ must be greater than atmospheric pressure and therefore greater than P_{inside}, F_{net} pushes in on the hatch. So the force necessary to open the hatch F_{open} must be directed outward:

$$F_{open} > (P_{outside} - P_{inside})A$$

Use Equation 11-6 to find the pressure at a depth d:

$$F > F_{net} = [(P_0 + \rho g d) - P_{inside}]A$$

Pressure at the surface of the lake P_0 is 1 atm, and we have been given P_{inside} equal to 1 atm as well, so these two terms cancel, leaving

$$F_{open} > \rho g d A = 1000\ \frac{kg}{m^3}\left(9.8\ \frac{m}{s^2}\right)100\ m\,(2.00\ m^2) = 2 \times 10^6\ N$$

REFLECT

The outward force required to open the hatch is about 2×10^6 N. That's a lot, more than 200 tons in English units!

Practice Problem 11-3 The crew compartment in a blimp is maintained at a pressure of 1 atm. When such an airship hovers at an altitude of 3000 m, where the air pressure is 0.70×10^5 Pa, what is the direction and magnitude of the force on a square window 0.55 m on a side?

Pressure differences drive air into and out of the lungs. Frogs, for example, inflate their lungs by swallowing mouthfuls of air. In contrast, we suck air into our lungs by contracting the diaphragm and the muscles of the chest wall to increase the volume of our chest cavity. This pulls the lungs open, increasing their volume and therefore decreasing the pressure inside them. This pressure difference pulls air into the expanding lungs.

It takes energy on our part to contract the muscles required to inhale. However, normal exhalation requires no work. Relaxing the muscles allows the diaphragm to rise and the ribs to fall, decreasing the volume of the chest cavity. This causes pressure in the chest to rise above atmospheric pressure and forces air out of the lungs. Because atmospheric pressure is constant, changes in lung pressure determine the direction of airflow.

Example 11-4 Breathing through a Tube

Super Agent Steele Branson dives into a shallow pond to avoid capture by his nemesis. Branson knows that his lungs cannot function when the force pushing in on his chest results in a difference in pressure (pushing in compared to pushing out) greater than about 0.10 atm. He intends to lie flat on his back on the bottom of the pond and breathe through a hollow reed. What is the length L of the longest reed through which he can breathe underwater?

SET UP
Because of the hollow reed, the air pressure in Branson's lungs is essentially that of the air at the surface of the pond. To be able to breathe, then, Branson needs to ensure that the difference between the pressure of the water on his chest and the pressure at the surface is no greater than 0.10 atm. If the difference were greater, he wouldn't be able to expand his chest cavity in order to inhale. The pressure of the water increases with depth, so our goal is to find the depth L so that the difference in pressure between the surface and his depth doesn't exceed 0.10 atm.

SOLVE
The difference in pressure ΔP between the surface and depth L is

$$\Delta P = P_{depth} = L - P_0$$

The pressure at depth L is given by the pressure versus depth equation, Equation 11-6:

$$\Delta P = (P_0 + \rho_{water}gL) - P_0 = \rho_{water}gL$$

Notice that we have two P_0 terms, which cancel. Now solve for L, and substitute the given values. Be careful to convert 0.10 atm to pascals, that is, $\Delta P = 0.10 \text{ atm} \times 1.01 \times 10^5 \text{ Pa/atm} = 1.01 \times 10^4 \text{ Pa}$:

$$L = \frac{\Delta P}{\rho_{water}g} = \frac{1.01 \times 10^4 \text{ Pa}}{1000 \text{ kg/m}^3 (9.8 \text{ m/s}^2)} = 1.03 \text{ m}$$

REFLECT
The longest hollow reed through which Super Agent Steele can breathe underwater is about 1 m in length—not that long. But perhaps this result shouldn't be surprising, when you consider that the weight of water which is a meter deep and on an area the size of someone's chest is easily more than 300 lb. If you wonder how people can breathe at great depths underwater when scuba diving, remember that the air in a scuba tank is pressurized!

Practice Problem 11-4 Over the range of altitudes below sea level found on Earth, the density of air is relatively constant, so Equation 11-6 holds. Imagine Super Agent Steele Branson has traveled to a place where the altitude is below sea level and he is attempting to breathe from a straw long enough so that the other end is at sea level. (He is not underwater.) How far below sea level could he breathe, assuming he could actually move the volume of air in the tube into his lungs?

> ✳ **What's Important 11-5**
> The net force exerted on an object due to fluid pressure is proportional to the difference in pressure inside the object and the pressure outside.

11-6 Pascal's Principle

When you squeeze one end of a tube of toothpaste, the pressure you apply is transmitted throughout the tube, and toothpaste comes out the other end. This simple example is representative of a more general principle first proposed by Pascal:

> Pressure applied to a confined, static fluid is transmitted undiminished to every part of the fluid as well as to the walls of the container.

Hydraulic jacks exploit **Pascal's principle**, allowing you to lift heavy objects such as your car. Figure 11-5 shows the construction of a simplified version that demonstrates the physics in action. Start with a tube bent into a U-shape. One side of the tube, labeled 2, is much wider than the other side, labeled 1; we'll call the cross-sectional areas of the tube ends A_1 and A_2. We partially fill the tube with an incompressible liquid such as water or oil. (This is, of course, an idealized or not-quite-possible case.) The liquid rises to the same height on both sides of the tube. (Why?) We'll put a moveable cap at both ends of the tube to keep the fluid from leaking out.

How does the system react when we apply a downward force F_1 on the cap on side 1? Certainly the force causes pressure under the cap to increase by $\Delta P_1 = F_1/A_1$, and according to Pascal's principle, this change in pressure is transmitted throughout the tube. We will therefore see this same pressure increase below the cap on side 2; that is, $\Delta P_2 = \Delta P_1$. We can then relate the force F_2 applied upward on the cap on side 2 due to the force F_1:

$$\frac{F_2}{A_2} = \frac{F_1}{A_1}$$

or

$$F_2 = \frac{A_2}{A_1}F_1 \qquad \textbf{(11-8)}$$

From this equation we can see the mechanical advantage of the hydraulic jack: Because A_2 is greater than A_1, F_2 is greater than F_1. A small force applied to side 1 results in a larger force applied to the cap on side 2. Do you need to lift up your car in order to change a flat tire? The physics of Pascal's principle will assist you, converting the relatively small force you are able to apply into a force large enough to raise your car off the ground.

How far can we move the cap on side 2? When F_1 is applied, the cap on side 1 moves down a distance d_1, displacing a volume of liquid equal to $V_1 = d_1A_1$. The increase in pressure throughout the tube causes liquid to rise on side 2, pushing the cap up a distance d_2. The increase in volume under cap 2 is $V_2 = d_2A_2$. Because we required the liquid to be incompressible, these two volumes of liquid must be equal:

$$d_2A_2 = d_1A_1$$

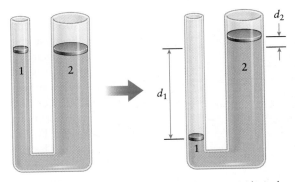

When a downward force is applied to the cap on side 1, the increase in pressure is transmitted undiminished throughout the fluid, forcing the cap on side 2 to rise.

Figure 11-5 A simple hydraulic jack demonstrates Pascal's principle.

or

$$d_2 = \frac{A_1}{A_2}d_1 \qquad (11\text{-}9)$$

Look carefully; the multiplicative factor is the inverse of the factor that relates F_1 and F_2. Although the applied force on side 2 is larger than the force on side 1, the cap on side 2 moves a shorter distance than the cap on side 1. So there's a price to be paid for the mechanical advantage gained from the hydraulic jack! If you've ever lifted a car up with a manual jack, you know that to get the car only a few inches off the ground requires many, many pumps on the jack handle. Each application of force can lift the car, but only a very short distance.

To put the hydraulic lift in perspective, consider the work done on each side of the tube. Our simple definition of work is $W = Fd$, thus $W_1 = F_1d_1$ and $W_2 = F_2d_2$. We can write the second equation in terms of F_1 and d_1 using Equations 11-8 and 11-9:

$$W_2 = \frac{A_2}{A_1}F_1\frac{A_1}{A_2}d_1 = F_1d_1$$

The work done on side 2 equals the work done on side 1. You can't get something for nothing—whatever work you do on side 1 is translated into an equal amount of work on side 2. (And of course, this example is idealized. Any real-world jack would be less efficient.)

★ What's Important 11-6

As described by Pascal's principle, when applied to a confined, static fluid, pressure is transmitted to every part of the fluid as well as to the walls of the container holding it.

11-7 Buoyancy—Archimedes' Principle

When you drop a plastic toy into a pond, you're not surprised when it pops back up to the surface. Yet while a container ship transporting a massive load also floats, its anchor sinks immediately when dropped overboard. Why does the ship float on the surface even though it weighs much more than the anchor, which sinks?

Because the pressure in a body of fluid changes with depth, the top of an object submerged in the fluid will feel a different pressure than the bottom of the object. Because the bottom is farther from the surface than the top, the pressure there is necessarily greater than at the top, resulting in an upward force on the object, which is called the **buoyant force**. If that buoyant force is greater than the weight of an object, as it usually is in the case of a ship, the object floats.

Let's submerge an object in a tank of still fluid as shown in **Figure 11-6**. The object in the figure is shaped liked a box, but it could be any shape for this discussion. The area of the top and bottom of the box is A.

The pressure of the water on the top of the object, due to the weight of the water (and air) above it, exerts a force F_d downward. The pressure of the water below the object pushes upward on the object with force F_u. Of course, the weight of the object itself, shown as W in the figure, exerts a downward force on the object. So the net downward force is $F_d + W$ and the net upward force is F_u. The buoyant force F_b is the difference between the forces due to the pressures:

$$F_b = F_u - F_d$$

An object is submerged in a stationary fluid.

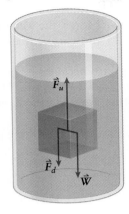

The upward force $\vec{F}_u$ is due to the pressure below the object.

The downward force $\vec{F}_d$ is due to the pressure above the object.

If $\vec{F}_d$ plus the object's weight is less than $\vec{F}_u$, the object floats to the surface.

Figure 11-6 An object is submerged in a fluid. Whether it floats to the surface, sinks, or remains stationary depends on how its weight compares to the net force due to the pressure in the fluid.

There are three possibilities for the relationship between the upward and downward forces. When

$$F_d + W > F_u \quad \text{the object sinks}$$

$$F_d + W = F_u \quad \text{the object neither sinks nor rises}$$

$$F_d + W < F_u \quad \text{the object rises}$$

Perhaps most interesting is the second condition, in which the upward and downward forces exactly balance and are therefore in equilibrium. Intuition tells us that this condition will only be met when the object has the same density as the liquid; if the object is denser than the liquid it should sink, and if it is less dense than the liquid it should rise to the surface. In the special case where the upward and downward forces balance, then, the density of the object must be the same as the density of the liquid, which means that the weight W of the object is equal to the weight of the volume of liquid displaced by the object.

We can write the equilibrium statement in terms of the buoyant force; from $F_d + W = F_u$,

$$F_b = F_u - F_d = W$$

As we have seen, in this case the weight W of the object must equal $W_{displaced}$, the weight of the fluid displaced by the object; thus

$$F_b = W_{displaced}$$

We can write this equation in a more useful form:

$$F_b = M_{displaced}g = \rho_{displaced} V_{displaced}g \qquad (11\text{-}10)$$

This equation expresses **Archimedes' principle,**

> The buoyant force on an object immersed in a fluid is equal to the weight of the fluid displaced by the object.

 Go to Picture It 11-1 for more practice dealing with buoyant force

Example 11-5 Ball and Chain

A solid plastic ball with a radius R equal to 2.0 cm is attached by a massless chain to the bottom of an aquarium filled with fresh water. The density of the plastic is $\rho_{plastic} = 600 \text{ kg/m}^3$. Find the tension in the chain.

 Go to Interactive Exercise 11-1 for more practice dealing with forces including the buoyant force

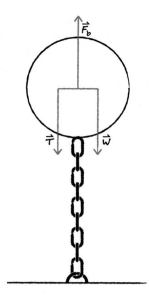

Figure 11-7

See the Math Tutorial for more information on Geometry

SET UP

To solve this problem we will need to understand the relationship between three forces that act on the ball: the buoyant force, the weight of the ball, and the tension in the chain. We therefore start by drawing a free-body diagram, as shown in **Figure 11-7**.

The buoyant force must be upward because the plastic is less dense than water, and the other two forces are directed downward. The system is in equilibrium because the sum of the forces acting upward exactly equals the sum of the forces acting downward.

The mathematical statement that describes the equilibrium condition in this case is

$$F_b = W + T$$

where F_b is the buoyant force, W is the weight of the ball, and T is the tension in the chain. Our goal is to find the tension:

$$T = F_b - W$$

SOLVE

Using Archimedes' principle (Equation 11-10), the buoyant force is equal to the weight of the water displaced by the ball:

$$F_b = W_{\text{water displaced}} = M_{\text{water displaced}}g$$

The mass of the water displaced by the ball is equal to the product of the density of water and the volume of water displaced, so

$$F_b = M_{\text{water displaced}}g = \rho_{\text{water}} V_{\text{water displaced}}g$$

In a similar way, the weight of the ball can be written

$$W = M_{\text{ball}}g = \rho_{\text{plastic}} V_{\text{ball}}g$$

Because the ball is completely submerged, the volume of water it displaces is equal to the volume of the ball, $V_{\text{water displaced}} = V_{\text{ball}} = \frac{4}{3}\pi R^3$. Remember $T = F_b - W = \rho_{\text{water}} V_{\text{water displaced}}g - \rho_{\text{plastic}} V_{\text{ball}}g$. Substituting $\frac{4}{3}\pi R^3$ into the equation for T yields

$$T = \rho_{\text{water}}\left(\frac{4}{3}\pi R^3\right)g - \rho_{\text{plastic}}\left(\frac{4}{3}\pi R^3\right)g = (\rho_{\text{water}} - \rho_{\text{plastic}})\frac{4}{3}\pi R^3 g$$

The numeric value of the tension is therefore

$$T = (1000\,\text{kg/m}^3 - 600\,\text{kg/m}^3)\frac{4}{3}\pi(0.020\,\text{m})^3(9.8\,\text{m/s}^2) = 0.13\,\text{N}$$

REFLECT

Is there a range of values we can reasonably expect for the tension? Certainly the tension must be greater than zero, that is, the buoyant force must be greater than the weight of the ball, otherwise the ball would sink. However, the maximum possible tension depends on the breaking strength of the chain. A large, low-density ball could potentially exert a large upward force on the chain, and as long as the chain remained intact, the tension would be large as well.

Practice Problem 11-5 A balloon of radius 20 m is filled with helium and tethered to the ground by means of a light cord. What is the tension in the cord if the weight of the empty balloon is 3200 N?

Got the Concept 11-2
Floating, Two Ways

A plastic cube with a coin taped to its top surface floats partially submerged in water. Mark the level of the water on the cube, then remove the coin and tape it to the *bottom* of the cube. Is the water level on the cube higher, lower, or the same as it was when the coin was on the top of the cube? If you used the cork from a wine bottle instead of a coin would your answer be higher, lower, or the same? Remember that the buoyant force on an object is the weight of the fluid which it displaces, and in this problem, water is being displaced by whatever is below the surface.

Got the Concept 11-3
Sinking Bottles

An open, glass soda bottle will float in a tub of water as long as it is empty. Place an empty, open glass bottle into a tub of water so that it floats, and mark the level of the water on the wall of the tub. Now submerge the bottle so that it fills with water and sinks to the bottom. Is the level of the water higher, lower, or the same as it was when the bottle was floating?

An object only partially submerged in a liquid feels a buoyant force. According to Archimedes' principle, the magnitude of the force depends on the volume of liquid displaced; in other words, the buoyant force depends on the volume of that portion of the object that is submerged.

Example 11-6 Ice Cubes

At 0 °C, ice has density ρ_{ice} equal to 917 kg/m³. What fraction of an ice cube (**Figure 11-8**) floats below the surface in a glass of water at room temperature?

 Go to Interactive Exercise 11-2 for more practice dealing with the buoyant force

SET UP

The upward, buoyant force on the ice cube must equal its weight because it is floating and in equilibrium. We can therefore solve the problem by setting the two forces equal.

According to Archimedes' principle the buoyant force is equal to the weight of the water displaced by the ice cube. There is no requirement that the ice cube be completely submerged, however, and only the volume of ice below the water's surface contributes to the buoyant force. We must take this into account when setting the forces equal.

SOLVE

Once again, express the equivalence of the buoyant force and the weight of the displaced water:

$$F_b = W_{\text{water displaced}} = M_{\text{water displaced}}g = \rho_{\text{water}}V_{\text{water displaced}}g$$

Because the ice cube floats, the buoyant force F_b must equal the weight of the ice cube:

$$F_b = M_{\text{ice}}g = \rho_{\text{ice}}V_{\text{ice}}g$$

Figure 11-8

so

$$\rho_{\text{water}} V_{\text{water displaced}} g = \rho_{\text{ice}} V_{\text{ice}} g$$

But here, unlike in Example 11-5, the volume of water displaced is not equal to the volume of the whole ice cube because part of it sits above the water. Instead, the volume of displaced water is a fraction of the total volume of the ice cube, and it's that fraction that we want to find. Using f to represent the fraction, $V_{\text{water displaced}} = f V_{\text{ice}}$, so

$$\rho_{\text{water}} f V_{\text{ice}} g = \rho_{\text{ice}} V_{\text{ice}} g$$

or

$$f = \frac{\rho_{\text{ice}}}{\rho_{\text{water}}}$$

Substitute in our known values to calculate the fraction of the ice cube that is submerged:

$$f = \frac{917 \text{ kg/m}^3}{998 \text{ kg/m}^3} = 0.919$$

REFLECT

You may have heard it said that about 90% of an iceberg floats below the water surface; our calculation certainly supports that claim. (The percent of the ice cube submerged is 91.9%.) An iceberg in the ocean would actually float a bit higher because the density of sea water is higher than that of freshwater.

Practice Problem 11-6 At $0\,°C$, ice has density $\rho_{\text{ice}} = 917 \text{ kg/m}^3$. What fraction of an iceberg floats below the surface in seawater at a temperature of $4\,°C$ and a density of 1025 kg/m^3?

Figure 11-9a shows two identical objects hanging from spring balances. The object on the left is submerged in water, while the one on the right is not. Although both objects have the same mass M and the same volume V, the weights registered on the spring balances are not the same. Why? The buoyant force due to the displaced water opposes the force of gravity for the object on the left.

An object submerged in a fluid (that is denser than air) has an *apparent weight* less than its weight when submerged in air.

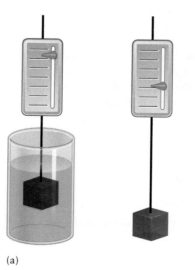

The bouyant force $\vec{F}_B$ in this free-body diagram opposes the weight force $\vec{w}$, reducing the net force. Net force is what the scale measures.

Figure 11-9 The apparent weight of an object hung from a spring balance depends on the fluid in which it is submerged.

(a)

(b)

A free-body diagram for the submerged object is shown in **Figure 11-9b**, where $\vec{W}$ is the weight of the object and $\vec{F}_b$ is the buoyant force. The spring balance registers the difference between these two; the magnitude of the net force on the object is its **apparent weight** W_{app}:

$$W_{app} = W - F_b = W - \rho_{water} V g$$

So by measuring the weight of an object while both submerged and not submerged, we can determine its volume:

$$V = \frac{W - W_{app}}{\rho_{water} g} \qquad (11\text{-}11)$$

We can also write the apparent weight in terms of the density of the object using $W = Mg = \rho_{obj} V g$. Using this expression, the equation for apparent weight becomes

$$W_{app} = W - F_b = \rho_{obj} V g - \rho_{water} V g \qquad (11\text{-}12)$$

Note that any of these expressions can be generalized for the case when an object is submerged in a fluid other than water; for example, Equation 11-11 becomes

$$V = \frac{W - W_{app}}{\rho_{fluid} g}$$

For an object of known volume (which can be determined using the weight and the apparent weight in a known fluid using Equation 11-11), we can also determine the density of the object. It was supposedly this realization by Archimedes (while submerged in a bath!) that led him to his understanding of buoyancy. More interesting, perhaps, is that if an object is composed of two materials of known density, it is possible to determine the volume of each by comparing the weight of the object to its apparent weight. This is the physics behind measurements of body fat, in which a person's weight in air and while submerged in water are used to determine the ratio of lean tissue to fat tissue in the body.

Example 11-7   Body Fat

In the human body, the density of lean tissue is often taken to be $\rho_{lean} = 1100 \text{ kg/m}^3$, and that of fat tissue to be $\rho_{fat} = 900 \text{ kg/m}^3$. If a person (perhaps one of the authors) who weighs $W = 940 \text{ N}$ (about 210 lb) in air has an apparent weight when submerged in water of $W_{app} = 60 \text{ N}$, what percent of his body is fat? Assume (somewhat unrealistically) that the person is composed only of lean and fat tissue.

 Go to Bio Animation 11-1 to explore body fat and buoyancy

SET UP

We observed that when an object is weighed while submerged in a fluid, the buoyant force acts against gravity. As a result the apparent weight is less than would be measured if the object were weighed while not submerged. From Equation 11-12:

$$W_{app} = W - \rho_{water} V g$$

To connect the weights and volumes in this problem, notice the assumption that the person is composed of lean and fat tissue only. Therefore, his total weight W is the sum of the weight of lean tissue and the weight of fat tissue:

$$W = W_{fat} + W_{lean} \qquad (11\text{-}13)$$

Likewise, the person's total volume V is the sum of the volumes of lean and fat tissues:

$$V = V_{lean} + V_{fat} \qquad (11\text{-}14)$$

Our goal is to find the body fat ratio W_{fat}/W. We will determine W_{fat} by first finding the volume of fat tissue, because we know the density of fat,

$$W_{fat} = \rho_{fat} V_{fat} g$$

so

$$\frac{W_{fat}}{W} = \frac{\rho_{fat} V_{fat} g}{W} \tag{11-15}$$

SOLVE

To connect the given densities to weight, in particular to the apparent weight in water, we express the weights in Equation 11-13 in terms of density and volume:

$$W = \rho_{fat} V_{fat} g + \rho_{lean} V_{lean} g$$

The volume of lean tissue isn't given and doesn't appear in the ratio W_{fat}/W, so we need to eliminate V_{lean}. We can do that using Equation 11-14:

$$W = \rho_{fat} V_{fat} g + \rho_{lean} (V - V_{fat}) g$$

Finally, we can introduce the apparent weight using Equation 11-11, which shows how apparent weight is related to total volume and total weight:

$$W = \rho_{fat} V_{fat} g + \rho_{lean} \frac{(W - W_{app})}{(\rho_{water} g)} g - \rho_{lean} V_{fat} g$$

A bit cumbersome, perhaps, but notice that the only unknown quantity in the expression is V_{fat}, which is what we need in order to find W_{fat}. Rearrange to get

$$V_{fat} = \frac{W - \rho_{lean} \dfrac{W - W_{app}}{\rho_{water}}}{(\rho_{fat} - \rho_{lean}) g}$$

So to find the percent body fat, we find the ratio given in Equation 11-15:

$$\frac{W_{fat}}{W} = \frac{\rho_{fat} V_{fat} g}{W} = \frac{\rho_{fat} g}{W} V_{fat} = \frac{\rho_{fat} g}{W} \left(\frac{W - \rho_{lean} \dfrac{W - W_{app}}{\rho_{water}}}{(\rho_{fat} - \rho_{lean}) g} \right)$$

A bit of simplification makes the expression look a little clearer:

$$\frac{W_{fat}}{W} = \frac{\rho_{fat}}{\rho_{fat} - \rho_{lean}} \left[1 - \frac{W - W_{app}}{W} \left(\frac{\rho_{lean}}{\rho_{water}} \right) \right]$$

In particular, this simplification makes it easier to see that our expression for the ratio W_{fat}/W is dimensionless because each fraction on the right-hand side is dimensionless. Using the known values, we find the ratio to be

$$\frac{W_{fat}}{W} = \frac{900 \text{ kg/m}^3}{900 \text{ kg/m}^3 - 1100 \text{ kg/m}^3} \left[1 - \frac{940 \text{ N} - 60 \text{ N}}{940 \text{ N}} \left(\frac{1100 \text{ kg/m}^3}{1000 \text{ kg/m}^3} \right) \right] = 0.13$$

The body fat ratio is 0.13 or 13%.

REFLECT

A healthy percentage of body fat for adult men is between 15 and 18%, so our subject clearly is not one of the authors! Note, though, that the method outlined above is only an approximation of an actual body fat calculation; we ignored the volumes of the lungs and bones in order to simplify problem.

Estimate It! 11-2 Lighter in Air

In discussing apparent weight we assumed that the weight of an object in air is its actual weight. Is that reasonable? After all, an object weighed in air experiences a buoyant force due to being submerged in air. Let's do a quick calculation to see if the buoyant force due to displaced air can be neglected when weighing, say, you. Estimate the buoyant force on you due to your being immersed in air.

SET UP
The buoyant force is described by Archimedes' principle:

$$F_b = \rho_{displaced} V_{displaced} g \qquad (11\text{-}10)$$

What volume of air do you displace? For an estimate, it's reasonable to imagine you to be cylindrical, so

$$V = \pi R^2 H$$

where R and H are the radius and height of a cylindrical you. Big, round—but reasonable—values could be $R = 0.2$ m and $H = 2$ m. The displaced fluid is air, so $\rho_{displaced} = \rho_{air}$. For the purposes of estimation, we'll take the density of air to be 1 kg/m^3, and let g be about 10 m/s^2.

SOLVE

$$F_b = \rho_{displaced} V_{displaced} g = \rho_{air} \pi R^2 H g$$

or

$$F_b \approx (1 \text{ kg/m}^3) \pi (0.2 \text{ m})^2 (2 \text{ m}) (10 \text{ m/s}^2) = 2.5 \text{ N}$$

To one significant digit, our estimate of the buoyant force on you in air is 3 N.

REFLECT
Is it important to consider a buoyant force on you of about 3 N? Thinking of weight as mass, that's about 0.3 kg, which is probably just a few tenths of a percent of your total weight. So it's not worth worrying about.

★ What's Important 11-7

An object surrounded by a fluid experiences a buoyant force equal to the weight of the fluid displaced by the object. When an object is weighed while immersed in a fluid, the apparent weight is its weight when measured in air reduced by the buoyant force. Because the buoyant force is proportional to the object's volume, measuring apparent weight in a known fluid allows a determination of an object's density.

11-8 Fluids in Motion and the Equation of Continuity

We have considered various properties of fluids at rest. Now let's consider the physics of fluids in motion. Examples of fluids in motion include blood circulating through the cardiovascular system of animals and water squirting out through a sprinkler head. Why does the water leaving the sprinkler shoot so far? How does the speed of blood change as it flows through our bodies? Our study of fluids in motion will answer these and other questions.

A thorough treatment of fluid dynamics is complex. Fluids can exhibit *turbulent flow*, which is characterized by chaotic whirlpools or eddy currents in some or all regions of the fluid. The eddies make it difficult to describe fluid flow without complex mathematics. But to understand general properties of moving fluids, we won't be too inaccurate if we consider only the special case of smooth and uniform flow. Under these conditions, the speed of the fluid is the same everywhere in a small region perpendicular to the direction of flow. We'll also require that the fluid flow is *irrotational*, which means that an object placed in the fluid does not rotate as it moves with the fluid, even if the net motion of the object is smooth and constant. Finally, as we did in our earlier consideration of static fluids, we'll consider only the special case of *incompressible* fluids; the density of these fluids is constant throughout their entire volume. Although all of these conditions are often not true for real fluids, these simplifying assumptions make the fundamental physical principles easier to understand.

Okay! Let's go back to the picture of the human circulatory system, in which blood leaves the heart through the aorta and eventually flows through billions of tiny capillaries. Although each capillary has a small diameter, the total cross-sectional area of all of the capillaries taken together is enormous. Schematically, we can imagine the arrangement of vessels to be as shown in Figure 11-10. In the figure

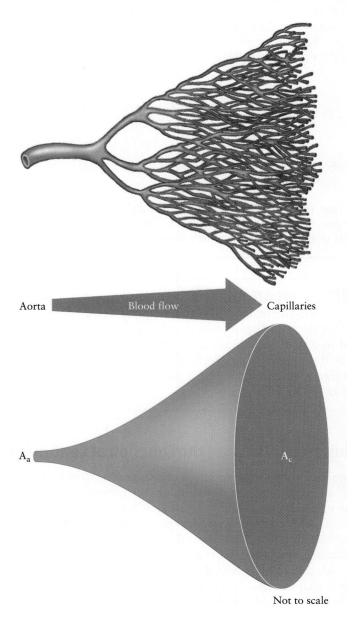

Aorta　　　Blood flow　　　Capillaries

A_a　　　　　　　A_c

Not to scale

Figure 11-10 Blood leaving the heart flows first through a single vessel, the aorta; branches off the aorta eventually lead to billions of capillaries. Although each capillary is very small, the total cross-sectional area of all capillaries taken together is large.

we see blood flowing through the aorta from the left and into a larger vessel on the right that represents all of the capillaries taken together. The cross-sectional area of the aorta is A_a and the cross-sectional area of all the capillaries taken together is A_c.

We'll consider the blood as the idealized fluid described above, flowing smoothly through from aorta through the capillaries. For blood in a circulatory system, this assumption will never be correct, because the flow of blood changes with the rhythm of the cycle of the heart. We will neglect the changes in flow, however, in order to uncover a fundamental property of fluids in motion.

We previously defined density as the quantity of mass in a region divided by the volume of that region (Equation 11-1). Because we have declared that blood is incompressible, the density of blood in the aorta (ρ_a) must be the same as the density of blood flowing through the capillaries (ρ_c), that is, $\rho_a = \rho_c$. The mass of blood in a given volume of the aorta (m_a) must, in turn, be the same as the mass of blood in the same volume of the capillaries (m_c), that is, $m_a = m_c$.

In any short period of time Δt, the volume of blood flowing past a point in the aorta (V_a) will equal the volume of blood flowing past a point in the capillaries (V_c):

$$V_a = V_c \qquad \text{(11-16)}$$

If we consider the blood vessels to be cylindrical, as shown in **Figure 11-11** the volume of a region is determined by the cross-sectional area of the vessel multiplied by the length of the region along the vessel:

$$V = A\,\Delta x$$

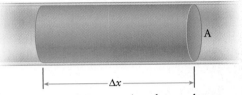

Figure 11-11 A volume element in a flowing fluid is the product of the cross-sectional area A of the tube and the distance Δx along the tube.

We can now relate the distance Δx that a small quantity of the fluid travels in a time Δt with the speed of the flow. The definition of speed is the time rate of change of position:

$$v = \lim_{\Delta t \to 0} \frac{\Delta x}{\Delta t}$$

Because our Δt is small we can eliminate the limit and write $\Delta x = v\,\Delta t$. The volume through which a small quantity of fluid flows in time Δt is then

$$V = Av\,\Delta t$$

So we can rewrite Equation 11-16, which states that the volume of blood flowing past a point in the aorta is equal to the volume of blood flowing past a point in the capillaries:

$$A_a v_a \Delta t = A_c v_c \Delta t$$

Here v_a and v_c represent the speed of blood through the aorta and through the capillaries, respectively. Because we considered the volume flowing in the two parts of the system during the same time Δt, we can reduce the equation to

$$A_a v_a = A_c v_c$$

This brings us to a general, powerful statement: If we measure the cross-sectional areas and flow velocities in two regions of a moving, ideal fluid,

$$A_1 v_1 = A_2 v_2 \qquad \text{(11-17)}$$

Said another way,

$$Av = \text{constant}$$

Because the product of the cross-sectional area and the speed of flow anywhere in a closed system of moving fluid (one to which no fluid is added and one from which no fluid leaves) is always the same, as the cross-sectional area goes up, the speed must go down and vice versa. If you've ever covered part of the end of a garden hose with your thumb in order to create a jet of water, you've experimentally tested this **equation of continuity**!

Notice that the equation of continuity describes the **volume flow** of the fluid, that is, the rate at which a given volume of fluid moves past a certain point. **Figure 11-12** makes this point clear; a volume of fluid moves through a pipe of changing cross-sectional area. On the left side, where the cross-sectional area of the tube is A_1, the fluid volume extends for a distance d_1. As we saw above, because we treat the fluid as incompressible, the volume of the fluid must remain the same as it moves to the right. Because the cross-sectional area there (A_2) is greater than on the left, the extent of the volume along the tube d_2 must be smaller than d_1. The speed of the fluid flow is determined by these distances; because d_2 is less than d_1, v_2 must be less than v_1. The higher cross-sectional area results in lower speed.

You can also see the relationship between the equation of continuity and volume flow by multiplying the units of area A (m²) with those of speed v (m/s) to get m³/s, or volume per time interval. As we noted early on in this chapter, most often we study the bulk properties of fluids. So here we concern ourselves not with the speed of a single molecule of a fluid, but rather the bulk or volume flow of the fluid.

? Got the Concept 11-4
Gushing Water

While walking past a construction site, you notice a pipe with water rushing out of it sticking out of a fifth floor window. Considering that as the water flows (falls) to the ground, it must speed up (conservation of energy applies to fluids, of course), what changes, if any, would you expect in the appearance of the flow as it nears the ground?

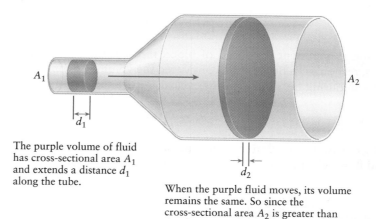

The purple volume of fluid has cross-sectional area A_1 and extends a distance d_1 along the tube.

When the purple fluid moves, its volume remains the same. So since the cross-sectional area A_2 is greater than A_1, the extent of the volume along the tube d_2 must be smaller than d_1.

Figure 11-12 A volume of fluid is pushed through a tube of varying cross-sectional area.

Example 11-8 Blood Flow

The typical diameter D_a of a human aorta is equal to 2.5 cm and the mean linear speed of blood through the aorta v_a is equal to 20 cm/s. If the capillaries are considered a single vessel their equivalent total cross-sectional area A_c would be about equal to 4000 cm². At what linear speed does blood flow through the capillaries?

SET UP

The equation of continuity describes the speed of flow through a tube with varying cross-sectional area as

$$A_1 v_1 = A_2 v_2 \qquad\qquad (11\text{-}17)$$

We can apply this equation to the problem at hand.

SOLVE

Rewriting the Equation 11-17 for the circulatory system gives

$$A_a v_a = A_c v_c$$

where the subscripts a and c represent the aorta and the capillaries. A_a can be expressed in terms of the given diameter of the aorta:

$$\pi \left(\frac{D_a}{2} \right)^2 v_a = A_c v_c$$

or

$$v_c = \frac{\pi}{A_c} \left(\frac{D_a}{2} \right)^2 v_a$$

Using the numbers provided in the problem statement,

$$v_c = \frac{\pi}{4000\ \text{cm}^2} \left(\frac{2.5\ \text{cm}}{2} \right)^2 20\ \text{cm/s} = 2.5 \times 10^{-2}\ \text{cm/s}$$

REFLECT

Our result is in good agreement with measured values. The very low speed of blood—a few cell diameters per second—through the capillaries also allows more time for exchange of nutrients, gases, and waste products between the blood and the surrounding tissues. This is a spectacular example of the interplay between physics and physiology!

Practice Problem 11-8 The mean linear speed v of water in a garden hose of diameter 1.3 cm is equal to 4.0 m/s. The hose is connected to a sprinkler head which has 75 nozzles, each 1 mm in diameter. At what linear speed does water leave each nozzle?

 Watch Out

The speed of blood increases as it moves from capillaries back to the heart.

Students sometimes expect that blood is moving slower when it gets to the heart than when it leaves the capillaries, a confusion that arises because of an expected relationship between pressure and speed. We'll see in the next example that as blood approaches the heart, pressure drops but speed actually increases because the total cross-sectional area of the vessels decreases.

Example 11-9 From Capillaries to the Vena Cavae

How fast will blood be moving when it returns to the heart from the capillaries through the vessels known as the vena cavae? The vena cavae have a cross-sectional area of about 10 cm².

SET UP

Once again, apply the equation of continuity (Equation 11-17), now between the vena cavae (subscript vc) and the capillaries (subscript c):

$$A_{vc} v_{vc} = A_c v_c$$

SOLVE

Solving the equation for v_{vc} yields

$$v_{vc} = \frac{A_c v_c}{A_{vc}}$$

Using A_c and v_c values from Example 11-8 and $A_{vc} = 10$ cm² from the problem statement,

$$v_{vc} = \frac{(4000 \text{ cm}^2)(2.5 \times 10^{-2} \text{ cm/s})}{10 \text{ cm}^2} = 10 \text{ cm/s}$$

REFLECT

Comparing our results with Example 11-8, clearly speed increases as blood returns to the heart.

 What's Important 11-8

The product of the cross-sectional area of a fluid flow and the linear speed at which it flows is constant. As the cross-sectional area increases, the speed decreases, and vice versa.

11-9 Fluid Flow—Bernoulli's Equation

Daniel Bernoulli, a brilliant and colorful 18th century mathematician, discovered that when fluid flows, the speed of the flow and the pressure at any given point are connected. More specifically, **Bernoulli's equation** tells us that in a fluid flowing horizontally, when speed is high, pressure is low, and when speed is low, pressure is high. In this section, we'll examine a number of curious effects that result from this simple relationship, including what keeps an airplane up in the air and how a thrown baseball can be made to curve. Bernoulli himself explored the math and physics behind this principle because of his interest in measuring blood pressure.

You can easily demonstrate Bernoulli's equation for yourself as shown in Figure 11-13. Drop a clear straw into a glass of water and blow across the top, open end. Without much effort, you should see water being drawn up into the straw from the glass. But let's be clear about the fluid to which we're applying Bernoulli's equation: it's the air, not the water! Air is the fluid in flow in this example. We consider two regions in the air; one is inside the straw and the other is right above the open end. Inside the straw the air speed is zero, while above the straw your breath makes the speed greater than zero. Because the air's speed is higher above the straw than inside it, the pressure above the straw must be lower above the straw than inside it. As we've seen, a pressure difference results in a net force, and in this case, the net force on the water in the straw is up. That net force is what pulls water up into the straw.

Although not exactly correct, it might be helpful to think of Bernoulli's equation working this way: As you blow across the straw, causing the speed in that region to be high, air molecules are drawn away from that region more quickly, lowering the pressure.

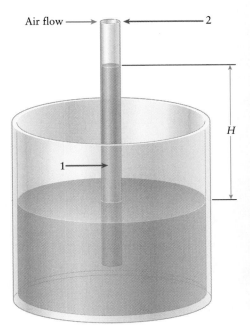

Figure 11-13

> ### ? Got the Concept 11-5
> #### Closing the Gap
>
> Imagine holding two pieces of paper vertically, with a small gap between them, and then blowing gently into the gap. How would you expect the pieces of paper to move? Next imagine tearing off a narrow strip of paper, holding one end of it below your mouth and then blowing gently across the top of it. How would you expect the strip of paper to move? Now go try these two experiments!

The exact relationship between speed and pressure in a moving fluid can be obtained by examining the work required to move a volume of the fluid, and the resulting change in energy. (Said another way, we want to apply a statement of energy conservation to a moving fluid.) Consider the volume of fluid shown flowing through a tube of varying cross-sectional area in Figure 11-14. Only a section of the tube is shown; the tube extends to both the right and left of the section shown in the figure.

Fluid flows from left to right in the figure; this motion occurs because the fluid to the left of the section of tube shown (in other words, fluid not shown in the figure) in the top part of the figure exerts a pressure P_1 and a related force F_1 on the fluid in this section of the tube. Fluid in the right side of the tube in turn feels a pressure P_2 and a related force F_2. In addition, it's important to note that when the left boundary of the volume of fluid shown has moved a distance d_1, as shown in the bottom part of the figure, the right boundary of the volume will have moved a smaller distance d_2 because the cross-sectional area of the tube is bigger on the right side than on the left side.

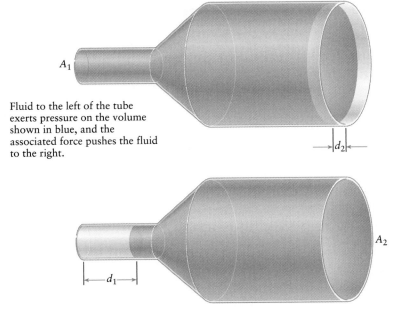

Fluid to the left of the tube exerts pressure on the volume shown in blue, and the associated force pushes the fluid to the right.

As the fluid moves through the tube, it in turn exerts a pressure on the fluid to the right. The distance d_2 is less than d_1 because the cross-sectional area A_2 is bigger than A_1.

Figure 11-14 Fluid flows from left to right because of the pressure difference between the left and right boundaries of this section of tube. Note that when the left boundary of the volume of fluid shown has moved a distance d_1 the right boundary of the volume moves a smaller distance d_2 because the cross-sectional area of the tube is bigger on the right side than on the left side.

When force and displacement are constant, we compute the work from $W = \vec{F} \cdot \vec{d}$, so the work done on the volume of interest due to P_1 is

$$W_1 = F_1 d_1$$

Likewise, the work done on the volume of fluid due to P_2 is

$$W_2 = -F_2 d_2$$

The minus sign comes from the dot product in the definition of work. Said another way, the force F_1 due to pressure P_1 is doing work on our volume of interest, but the force related to P_2 is due to the volume of interest itself doing work on the fluid to the left of it.

The work–energy theorem tells us that the total work done on the fluid must equal the change in its kinetic energy. That is, $W = \Delta K$, where

$$W = W_1 + W_2 = F_1 d_1 - F_2 d_2 \tag{11-18}$$

and

$$\Delta K = K_2 - K_1 = \frac{1}{2}mv_2{}^2 - \frac{1}{2}mv_1{}^2$$

You might be scratching your head about the mass term that appears in the last equation, because it's the first time in this chapter on fluids that we've used mass. Indeed, because we usually describe fluids by treating them as an average over many particles, we have been using density rather than mass when that quantity is required. Let's do the same here, by writing the mass in terms of density, that is $m = \rho V$. Also, because we are treating the fluid as incompressible, the volume of fluid that moves in regions 1 and 2 is the same, as is the density of the fluid in the two regions, so we can leave both variables without subscripts.

Putting the work–energy theorem all together, then, we have

$$\frac{1}{2}\rho V v_2{}^2 - \frac{1}{2}\rho V v_1{}^2 = F_1 d_1 - F_2 d_2 \tag{11-19}$$

The volume of fluid that moves is a constant, and as we saw in Section 11-8 when the cross-sectional area of the flow changes, this implies that

$$A_1 d_1 = A_2 d_2$$

Dividing through by the volume term gives

$$\frac{1}{2}\rho v_2{}^2 - \frac{1}{2}\rho v_1{}^2 = \frac{F_1 d_1}{A_1 d_1} - \frac{F_2 d_2}{A_2 d_2}$$

By now you recognize the definition of pressure that has appeared in the right-hand terms, and we can rewrite this as a simplified version of Bernoulli's equation:

$$\frac{1}{2}\rho v_2{}^2 - \frac{1}{2}\rho v_1{}^2 = P_1 - P_2$$

A more standard form of the equation shows the relationship between pressure and speed a bit more clearly:

$$\frac{1}{2}\rho v_2{}^2 + P_2 = \frac{1}{2}\rho v_1{}^2 + P_1 \tag{11-20}$$

or

$$\frac{1}{2}\rho v^2 + P = \text{constant}$$

You can see that in any region of a flowing fluid, the speed term and the pressure term must always produce the same sum. The result brings us back to Bernoulli's equation: If the speed in the flow gets higher, pressure must become lower to compensate.

Although it's been slightly hidden in the language of fluids, don't lose sight of the physics which underlies Bernoulli's equation, that of the conservation of energy.

Example 11-10 Water in a Straw

When you blow directly across the top, open end of a straw held vertically in a cup of water, the water level in the straw will rise. You can probably get water up to a height H equal to 2.5 cm in a straw in this way. (Try it!) With what speed are you able to blow across the straw?

SET UP
Let's think through a plan. A simple picture, such as the one in Figure 11-13, might help. Because air flows across the top of the straw but air in the straw is not moving, the speed of air must be higher above the straw than inside it. Bernoulli's equation tells us the pressure above the straw must therefore be lower than the pressure inside it, and this pressure difference results in a net force upward on the surface of the water in the straw. That net force will push an equivalent weight of water up into the straw.

SOLVE
Let's get quantitative. According to Bernoulli's equation,

$$\frac{1}{2}\rho v_2{}^2 + P_2 = \frac{1}{2}\rho v_1{}^2 + P_1 \tag{11-20}$$

The fluid in motion is air, so the relevant density is *not* that of water. Let's label region 1 as inside the straw and region 2 as directly above the open end.

Our goal is to determine v_2, the speed of the air you are forcing across the end of the straw, so let's solve for v_2:

$$v_2 = \sqrt{\frac{2\left(\frac{1}{2}\rho_{air}v_1{}^2 + P_1 - P_2\right)}{\rho_{air}}}$$

First, because the air in the tube is not flowing, $v_1 = 0$, so

$$v_2 = \sqrt{\frac{2\left(0 + P_1 - P_2\right)}{\rho_{air}}}$$

Also, note that the pressure difference $P_1 - P_2$ results in a net force F_{net} on the surface area of the water in the straw:

$$P_1 - P_2 = \frac{F_{net}}{A}$$

The net force F_{net} will balance the weight W of the water drawn up into the straw, so

$$v_2 = \sqrt{\frac{\dfrac{2F_{net}}{A}}{\rho_{air}}} = \sqrt{\frac{\dfrac{2W}{A}}{\rho_{air}}}$$

Finally, the weight of the volume of water pulled up into the straw is determined by $W = \rho_{water} AHg$, so

$$v_2 = \sqrt{\frac{\left(\dfrac{2\rho_{water}AHg}{A}\right)}{\rho_{air}}} = \sqrt{\frac{2\rho_{water}Hg}{\rho_{air}}}$$

Now we're ready for the numbers! Using known data and the information in the problem statement, the velocity is

$$v_2 = \sqrt{\frac{2\,(1000\ \text{kg/m}^3)\,(0.025\ \text{m})\,(9.8\ \text{m/s}^2)}{1.217\ \text{kg/m}^3}} = 20\ \text{m/s}$$

REFLECT

Is this a reasonable answer? Why not check it by conducting a simple experiment? Hold a piece of tissue at arm's length from your mouth, time approximately how long it takes for the tissue to move after you blow toward it, and compute the speed. Certain effects, such as air resistance, will make your result differ from the true speed with which you can blow air, but you should get close.

Practice Problem 11-10 You use a compressor to blow air at 25 m/s across the open end of a tube held vertically in a bucket of water. How high up into the tube will water rise?

The previous example is a simple form of a *venturi meter*, a device for measuring fluid speed, typically for a gas. A schematic of a venturi meter, or flowmeter, is shown in Figure 11-15. The cross-sectional areas in the regions marked 1 and 2 in the flow are A and a, respectively. Because A is greater than a, and from the equation of continuity ($Av = $ constant), the speed in region 2 must be higher than in region 1. Bernoulli's equation tells us that the pressure in region 2 must be less than the pressure in region 1; the pressure difference draws fluid up higher in the tube on the region 2 side. As in the previous example, the height of the fluid is directly related to the speed of the flow.

The Complete Form of Bernoulli's Equation

In Equation 11-18, the work done on the fluid is due only to the pressure of the fluid. If the ends of the tube are at different heights relative to some reference, however, we must account for the work that gravity does on the fluid as well.

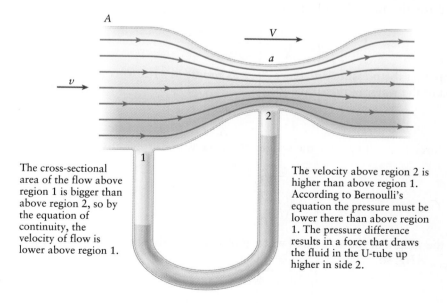

Figure 11-15 A simple form of a venturi meter, a device for measuring fluid speed. A venturi meter, or flowmeter, is typically used to measure the velocity of a gas.

The cross-sectional area of the flow above region 1 is bigger than above region 2, so by the equation of continuity, the velocity of flow is lower above region 1.

The velocity above region 2 is higher than above region 1. According to Bernoulli's equation the pressure must be lower there than above region 1. The pressure difference results in a force that draws the fluid in the U-tube up higher in side 2.

Consider the heights of the ends of the tube to be y_1 and y_2 as shown in Figure 11-16. Work must be done on the fluid by gravity as the fluid rises from y_1 to y_2. This work is

$$W_{\text{gravity}} = -mg(y_2 - y_1)$$

where the minus sign indicates that work is being done against the force of gravity. We can now add this term to Equation 11-18:

$$W = W_1 + W_2 + W_{\text{gravity}} = F_1 d_1 - F_2 d_2 - mg(y_2 - y_1)$$

Using the definition for mass in terms of density and volume, the work–energy theorem (Equation 11-19) then becomes

$$\frac{1}{2}\rho V v_2{}^2 - \frac{1}{2}\rho V v_1{}^2 = F_1 d_1 - F_2 d_2 - \rho V g(y_2 - y_1)$$

and again using the definition $V = Ad$, we can divide through by the (constant) volume:

$$\frac{1}{2}\rho v_2{}^2 - \frac{1}{2}\rho v_1{}^2 = P_1 - P_2 - \rho g y_2 + \rho g y_1$$

This expression is the complete form of **Bernoulli's equation**, which is often written as

$$\frac{1}{2}\rho v_2{}^2 + P_2 + \rho g y_2 = \frac{1}{2}\rho v_1{}^2 + P_1 + \rho g y_1 \qquad \text{(11-21)}$$

or

$$\frac{1}{2}\rho v^2 + P + \rho g y = \text{constant}$$

This equation is, again, a statement related to conservation of energy.

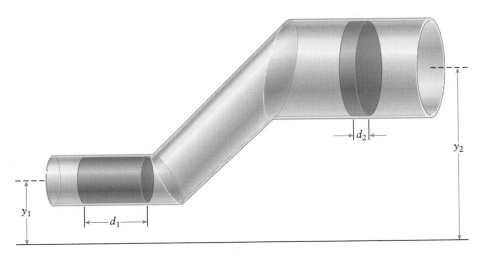

Figure 11-16 The work done on a fluid moving through a tube depends on the difference in pressure, but also in height, from region of the tube to another. The complete form of Bernoulli's equation takes into account the difference in height.

Figure 11-17 The speed of fluid flow around an airfoil is higher above it than below.

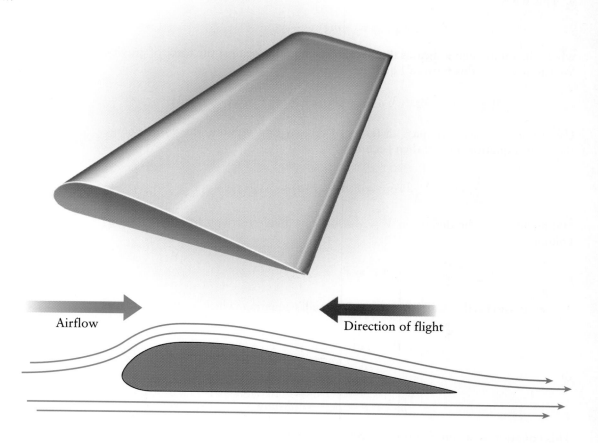

Airflow

Direction of flight

Airfoil Lift

While rolling down the runway in preparation for takeoff, a Boeing 747 jet is about 400,000 kg (400 tons) of metal, fuel, people, and luggage. We're so accustomed to the concept of flying, however, that you probably don't give too much thought to how it's possible to get all of that mass up into the air, and to keep it up until it reaches, say, Toronto. A big part of the answer lies in Bernoulli's principle.

The wings of an aircraft have a particular shape, often called an "airfoil." As you can see from the cross section of a typical airfoil shown in **Figure 11-17**, the bottom surface is relatively flat while the top is curved. The rounded edge of the airfoil is the front of the wing. On the top side of the wing air flows up over the rounded edge and back down to the tapered edge as the plane moves forward. In contrast, air moves straight back along the flat, bottom side of the wing. The difference in the two paths, and the difference in the shape of the top and bottom of the airfoil, causes air to move more quickly past the top of the airfoil than the bottom. This is where Bernoulli's equation comes in: Because the air speed is higher above compared to below the wing, the pressure above it is lower than the pressure below it. The pressure difference results in a net upward force on the wing. We call that upward force *lift*.

Often the wings of an airplane are tilted slightly upward. Because of this tilt, the force imparted to the bottom surfaces of the wings imparted by the air striking them has a vertical component, and this contributes to the upward force that enables an airplane to fly. This effect is not related to the physics of fluids, on which we've concentrated in this chapter, but no discussion of aerodynamic lift would be complete without including it.

Curveballs

The major league baseball pitcher goes into his windup and throws the ball at high speed toward the batter. It sure looks like the ball doesn't follow a straight path, or anything close. Did you imagine this, or did the ball really curve? In the early days of baseball, most people thought a curveball was just an optical illusion. It wasn't until 1941, when *Life* magazine published photographs of a curveball taken with a strobe light, that baseball fans (and everyone else) had proof that, thrown properly, a baseball can be made to curve.

The physics of the curveball is a bit complicated, as so many real-world situations are. A simple treatment, however, will get us relatively close to the way a real baseball moves. Imagine a baseball thrown without any spin, what ball players call a knuckleball. As the ball flies, the air rushes past uniformly on all sides. In the absence of other effects, then, since the speed of the air is the same on all sides, according to Bernoulli's equation the pressure in the airflow is the same as well. With pressure the same in every direction, there is no net force on the ball due to the air and it will go straight.

Now imagine the ball spinning as it moves. In **Figure 11-18**, we're looking at the ball from above, and it is rotating counterclockwise as seen from our vantage point. The ball is moving in the direction shown by the green arrow, and air rushes past it in the opposite direction. Because the surface of the ball is rough, and because the baseball has raised seams that hold the leather cover of the ball together (**Figure 11-19**), the spinning motion drags a layer of air around in the same direction of the spin of the ball.

The air being dragged along with the rotation of the ball moves in the same direction as the air flowing past the ball in the region to the left of the ball. But to the right of the ball the boundary layer of air is moving in opposition to the overall flow. As a result, the net speed of air is lower to the right of the ball than to the left. For the rotating ball, Bernoulli's equation suggests that the fluid velocities in different regions around the ball are not the same, so neither are the pressures. In this case, the lower speed to the right of the ball results in higher pressure, and the net force pushes the ball to the left. Viewed from above, the baseball will curve to the left!

It's interesting to note a few other effects that come into play for the pitched baseball. One effect is caused by the way that air is disturbed by the baseball moving through it. Any object moving through a fluid leaves a turbulent wake behind it. A symmetric, nonspinning object would create a uniform, symmetric wake as it passes

As the ball rotates counterclockwise (in this picture) a layer of air close to the ball is also dragged around counterclockwise. This boundary layer moves in the same direction as the air rushing past the ball on the left, but opposes the air flow on the right side of the bau.

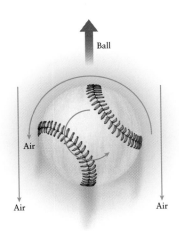

According to Bernoulli's equation, higher net air velocity on the left results in lower pressure. The pressure difference between the right and left sides of the ball results in a force, so this ball will curve to the left!

Figure 11-18 Airflow around a spinning, thrown baseball. The spinning motion drags a layer of air around in the same direction as the rotation of the ball.

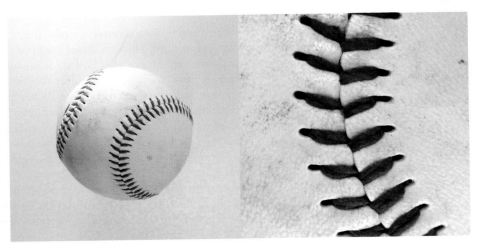

Figure 11-19 The surface of a baseball is rough and the seams that hold the leather cover of the ball together are raised above the surface of the ball. *(Courtesy David Tauck.)*

through a fluid. In contrast, a baseball drags a layer of air around as it spins. The boundary between that layer of air spinning with the ball and the stationary air through which it is moving separates into a turbulent wake, and because of the spinning motion, the separation occurs closer to one side of the ball than the other. You can see in Figure 11-19 that the separation occurs closer to the ball on the side where the boundary layer of air is moving in the direction opposite to the flow of air caused by the ball's motion. The net effect is a deflection of the ball; the associated force is known as the *Magnus force*. The deflection is in the same direction as the force due to the pressure difference predicted by Bernoulli's equation.

As any baseball pitcher will tell you, the orientation of the baseball's seams when the ball is released plays a major role in determining the path the ball takes. Given the discussion above that's not too surprising, since the very presence of seams must contribute to the effect that the ball has on the layer of air which surrounds it. But because of the distinctive pattern of the seams (**Figure 11-20**), a clever pitcher can release a baseball in a variety of orientations. For example, if a ball is released so that four regions of the seams are presented to the air rushing past it as it spins (a "four-seamer"), it approaches the batter at high speed and along a relatively straight line. A ball released so that only two regions of the seams are presented to the air (a "two-seamer") curves a lot and approaches the batter more slowly.

✳ What's Important 11-9

Bernoulli's equation describes the relationship between the pressure and the linear speed in a fluid flow. As speed increases, pressures decreases, and vice versa.

11-10 Viscous Fluid Flow

If you live near sea level, whenever you visit a place at high altitude you immediately notice that it's harder to exercise or even to catch your breath. Over time, however, one of the most important ways your body acclimates to high altitude is by producing more oxygen-transporting red blood cells. As the concentration of red blood cells increases, however, your blood becomes more viscous and flows a little less easily through your blood vessels. This phenomenon leads some people who stay at high altitude for too long to develop chronic mountain sickness, a potentially fatal disease in which the decrease in blood flow results in reduced oxygen delivery to their tissues. (Happily, most people will recover if they are evacuated to a lower elevation.) In this section, we explore viscosity and its effect on fluid flow through tubes.

We have seen that a fluid flows to conform to the shape of the container in which it is placed. This is a direct result of the fluid's inability to sustain a shear force, that is, a force pushing parallel to the surface. A fluid does, however, offer resistance to shear, a tendency referred to as **viscosity**. The higher the viscosity, the more resistive a fluid is to flow. We often refer to a higher viscosity fluid as being "thicker" than a less viscous one. Honey for example, has a higher viscosity than water.

Viscosity arises from friction between adjacent regions in a fluid a force that resists the motion of one region relative to the other. In the case of **laminar flow**, in which the fluid moves in parallel layers, viscosity is a manifestation of the frictional forces between layers of fluid sliding past one another.

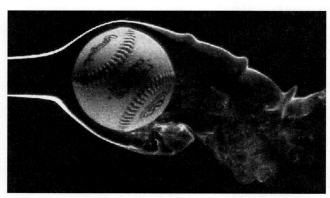

Figure 11-20 The separation between a thrown, spinning baseball and the boundary layer of air being dragged by the rotating surface. *(James J. Pallis, courtesy of Cislunar Aerospace Inc. and NASA Ames Research Center, California, USA.)*

Have you ever heated some honey in a microwave oven, in order to make it easier to pour from the container? Perhaps you've experienced poor performance of your car's engine on a winter morning when the oil in the pistons is cold. Both phenomena are due, in part, to the temperature dependence of the viscosity of a fluid. Viscosity decreases as temperature increases. Fluids expand when the fluid temperature increases, increasing the average distance between molecules and resulting in fewer interactions between them. Fewer interactions results in a lower average force between molecules and therefore a lower viscosity. Both honey and motor oil flow more easily as temperature increases.

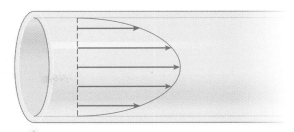

Figure 11-21 The speed of the fluid flowing in a tube is zero immediately adjacent to the walls of a straight tube, and the speed gets higher closer to the central axis.

Fluid does not flow at the same speed across the width of a tube. Both because of interactions between the fluid and the walls of the tube and because of the fluid's viscosity, the speed of the fluid immediately adjacent to the walls of a straight tube is zero, and speed gets higher closer to the central axis. A typical speed profile for a fluid in a straight tube is shown in **Figure 11-21**.

The drag force F that resists the motion of one layer in a fluid relative to the adjacent ones is proportional to the speed of flow v; the higher the speed, the higher the force. (The relationship changes when the speed of flow is so high that it is no longer laminar.) The resistive force also increases with the contact area A between regions of fluid, and decreases with the length L of the flow, for example, the length of a tube in which fluid is flowing. In other words,

$$F \propto \frac{vA}{L}$$

or

$$F = \eta \frac{vA}{L} \qquad \textbf{(11-22)}$$

where the proportionality constant η is viscosity. Notice that by arranging Equation 11-22 for η,

$$\eta = \frac{FL}{vA}$$

we see that the SI units of viscosity are

$$[\eta] = \frac{[F][L]}{[v][A]} = \frac{\text{N m}}{(\text{m/s}) \, \text{m}^2} = \text{N} \cdot \text{s/m}^2$$

The flow of fluid in a tube is a direct result of a pressure difference between two points in the flow. It's not surprising, then, that the units of pressure, N/m^2, appear in the units of viscosity. The units of viscosity can also be written in terms of pressure as pascal-seconds ($\text{Pa} \cdot \text{s}$).

At room temperature (20 °C), motor oil has a viscosity of about $0.1 \, \text{N} \cdot \text{s/m}^2$. At the same temperature honey is rather thicker, with a viscosity around $10 \, \text{N} \cdot \text{s/m}^2$, and tar, which is much thicker, has a viscosity of $30,000 \, \text{N} \cdot \text{s/m}^2$. For comparison, water has viscosity equal to $0.001 \, \text{N} \cdot \text{s/m}^2$ at room temperature. To give an idea of the temperature dependence of viscosity, at 40 °C the viscosity of water decreases to $0.0006 \, \text{N} \cdot \text{s/m}^2$ and at 100 °C, the viscosity of water is $0.00028 \, \text{N} \cdot \text{s/m}^2$. The viscosity of blood at human body temperature is between 0.003 and $0.004 \, \text{N} \cdot \text{s/m}^2$. The viscosity of other fluids is listed in **Table 11-2**.

We defined the volume flow rate (Equation 11-17) as the product of the cross-sectional area of a fluid flow A and the speed of flow v. Taking Q to represent the volume of fluid moving past a fixed point,

$$Q = Av \qquad \textbf{(11-23)}$$

Table 11-2 Viscosities of Selected Substances	
Substance	**Viscosity (N · s/m²)**
blood (37 °C)	0.003 – 0.004
chocolate syrup (20 °C)	1.5
honey (20 °C)	10
ketchup (20 °C)	5.1
motor oil (20 °C)	0.1
peanut butter (20 °C)	25.5
tar (20 °C)	3×10^4
seawater (20 °C)	1.0×10^{-3}
water (20 °C)	1.0×10^{-3}
water (40 °C)	6.0×10^{-4}
water (100 °C)	2.8×10^{-4}

The speed at which fluid flows depends on the pressure difference along the flow:

$$v = \frac{\Delta P A}{8\pi\eta L} = \frac{\Delta P \pi r^2}{8\pi\eta L} \tag{11-24}$$

where ΔP is the pressure difference between two points separated by distance L along a flow of cross-sectional area A. The volume flow rate is then

$$Q = \frac{\Delta P (\pi r^2)^2}{8\pi\eta L} = \frac{\Delta P \pi r^4}{8\eta L} \tag{11-25}$$

This equation is **Poiseuille's law** for laminar flow. Jean Louis Marie Poiseuille, a nineteenth century French physician, arrived at this relationship between pressure difference and the volume flow rate in his study of blood flow. Blood flow is often turbulent, so Poiseuille's law is at best an approximation to blood flow in the body. However, Poiseuille's law does describe qualitatively many aspects of blood flow. For example, changes in both blood vessel size and blood pressure regulate blood flow. Because healthy arteries are elastic and not hard-walled pipes, when blood pressure in an artery increases, its walls stretch, causing an increase in the radius of the vessel. Notice that in Poiseuille's law, flow rate is proportional to the fourth power of vessel radius. So, even a small change in pressure can have a big effect on blood flow because not only does it affect the driving force, it also changes the radius of the artery. When your muscles demand higher blood flow during exercise, more blood vessels open in the muscles, heart rate goes up, blood pressure increases, and the increased blood pressure causes an increase in the radius of the arteries, dramatically increasing the flow rate to the muscles.

 Go to Bio Animation 11-2 to explore ischemic heart disease and Poiseuille's Law

? Got the Concept 11-6
Plugged with Plaque

Inadequate blood flow to the heart, *ischemic heart disease*, affects millions of people and is the most common cause of death in the developed world. The disease begins when fat and cholesterol accumulate in the walls of arteries supplying the heart muscle. This leads to the formation of plaque that protrudes into the lumen of the vessels. How does Poiseuille's law explain the effect of plaque on blood flow to the heart?

> **? Got the Concept 11-7**
> **Hosed**
>
> Water flows through a garden hose at a volume flow rate of 8 cm^3/s as the water pressure drops to zero at the open end of the hose. What flow rate would you expect if the length of the hose were doubled?

> **? Got the Concept 11-8**
> **Hung from a Pole**
>
> Intravenous fluids are often administered to patients from a bag hung high on a pole. What advantage does placing the fluid container up above the patient provide?

Example 11-11 IV Fluids

A patient receives fluids intravenously (IV) at a rate of 1.0×10^{-7} m^3/s (360 mL/h) through a needle 2.0 cm long and 0.30 mm in diameter. The bag of saline solution hangs from a pole and is connected to the needle by a tube. Blood pressure in the patient's vein is 7.0 mmHg (933 Pa). (a) What pressure must be supplied at the end of the tube (where it connects to the needle)? (b) How high above the end of the tube should the surface of the IV solution be hung? Take the viscosity and density of the solution to be the same as seawater at room temperature, about 0.0010 Pa · s (from Table 11-2) and about 1025 kg/m^3, respectively. Assume laminar flow in the tube.

SET UP

Poiseuille's law provides the relationship between the volume flow rate from the output of the needle as a function of the pressure difference between the ends of the needle as well as other parameters in the problem. So first, rearrange Poiseuille's law (Equation 11-25) for pressure difference:

$$\Delta P = \frac{8\eta L Q}{\pi r^4}$$

Here, L is the length and r is the inner radius of the needle. Because the outlet of the needle is inside the vein, the pressure at the outlet of the needle is equal to the blood pressure P_{blood}. For that reason ΔP is the difference between the pressure in the vein and the pressure at the inlet of the needle, so

$$P_{\text{needle inlet}} - P_{\text{blood}} = \frac{8\eta L Q}{\pi r^4}$$

or

$$P_{\text{needle inlet}} = \frac{8\eta L Q}{\pi r^4} + P_{\text{blood}} \qquad (11\text{-}26)$$

In addition, elevating the bag of fluid above the patient, and therefore above the inlet of the needle, results in an increase in fluid pressure. The pressure in a column of fluid of density ρ is

$$P = P_0 + \rho g d \qquad (11\text{-}6)$$

where in this case P is the pressure at the inlet of the needle, P_0 is the pressure at the surface of the fluid (1 atm), and d is the height of the surface of the saline solution above the point at which the tube connects to the needle. The pressure at the needle is 1 atm in the absence of the column of fluid, so the difference in pressure, that is, the pressure that drives fluid through the needle is the difference between P in Equation 11-6 and 1 atm:

$$P_{\text{needle inlet}} - 1 \text{ atm} = (1 \text{ atm} + \rho_{\text{blood}}gd) - 1 \text{ atm} = \rho_{\text{blood}}gd \quad (11\text{-}27)$$

SOLVE

(a) The pressure at the inlet of the needle required to achieve a volume flow rate Q of $1.0 \times 10^{-7} \text{ m}^3/\text{s}$ is found using Equation 11-26, with η equal to 0.0010 Pa·s, L equal to 2.0 cm (0.020 m), r equal to 0.00015 m (one half the needle's diameter of 0.30 mm), and P_{blood} equal to 933 Pa:

$$P_{\text{needle inlet}} = \frac{8(0.0010 \text{ Pa} \cdot \text{s})(0.020 \text{ m})(1.0 \times 10^{-7} \text{ m}^3/\text{s})}{\pi(0.00015 \text{ m})^4} + 933 \text{ Pa}$$

or

$$P_{\text{needle inlet}} = 1.1 \times 10^4 \text{ Pa}$$

(b) We convert this to a height above the needle (and patient) using Equation 11-27:

$$P_{\text{needle inlet}} = \rho_{\text{blood}}gd$$

or

$$d = \frac{P_{\text{needle inlet}}}{\rho_{\text{blood}}g} = \frac{1.1 \times 10^4 \text{ Pa}}{(1025 \text{ kg/m}^3)(9.8 \text{ m/s}^2)} = 1.1 \text{ m}$$

where we used the density of seawater at room temperature, about 1030 kg/m^3, as the density of the IV fluid.

REFLECT

The bag of IV saline solution must be hung about 1 m above the patient, a reasonable height.

Practice Problem 11-11 Health codes often require that the volume of air in a classroom be changed 8 to 12 times per hour. If the ventilation system can exert a net pressure of 0.70 atm on the air in a 30-m-long duct, what is the radius of the duct running to 10 classrooms each of which is 8 m wide, 15 m deep, and 4 m high? Assume 10 volume changes per hour. The viscosity of air at room temperature is 1.8×10^{-5} Pa·s.

✱ What's Important 11-10

Viscosity characterizes a fluid's resistance to flow. The higher the viscosity, the more resistive a fluid is to flow. Poiseuille's law describes the relationship between pressure difference and the volume flow rate of a fluid in laminar flow.

Answers to Practice Problems

11-1 0.068 m³

11-2 1.4 atm

11-3 9.4 × 10³ N outward

11-4 850 m below sea level. (The lowest point near the Dead Sea is 423 m below sea level.)

11-5 3.4 × 10⁵ N

11-6 89.5%

11-8 9.0 m/s

11-10 3.9 cm

11-11 2.3 cm

Answers to Got the Concept Questions

11-1 Air is a fluid, so higher up is less "deep" in the fluid. That means that the ambient air pressure is lower at high altitudes. Even though the cabins of commercial planes are pressurized, the pressure you experience while flying is lower than when you're on the ground. Because you had your water bottle open at that low pressure and then sealed it, the inside of the bottle remains at low pressure when you land. Because the air pressure outside the bottle is now higher than inside, the bottle will be crushed inward. When you open the cap, the pressure difference drives air into the bottle; you'll hear a rush of air, and the plastic bottle will be restored to its normal shape.

11-2 Regardless of whether the coin is on the top or bottom of the cube, because the cube and coin are floating, the buoyant force of the water displaced must equal the weight of the cube plus the weight of the coin. (If that weren't true, the cube and coin would sink.) So the amount of water displaced is the same in both cases. However, when the coin is on the top, only the cube is displacing water, but when the coin is on the bottom, both the cube and the coin displace water. That means that in the second case, the volume of the coin is contributing to the buoyant force, which means that the cube won't displace as much water as when the buoyant force was due totally to the amount of the cube submerged. For that reason, more of the cube will be above the surface of the water when the coin is taped to the bottom. The cube would float higher if a wine bottle cork were substituted for the coin.

11-3 When the bottle was floating, it displaced a volume of water equal to its weight. That's because in order to float, the buoyant force must equal the bottle's weight, and according to Archimedes' principle, the buoyant force is also equal to the weight of the displaced water. However, when the bottle has sunk to the bottom of the tub, the volume of water it displaces is equal only to the volume of the glass which forms the bottle. Since the bottle sank, we know that the glass must be more dense than water, which means the glass itself takes up less volume than the equivalent weight of water. The net result is that when the bottle sinks, the water level goes down!

11-4 According to the equation of continuity, the product of the cross-sectional area of a flow of fluid and the speed of flow is constant. For that reason, as the speed of the falling water increases, the cross-sectional area must decrease. The "neck-down" effect is visible even for height differences of about a meter and can be quite noticeable, for example, in waterfalls at Yosemite National Park.

11-5 According to Bernoulli's equation, when two regions in a flowing fluid have different velocities, the pressure must be lower in the region where the speed is higher. When you blow between two papers, because air flows faster between the papers than in the region surrounding the papers, the pressure between them is lower than in the surrounding region, causing the papers to be drawn together. (Not pushed apart!) When you blow across the top of a strip of paper, for the same reason, the pressure above the strip will be lower than below the strip, which causes the strip to rise up.

11-6 According to Poiseuille's law, as the radius of an artery shrinks, blood flow through it decreases dramatically because the volume flow rate depends on the fourth power of radius. The only way to maintain flow under such conditions is to increase blood pressure. The heart needs even more nutrients to generate higher pressure. Eventually blood flow to the heart fails to provide the muscle with the nutrients it needs, and cardiac cells begin to die.

11-7 Because the pressure difference between the ends of the hose is the same in both cases, the only difference between the two cases is the length of the hose. Volume flow rate in Poiseuille's law (Equation 11-24) is inversely proportional to the length of the tube in which fluid flows, so doubling the length results in the flow dropping by a factor of 2. The flow rate in the longer hose is 4 cm³/s.

11-8 Elevating the container with respect to the patient results in a higher pressure at the outlet of the tube, and therefore where the needle is connected to the patient's vein. The pressure difference is necessary to overcome the (blood) pressure in the vein and enable flow of the IV fluid into the vein.

SUMMARY

Topic	Summary	Equation or Symbol
absolute pressure	Absolute pressure is the actual pressure in a system. For example, the absolute pressure in a deflated car tire is 1 atm, in contrast to the gauge pressure, which is 0 atm.	
apparent weight	When an object is weighed while suspended in a fluid, the measured weight, called the apparent weight, is its actual weight reduced by the buoyant force.	W_{app}
Archimedes' principle	Archimedes' principle describes buoyancy in a fluid. The buoyant force F_b on an object immersed in a fluid is equal to the weight of the fluid displaced by the object.	$F_b = \rho_{displaced} V_{displaced} g$ $\qquad$ (11-10)
Bernoulli's equation	Bernoulli's equation relates the speed of flow v, the pressure P, the density ρ, and the height y above a reference height of a moving fluid. According to Bernoulli's equation, as the speed of a fluid flowing horizontally increases, its pressure decreases, and vice versa.	$\frac{1}{2}\rho v_2{}^2 + P_2 + \rho g y_2 = \frac{1}{2}\rho v_1{}^2 + P_1 + \rho g y_1$ $\qquad$ (11-21)
buoyant force	An object partially or completely submerged in a fluid feels an upward, buoyant force due to the fluid it displaces. Archimedes' principle describes the magnitude of the buoyant force.	F_b
density	Density ρ is defined as mass M per unit volume V. An object's density does not directly indicate its weight—a small, dense object can weigh more than a large, less dense one.	$\rho = \dfrac{M}{V}$ $\qquad$ (11-1)
equation of continuity	For an ideal fluid, one that flows uniformly and is incompressible, the product of the flow speed v and cross-sectional area A of the flow is a constant.	$A_1 v_1 = A_2 v_2$ $\qquad$ (11-17)
fluid	The molecules in a fluid are not arranged in any organized structure. A fluid does not have any shape of its own; it flows to conform to the shape of the container that holds it.	
gauge pressure	Gauge pressure is equivalent to the pressure in a system above 1 atm. For example, the gauge pressure in a deflated car tire is zero, in contrast to the absolute pressure, which is 1 atm.	
laminar flow	In laminar flow, fluid moves in parallel layers.	
Pascal's principle	According to Pascal's principle, when pressure is applied to a confined, static fluid, it is transmitted to every part of the fluid as well as to the walls of the container holding it.	

Poiseuille's law	Poiseuille's law describes the relationship between pressure difference and the volume flow rate for laminar flow in a fluid.	$Q = \dfrac{\Delta P (\pi r^2)^2}{8\pi \eta L} = \dfrac{\Delta P \pi r^4}{8\eta L}$	(11-25)
pressure	Because a fluid flows to conform to the shape of its container, it exerts a force F on the surface area A of the container and perpendicular to it. The proportionality constant between force and area is pressure.	$\vec{F} = P\vec{A}$	(11-2)
pressure difference	The pressure difference ΔP between two regions results in a net force $\vec{F}_{net}$ on a surface of area A. The area vector $\vec{A}$ points perpendicular to the surface and has magnitude equal to the area.	$\vec{F}_{net} = \Delta P \vec{A}$	(11-7)
pressure versus depth	Pressure in a fluid depends on depth. Pressure varies with depth because pressure is a measure of the weight of all the fluid directly above the point of interest.	$P = P_0 + \rho g d$	(11-6)
specific gravity	The specific gravity of a substance is the ratio of the density of the substance to the density of water at the standard temperature of $4\,°C$ (277.15 K).		
viscosity	Viscosity is a manifestation of the frictional forces between layers of fluid sliding past one another. The SI units of viscosity are $N \cdot s/m^2$.	η	
volume flow	Volume flow, or volume flow rate, describes the rate at which a volume of fluid moves.	Q	

QUESTIONS AND PROBLEMS

In a few problems, you are given more data than you actually need; in a few other problems, you are required to supply data from your general knowledge, outside sources, or informed estimate.

Interpret as significant all digits in numerical values that have trailing zeros and no decimal points.

For all problems, use $g = 9.8\ m/s^2$ for the free-fall acceleration due to gravity. Neglect friction and air resistance unless instructed to do otherwise.

- Basic, single-concept problem
- •• Intermediate-level problem, may require synthesis of concepts and multiple steps
- ••• Challenging problem

SSM *Solution is in Student Solutions Manual*

Conceptual Questions

1. •Aluminum is more dense than plastic. You are given two identical, closed, and opaque boxes. One con-

tains a piece of aluminum and the other contains a piece of plastic. Without opening the boxes, is it possible to tell which box contains the plastic and which box contains the aluminum?

2. •Two identically shaped containers in the shape of a truncated cone are placed on a table, but one is inverted such that the small end is resting on the table. The containers are filled with the same height of water. The pressure at the bottom of each container is the same. However, the weight of the water in each container is different. Explain why this statement is correct.

3. •A dam will be built across a river to create a reservoir. Does the pressure in the reservoir at the base of the dam depend on the shape of the dam, given that the depth of the reservoir will be the same regardless of the choice of dam shape? Explain your answer. SSM

4. •**Medical** After sitting for many hours during a trans-Pacific flight, a passenger jumps up quickly as soon as the "fasten seat belt" light is turned off, and immediately falls to the floor, unconscious! Within seconds of being in the horizontal position, however, consciousness is restored. What happened?

5. •**Medical** Usually blood pressure is measured on the arm at the same level as the heart. How would the results differ if the measurement were made on the leg instead?

6. •**Medical** When you cut your finger badly, why might it be wise to hold it high above your head?

7. •**Medical** When you donate blood, is the collection bag held below or above your body? Why?

8. •An ice cube floats in a glass of water so that the water level is exactly at the rim. After the ice cube melts, will all the water still be in the glass? Explain your answer.

9. •The salinity of the Great Salt Lake varies from place to place, and from about two to as much as eight times the salinity of ocean water. Is it easier or harder for a person to float in the Great Salt Lake compared to floating in ocean water? Why? SSM

10. •A wooden boat floats in a small pond, and the level of the water at the edge of the pond is marked. Will the level of the water rise, fall, or stay the same when the boat is removed from the pond?

11. •You are given two objects of identical size, one made of aluminum and the other of lead. You hang each object separately from a spring balance. Because lead is more dense than aluminum, the lead object weighs more. Now you weigh each object while it is submerged in water. Will the difference between the weight of the aluminum object and weight of the lead object be greater than, less than, or the same as it was when the objects were weighed in air?

12. •A river runs through a wide valley and then through a narrow channel. How do the velocities of the flows of water compare between the wide valley and the narrow channel?

13. •A dam holds back a very long, deep lake. If the lake were twice as long, but still just as deep, how much thicker would the dam need to be?

14. •The wind is blowing from west to east. Should landing airplanes approach the runway from the west or the east? Explain your answer.

15. •Use Bernoulli's equation to explain why a house roof is easily blown off during a tornado or hurricane. SSM

16. •Why does the stream of water from a faucet become narrower as it falls?

17. •A cylindrical container is filled with water. If a hole is cut on the side of the container so that the water shoots out, what is the direction of the water flow the instant it leaves the container?

Multiple-Choice Questions

18. •Object A has density ρ_1. Object B has the same shape and dimensions as object A, but it is three times as massive. Object B has density ρ_2 such that
  A. $\rho_2 = 3\rho_1$.
  B. $\rho_2 = \dfrac{\rho_1}{3}$
  C. $\rho_2 = \rho_1$.
  D. $\rho_2 = \dfrac{\rho_1}{3}$.
  E. $\rho_2 = \dfrac{\rho_1}{2}$.

19. •If the gauge pressure is doubled, the absolute pressure will
  A. be halved.
  B. be doubled.
  C. be unchanged.
  D. be increased, but not necessarily doubled.
  E. be decreased, but not necessarily halved. SSM

20. •A toy floats in a swimming pool. The buoyant force exerted on the toy depends on the volume of
  A. water in the pool.
  B. the pool.
  C. the toy under water.
  D. the toy above water.
  E. none of the above choices.

21. •An object floats in water with 5/8 of its volume submerged. The ratio of the density of the object to that of water is
  A. 8/5.
  B. 5/8.
  C. 1/2.
  D. 2/1.
  E. 3/8.

22. •An ice cube floats in a glass of water. As the ice melts, what happens to the water level?
  A. It rises.
  B. It remains the same.
  C. It falls by an amount that cannot be determined from the information given.
  D. It falls by an amount proportional to the volume of the ice cube.
  E. It falls by an amount proportional to the volume of the ice cube that was initially above the water line.

23. •Water flows through a 0.5-cm-diameter pipe connected to a 1-cm-diameter pipe. Compared to the speed of the water in the 0.5-cm pipe, the speed in the 1-cm pipe is
  A. one-quarter the speed in the 0.5-cm pipe.
  B. one-half the speed in the 0.5-cm pipe.
  C. the same as the speed in the 0.5-cm pipe.

This page is intentionally left blank.

For complete end of chapter problem sets, please go to
www.whfreeman.com/kestentauck

45. •**Astro** When a star reaches the end of its life it is possible for a supernova to occur. This may result in the formation of a very small, but very dense, neutron star. The density of such a stellar remnant is about the same as a neutron. Determine the radius of a neutron star that has the mass of our Sun and the same density as a neutron. A neutron has a mass of 1.7×10^{-27} kg and an approximate radius of 1.2×10^{-15} m. Assume that both the neutron star and the neutron itself are spherical in shape. SSM

11-2: Pressure

46. •(a) An object of mass m_1 is supported by a square surface area of side s_1. Find the equation for the mass m_2 of a second object that will produce the same pressure when supported by a square surface area of side s_2. (b) What is the ratio of m_2 to m_1 if $s_2 = 5s_1$?

47. ••**Calc** Assuming that Earth is a sphere of radius 6380 km and that the pressure is 1 atm at the surface, give a maximum estimate of the mass of the atmosphere.

48. •An elephant that has a mass of 6000 kg evenly distributes her weight on all four feet. (a) If her feet are approximately circular and each has a diameter of 50 cm, estimate the pressure on each foot. (b) Compare the answer in part (a) with the pressure on each of your feet when you are standing up. Make some rough but reasonable assumptions about the area of your feet.

49. •The head of a nail is 0.32 cm in diameter. You hit it with a hammer with a force of 25 N. (a) What is the pressure on the head of the nail? (b) If the pointed end of the nail, opposite to the head, is 0.032 cm in diameter, what is the pressure on that end? SSM

11-3: Pressure versus Depth in a Fluid

50. •Find the pressure at the bottom of a graduated cylinder that is half full of mercury and half full of water. Assume the height of the cylinder is 0.25 m and don't forget about atmospheric pressure!

51. •A swimming pool 5.0 m wide by 10 m long is filled to a depth of 10 m. What is the gauge pressure on the bottom?

52. •A diver is 10.0 m below the surface of the ocean. The surface pressure is 1 atm. What is the absolute pressure and gauge pressure he experiences? The density of seawater is 1.025×10^3 kg/m³.

53. •What is the difference in blood pressure (mmHg) between the top of the head and bottom of the feet of a 1.75-m-tall person standing vertically? The density of blood is 1.06×10^3 kg/m³. SSM

54. •Determine the pressure at the point P in the U-tube manometer in each of the cases shown **Figure 11-22**. In all cases, the distance between points O and P is 37 cm.

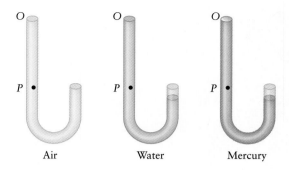

Figure 11-22 Problem 54

55. • At 25 °C the density of ether is 72.7 kg/m³ and the density of iodine is 4930 kg/m³. A cylinder is filled with iodine to a depth of 1.5 m. How tall would a cylinder filled with ether need to be so that the pressure at the bottom is the same as the pressure at the bottom of the cylinder filled with iodine?

11-4: Atmospheric Pressure and Common Pressure Units

56. •Convert the following units of pressure to the SI unit of pascals (Pa), where 1 Pa = 1 N/m².
    (a) 1500 kPa = _____ Pa
    (b) 35 psi = _____ Pa
    (c) 2.85 atm = _____ Pa
    (d) 883 torr = _____ Pa

57. •**Medical** Suppose that your pressure gauge for determining the blood pressure of a patient measured absolute pressure instead of gauge pressure. How would you write the normal value of systolic blood pressure, 120 mmHg, in such a case? SSM

58. •Assuming the atmospheric pressure is 1 atm at sea level, determine the atmospheric pressure at Badwater (in Death Valley, CA) where the elevation is 85 m below sea level.

59. •Elaine wears her wide-brimmed hat at the beach. If the atmospheric pressure at the beach is exactly 1 atm, determine the weight of the imaginary column of air that "rests" on her hat if its diameter is 45 cm.

60. •What is the *absolute* pressure of the air inside a bicycle tire that is inflated with a hand pump to 65 psi? Give your answer in pascals (Pa).

11-5: Pressure Difference and Net Force

61. •What is the net force on an airplane window of area 1000 cm² if the pressure inside the cabin is 0.95 atm and the pressure outside is 0.85 atm? SSM

62. •A rectangular swimming pool is 8 m × 35 m in area. The depth varies uniformly from 1 m in the shallow to 2 m in the deep end. (a) Determine the pressure at the deep end of the pool and at the shallow end. (b) What is the net force on the bottom of the pool due to the water in the pool? (Ignore the effects of the air above the pool for this part.)

63. •What force must the surface of a basketball withstand if it is inflated to a pressure of 8.5 psi? Assume the ball has a diameter of 23 cm.

64. •What is the net force on the walls of a 55-gal drum that is on the bottom of the ocean (the specific gravity of the ocean is 1.025) at a depth of 250 m. Assume the drum is a cylinder that has a height of $34\frac{1}{2}$ in. and a diameter of $21\frac{5}{8}$ in. (Remember to convert inches to meters!) Consider the interior pressure of the drum to be 1 atm.

65. •Suppose that the hatch on the side of a Mars lander is built and tested on Earth so that the internal pressure just balances the external pressure. The hatch is a disk 50.0 cm in diameter. When the lander goes to Mars, where the external pressure is 650 N/m², what will be the net force (in newtons and pounds) on the hatch, assuming that the internal pressure is the same in both cases? Will it be an inward or an outward force? SSM

11-6: Pascal's Principle

66. •What is the maximum weight that can be raised by the hydraulic lift shown in **Figure 11-23**?

Figure 11-23 Problem 66

67. •A hydraulic lift is designed to raise a 900-kg car. If the "large" piston has a radius of 35 cm and the "small" piston has a radius of 2 cm, determine the minimum force exerted on the small piston to accomplish the task.

68. •A hydraulic lift has a leak so it is only 75% efficient in raising its load. If the large piston exerts a force of 150 N when the small piston is depressed with a force of 15 N and the radius of the small piston is 0.05 m, what is the radius of the large piston?

69. •(a) What force will the large piston provide if the small piston in a hydraulic lift is moved down as shown in **Figure 11-24**? (b) If the small piston is depressed a distance of Δy_1, by how much will the large piston rise? (c) How much work is done in slowly pushing down the small piston compared to the work done in raising the large piston if $\Delta y_1 = 0.20$ m? SSM

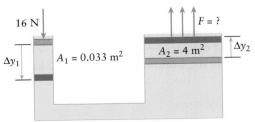

Figure 11-24 Problem 69

11-7: Buoyancy—Archimedes' Principle

70. •A rectangular block of wood, 10 cm × 15 cm × 40 cm, has a specific gravity of 0.6. (a) Determine the buoyant force that acts on the block when it is placed in a pool of freshwater. *Hint:* Draw a free-body diagram labeling all of the forces on the block. (b) What fraction of the block is submerged? (c) Determine the weight of the water that is displaced by the block.

71. •A block of wood will sink 10 cm in freshwater. How far will it sink in seawater? (The specific gravity of seawater is 1.025).

72. ••A cube of side s is completely submerged in a pool of freshwater. (a) Derive an expression for the pressure difference between the bottom and top of the cube. (b) After drawing a free-body diagram, derive an algebraic expression for the net force on the cube. (c) What is the weight of the displaced water when the cube is submerged? Your expressions may include some or all of the following quantities: P_{atm}, ρ_{fluid}, s, m_{cube}, and g.

73. •A crown that is supposed to be made of solid gold is under suspicion. When the crown is weighed in air it has a weight of 5.15 N. When it is suspended from a digital balance and lowered into water, its apparent weight is measured to be 4.88 N. Given that the specific gravity of gold is 19.3, comment on the authenticity of the crown. SSM

74. •Tom and Huck fashion a river raft out of logs. The raft is 3 m × 4 m × 0.15 m and is made from trees that have an average density of 700 kg/m³. How many people can stand on the raft and keep their feet dry, assuming an average person has a mass of 70 kg?

75. ••(a) A spherically shaped buoy is made from wood, which has a specific gravity of 0.6. Find the acceleration of the buoy when released from rest at the bottom of a freshwater lake. (b) If the buoy starts out 10 m below the surface of the water, determine the height above the water that it will rise after it shoots out of the water.

76. •A woman floats in a region of the Great Salt Lake where the water is about 4 times saltier than the ocean and has a density of about 1130 kg/m³. The woman has a mass of 55 kg and her density is 985 kg/m³ after exhaling as much air as possible from her lungs. Determine the percentage of her volume that will be above the waterline of the Great Salt Lake.

11-8: Fluids in Motion and the Equation of Continuity

77. •(a) A hose is connected to a faucet and used to fill a 5-L container in a time of 45 s. Determine the volume flow rate in m³/s. (b) Determine the velocity of the water in the hose in part (a) if it has a radius of 1 cm. SSM

78. •How fast is the water leaving the nozzle of a hose with a volume flow rate of 0.45 m³/s? Assume there are no leaks and the nozzle has a circular opening with diameter of 7.5 mm.

79. •Determine the time required for a 50-L container to be filled with water when the speed of the incoming water is 25 cm/s and the cross-sectional area of the hose carrying the water is 3 cm².

80. •**Medical** A cylindrical blood vessel is partially blocked by the build up of plaque. At one point, the plaque decreases the diameter of the vessel by 60.0%. The blood approaching the blocked portion has speed v_0. Just as the blood enters the blocked portion of the vessel, what will be its speed in terms of v_0?

81. •**Medical** You inject your patient with 2.5-mL of medicine. If the inside diameter of the 31-gauge needle is 0.114 mm and the injection lasts for 0.65 s, determine the speed of the fluid as it leaves the needle. SSM

82. •In July of 1995, a spillway gate broke at the Folsom Dam in California. During the uncontrolled release, the flow rate through the gate peaked at 40,000 ft³/s and about 1.35 billion gallons of water were lost (nearly 40% of the reservoir). Estimate the time that the gate was open.

83. •The return-air ventilation duct in a home has a cross-sectional area of 900 cm². The air in a room that has dimensions 7 m × 10 m × 2.4 m is to be completely circulated in a 30-min cycle. What is the speed of the air in the duct?

11-9: Fluid Flow—Bernoulli's Equation

84. ••Water flows from a fire truck through a hose that is 11.7 cm in diameter and has a nozzle that is 2 cm in diameter. The firemen stand on a hill 5 m above the level of the truck. When the water leaves the nozzle, it has a speed of 20 m/s. Determine the minimum gauge pressure in the truck's water tank.

85. •At one point, Hurricane Katrina had maximum sustained winds of 175 mi/h (240 km/h) and a low pressure in the eye of 666.52 mmHg (0.877 atm). Using the given air speed, compare the atmospheric pressure that would be predicted by Bernoulli's equation to the measured value. Comment on any discrepancies. Assume the pressure of the air is normally 1 atm. SSM

86. •When the atmospheric pressure is 1 atm, a water fountain ejects a stream of water that rises to a height of 5 m (Figure 11-25). There is a 1-cm-radius pipe that leads from a pressurized tank to the opening that ejects the water. What would happen if the fountain were operating when the eye of a hurricane passes through? Assume that the atmospheric pressure in the eye is 0.877 atm and the tank's pressure remains the same.

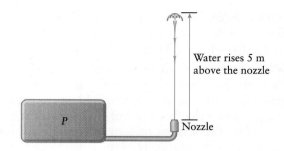

Figure 11-25 Problem 86

87. ••A cylinder that is 20 cm tall is filled with water (Figure 11-26). If a hole is made in the side of the cylinder, 5 cm below the top level, how far from the base of the cylinder will the stream land? Assume that the cylinder is large enough so that the level of the water in the cylinder does not drop significantly.

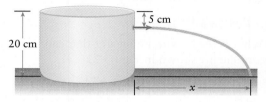

Figure 11-26 Problem 87

11-10: Viscous Fluid Flow

88. •Water flows through a horizontal tube that is 30.0-cm-long and has an inside diameter of 1.50 mm at 0.500 mL/s. Find the pressure difference required to drive this flow if the viscosity of water is 1.00 mPa · s. Assume laminar flow.

89. •**Biology** Blood takes about 1.50 s to pass through a 2.00-mm-long capillary in the human circulatory system. If the diameter of the capillary is 5.00 μm and the pressure drop is 2.60 kPa, calculate the viscosity of blood. Assume laminar flow.

90. ••A very large tank is filled to a depth of 250 cm with oil that has a density of 860 kg/m3 and a viscosity of 180 mPa · s. If the container walls are 5.00 cm thick, and a cylindrical hole of radius 0.750 cm has been bored through the base of the container, what is the initial volume flow rate (in L/s) of the oil through the hole?

91. •Water is supplied to an outlet from a pumping station 5.00 km away. From the pumping station to the outlet there is a net vertical rise of 19 m. Take the coefficient of viscosity of water to be 0.0100 poise. The pipe leading from the pumping station to the outelt is 1.00 cm in diameter, and the gauge preassure in the pipe at the point where it exits the pumping station is 520 kPa. At what volume flow rate does water flow from the outlet?

General Problems

92. •**Medical** The human body contains about 5.0 L of blood that has a density of 1060 kg/m³. Approximately

45% (by mass) of the blood is cells and the rest is plasma. The density of blood cells is approximately 1125 kg/m^3, and about 1% of the cells are white blood cells, the rest being red blood cells. The red blood cells are about 7.5 μm across (modeled as spheres). (a) What is the mass of the blood in a typical human? (b) Approximately how many blood cells (of both types) are in the blood?

93. •In August 2009 a person in Bagnone, Italy, won the largest lottery prize in European history, 146.9 million Euros, which at the time was worth 211.8 million U.S. dollars. Gold at the time was worth $953 per troy ounce, and silver was worth $14.16 per troy ounce. A troy ounce is 31.1035 g, the density of gold is 19.3 g/cm^3, and the density of silver is 10.5 g/cm^3. (a) If the lucky lottery winner opted to be paid in a single cube of pure gold, how high would the cube be? Could he carry it home? (b) What would be the height of a silver cube of the same value? Try to solve this part by using proportional reasoning with the results of part (a) instead of just repeating the same procedure you used in (a). SSM

94. •A hybrid car travels about 50 mi/gal of gasoline. The density of gasoline is 737 kg/m^3 and 1 gal equals 3.788 L. Express the car's mileage in miles per kilogram (mi/kg) of gas.

95. •Medical Blood pressure is normally expressed as the ratio of the *systolic* pressure (when the heart just ejects blood) to the *diastolic* pressure (when the heart is relaxed). The measurement is made at the level of the heart (usually at the middle of the upper arm), and the pressures are given in millimeters of mercury, although the units are not usually written. Normal blood pressure is typically 120/80. How would you write normal blood pressure if the units of pressure used were (a) pascals, (b) atmospheres, or (c) pounds per square inch (lb/in.², psi)? (d) Is the blood pressure, as typically stated, the absolute pressure or the gauge pressure? Explain your answer.

96. •Astro Evidence from the Mars rover *Discovery* suggests that oceans as deep as 0.50 km once may have existed on Mars. The acceleration due to gravity on Mars is $0.379g$ and the current atmospheric pressure is 650 N/m^2. (a) If there were any organisms in the Martian ocean in the distant past, what pressure (absolute and gauge) would they have experienced at the bottom, assuming the surface pressure was the same as today? Assume that the salinity of Martian oceans was the same as oceans on Earth. (b) If the bottom-dwelling organisms in part (a) were brought from Mars to Earth, how deep could they go in our ocean without exceeding the maximum pressure they experienced on Mars?

97. ••Calc Determine the net force on a freshwater dam that is 100 m wide by 50 m high.

98. •Medical Blood pressure is normally taken on the upper arm at the level of the heart. Suppose, however, that a patient has his arms in a cast so you cannot take his blood pressure in the usual way. If you have him stand up and take the blood pressure at his calf, which is 95.0 cm below his heart, what would normal blood pressure be? The density of blood is 1060 kg/m^3.

99. •Medical A syringe which has an inner diameter of 0.6 mm is attached to a needle which has an inner diameter of 0.25 mm. A nurse uses the syringe to inject fluid into a patient's artery where blood pressure is 140/100. Assume the liquid is an ideal fluid. What is the minimum force the nurse needs to apply to the syringe? SSM

100. •A spherical water tank is being filled by a 1-cm-diameter hose. The water in the hose has a uniform speed of 15 cm/s. Meanwhile, the tank springs a leak at the bottom of the sphere. The hole has a diameter of 0.5 cm. Determine the equilibrium level of the water in the tank if water continues flowing through the hose and into the tank at the same rate.

101. •A large water tank is 18 m above the ground (Figure 11-27). Suppose a pipe with a diameter of 8 cm is connected to the tank and leads down to the ground. How fast does the water rush out of the pipe at ground level? Assume that the tank is open to the atmosphere.

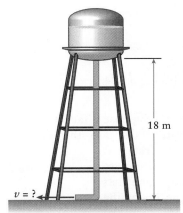

Figure 11-27 Problem 101

102. •Medical The aorta is approximately 25 mm in diameter. The mean pressure there is about 100 mmHg and the blood flows through the aorta at approximately 60 cm/s. Suppose that at a certain point a portion of the aorta is blocked so that the cross-sectional area is reduced to 3/4 of its original area. The density of blood is 1060 kg/m^3. (a) How fast is the blood moving just as it enters the blocked portion of the aorta? (b) What is the gauge pressure (in mmHg) of the blood just as it has entered the blocked portion of the aorta?

103. ••A diver is located 60 m below the surface of the ocean (the specific gravity of seawater is 1.025). To raise a treasure chest that she discovered, the diver inflates a plastic, spherical buoy with her compressed air tanks.

The radius of the buoy is inflated to 40 cm. (a) If the treasure chest has a mass of 200 kg, what is the acceleration of the buoy and treasure chest when they are tethered together and released? Assume the chest is 20 cm × 40 cm × 10 cm in size, and neglect the tether's mass and volume. (b) What happens to the radius of the buoy as it ascends in the water? If there are any changes, determine them numerically. How does this change your answer in part (a)?

104. ••**Sports** The air around the pinning baseball in **Figure 11-28** (top view) experiences a faster speed on the left than the right side and hence a smaller pressure on the left than the right. Suppose the ball is traveling forward in the x direction at 38 m/s and it breaks 15 cm to the left in the −y direction (neglect the pull of gravity in the z direction). Determine the pressure difference between the right and left sides of the ball. Assume the ball has a mass of 142 g and a radius of 3.55 cm. In addition, assume it travels at a constant speed of 38 m/s in the x direction from the pitcher's mound to home plate (60.5 ft = 18.44 m). *Hint:* $\Delta P = F_y/\pi r^2$.

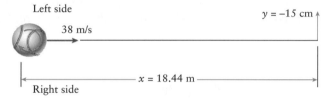

Figure 11-28 Problem 104

105. •••**Calc** (a) Suppose a faucet is turned on and the stream of water falls into the sink (**Figure 11-29**). Determine the rate at which the radius decreases in terms of how far it has fallen from the lip of the faucet. Assume that the radius of the stream of water is r_0 when it leaves the faucet at speed v_0. *Hint:* First determine a formula for the radius of the stream after it has fallen through a vertical distance of "y"; call this

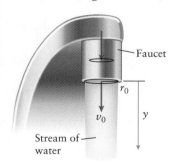

Figure 11-29 Problem 105

$r(y)$. Then, differentiate the function with respect to t using the chain rule. Notice that $v(y) = dy/dt$. (b) If the initial values are $v_0 = 1.5$ m/s and $r_0 = 1$ cm, determine the radius of the stream (r) and the rate that it is changing (dr/dt) when $y = 10$ cm.

106. •••**Calc** The pressure P is constant over the entire surface of a hemisphere of radius R (**Figure 11-30**). Using

calculus, integrate to find the net force acting on the outer surface. Recall $F = \int P\,dA$. *Hint:* The pressure is perpendicular to the outer surface of the hemisphere at every point. Be sure to give the direction of the net force, too!

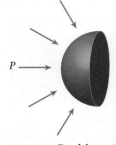

Figure 11-30 Problem 106

107. •••An equilateral prism of wood (the specific gravity is 0.6) is placed into a fresh-water pool (**Figure 11-31**). Determine the ratio y_d/y_u of the depth of submersion when the prism is pointed down to the depth of submersion when the prism is pointed up. Assume the prism has a side of s.

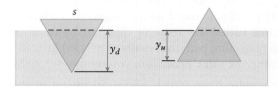

Figure 11-31 Problem 107

MCAT® Section

The section that follows includes material from previously administered MCAT® items and is reprinted with permission of the Association of American Medical Colleges (AAMC).

Passage 12 (Questions 73–80)

A pressure difference between the ends of a horizontal, uniform pipe is required to maintain the steady flow of a viscous fluid through the pipe. The fluid pressure at the downstream end of the pipe (P_2) depends on the tube length (L), inside radius of the pipe (r), viscosity (η), volume flow rate (f), and the pressure at the beginning of the pipe (P_1). These quantities are related by Poiseuille's equation $\Delta P = P_2 - P_1 = \dfrac{(8\eta Lf)}{\pi r^4}$, where ΔP is the pressure differential.

Two experiments were conducted to investigate the flow of distilled water and glycerin through various pipes.

Experiment 1
Distilled water flowed through pipes that were 5.0 m long and had different radii. The pressure differential between the beginning and end of the pipes was measured while the flow rate was held constant. The results are listed in Table 1. (Note: At 20 °C, distilled water has a viscosity of 1×10^{-3} kg/m·s and a density of 1×10^3 kg/m^3.)

Table 1

Trial	L(m)	r(m)	f (m³/s)	ΔP (N/m²)
1	5.0	0.05	0.003	6.0
2	5.0	0.04	0.003	15.0
3	5.0	0.03	0.003	47.0
4	5.0	0.02	0.003	240
5	5.0	0.01	0.003	3820

Experiment 2
Glycerin flowed through pipes that had different lengths but the same radius. The flow rates were measured while the pressure differential between the beginning and end of the pipes was held constant. The results are listed in Table 2. (Note: At 20°C, glycerin has a viscosity of 1.5×10^{-3} kg/m·s and a density of 1.3×10^{3} kg/m³.)

Table 2

Trial	L(m)	r(m)	ΔP (N/m²)	f (m³/s)
1	1.0	0.01	100	2.6×10^{-7}
2	1.5	0.01	100	1.7×10^{-7}
3	2.0	0.01	100	1.3×10^{-7}
4	2.5	0.01	100	1.0×10^{-7}
5	3.0	0.01	100	8.7×10^{-8}

73. When viscosity increases, an increase in pressure differential is required to maintain the same volume flow rate. This is because an increase in viscosity is related to an increase in which of the following factors?
  A. Pipe radius
  B. Friction
  C. Fluid density
  D. Turbulence

74. When fluid flow is compared to electricity, which of the following fluid characteristics is analogous to electrical current?
  A. Pressure differential
  B. Volume flow rate
  C. Viscosity
  D. Flow speed

75. In which trial of Experiment 2 did a unit length of glycerin have the greatest kinetic energy?
  A. Trial 1
  B. Trial 2

  C. Trial 4
  D. Trial 5

76. If an additional trial of Experiment 1 were conducted using a pipe with a radius of 0.06 m, what would be the expected pressure differential?
  A. 0.4 N/m²
  B. 1.0 N/m²
  C. 1.5 N/m²
  D. 2.9 N/m²

77. If water and glycerin flowed through identical pipes and had the same pressure differential, which fluid would have the greater flow rate?
  A. Water, because it has a lower viscosity
  B. Water, because it has a lower density
  C. Glycerin, because it has a higher viscosity
  D. Glycerin, because it has a higher density

78. If ΔP is held constant and *r* is reduced by 33%, by what percentage does *f* decrease?
  A. 33%
  B. 56%
  C. 80%
  D. 98%

79. If an additional trial of Experiment 2 were conducted using a pipe that was 4.0 m long, what would be the expected flow rate?
  A. 2.6×10^{-8} m³/s
  B. 4.3×10^{-8} m³/s
  C. 6.5×10^{-8} m³/s
  D. 7.8×10^{-8} m³/s

80. Which of the following graphs best shows how the volume flow rate of a fluid flowing out of a uniform pipe varies as the pressure differential between the beginning and end of the pipe is changed?

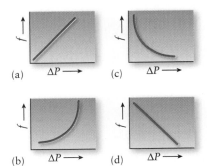

12 Oscillations

(Courtesy David Tauck.)

Who hasn't watched the hypnotic back-and-forth swing of a pendulum? The bob slows as it rises and then accelerates as it returns to its lowest position. Gravitational potential energy is stored on the upward part of the swing, then converted to kinetic energy as the bob swings back down. As you walk, your body employs a similar mechanism to store and release energy. When viewed upside down, your center of mass, marked with a red dot in the photograph, moves like the bob at the end of a rod that can swing around a pivot in your foot. With each step, the bob rises and falls like a pendulum, a pendulum that allows you to conserve about 60% of the energy needed to raise your center of mass on each stride.

When a doctor taps your bended knee with a rubber hammer your leg kicks up. If your muscles are relaxed, what happens next is physics; gravity tugs on your leg causing it to swing back down. As your leg swings back it accelerates, and the nonzero momentum it acquires carries it past the vertical (straight down) position. Your leg continues to swing, up and backward, then stops momentarily before swinging back down toward the vertical again. If there were no forces other than gravity, the swinging would repeat over and over. In this chapter, we explore this kind of cyclic motion, which also appears in the up-and-down oscillation of a spring scale in the grocery store, the vibrations of an eardrum, the strings of a musical instrument, the tail feathers of a hummingbird as it pulls up out of a vertical dive, and the thorax of a flying bee.

12-1 Simple Harmonic Motion

As you breathe, the partial pressure of carbon dioxide retained in your lungs fluctuates with every cycle of inhalation and exhalation. As your heart pushes blood into your arteries, blood pressure increases and then falls between heartbeats. Many other physiological variables including body temperature, hormone levels and urine osmolarity also exhibit circadian rhythms, oscillating on a 24-h cycle. In this section we consider the physics of cyclic variations.

Motion that repeats itself, cyclic motion, is characterized by a **restoring force** that tends to return the system to the position in which it experiences no net force. A system that experiences no net force and is stationary is in **equilibrium**. When a system that is in motion approaches a position at which no net force is exerted, it may continue to move; we would say that the system has moved through an *equilibrium position*. As you sit with your knee over the edge of the doctor's exam table, your lower leg is in equilibrium when motionless and hanging straight down. A child sitting on a motionless swing is in equilibrium until someone gives her a push. For an object attached to one end of a horizontal spring with the other end fixed, as in Figure 12-1, the equilibrium position is the point at which the spring is neither stretched nor compressed. When a system, such as your leg, a swing, or the object and spring, is displaced from the equilibrium position, the restoring force brings it back; however, under most circumstances momentum carries it past the equilibrium position. Oscillatory motion results as the process of returning to the equilibrium position overshoots it and then repeats.

We encountered the kind of force responsible for the spring's cyclic motion in Chapter 6. Hooke's law describes a force proportional to the displacement from equilibrium and directed opposite to the displacement vector:

$$\vec{F} = -k\vec{x} \qquad (6\text{-}16)$$

The minus sign is a clue that this is a restoring force. When the displacement x is positive, the force points in the negative direction, back toward the equilibrium position. When the displacement is negative, the force points in the positive direction, again toward equilibrium.

Let's use Hooke's law to find and solve an equation that describes the motion of the object and spring system. We need an equation that includes position and

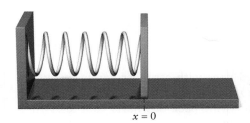

The mass at the free end of the spring is in the equilibrium position when the spring is neither stretched nor compressed.

$x = 0$

When the spring is stretched, the spring force pulls the object back to equilibrium – positive $\vec{x}$, negative $\vec{F}$.

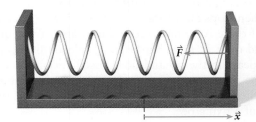

$\vec{x}$

Figure 12-1 An object attached to one end of a horizontal spring with the other end fixed is in its equilibrium position when the spring is neither stretched nor compressed. When displaced from the equilibrium point, the spring exerts a restoring force that brings the object back toward the equilibrium point.

When the spring is compressed, the spring force pushes the object back to equilibrium – negative $\vec{x}$, positive $\vec{F}$.

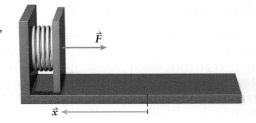

$\vec{x}$

time, as well as one that makes use of the force that drives the system's motion. Newton's second law connects force and motion:

$$\sum \vec{F}_{\text{dir}} = m\vec{a}_{\text{dir}} \qquad (4\text{-}1)$$

Along the axis of the spring, the only force is $-kx$:

$$-kx = ma$$

To see the time dependence more clearly, we replace acceleration a with its definition, the second derivative of position with respect to time:

$$-kx = m\frac{d^2x}{dt^2}$$

 See the Math Tutorial for more information on Calculus

Recall that Hooke's law applies only when the displacement from the equilibrium position is small, so this equation describes the position of an object that has been displaced slightly from equilibrium (and is acted on by a restoring force). By convention we write the **simple harmonic oscillator equation** as

$$\frac{d^2x}{dt^2} + \frac{k}{m}x = 0 \qquad (12\text{-}1)$$

All solutions of Equation 12-1 are expressions for x such that the second time derivative added to k/m multiplied by the function x results in zero. By knowing x as a function of time, that is, the displacement from equilibrium at any time, we can understand the motion of the system. It is standard to include the time variable in parentheses, writing $x(t)$, to make it more obvious that the solution is a function of time.

What form can the solution to the simple harmonic oscillator equation take? To satisfy the equation, we need a function such that its second derivative with respect to time is proportional to the negative of the function itself—otherwise the two could not cancel to give zero. The sine and cosine satisfy this requirement; for example, when $x(t)$ equals $A \cos(\omega_0 t)$,

$$\frac{dx}{dt} = -A\omega_0 \sin(\omega_0 t)$$

and

$$\frac{d^2x}{dt^2} = -A\omega_0^2 \cos(\omega_0 t)$$

Using this generic sinusoidal form of $x(t)$, the simple harmonic oscillator equation becomes

$$\frac{d^2}{dt^2} A \cos(\omega_0 t) + \frac{k}{m} A \cos(\omega_0 t) = 0$$

or

$$-\omega_0^2 A \cos(\omega_0 t) + \frac{k}{m} A \cos(\omega_0 t) = 0$$

This statement *can* be made true, as long as

$$\omega_0{}^2 = \frac{k}{m}$$

or

$$\omega_0 = \sqrt{\frac{k}{m}} \qquad (12\text{-}2)$$

So a solution to Equation 12-1 is

$$x(t) = A \cos\left(\sqrt{\frac{k}{m}}t\right) \tag{12-3}$$

The simple harmonic motion equation (Equation 12-1) defines the relationship between time and the position of an oscillating quantity, such as the displacement of a spring from equilibrium or the concentration of a hormone in the blood. The *solution* to this equation, Equation 12-3, gives the actual relationship between the variables. For an object on the end of a spring, for example, the solution to the simple harmonic motion equation is the specific equation of position as a function of time that applies to that object and spring. The solution describes how the state of the system changes over time.

The units of ω_0 are radians per second (rad/s). We can deduce this by considering the argument of the cosine function in Equation 12-3, which must carry units of radians:

$$\left[\sqrt{\frac{k}{m}}t\right] = \left[\sqrt{\frac{k}{m}}\right]s = \text{rad}$$

√x̄ **Go to Interactive Exercise 12-1**
*for more practice dealing with
simple harmonic motion*

or

$$\left[\sqrt{\frac{k}{m}}\right] = [\omega_0] = \frac{\text{rad}}{\text{s}}$$

This ω_0 is the same **angular velocity** we encountered in circular motion. We will explore the relationship between simple harmonic motion and circular motion in the next section.

A graph of the position versus time of an object attached to a spring is shown in **Figure 12-2**. For this figure we have imagined a vertical spring, but the physics is the same for a horizontal one. At the equilibrium position, shown on the left, the object hangs motionless. To set the object into motion, we lift it to an initial position $+A$ above equilibrium and release it. Because the restoring force pulls the object down from the moment of release until it reaches equilibrium, the speed of the object continues to increase. Speed is greatest as the object passes through equilibrium, and the associated momentum causes the object to overshoot the equilibrium position. The restoring force now pulls upward, causing the object to decelerate and momentarily come to rest, before accelerating back upward. The object passes through equilibrium twice, once on the way down and again on the way back up, before returning to the initial position.

Midway through the cycle, the object has stretched the spring so that it reaches a displacement $-A$ below equilibrium; because displacement above equilibrium is

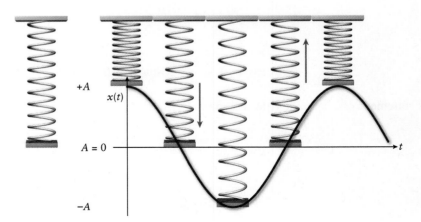

Figure 12-2 The image on the left shows a vertical spring at its equilibrium position. The graph on the right shows the position versus time of an object attached to the spring as it oscillates between $+A$ and $-A$.

taken as positive, the midway point is a negative displacement. Once the object returns to its initial position, the cycle repeats.

✷ **What's Important 12-1**

Cyclic motion results from a restoring force that tends to return the system to its equilibrium position, the position in which it experiences no net force. When a system is displaced from its equilibrium position, the restoring force brings it back; however, momentum carries it past the equilibrium position under most circumstances. Oscillatory motion results as the process of returning to equilibrium overshoots it and then repeats. The simple harmonic motion equation (Equation 12-1) defines the relationship between time and the position of an oscillating quantity.

12-2 Oscillations Described

We've seen that not all oscillations in nature are mechanical. In some cells, for example, small changes in the extracellular concentration of calcium ions (Ca^{2+}) cause sinusoidal oscillations of the intracellular Ca^{2+} concentration, which can regulate enzyme activity, mitochondrial metabolism, hormone release, or even gene expression. One such example of intracellular Ca^{2+} oscillations is shown in **Figure 12-3a**; a sinusoidal curve drawn on top of the data matches the oscillations well. In **Figure 12-3b**, the volume of air in a person's lungs is shown as a function of time. A sinusoidal curve has been superimposed on this data as well; in both cases the oscillations exhibit the same time variation as simple harmonic motion.

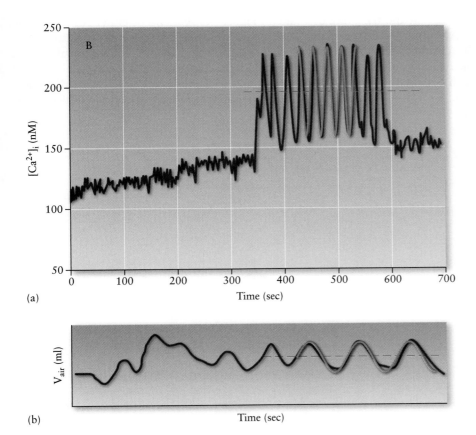

(a)

(b)

Figure 12-3 (a) Amino acids and Ca^{2+} stimulate different patterns of Ca^{2+} oscillations through the Ca^{2+}-sensing receptor. (After Steven H. Young and Enrique Rozengurt, (2002) *Am. J. Physiol. Cell. Physiol.* 282, pp. 1414–1422.) (b) The volume of air in a person's lungs is shown as a function of time during and after coughing. A sinusoidal curve has been superimposed on the data. The dashed red line shows the equilibrium position of the lungs. (After Michael A Coyle, et al., (2005) *Cough* 1, p. 3.)

In the last section, we described simple harmonic motion with an equation (Equation 12-1) based on a restoring force that is proportional to a negative constant multiplied by displacement from equilibrium. We developed a solution that gives position as a function of time; from Equation 12-3,

$$x(t) = A \cos(\omega_0 t)$$

where

$$\omega_0 = \sqrt{\frac{k}{m}} \tag{12-2}$$

Did you wonder why we chose the cosine and not sine for the solution? We can verify that sine also satisfies Equation 12-1 by substituting $x(t) = A \sin(\omega_0 t)$ into it:

$$\frac{d^2 A \sin(\omega_0 t)}{dt^2} + \frac{k}{m} A \sin(\omega_0 t) = 0$$

or

$$-\omega_0^2 A \sin(\omega_0 t) + \frac{k}{m} A \sin(\omega_0 t) = 0$$

This is true when $\omega_0 = \sqrt{k/m}$. So both $x(t) = A \cos(\omega_0 t)$ and $x(t) = A \sin(\omega_0 t)$ satisfy Equation 12-1 and either expression can therefore be used to give the position (ion concentration, lung volume, body temperature or other quantities) as a function of time of a system that displays simple harmonic oscillation. In Figure 12-3a the curve plotted on top of the data appears to be cosine, because we have drawn the curve starting from a peak of the oscillations. In the same way the overlaid curve in Figure 12-3b appears to be sine. But cosine and sine have exactly the same shape, and differ only by the position along the horizontal axis that we choose to label as zero, as you can see from **Figure 12-4**. So both the curve overlaid on Figure 12-3a and the curve overlaid on Figure 12-3b could be either cosine or sine.

Although both sine and cosine satisfy the simple harmonic oscillator equation, cosine is often the more convenient and therefore the conventional choice. When position is defined as $x(t) = A \cos(\omega_0 t)$ the displacement at $t = 0$ equals $A \cos 0$ which equals A. For an object on a spring, this corresponds to releasing the object from its maximum displacement from equilibrium. We will always take *the* **amplitude A** to be positive.

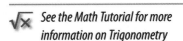 *See the Math Tutorial for more information on Trigonometry*

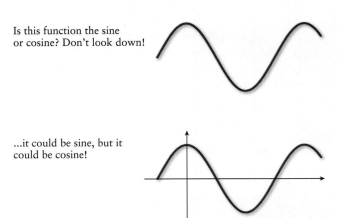

Is this function the sine or cosine? Don't look down!

...it could be sine, but it could be cosine!

Sine and cosine have the same shape. They differ only by our choice of where to label zero on the horizontal axis.

Figure 12-4 The cosine and sine functions have exactly the same shape, and differ only by the position along the horizontal axis that we choose to label as zero.

> ## ❗ Watch Out
> ### The amplitude of oscillation is the maximum displacement from equilibrium.
>
> It is tempting to think of amplitude as the difference between the minimum and maximum displacement from equilibrium. This peak-to-peak variation can be a useful parameter when studying an oscillation. However, the amplitude serves as a multiplier that enables deviations from the equilibrium position to be scaled up or down from 1, the maximum value of sine or cosine. When an object on a spring is pulled 0.05 m from the position at which it experiences no net force, where $x(t) = A \cos(\omega_0 t)$, we ensure that the maximum value of $x(t)$ is 0.05 by setting $A = 0.05$ m.

Lung volume versus time is sinusoidal (unless you're holding your breath!), as shown in Figure 12-3b. We plotted a sine curve on top of the data in that figure. Should we then use $x(t) = A \sin(\omega_0 t)$ (let x be the lung volume) to describe the time variation of lung volume? We could, but for consistency we prefer to use cosine, and to include an additional parameter in the argument of cosine to align the cosine function with the data:

$$x(t) = A \cos(\omega_0 t + \phi) \tag{12-4}$$

The parameter ϕ, the **phase angle**, allows us to set $t = 0$ at whatever value of displacement from equilibrium we choose. Just as $\omega_0 t$ has units of radians, so must ϕ. For the lung volume versus time data, for example, we want displacement from equilibrium, V_{air} in the figure, to be equal to the value along the dashed line at $t = 0$ and for it to become more and more positive immediately afterward. We therefore require $x(0)$ to be zero in Equation 12-4:

$$0 = A \cos(0 + \phi) = A \cos \phi$$

For this equation to be accurate, $\cos \phi$ must equal 0; so, either $\phi = \pi/2$ rad or $\phi = 3\pi/2$ rad. By comparing Figure 12-3b and the curves in **Figure 12-5**, for $x(t) = A \cos(\omega_0 t + \pi/2)$ and $x(t) = A \cos(\omega_0 t + 3\pi/2)$, we see that when $\phi = 3\pi/2$ rad, $x(t)$ initially becomes more and more positive.

You can think of the phase angle as shifting the location of the vertical axis (at which time equals zero) to the right or left on a graph of position versus time (Equation 12-4). In the top graph in **Figure 12-6**, one full cycle of $\cos \theta$ is shown in

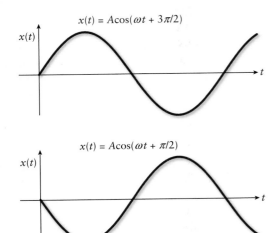

$$x(t) = A\cos(\omega t + 3\pi/2)$$

$$x(t) = A\cos(\omega t + \pi/2)$$

Figure 12-5 The two plots show cosine functions that differ by an integral multiple of $\pi/2$ rad.

Figure 12-6 The two cycles shown in blue differ by a phase shift ϕ. This is equivalent to shifting the vertical axis that marks zero to the right by ϕ in the horizontal direction.

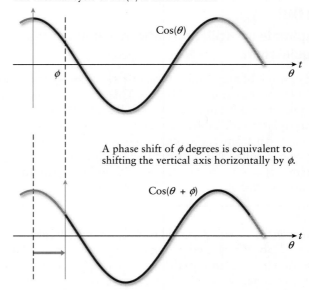

The first full cycle of $\cos(\theta)$ is traced in blue.

$\text{Cos}(\theta)$

A phase shift of ϕ degrees is equivalent to shifting the vertical axis horizontally by ϕ.

$\text{Cos}(\theta + \phi)$

blue. In the lower graph, one full cycle of $\cos(\theta + \phi)$ is shown in blue; the position of the vertical axis, which marks the zero point of the horizontal axis, has been shifted by phase angle ϕ. So for a simple harmonic oscillator described by $x(t) = A\cos(\omega_0 t + \phi)$, the choice of ϕ slides the position of $t = 0$ right or left, allowing any value of displacement from equilibrium to be used as the initial ($t = 0$) position.

? Got the Concept 12-1
Air in a Breath

In physiology lab, Kharissia measures the amount of air that enters and leaves Blythe's chest with every breath. The volume of air versus time obeys the description $V(t) = V_0 \cos(\omega_0 t + \phi)$, an adaptation of Equation 12-4. If the measurements start ($t = 0$) at the end of one exhalation, what is the value of ϕ?

Example 12-1 Finding the Phase

Sound causes a frog's eardrum to move back and forth as shown in the plot of displacement versus time in **Figure 12-7**. In this case, displacement is given as a percentage of the amplitude of the oscillations. At the start of the measurements, the displacement is about 12% of the maximum in the negative direction. Find the phase angle ϕ if the displacement obeys $x(t) = A\cos(\omega_0 t + \phi)$ (Equation 12-4). (We will address the fact that the amplitude of the oscillations changes from one cycle to the next in Section 12-7.)

SET UP
Phase angle provides the flexibility to adjust the mathematical description of the cyclic motion to allow any initial value of displacement. Here, we demand that $x(0) = -12$. From Equation 12-4, $x(0)$ is

$$x(0) = A\cos(\omega_0(0) + \phi) = A\cos\phi$$

where the amplitude A is 100 (percent).

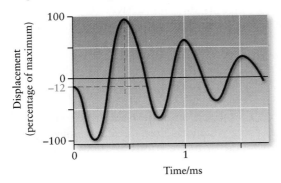

Displacement from equilibrium of eardrum in a frog...

Figure 12-7 Sound causes a frog's eardrum to move back and forth, in (approximately) harmonic motion. The vertical dashed red line marks the beginning of a cosine cycle; the horizontal dashed red line marks −12% of the maximum displacement from equilibrium. The time interval between the two red arrows is just slightly more than one-quarter of a full cycle. (After Chung, Pettigrew, and Anson, (1981) *Proc. Royal Soc. London Ser. B Biol. Sci.* 212, No. 1189, pp. 459–485.)

SOLVE

Set the two versions of $x(0)$ equal:

$$-12 = 100 \cos \phi$$

or

$$\cos \phi = -0.12$$

We take the first positive solution for ϕ:

$$\phi = 1.7 \text{ rad}$$

REFLECT

The cosine function completes one full cycle in 2π rad, so a phase angle $\phi = 1.7$ rad amounts to a bit more than one-quarter of a full cycle. You can verify that the result is reasonable by inspecting Figure 12-7. The vertical dashed line marks the start of a cosine cycle; the separation from that value of time and the time at which the oscillation reaches −12% is just slightly more than one-quarter of a full cycle. Remember that because harmonic motion is repetitive, we can use any cycle or any part of a cycle to measure ϕ.

Practice Problem 12-1 Two sound waves of the same frequency are out of phase, so that when one is at its maximum amplitude, the other is at its minimum. Find the phase angle between the two waves.

We first encountered angular velocity ω, one of the parameters with which we describe simple harmonic oscillation, in the context of uniform circular motion. The relationship between oscillatory and circular motion is illustrated in **Figure 12-8**.

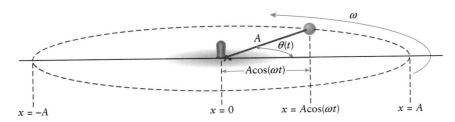

Figure 12-8 A red ball tied to a green peg by a string moves in uniform circular motion. Its position projected onto the x axis is identical to that of an object attached to a spring undergoing simple harmonic motion.

A ball tied to a peg by a string of length A rotates at a constant angular velocity ω. Angular velocity is defined as the time derivative of angular position, so at any moment the angle θ between the string and the positive x axis is

$$\theta = \omega t$$

The x component of the ball's position at any time is then

$$x(t) = A \cos \theta = A \cos(\omega t)$$

The motion of the ball projected onto the x axis is therefore identical to that of an object attached to a spring which has been stretched a distance A from equilibrium.

We can take advantage of the relationship between the uniform motion of an object in a circular path and the movement of an object attached to a spring to determine the time it takes for one full oscillation of the object. The ball on the string rotates at angular velocity ω rad/s, so the time T for one full revolution is 2π rad (one full circle) divided by ω:

$$T = \frac{2\pi}{\omega} \tag{12-5}$$

The units of T are, naturally, seconds:

$$[T] = \frac{[2\pi]}{[\omega]} = \frac{\text{rad}}{\text{rad/s}} = \text{s}$$

The time T is both the **period** of rotation of the ball and the period of one full oscillation of the object attached to the spring.

If an object undergoing simple harmonic motion has a period of T, then it has a **frequency** of one cycle every T s. We use f to represent frequency, so

$$f = \frac{1}{T} \tag{12-6}$$

The units of frequency are cycles per second; in the SI system,

$$1 \frac{\text{cycle}}{\text{s}} = 1 \text{ hertz} = 1 \text{ Hz}$$

We can combine Equations 12-5 and 12-6 to obtain another useful relationship:

$$f = \frac{1}{T} = \frac{\omega}{2\pi}$$

or

$$\omega = 2\pi f \tag{12-7}$$

Both ω and f are frequencies; the factor of 2π makes ω an angular quantity.

? Got the Concept 12-2
Heartbeats

Jess has a resting pulse of 45 heartbeats per minute. What is the frequency in hertz and the period in seconds of this cycle? Her respiratory rate is 0.20 Hz. How many breaths does she take in 1 min? What is the period of her breathing?

Example 12-2 Four Times the Mass

A small object of mass m equal to 0.80 kg is attached to a horizontal spring of constant k equal to 180 N/m and set into simple harmonic motion. (a) Find the angular velocity, frequency, and period of the object. (b) Find the angular velocity, frequency, and period of simple harmonic motion if the mass of the object is increased by a factor of 4 to 3.2 kg.

SET UP

The angular velocity can be found directly from the information given in the problem statement, according to

$$\omega_0 = \sqrt{\frac{k}{m}} \qquad (12\text{-}2)$$

Once the angular velocity is known we can determine the frequency using $\omega_0 = 2\pi f$ (from Equation 12-7). The period can be determined from ω_0 as well, using $T = \dfrac{2\pi}{\omega_0}$ (Equation 12-5).

SOLVE

(a) The angular velocity is

$$\omega_0 = \sqrt{\frac{k}{m}} = \sqrt{\frac{180 \text{ N/m}}{0.80 \text{ kg}}} = 15 \, \frac{\text{rad}}{\text{s}}$$

Using Equation 12-7, the frequency is

$$f = \frac{\omega_0}{2\pi} = \frac{15 \text{ rad/s}}{2\pi \text{ rad}} = 2.4 \text{ Hz}$$

and using Equation 12-5, the period is

$$T = \frac{2\pi}{\omega_0} = \frac{2\pi \text{ rad}}{15 \text{ rad/s}} = 0.42 \text{ s}$$

(b) Before we plug in more numbers, let's consider how the angular velocity changes when mass increases by a factor of 4. When we have a new mass $m' = 4m$, the new angular velocity ω' is given by

$$\omega' = \sqrt{\frac{k}{m'}} = \sqrt{\frac{k}{4m}} = \frac{1}{2}\sqrt{\frac{k}{m}}$$

So when the mass is increased by a factor of 4, the angular velocity becomes one half of its original value, $\omega' = \frac{1}{2}\omega_0$. From Equation 12-7, we see that the frequency must also decrease by a factor of 2, and from Equation 12-5, we see that the period increases by the same factor. So the new angular velocity, frequency, and period are

$$\omega' = \frac{1}{2}\omega_0 = \frac{1}{2}15 \, \frac{\text{rad}}{\text{s}} = 7.5 \, \frac{\text{rad}}{\text{s}}$$

$$f' = \frac{1}{2}f = \frac{1}{2}2.4 \text{ Hz} = 1.2 \text{ Hz}$$

and

$$T' = 2T = 2(0.42 \text{ s}) = 0.84 \text{ s}$$

REFLECT

When mass increases, angular velocity and frequency decrease, so the period is longer.

Practice Problem 12-2 A small object of mass m equal to 0.80 kg is attached to a horizontal spring of constant k equal to 720 N/m and set into simple harmonic motion. Find the angular velocity, frequency, and period of the object. Note that the spring constant is four times that in the original problem.

> ### ✴ What's Important 12-2
> The amplitude of oscillation is the maximum displacement of a system from equilibrium. We will always take the amplitude to be positive. Phase angle allows us to set time equal to zero at whatever value of displacement from equilibrium we choose. Phase angle shifts the location of the vertical axis (where time equals zero) to the right or left in a graph of position versus time.

12-3 Energy Considerations

In Chapter 6, we considered the kangaroo, whose tendons act like springs to store potential energy which then gets transformed into kinetic energy as the animal hops (Figure 12-9); this increases the efficiency of movement. A similar mechanism propels dolphins through the water. Even when a dolphin increases its speed by beating its tail faster, the amount of oxygen it consumes, an indication of the amount of energy used, hardly changes. Tissue in a dolphin's tail acts much like a spring, storing potential energy as the tail flips either up or down and then transforming the potential energy into kinetic energy. As a result, the dolphin can swim efficiently at high speeds. Systems which contain mechanical or biological springs share the common property that energy is transformed back and forth between potential and kinetic forms. In this section, we study the energy of an oscillating system that consists of an object attached to the free end of a spring.

To follow the energy flow in an oscillating system more clearly, we start by idealizing the system to include no frictional or other dissipative forces. In this ideal case, we only need to account for potential energy and kinetic energy. Qualitatively, when an object attached to a spring is pulled away from equilibrium, the potential energy increases. When the object is released, it accelerates back toward equilibrium due to the restoring force. It reaches maximum velocity (and therefore maximum kinetic energy) as it passes through equilibrium, because after that the restoring force opposes the motion and the object decelerates.

To examine the energy of a system in simple harmonic oscillation quantitatively, we start from the total energy E,

$$E = K + U = \frac{1}{2}mv^2 + \frac{1}{2}kx^2$$

where we have used the form of spring potential energy $U = \frac{1}{2}kx^2$ that derives from a Hooke's law restoring force $F = -kx$. We also know the position of the object as a function of time:

$$x(t) = A\cos(\omega_0 t + \phi) \tag{12-4}$$

So the potential energy U as a function of time is

$$U = \frac{1}{2}kA^2\cos^2(\omega_0 t + \phi) \tag{12-8}$$

Figure 12-9 Tendons in a kangaroo's legs act like springs to store potential energy which then gets transformed back into kinetic energy as the animal hops. *(Courtesy David Tauck.)*

To determine the kinetic energy, we first find velocity by taking the time derivative of position as a function of time (Equation 12-4).

$$v(t) = \frac{dx(t)}{dt} = \frac{d}{dt} A \cos(\omega_0 t + \phi) = -A\omega_0 \sin(\omega_0 t + \phi)$$

This is a good example of the value of applying calculus in physics; the velocity of an object at the end of the spring is continually changing, so we can't simply take the difference between its position at two different times to determine the velocity.

Knowing the velocity of the object, we can now find the kinetic energy:

$$K = \frac{1}{2} m A^2 \omega_0^2 \sin^2(\omega_0 t + \phi)$$

The total energy E is the sum of the kinetic energy and the potential energy of the system

$$E = \frac{1}{2} m A^2 \omega_0^2 \sin^2(\omega_0 t + \phi) + \frac{1}{2} k A^2 \cos^2(\omega_0 t + \phi)$$

 See the Math Tutorial for more information on Trigonometry

Whenever you encounter a $\sin^2 \alpha$ term added to a $\cos^2 \alpha$ term, think first of applying the trigonometric identity $\sin^2 \theta + \cos^2 \theta = 1$. In order for the two energy terms to be combined using this identity, we need the coefficients that multiply the sinusoidal terms to be the same. We accomplish this by rearranging the relationship for ω_0:

$$\omega_0 = \sqrt{\frac{k}{m}} \qquad (12\text{-}2)$$

so

$$m\omega_0^2 = k$$

The kinetic energy of the simple harmonic oscillator becomes

$$K = \frac{1}{2} k A^2 \sin^2(\omega_0 t + \phi) \qquad (12\text{-}9)$$

and the total energy is then

$$E = \frac{1}{2} k A^2 \sin^2(\omega_0 t + \phi) + \frac{1}{2} k A^2 \cos^2(\omega_0 t + \phi)$$

We pull out the common $\frac{1}{2} k A^2$ term to make use of the trigonometric identity:

$$E = \frac{1}{2} k A^2 \left[\sin^2(\omega_0 t + \phi) + \cos^2(\omega_0 t + \phi) \right]$$

or

$$E = \frac{1}{2} k A^2 \qquad (12\text{-}10)$$

Because both the spring constant k and the amplitude A are constant for a specific motion of a given oscillator, the total energy is constant. This is a statement of conservation of energy. In the absence of dissipative forces (we haven't introduced any), total energy is constant.

Equations 12-9 and 12-8 give kinetic energy and potential energy as functions of time. It is helpful to write the relationships as functions of displacement, so that we can make plots of the two and consider them graphically. The potential energy is already expressed in terms of x,

$$U(x) = \frac{1}{2}kx^2 \tag{12-11}$$

We rearrange the trigonometric identity $\sin^2 \theta + \cos^2 \theta = 1$ to write kinetic energy K in terms of x:

$$K = \frac{1}{2}kA^2 \sin^2(\omega_0 t + \phi)$$

$$= \frac{1}{2}kA^2[1 - \cos^2(\omega_0 t + \phi)]$$

$$= \frac{1}{2}kA^2 - \frac{1}{2}kA^2 \cos^2(\omega_0 t + \phi)$$

Using $x(t) = A \cos(\omega_0 t + \phi)$ (Equation 12-4), kinetic energy as a function of displacement is then

$$K(x) = \frac{1}{2}kA^2 - \frac{1}{2}kx^2 \tag{12-12}$$

As a check, note that the expressions for U and K immediately lead to the relationship we just uncovered for total energy:

$$E = K + U = \left(\frac{1}{2}kA^2 - \frac{1}{2}kx^2\right) + \frac{1}{2}kx^2 = \frac{1}{2}kA^2$$

Plots of $K(x)$ (Equation 12-12) and $U(x)$ (Equation 12-11) versus displacement are shown in **Figure 12-10**. (The specific value of energy doesn't matter here, so we plotted the energies in terms of percentage of total energy to make the curves easy to read.) As expected, $U(x)$ is largest at the two extremes of displacement ($x = +A$ and $x = -A$) and zero at equilibrium ($x = 0$). $K(x)$ is zero at the two extremes, where the object must momentarily come to a stop to reverse direction, and largest as the object passes through $x = 0$. Then at $x = \pm A$, the total energy $E = 0 + U_{max}$ because K is zero, and at $x = 0$ the total energy $E = K_{max} + 0$ because U is zero.

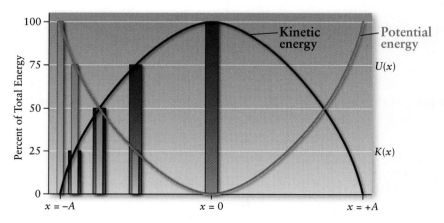

Figure 12-10 The sum of the kinetic energy and potential energy associated with an object in simple harmonic motion is constant. Red bars indicate the relative magnitude of the potential energy at different displacements from equilibrium. Blue bars show the relative magnitude of kinetic energy at different displacements from equilibrium. The red and blue curves show the contribution in percent of the potential energy and kinetic energy, respectively, to the total energy during simple harmonic motion.

The total energy is therefore equal to the maximum value of K, and also equal to the maximum value of U.

At any value of displacement, the total energy $E = K + U$. For example, when K is 25% of E (the leftmost blue bar in Figure 12-10), U is 75% of E. The sum of 25% of the total energy and 75% of the total energy gives, of course, the total energy. Notice also that at the point where the two curves cross, the potential energy and kinetic energy both equal half of the maximum of K or U, and also half of the total energy. At any value of displacement x on the graph, the sum of K and U equals the total energy. The energy of the simple harmonic oscillator is indeed constant.

Example 12-3 Where Does Kinetic Energy Equal Potential Energy?

In the discussion of Figure 12-10, we considered the point at which both kinetic energy and potential energy equal one-half of the total energy. What fraction of the amplitude of the motion is the displacement at that point?

SET UP
Because kinetic energy and potential energy are equal, you might be tempted to set $\frac{1}{2}mv^2$ equal to $\frac{1}{2}kx^2$ to solve for the particular value of x in which we're interested. Such a calculation would be unproductive because we don't know the speed of the object at this particular displacement. However, because K and U are equal, it must also be true that the potential energy is one-half of the total energy:

$$U = \frac{1}{2}E$$

So by substituting the expression in Equation 12-10 for total energy E and the expression in Equation 12-11 for potential energy, we find that

$$\frac{1}{2}kx^2 = \frac{1}{2}\left(\frac{1}{2}kA^2\right)$$

We can solve this equality for x!

SOLVE
Using the equation to solve for x yields

$$x^2 = \frac{1}{2}A^2$$

or

$$x = 0.707A$$

At a displacement of about 70% of the maximum amplitude (the maximum displacement), the kinetic energy equals the potential energy.

REFLECT
Convince yourself that this is reasonable by estimating the displacement at which K equals U in Figure 12-10.

Practice Problem 12-3 A system undergoing simple harmonic motion reaches a displacement from equilibrium at which the kinetic energy is one-third of the potential energy. What fraction of the amplitude of the motion is the displacement at that point?

> ### ✳ What's Important 12-3
> Energy is transformed back and forth between potential and kinetic forms in systems which contain mechanical or biological springs. When an object attached to a spring is pulled away from equilibrium, the potential energy increases; when the object is released it accelerates back toward equilibrium due to the restoring force. It reaches maximum velocity (and therefore maximum kinetic energy) as it passes through equilibrium because after that the restoring force opposes the motion and the object decelerates. Throughout the process, the sum of the kinetic energy and potential energy of a simple harmonic oscillator, and therefore the total energy, remains the same.

12-4 The Simple Pendulum

Your legs swing back and forth as you walk, in much the same way that a ball swings when hanging from the end of a string. The motion of your leg and the ball on a string share certain characteristics. When allowed to move freely, both would hang straight down under the influence only of gravity. This is the definition of an equilibrium position. Also, when either the leg or ball is displaced from equilibrium, released, and allowed to move freely, the force of gravity pulls it back toward equilibrium. Gravity, then, serves as a restoring force. The motions of such systems are characterized both by an equilibrium position and a restoring force; we should expect the motion of both to be oscillatory. In this section, we explore the connections between simple harmonic motion and the motion of a pendulum.

Let's consider an idealized or **simple pendulum**, in which all of the mass is concentrated a fixed distance from a rotation point. (We will examine a nonidealized pendulum in Section 12-6.) A simple pendulum could be a tiny object that has a mass m and is attached to a thread that has a negligible mass and a length L, as shown in Figure 12-11a. In the figure, the system has been displaced by an angle θ from the vertical. To analyze the simple harmonic motion of this pendulum, we need to identify the restoring force.

The free-body diagram for the object is shown in Figure 12-11b. In the radial direction, the tension force $\vec{T}$ is balanced by $m\vec{g}\cos\theta$, the radial component of the

Figure 12-11 (a) A simple pendulum consists of a small object connected to a pivot point by a thread of negligible mass. (b) The free-body diagram for the object includes two forces, the tension in the thread and the force due to gravity. The components of the force due to gravity in the radial and tangential directions are shown as dashed red arrows.

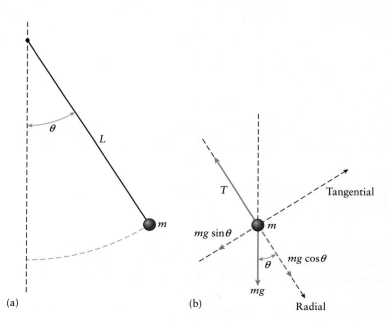

(a)   (b)

gravitational force. The net force in the radial direction is therefore zero. A zero net force is not the case in the tangential direction, however. Because $\vec{T}$ is radial, it has no tangential component. So, the only tangential force is $m\vec{g} \sin \theta$, the component of $m\vec{g}$ in the direction tangential to the path of the object. Because the net radial force is zero, $m\vec{g} \sin \theta$ is therefore not only the tangential force but also the net force on the object. When the object is displaced from hanging straight down, the force

$$\vec{F} = -m\vec{g} \sin \theta \qquad (12\text{-}13)$$

acts to restore the object to its equilibrium position. We included the minus sign to indicate that it is a restoring force.

Does $\vec{F} = -m\vec{g} \sin \theta$ have the same form as Hooke's law? If so, we can simply write down the motion as a function of time because we've already solved the problem of an object that experiences a Hooke's law force. A Hooke's law force has the form

$$\vec{F} = -k\vec{x} \qquad (6\text{-}16)$$

where the variable k is a constant and $\vec{x}$ is the displacement from the equilibrium position. Also, note that for the term on the left to equal the term on the right, the $\vec{x}$ and $\vec{F}$ vectors must point in the same direction.

The displacement of the object from its equilibrium position is the length of the arc of radius L subtended by angle θ,

$$x = L\theta \qquad (12\text{-}14)$$

In this equation the variable x represents the distance through which the object must swing to return to equilibrium; it does not imply a horizontal distance. Because $\sin \theta$, the term that follows the constant mg in $\vec{F} = -m\vec{g} \sin \theta$, is not the displacement of the object from its equilibrium position, we must therefore conclude that Equation 12-13 is *not* a form of Hooke's law. Moreover, notice that the dimensions of displacement should be distance, which is not the case for $\sin \theta$. In addition, because the distance is an arc length, it cannot be aligned with the force vector $\vec{F}$. We could, however, allow ourselves to deal only with cases in which the angle θ is relatively small. When the arc is short enough, we can treat it as a straight line, that is, a vector perpendicular to the radial direction and parallel to $\vec{F} = -m\vec{g} \sin \theta$. When the angle, measured in radians, is small we can also employ another simplification:

$$\sin \theta \approx \theta \qquad \text{when } \theta \text{ is small} \qquad (12\text{-}15)$$

(A graphical proof of this approximation is shown in **Figure** 12-12). So the magnitude of the force given by Equation 12-13 becomes

$$F \approx -mg\theta$$

Using Equation 12-14 to write θ in terms of x, the restoring force on the object attached to the free end of the pendulum can be written

$$F \approx -mg\frac{x}{L}$$

or

$$F \approx -\frac{mg}{L}x$$

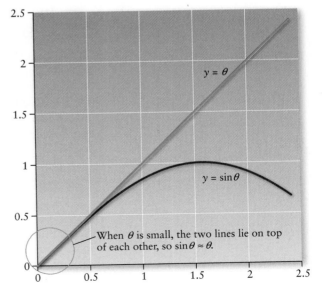

When θ is small, the two lines lie on top of each other, so $\sin \theta \approx \theta$.

Figure 12-12 For small values of θ, $\sin \theta$ is approximately equal to θ.

This equation *is* Hooke's law! The constant is $k = mg/L$, so we can immediately use the previous results for simple harmonic motion to describe the motion of the pendulum. We found that the displacement from equilibrium as a function of time can be written as

$$x(t) = A\cos(\omega_0 t + \phi) \tag{12-4}$$

where

$$\omega_0 = \sqrt{\frac{k}{m}} \tag{12-2}$$

Equation 12-4 applies equally well to the pendulum as to an object attached to a spring; to express angular velocity in terms of the parameters of the pendulum, we substitute $k = mg/L$ into Equation 12-2:

$$\omega_0 = \sqrt{\frac{mg/L}{m}} = \sqrt{\frac{g}{L}}$$

 Go to Picture It 12-1 *for more practice dealing with simple pendulums*

The period, using Equation 12-5, is then

$$T = \frac{2\pi}{\omega_0} = \frac{2\pi}{\sqrt{g/L}} = 2\pi\sqrt{\frac{L}{g}} \tag{12-16}$$

Notice that neither the angular velocity ω_0 nor the period T of the simple pendulum depends on the mass of the object hanging from the support. (Remember that a simple pendulum is idealized as an object attached to a rotation axis by a support of negligible mass.) Two simple pendula of the same length but of different masses nevertheless swing back and forth at the same rate and with the same period.

Equation 12-16 only applies when the angle of the initial displacement is so small that we are justified in using θ instead of $\sin\theta$ to express the restoring force. How small is enough? Is there some specific angle smaller than which the results hold and larger than which they don't? No. Rather, the smaller the angle, the better the approximation works. When θ is about 15°, the percent error in the difference between $\sin\theta$ and θ is about 1%, but when θ is about 5° the percent error is about 0.1%.

? Got the Concept 12-3
Period of a Pendulum I

A small marble is attached to a thread that has a negligible mass and is hung from a support. When the marble is pulled back a small distance and released, it swings in simple harmonic motion with a period of 3 s. What is the period of the pendulum after the length of the thread is increased by a factor of 4? The marble is released in the same way. Assume that the thread does not stretch as the marble swings down.

? Got the Concept 12-4
Period of a Pendulum II

A small marble is attached to a thread that has a negligible mass and is hung from a support. When the marble is pulled back a small distance and released, it swings in simple harmonic motion with a period of 3 s. What is the period of the pendulum after the marble is replaced by one that is 4 times its mass? The second marble is released in the same way as the first was and as before, the thread does not stretch.

Got the Concept 12-5
Period of a Pendulum III

A small marble is attached to a thread that has a negligible mass and is hung from a support. When the marble is pulled back a small distance and released, it swings in simple harmonic motion with a period of 3 s. What is the period of the pendulum when pulled back twice as far from equilibrium, assuming that it swings in simple harmonic motion?

 Got the Concept 12-6
Pendulum on the Moon

A small marble is attached to a thread that has a negligible mass and is hung from a support. When the marble is pulled back a small distance and released, it swings in simple harmonic motion with a period of 3 s. What would the period of this pendulum if it were on the surface of the Moon, where the acceleration due to gravity is approximately one-sixth that on the surface of Earth?

Example 12-4 Changing Pendulum

A simple pendulum is created by hanging a small 60-g sphere from an elastic thread of negligible mass. With the sphere attached, the thread is 40 cm long. (a) Find the period when the sphere is pulled back a small distance and released. (b) When the sphere is replaced by another one of mass 260 g, the thread stretches an additional 10 cm. Find the new period.

SET UP
We can rely on Equation 12-16 for the period of a simple pendulum for both parts of the problem

$$T = 2\pi\sqrt{\frac{L}{g}} \qquad (12\text{-}16)$$

Based on the previous Got the Concept questions, we should not expect the difference in mass to affect the period.

SOLVE
For part (a), the length of the pendulum is $L = 40$ cm $= 0.40$ m, so

$$T = 2\pi\sqrt{\frac{0.40\text{ m}}{9.8\text{ m/s}^2}} = 1.3\text{ s}$$

For part (b), the change in length affects the period but the change in mass does not. We can therefore directly apply Equation 12-16 again:

$$T = 2\pi\sqrt{\frac{0.50\text{ m}}{9.8\text{ m/s}^2}} = 1.4\text{ s}$$

REFLECT
The period of a simple pendulum depends on its length but not on the mass of the object that swings.

Practice Problem 12-4 A simple pendulum is created by hanging a small 60-g sphere from an elastic thread of negligible mass. When the sphere is attached, the thread is 40 cm long. When the sphere is replaced by a smaller one that has a mass of 20 g, the thread is 38 cm long. Find the period of the new pendulum.

Estimate It! 12-1   Harry Potter and the Big Pendulum

In a scene from the movie *Harry Potter and the Order of the Phoenix*, a large, long pendulum apparently swings from the top of a tall tower. While watching the movie, one of the authors (he's somewhat of a geek) estimated the period of the pendulum in order to compute its length. He crudely measured the time for one-half of a full swing (half the period) to be 2 s. Estimate the length of the pendulum, assuming it can be considered simple. (The pendulum's support appears relatively small, far less massive than the disk at the end of the pendulum, so this assumption is not too bad.) Are you comfortable with the filmmaker's grasp of physics?

SET UP

We can estimate the length of the pendulum using the relationship between the period and length of a simple pendulum:

$$T = 2\pi\sqrt{\frac{L}{g}} \tag{12-16}$$

or

$$L = \frac{gT^2}{4\pi^2}$$

SOLVE

We have the author's estimate that the pendulum swings through half a period in 2 s, so the period T is approximately 4 s. To the same number of significant figures, g is 10 m/s^2. All together, then

$$L = \frac{(10 \text{ m/s}^2)(4\text{s})^2}{4\pi^2} = 4 \text{ m}$$

REFLECT

The swing of the pendulum caught the author's attention because it was too rapid for the pendulum to be as long as the size of the disk and the tower suggested. Our estimate of 4 m confirms that. A pendulum that has the length of a castle tower's height (say, 30 m) would have a period of about 11 s (using Equation 12-16), more than twice as long as suggested in the movie. It appears that the filmmaker did not properly account for physics in constructing the special effects for the pendulum.

★ What's Important 12-4

In a simple pendulum (for example, a tiny object hanging from a support of negligible mass), all of the mass of the pendulum is concentrated a fixed distance from a rotation point. Neither the angular velocity ω_0 nor the period T of a simple pendulum depends on the mass of the object hanging from the support. Two simple pendula that have the same length but different masses nevertheless swing back and forth at the same rate and with the same period.

12-5 Physical Oscillators

We have considered a number of oscillating systems, for example, a system that consists of an object attached to a spring, a frog's eardrum, and a simple pendulum. In studying the mathematical underpinnings of oscillating systems, we neglected some of the real-world details in order to make the physics a bit easier to understand. In the case of the object and spring, we neglected resistance and other forces that would damp out the motion as well as nonconservative forces that convert energy in the spring to heat. In examining the pendulum, we assumed that all of its mass is concentrated at the end of the pendulum's support. Such an approximation is reasonable for a small ball swinging from a thin thread, but not as reasonable for, say, your leg as it swings when you walk. To better understand oscillations that occur in the real world, we will examine more carefully the pendulum and the object–spring systems under conditions that more closely resemble those likely to be present in observable phenomena.

An oscillatory system that you've probably experienced is the combination of springs and shock absorbers that comprise the suspension of most cars. This example illustrates two important real-world complications we must consider. To isolate the passengers from bumps in the road, the wheels and axles are separated from the frame of a car by a set of springs. If you push down on the front end of a car, it will bounce up and down, once or twice if your suspension is in good working order, and a few more times if it's not. This harmonic motion is due to the springs. Unlike the object and spring we considered in Section 12-1, however, these oscillations don't have constant amplitude. The shock absorbers, also part of the connection between the axles and the frame, exert a **damping force** that dissipates the energy stored in the springs. We will consider the damped harmonic oscillator in Section 12-7.

As a car drives along a road, even a smooth one, irregularities in the surface cause the wheels to bounce up and down. The bumps in the road exert a time-varying force on the springs, which changes the motion of the object (the car's frame) attached to the other end. When we considered the object and spring earlier, we displaced one end of the spring from equilibrium and then applied no other force to the system; the resulting oscillatory motion was due to the cyclic transfer of the spring's energy between potential energy and kinetic energy. It is more often the case in nature, however, that the force which displaces the system from equilibrium has time dependence. The bumps in a road repeatedly bounce a car's frame and springs, in the same way that you might wiggle your hand up and down while holding a spring with an object attached to the other end. We will discuss forced oscillations, in which a system experiences a time-varying **driving force**, in Section 12-8.

Our simple pendulum, as the name implies, is also an idealized system. For a light enough thread and a small enough mass, the requirement that all of the pendulum's mass be concentrated at the end of the support is a reasonable approximation. It is rare, however, to find such a pendulum in nature. Your leg, for example, which can swing in harmonic motion, has a relatively uniform distribution of mass from the pivot in your hip down to your foot. In the next section, we will consider the *physical pendulum*, one in which the mass is not concentrated at a single point at the end of a support that can rotate around a pivot point.

★ What's Important 12-5

A damping force dissipates energy, such as that stored in an oscillating object–spring system. A driving force is an external force acting on a system. Forced oscillations occur when a system that can undergo harmonic motion experiences a time-varying driving force.

12-6 The Physical Pendulum

During a test of the knee-jerk reflex, the lower leg will first kick upward and then swing back and forth like a pendulum. Up to this point we have only discussed a simple pendulum, one in which the entire mass is concentrated at a point attached to the free end of a support of negligible mass. The leg is clearly not a simple pendulum because its entire mass is not concentrated in the foot. To study the periodic motion of a knee jerk and obtain equations for the period and for the position of the leg as a function of time, we need to reexamine our analysis of the pendulum.

Figure 12-13 shows a generic **physical pendulum**, one in which the pendulum's mass is not concentrated at a point at the end of a support of negligible mass. As we determined in Section 7-6, we can analyze the motion of an extended object by considering the net force as acting on the center of mass. For the physical pendulum of mass m, the net force is the pendulum's weight $m\vec{g}$, which points vertically down. And as we determined in Section 8-6, when an object that experiences a force is free to rotate, we must consider the torque exerted by that force. The torque around the axis which passes through the pivot point, from Equation 8-23, is

$$\tau = hmg \sin \phi \qquad (12\text{-}17)$$

We follow the convention here of using the variable h rather than r to represent the distance from the center of mass to the pivot point.

The torque in Equation 12-17 restores the physical pendulum to its equilibrium position after it has been displaced. To make it clear that this is a restoring torque, we can introduce a minus sign by using θ, as shown in the figure, instead of ϕ, in Equation 12-17. Because $\sin \theta = -\sin \phi$ when $\theta = \phi - 180$

$$\tau = -hmg \sin \theta \qquad (12\text{-}18)$$

When the pendulum is displaced from equilibrium, the torque causes it to return to that position, hanging straight down. Because the torque on the physical pendulum is restoring, it is natural for us to ask whether it fits the form of Hooke's law. If it does, we can apply the technique we used to analyze the simple pendulum.

The restoring torque described in Equation 12-18 is not in the form of Hooke's law because it is not proportional to θ, the displacement from equilibrium. However, recall that $\sin \theta \approx \theta$ when θ is small (Equation 12-15). So if we limit ourselves to small displacements from equilibrium, then with minor rearranging Equation 12-18 becomes

$$\tau \approx -mgh \, \theta \qquad (12\text{-}19)$$

This expression *is* in Hooke's law form, so we can apply the same analytical techniques we used to describe the simple pendulum (Section 12-4). The rotational equivalent of Newton's second law connects torque and motion:

$$\sum \vec{\tau} = I \vec{\alpha} \qquad (8\text{-}24)$$

where the moment of inertia I and angular acceleration α describe the motion. There is only one torque acting on the physical pendulum, so the magnitude of the torque is

$$\tau = I\alpha$$

and then by entering the value of τ from Equation 12-19, we find that

$$-mgh \, \theta = I\alpha$$

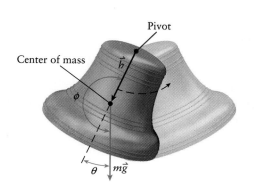

Center of mass

Pivot

$\vec{h}$

ϕ

θ $m\vec{g}$

Figure 12-13 The mass of a physical pendulum is distributed, perhaps not uniformly, around a pivot point.

for small displacements from equilibrium. We make the time dependence explicit by including the definition of angular acceleration α:

$$-mgh\,\theta = I\frac{d^2\theta}{dt^2}$$

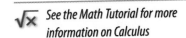 See the Math Tutorial for more information on Calculus

We can write this equation in a form that is analogous to the simple harmonic oscillator equation, $d^2x/dt^2 + (k/m)x = 0$, Equation 12-1):

$$\frac{d^2\theta}{dt^2} + \frac{mgh}{I}\theta = 0 \qquad (12\text{-}20)$$

In the same way that Equation 12-1 results in a relationship for linear displacement as a function of time, Equation 12-20 gives angular displacement as a function of time:

$$\theta(t) = \theta_{max}\cos(\omega_0 t + \phi) \qquad (12\text{-}21)$$

Also analogous to our solution to Equation 12-1, the angular velocity ω_0 is defined as the square root of the coefficient of the linear term in Equation 12-20:

$$\omega_0 = \sqrt{\frac{mgh}{I}} \qquad (12\text{-}22)$$

Finally, the period T of the physical pendulum is obtained by combining this relationship for ω_0 with the definition of the period:

$$T = \frac{2\pi}{\omega_0} \qquad (12\text{-}5)$$

This leads to a relationship describing the period of a physical pendulum of mass m and moment of inertia I around a rotation axis a distance h from the center of mass

$$T = 2\pi\sqrt{\frac{I}{mgh}} \qquad (12\text{-}23)$$

? Got the Concept 12-7
Period of a Pendulum

A physical pendulum is supported so that it rotates around its center of mass. Is the period large or small? Interpret your answer by describing the oscillations of the pendulum.

Example 12-5 Simple Pendulum from a Physical One

The simple pendulum that we covered previously is a special case of the physical pendulum we've discussed in this section. Show that the period T we found for a simple pendulum (Equation 12-16) of length L and mass m can be obtained from the physics that describes the physical pendulum.

SET UP
Our goal is to obtain Equation 12-16:

$$T = 2\pi\sqrt{\frac{L}{g}}$$

from the period of a physical pendulum (Equation 12-23):

$$T = 2\pi\sqrt{\frac{I}{mgh}}$$

To do so, we need to find h, the distance from the pivot to the center of mass, and also the moment of inertia I for a simple pendulum of length L and mass m.

SOLVE
Because the only mass in the system is at a distance L from the support, the center of mass is also there, so $h = L$. In addition, according to the definition of moment of inertia $I = \sum m_i r_i^2$ (Equation 8-7), for the single object of mass m at a distance L from the rotation axis, $I = mL^2$. So using the period of the physical pendulum,

$$T = 2\pi\sqrt{\frac{mL^2}{mgL}} = 2\pi\sqrt{\frac{L}{g}}$$

REFLECT
We have reproduced the relationship we developed earlier for the period of a simple pendulum by treating it as a physical one.

Example 12-6 A Swinging Rod

A uniform rod of length L and mass M is supported so that it can swing freely from one end. Find the period of oscillation when the rod is pulled slightly from vertical and released.

SET UP
The period of a physical pendulum is

$$T = 2\pi\sqrt{\frac{I}{Mgh}} \qquad \text{(12-23)}$$

To apply this expression to any particular pendulum, we need to determine the moment of inertia I around the rotation axis. We must also identify the location of the center of mass to determine h.

   We found the moment of inertia of a uniform rod rotating around one end in Chapter 8. As summarized in Table 8-1,

$$I = \frac{ML^2}{3}$$

Also, by symmetry, the center of mass of the rod is at its center, a distance $L/2$ from the pivot, so $h = L/2$.

SOLVE
Using I and h for the rod, the period T of the physical pendulum becomes

$$T = 2\pi\sqrt{\frac{ML^2/3}{Mg(L/2)}} = 2\pi\sqrt{\frac{2L}{3g}} \qquad \text{(12-24)}$$

REFLECT
We can write the result as

$$T = 2\pi\sqrt{\frac{\left(\frac{2}{3}\right)L}{g}}$$

 Go to Interactive Exercise 12-2 for more practice dealing with physical pendulums

So a uniform rod allowed to swing freely from one end has the same period of oscillation as a simple pendulum of two-thirds the length of the rod.

Practice Problem 12-6 A uniform rod of length L and mass M is supported so that it can swing freely from a point halfway between the center and one end of the rod. Find the period of oscillation when the rod is pulled slightly from vertical and released. Express your answer in terms of the period T_0 of a simple pendulum of the same length.

Example 12-7 Moment of Interia of a Human Leg

Your gait is affected in part by how your legs swing around the rotation axis created by your hip joints. As with most irregular, nonuniform objects, it is not possible to calculate the moment of inertia of a leg around the hip using the definition of I (Equation 8-9). I can be measured, however. A common method to measure I for an irregular object makes use of the period of the object when allowed to swing freely. In a clinical study, a person's leg of length 0.88 m is estimated to have a mass of 6.5 kg and a center of mass 0.37 m from the rotation axis through the hip. When allowed to swing freely, it oscillates with a period of 1.2 s. Find the moment of inertia of the leg in rotation around the axis through the hip. Compare your answer to a uniform rod that has the same mass and length.

SET UP
The period of a physical pendulum (Equation 12-23) can be rearranged as an expression for I:

$$I = mgh\left(\frac{T}{2\pi}\right)^2$$

We know h, the distance from the rotation axis to the center of mass, as well as the mass and the period.

SOLVE
Substituting the known values gives

$$I = (6.5 \text{ kg})(9.8 \text{ m/s}^2)(0.37 \text{ m})\left(\frac{1.2 \text{ s}}{2\pi}\right)^2 = 0.86 \text{ kg} \cdot \text{m}^2$$

We can compare the moment of inertia of the leg rotating around the hip joint to the moment of inertia of a uniform rod rotating around its end (from Table 8-1):

$$I = \frac{ML^2}{3}$$

Using the length and mass given for the human leg,

$$I = \frac{(6.5 \text{ kg})(0.88 \text{ m})^2}{3} = 1.7 \text{ kg} \cdot \text{m}^2$$

REFLECT
The moment of inertia of a uniform rod of the length and mass of a leg is twice as large as the moment of inertia we found for a leg. Consider that more of the leg's mass is above the knee, as suggested by **Figure 12-14**, so the center of mass is closer to the rotation axis than a point halfway down the leg. As you recall from Chapter 8 and can see from $I = \int r^2 dm$ (Equation 8-9), the moment of inertia increases as the square of the distance of the mass from the rotation axis. So we would expect that the leg has a smaller moment of inertia than a uniform rod because the center of mass of the rod *is* halfway along its length.

Figure 12-14 The tapered shape superimposed on a leg suggests that more of the leg's mass is above the knee.

 Got the Concept 12-8
Period of a Rod

A uniform rod has the same length and mass as the leg in the previous example. When supported at one end and allowed to rotate, would the period of the rod be greater than, the same as, or smaller than the period of the leg?

* What's Important 12-6
A physical pendulum is a nonidealized, real pendulum. Its mass is distributed so that, unlike a simple pendulum, it cannot be treated as a point object suspended at the end of a support of negligible mass. We analyze a physical pendulum by considering the torque that arises when the pendulum is displaced from equilibrium. The restoring torque causes the pendulum to return to its equilibrium position.

12-7 The Damped Oscillator

When you push down on and then release one of the fenders of a car, the car springs back up but it doesn't continue to bounce up and down forever. When your knee-jerk reflex is tested, after one or two swings your leg stops moving. Similarly, as we saw in Example 12-1, the oscillations of a frog's eardrum diminish over time. In all of these cases, some mechanism is exerting a damping force that dissipates the energy required for the oscillations to continue. In this section, we discuss the damped oscillator.

We calculated the magnitude of the motion of the simple object and spring system by setting the sum of the forces equal to mass multiplied by acceleration (Newton's second law). Using acceleration $a = d^2x/dt^2$, we obtained

$$-kx = m\frac{d^2x}{dt^2}$$

which led directly to the simple harmonic oscillator equation:

$$\frac{d^2x}{dt^2} + \frac{k}{m}x = 0 \tag{12-1}$$

In developing this equation and its solution, we considered only one force on the object, a force that can be described by Hooke's law, $F = -kx$. Because we did not include a damping force, however, the equation does not predict a motion which decreases over time. Damping forces, such as the drag forces we considered in Chapter 5 (Equation 5-22), are generally proportional to the velocity of a moving object raised to some power. The simplest damping force is

$$F_{damping} = -bv$$

where the value of the damping coefficient b depends on various physical characteristics of a particular system. By adding this force term to the simple harmonic oscillator equation and using $v = dx/dt$, we account for a damping force on the object as well as the restoring force (described by Hooke's law):

$$m\frac{d^2x}{dt^2} + b\frac{dx}{dt} + kx = 0 \tag{12-25}$$

 Go to Picture It 12-2 *for more practice dealing with damped oscillations*

Solutions of the damped harmonic oscillator equation take a somewhat different form depending on the relative magnitude of the damping force. We will consider only the most common situation, in which the damping force is relatively small. In this case the displacement from equilibrium of the oscillator as a function of time can be expressed as

$$x(t) = Ae^{-(b/2m)t}\cos(\omega_1 t + \phi) \tag{12-26}$$

where the frequency of oscillation ω_1 is

$$\omega_1 = \sqrt{\omega_0^2 - (b/2m)^2}$$

The angular frequency of the oscillator if no damping were present, ω_0, is referred to as the **natural frequency** in order to differentiate it from the oscillation frequency ω_1.

Equation 12-26 describes general characteristics of the relationship between displacement and time in damped oscillators. First, notice that as in the case of the undamped oscillator, displacement depends on a cosine function, so the motion of a damped oscillator is periodic. In fact, if the value of the damping coefficient b is sufficiently small, the frequency of oscillation ω_1 is approximately equal to the natural frequency ω_0, making the cosine term in both the damped and undamped oscillator solutions the same.

The significant difference between the damped and undamped solutions appears in the amplitude of the motion. Whereas the amplitude of the undamped motion is a constant A, the amplitude for the damped oscillator decreases as a function of time:

$$A(t) = Ae^{-(b/2m)t} \tag{12-27}$$

As a result of the decreasing exponential function, the oscillator deviates less and less from the equilibrium position over time, as we would expect. A graph of $A(t)$ versus time is shown in **Figure 12-15**.

In Example 12-1 we noted that the oscillations of a frog's eardrum change from one cycle to the next. We now know that the changing amplitude results from a damping force in the system. The graph of displacement from equilibrium of the

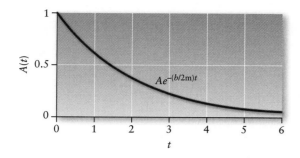

Figure 12-15 A damped harmonic oscillator deviates less and less from the equilibrium position over time.

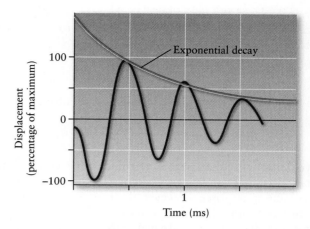

Figure 12-16 Sound causes a frog's eardrum to move back and forth, in (approximately) damped harmonic motion. The red curve that follows the peaks of the periodic displacement curve is described by an exponential decay as in Equation 12-27. (After Chung, Pettigrew, and Anson (1981) *Proc. Royal Soc. London Ser. B Biol. Sci.* 212, No. 1189, pp. 459–485.)

eardrum with time is shown again in Figure 12-16. This time, we added the red curve that follows the peaks of the periodic displacement curve. Equation 12-27 exactly defines the red curve.

Example 12-8 Balance

Special sensory cells in your inner ear play a role in balance. (See Section 3-5 for a discussion of the vestibular system.) In response to a prolonged stimulus, the cells send out an oscillating signal which can look like the data in Figure 12-17. The red dashed line follows Equation 12-27; for the data shown, $b/2m = 42$ s^{-1}. Find the time it takes for the amplitude of the oscillations to fall to 20% of the initial amplitude. Check your answer against the figure. (The period of the oscillations is about 3.2×10^{-3} s.)

SET UP

Our goal is to find the time t_1 when $A(t_1)$ is equal to 0.20 times the initial amplitude, that is, when

$$A(t_1) = 0.20A$$

Figure 12-17 In response to a prolonged stimulus, special sensory cells in the inner ear send out an oscillating signal. The red dashed line follows Equation 12-27; for the data shown, $b/2m = 42$ s^{-1}. (After, Armstrong and Roberts (1998) *J Neurosci*, 18(8):2962–2973.)

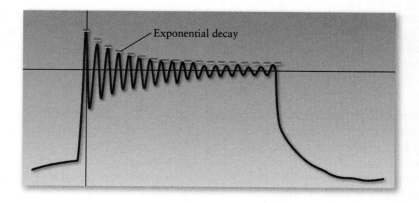

From Equation 12-27, we know that the amplitude at time t_1 is

$$A(t_1) = A\,e^{-(b/2m)t_1}$$

We can combine these two equations to solve for t_1.

SOLVE

So,

$$0.20A = A\,e^{-(b/2m)t_1}$$

or

$$0.20 = e^{-(b/2m)t_1} \tag{12-28}$$

Although we have a bit more work to do, you can see that because the value of $b/2m$ is known, this expression uniquely determines t_1.

To isolate t_1 we employ the definition of the logarithm. The definition of a logarithm to the base n is such that for $\log_n a = x$, then $n^x = a$. So $\log_e e^x = \ln e^x = x$. We therefore take the natural logarithm of both sides of Equation 12-28:

$$\ln(0.20) = \ln(e^{-(b/2m)t_1}) = -(b/2m)t_1$$

All that remains is to substitute known values for the variables and calculate an answer

$$t_1 = \frac{-\ln(0.20)}{b/2m} = \frac{-(-1.6)}{42\ \text{s}^{-1}} = 3.8 \times 10^{-2}\ \text{s}$$

REFLECT

Given $T = 3.2 \times 10^{-3}$ s, the number of cycles corresponding to our value of t_1 is

$$\#\ \text{cycles} = \frac{t_1}{T} = \frac{3.8 \times 10^{-2}\ \text{s}}{3.2 \times 10^{-3}\ \text{s}} = 12\ \text{cycles}$$

Using a ruler to estimate the amplitude of the oscillations, you can verify that the height of the peak of the twelfth cycle in Figure 12-17 is close to 20% of the height of the first peak. As we have noted, amplitude, and therefore the height, is measured from the middle of the waveform.

Practice Problem 12-8 Special sensory cells in the inner ear play a role in balance. In response to a prolonged stimulus, the cells send out an oscillating signal which can look like the data in Figure 12-17. The red dashed line follows Equation 12-27; for the data shown, $b/2m = 42\ \text{s}^{-1}$. Find the time it takes for the amplitude of the oscillations to fall to 2% of the initial amplitude, assuming that the stimulus lasts that long.

> ✳ **What's Important 12-7**
> The motion of a damped oscillator is periodic with an amplitude that decreases as a function of time. If no damping is present, the angular frequency of an oscillator is referred to as the natural frequency.

12-8 The Forced Oscillator

Up to this point in our study of oscillating systems, we have only considered the effects of restoring forces and damping forces. The motion of a child on a playground swing is not unlike that of a physical pendulum; the child and swing exhibit something close to simple harmonic motion when allowed to move freely. But when someone pushes on the swing, the motion can be quite different. If pushes are timed so that they correspond in frequency to the frequency at which the swing moves naturally, the child will swing higher and higher. The same is true of a car on a bumpy road; the bumps exert a time-varying force on the car's springs, and if the timing of the frequency of the forces matches the frequency at which the frame of the car bounces up and down naturally, the amplitude of the motion could get large.

We obtained Equation 12-25 by setting the sum of the restoring force and damping forces equal to mass multiplied by acceleration:

$$-b\frac{dx}{dt} - kx = m\frac{d^2x}{dt^2}$$

or

$$m\frac{d^2x}{dt^2} + b\frac{dx}{dt} + kx = 0 \tag{12-25}$$

To study situations like those mentioned earlier in this section, we introduce a third kind of force, a time-varying driving force $F(t)$:

$$F(t) - b\frac{dx}{dt} - kx = m\frac{d^2x}{dt^2}$$

or

$$m\frac{d^2x}{dt^2} + b\frac{dx}{dt} + kx = F(t) \tag{12-29}$$

This equation is a complete description of an oscillating system. The force kx restores the system to equilibrium. The bv term ($b\,dx/dt$ in the equation) is a damping force which dissipates energy from the system. The force $F(t)$, which would correspond to the force exerted by someone pushing on a swing or by bumps in the road on a car, is the driving force.

Solutions of Equation 12-29 are not straightforward to obtain, but we can gain insight into the nature of driven oscillations by making two simplifications. First, we initially consider the system in the absence of a damping force; that is, we remove the $b\,dx/dt$ term from Equation 12-29. We also generalize the driving force as a periodic function of time, which makes the math more straightforward but still allows us to see the physics clearly. We could pick either sine or cosine as the periodic function; because we have been using a cosine function to describe harmonic oscillations, we'll use a driving force of the form

$$F(t) = F_0 \cos(\omega t)$$

The angular frequency ω in this expression is the frequency at which the system is driven. Unlike the natural frequency ω_0, the driving frequency ω is not a characteristic of the oscillating system but rather is determined externally. If you hold the end of a spring and let an object attached to the other end bounce up and down, the natural frequency you observe is determined by the spring constant and the mass of the object. Once you've chosen the spring and the object, the natural frequency ω_0 is fixed. You are free, however, to wiggle the end of the spring in your hand at any frequency you pick; the wiggle frequency is the driving frequency ω.

Rewriting Equation 12-29 without the damping term and using $F(t) = F_0 \cos(\omega t)$:

$$m\frac{d^2x}{dt^2} + kx = F_0 \cos(\omega t)$$

Noting that $\omega_0^2 = k/m$, divide the equation by m to get

$$\frac{d^2x}{dt^2} + \omega_0^2 x = \frac{F_0}{m}\cos(\omega t)$$

For this expression to be a true statement for all values of time t, $x(t)$ must contain a cos ωt term, that is,

$$x(t) = A \cos(\omega t)$$

Then, by substitution

$$\frac{d^2 A \cos(\omega t)}{dt^2} + \omega_0^2 A \cos(\omega t) = \frac{F_0}{m} \cos(\omega t)$$

or

$$-\omega^2 A \cos(\omega t) + \omega_0^2 A \cos(\omega t) = \frac{F_0}{m} \cos(\omega t)$$

This equation is true as long as

$$A(\omega_0^2 - \omega^2) = \frac{F_0}{m}$$

or

$$A = \frac{F_0}{m(\omega_0^2 - \omega^2)}$$

This expression is the amplitude of oscillations when the system is driven with the periodic force $F(t) = F_0 \cos(\omega t)$. A most peculiar phenomenon occurs when the driving frequency is the same as the natural frequency of the system: the denominator becomes zero and the amplitude becomes infinitely large! This is **resonance**, in which each peak in the driving force occurs at the peak of the oscillatory motion, reinforcing it and causing the system to be displaced even farther from equilibrium. When damping is included, the amplitude gets large but not infinite. In the case when the damping is small (such as the damping we considered in Section 12-7), the amplitude is

$$A = \frac{F_0}{m\sqrt{(\omega_0^2 - \omega^2)^2 + b^2\omega^2/m^2}} \tag{12-30}$$

Although the denominator is not zero when ω equals ω_0, it is nevertheless as small as it can possibly be, resulting in the largest possible oscillation amplitude.

Resonance in a driven oscillator system occurs in physiological systems as well as mechanical systems, such as a swing or an object and spring. For example, insects such as flies, wasps, and bees have special flight muscles that allow them to beat their wings at a frequency as high as 1000 Hz or more and thereby exploit the phenomenon of resonance. When the flight muscles contract and relax repetitively at the resonant frequency of the thorax and wings, a rhythmic oscillation is established in the thorax, sending the wings into large amplitude resonant vibration.

? Got the Concept 12-9
Pushing, Bumping, and Vibrating

What kinds of forces can drive an oscillating system? You can push a child on a swing. The bumps on a road can push the springs in your car up and down. Vibrations associated with sound can also act as a driving force. **Figure 12-18** shows measurements of a quantity related to the amplitude of vibration versus frequency of sound impinging on a frog's eardrum. What is the natural frequency of the eardrum?

Figure 12-18 The amplitude of vibration of a frog's eardrum versus the frequency of sound impinging on it shows a clear resonance peak. (From Anson et al. (1985) *J Acoust Soc Am* 78(3):916.)

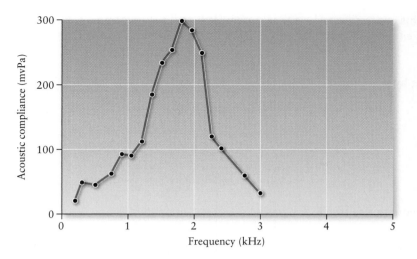

? Got the Concept 12-10
Buzzzzzz . . .

Male and female mosquitoes buzz at different frequencies. The flight sounds of male mosquitoes are in the range of 650 Hz while those of females are close to 400 Hz. Curiously, the antennae of mosquitoes differ from male to female, as seen in **Figure 12-19**, so that the natural frequency of a male's antennae is about 400 Hz while that of the female's is about 200 Hz. Some mosquitoes can detect the presence of others based on nerve signals generated when their antennae vibrate. Would you expect male mosquitoes to be able to detect other males? Females? Would you expect female mosquitoes to detect other females? Males?

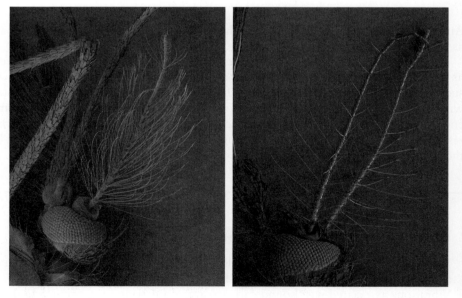

Figure 12-19 The antenna of a male mosquito (left) differs from that of a female (right); the natural frequency of a male's antennae is different from that of a female. *(Dennis Kunkel.)*

What's Important 12-8

Driving frequency is not a characteristic of an oscillating system but rather is determined externally. When the driving frequency is the same as the natural frequency of the system, the amplitude of oscillations becomes infinitely large unless a damping force is also present. This is resonance, in which each peak in the driving force occurs at the peak of the oscillatory motion, reinforcing it and causing the system to be displaced even farther from equilibrium.

Answers to Practice Problems

12-1 π rad

12-2   30 rad/s, 4.8 Hz, 0.21 s

12-3 $\sqrt{\dfrac{3}{4}}A$

12-4   1.2 s

12-6 $\sqrt{\dfrac{7}{12}}T_0$

12-8 9.3×10^{-2} s

Answers to Got the Concept Questions

12-1   Using the approach of sliding the horizontal zero as described in Figure 12-6, we need to move the vertical axis by π rad in order to set the minimum at $t = 0$. This result is shown in Figure 12-20.

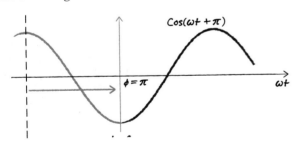

Figure 12-20 A phase shift of π rad is required to set the time at which the minimum occurs as $t = 0$.

12-2   Frequency is cycles per second or beats per second in the case of the heart:

$$f_{heart} = 45\ \text{beats}/60\ \text{s} = 0.75\ \text{beats/s} = 0.75\ \text{Hz}$$

Period and frequency are inversely related, as in Equation 12-6. Inverting Equation 12-6,

$$T_{heart} = \frac{1}{f_{heart}} = \frac{1}{0.75\ \text{Hz}}$$

$$= \frac{1}{0.75\ \text{beats/s}} = 1.3\ \text{s (per beat)}$$

Her breathing rate is 0.20 Hz, that is, 0.20 breaths per second. So in 1 min, she takes

$$f_{resp} = 0.20\ \frac{\text{breaths}}{\text{s}}$$

$$= \left(0.20\ \frac{\text{breaths}}{\text{s}}\right)\left(60\ \frac{\text{s}}{\text{min}}\right) = 12\ \frac{\text{breaths}}{\text{min}}$$

The period is

$$T_{resp} = \frac{1}{f_{resp}} = \frac{1}{0.20\ \text{Hz}}$$

$$= \frac{1}{0.20\ \text{breaths/s}} = 5.0\ \text{s (per breath)}$$

The heart rate is faster than the respiratory rate so the period is shorter.

12-3   The period is proportional to the square root of the length of the pendulum. If the length is increased by a factor of 4, the period increases by a factor of $\sqrt{4}$. So the period doubles when the length of the pendulum increases by a factor of 4. The new period is 6 s.

12-4   The period is independent of the mass of the pendulum. Therefore, the period remains the same regardless of the mass that swings at the end of the pendulum. The period of the new pendulum is 3 s.

12-5   Notice that as long as Hooke's law holds (that is, for small displacements from equilibrium), the period doesn't depend on the release point. Because we stipulated that the pendulum exhibits simple harmonic motion, even when released from the greater distance from equilibrium Hooke's law still holds, so the period remains 3 s.

12-6 The period is proportional to the square root of the inverse of the acceleration due to gravity. Using g_{Moon} equal to $1/6 \, g_{Earth}$, the fraction under the square root in Equation 12-16 increases by a factor of 6, so the period increases by a factor of $\sqrt{6}$. The new period on the surface of the Moon is $3\sqrt{6}$ s, or about 7 s.

12-7 A physical pendulum supported at its center of mass is in equilibrium no matter what its orientation. It therefore experiences no net force and does not oscillate. We can see this result from the definition of the period of a physical pendulum as well. The denominator in the square root term in Equation 12-23 is zero when the object rotates around the center of mass. The numerator, however, cannot be zero. From the parallel-axis theorem, $I = I_{CM} + Mh^2$ (Equation 8-16), so when $h = 0$, the moment of inertia is minimized but not zero. When the denominator of the period equation is zero but the numerator is not, the fraction (and therefore the period) is infinite. It therefore takes an infinite time for the pendulum to swing once; in other words, we observe no oscillation.

12-8 We know from Equation 12-23 that the period of a physical pendulum is proportional to the square root of I/mgh. At the end of the last example, we found the moment of inertia I of the rod is twice as large as that for the leg when both rotate around one end. The distance h between the center of mass and the rotation axis, however, is about the same for both, 0.37 m for the leg and 0.44 m (half the length) for the uniform rod. So when determining I/mgh, while the denominator is about the same for both, the moment of inertia is larger for the rod. Therefore, the period of the uniform rod must be greater than that of the leg.

12-9 By eye, the peak of the curve appears to be at about 1.8 kHz or 1800 Hz. The peak, according to Equation 12-30, corresponds to the time when the frequency of the driving force, sound in this case, equals the natural frequency of the frog's eardrum. We conclude that the natural frequency of the eardrum is 1800 Hz. The natural frequency of a human eardrum is similar, measured to be in the range of 1700 to 4000 Hz.

12-10 Male mosquitoes evolved to detect female mosquitoes. The frequency of sound made by the female in flight is well tuned to the natural frequency of a male's antennae, so his antennae will exhibit large amplitude vibrations (resonance) in the presence of a female mosquito. This matching of natural frequency of antennae and frequency of flight sounds does not occur in females or between pairs of male mosquitoes.

SUMMARY

Topic	Summary	Equation or Symbol
amplitude	The amplitude of an oscillation is the maximum displacement of a system from equilibrium. We will always take amplitude to be positive.	A
angular velocity	For a system undergoing harmonic motion, angular velocity is the rate, in radians per second (rad/s), at which the object moves. Angular velocity is also referred to as angular frequency.	$\omega_0 = \sqrt{\dfrac{k}{m}}$ (12-2)
damping force	A damping force is one that acts to resist motion in an oscillating system.	
driving force	A driving force is an external force acting on a system. Forced oscillations occur when a system that can undergo harmonic motion experiences a time-varying driving force.	
equilibrium	A system that experiences no net force and is also stationary is said to be in equilibrium. It is possible for a system to move through a point at which equilibrium could occur; such a location is an equilibrium position.	

frequency	The frequency f of a system undergoing simple harmonic motion is the number of cycles the system makes per second. If an object undergoing simple harmonic motion has a period of T, then it has a frequency of one cycle every T s. Frequency and angular frequency (or angular velocity) are related by a factor of 2π.	$f = \dfrac{1}{T}$ (12-6) $\omega = 2\pi f$ (12-7)
natural frequency	The angular frequency of the oscillator if no damping were present, ω_0, is referred to as the natural frequency. For an object attached to a spring, the natural frequency ω_0 is determined by the spring constant k and the mass m of the object.	$\omega_0 = \sqrt{\dfrac{k}{m}}$ (12-2)
period	The period T of a system undergoing simple harmonic motion is the time required for one full oscillation. The period of a system is the inverse of its frequency f.	$T = \dfrac{1}{f} = \dfrac{2\pi}{\omega}$ (12-5 and 12-6)
phase angle	Phase angle allows us to set time equal to zero at any convenient choice of displacement from equilibrium. Phase angle shifts the location of the vertical axis (where time equals zero) to the right or left when position versus time is graphed. The units of phase angle are radians (rad).	ϕ
physical pendulum	In a physical pendulum, the mass is not concentrated at a single point at the end of a support that can rotate around a pivot point.	
resonance	Resonance occurs when a system that can oscillate is driven at its natural frequency. At resonance, the amplitude of oscillations becomes large, tending toward infinity in the absence of a damping force.	
restoring force	Cyclic motion is characterized by a restoring force that tends to return the system to the position in which it experiences no net force.	
simple harmonic oscillator equation	The simple harmonic oscillator equation describes the displacement from equilibrium x as a function of time in a system undergoing simple harmonic motion. The specific solution to the simple harmonic oscillator equation depends on the mass m of the system and k, the effective spring constant.	$\dfrac{d^2 x}{dt^2} + \dfrac{k}{m}x = 0$ (12-1)
simple pendulum	A simple pendulum is an idealized pendulum, treated as if all of the mass were concentrated a fixed distance from a rotation point. A simple pendulum would be a pointlike object hanging from a thread of negligible mass.	

QUESTIONS AND PROBLEMS

In a few problems, you are given more data than you actually need; in a few other problems, you are required to supply data from your general knowledge, outside sources, or informed estimate.

Interpret as significant all digits in numerical values that have trailing zeros and no decimal points.

For all problems, use $g = 9.8 \text{ m/s}^2$ for the free-fall acceleration due to gravity. Neglect friction and air resistance unless instructed to do otherwise.

• Basic, single-concept problem

•• Intermediate-level problem, may require synthesis of concepts and multiple steps

••• Challenging problem
SSM *Solution is in Student Solutions Manual*

Conceptual Questions

1. •Not all oscillatory motion is simple harmonic, but simple harmonic motion is always oscillatory. Explain this statement and give an example to support your explanation.

2. •The fundamental premise of simple harmonic motion is that a force must be proportional to an object's displacement. Is anything else required?

3. •List several examples of simple harmonic motion that you have observed in everyday life.

4. •Explain the difference between a simple pendulum and a physical pendulum.

5. •If the rise and fall of your lungs is considered to be simple harmonic motion, how would you relate the period of the motion to your breathing rate (breaths per minute)? SSM

6. •Explain how *either* a cosine or a sine function will satisfy the force equation for simple harmonic motion.

7. •(a) What are the units of ω? (b) What are the units of ωt?

8. •Galileo was one of the first scientists to observe that the period of a simple harmonic oscillator is independent of its amplitude. Explain what it means that the period is independent of the amplitude. Be sure to mention how the requirement that simple harmonic motion undergo small oscillations is affected by this supposition.

9. •Compare $x(t) = A \cos \omega t$ to $x(t) = A \cos(\omega t + \phi)$. What is the phase angle ϕ and how does it change the solution to simple harmonic motion?

10. •What are three factors that can help you distinguish between a simple pendulum and a physical pendulum?

11. •Explain how you could do an experiment to measure the elevation of your location through the use of a simple pendulum. SSM

12. •In the case of the damped harmonic oscillator, what are the units of the damping constant, b?

13. •The application of an external force on a simple pendulum can create many different outcomes, depending on how frequently the force is applied. Explain what will happen to the amplitude of the motion if an external force is applied to a simple pendulum at the same frequency as the natural frequency of the pendulum.

14. •Starting from the full description of an oscillating system,

$$m\frac{d^2x}{dt^2} + b\frac{dx}{dt} + kx = F(t)$$

under what physical and mathematical circumstances will you arrive at the expression describing the basic case of simple harmonic motion?

15. •Explain the difference between the frequency of the driving force and the natural frequency of an oscillator. SSM

Multiple-Choice Questions

16. •Replacing an object on a spring with an object having one-quarter the original mass will have the result of changing the frequency of the vibrating spring by a factor of
  A. ¼.   D. 2.
  B. ½.   E. 4.
  C. 1 (no change).

17. •A block is attached to a horizontal spring and set in simple harmonic motion, shown in **Figure 12-21**. At what point in the motion is the speed of the block at its maximum? SSM

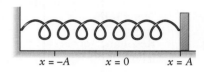

$x = -A$ $x = 0$ $x = A$

Figure 12-21 Problems 17 and 18

  A. $x = A$ and $x = -A$
  B. $x = 0$
  C. $x = 0$ and $x = A$
  D. $x = 0$ and $x = -A$
  E. $x = 0$, $x = -A$, and $x = A$

18. •A block is attached to a horizontal spring and set in simple harmonic motion, as shown in Figure 12-21. At what point in the motion is the acceleration of the block at its maximum?
  A. $x = A$ and $x = -A$
  B. $x = 0$
  C. $x = 0$ and $x = A$
  D. $x = 0$ and $x = -A$
  E. $x = 0$, $x = -A$, and $x = A$

19. •A small object is attached to a horizontal spring, pulled to position $x = -A$, and released. In one full cycle of its motion, the total distance traveled by the object is
  A. $A/2$.   D. $2A$.
  B. $A/4$.   E. $4A$.
  C. A.

20. •A small object is attached to a horizontal spring and set in simple harmonic motion with amplitude A and period T. How long does it take for the object to travel a total distance of $6A$?
  A. $T/2$ D. $3T/2$
  B. $3T/4$ E. $2T$
  C. T

This page is intentionally left blank.

For complete end of chapter problem sets, please go to
www.whfreeman.com/kestentauck

40. •A simple pendulum that has a length of 1.25 m and a bob with a mass of 4 kg is pulled an angle θ from its natural equilibrium position (hanging straight down). Use a spreadsheet or programmable calculator to calculate the maximum potential energy of the Earth–pendulum system and complete the table below. Using the small-angle approximation, determine the largest angle that applies. Note: This formula only works for angles in radians!

Maximum angle from vertical	Maximum potential energy of bob (J)	Maximum speed of bob (m/s)
5°		
10°		
15°		
20°		
25°		
Largest "valid" angle		

Problems

12-1: Simple Harmonic Motion

41. •Calc The second-order differential equation that describes simple harmonic motion can be written as follows:

$$\frac{d^2x}{dt^2} + \omega^2 x = 0$$

Determine, by differentiating, which of the following functions will satisfy the equation (assume A and ω are constants): (a) $x(t) = A \cos(\omega t)$, (b) $x(t) = A \sin(\omega t)$, (c) $x(t) = A \cos(\omega t) + A \sin(\omega t)$, (d) $x(t) = A e^{\omega t}$, (e) $x(t) = A e^{+\omega t} + A e^{-\omega t}$, (f) $x(t) = A e^{i\omega t}$, (g) $x(t) = A e^{+i\omega t} + A e^{-i\omega t}$.

42. •Calc Starting with Hooke's law ($F = -kx$) and Newton's second law of motion ($F = ma$), derive the general formula for simple harmonic motion when an object that has a mass M is attached to a spring that has a constant k.

43. •Calc Suppose an object of mass M attached to a spring that has a force constant k is suspended *vertically* before being slightly displaced from equilibrium. Use the techniques of Section 12-1 to predict the motion of the object. SSM

44. •Calc A friend of yours suggests that the solutions to the simple harmonic motion equation can be multiplied together to form another, valid solution. Test the suggestion for $x_1(t) = A_1 \cos \omega t$ and $x_2(t) = A_2 \sin \omega t$. Does $x = x_1 x_2$ satisfy simple harmonic motion?

45. •Calc Euler's formula states $e^{i\omega t} = \cos(\omega t) + i \sin(\omega t)$ (where $i = \sqrt{-1}$). (a) Show that $x(t) = A e^{i\omega t}$ solves the force equation for simple harmonic motion. (b) What are the units of the constant A?

46. •Calc Determine if the function $x(t) = A \cos(\alpha t^2)$ satisfies simple harmonic motion (where α is constant).

47. •Plot the following functions for two periods: (a) $x(t) = A \cos(2\pi t)$, (b) $x(t) = A \cos(2\pi t + \pi/2)$, and (c) $x(t) = A \cos(\pi t + \pi/4)$.

48. •A simple harmonic oscillator is observed to start its oscillations at the maximum amplitude when $t = 0$. Devise an appropriate solution that is consistent with this initial condition. Repeat when the oscillations start at the equilibrium position when $t = 0$.

49. •Calc The solution for a particular simple harmonic oscillator is $x(t) = (0.15 \text{ m}) \cos(\pi t + \pi/3)$. Calculate (a) the velocity of the oscillator at $t = 1$ s and (b) its acceleration at $t = 2$ s. SSM

50. ••Calc A 2-kg object is attached to a spring and undergoes simple harmonic motion. At $t = 0$, the object starts from rest, 10 cm from the equilibrium position. If the force constant of the spring k is equal to 75 N/m, calculate (a) the maximum speed and (b) the maximum acceleration of the object. Find the velocity of the object at $t = 5$ s.

12-2: Oscillations Described

51. •The period of a simple harmonic oscillator is 0.0125 s. What is the frequency?

52. •A simple harmonic oscillator completes 1250 cycles in 20 min. Calculate (a) the period and (b) the frequency of the motion.

53. •An object on the end of a spring oscillates with a frequency of 15 Hz. Calculate (a) the period of the motion and (b) the number of oscillations that the object undergoes in 2 min. SSM

54. ••The amplitude of an object undergoing simple harmonic motion is 20 cm. At $t = 0$, the oscillating object has a position of -12 cm. (a) What is the phase angle ϕ? (b) If the period is 2 s, write out the mathematical function that describes the position of the object as a function of time.

55. •A 200-g object is attached to the end of a 55 N/m spring. It is displaced 10 cm to the right of equilibrium and released on a horizontal, frictionless surface. Calculate the period of the motion.

56. •What is the mass of an object that is attached to a spring with a force constant of 200 N/m if 14 oscillations occur each 16 s?

57. ••Two springs are attached end to end to a green box of mass M as shown in **Figure 12-23**. Determine an expression that describes the period of motion for the box attached to this combination of springs. The two individual springs have spring constants of k_1 and k_2, respectively. SSM

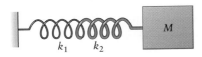

Figure 12-23 Problem 57

58. ••Two springs are attached side by side to a green box of mass M as shown in **Figure 12-24**. Determine an expression that describes the period of the motion of the box attached to this combination of springs. The two individual springs have spring constants of k_1 and k_2, respectively.

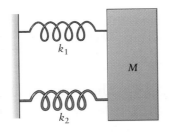

Figure 12-24 Problem 58

59. •When the mass of a spring (m_s) is appreciably large, the period of an oscillating object of mass m is given by

$$T_{corrected} = 2\pi\sqrt{\frac{(m + \frac{1}{3}m_s)}{k}}$$

Calculate the ratio of the corrected period to the standard period when the mass of the spring is 50 g, the mass on the end of the spring is 375 g, and the spring constant is 44 N/m.

60. ••High tide occurs at 8:00 A.M. and is 1 m above sea level. Six hours later, low tide is 1 m below sea level. After another 6 h, high tide occurs (again 1 m above sea level), then finally one last low tide (6 h later, 1 m below sea level). (a) Write a mathematical expression that would predict the level of the ocean at this beach at any time of day. (b) Find the times in the day when the ocean level is exactly at sea level.

12-3: Energy Considerations

61. ••Calc If the total energy of a simple harmonic oscillator is given by $E = \frac{1}{2}mv^2 + \frac{1}{2}kx^2$, show that the total energy is constant, for any sinusoidal solution of the form $x(t) = A\cos(\omega t + \phi)$.

62. •An object on a spring slides on a horizontal frictionless surface with simple harmonic motion. Determine where the object's kinetic energy and potential energy are the same. Assume the maximum displacement of the object is A so it oscillates between $+A$ and $-A$.

63. ••Calc The position as a function of time for an object that has a mass m, is attached to a spring that has a force constant k, and is sliding on a horizontal frictionless table is given by $x(t) = A\cos(\omega t + \phi)$ where $\omega = \sqrt{k/m}$. As a function of time, determine an expression for (a) the potential energy of the object–spring system and (b) the kinetic energy of the object–spring system. (c) Show that the total energy of the object–spring system is conserved. SSM

64. •Calc Starting with the formula $x(t) = A\cos(\omega t + \phi)$, (a) use direct differentiation to find the velocity at $t = 0$ and $t = T/2$. (b) Verify these results using conservation of energy for this simple harmonic oscillator.

65. ••A 250-g object attached to a spring oscillates on a frictionless horizontal table with a frequency of 4 Hz and an amplitude of 20 cm. Calculate (a) the maximum potential energy of the system, (b) the displacement of the object when the potential energy is one-half of the maximum, and (c) the potential energy when the displacement is 10 cm. SSM

66. ••The potential energy of an object on a spring is 2.4 J at a location where the kinetic energy is 1.6 J. If the amplitude of the simple harmonic motion is 20 cm, (a) calculate the spring constant and (b) find the largest force that it experiences.

67. ••The potential energy of a simple harmonic oscillator is given by $U = \frac{1}{2}kx^2$. (a) If $x(t) = A\sin\omega t$, plot the potential energy versus time for three full periods of motion. (b) Derive an expression for the velocity, $v(t)$, and (c) add the plot of the kinetic energy, $K = \frac{1}{2}mv^2$, to your graph. SSM

12-4: The Simple Pendulum

68. •A simple pendulum is 1.24 m long. What is the period of its oscillation?

69. •The period of a simple pendulum is 2.25 s. Determine its length.

70. •In 1851, Jean Bernard Léon Foucault suspended a pendulum (later named the Foucault pendulum) from the dome of the Panthéon in Paris. The mass of the pendulum was 28 kg and the length of the rope was 67.00 m. The acceleration due to gravity in Paris is 9.809 m/s². Calculate the period of the pendulum.

71. •Geoff counts the number of oscillations of a simple pendulum at a location where the acceleration due to gravity is 9.8 m/s², and finds that it takes 25 s for 14 complete cycles. Calculate the length of the pendulum. SSM

72. •Astro What is the period of a 1-m-long simple pendulum on each of the planets in our solar system? You will need to look up the acceleration due to gravity on each planet.

73. •(a) What is the period of a simple pendulum of length 1 m at the top of Mt. Everest, 8848 m above sea level. (b) Express your answer as a number times T_0, the period at sea level where h equals 0. The acceleration due to gravity in terms of elevation is

$$g = g_0 \left(\frac{R_E}{R_E + h} \right)^2$$

where g_0 is the average acceleration due to gravity at sea level, R_E is Earth's radius, and h is elevation above sea level. Take g_0 to be 9.800 m/s² and Earth's radius to be R_E is 6.380×10^6 m.

74. ••A simple pendulum of length 0.350 m starts from rest at a maximum displacement of 10° from the equilibrium position. (a) At what time will the pendulum be located at an angle of displacement of 8°? (b) What about 5°? (c) When will the pendulum return to its starting position?

75. ••A simple pendulum oscillates between ±8° (as measured from the vertical). The length of the pendulum is 0.50 m. Compare the time intervals between ±8° and ±4°. SSM

12-6: The Physical Pendulum

76. •A rod pendulum with a length of 30 cm is set into harmonic motion about one end. Calculate the period of the motion.

77. •A physical pendulum consists of a uniform spherical bob that has a mass M of 1.0 kg and a radius R of 0.50 m suspended from a massless string that has a length L of 1.0 m. What is the period T of small oscillations of the pendulum?

78. •A thin, round disk made of acrylic plastic (density is 1.1 g/cm³) is 20 cm in diameter and 1 cm thick (**Figure 12-25**). A very small hole is drilled through the disk at a point 8 cm from the center. The disk is hung from the hole on a nail and set into simple harmonic motion with a maximum angular displacement (measured from vertical) of 7°. Calculate the period of the motion.

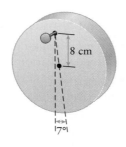

Figure 12-25 Problem 78

79. •A solid sphere, made of acrylic plastic (density is 1.1 g/cm³) has a radius of 5 cm (**Figure 12-26**). A very small "eyelet" is screwed into the surface of the sphere and a horizontal support rod is passed through the eyelet, allowing the sphere to pivot

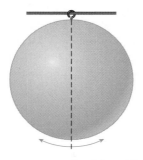

Figure 12-26 Problem 79

around this fixed axis. If the sphere is displaced slightly from equilibrium, find the period of its harmonic motion when it is released. SSM

80. ••**Calc** A thin, rectangular piece of aluminum (density is 2700 kg/m³) is designed to pivot about a pin that smoothly slides into a small hole (**Figure 12-27**). Where should the hole be drilled in order to minimize the period of harmonic motion? The aluminum has dimensions of 2 cm × 40 cm × 0.1 cm.

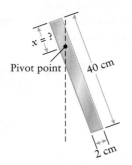

Note: This is not drawn to scale.
Figure 12-27 Problem 80

12-7: The Damped Oscillator

81. ••**Calc** Starting with the force equation for a damped harmonic oscillator, show that a solution of the form $x(t) = A\, e^{-(b/2m)t} \sin \omega_1 t$ works. The differential equation and the lightly damped oscillation frequency are

$$m\frac{d^2x}{dt^2} + b\frac{dx}{dt} + kx = 0 \quad \text{and} \quad \omega_1 = \sqrt{\omega_0{}^2 - \frac{b^2}{4m^2}}.$$

82. •An oscillator that has a mass of 100 g experiences damped harmonic oscillation. The amplitude decreases to 36.8% of its initial value in 10.0 s. What is the value of the damping coefficient b?

83. •A 110-g pendulum bob swings at the end of a 15.0-m-long wire. The pendulum's damping coefficient is 0.010 kg/s. What is the amplitude of the pendulum after 3 oscillations if the initial amplitude is 1.5 m? SSM

84. •A 500-g object is attached to a spring with a force constant of 2.5 N/m. The object rests on a horizontal surface that has a viscous, oily substance spread evenly on it. The object is pulled 15 cm to the right of the equilibrium position and set into harmonic motion. After 3 s, the amplitude has fallen to 7 cm due to frictional losses in the oil. (a) Calculate the natural frequency of the system, (b) the damping constant for the oil, and (c) the frequency of oscillation that will be observed for the motion. (d) How much time will it take before the oscillations have died down to one-tenth of the original amplitude (1.5 cm)?

85. •A damped pendulum of length L is described with the following equations:

$$x(t) = A\, e^{-(a/2L)t} \cos(\omega_1 t)$$

$$\omega_1 = \sqrt{\frac{g}{L} - \frac{a^2}{4L^2}}$$

Suppose a 1.25-m-long pendulum oscillates in a light fluid with a damping constant $a = 5$ m/s. (a) Find the percent difference between the natural frequency and the frequency of oscillation. (b) By what fraction will the amplitude be reduced after 2 s elapse?

12-8: The Forced Oscillator

86. •A forced oscillator is driven at a frequency of 30 Hz with a peak force of 16.5 N. The natural frequency of the physical system is 28 Hz. If the damping constant is 1.25 kg/s and the mass of the oscillating object is 0.75 kg, calculate the amplitude of the motion.

87. ••An oscillating system has a natural frequency of 50 rad/s. The damping coefficient is 2.0 kg/s. The system is driven by a force $F(t) = (100$ N$) \cos((50$ rad/s$)t)$. What is the amplitude of the oscillations? SSM

88. •A 5.0-kg object oscillates on a spring with a force constant of 180 N/m. The damping coefficient is 0.20 kg/s. The system is driven by a sinusoidal force of maximum value 50 N and the angular frequency is 20 rad/s. (a) What is the amplitude of the oscillations? (b) If the driving frequency is varied, at what frequency will resonance occur?

89. ••Calc (a) By taking derivatives, show that the following function for $x(t)$ satisfies the complete differential equation for oscillating systems:

$$m\frac{d^2x}{dt^2} + b\frac{dx}{dt} + kx = F(t)$$

$$F(t) = F_0 \cos(\omega t); \quad x(t) = A \sin(\omega t + \phi)$$

(b) Find the values of A and ϕ.

90. ••The *quality factor* Q is a parameter that specifies the width and height of the resonant peak when an oscillator is driven by a sinusoidal external force. It is defined as follows:

$$Q = \frac{m\omega_0}{b}$$

The "width" of the resonant peak is $\Delta\omega = \omega_0/Q$. Although different definitions exist, we'll use the FWHM ("full-width half-max") concept for this width. FWHM is basically the width of the peak at the point that is one-half of the maximum value. For a driven oscillator that has a mass of 100 g, a damping constant of 0.2 kg/s, a peak force of 7.5 N, and a spring constant 25 N/m, calculate the quality factor Q and the FWHM of the resonant peak.

General Problems

91. •Calc The acceleration of an object that has a mass of 0.025 kg and exhibits simple harmonic motion is

given by $a(t) = (10$ m/s$^2) \cos(\pi t + \pi/2)$. Calculate its velocity at $t = 2$ s, assuming the object starts from rest at $t = 0$. SSM

92. •Show that the formulas for the period of an object on a spring $(T = 2\pi\sqrt{m/k})$ and a simple pendulum $(T = 2\pi\sqrt{L/g})$ are dimensionally correct.

93. ••A 200-g object is attached to a spring that has a force constant of 75 N/m. The object is pulled 8 cm to the right of equilibrium and released from rest to slide on a horizontal, frictionless table. (a) Calculate the maximum speed of the object. (b) Find the location of the object when its velocity is one-third of the maximum speed heading toward the right.

94. ••A 100-g object is fixed to the end of a spring that has a spring constant of 15 N/m. The object is displaced 15 cm to the right and released from rest at $t = 0$ to slide on a horizontal, frictionless table. (a) Find the first three times when the object is at the equilibrium position. (b) Find the first three times when the object is 10 cm to the left of equilibrium. (c) What is the first time that the object is 5 cm to the right of equilibrium, moving toward the left?

95. •A damped oscillator with a period of 30 s shows a reduction of 30% in amplitude after 1 min. Calculate the percent loss in mechanical energy to the damping material per cycle. SSM

96. •A system consisting of a small 1.20-kg object attached to a light spring oscillates on a smooth, horizontal surface. A graph of the position x of the object as a function of time is shown in **Figure 12-28**. Use the graph to answer the following questions. (a) What are the angular velocity ω_0 and the frequency of the motion? (b) What is the spring constant of the spring? (c) What is the maximum speed of the object? (d) What is the maximum acceleration of the object?

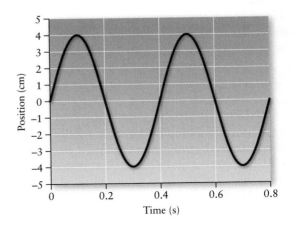

Figure 12-28 Problem 96

97. ••A rod pendulum that has a length L_0 of 85 cm hangs from one end and is allowed to oscillate (**Figure 12-29**). (a) What length of a simple pendulum L_1 will have the same period of simple harmonic motion? (b) A second rod pendulum is hanging from a point that is 5 cm from its end. What total length L_2 should it have to give it the same period as the first rod pendulum?

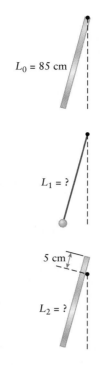

Figure 12-29 Problem 97

98. •••A block of wood floats in a basin of water. If it is pushed down slightly and released, the subsequent motion is oscillatory. Consider the buoyant force and gravitational force acting on the block to derive an algebraic expression for the period of this simple harmonic motion in terms of the surface area on which the block floats, the density of the water, the gravitational constant, the mass of the block, and constants.

99. ••**Biology, Calc** Ruby-throated hummingbird wing flaps have been timed at 53 flaps each second. A typical wing is 4.5 cm long, and each wing rotates through approximately a 90° angle. Assuming that the motion of the wing is simple harmonic, find (a) the period of the wing motion, (b) the frequency of the wing motion, (c) the angular velocity ω_0 of the wing motion, and (d) the maximum speed (in m/s and mph) of the tip of the wing.

100. •(a) Using conservation of energy, derive a formula for the speed of an object that has a mass M, is on a spring that has a force constant k, and is oscillating with an amplitude of A as a function of position $v(x)$. (b) If M has a value of 250 g, the spring constant is 85 N/m, and the amplitude is 10 cm, use the formula to calculate the speed of the object at $x = 0$ cm, 2 cm, 5 cm, 8 cm, and 10 cm.

101. ••A spinning golf ball of radius R can be suspended in a stream of high velocity air. The ball is in equilibrium at the vertical center of the air stream ($y = 0$), and not moving except for its rotation at velocity v_{Rot}. Show that the ball undergoes simple harmonic motion when it is released a small vertical distance y below the equilibrium position. Assume that for small y there is no net horizontal force, and that for small values of y the speed of the air drops off linearly from the center of the air stream. (For small y, the speed varies as $v(y) = v_0 - b|y|$, where v_0

is the speed of the air in the center of the stream and b is a constant.) SSM

102. ••Devise a "correction" to the formula for the period of a pendulum due to a change in altitude (and the corresponding change in the gravitational field). Your answer should provide the percent change in the period for each 1000 m increase (or decrease) in elevation.

103. ••When a small ball swings at the end of a very light, uniform bar, the period of the pendulum is 2.00 s. What will be the period of the pendulum if (a) the bar has the same mass as the ball and (b) the ball is removed and only the bar swings?

104. •••A bullet that has a mass of 10 g is fired at a speed of 150 m/s into a 2.2-kg block that rests on a frictionless tabletop and is attached to a spring (k equal to 80 N/m) as shown in **Figure 12-30**. Calculate the period and amplitude of the resulting simple harmonic motion once the bullet becomes embedded in the block.

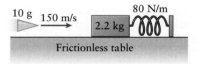

Figure 12-30 Problem 104

105. •••Two identical objects are released from rest at the same height on either side of a symmetric, frictionless ramp. They collide elastically as shown in **Figure 12-31**. (a) Find the period of the cyclic process in terms of g, the angle θ, the distance x_0, and the starting height y_0. (b) Calculate the numerical value for the period if $y_0 = 2$ m, $\theta = 30°$, and $x_0 = 5$ m.

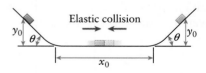

Figure 12-31 Problem 105

106. •**Astro** Your spaceship lands on a satellite of a planet around a distant star. As you initially circled the satellite, you measured its diameter to be 5480 km. After landing you observe that a simple pendulum that had a frequency of 3.50 Hz on Earth now has a frequency of 1.82 Hz. (a) What is the mass of the satellite? Express your answer in kilograms and as a multiple of our Moon's mass. (b) Could you have used the vibrations of a spring–object system to determine the satellite's mass? (c) Explain your reasoning.

107. •A bungee jumper who has a mass of 80 kg leaps off a very high platform. A crowd excitedly watches as the jumper free-falls, reaches the end of the bungee cord, then gets "yanked" up by the elastic cord, again and again. One observer measures the time between the low points for the jumper to be 9.5 s. Another observer realizes that simple harmonic motion can be used to describe the process because several of the subsequent bounces for the jumper require 9.5 s also. Finally, the jumper comes to rest a distance of 40 m below the jump point. Calculate (a) the effective spring constant for the elastic bungee cord and (b) its unstretched length.

108. •Biology Spiderwebs are quite elastic, so when an insect gets caught in the web, its struggles cause the web to vibrate. This alerts the spider to a potential meal. The frequency of vibration of the web gives the spider an indication of the mass of the insect. (a) Would a rapidly vibrating web indicate a large (massive) or a small insect? Explain your reasoning. (b) Suppose that a 15-mg insect lands on a horizontal web and depresses it 4.5 mm. If we model the web as a spring, what would be its effective spring constant? (c) At what rate would the web in part (b) vibrate, assuming that its mass is negligible compared to that of the insect? (d) Would the vibration rate differ if the web were not horizontal?

109. •••Biology A 20.0-cm string is made by combining ten parallel cylindrical strands of human hair, each 20.0 cm long. Young's modulus for the hair is 4.50 GN/m², and each strand is 125 μm thick. A 175-g utensil is tied to one end of the string and the other end is fastened to the ceiling. (a) By how much does the string stretch beyond its original 20.0 cm length when the utensil is attached? (b) If the utensil is now pulled down an additional 1.40 mm and then released, how long will it take for the utensil to first return to the position from which it was released? SSM

110. •••A 475-kg piece of delicate electronic equipment is to be hung by a 2.80-m-long steel cable. If the equipment is disturbed (such as by being bumped), vertical vibrational frequency must not exceed 25.0 Hz. (a) What is the maximum diameter the cable can have? (b) Would the piece of equipment cause the cable to stretch appreciably?

111. •••Imagine that a narrow tunnel, just large enough for a baseball to fit, is drilled through the center of Earth. Suppose a standard baseball is dropped into the tunnel from the surface of Earth. Calculate the period of the subsequent motion of the baseball, assuming that Earth is a solid, uniform sphere that does not spin. *Hint*: Draw the free-body diagram for the ball and recall that the gravitational field inside a sphere is $g(r) = -g_0(r/R)$,

where g_0 equals 9.8 m/s², R is the radius of Earth and equals 6380 km, and r is the distance from the center of Earth to the center of the ball. Newton's second law will lead you to the solution. SSM

112. ••A 60-cm-long thin rod of copper has a radius r of 0.4 cm (the density of copper is 8.92 g/cm³). The rod is suspended from a thin wire that is welded to the exact center of the copper rod. The wire is also made of copper and has a length of 20 cm and a cross-sectional diameter of 1 mm. The rod is displaced from the equilibrium position and the torque on the thin wire causes it to twist the rod back and forth in harmonic motion. This is a *torsion pendulum* (**Figure 12-32**). The torque acts on the rod according to following equation:

$$\tau = -K\theta$$

where τ is the torque, K is the torsional constant for wire and equal to $\pi Gr^4/2l$, G is the modulus of rigidity for copper and equal to 45 GPa, l is the length, and θ is the angular displacement from equilibrium. Calculate the period of the harmonic motion.

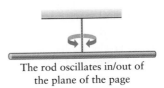

The rod oscillates in/out of
the plane of the page

Figure 12-32 Problem 112

113. •••Calc In addition to sine and cosine, many other mathematical functions are periodic and satisfy the basic differential equation for simple harmonic motion. Show that the function $x(t) = A(\sin \omega t + \cos \omega t)$ satisfies the simple harmonic motion equation.

ANSWER for odd answers in the back of the book:

$$\frac{dx}{dt} = A(-\omega \sin \omega t + \omega \cos \omega t) \text{ and}$$

$$\frac{d^2x}{dt^2} = A(-\omega^2 \cos \omega t - \omega^2 \sin \omega t), \text{ so } \frac{d^2x}{dt^2} + \omega^2 x = 0.$$

114. •••In addition to the definition provided in problem 12-90, for a damped, driven oscillator the quality factor, Q can also be shown to be

$$Q = 2\pi \frac{\text{total energy}}{\text{energy loss per period}}$$

Using the two definitions for the quality factor, calculate the energy loss per period for a driven oscillator where $m = 0.100$ kg, $b = 0.2$ kg/s, $k = 25$ N/m, $\omega = \omega_0$, and $F_0 = 7.5$ N.

MCAT® SECTION

The section that follows includes material from previously administered MCAT® items and is reprinted with permission of the Association of American Medical Colleges (AAMC).

Passage 7 (Questions 41–48)

Three experiments were conducted to examine collisions between masses suspended at the ends of simple pendula. Each pendulum consisted of a mass, or bob, attached to the bottom of a rod of negligible mass. The top of the rod was anchored to a solid surface from which it could swing free of friction. For sufficiently small amplitudes, each pendulum executed simple harmonic motion with a period $T = 2\sqrt{\frac{l}{g}}$, where l is the length of the pendulum, and g is the acceleration due to gravity. (Note: Assume air resistance is negligible in the experiments.)

Experiment 1
Two identical pendula with steel balls for bobs were hung next to each other so that their bobs barely touched when they hung vertically. The left pendulum bob was moved to the side and then released. The bobs collided, and immediately after the collision, the left bob was at rest while the right bob moved with the same speed that the left bob had before the collision.

Experiment 2
The steel bobs were replaced with nonidentical balls made of an unknown substance. The left pendulum bob was moved to the same position as in Experiment 1 and then released. After the collision, the right bob moved to the right while the left bob rebounded to the left.

Experiment 3
The bobs were replaced with identical balls made of a deformable, sticky material. The left pendulum bob was moved to the same position as in Experiment 1 and then released. After the collision, the bobs stuck together, and the final speed of the combined bobs was less than the speed of the left bob before the collision.

41. Which of the following diagrams best describes the subsequent motion of a bob that detaches from a simple pendulum When the bob moves through the lowest point of its arc?

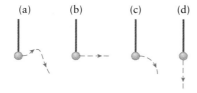

42. Which of the following quantities is NOT conserved in the collision in Experiment 3?
    A. Linear momentum
    B. Angular momentum
    C. Potential energy
    D. Kinetic energy

43. Which of the following would occur if the collision in Experiment 1 was perfectly elastic?
    A. The right bob would rise to the same height that the left bob started from.
    B. The support rod would stretch enough that the right bob would move in a straight line.
    C. Both bobs would move away from the impact with equal and opposite velocities.
    D. Both bobs would 'move to the right after the collision.

44. Which of the following accurately describes the difference between the pendulum bobs used in Experiment 2?
    A. The left bob was more elastic than the right bob.
    B. The left bob was less elastic than the right bob.
    C. The left bob weighed less than the right bob.
    D. The left hob weighed more than the right bob.

45. If the speed of the left bob in Experiment 3 was v immediately before the collision, what was its speed immediately after the collision?
    A. $\dfrac{v}{4}$ B. $\dfrac{v}{2}$ C. $\dfrac{v}{\sqrt{2}}$
    D. Approximately equal to but slightly less than v

46. If the maximum height of the left bob in Experiment 1 was increased by a factor of 2, by what factor would the maximum height of the right bob increase?
    A. $\sqrt{2}$
    B. $\dfrac{3}{2}$
    C. 2
    D. 4

47. The speed with which the left bob rebounds in Experiment 2 is
    A. less than its speed before the collision.
    B. equal to its speed before the collision.
    C. greater than its speed before the collision.
    D. equal to the speed of the right bob before the collision.

48. A pendulum bob is replaced with one that weighs 2 times as much. How does the time it takes for the new bob to move from a small angle of displacement to the lowest point of its arc compare to the time it took the old bob?
    A. It takes 1/2 as long.
    B. It takes the same amount of time.
    C. It takes $\sqrt{2}$ times as long.
    D. It takes 2 times as long.

13 Waves

(Horia Vlad Bogdan/Dreamstime.com.)

As a mouse scurries across the desert floor, its footsteps generate surface waves in the sand that spread like the ripples a dropped stone creates on the surface of a pond. The horned desert viper (*Cerastes cerastes*) detects these tiny waves by resting its lower jaw gently on the sand. Bones in the snake's jaw loosely connect to the inner ear, so the jaw acts like our eardrum—vibrations of the jaw cause vibrations in the inner ear which in turn stimulates the brain. Because the jawbones on each side of the head move independently, the snake collects stereo information in the same way we do with our two ears. This allows the snake to determine both the distance and direction of its next meal.

A visible disturbance propagates around a crowded soccer stadium when fans, section by section, jump up and then sit back down (Figure 13-1). Although not a wave in the strict sense, the phenomenon shares many features with its physics analog. Perhaps most important, the human "wave" temporarily displaces people from their seats as it travels around the stadium, but notice that the individual people don't move in the direction of that propagation. The disturbance moves through the crowd, but the fans don't move along with it.

If you've ever shaken one end of a taut rope tied down at the other end, you've likely seen a wave traveling along the rope. When two children talk to each other using a string stretched between two paper cups, sound travels from one to the other by means of a wave. Ripples on the surface of a pond spread out from a stone thrown into the water (Figure 13-2). In each case, as the wave travels, it briefly displaces elements of the medium through which it propagates—the rope, the string, the air, and the water. However, the elements of the medium then return to their original position; the medium experiences no net displacement. After the sound travels from one paper cup to the other, for example, the string returns to its initial position.

In this chapter we examine physical properties of waves and their mathematical description, as well as applications to sound and hearing.

Figure 13-1 Fans in a crowded soccer stadium "do the wave" by jumping up and throwing their hands into the air section by section and then sitting back down. (*Courtesy of Elijah Light.*)

Figure 13-2 Ripples on the surface of a pond spread out from a stone thrown into the water. *(Aspireimages/Inmagine.)*

13-1 Types of Waves

A **mechanical wave** moving through a medium displaces particles or elements of the medium as it propagates. (In Chapter 22, we will examine the nature of electromagnetic waves such as light, which do not require a medium to propagate.) We will concentrate on periodic waves that follow a cyclic, repeating pattern. Based on our experience with oscillating systems, we will employ sinusoidal functions to describe the waves mathematically.

We will consider the three most commonly experienced waves. A **transverse wave** is one in which the elements of the medium are disturbed in the direction perpendicular to the direction in which the wave propagates. **Longitudinal waves** cause elements of the medium to move back and forth along the direction of wave propagation. Propagating on the surface of a medium, surface waves are a combination of a transverse and longitudinal waves.

Imagine a taut rope tied to a post at one end. Now imagine creating a mechanical wave pulse by quickly snapping the free end of the rope up and then down just once. As your hand moves up, the bit of rope at your end moves up as well. That bit of rope tugs on the bit of rope immediately adjacent to it, which in turn tugs on the next tiny element of the rope. Because each tiny segment of the rope connects to the adjacent one, this tugging of one bit on the next bit travels along the rope. But in your quick motion, you pulled the end of the rope down as well as up, so with the same timing as the motion of your hand, the bits of rope tug down one to the next; the resulting wave pulse propagates along the rope as shown in **Figure 13-3**. Notice that the elements of the rope move in the direction perpendicular, or transverse, to the motion of the pulse, not in the direction the pulse itself propagates. We call this a *transverse wave pulse*.

The motion just described, a single up and down snap of the end of the rope, produces a transverse wave pulse. If you instead move your hand and the end of the rope up and down in oscillatory motion, you would create a series of pulses, resulting in the periodic transverse wave shown in **Figure 13-4**. Note that the passage of the transverse wave causes no net motion of the rope.

With the passage of a longitudinal wave, the elements of the medium move in the same direction as the wave propagates. Imagine clapping your hands together

Figure 13-3 A transverse wave pulse travels down a stretched rope as the result of a quick up–down motion of its free end.

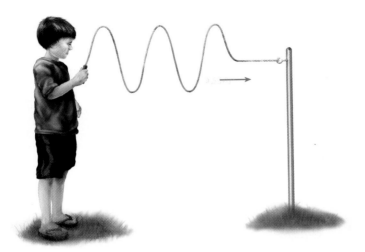

Figure 13-4 A transverse wave pulse travels down a stretched rope as the result of an oscillatory up–down motion of its free end.

sharply at the opening of a tube, as shown in Figure 13-5. As your hands squeeze on the air in between them, the air pressure increases to a higher level than the pressure of the air that surrounds your hands. As the pulse of air at higher pressure moves outward from your clapping hands, a region of air that has lower pressure (in the volume from which air molecules have been pushed out) follows behind. If you clap your hands rhythmically, each region of air that has higher pressure pushes on the air molecules at the opening of the tube, which in turn exert an increased force on the air molecules slightly farther down the tube. Each region of lower pressure pulls the air molecules back toward their equilibrium position. In this way each region of increased air pressure propagates down the tube, but just as in the case of a transverse wave, individual air molecules experience no *net* motion. Air molecules simply slosh back and forth as a longitudinal wave passes, as shown in Figure 13-6.

A longitudinal wave such as sound in a tube consists of regions of higher pressure (compressions) surrounded by regions of lower pressure (rarefactions). For this reason, physicists sometimes refer to sound and other longitudinal waves as **pressure waves.**

A wave that propagates along the surface of a material, an ocean wave, for example, is a combination of a transverse wave and a longitudinal wave. As a surface wave passes, particles in a medium move in circular or elliptical paths; in other words, their motion is both transverse and longitudinal to the direction of wave

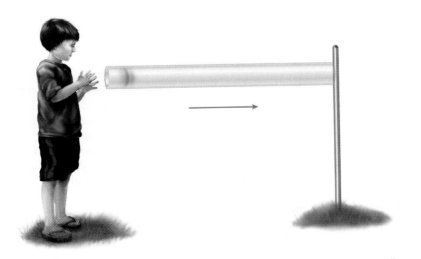

Figure 13-5 A longitudinal wave pulse travels down a tube filled with air as the result of a single hand clap.

Figure 13-6 Clapping your hands rhythmically at the opening of a tube filled with air results in a longitudinal wave traveling down the tube.

propagation. In deep water, water molecules near the surface move in circular paths, while in shallow water the paths get compressed into ellipses.

 Got the Concept 13-1
Types of Waves

Earlier in this section, we considered waves generated by shaking one end of a taut rope, talking into a paper cup attached to another cup by a stretched string, and tossing a stone into a pond. What type of wave does each example represent?

✳ **What's Important 13-1**

A mechanical wave moving through a medium displaces the medium as it propagates. As a transverse wave moves, the elements of the medium are disturbed in the direction perpendicular to the direction the wave propagates. A longitudinal wave, sometimes called a pressure wave, causes elements of the medium to move back and forth along the direction of wave propagation. A combination of a transverse and longitudinal wave, surface waves propagate on the surface of a medium.

13-2 Mathematical Description of a Wave

Transverse waves typically move along linear media, such as a stretched rope. (Electromagnetic waves, which we'll discuss in Chapter 22, are transverse waves that can propagate in free space.) Longitudinal waves travel through a volume of medium, for example, sound moving through air. Surface waves, as the name implies, propagate on the surface of a material. Despite the differences, however, the mathematical descriptions of transverse, longitudinal, and surface waves have common parameters, for example, wavelength λ, period T, frequency f, and amplitude A.

Wavelength λ is the distance the disturbance travels over one full cycle of the wave. **Figure 13-7** shows the displacement of the medium versus position along a generic, sinusoidal periodic wave. In Figure 13-7, time has been fixed, so the figure

Time is fixed in this snapshot of the wave.

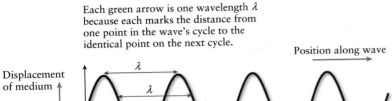

Each green arrow is one wavelength λ because each marks the distance from one point in the wave's cycle to the identical point on the next cycle.

Position along wave

Displacement of medium

λ

λ

λ

The red arrow does not mark a full wavelength. The magnitude of the disturbance caused by the wave is the same at both ends, but these endpoints are not identical points in the cycle. On the left side the magnitude of the disturbance is decreasing while on the right it is increasing.

Figure 13-7 The wavelength λ of a wave is the distance the disturbance travels over one full cycle. Time is fixed in this snapshot of a wave.

is essentially a snapshot of a region of the wave. You can measure λ from one crest of the wave to the next crest, one trough to the next trough, or any specific point in the cycle to the analogous point in the next cycle. In the SI system, the units of wavelength are meters (m).

Watch Out

The wavelength is the distance the wave travels over one full cycle.

The red arrow in Figure 13-7 marks two points along a transverse wave where the medium is displaced by the same amount. Although you might be tempted to use these two locations to measure the wavelength, that would not be correct because the two points are not equivalent moments in the cycle of the wave. At the point along the wave on the left side of the arrow, the elements of the medium are returning to a position of zero displacement, while on the right they are moving farther away. The two points do not measure the wavelength because they do not span a full cycle of the displacement of the medium.

Got the Concept 13-2

One Full Wavelength

The red arrow in Figure 13-7 does not mark a wavelength of the transverse wave shown. How would you extend the arrow, either to the right or to the left, in order for the arrow to correctly identify one full wavelength?

The **period** T of a wave is the time it takes for the disturbance to go through one full cycle as it moves through the medium. Whether we're talking about a small section of a vibrating string or a volume of air next to a barking dog, the specific element of the medium through which the wave propagates moves in a periodic way. The bit of air, say, gets farther from its initial or equilibrium position in one direction, reaches a maximum displacement, then passes through the initial position on its way to a maximum displacement in the other direction before returning again to equilibrium. In this way, the pressure variations exert a restoring force on the air molecules; their motions are analogous to the simple harmonic motion of an object attached to a spring. The time for that complete cycle is the period T. The units of T in the SI system are seconds (s).

You probably notice the similarities between harmonic, oscillatory motion and the sinusoidal nature of waves. In both cases, some element of a system undergoes periodic motion. The period is the same physical quantity for both kinds of motion, so as we did for harmonic motion, we can therefore represent the **frequency f** of a periodic wave as the inverse of its period:

$$f = \frac{1}{T} \tag{12-6}$$

The units of f are reciprocal seconds, or hertz (s^{-1}, or Hz). Often, but not always, physicists use the term *tone* to describe a sound of a specific frequency.

? Got the Concept 13-3
Sound Waves

Humans hear sound waves that range from approximately 20 Hz to 20 kHz. At which end of this range does a sound wave take longer to cycle through one period?

The **amplitude A** of a wave is the maximum displacement from equilibrium that any element of the wave experiences as the disturbance propagates through a medium. Because amplitude is measured from equilibrium, it is not the difference between the maximum positive and maximum negative displacement. Both green arrows in **Figure 13-8** mark the amplitude of the wave, while the red arrow does not.

To study the details of wave phenomena, we need to describe mathematically the parameters that define waves. We will explore this description from the perspective of a transverse wave, but the result will apply equally well to any wave.

Consider a disturbance moving along a taut rope. The displacement of a small bit of rope at any specific position x along the rope can be described by the variable

Figure 13-8 The amplitude A of a wave is the maximum displacement from equilibrium that any element of the wave experiences as the disturbance propagates through a medium. Green arrows indicate the amplitude here. Amplitude is *not* the difference between the maximum positive and maximum negative displacement, indicated by the red arrow. Time is fixed in this snapshot of a wave.

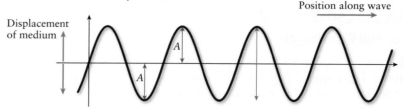

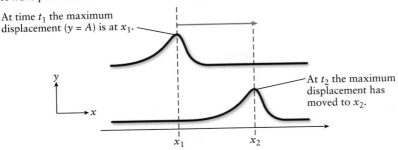

A wave pulse moves down a stretched string.

At time t_1 the maximum displacement ($y = A$) is at x_1.

At t_2 the maximum displacement has moved to x_2.

The change in distance is related to the time interval by the propagation velocity v_p: $x_2 - x_1 = v_p\,(t_2 - t_1)$.

Figure 13-9 The speed at which a longitudinal wave pulse propagates is the displacement divided by the corresponding time interval.

y—that is, a displacement perpendicular to the x direction. As a wave moves along the rope, the value of y at any specific x varies as a function of time. To describe the wave mathematically, we need to determine y as a function of both position x and time t; in other words, we need to find $y(x, t)$.

Figure 13-9 shows a wave pulse at two times during its motion down a taut rope. The location of the maximum displacement ($y = A$) is $x = x_1$ when $t = t_1$. The peak of the pulse will have moved to a new location $x = x_2$ at time t_2. The constant propagation speed v_p is simply the change in distance divided by the time interval:

$$v_p = \frac{x_2 - x_1}{t_2 - t_1}$$

or

$$x_2 - x_1 = v_p(t_2 - t_1)$$

Gathering the x_1 and t_1 terms on one side of the equation and the x_2 and t_2 terms on the other

$$x_2 - v_p t_2 = x_1 - v_p t_1$$

The only way that the quantity $x - v_p t$ can be the same for some point on the pulse at two arbitrary times is if it is the same for all times. We conclude that $x - v_p t$ remains constant over time for any fixed point on the pulse, for example, the peak. For this reason we can use this expression to define the time evolution of the wave. We will therefore look for $y(x, t)$ as a function of $x - v_p t$:

$$y(x, t) = f(x - v_p t)$$

Recall that we anticipated using a sinusoidal function to represent a periodic wave. Would it be correct to use $x - v_p t$ as the argument of a sine function? Not quite, because the argument of sine must be an angle, and both x and $v_p t$ are distances. We can, however, multiply $x - v_p t$ by a constant k that has units of angle divided by distance, for example, rad/m. The result will be an angle and therefore acceptable as an argument of a sinusoidal function. We can then represent a periodic wave as

$$y(x, t) = A \sin(k(x - v_p t)) = A \sin(kx - kv_p t) \qquad (13\text{-}1)$$

Because the maximum value of sine is 1, we multiplied $\sin(kx - kv_p t)$ by amplitude A so that the maximum of $y(x, t)$ can be any value.

A drawing of the periodic wave at one instant of time t_0 reveals the physical significance of k, as shown in Figure 13-10. (Imagine taking a photograph of a wave on a taut rope.) The positions x_1 and x_2 differ by the wavelength λ. Because

Figure 13-10 The positions x_1 and x_2 differ by the wavelength λ.

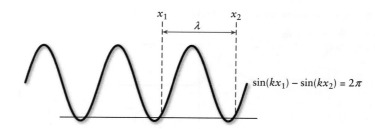

$$\sin(kx_1) - \sin(kx_2) = 2\pi$$

$\sqrt{x}$ See the Math Tutorial for more information on Trigonometry

the wave repeats itself over this distance, the values of $y(x, t)$ must be the same at x_1 and x_2:

$$A \sin(kx_2 - kv_p t_0) = A \sin(kx_1 - kv_p t_0)$$

Sine repeats every 2π radians, which requires that

$$kx_2 - kv_p t_0 = 2\pi + (kx_1 - kv_p t_0)$$

or

$$kx_2 = 2\pi + kx_1$$

Thus,

$$k(x_2 - x_1) = 2\pi$$

The term $x_2 - x_1$ is the wavelength λ, so

$$k = \frac{2\pi}{\lambda} \tag{13-2}$$

The **angular wave number** (or simply **wave number**) k describes the number of cycles of a wave that fit into one full cycle of a sine function (2π radians or $360°$). The angular wave number is a common way to include the wavelength in mathematical descriptions of wave phenomena.

We studied the first term in the argument of the sine in Equation 13-1 by looking at the variations of $y(x, t)$ at one fixed time. In a similar way, we can examine the second term by looking at the variations of $y(x, t)$ at one fixed position x_0 as a function of time, as shown in **Figure 13-11**. Note that the horizontal axis in this figure is time, not position; for this reason we have not included an arrow to indicate motion. The times have been selected so that t_2 is exactly one cycle later than t_1, so they differ by the period T. (So $t_2 - t_1 = T$.) Because the wave repeats itself over the time interval from t_1 to t_2, the values of $y(x, t)$ must be the same at t_1 and t_2:

$$A \sin(kx_0 - kv_p t_2) = A \sin(kx_0 - kv_p t_1)$$

As we saw in our discussion of variations of $y(x, t)$ at one fixed time, this expression requires that the arguments of the two sines differ by 2π. Because $-t_2$ is less than $-t_1$, we add 2π to the argument on the *left* side of the expression to require that

$$(kx_0 - kv_p t_2) + 2\pi = kx_0 - kv_p t_1$$

or

$$-kv_p t_2 + 2\pi = -kv_p t_1$$

Thus,

$$kv_p(t_2 - t_1) = 2\pi$$

The term $t_2 - t_1$ is equal to the period T, so

$$kv_p = \frac{2\pi}{T}$$

This curve shows the displacement of one fixed position along the wave as a function of time.

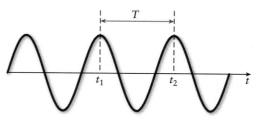

Figure 13-11 The period of a wave is the time it takes to complete one cycle.

Recall that period and angular velocity are related (Equation 12-5):

$$\omega = \frac{2\pi}{T}$$

so

$$k v_p = \omega \tag{13-3}$$

This equation immediately gives us two important results. First, we can obtain a standard mathematical description of a periodic wave by substituting this expression into Equation 13-1:

$$y(x, t) = A \sin(kx - kv_p t) = A \sin(kx - \omega t) \tag{13-4}$$

Second, Equation 13-3 gives us a relationship for the propagation speed of a periodic wave:

$$v_p = \frac{\omega}{k} \tag{13-5}$$

The angular velocity ω is related to frequency f by $\omega = 2\pi f$ (Equation 12-7). The angular wave number k is related to the wavelength by $k = 2\pi/\lambda$ (Equation 13-2). So Equation 13-5 is a compact description of wave motion that brings together most of the fundamental quantities associated with waves. Equation 13-5 also provides a glimpse into nature at a fundamental level. Because the propagation speed of a wave in any particular medium is a constant, as the angular velocity of waves increases, the angular wave number also increases. In other words, as the frequency of waves increases, wavelength decreases.

> **? Got the Concept 13-4**
> **Propagation Speed**
>
> The propagation speed of sound in air does not depend on the wavelength or frequency of the waves. (a) When the frequency of a tone produced by an audio system is increased, the wave number also increases. How does the angular velocity change? (b) When the wavelength of the sound wave is increased, how does the period change?

The mathematical description we've developed for a transverse wave applies equally well to a longitudinal wave (for example, a sound wave propagating down a straight tube). As the sound wave moves, molecules of the air are disturbed along the propagation direction of the wave so that they move forward and then backward sinusoidally along the tube. So we can define a relationship between displacement s of the medium, in terms of position x along the tube and time t, that has the same form as the general description of a periodic transverse wave (Equation 13-4):

$$s(x, t) = s_{max} \sin(kx - \omega t) \tag{13-6}$$

where s_{max} is the maximum displacement, or displacement amplitude, of the disturbance.

As a sound wave propagates, it squeezes together air molecules at some places along the wave, resulting in regions of air that have higher pressure. Adjacent regions of air are slightly depleted of air molecules and therefore have relatively lower pressure. Because the pressure variations are related to the displacement of air molecules, we can represent the pressure along the wave as

$$p(x, t) = p_{max} \cos(kx - \omega t) \tag{13-7}$$

where p_{max} is the maximum pressure, or pressure amplitude, of the wave. A positive value of $p(x, t)$ corresponds to regions of compression, while negative values indicate regions of expansion. Notice that we have written $s(x, t)$ as a sine but $p(x, t)$ as a cosine. Displacement and pressure are 90° out of phase as a longitudinal wave propagates.

Because the displacement and pressure along the propagation direction of a longitudinal wave can be represented by sinusoidal functions, it is common to draw such waves in the same way we have for transverse waves, for example, in Figures 13-7 and 13-8. This result reinforces our conclusion that the mathematical descriptions of longitudinal and transverse waves are the same. The concepts and relationships for the wavelength and frequency, for example, are the same for longitudinal and transverse waves.

★ What's Important 13-2

Wavelength λ is the distance the disturbance in a medium travels over one full cycle of the wave. The period T of a wave is the time it takes for the disturbance to go through one full cycle as it moves through the medium. The amplitude A of a wave is the maximum displacement from equilibrium that any element of the wave experiences as the disturbance propagates through a medium. The angular wave number k describes the number of cycles of a wave that fit into one full cycle of a sine function (2π rad or 360°).

13-3 Wave Speed

The hero in an action movie signals his sidekick, who is a few kilometers away, by pounding the steel rail of a train track with a rock. Having studied physics, our hero knows that sound travels far more quickly in steel than in air. We have already uncovered one relationship for the propagation speed of a wave. Equation 13-5, which applies to both transverse and longitudinal waves, relates angular velocity ω and wave number k:

$$v_p = \frac{\omega}{k} \tag{13-5}$$

In this section, we further explore the speed at which waves propagate.

First, note that Equation 13-5 can be expressed in terms of frequency f and wavelength λ. Using $\omega = 2\pi f$ (Equation 12-7) and $k = 2\pi/\lambda$ (Equation 13-2), Equation 13-5 becomes

$$v_p = \frac{2\pi f}{2\pi/\lambda} = f\lambda \tag{13-8}$$

This equation states that propagation speed is the product of frequency and wavelength. Again, this applies to both transverse and longitudinal waves.

Estimate It! 13-1 Wave Speed on a Sperm's Flagellum

Spermatozoa move by beating a long, tail-like flagellum against the surrounding fluid, as seen in the sequence of images of a swimming sea urchin spermatozoon in **Figure 13-12.** Although some aspects of the motor driving the sinusoidal motion are distributed along the length of the flagellum, we can approximate the motion

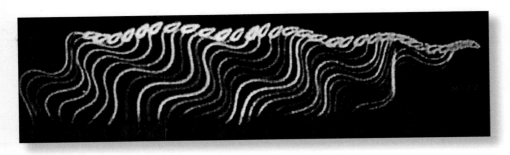

Figure 13-12 A sequence of images of a swimming sea urchin spermatozoon shows how it moves by beating its long, tail-like flagellum against the surrounding fluid. *(Cosson, J. 1996. A moving image of flagella: news and views on the mechanisms involved in axonemal beating.* Cell Biol. Int. *20:83–94.)*

of the flagellum as a pure transverse wave moving along it. The spermatozoon is 40 μm long, and the images are separated by approximately 0.01 s. Estimate the speed of the transverse wave as it propagates along the flagellum.

SET UP
We can estimate the speed at which the displacement moves along the sperm's flagellum by treating the motion as a pure transverse wave moving along it.

$$v_p = f\lambda \qquad (13-8)$$

We can estimate the wavelength of one cycle of displacement along the flagellum by making a measurement directly from Figure 13-12. Using the spermatazoon's 5-μm head as a guide, it's about 20 μm from one crest to the next in the first image in the figure.

We can estimate the frequency of the transverse wave by determining how often the wave pattern repeats. Notice that a peak of motion is directly behind the head of the sperm in the first and tenth images. The images were recorded 1/300 s apart, so the time interval is 9/300 s or 0.03 s. The period is therefore 0.03 s, and the frequency is the inverse of this.

SOLVE
The wave propagation speed is

$$v_p = \frac{1}{0.03 \text{ s}} 20 \ \mu\text{m} = 667 \ \mu\text{m/s}$$

or about 700 μm/s to one significant digit.

REFLECT
To put the speed in context, consider that the world record holder in the World Snail Racing Championships (held annually in Congham, England) covered the 13-in. course in 2 min, which is just under 10 m/h. The propagation speed we found of 700 μm/s is about 2 m/h. The value is only a factor of 5 slower—not bad for something 1000 times smaller than a snail.

To further explore the speed at which waves propagate, we consider a transverse wave moving down a stretched string. The propagation speed depends on how quickly each infinitesimally small piece of the string can be accelerated; the force that causes the acceleration is the tension exerted on the string. Lower tension in the string results in less acceleration, which leads to slower overall wave motion. The reverse is also true. You can probably guess that if you pluck a somewhat saggy string, one under relatively low tension, the wave you create will move sluggishly. If you pull on the end of the string to increase the tension, the wave will move more quickly.

Wave speed also depends on how much mass the wave must move as it propagates. Mass and inertia, the tendency of an object to resist a change in motion, are related; so, the more mass in any stretch of the string, the more sluggishly it will respond to a disturbance. We conclude that as the mass per unit length of a string (under constant tension) increases, the wave propagation speed decreases. Taking both effects into account, and applying Newton's second law to a stretched string, the propagation speed v_p of a transverse wave is found to be

$$v_p = \sqrt{\frac{T}{\mu}} \qquad (13\text{-}9)$$

where T is the tension in the rope or other linear, stretchable medium and μ is the mass of the medium per unit length or **linear mass density**. The SI units of μ are kilograms per meter (kg/m). Notice that the relationship accounts for both observations above, that either lower tension or higher mass per unit length slows the wave propagation speed.

Example 13-1 Correct Units for v_p

Show that the units of propagation speed v_p as defined by Equation 13-9 are meters per second (m/s).

SET UP

This problem requires us to manipulate the units of tension (force) and mass density.

SOLVE

We will use SI units, that is,

$$[v_p] = \left[\sqrt{\frac{T}{\mu}}\right] = \sqrt{\frac{[T]}{[\mu]}} = \sqrt{\frac{\text{N}}{\text{kg/m}}}$$

If you can't recall the definition of a newton (N) in terms of other SI units, think of Newton's second law, $\sum \vec{F}_{\text{dir}} = m\vec{a}_{\text{dir}}$ (Equation 4-1). Because the units of force must equal the units of mass multiplied by the units of acceleration,

$$1\,\text{N} = 1\,\text{kg}\,\frac{\text{m}}{\text{s}^2}$$

so

$$[v_p] = \sqrt{\frac{\text{kg m/s}^2}{\text{kg/m}}} = \sqrt{\frac{\text{m}^2}{\text{s}^2}} = \frac{\text{m}}{\text{s}}$$

REFLECT

The relationship for the propagation speed of a transverse wave defined by Equation 13-9 has the correct units, distance (m) per unit time (s).

We arrived at Equation 13-9 by considering a transverse wave moving down a stretched string. According to the equation, once we select a particular string, if we know the value of its mass per unit length μ, and if we put the string under some fixed tension T, the propagation speed v_p has a known, unchanging value. Equation 13-8 ($v_p = f\lambda$) therefore requires that the product of the wave's frequency and wavelength be constant. Therefore, the wavelength depends on the frequency with which we wiggle one end of the string to start the series of transverse waves. We cannot select both the frequency and the wavelength for a transverse wave. Once we select either the frequency or wavelength, the other is determined because the product is a constant set by the tension and linear mass density of the medium.

Watch Out

The propagation speed is not the time derivative of *y(x, t)*.

You might have wondered why we didn't take the time derivative of the general expression for a wave, $y(x, t) = A \sin(kx - \omega t)$ (Equation 13-4), to find the propagation speed. Certainly the time derivative of $y(x, t)$ yields *some* speed associated with the wave. It's just not the propagation speed. The variable $y(x, t)$ describes not the wave position as a function of time but rather the transverse displacement of any small element of the medium along which the wave propagates. The time derivative of $y(x, t)$ therefore gives the speed of elements of the medium in the direction perpendicular to the motion of the wave. To be precise, the speed of an element of the medium at some fixed position x_0 is

$$v(x_0, t) = \frac{d}{dt}y(x_0, t) = \frac{d}{dt}A \sin(kx_0 - \omega t)$$

or

$$v(x_0, t) = -A\omega \cos(kx_0 - \omega t) \tag{13-10}$$

Notice that this **transverse speed** of a transverse wave is periodic, not constant. Also, as with simple harmonic motion, when the magnitude of the displacement of an element of the medium is maximal, the transverse speed is zero. The transverse speed is at its maximum when the magnitude of the displacement is zero. Indeed, each small bit of the medium is exhibiting simple harmonic motion in the transverse direction.

We want to extend the relationship for propagation speed (Equation 13-9) to apply to longitudinal waves. Longitudinal waves propagate through a volume while transverse waves move along linear media. We should expect, then, that while the propagation speed of a transverse wave (Equation 13-9) involves linear quantities such as linear mass density (mass per unit length μ) and tension T, propagation speed of a longitudinal wave will depend on analogous variables that apply to a three-dimensional medium, such as a fluid or a metal rod.

Linear mass density μ (which is a two-dimensional quantity) corresponds to volume density ρ (which is the mass per unit volume and a three-dimensional quantity). By changing the denominator in Equation 13-9 from a quantity that has dimensions of mass divided by length to one that has dimensions of mass divided by volume (length multiplied by length multiplied by length), we have added area, or two powers of length, to the equation. Thus, the quantity that we substitute for T in the numerator of Equation 13-9 must have dimensions of force divided by area.

For a wave on a linear medium such as a string, wave speed depends on the ability of the medium to stretch, whereas for a wave in a three-dimensional volume such as the air in a tube, the wave speed depends on the ability of the medium to expand. As we saw in Chapter 9, the bulk modulus B and Young's modulus Y define the compressibility of a liquid or solid, respectively. For a wave propagating in a three-dimensional medium, we replace the tension T with B, and the one-dimensional linear mass density μ with volume density ρ. The propagation speed for a longitudinal wave in a fluid is therefore

$$v_p = \sqrt{\frac{B}{\rho}} \tag{13-11}$$

Likewise, the propagation speed for a longitudinal wave in a solid is

$$v_{\mathrm{p}} = \sqrt{\frac{Y}{\rho}} \qquad (13\text{-}12)$$

Notice that because both B and Y have dimensions of force per area (or units of $\mathrm{N/m^2}$), the right-hand sides of both Equation 13-11 and Equation 13-12 have the same units:

$$\left[\sqrt{\frac{B}{\rho}}\right] = \left[\sqrt{\frac{Y}{\rho}}\right] = \sqrt{\frac{\mathrm{N/m^2}}{\mathrm{kg/m^3}}} = \sqrt{\frac{\mathrm{kg\,m/s^2\,m}}{\mathrm{kg}}} = \sqrt{\frac{\mathrm{m^2}}{\mathrm{s^2}}} = \frac{\mathrm{m}}{\mathrm{s}}$$

In other words, the quantity defined in both Equations 13-11 and 13-12 has units of velocity.

The speed of sound in air is not a constant value, but instead depends on the pressure, density, and temperature of the air, as well as other parameters. For our purposes, however, unless otherwise stated, we will use

$$v_{\mathrm{sound}} = 343\ \mathrm{m/s} \qquad (13\text{-}13)$$

as the standard value for the speed of sound in dry air at sea level and at 20 °C.

Example 13-2 Propagation Speeds

Compare the speed of a sound wave in two fluids, air and water. Use ρ_{air} equal to $1.2\ \mathrm{kg/m^3}$, B_{air} equal to $1.4 \times 10^5\ \mathrm{N/m^2}$, ρ_{water} equal to $1.0 \times 10^3\ \mathrm{kg/m^3}$, and B_{water} equal to $2.0 \times 10^9\ \mathrm{N/m^2}$ at room temperature and 1 atm of pressure. Note that the value of the bulk modulus of air given here is not the value found in Table 9.1; the value given here is the so-called "adiabatic bulk modulus" which is used to characterize the elastic properties of a fluid when under pressure from all sides.

SET UP

The compressibility and density of the media through which a wave travels determine the wave's propagation speed. We will substitute the given numbers into Equation 13-11:

$$v_{\mathrm{p}} = \sqrt{\frac{B}{\rho}}$$

SOLVE

For air of the density and compressibility given, we find the speed of sound to be

$$v_{\mathrm{p}} = \sqrt{\frac{1.4 \times 10^5\ \mathrm{N/m^2}}{1.2\ \mathrm{kg/m^3}}} = 3.4 \times 10^2\ \mathrm{m/s}$$

For water that has the density and compressibility given, we find the speed of sound to be

$$v_{\mathrm{p}} = \sqrt{\frac{2.0 \times 10^9\ \mathrm{N/m^2}}{1.0 \times 10^3\ \mathrm{kg/m^3}}} = 1.4 \times 10^3\ \mathrm{m/s}$$

REFLECT

The value we found for the speed of sound in air is only approximate, based on the values given in the problem statement. We will generally use the accepted value of 343 m/s for the speed of sound in air at room temperature and 1 atm of pressure.

Of more interest than the values themselves are their relative magnitudes. We argued earlier that the speed of a wave depends in part on the inertia of the medium—the more massive the medium (by volume), the slower the wave. The mass density of water is far greater than that of air; so, on the basis of density alone, we might expect the speed of sound in water to be much lower than its speed in air. Yet by our calculation the speed of sound in water is more than 4 times *larger* than the speed of sound in air. The reason, as you can see, is that although the density of water ρ_{water} is about a thousand times greater than the density of air ρ_{air}, water is about 10,000 times less compressible than air. These two effects together combine to give the relative magnitudes of speeds that we found.

Finally, notice that because the *ratio* of the bulk modulus B and the density ρ determines v_p, it is unwise to guess the speed of a wave based on B or ρ. For example, just because the density of water is larger than the density of air, it is not the case that the speed of sound is lower in water than air.

Practice Problem 13-2 Find the speed of a sound wave in ethyl alcohol at room temperature and 1 atm of pressure. Take the density of ethyl alcohol to be ρ equal to 0.81×10^3 kg/m^3 and the bulk modulus to be B equal to 1.0×10^9 N/m^2.

Example 13-3   Action Hero

At the beginning of this section, we described an action hero in a movie who signals his sidekick by pounding on a steel rail of a train track. Compare the speed of sound along a steel rail to the speed in air. Use ρ_{steel} equal to 8.0×10^3 kg/m^3 and Y_{steel} equal to 210×10^9 N/m^2.

SET UP
Given values of Y and ρ, Equation 13-12

$$v_p = \sqrt{\frac{Y}{\rho}}$$

can be applied directly.

SOLVE
For steel, the speed of sound is

$$v_p = \sqrt{\frac{210 \times 10^9 \text{ N/m}^2}{8.0 \times 10^3 \text{ kg/m}^3}} = 5.1 \times 10^3 \text{ m/s}$$

The speed of sound in air is 343 m/s, so speed travels approximately 15 times faster in steel than in air.

REFLECT
The speed with which sound travels down the steel rail is nearly 15 times faster than the speed of sound in air, because although steel is about 8000 times denser than air, the compressibility of air is more than 2 million times greater than that of steel.

Practice Problem 13-3 Determine the speed of sound along a rail made of aluminum. Use $\rho = 2.7 \times 10^3$ kg/m^3 and $Y = 70 \times 10^9$ N/m^2 as the density and Young's modulus of aluminum.

? **Got the Concept 13-5**
Crack of a Bat

While sitting in the bleachers at a baseball game (far from home plate), you hear the crack of the bat hitting the ball almost a half second after you see the batter swing. What's going on?

? Got the Concept 13-6
Lightning

You may know this rule for estimating the distance of a lightning strike. After you see the lightning, count the number of seconds until you hear the thunder clap. For every 5 s you count, the lightning strike is 1 mi away (or for every 3 s, the lightning is 1 km away). What is the basis for the rule?

✱ What's Important 13-3

The propagation speed of both transverse and longitudinal waves is the product of the wavelength and the frequency of the wave. Propagation speed in a one-dimensional medium (for example, a stretched string) is also a function of the tension in the medium and the linear mass density. In air and in other compressible, three-dimensional media, the propagation speed is also a function of the mass density and bulk modulus of the medium. The transverse speed of a transverse wave varies periodically along the length of the wave and it is at its maximum when the magnitude of the displacement is zero.

13-4 Superposition and Interference

Ripples spread away from the point where a stone thrown into a pond enters the water (Figure 13-13a). As surface waves propagate outward, the water height varies sinusoidally along radial lines. At the instant captured in the photograph, the peaks appear as bright bands (marked with blue dots) while the troughs appear as darker bands (marked by red dots). Figure 13-13b shows the height of the water versus radial distance from the entry point.

Now consider the situation when two stones are thrown simultaneously into a pond (Figure 13-14). At the instant captured in the sketch, the ripples spreading from each stone have begun to overlap. Notice that the circles which mark the peaks and troughs of the surface waves remain clearly visible as the two waves pass through each other. Two waves moving along the same line in opposite directions, such as the waves shown in Figure 13-15, pass through each other and emerge from the interaction completely intact and unchanged. This property characterizes the surface waves on the pond in Figure 13-14 as well as transverse and longitudinal waves.

What happens during the time when the two waves overlap? Waves obey the principle of **superposition**: at every point where more than one wave passes simultaneously, the net disturbance of the medium equals the sum of the displacements that each wave would have caused individually. You can understand this as a result of the *independence* of waves. When one wave would disturb the medium in one direction at some location while another wave would disturb it in the opposite direction, some amount of cancellation occurs. If two waves both disturb the medium in the same direction at some location, the two waves *reinforce* each other. Two waves can either completely reinforce each other or completely cancel each other out, because each one disturbs the medium independently of the other. Any intermediate state is also possible. We refer to the combination of two or more waves as **interference**.

Figure 13-16 shows three moments in time as the blue wave moving to the left passes through the red wave moving to the right. To make it easier to study the process by which this occurs, we selected two waves of the same amplitude and

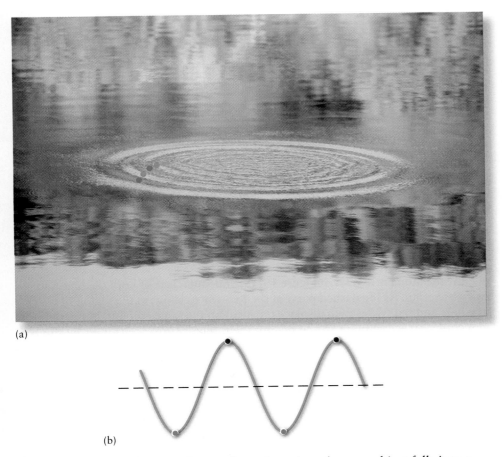

(a)

(b)

Figure 13-13 (a) Ripples spread away from the point where an object falls into a pool of water. Red dots mark troughs and blue dots mark crests. (b) The height of the water varies sinusoidally as the ripples propagate away from the point at which the object entered the water.

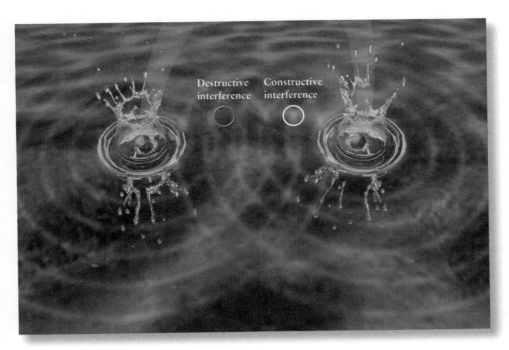

Figure 13-14 The ripples caused by two objects that fall simultaneously into a pool of water overlap as they move away from the points where the objects entered the water. The two waves are nevertheless independent; the crests and troughs of each wave remain clearly visible. The white circle marks a point of constructive interference. The red circle marks a point of destructive interference.

Figure 13-15 Waves moving along the same line in opposite directions (top) pass through each other and emerge from the interaction completely intact and unchanged (bottom).

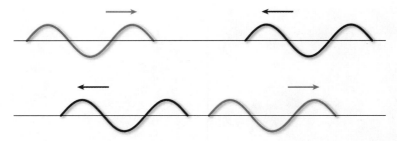

wavelength. In Figure 13-16a, the waves overlap by exactly one-half cycle (one-half wavelength). In this region, thin dashed red and blue lines show each wave as it would be if the other were not present. In the case shown in Figure 13-16a, both waves cause a maximal disturbance of the medium in the same direction at the same place so that their sum, shown in purple, equals twice the amplitude of either one separately. This result is sometimes referred to as complete or fully **constructive interference**. Partially constructive interference occurs when two waves reinforce each other, but not fully.

At the moment shown in Figure 13-16b, one full cycle of both the blue and red waves overlap. As the dashed parts of the two waves show, in the region of overlap the blue and red waves each displace the medium the same amount, but in opposite directions. At every point in the region, then, the disturbance caused by the blue wave cancels the disturbance caused by the red wave. As a result, no net disturbance of the medium occurs, a phenomenon known as **destructive interference**. This result, sometimes referred to as complete or fully destructive interference, is shown by the horizontal purple line in Figure 13-16b. Partially destructive interference occurs when two waves cancel each other, but not fully.

Even if two waves interfere destructively as they pass through each other, the waves remain unchanged by the interaction. At the point shown in Figure 13-16c the two waves begin to emerge from the overlap region, the blue wave to the left and the red wave to the right. At the moment shown, one-half of a full wavelength of each wave overlaps, so the waves interfere constructively in that region. Outside the region of overlap the two waves remain as they were before the interaction.

The **phase difference** φ between two waves determines how they interfere with each other. Phase difference, commonly measured in degrees or radians, quantifies the relationship between two waves relative to the periodic cycle of each. Waves in phase with each other rise and fall together; the phase difference $\varphi = 0$ rad. Waves that do not rise and fall together are out of phase. Suppose two waves pass through each other and one wave peaks when the other wave crosses the zero mark as its

Figure 13-16 Two waves traveling in opposite directions pass through each other and interfere. Regions of interference are shown in purple. Dashed lines indicate what each wave would be in the absence of the other. (a) One half cycle (one-half wavelength) of each wave overlap, resulting in constructive interference. (b) One full cycle of both the blue and red waves overlap, resulting in destructive interference. (c) The two waves begin to emerge from the overlap region; at the moment shown, one-half of a full wavelength of each wave overlap, so the waves again interfere constructively.

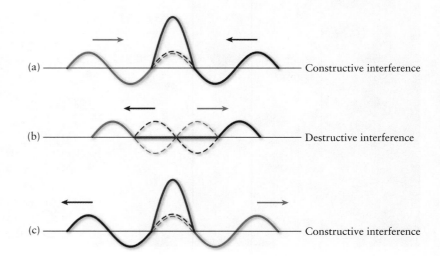

(a) — Constructive interference

(b) — Destructive interference

(c) — Constructive interference

A full cycle of a sine wave is considered to start where the sine equals zero and is increasing.

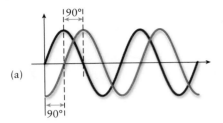

(a)

The first full cycle of the red sine curve starts at 90°, where the blue sine curve is at its peak. The red sine curve therefore leads the blue sine curve by 90°, so the phase difference equals 90° or π/2 rad.

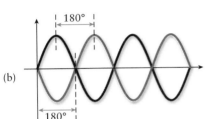

(b)

The first full cycle of the red sine curve starts at 180°, where the blue sine curve has returned to zero. The red sine curve therefore leads the blue sine curve by 180°, so the phase difference φ equals 180° or π rad.

Figure 13-17 (a) Phase difference φ quantifies the relationship between two waves relative to the periodic cycle of each. (b) Destructive interference occurs when the phase difference between two waves is 180°.

amplitude is increasing; the phase difference $\varphi = \pi/2$ rad (or $\varphi = 90°$) (**Figure 13-17a**). Treating both waves as sine functions, the peak of the blue curve occurs at an angle of $\pi/2$ rad; because the red curve begins a new cycle at that point, the two curves differ by a phase of $\pi/2$ rad. **Figure 13-17b** shows two curves out of phase by π rad (180°). The red curve is shifted by π rad relative to the blue one.

The combination of the two waves across a region of overlap can result in a complex but regular **interference pattern,** in which the waves interfere constructively at some places and destructively at others. In Figure 13-14, for example, a peak of one ripple coincides with a peak of the other at certain locations, resulting in full constructive interference. One such place is indicated by the white circle; at that location, the combined wave is twice as high, which results in a region on the image that is brighter than the peaks of undisturbed waves. The red circle marks a region in which a peak of one wave interferes destructively with a trough of another wave. Destructive interference results in zero net disturbance of the water, so here the image is less bright than at the peak of an undisturbed wave but not as dark as at the wave troughs. Regions of constructive interference and destructive interference form a complex but regular pattern across the whole region of overlap.

> **? Got the Concept 13-7**
> **Sound Generator**
>
> Your physics instructor uses a sound generator attached to a speaker to create a tone of a single frequency in the classroom. When you cover one ear and move your head from side to side you notice that the sound gets louder and softer. What's going on?

To determine how two waves interfere at any specific point in space, we need to know how they overlap at that point. Let's consider how the cycles of two sound waves are aligned at point P, which is a distance D_1 from the source of one wave (source 1) and is a distance D_2 from the source of the other wave (source 2). In **Figure 13-18a**, we represent the sound waves as sine functions to indicate the magnitude of the disturbance of the air along the direction of propagation. To simplify the math, we take both sounds to be of the same amplitude and wavelength. We will also let the waves be in phase; that is, their cycles rise and fall together at the two sources, as shown in the figure.

Figure 13-18 (a) Two sound waves of the same amplitude and wavelength, represented here as sine functions that indicate the magnitude of the disturbance of the air, interfere at point P. (b) The specific way the two waves overlap depends on how many cycles fit into the difference in the length of the paths each travels to reach P.

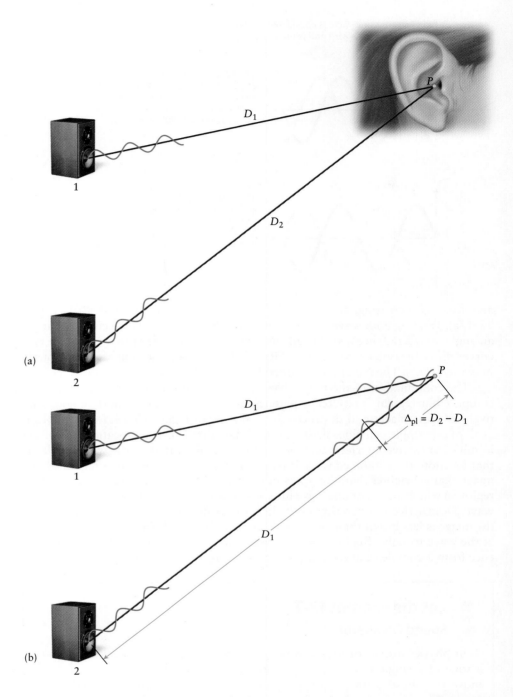

In **Figure 13-18b**, wave 1 reaches a peak of a cycle at point P. Notice that wave 2 travels farther to reach P; the additional distance is the path length difference $\Delta_{pl} = D_2 - D_1$. Because the two waves are in phase and of the same wavelength and frequency, both waves hit a peak a distance D_1 from its source. (Point P is D_1 from the source of wave 1.) So, to determine how the two waves overlap at P we only need to ask how many cycles of wave 2 fit into the extra distance Δ_{pl}.

If one wavelength exactly fits into Δ_{pl} then wave 2 will be at a peak at point P, just as if it were at distance D_1 from the source. The same is true if two full wavelengths fit into Δ_{pl}, or three full wavelengths, or *any* integer number of wavelengths fit into Δ_{pl}. And because we've already established that wave 1 is at the same point in its cycle at P as wave 2 is at D_1 from source 2, we can make the comparison regardless of where in the cycles the waves are at P. If an integer number of wavelengths fit into Δ_{pl}, the two waves will rise and fall together as they arrive at point

P. The two waves are therefore in phase and interfere constructively. If two sources emit waves of wavelength λ in phase with each other, constructive interference occurs at a point D_1 from one source and D_2 from the other when

$$\Delta_{\mathrm{pl}} = n\lambda, \qquad n = 1, 2, 3, \ldots \quad \text{(constructive interference)} \qquad \textbf{(13-14)}$$

where $\Delta_{\mathrm{pl}} = D_2 - D_1$. Equation 13-14 is the mathematical equivalent of requiring that an integer number of wavelengths fit into Δ_{pl}.

If exactly one-half of a wavelength, or an odd number of half wavelengths, fits into Δ_{pl}, then when wave 2 is at a peak at distance D_1 from the source it will be at a trough in its cycle at point P. So waves 1 and 2 will be π rad (180°) out of phase at point P, resulting in destructive interference. If two sources emit waves of wavelength λ in phase with each other, destructive interference occurs at a point D_1 from one source and D_2 from the other when

$$\Delta_{\mathrm{pl}} = m\lambda, \qquad m = \frac{1}{2}, \frac{3}{2}, \frac{5}{2}, \ldots \quad \text{(destructive interference)} \qquad \textbf{(13-15)}$$

where $\Delta_{\mathrm{pl}} = D_2 - D_1$. Equation 13-15 is the mathematical equivalent of requiring that an odd number of half wavelengths fit into Δ_{pl}.

Example 13-4 Two Elephants

While at a zoo, you stand 5 m from one elephant and 12 m from another. At the same instant, both elephants begin to make low frequency calls, in which most of the sound's power is at a frequency of 24.5 Hz (just above the threshold of human hearing). At your location, would you expect to observe constructive or destructive interference, or something in between?

SET UP

To determine how the two sound waves of the same frequency (and wavelength) interfere, we need to know how they overlap. Constructive interference occurs when the peaks and troughs align exactly. Destructive interference occurs when the peaks of one align exactly with the troughs of the other. To solve this problem then, we need to determine how many wavelengths fit into the difference of the lengths of the paths that the two waves take before arriving at your ears.

SOLVE

The frequency of the elephants' call is $f = 24.5$ Hz, so according to Equation 13-8, the wavelength is

$$\lambda = \frac{v_{\mathrm{p}}}{f} = \frac{343 \text{ m/s}}{24.5 \text{ s}^{-1}} = 14 \text{ m}$$

Here the speed of sound in air v_{p} is 343 m/s. The path length difference is

$$\Delta_{\mathrm{pl}} = D_2 - D_1 = 12 \text{ m} - 5 \text{ m} = 7 \text{ m}$$

One-half of a wave of $\lambda = 14$ m fits into Δ_{pl}, that is,

$$\Delta_{\mathrm{pl}} = \frac{1}{2}\lambda$$

So according to Equation 13-15, you observe destructive interference.

REFLECT

Notice that we did not consider the number of cycles that fit into the distance between you and either elephant. The phase relationship between waves determines how they interfere with each other. The specific number of cycles a wave undergoes between the source and point of interest has no bearing on the conclusion. If one wave is, say, at a peak at the point of interest, it doesn't matter whether the wave has undergone 1 or 100 or 1000 cycles before arriving at the point.

Practice Problem 13-4 While at a zoo, you stand 6 m from one elephant and 20 m from another. At the same instant, both elephants begin to make low frequency calls. Most of the sound's power is at a frequency of 24.5 Hz (just above the threshold of human hearing). At your location, would you expect to observe constructive or destructive interference, or something in between?

★ What's Important 13-4

Waves obey the principle of superposition. At every point where more than one wave passes simultaneously, the net disturbance of the medium equals the sum of the displacements that each wave would have caused individually. If two waves disturb the medium in the same direction at some location, the two waves *reinforce* and interfere constructively. When two waves each displace the medium by the same amount but in opposite directions, the disturbance caused by one wave cancels the disturbance caused by the other one, resulting in destructive interference.

13-5 Transverse Standing Waves

Two tuning forks held relatively close together and struck simultaneously produce sound waves that interfere with each other. You might be able to detect the interference pattern by moving your head around the tuning forks and listening for regions where the sound is louder or softer. An easier, more direct way to observe this kind of phenomenon is to watch two transverse waves on a string. Their interference will be immediately visible in the string's motion.

The person starts by initiating a single pulse on a string stretched between his hand and a pole (**Figure 13-19a**). In this case, when the pulse reaches the fixed end, the string exerts an upward force on the pole which must exert a downward force on the string as required by Newton's third law. The action reflects an inverted pulse back along the string (**Figure 13-19b**). If he were to wiggle the end of the string in his hand periodically, creating a series of waves on the string, the reflection would be a train of inverted waves traveling back toward us with the same amplitude, wavelength, and frequency as the incoming waves.

Remember from the previous section that two waves in one region of a medium result in interference. We studied the interference pattern by using the mathematical description of a transverse wave. In general,

$$y(x, t) = A \sin(kx - \omega t) \tag{13-4}$$

where $y(x, t)$ is the transverse displacement of the string at any position x and time t, A is the amplitude of the displacement, and k and ω are the angular wave number and angular velocity, respectively. The angular wave number k is related to the wavelength λ ($k = 2\pi/\lambda$, Equation 13-2) and angular velocity ω is related to the frequency f ($\omega = 2\pi f$, Equation 12-7). By convention, Equation 13-4 represents a wave traveling from left to right on the string. How can we use this equation to represent the reflected wave (that is, a wave moving from right to left)?

The sign of the propagation speed describes the direction a wave propagates. Although the speed is not explicit in Equation 13-4, we can get it from the equation by taking a derivative with respect to time. To get the propagation speed, we need the time derivative of the position x, but not the time derivative of the transverse displacement of the string. We compute this derivative by selecting a

As the pulse arrives at the pole, the string exerts an upward force on the pole...

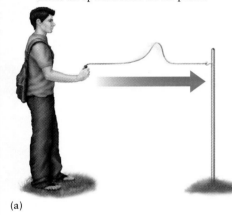

(a)

...so the pole exerts a downward force on the string. This causes the pulse to become inverted as it is reflected.

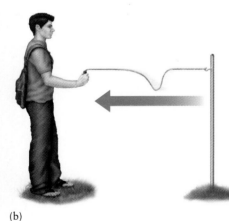

(b)

Figure 13-19 (a) A single pulse moves down a stretched string tied to a pole. (b) When the pulse reaches the fixed end the string exerts an upward force on the pole which therefore exerts a downward force on the string; the pulse is inverted as it reflects back from the pole.

specific position on the wave (a position of constant transverse displacement), and watching how it moves with time. For example, we could follow the motion over time of one of the peaks of the wave. Any fixed position on the wave will work, so we accomplish this by setting $y(x, t)$ in Equation 13-4 equal to a constant and then taking the time derivative. When the transverse displacement of the string (the value of y) is a constant

$$kx - \omega t = \text{constant} \tag{13-16}$$

taking the time derivative results in a relationship for propagation speed dx/dt:

$$\frac{d}{dt}(kx - \omega t) = \frac{d}{dt}(\text{constant})$$

so

$$k\frac{dx}{dt} - \omega = 0$$

or

$$\frac{dx}{dt} = \frac{\omega}{k} = v_p$$

√x̄ *See the Math Tutorial for more information on Calculus*

We recognize the fraction ω/k as the propagation speed v_p, but what interests us here is that the speed is positive. By convention a positive speed indicates motion from left to right, which corresponds to a wave approaching the pole in Figure 13-19a. In Equation 13-4,

$$y(x, t) = A\sin(kx - \omega t)$$

and the minus sign in front of the ωt term indicates positive speed; this equation represents a wave traveling from left to right on the string. The reflected wave travels in the opposite direction, so we need the sign of the ωt term to be positive; the reflected wave is described by

$$y(x, t) = A\sin(kx + \omega t) \tag{13-17}$$

The wave has the same amplitude, wavelength, and frequency as the wave before reflection, but it moves from right to left, the direction opposite to the propagation direction of the original wave.

Interference results from adding together two or more waves. We can represent the interference pattern formed by the reflected wave and the original one by adding the mathematical descriptions, Equation 13-17 and Equation 13-4, respectively:

$$y_{\text{int}}(x, t) = A\sin(kx + \omega t) + A\sin(kx - \omega t)$$

We can simplify the sum using the trigonometric identity

$$\sin a + \sin b = 2\sin\left(\frac{a + b}{2}\right)\cos\left(\frac{a - b}{2}\right)$$

which gives

$$y_{\text{int}}(x, t) = 2A\sin\left(\frac{(kx + \omega t) + (kx - \omega t)}{2}\right)\cos\left(\frac{(kx + \omega t) - (kx - \omega t)}{2}\right)$$

Notice how the terms in the arguments of sine and cosine add or cancel, leaving

$$y_{\text{int}}(x, t) = 2A\sin(kx)\cos(\omega t) \tag{13-18}$$

This mathematical construct describes a wave pattern unlike any we've seen yet. Perhaps most important, the x and t terms no longer appear in the same sinusoidal function. As we have seen, the derivative with respect to time of $kx - \omega t$ (Equation 13-16) yields velocity. A wave of the form $y(x, t) = A \sin(kx - \omega t)$ therefore has a velocity, in other words, it moves. We refer to such a wave as a **traveling wave**. Equation 13-18 does not contain a $kx - \omega t$ term, however, so the wave it describes does not propagate. Equation 13-18 describes a **standing wave**. The interference of a traveling wave and its reflection forms a standing wave, but although the traveling waves move along the direction of propagation and although elements of the medium move transverse to that direction, the *pattern* created by the interference, the standing wave, remains stationary.

At the end of the string tied to the pole ($x = L$, where L is the length of the string) the transverse displacement of the string must be zero at all times. The sin kx term is therefore equal to zero at $x = L$ in Equation 13-18. A sine function equals zero when the argument is an integer multiple of π, so it must be true that

$$kL = n\pi$$

where n is allowed to take on only integer values, that is, $n = 1, 2, 3$, and so on. So

$$L = \frac{n\pi}{k}$$

or, using the definition of the angular wave number $k = 2\pi/\lambda$ (Equation 13-2),

$$L = \frac{n\pi}{2\pi/\lambda} = \frac{n\lambda}{2}$$

To make the physics clear, we rewrite the equation separating the wavelength from the fraction:

$$L = \frac{n}{2}\lambda, \qquad n = 1, 2, 3 \dots \tag{13-19}$$

Here λ represents the wavelength of both the standing wave and also the original wave we created on the string of length L. So according to Equation 13-19, only waves such that half integer multiples of the wavelength fit onto the string generate the standing wave interference pattern described by Equation 13-18. Each value of n identifies a different interference pattern or resonant **mode**. For $n = 1$, one-half wavelength ($\frac{1}{2}\lambda$) fits on the string. For $n = 2$, one full wavelength ($\frac{2}{2}\lambda$) fits, and for $n = 3$, one and a half wavelengths ($\frac{3}{2}\lambda$) fit on the string. **Figure 13-20** shows the three modes; the dark orange line is the sine function itself, and the other lines show the string at various times in the standing wave vibration. Photographs of the $n = 1$, 2, and 3 modes of standing waves on a real string are shown in **Figure 13-21**. In the photographs, as well as in the drawings of the standing wave patterns in Figure 13-20, the movement of the end of the string that creates the standing waves is extremely slight. We therefore treat that end of the string as fixed, as is required if some number of half cycles of a sine function fits on the string.

Take another look at Equation 13-18, our expression for a standing wave. The properties of the sine function dictate that there can be positions along the string, called **nodes**, at which transverse displacement is zero. For the mode labeled $n = 2$ in Figure 13-20, there are three nodes, one at each end and one in between. There are four nodes for the $n = 3$ vibrational mode. A node occurs at any point such that kx is an integer (n) multiple of π because the sine of any integer multiple of π is zero. Positions along the standing wave at which the vibration of the string is maximal are **antinodes**.

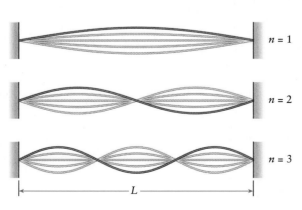

Figure 13-20 A string connected to two supports is made to vibrate in its first three standing wave modes. The mode number n counts the number of half cycles that fit onto the string.

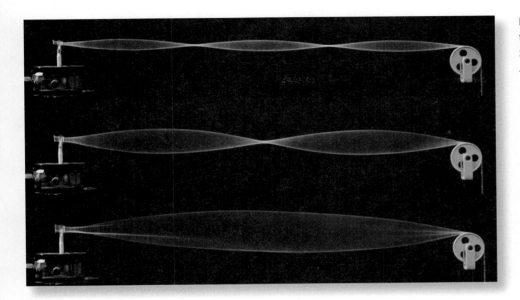

Figure 13-21 Photographs show the first three standing wave modes on a string. (Courtesy David Tauck.)

? Got the Concept 13-8
Elastic Band I

When you pluck an elastic band stretched to a length of 15 cm between your friend's fingers, it vibrates in its lowest (*n* equal to 1) resonant mode. What is the wavelength of this mode of the elastic band?

? Got the Concept 13-9
Elastic Band II

When you pluck an elastic band stretched to a length of 15 cm between your friend's fingers, it vibrates in its lowest vibrational mode, with a frequency of 110 Hz. What is the wavelength of the sound produced?

Recall that because the linear mass density μ of a string and the tension T applied to it determine wave propagation speed ($v_p = \sqrt{T/\mu}$, Equation 13-9), for a given string and tension there is some specific value of v_p. Because v_p is fixed, the relationship between v_p, λ, and f ($v_p = f\lambda$, Equation 13-8) determines f for any specific value of λ and vice versa. In other words, you can't choose both f and λ for a wave on a string; once you select either one, the other is also determined. So for example, we can combine Equations 13-8 and 13-19 to give a relationship for the frequencies f_n in terms of n and the length of the string:

$$f = \frac{v_p}{\lambda} = \frac{v_p}{2L/n}$$

or

$$f_n = n\frac{v_p}{2L}, \quad n = 1, 2, 3 \ldots \tag{13-20}$$

Each integer value of n sets a value of λ_n, which in turn determines a **resonant frequency** f_n of the string. The stretched string will vibrate in one of its distinctive resonant modes only when waves of one of the resonant frequencies are applied, say by wiggling one end of the string slightly. The frequency that corresponds to $n = 1$ is called the natural or **fundamental frequency**, or just "the fundamental." The $n = 2$ mode is the first overtone or second harmonic, the $n = 3$ mode is the second overtone or third harmonic, and so on. Notice that the overtones are multiples of the fundamental:

$$f_n = nf_1, \quad n = 1, 2, 3 \ldots$$

√x **Go to Picture It 13-1** for more practice dealing with overtones

where

$$f_1 = \frac{v_p}{2L} \tag{13-21}$$

is the fundamental frequency of the string. For stringed instruments like a guitar, plucking a string normally causes it to vibrate in its fundamental. The vibration sets up sound waves in the air that surrounds the string.

If a vibration of a frequency that is not the fundamental or one of the harmonics is applied to a string, interference of the waves traveling in opposite directions along the string does not produce a standing wave pattern. Instead, a jumble of wiggles will appear on the string, or perhaps an interference pattern which is not a standing wave.

√x **Go to Interactive Exercise 13-1** for more practice dealing with waves, tension, and linear mass density

? Got the Concept 13-10
Strings of a Guitar

Why are strings that produce the low (frequency) notes on a guitar thicker than those that produce the high notes?

? Got the Concept 13-11
Tension of a String

Does the fundamental frequency of a stretched string rise or fall when the tension in the string is increased by a factor of 2? By what factor does it change?

Example 13-5 Fundamental

If you tune a 0.65-m-long guitar string so that it vibrates (in its fundamental) at 440 Hz, what is the corresponding λ? Determine the frequency and wavelength of the sound produced when the string is plucked. The speed of sound in air is 343 m/s.

SET UP
A stretched string only resonates with frequencies such that the corresponding wavelengths are related to the length of the string according to Equation 13-19, so

$$\lambda = \frac{2L}{n}, \quad n = 1, 2, 3 \ldots$$

When the string vibrates in the fundamental, for which $n = 1$, the wavelength is therefore twice the length of the string.

The vibration of the string causes the surrounding air to move back and forth. The frequency at which the air molecules are disturbed from their equilibrium position matches the vibrational frequency of the string. From Equation 13-8, the wavelength of the sound is then

$$\lambda = \frac{v_{sound}}{f_{sound}} \tag{13-22}$$

where v_{sound} is the velocity of sound in air and f_{sound} is the fundamental frequency of the string and also the frequency of the sound wave.

SOLVE

From Equation 13-19, the wavelength that corresponds to the fundamental resonant frequency is twice the length of the string, so

$$\lambda = 2(0.65 \text{ m}) = 1.3 \text{ m}$$

The frequency of the sound wave matches the given frequency of vibration of the string, so

$$f_{sound} = f_{string} = 440 \text{ Hz}$$

The wavelength of the sound is given by Equation 13-22:

$$\lambda = \frac{v_{sound}}{f_{sound}} = \frac{343 \text{ m/s}}{440 \text{ Hz}} = 0.78 \text{ m}$$

REFLECT

The frequency of the standing wave on the string is the same as the frequency of the sound wave, because it is the vibration of the string that generates the motion of the air molecules. The wavelength of the sound is not, however, equal to the wavelength of the standing wave on the string. The product of frequency and wavelength is propagation speed ($v_p = f\lambda$, Equation 13-8). Because the propagation speed on the string is different from, and unrelated to, the speed of sound in air, the wavelengths of the string vibration and the sounds that it creates will be different. (The two wavelengths would only be the same if the propagation speed of waves on the string was equal to the speed of sound in air.)

Also, note that our answer would be different if we took the guitar to the top of a mountain where the lower air pressure results in a lower speed of sound. At the top of Mt. Everest, for example, the speed of sound in air is about 310 m/s under typical atmospheric conditions.

Practice Problem 13-5 If you tune a 0.55-m-long guitar string so that it vibrates in its second vibrational mode at 520 Hz, what is the corresponding λ? Determine the frequency and wavelength of the sound produced when the string is plucked.

★ What's Important 13-5

The interference of two identical waves traveling in opposite directions, for example, the interference of a traveling wave and its reflection, can result in a standing wave. A stretched string will vibrate in a standing wave only when waves of one of its resonant frequencies are applied.

13-6 Longitudinal Standing Waves

A male tree-hole frog attracts females by croaking out a simple call dominated by a tone of a single frequency. To amplify his call and enhance his attractiveness to potential mates, he adjusts the frequency of his call to match the fundamental resonant frequency of the cavity. Because the tree hole cavity is essentially a cylinder, the interference of sound waves reverberating back and forth results in standing waves, so the resonant frequencies of the cavity depend on its height in much the same way that the resonant frequencies of the stretched string depends on its length.

Sound and other longitudinal waves confined to a particular region, such as a tube (or tree hole!), can interfere to produce three-dimensional standing waves. These longitudinal disturbances of the medium create regions (volumes) of higher and lower pressure and density. During the 1860s, the German physicist August Kundt invented a device to see such standing waves. The apparatus that bears his name consists of a horizontal glass tube sealed at both ends in which small bits of lightweight material (Kundt used a fine powder) are spread along the bottom. Small styrene balls were used in the Kundt's tube shown in **Figure 13-22**. When sound waves are created in the tube, say, by placing a speaker at one end, they bounce back and forth between the ends of the tube setting the air into a standing wave. Longitudinal motion of air molecules is greatest at the antinodes of the standing wave. This causes the styrene balls to gather in the region of the antinodes and tends to lift them up. Figure 13-22 clearly shows the second harmonic (two half cycles) of a standing pressure wave in a Kundt's tube.

The magnitude of the longitudinal motion of air molecules in the tube can be represented by the same figures we used to describe a standing wave on a string. The standing wave in the Kundt's tube shown in Figure 13-22, for example, could be represented by the $n = 2$ mode in Figure 13-20. The farther points on the outer lines are from the center line, the greater the magnitude of the displacement of the molecules from their equilibrium position. As we saw with transverse standing waves, the wavelength of the standing wave and the waves that create it are identical. So the drawings in Figure 13-20 also represent the sound waves that interfere to create the standing wave. However, although we originally used the curves in Figure 13-20 to

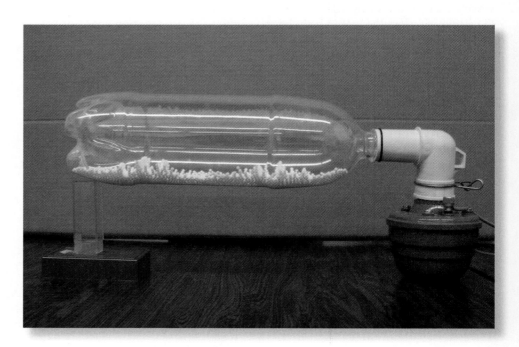

Figure 13-22 Sound waves moving back and forth in a Kundt's tube form a standing wave. Small styrene balls move in response to the standing wave. The $n = 2$ mode is shown. *(Courtesy of Tatsuya Kitamura.)*

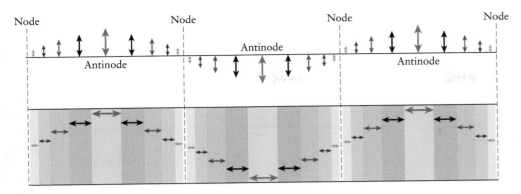

Figure 13-23 The magnitude of the longitudinal oscillations of air molecules vibrating in the $n = 3$ standing wave in a Kundt's tube varies along the length of the tube in the lower sketch. The magnitude of the oscillations is represented by a sinusoidal function in the upper part of the figure. The length of the arrows in the lower part of the figure indicate the magnitude of the displacement of air molecules in each region along the tube.

represent transverse waves on a stretched string, neither the sound waves in the tube nor the standing wave interference pattern that forms is transverse. Both are longitudinal waves; the air molecules oscillate back and forth along the length of the tube. To make this clearer, in **Figure 13-23** we have sketched the magnitude of the oscillation of air molecules for the third harmonic of a standing pressure wave in a Kundt's tube. In the lower half of the figure, colored regions in the tube indicate the magnitude of the *longitudinal* oscillation of air molecules in various regions along the tube. No air moves at the ends nor at the two nodes in between. At the three antinodes, indicated by orange shading, the displacement of the air molecules is greatest. You can see how a longitudinal wave can be represented by a sinusoidal function by comparing the upper and lower parts of the figure.

Longitudinal waves in a tube cause air molecules to move back and forth along the direction of the axis of the tube. For this reason, molecules of air at a closed end of a tube do not move. There must therefore be a node at both ends when a standing wave forms in a Kundt's tube. Nodes are plainly visible at the ends of the tube in Figure 13-22. In contrast, when the air in a tube not sealed at both ends is disturbed by the propagation of a longitudinal wave, the air at the open end can move back and forth freely. So when a standing wave forms in a tube with an open end, the amplitude of the motion of air at the open end is not limited; *this is the definition of an antinode.* An antinode exists, for example, at the opening of the tree cavity used by a tree-hole frog to amplify his mating call. We will consider the resonant frequencies of a tube that is open at one end and closed at the other, and a tube that is open at both ends.

Figure 13-24 shows the first three standing wave modes in a tube closed at one end. To form the waves, we send a sound wave into the open end on the left; as the waves reflect from the closed end on the right, waves traveling in opposite directions in the tube interfere to form the standing waves. Again, we represent the magnitude of the displacement of air molecules from equilibrium by the separation of the blue curves from the dashed center line. Regardless of the harmonic, a node always appears at the closed end of the tube and an antinode at the open end.

The sound wave that produces the first harmonic of a tube closed at one end is longer than the tube. As shown in Figure 13-24a only one-quarter of the wave fits. To make this clear, in **Figure 13-25a** we

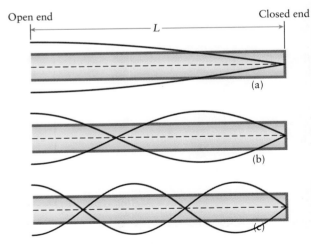

Figure 13-24 Standing wave modes in a tube closed at one end and open at the other have a node at the closed end and an antinode at the open end. (a), (b), and (c) show the first three standing wave modes.

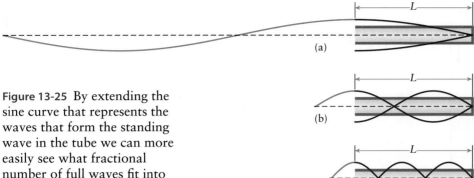

Figure 13-25 By extending the sine curve that represents the waves that form the standing wave in the tube we can more easily see what fractional number of full waves fit into the tube.

extend the sinusoidal representation of the sound wave out beyond the open end to show one full wave cycle. The extension does not represent the standing wave, which exists only inside the tube, but it allows you to compare the wavelength of the sound with the length of the tube. For the tube closed at one end, then, the relationship between the length of the tube L and the wavelength λ of the sound wave that sets up the standing wave interference is

$$L = \frac{1}{4}\lambda$$

Figures 13-24b and 13-24c show the standing waves formed when the tube open at one end and closed at the other is excited in the second and third resonant modes. Figures 13-25b and 13-25c show one full period of the sound waves that excite the resonant standing waves. Three-quarters (3/4) and five-quarters (5/4) of a wavelength fit into the tube, respectively, for the modes. The general relationship between L and λ for a tube closed at one end is therefore

$$L = \frac{n}{4}\lambda, \quad n = 1, 3, 5 \dots \tag{13-23}$$

Notice that only the odd harmonics can exist in a tube closed at one end (mode numbers 1, 3, 5, ...). From $v_p = f\lambda$ (Equation 13-8), the corresponding resonant frequencies are

$$f = \frac{v_p}{\lambda} = \frac{v_p}{4L/n}$$

or

$$f_n = n\frac{v_p}{4L}, \quad n = 1, 3, 5 \dots \tag{13-24}$$

for a tube closed at one end.

? Got the Concept 13-12
Ruler

A physics student presses one end of a plastic ruler tightly against a desk so that the other end extends over the edge. Pushing down and then releasing the free end causes the ruler to vibrate and produce a tone. As she slides the ruler so that less and less sticks out over the edge of the desk, does the frequency of the sound produced increase, stay the same, or decrease? *Hint:* As the ruler vibrates, one end is a node and the other an antinode.

? Got the Concept 13-13
Water Bottle

Gently tap the side of water bottle as you fill it. Why does the tone you hear go up in pitch as the water rises?

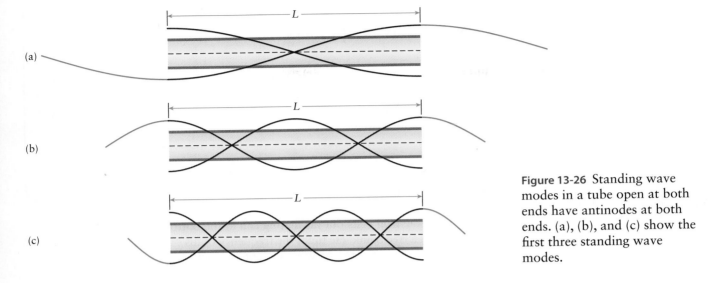

Figure 13-26 Standing wave modes in a tube open at both ends have antinodes at both ends. (a), (b), and (c) show the first three standing wave modes.

Figure 13-26 shows the first three standing wave modes in a tube open at both ends. To form the waves, we send a sound wave into either open end; although some of the sound passes out of the tube at the ends, the waves are partly reflected as they approach the interfaces between the inside and outside of the tube. The interfaces generate waves traveling in opposite directions in the tube, which can interfere to form standing waves. Regardless of the harmonic, when a standing wave is created in an open tube, an antinode occurs at both ends because air moves freely at both ends.

One-half of the sound wave which produces the first harmonic of a tube open at both ends fits inside the length of the tube. This is evident in Figure 13-26a, in which we extend the sinusoidal representation of the sound wave out beyond the ends of the tube to show one full wave cycle. The extension does not represent the standing wave, which only exists inside the tube. As in the case of the tube open at one end and closed at the other, the extension allows you to compare the wavelength of the sound with the length of the tube. For the tube open at both ends, the wavelength of the sound wave that sets up the first harmonic standing wave equals twice the length of the tube:

$$L = \frac{1}{2}\lambda$$

Figures 13-26b and 13-26c show one full period of the sound waves that excite the next two resonant standing waves in the tube open at both ends. One full (two-halves) and one and a half (three-halves) of a wavelength fit into the tube, respectively, for the modes. The general relationship between L and λ for a tube open at both ends is therefore

$$L = \frac{n}{2}\lambda, \quad n = 1, 2, 3 \ldots \tag{13-25}$$

From $v_p = f\lambda$ (Equation 13-8), the corresponding resonant frequencies are

$$f = \frac{v_p}{\lambda} = \frac{v_p}{2L/n}$$

or

$$f_n = n\frac{v_p}{2L}, \quad n = 1, 2, 3 \ldots \tag{13-26}$$

for an open tube.

? Got the Concept 13-14
Horn

By blowing into the mouthpiece, a musician causes air in the tube of a horn to vibrate. The waves set up by the vibrations reflect back and forth in the horn to create standing waves. Which horn has a lower fundamental frequency, a longer or a shorter one?

Example 13-6 Tree-hole Frog

A male tree-hole frog belts out his mating call at the bottom of a cylindrical, 11-cm-deep cavity in a tree trunk. He tunes the frequency of his call to match the fundamental resonant mode of the cavity. What is the frequency of the frog's mating call?

SET UP

The cavity in the tree trunk acts like a tube closed at one end (the bottom) and open at the other. The fundamental frequency of the cavity is given by Equation 13-24, where $n = 1$:

$$f = \frac{v_p}{4L}$$

SOLVE

Sound travels through air at a velocity $v_p = 343$ m/s, so the fundamental frequency of a cavity of depth $L = 0.11$ m is

$$f = \frac{343 \text{ m/s}}{4(0.11 \text{ m})} = 780 \text{ Hz}$$

REFLECT

Tree-hole frogs have been observed to match the resonant frequency of a tree cavity over a cavity depth of between 10 and 15 cm. This calculation is close to the measured value of the frequency of a male tree-hole frog's mating call in a cavity 11 cm deep.

Practice Problem 13-6 A male tree-hole frog belts out his mating call at the bottom of a cylindrical, 13-cm deep cavity in a tree trunk. He tunes the frequency of his call to match the fundamental resonant mode of the cavity. What is the frequency of the frog's mating call?

Estimate It! 13-2 A Flute and a Clarinet

Estimate the lowest fundamental frequency of a clarinet and a flute.

SET UP

The clarinet is a tube closed at one end and open at the other. The flute acts like a tube open at both ends. We'll therefore want to use Equation 13-23 for the clarinet and Equation 13-25 for the flute. For both we need the length of the instrument; from having seen a flute and clarinet, we estimate that both are about 0.5 m long. (Yes, you could look up the exact length of each, but remember this is an estimate. We only need to use big, round numbers, as long as they're reasonable.)

SOLVE

The clarinet behaves like a tube open at one end and closed at the other. From Equation 13-23, only one-quarter of a wavelength fits inside the clarinet when it resonates in the fundamental. So,

$$\lambda_{clarinet} = 4(0.5) = 2 \text{ m}$$

The fundamental frequency is found from $f = v_p/\lambda$ (or Equation 13-24, where $n = 1$):

$$f_{clarinet} = 343 \text{ m/s}/2 \text{ m} = 171.5 \text{ Hz}$$

To one significant figure, our estimate for the fundamental frequency of the clarinet is about 200 Hz.

The flute is a tube open at both ends, so its longest (fundamental) wavelength is twice the length according to Equation 13-25:

$$\lambda_{flute} = 2(0.5) = 1 \text{ m}$$

The fundamental frequency, from $f = v_p/\lambda$ (or Equation 13-26, where $n = 1$), is

$$f_{flute} = 343 \text{ m/s}/1 \text{ m} = 343 \text{ Hz}$$

So our estimate for the fundamental frequency of the flute is about 300 Hz. Again, our rough approximation for the length of the flute requires us to state the answer to one significant figure.

REFLECT

The actual sounding length of a flute is about 0.67 m long, which gives a fundamental frequency of $f = 343 \text{ m/s}/2(0.67 \text{ m}) = 256 \text{ Hz}$. This is therefore the lowest note that can be produced on a flute. A (B-flat) clarinet is a bit shorter, about 0.58 m in length. The clarinet therefore has an approximate fundamental frequency of $f = 343 \text{ m/s}/4(0.58 \text{ m}) = 150 \text{ Hz}$. Both of our estimates are reasonable.

★ What's Important 13-6

Sound and other longitudinal waves confined to a tube (or some particular region) can interfere to produce three-dimensional standing waves. The longitudinal disturbances of the medium create regions of higher and lower pressure and density. A node always appears at a closed end of the tube and an antinode always appears at an open end.

13-7 Beats

You might have heard a sound something like "wahwahwah" when a guitar player tunes his instrument by plucking two strings at the same time. As he adjusts the tension on one string, the "wahs" and also the time between them gets longer and longer until they can no longer be distinguished. This phenomenon, known as **beats**, arises from the interference between sound waves. Beats are most pronounced when two waves of nearly identical frequencies and wavelengths interfere.

Consider two waves that have wave numbers k_1 and k_2 and angular velocities ω_1 and ω_2, where k_1 and k_2 are almost equal and ω_1 and ω_2 are almost equal. We let the waves have the same amplitude A and represent them as

$$u_1(x, t) = A \sin(k_1 x - \omega_1 t)$$

and

$$u_2(x, t) = A \sin(k_2 x - \omega_2 t)$$

Although the phenomenon of beats occurs in both transverse and longitudinal waves, here we focus on the beats formed when sound waves interfere, so the variable $u(x, t)$ represents longitudinal displacement of air molecules. The magnitude of the displacement is directly related to the pressure along the wave as well as the loudness of the sound we hear.

Interference occurs when two waves overlap. To explore the pattern formed, we add waves $u_1(x, t)$ and $u_2(x, t)$ together:

$$u(x, t) = u_1(x, t) + u_2(x, t)$$
$$= A \sin(k_1 x - \omega_1 t) + A \sin(k_2 x - \omega_2 t)$$

Using the trigonometric identity

$$\sin a + \sin b = 2 \sin\left(\frac{a + b}{2}\right)\cos\left(\frac{a - b}{2}\right)$$

the expression becomes

 See the Math Tutorial for more information on Trigonometry

$$u(x, t) = 2A \sin\left(\frac{(k_1 x - \omega_1 t) + (k_2 x - \omega_2 t)}{2}\right)\cos\left(\frac{(k_1 x - \omega_1 t) - (k_2 x - \omega_2 t)}{2}\right)$$

Don't glaze over—the equation can be simplified! Because $k_1 \approx k_2$ we can let both k_1 and k_2 be approximately equal to the same value k. Likewise, because ω_1 is approximately ω_2, we let $\omega_1 \approx \omega_2 \approx \omega$. In the sine term, then,

$$\frac{k_1 x + k_2 x - \omega_1 t - \omega_2 t}{2} = \frac{(k_1 + k_2)x - (\omega_1 + \omega_2)t}{2} = \frac{2kx - 2\omega t}{2} = kx - \omega t$$

This expression is the general form of the argument of a traveling wave, as in Equation 13-4. So, the sine term in the interference pattern is a traveling wave, and the wavelength and frequency of the wave are approximately the same as for the two interfering waves $u_1(x, t)$ and $u_2(x, t)$. The sine term of wave $u(x, t)$ is shown in **Figure 13-27a**.

We simplified the argument of the sine piece by combining the two k terms into one and also the two ω terms into one, which made it more apparent that the interference pattern is a traveling wave. By combining the k and ω terms in the argument of cosine in $u(x, t)$, it becomes clear that the cosine piece is also a traveling wave. We define Δk to be the difference between k_1 and k_2 and $\Delta\omega$ to be the difference between ω_1 and ω_2; that is, $\Delta k = k_1 - k_2$ and $\Delta\omega = \omega_1 - \omega_2$. So then $u(x, t)$ becomes

$$u(x, t) = 2A \sin(kx - \omega t)\cos\left(\frac{\Delta k}{2}x - \frac{\Delta\omega}{2}t\right) \tag{13-27}$$

The arguments of the cosine and sine terms are of the same form, but the wave number and angular velocity of the second wave are $\Delta k/2$ and $\Delta\omega/2$, respectively. Notice that because $k_1 \approx k_2$ and $\omega_1 \approx \omega_2$ the differences Δk and $\Delta\omega$ are small. A wave that has a small wave number has a long wavelength, because the two quantities are inversely proportional to one another ($\lambda_{\text{cosine}} = 2\pi/\Delta k$, using Equation 13-2). In the same way, because period and angular velocity are inversely proportional ($T_{\text{cosine}} = 2\pi/\Delta\omega$, using Equation 12-5), the period associated with the cosine term is large because $\Delta\omega$ is small. Because the cosine term has a relatively longer wavelength and larger period, it varies much more slowly than the sine term, as shown in **Figure 13-27b**.

So the interference of two similar waves is the product of two traveling waves, one varying more slowly than the other (Equation 13-27). The result is a traveling wave that oscillates at (approximately) the frequency of the two interfering waves, but with an amplitude limited by the slowly varying cosine term.

Beats: The distinctive interference pattern formed by the superposition of two similar waves is the product of two traveling waves, $\sin(kx - \omega t)$ and $\cos(\frac{\Delta k}{2}x - \frac{\Delta\omega}{2}t)$, which varies more slowly.

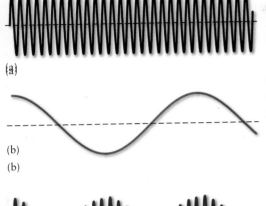

(a)

$\sin(kx - \omega t)$
Because $k_1 \approx k_2 \approx k$ and $\omega_1 \approx \omega_2 \approx \omega$, this wave is approximately the same as both of the original waves.

(b)

(b)

$\cos(\frac{\Delta k}{2}x - \frac{\Delta\omega}{2}t)$
$\Delta k = k_1 - k_2$, so Δk is small.
$\Delta\omega = \omega_1 - \omega_2$, so $\Delta\omega$ is small.
Because $\lambda = 2\pi/k$ and $T = 2\pi/\omega$ the wavelength and period of this wave are therefore relatively large; the cosine term varies far more slowly than the sine term.

(c)

$\sin(kx - \omega t)\cos(\frac{\Delta k}{2}x - \frac{\Delta\omega}{2}t)$
The product oscillates with the frequency of the sine term but with amplitude limited by the slowly varying cosine term. The interference pattern, a traveling wave of varying amplitude, gives rise to the phenomenon of beats.

Figure 13-27 The product of (a) a rapidly varying sinusoidal function and (b) a slowly varying one results in (c) the distinctive interference pattern known as beats.

The distinctive interference pattern which arises, shown in **Figure 13-27c**, represents the phenomenon known as beats. For sound, beats are the periodic variation of intensity that can result, for example, when two strings on a guitar, fingered to produce the same note, are plucked simultaneously on an instrument that is out of tune.

The sine term in $u(x, t)$ (Equation 13-27) is approximately the same as the two separate sound waves causing the beats, so it is through this term that we hear the pitch of the sound. The rate at which beats are heard is determined by the cycle of the cosine term in $u(x, t)$. As shown in **Figure 13-28**, two beats occur in every cycle of $\cos(\Delta k\, x/2 - \Delta\omega\, t/2)$, so the angular velocity of the beats is twice the angular velocity of the cosine. The angular velocity of beats is therefore the difference in angular velocity of the two waves that are interfering with each other:

$$\omega_{\text{beats}} = 2\left(\frac{\Delta\omega}{2}\right) = \Delta\omega$$

There are two beats in every cycle of the cosine term.

Beat

Beat

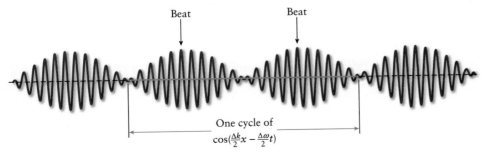

One cycle of
$\cos(\frac{\Delta k}{2}x - \frac{\Delta\omega}{2}t)$

Figure 13-28 Based on our mathematical description of beats, two beats occur in every cycle of the cosine term.

Using Equation 12-7, the relationship between frequency and angular velocity ($\omega = 2\pi f$), we can express the beat frequency in terms of the frequencies f_1 and f_2 of the original waves:

$$f_{\text{beats}} = \frac{\omega_{\text{beats}}}{2\pi} = \frac{\Delta\omega}{2\pi} = \frac{\omega_1 - \omega_2}{2\pi} = \frac{\omega_1}{2\pi} - \frac{\omega_2}{2\pi} = f_1 - f_2$$

The frequency at which beats occur equals the difference in frequencies of the original two waves. It doesn't matter which of the frequencies of the original waves is greater, and in addition, a negative frequency has no meaning. So to be strictly correct, the beat frequency is the absolute value of the difference between f_1 and f_2:

$$f_{\text{beats}} = |f_1 - f_2| \tag{13-28}$$

 Got the Concept 13-15
Tuning Forks I

Two tuning forks of frequencies 440 Hz and 438 Hz are struck simultaneously. What frequency of beats is heard?

 Got the Concept 13-16
Tuning Forks II

When two tuning forks are struck simultaneously, 5 beats per second are heard. The frequency of the first tuning fork is 440 Hz. What is the frequency of the second tuning fork?

★ **What's Important 13-7**

Beats are the periodic variation of intensity that results when two waves of nearly identical frequencies and wavelengths interfere. Beats occur at a frequency equal to the difference in frequencies of the original two waves.

13-8 Volume, Intensity, and Sound Level

Humans hear sounds over a wide range of volumes, from faint sounds such as leaves rustling in a gentle breeze to sounds so loud, say, the roar of the engines on a jet about to take off, that they might cause pain. We have a good sense of faint and loud, but how can we quantify the loudness of a sound in a physically meaningful way?

We defined a wave as a propagating displacement of the elements of a medium. As a sound wave travels through air, for example, any individual air molecule oscillates back and forth along the direction of the wave propagation. The air molecules therefore possess kinetic energy; the transfer of energy from air molecule to air molecule allows the wave to propagate. The energy of any oscillating object increases with the square of the amplitude of the motion ($E = \frac{1}{2}kA^2$, Equation 12-10), so the farther the air molecules move back and forth from equilibrium, the greater the energy of the disturbance that creates a sound.

The process of hearing involves translating the kinetic energy carried by a sound wave into electrical signals in the brain. As part of that process the outer ear gathers sound and funnels it down to the eardrum which vibrates as the energy of the sound wave impinges on it. But as we saw in the last chapter, when an oscillating system vibrates close to its natural frequency, it resonates with large amplitude motion. Thus, sounds in the range of the natural frequency of the human eardrum, 700 to 4000 Hz, require less energy to cause large amplitude vibrations of the eardrum than sounds at other frequencies. We perceive sounds that cause large amplitude vibrations of the eardrum as loud. So even if it carries more energy, a 100-Hz tone might sound fainter to you than a 2000-Hz tone. Clearly, our perception of loudness correlates poorly with the physical properties of sound.

The amount of energy that a wave transmits in a given period of time is a better measure of the impact of a sound than our perception of its loudness. We define the quantity **power P** as the energy E delivered per unit time t:

$$P = \frac{E}{t} \tag{13-29}$$

The SI units of power are watts (W):

$$1\,\text{W} = 1\frac{\text{J}}{\text{s}}$$

The amount of power transmitted to a specific object by a sound depends on the surface area of the object. For example, sound delivers far less energy per second when striking your relatively small eardrum than when it hits a relatively large parabolic microphone like the ones used on the sideline of a football game. To eliminate the effect of the detector surface area on our quantitative description of sound, we define **intensity** as power per unit area A:

$$I = \frac{P}{A} \tag{13-30}$$

Said another way, power per unit area (intensity) multiplied by a specific area gives the power delivered to that area. But more important is the reverse statement, that while the power delivered by a sound depends on the surface area of the detector, the intensity is the same regardless of whether the sound strikes your eardrum or a parabolic microphone. Intensity is therefore a good way to characterize sound.

The intensity of a sound wave of frequency f depends on the mass density ρ of air (or of the medium through which the wave propagates), the propagation speed v_p, and the amplitude of the displacement of air molecules s_max (Equation 13-6):

$$I = 2\pi^2 \rho v_\text{p} f^2 s_\text{max}^2 \tag{13-31}$$

Intensity depends on energy. The kinetic energy of the oscillating air molecules is proportional to s_max^2, the square of the amplitude of the motion of air molecules; therefore, intensity is also proportional to s_max^2.

The human ear is sensitive to a wide range of sound intensities. The threshold of hearing is commonly taken to be a sound of $I = 10^{-12}\,\text{W/m}^2$, a factor of 10^{12} lower than the intensity of sound that could cause the listener pain. Intensity $I = 1\,\text{W/m}^2$ is sometimes referred to as the "threshold of pain." (It's not very scientific, but we like to say it.)

Example 13-7 Faint and Loud

The human ear detects sound over a wide range of intensities. The amplitude of the motion of your eardrum in response to a sound wave more or less matches the amplitude of the motion of nearby air molecules. Through what distance does your eardrum move when you hear a sound that has a frequency of 1000 Hz and an intensity close to the threshold of pain, $I = 1 \text{ W/m}^2$? What about when you hear a barely audible tone of that frequency? Use $\rho = 1.2 \text{ kg/m}^3$ for air density and $v = 343 \text{ m/s}$ for the speed of sound in air.

SET UP

The displacement amplitude that results from a passing sound wave is proportional to the square root of intensity. Using Equation 13-31,

$$s_{max} = \sqrt{\frac{I}{2\pi^2 \rho v_p f^2}}$$

As your eardrum oscillates in response to a sound wave, the peak-to-peak amplitude of the oscillation is equal to $2s_{max}$.

SOLVE

We substitute known values to obtain numeric answers. For painfully loud sounds, the intensity I is 1 W/m^2,

$$s_{max} = \sqrt{\frac{1 \text{ W/m}^2}{2\pi^2 (1.2 \text{ kg/m}^3) \, (343 \text{ m/s}) \, (1000 \text{ Hz})^2}} = 1.1 \times 10^{-5} \text{ m}$$

So the distance through which the eardrum moves is $2s_{max} = 2.2 \times 10^{-5} \text{ m}$.

For the barely audible sound we set $I = 10^{-12} \text{ W/m}^2$, the threshold of hearing, so

$$s_{max} = \sqrt{\frac{10^{-12} \text{ W/m}^2}{2\pi^2 (1.2 \text{ kg/m}^3) (343 \text{ m/s}) (1000 \text{ Hz})^2}} = 1.1 \times 10^{-11} \text{ m}$$

and $2s_{max} = 2.2 \times 10^{-11} \text{ m}$.

REFLECT

For the painfully loud sound, your eardrum moves back and forth $2.2 \times 10^{-5} \text{ m}$ or about 20 μm. On a human scale, this distance is rather small; 20 μm is the about three times the diameter of a human red blood cell.

In response to a faint sound, the eardrum moves only about 2×10^{-11} m, a distance approximately one thousand times smaller than the smallest virus and one-tenth the diameter of a hydrogen atom!

The energy radiated per second through sound depends only on the source of the sound. The intensity of the sound, however, also depends on how far it travels from the source, perhaps a speaker connected to an audio system, to the detector or listener. As the sound radiates away from its source, the total power spreads out over an ever-expanding surface. If a source radiates sound of power P_0 in *all* directions uniformly, the power P_0 spreads out over the surface area of an expanding sphere of radius r. So by substituting the surface area of a sphere into Equation 13-30, at a distance r the intensity equals

$$I = \frac{P_0}{4\pi r^2} \tag{13-32}$$

> **? Got the Concept 13-17**
> **Spherical Speakers**
>
> So-called spherical speakers radiate sound isotropically, that is, uniformly in all directions. When you double your distance from a spherical speaker, how does the intensity of the sound you hear change? Assume an initial distance large enough to neglect the speaker's size. Neglect reflections of the sound from the floor, ceiling, and walls, too.

Example 13-8 Energy on an Eardrum

If you stand 20 m from a speaker tower generating 2500 W of sound power at an outdoor concert, how much energy per second impinges on each of your eardrums? Assume the surface area of a typical human eardrum is 55 mm² and that the speaker generates sound uniformly over a hemisphere in the direction of the audience.

SET UP
As the sound generated radiates away from the speakers, the total power, which remains constant, spreads out over the surface of an ever-expanding, imaginary sphere. The power intercepted by an eardrum depends on the product of the area of the eardrum and the power P_0 of the sound wave per unit area:

$$P_{eardrum} = \frac{P_0}{area} A_{eardrum}$$

At a distance r, the power P_0 spreads out over the surface area of a hemisphere (*half* of a sphere) that has a radius r. So, $A = 2\pi r^2$ and

$$P_{eardrum} = \frac{P_0}{2\pi r^2} A_{eardrum}$$

SOLVE
Using the values given, the energy that impinges on each eardrum per second is found to be

$$P_{eardrum} = \frac{2500\ W}{2\pi(20\ m)^2} 55\ mm^2 \left(\frac{1\ m}{1000\ mm}\right)^2 = 5.5 \times 10^{-5}\ W$$

REFLECT
Power of $P_{eardrum} = 5.5 \times 10^{-5}$ W means that 5.5×10^{-5} J of energy hits the eardrum every second. Remember that it takes 9.8 J to lift a 1-kg object through a vertical distance of 1 m near the surface of Earth; this answer is a relatively small amount of energy. But consider the intensity of the sound:

$$I = \frac{P_0}{2\pi r^2} = \frac{2500\ W}{2\pi(20\ m)^2} = 0.99\ \frac{W}{m^2}$$

This comes from Equation 13-32, but we have used $2\pi r^2$ rather than $4\pi r^2$ because the sound in this problem is distributed over a hemisphere. The value of I we obtain is effectively the so-called threshold of pain. Although the sound delivers a relatively small amount of energy to the eardrum, were you to stand 20 m from a speaker tower generating 2500 W, you would experience sound so loud that it would be painful.

Practice Problem 13-8 If you stand 20 m from a speaker tower generating 500 W of sound power at an outdoor concert, how much energy per second impinges on each of your eardrums? What is the intensity of the sound? Assume the typical human eardrum surface area to be 55 mm² and that the speaker generates sound uniformly over a hemisphere in the direction of the audience.

The range of sound intensities that we can hear spans the range from 10^{-12} W/m² to 1 W/m². A range so large makes it difficult to communicate levels of loudness to other people. Imagine if we measured temperature over such a large range—on a cold day it might be 12 °C, but on a warm day it might be 327,487,300,578 °C! This wouldn't be convenient—in our everyday experiences we aren't comfortable with numbers distributed over 12 orders of magnitude. We can compress a large range, however, by using a logarithm. The logarithm of x to base b is defined to be equal to y when $b^y = x$. By convention, we assume that $b = 10$ unless specified otherwise; for example, $\log 1000 = 3$ because $10^3 = 1000$. So for numbers ranging from 1 to 1000, the log of those numbers ranges only from 1 to 3. In the same way, taking the log of numbers that range from 10^{-12} to 1 results in a range from $\log 10^{-12} = -12$ to $\log 1 = 0$. (Any number to the power of zero equals 1, so $\log 1$ equals 0.)

The argument of the logarithm must be a pure number; that is, it cannot have dimensions or units. We satisfy this requirement by defining the **sound level β** of a sound that has intensity I as

$$\beta = 10 \log \left(\frac{I}{10^{-12}\,\text{W/m}^2} \right) \tag{13-33}$$

The units of sound level are the decibel (dB), named after the American scientist and inventor Alexander Graham Bell. (Note that in order for the units to work out correctly, the constant value 10 in Equation 13-33 must have units of dB. (In calculating numeric values for problems that involve β, remember to include these units!) The sound level at the threshold of hearing is then

$$\beta_{\text{faint}} = (10\,\text{dB}) \log \left(\frac{10^{-12}\,\text{W/m}^2}{10^{-12}\,\text{W/m}^2} \right) = (10\,\text{dB}) \log 1 = 0\,\text{dB}$$

The sound level at the threshold of pain is

$$\beta_{\text{pain}} = (10\,\text{dB}) \log \left(\frac{1\,\text{W/m}^2}{10^{-12}\,\text{W/m}^2} \right) = (10\,\text{dB}) \log 10^{12} = 120\,\text{dB}$$

Note that you don't need a calculator to find $\log 10^{12}$. The power to which 10 must be raised to give 10^{12} is 12! So by using sound level we compress the range of values of human hearing to 2 orders of magnitude instead of 12. We list the sound levels of a range of sounds in **Table 13-1**.

See the Math Tutorial for more information on Logarithms

Table 13-1
Sound Level

Sound	Sound level (dB)
jet engine at 25 m	150
live rock music	120
car horn at 1 m	110
jackhammer	100
city street	90
vacuum cleaner	70
quiet conversation	50
rustling leaves	20
breathing	10

> ## ❗ Watch Out
> ### A sound level of 0 dB is not the absence of sound
>
> We are used to thinking of something that has a value of zero is, well, nothing. If an object has a speed of 0 m/s, for example, it's not moving. However, a sound of sound level equal to 0 dB, while faint, is still a sound. Sound level is defined relative to humans' ability to hear; 0 dB corresponds to the faintest sound that *we* can hear. But a sound of sound level equal to 0 dB is nevertheless a propagating pressure wave that disturbs the medium through which it passes.

Example 13-9 Double the Distance

In Got the Concept 13-17, we determined that when we double the distance between a listener and an isotropic source of sound the intensity decreases by a factor of 4. How does the sound level change when the intensity decreases by the same factor?

SET UP

Let the initial intensity be $I_1 = I_0$, so that when the distance doubles, the intensity decreases to $I_2 = \frac{1}{4} I_0$. The sound levels at the two locations are

$$\beta_1 = (10\,\text{dB}) \log \left(\frac{I_1}{10^{-12}\,\text{W/m}^2} \right) = (10\,\text{dB}) \log \left(\frac{I_0}{10^{-12}\,\text{W/m}^2} \right)$$

and

$$\beta_2 = (10 \text{ dB}) \log \left(\frac{I_2}{10^{-12} \text{ W/m}^2} \right) = (10 \text{ dB}) \log \left(\frac{I_0/4}{10^{-12} \text{ W/m}^2} \right)$$

respectively. We can examine the change in sound level by finding the difference between β_1 and β_2.

SOLVE
The difference in sound level is

$$\beta_1 - \beta_2 = (10 \text{ dB}) \log \left(\frac{I_0}{10^{-12} \text{ W/m}^2} \right) - (10 \text{ dB}) \log \left(\frac{I_0/4}{10^{-12} \text{ W/m}^2} \right)$$

To simplify, we employ an important property of logarithms, that

$$\log a - \log b = \log (a/b)$$

So

$$\beta_1 - \beta_2 = (10 \text{ dB}) \log \left(\frac{I_0}{10^{-12} \text{ W/m}^2} \times \frac{10^{-12} \text{ W/m}^2}{I_0/4} \right)$$

or

$$\beta_1 - \beta_2 = (10 \text{ dB}) \log 4 = 6.0 \text{ dB}$$

REFLECT
When the intensity decreases by a factor of 4, the sound level decreases by 6 dB. Because $\log 2 = 0.301$, so that $10 \log 2 \approx 3$, every factor of 2 change in intensity corresponds to a change in sound level of about 3 dB. So, a factor of 4 in intensity corresponds to a 6 dB change in sound level. A factor of $2 \times 2 \times 2 = 8$ change in intensity corresponds to a difference in sound level of $3 \text{ dB} + 3 \text{ dB} + 3 \text{ dB} = 9 \text{ dB}$.

Practice Problem 13-9 If the sound level increases by 12 dB as you move toward the source of the sound, what is the ratio of your new distance to your original distance from the sound source?

 Got the Concept 13-18
Fan

You measure the sound level of an electric fan to be 60 dB. What is the sound level when a second identical fan is turned on, at the same distance from you as the first one?

 **What's Important 13-8**
The intensity of a sound depends on its energy, and not features of the device used to detect the sound. The power *delivered* by a sound, however, depends on the surface area of the detector.

13-9 Moving Sources and Observers of Waves

Until now, we have only considered waves generated by stationary emitters and observed by stationary observers. Everyday experience, however, suggests that something curious happens when a moving object creates a sound. As a police car with sirens blaring zooms by you, for example, the frequency of the sound increases as the car approaches and decreases after it passes. Even without a siren, a fast moving car generates a characteristic high–low frequency sound (something like "neee-urrrr") as it approaches and passes you. And although we don't often get the chance to experience the phenomenon in reverse, if you were to move at high speed toward a stationary police car with its sirens blaring, the sound of the siren would follow a similar high–low frequency shift as you approached and then passed it. This effect is known as the **Doppler effect**, named for the Austrian physicist Christian Doppler who first proposed the phenomenon associated with waves in 1842.

In **Figure 13-29a**, a police siren emits a periodic sound wave that has a fixed frequency and a fixed wavelength. Because the car is stationary, each region of highest pressure along the wave, shown as arcs of circles, is centered on the siren. On the left side of **Figure 13-29b** the car moves from left to right. The wave front associated with the largest circle originated first, when the car was somewhere to the left, and the circle of the smallest radius represents the most recently generated wave front. Because the car moves to the right as the sound wave propagates, the siren emits each new wave front closer to the previous one than if the car were stationary. Because the distance between peaks determines the wavelength heard by a stationary listener, someone standing in front of the car detects a shorter wavelength than the siren generates. The opposite is true on the right side of Figure 13-29b, in which the car moves away from the listener.

According to $f = v/\lambda$ (from Equation 13-8), a shorter wavelength results in a higher frequency, so the listener in Figure 13-29b hears a higher frequency than the approaching siren generates. Analogously, because the wave fronts spread farther apart behind the car, the observed wavelength lengthens, resulting in a lower observed frequency as the car moves away from the listener.

Wave crests spread outward from the siren. The most recently emitted crest is closest to the siren.

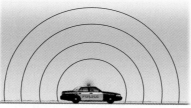

When the siren is not moving, a listener will hear a constant frequency.

(a)

Figure 13-29 A police car siren emits a periodic sound wave that has a fixed frequency and a fixed wavelength. (a) The police car is stationary, so each region of highest pressure along the wave, shown as arcs of circles, is centered on the siren. (b) The car moves from left to right. Because the source of the sound waves moves as the waves propagate, the wave crests get closer together as the car approaches an observer (the ear!), and farther apart as the car recedes.

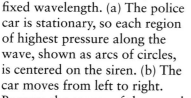

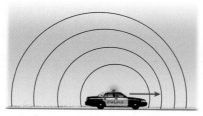

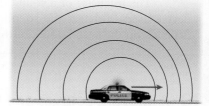

As the siren approaches the listener, the sound frequency will increase.

(b)

Listener

The frequency decreases as the siren moves away from the listener.

> ### ? Got the Concept 13-19
> ### Alarm
> As you run toward your parked car to turn off the blaring alarm system, will the frequency of the tone you hear be higher than, the same as, or lower than the frequency you hear when standing still? Explain your answer.

In **Figure 13-30**, we expanded the view of the moving police car to show the car at two instants in time separated by the period T of the tone created by the siren. At the top of the figure, the peak of the sound wave was created when the car was at the location shown; in the time T the peak has propagated as indicated by the red circle. That distance is given by the product of the speed of sound v_s and T, as shown. The bottom part of the figure shows that the distance the car moves in the time T is $v_{car}T$, where v_{car} is the velocity of the car. A stationary observer in front of the car detects λ_{obs} (the distance between two successive wave fronts) as the wavelength of the siren's tone. As seen in the figure, the three distances—the distance the sound moves, the distance the car moves, and the distance between two successive wave fronts at the place where the observer sits—are related by

$$v_{car}T + \lambda_{obs} = v_s T$$

or

$$\lambda_{obs} = v_s T - v_{car}T \qquad (13\text{-}34)$$

Although Equation 13-34 mathematically describes the Doppler effect, by convention we write the relationship as one between the actual and the observed frequency relative to the speed of the emitter rather than the speed of sound v. To write the expression, we use (from Equation 13-8)

$$\lambda_{obs} = \frac{v_s}{f_{obs}}$$

and also (from Equation 12-6)

$$T = \frac{1}{f}$$

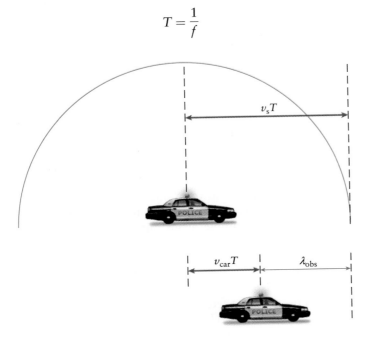

Figure 13-30 The wavelength of the sound heard by a stationary observer as a source of sound approaches depends on the distance the wave travels in a certain time and the distance the source travels in that same time.

Combining the equations with Equation 13-34 gives

$$\frac{v_s}{f_{obs}} = \frac{v_s}{f} - \frac{v_{car}}{f}$$

or

$$f_{obs} = \left(\frac{v_s}{v_s - v_{car}}\right)f$$

where f is the frequency of sound generated by the siren. Because v_{car} is a positive number, the fraction in parentheses must be greater than 1. So when the car is moving toward a stationary observer $f_{obs} > f$. We apply a similar approach to determine the observed frequency as the source moves away from a stationary observer, as well as for the cases in which the observer moves relative to a stationary source. We first write the cases separately, using "src" (source) instead of "car" as a more general way to indicate the source of the sound

$$f_{obs} = \left(\frac{v_s}{v_s - v_{src}}\right)f \quad \text{source approaching a stationary observer}$$

$$f_{obs} = \left(\frac{v_s}{v_s + v_{src}}\right)f \quad \text{source retreating from stationary observer}$$

$$f_{obs} = \left(\frac{v_s + v_{obs}}{v_s}\right)f \quad \text{observer approaching a stationary source}$$

$$f_{obs} = \left(\frac{v_s - v_{obs}}{v_s}\right)f \quad \text{observer retreating from stationary source}$$

We can use one equation to represent the preceding four equations:

$$f_{obs} = \left(\frac{v_s \pm v_{obs}}{v_s \mp v_{src}}\right)f \tag{13-35}$$

Notice that the plus sign is above the minus sign in the numerator but below it in the denominator. In Equation 13-35, we use the upper sign when the source or observer approaches the other and the lower sign when it retreats from the other. Equation 13-35 can be used for more than one case simultaneously. For example, if the source and observer are both approaching each other, the observed frequency is

$$f_{obs} = \left(\frac{v_s + v_{obs}}{v_s - v_{src}}\right)f$$

In this case, the numerator is larger than the speed of sound v_s and the denominator is smaller than v_s; both result in an observed frequency f_{obs} higher than the one generated f.

The speed of a moving object can be determined by measuring the shift in frequency associated with the Doppler effect. A common way to determine the initial frequency, which must be known in order to determine the frequency shift, is to create a wave of known frequency and then bounce it off the moving object to be studied. A radar gun, for example, sends out a wave of known frequency and compares it to the frequency of the wave after it is reflected by a baseball or a car. Bats and dolphins do the same to track the motion of prey and they don't know any physics! Measuring velocity using a Doppler shift has become important in many medical applications, too, for example, to measure blood flow and to generate real-time images of a moving fetus. Because the frequencies of sound used for such imaging, between 2 and 10 MHz, are well above the range of human hearing, the technique is called *ultrasonic imaging*.

Consider the process by which ultrasound is used to determine the speed of blood flow through a heart valve. (An obstructed valve is often characterized by an increase in the speed of the blood flow.) A probe that emits a low intensity sound wave of (high) frequency f_0 is placed over the chest and focused on a region localized around one heart valve. Because the blood is moving, it experiences a wave of frequency f' that has been shifted according to the Doppler effect for a moving observer. Using Equation 13-35, we write

$$f' = \left(\frac{v_s \pm v}{v_s}\right) f_0$$

where v is the speed of the blood, and the upper sign applies to motion toward the probe and the lower sign to motion away from the probe (Equation 13-35). The speed of sound in human tissue is about $v_s = 1540 \text{ m/s}$.

The sound wave is reflected back to the probe. The reflected wave is equivalent to blood emitting the sound, so we apply Equation 13-35 treating the wave as a moving source. The frequency f' of the reflected ultrasound detected by the probe is then

$$f'' = \left(\frac{v_s}{v_s \mp v}\right) f' = \left(\frac{v_s}{v_s \mp v}\right)\left(\frac{v_s \pm v}{v_s}\right) f_0$$

or

$$f'' = \left(\frac{v_s \pm v}{v_s \mp v}\right) f_0 \tag{13-36}$$

Equation 13-36 describes the frequency f'' observed when a wave of frequency f_0 hits an object moving with speed v and is then reflected back to the probe. Again, the upper signs apply to motion toward the probe and the lower signs to motion away from the probe.

Example 13-10 Ultrasound

A certain ultrasound machine can measure a fetal heart rate as low as 50 beats per minute, which corresponds to a speed of the surface of the heart of about $4 \times 10^{-4} \text{ m/s}$. If the transducer in the probe generates ultrasound that has a frequency of 2 MHz, what frequency shift must the machine be able to detect? Remember that the speed of sound in human tissue v_s is about 1540 m/s.

SET UP
The frequency shift Δf is the difference between the frequency of the reflected wave f'' and the one generated by the probe f_0 ($\Delta f = f'' - f_0$). Using Equation 13-36, we can express the reflected frequency f'' in terms of the generated frequency f_0, the speed of sound in tissue v_s, and the speed of the surface of the heart v_h. All three quantities are known. Therefore, by solving Equation 13-36 we can determine the required frequency shift Δf.

SOLVE
Using Equation 13-36 and considering a region of the heart moving toward the probe, the frequency shift is then

$$\Delta f = \left(\frac{v_s + v_h}{v_s - v_h}\right) f_0 - f_0 = \left(\frac{v_s + v_h}{v_s - v_h} - 1\right) f_0$$

The fraction $(v_s - v_h / v_s - v_h)$ as equal to 1, so

$$\Delta f = \left(\frac{v_s + v_h}{v_s - v_h} - \frac{v_s - v_h}{v_s - v_h}\right) f_0$$

or

$$\Delta f = \left(\frac{v_s + v_h - v_s + v_h}{v_s - v_h}\right)f_0 = \left(\frac{2v_h}{v_s - v_h}\right)f_0$$

Although not strictly necessary to finish the problem, note that because in this case v_h is so much smaller than v_s (4×10^{-4} m/s $\ll$ 1540 m/s) we can neglect v_s compared to v_h in the denominator. So, the machine must detect a frequency shift Δf of

$$\Delta f \approx \frac{2v_h}{v_s}f_0 \tag{13-37}$$

To calculate an answer we substitute in known values:

$$\Delta f \approx \frac{2(4 \times 10^{-4}\,\text{m/s})}{1540\,\text{m/s}}(2 \times 10^6\,\text{Hz}) = 1\,\text{Hz}$$

REFLECT

You can see from Equation 13-37 that the higher the frequency generated by the transducer, the larger the frequency shift will be for whatever speed we want to measure. Larger frequency shifts make it easier to detect smaller velocities, so a machine that operates at a higher frequency has a greater sensitivity. However, tissue absorption of the ultrasound increases with frequency, so higher frequencies cannot penetrate as deeply into the body.

Practice Problem 13-10 A certain ultrasound machine can measure a fetal heart rate as low as 50 beats per minute, which corresponds to a speed of the surface of the heart of about 4×10^{-4} m/s. If the transducer in the probe generates ultrasound that has a frequency of 4 MHz, what frequency shift must the machine be able to detect?

Something curious occurs when the speed of the source approaches and then exceeds the speed of sound. In **Figure 13-31a**, we see a jet plane sitting on the

Peaks of a sound wave are expanding spheres, drawn as circles.

(a) The plane is stationary on the tarmac.

Figure 13-31 (a) Concentric circles centered on a stationary jet plane represent the spherical wave fronts of the sound it generates. As the speed of the plane increases after takeoff, the plane catches up to the sound wave fronts. (b) The plane is not moving as fast as the sound waves propagate. (c) When the plane moves at the speed of the sound waves, it just exactly catches up to the wave fronts. (d) When the speed of the plane exceeds the speed of sound, the plane moves farther in any time interval than the corresponding sound wave front.

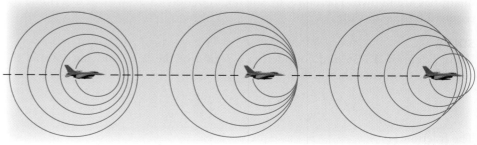

(b) As the plane flies it catches up to the sound peaks it created earlier.

(c) As the plane's speed increases it catches up with more sound peaks.

(d) When the plane exceeds the speed of sound, it overtakes previously created sound peaks.

tarmac before taking off; the concentric circles centered on the plane represent the spherical wave fronts of the sound it generates. The three drawings in **Figures 13-31b, 13-31c,** and **13-31d** show the plane while flying at increasingly higher speeds. As the plane moves, it catches up to the sound it generated earlier in the motion, squeezing the wave fronts closer together ahead of the plane. But notice that if the plane moves at the same speed as the sound, as in Figure 13-31c, the plane and the sound it generates travel the same distance in any time interval. In this case, the peaks of all of the sound waves bunch together. If the speed of the plane exceeds the speed of sound, as in Figure 13-31d, the plane moves farther in any time interval than the corresponding sound.

Figure 13-32a shows an expanded view of the case in which the plane exceeds the speed of sound. Each spherical wave front represents the spherical wave front of a sound generated when the plane was at the center of the spherical wave front. Notice that the fronts of all of the sound waves generated over the time Δt that the plane has moved from its initial position (the lighter-colored image to the left) to its final position (the image on the right) interfere constructively. This constructive interference takes place on the surface of a cone, known as the **Mach cone** after the Austrian physicist Ernst Mach who first explained the phenomenon in 1877. Because Figure 13-32a is two-dimensional, the cone is represented by the two gray lines beginning at the nose of the plane on the right.

The Mach cone, commonly known as a shock wave or a sonic boom, is a region of higher pressure relative to the surrounding air. The sonic boom is relatively narrow in extent, as suggested by **Figure 13-33**, a photograph of an X-15 aircraft in a wind tunnel with air moving at more than twice the speed of sound. In the figure, a number of separate shock waves, each originating

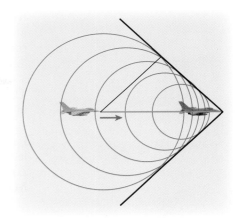

(a)

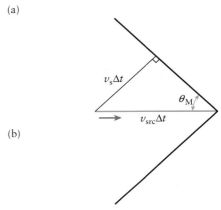

(b)

Figure 13-32 (a) In an expanded view of a plane exceeding the speed of sound, each circle represents the spherical wave front of a sound generated when the plane was at the center of the circle. (b) The fronts of all of the sound waves generated over the time that the plane has moved from its initial position to its final position interfere constructively on the surface of the Mach cone.

Figure 13-33 As air moves past an X-15 aircraft at more than twice the speed of sound in a supersonic wind tunnel, sonic shock waves originate from various discontinuities on its surface. The shock waves are narrow in extent. *(NASA/Science Photo Library.)*

from various discontinuities on the plane's surface, are made evident by a photographic technique that reveals regions of differing air density. Notice that each of the regions is only a fraction of the length of the plane. For this reason, if an X-15 aircraft were to fly overhead at a speed greater than that of sound, you would hear a series of sharp, explosive sounds, one from each shock wave as it passed by you. The sound of a sonic boom is not a long, low rumbling as is sometimes suggested in movies.

The distance the plane travels in Δt, the blue line in Figure 13-32a and Figure 13-32b, forms the hypotenuse of a right triangle and is labeled $v_{src}\Delta t$. The leg of the triangle labeled $v_s\Delta t$ is the distance the sound of the plane travels in time Δt. The other leg of the triangle lies along the region of constructive interference. **Mach angle θ_M** is labeled in Figure 13-32b and given by

$$\sin \theta_M = \frac{v_s}{v_{src}} \tag{13-38}$$

where v_s is the speed of sound in air and v_{src} is the speed of the plane. At Mach 1, when the speed of an object such as a jet plane equals the speed of sound, $\theta_M = 90°$ (Figure 13-32b). As an object's speed increases above v_s, θ_M decreases.

The right side of Equation 13-38 is the inverse of the **Mach number:**

$$\text{Mach number} = \frac{v_{src}}{v_s}$$

The Mach number describes the speed of an object relative to the speed of sound. For example, a jet flying at Mach 2 moves at twice the speed of sound in the air.

! Watch Out

The Mach cone and the object that creates it move at the same speed.

You certainly hear a loud noise when a sonic boom passes. However, the constructive interference which generates the phenomenon is not a sound wave, and does not propagate at the speed of sound. Rather, the Mach cone moves along with the object that creates it. So, for example, the sonic boom trailing a jet plane moving at Mach 2 also moves at twice the speed of sound.

Example 13-11 Sonic Boom

A supersonic jet flies over an air show at an altitude of 1200 m. You hear the explosive thunder clap of the sonic boom 1.5 s after the jet was directly overhead. What is the speed of the jet in terms of the speed of sound; that is, what is the Mach number of the jet?

SET UP

A good picture is particularly important to solve this problem. Figure 13-34 shows the plane at the moment that an observer on the ground hears the sonic boom. Notice that because the Mach cone moves with the plane, the time it takes for the sonic boom to be heard depends on the speed of the plane. We labeled the time Δt, so that the distance the plane travels between the time it was directly overhead and the time the sonic boom is heard equals $v_{jet}\Delta t$, where v_{jet} is the velocity of the jet. We can use the properties of the right triangle formed by the distances $v_{jet}\Delta t$ and H, together with Equation 13-38 that defines the Mach angle θ_M to determine the speed of the jet.

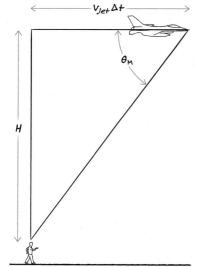

Figure 13-34 When a plane flies overhead at greater than the speed of sound, a line connecting an observer on the ground and the location of the plane at the instant the observer hears the sonic boom forms the hypotenuse of a triangle.

SOLVE
Using the triangle in Figure 13-34,

$$\tan \theta_M = \frac{H}{v_{jet} \Delta t} \tag{13-39}$$

See the Math Tutorial for more information on Trigonometry

We can solve the equation for v_{jet}:

$$v_{jet} = \frac{H}{(\tan \theta_M) \Delta t}$$

However, while H and Δt are known, θ_M is not. To solve for θ_M, we therefore need a second equation that includes v_{jet}, which is Equation 13-38:

$$\sin \theta_M = \frac{v_s}{v_{jet}} \tag{13-38}$$

By solving this equation for v_{jet}, we now have two different expressions for the unknown variable, θ_M. Setting them equal eliminates v_{jet} so that θ_M can be determined. And once we have θ_M we can find v_{jet}. So,

$$v_{jet} = \frac{v_s}{\sin \theta_M}$$

and therefore

$$\frac{H}{(\tan \theta_M) \Delta t} = \frac{v_s}{\sin \theta_M}$$

Because $\tan \theta = \sin \theta / \cos \theta$, the equation becomes

$$\frac{(\cos \theta_M) H}{(\sin \theta_M) \Delta t} = \frac{v_s}{\sin \theta_M}$$

or

$$\cos \theta_M = \frac{v_s \Delta t}{H}$$

All three variables on the right side are known, so we can find θ_M which allows us to determine v_{jet}:

$$\theta_M = \cos^{-1}\left(\frac{v_s \Delta t}{H}\right) = \cos^{-1}\left(\frac{343 \text{ m/s } 1.5 \text{ s}}{1200 \text{ m}}\right) = 64.6°$$

Then,

$$v_{jet} = \frac{v_s}{\sin \theta_M} = \frac{v_s}{\sin 64.6°} = 1.1 v_s$$

The jet is flying at Mach 1.1

REFLECT
The speed of the jet is Mach 1.1, just above the speed of sound. What about the Mach angle? You might expect that the Mach angle would be closer to 90° than the approximately 65° we found in this problem. But notice that the sine function is relatively close to 1 for angles of 60° or more. For example, $\sin 60° = 0.866$, so the inverse, $1/\sin 60°$, is 1.15. The Mach cone trailing a jet moving at 1.15 times the speed of sound would have a Mach angle 60°, so for the jet in this example, traveling at Mach 1.1, 65° is reasonable.

Practice Problem 13-11 A supersonic jet flies over an air show at Mach 1.5 and an altitude of 1500 m. How long after the jet was directly overhead do you hear the explosive thunder clap of the sonic boom?

> ## ✳ What's Important 13-9
>
> The frequency of a wave produced by a moving source increases as the source approaches the listener and decreases after it passes, a phenomenon called the Doppler effect. The speed of a moving object can be determined by measuring the shift in frequency associated with the Doppler effect. The Mach number describes the speed of an object relative to the speed of sound.

Answers to Practice Problems

13-2	1.1×10^3 m/s
13-3	5.1×10^3 m/s
13-4	constructive
13-5	0.55 m, 520 Hz, 0.66 m
13-6	660 Hz

13-8	1.1×10^{-5} W (or J/s), 0.20 W/m^2
13-9	1/4
13-10	2 Hz
13-11	3.3 s

Answers to Got the Concept Questions

13-1 The series of wiggles on the taut rope is a transverse wave. The sound wave that travels inside the paper cup is a longitudinal wave, while the vibration that travels along the string stretched between two paper cups is a transverse wave. The ripples created on the surface of the pond represent a surface wave.

13-2 See **Figure 13-35**. You can extend the arrow either to the left or to the right. Both solutions are shown.

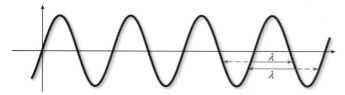

Figure 13-35 You can extend the red arrow either to the left or to the right to correctly identify one full wavelength.

13-3 The period T and the frequency f are inversely related, so the smaller the frequency, the longer the period. A sound of frequency 20 Hz has a much longer period than one of frequency 20,000 Hz.

13-4 (a) From Equation 13-5, when the wave number increases, the angular velocity must also increase in order for the propagation speed to remain the same. (b) From Equation 13-2, as the wavelength increases the angular wave number decreases. A decrease in angular wave number must result in a decrease in the angular velocity in order for the propagation speed to remain constant, according to Equation 13-5. Finally, from Equation 12-5, as angular velocity decreases, the period increases, because period and angular velocity are inversely related. Thus, as the wavelength of the sound wave is increased, the period also increases.

13-5 The speed of light is much faster than the speed of sound, so the light that brings the image of the batter swinging arrives at your eyes in an imperceptibly short time. The sound, however, travels at 343 m/s. Let's say you're sitting 130 m from home plate. At 343 m/s, the time it takes the sound of the crack of the bat to reach your ears is $t = 130$ m/(343 m/s) = 0.38 s. Although relatively short, the time is certainly long enough for you to notice the delay relative to when you see the batter swing.

13-6 For motion at a constant speed, distance equals the product of speed and time. Because the speed of light is so large, the time it takes to see the lightning is imperceptibly small. We can then say that the time for the sound of the thunder to travel to your ears is the time between seeing the lightning and hearing the sound. In 3 s, sound travels 343 m/s × 3 s = 1029 m, or about 1 km. In the English system of units sound travels 1125 ft/s × 5 s = 5625 ft in 5 s, which is just a shade under a mile.

13-7 The sound waves generated by the speaker can arrive at your ears directly or after bouncing off the walls, floor, or ceiling. You can imagine waves moving in a multitude of directions, setting up a complex interference pattern. The distance between locations of constructive and destructive interference is on the same order of magnitude as the wavelength of the sound, so if the wavelength is about 0.5 m, you might encounter one or more of the locations as you move your head. At any location where destructive interference occurs, the sound will be relatively soft. At locations where the waves interfere constructively, the sound will be relatively loud.

13-8 In order for the elastic band to vibrate in the lowest resonant mode, one-half of a wavelength must fit into the length of the elastic band. The wavelength of the fundamental is therefore twice the length of the elastic band, or 30 cm.

13-9 The vibration of the elastic band causes the surrounding air to vibrate at the same frequency. The wavelength of any wave is related to the frequency and the propagation speed according to

$$v_p = f\lambda \qquad (13\text{-}8)$$

So the wavelength of the sound is given by

$$\lambda = \frac{v_p}{f}$$

where in this case v_p is the speed of sound in air. So

$$\lambda = \frac{343 \text{ m/s}}{110 \text{ Hz}} = 3.1 \text{ m}$$

Notice that the wavelength is unrelated to the speed of a wave along the elastic band. Although the frequency of the disturbance of the surrounding air (the sound) is the same as the frequency of the vibration on the elastic band, the wavelength of the sound depends on the speed of sound *in air*, not the speed of the wave on the elastic band. The standing wave is created on the elastic band, but the sound it generates must travel through the surrounding air to reach your ears.

13-10 The fundamental frequency of a stretched string depends on the speed with which waves propagate along it. The lower the propagation speed, the lower the fundamental frequency, according to

$$f_1 = \frac{v_p}{2L} \qquad (13\text{-}21)$$

The propagation speed v_p depends on linear mass density μ:

$$v_p = \sqrt{\frac{T}{\mu}} \qquad (13\text{-}9)$$

A heavier, thicker string has a larger linear mass density, so v_p is lower than for a lighter, thinner one. The fundamental frequency, as well as the corresponding overtones, is therefore lower for a heavier, thicker string than for a lighter, thinner one.

13-11 Increasing the tension in the string increases the propagation speed (Equation 13-9). In addition, by using Equation 13-21, the higher the propagation speed is, the higher the fundamental is:

$$f_1 = \frac{v_p}{2L}$$

Using Equation 13-9, doubling the tension increases the propagation speed by $\sqrt{2}$:

$$v_p' = \sqrt{\frac{2T}{\mu}} = \sqrt{2}\sqrt{\frac{T}{\mu}} = \sqrt{2}\,v_p$$

So, the fundamental frequency also increases by a factor of $\sqrt{2}$.

13-12 The frequency increases. The vibration of the ruler behaves like the lowest frequency standing wave in a closed tube, with a node at one end (where the ruler is pressed against the edge of the desk) and an antinode at the other. As she decreases the length of the vibrating part of the ruler, the wavelength also decreases. Because wavelength and frequency are inversely related (Equation 13-8), the decrease in length causes the frequency to increase. Try it!

13-13 The bottle is effectively a tube that is open on one end and closed on the other. The frequency of sound that generates resonance in the air in the bottle is inversely proportional to the length of the air column in it. (Using Equation 13-24, the frequency of the fundamental is $f = v_p/4L$.) As the bottle fills with water L gets smaller, so f increases. A higher frequency is a higher pitch.

13-14 A longer horn has a lower fundamental frequency. The wavelength of vibrations that excite standing waves in the horn depends on the length of the horn; a long horn requires a longer wavelength of sound to create standing waves. Wavelength and frequency are inversely related (Equation 13-8), so the longer the wavelength (a bigger number), the lower the corresponding frequency (a smaller number).

13-15 The beat frequency is the difference in the frequencies of the two waves which interfere. When the two tuning forks are struck simultaneously, the beat frequency is $f_{\text{beats}} = |440 \text{ Hz} - 438 \text{ Hz}| = 2 \text{ Hz}$. Two beats are heard each second.

13-16 This question has two possible answers! Because the beat frequency is defined as the *absolute magnitude* of the difference between the frequency of the two waves (Equation 13-28), the frequency of the second tuning fork can be either higher or lower than the frequency of the first one. To create beats of frequency 5 Hz, the frequency of the second tuning fork must be either 435 Hz or 445 Hz. Without more information, however, it is not possible to know which of the two is correct.

13-17 Intensity decreases with distance from the speaker as the power generated spreads out over the surface area of an expanding sphere with the speaker at its center. We agreed to neglect the size of the speaker, so at a distance r the surface area of the sphere equals $4\pi r^2$. As the radius doubles, therefore, the surface goes up by a factor of 2^2, and the intensity drops by the same factor. That is, when you double your distance, the intensity of the sound goes down by a factor of 4, the square of the radius.

13-18 The sound waves from each fan carry the same energy and power, so at any distance from the fans the intensity is the same. When both fans are running, the total intensity doubles. A doubling of intensity corresponds to a change of 3 dB in sound level. The sound level due to both fans is 60 dB + 3 dB = 63 dB.

13-19 The frequency you hear is higher while you're running compared to when you're standing still. Because you are moving toward the source of the sound, you will intercept more wave peaks in a given amount of time than you would were you standing still. The qualitative approach we took using Figure 13-29 for a moving source of sound and a stationary observer applies equally well for a stationary source and a moving observer.

SUMMARY

Topic	Summary	Equation or Symbol
amplitude	The amplitude of a wave is the maximum displacement from equilibrium that any element of the wave experiences as the disturbance propagates through a medium.	A or s_{max}
angular wave number	The angular wave number (also called wave number) describes the number of cycles of a wave that fit into 2π radians, or one full cycle of a sine function.	k
antinode	Positions along a standing wave at which the disturbance of the medium is maximal are antinodes.	
beats	Beats arise from the interference between sound waves. Beats are most pronounced when two waves of nearly identical frequencies and wavelengths interfere. The beat frequency f_{beats} depends on the difference between the frequencies of the two waves f_1 and f_2.	$f_{beats} = \lvert f_1 - f_2 \rvert \qquad (13\text{-}28)$
constructive interference	When two waves overlap so that their peaks and troughs align, they reinforce each other. The result is constructive interference.	
destructive interference	When two overlapping waves each displace the medium the same amount at each point along the waves, but in opposite directions, the disturbance caused by one wave cancels the disturbance caused by the other. As a result, no net disturbance of the medium occurs, a phenomenon known as destructive interference.	
Doppler effect	When a source of sound moves relative to an observer, or an observer moves relative to a source of sound, or both, the observed wavelength and frequency are shifted according to the Doppler effect. For example, the observed frequency f_{obs} is higher than the emitted frequency f when an object creating sound moves toward a stationary observer, and lower when it moves away. The Doppler shifted frequency depends on the speed of the observer v_{obs}, the speed of the source of sound v_{src}, and the speed of sound v_s.	$f_{obs} = \left(\dfrac{v_s \pm v_{obs}}{v_s \mp v_{src}} \right) f \quad (13\text{-}35)$
frequency	The frequency of a wave is defined as the number of full cycles the wave makes per second. The frequency f of a periodic wave is the inverse of its period T.	$f = \dfrac{1}{T} \qquad (12\text{-}6)$

fundamental frequency	The fundamental frequency of a medium, for example, a stretched string, is the lowest frequency that results in a standing wave in or on the medium. For a stretched string, the fundamental frequency corresponds to a wavelength equal to one-half the length of the string, that is, a standing wave with a node at each end and an antinode in the middle.	
intensity	Intensity I is defined as power P delivered per unit area A.	$I = \dfrac{P}{A}$ (13-29)
interference	When two or more waves overlap, the net disturbance of the medium is the combination of each wave on the medium separately. This phenomenon is known as interference.	
interference pattern	The combination of two or more waves across a region of overlap can result in a complex but regular interference pattern, in which the waves reinforce each other at some places and partially or fully cancel each other at others.	
linear mass density	Linear mass density is defined as the mass of a string or other linear medium per unit length.	μ
longitudinal wave	A longitudinal wave causes elements of a medium to move back and forth along the direction of wave propagation. Sound in air propagates as a longitudinal wave.	
Mach angle	The sonic shock wave generated by an object traveling faster than the speed of sound extends behind the object as a cone-shaped surface. The angle between the line of motion and the surface of the cone, the Mach angle θ_M, depends on the speed of the object v_{src} and the speed of sound v_s.	$\sin \theta_M = \dfrac{v_s}{v_{src}}$ (13-37)
Mach cone	The Mach cone is a region of high pressure, in the shape of the surface of a cone, that results from the constructive interference of a series of sound waves generated by an object traveling faster than the speed of sound.	
Mach number	The Mach number describes the speed of an object v_{src} relative to the speed of sound v_s.	Mach number $= \dfrac{v_{src}}{v_s}$
mechanical wave	A mechanical wave is a disturbance that moves through a medium, displacing particles or elements of the medium as it propagates.	
mode	A system, for example, a stretched string or a column of air, can vibrate in a number of vibrational patterns. Each is a mode, identified by an integer counting number n.	
modes on a stretched string	The wavelengths λ and frequencies f of the vibrational modes on a stretched string depend on the length L of the string and the wave propagation speed v_p. The standing wave wavelengths are integer multiples of half waves.	$L = \dfrac{n}{2}\lambda, \quad n = 1, 2, 3 \ldots$ (13-19) $f_n = n\dfrac{v_p}{2L}, \quad n = 1, 2, 3 \ldots$ (13-20)
modes in a tube open on one end and closed on the other	The wavelengths and frequencies of the vibrational modes in a tube of fluid, such as air, that is open on one end and closed on the other depend on the length L of the tube and the speed of sound in that fluid. The standing wave wavelengths are odd multiples of quarter waves.	$L = \dfrac{n}{4}\lambda, \quad n = 1, 3, 5 \ldots$ (13-23) $f = n\dfrac{v_p}{4L}, \quad n = 1, 3, 5 \ldots$ (13-24)

modes in a tube open on both ends	The wavelengths of the vibrational modes in a tube of fluid, such as air, that is open on both ends depend on the length L of the tube and the speed of sound in that fluid. The standing wave wavelengths are integer multiples of half waves.	$L = \dfrac{n}{2}\lambda, \quad n = 1, 2, 3 \dots$ (13-25) $f = n\dfrac{v_{\mathrm{p}}}{2L}, \quad n = 1, 2, 3 \dots$ (13-26)
node	Nodes are positions along a standing wave where the displacement is zero at all times.	
period	The period of a wave is the time it takes for the disturbance to go through one full cycle as it moves through the medium.	T
phase difference	Phase difference quantifies the relationship between the start of a cycle of one wave relative to the start of a cycle of another wave. The phase difference between two waves determines how they interfere with each other.	φ
power	Power P is the energy E delivered per unit time t.	$P = \dfrac{E}{t}$ (13-29)
pressure wave	A longitudinal wave such as sound in a tube consists of regions of higher pressure (compressions) and lower pressure (rarefactions). Sound and other longitudinal waves are also called pressure waves.	
resonant frequency	In general, resonance is the phenomenon in which a system oscillates with larger amplitude at certain specific frequencies; in a medium of fixed extent, such as a string stretched between two posts or a column of air in a tube, only certain resonant frequencies will result in standing waves.	
sound level	Sound level β is used to characterize the intensity I of a sound wave. The SI units of sound level are decibels, or dB.	$\beta = 10 \log\left(\dfrac{I}{10^{-12}\ \mathrm{W/m^2}}\right)$ (13-33)
standing wave	A standing wave forms in or on a medium of fixed extent when traveling waves moving in different directions interfere. Only specific resonant frequencies of traveling waves result in a standing wave. As the name implies, standing waves do not move.	
superposition	At every point where more than one wave passes simultaneously, the net disturbance of the medium equals the sum of the displacements that each wave would have caused individually. In this way, waves obey the principle of superposition.	
surface wave	As a surface wave passes, particles in a medium move in circular or elliptical paths; in other words, their motion is both transverse and longitudinal to the direction of wave propagation.	
transverse speed	When a transverse wave propagates along a medium, the elements of the medium are disturbed in the direction transverse to (perpendicular to) the propagation direction. The speed at which elements of the medium move in this direction is the transverse speed. The transverse speed of any particular element of the medium varies in time, that is, it is not a constant of the motion.	

transverse wave	A transverse wave causes elements of a medium to be disturbed in the direction perpendicular to the direction of wave propagation. A wave that propagates along a stretched string is a transverse wave.	
traveling wave	Traveling wave is a general name for transverse, longitudinal, and surface waves. In contrast to standing waves, traveling waves move (propagate) through a medium.	
wavelength	Wavelength is the distance a disturbance in a medium travels over one full cycle of a wave.	λ

QUESTIONS AND PROBLEMS

In a few problems, you are given more data than you actually need; in a few other problems, you are required to supply data from your general knowledge, outside sources, or informed estimate.

Interpret as significant all digits in numerical values that have trailing zeros and no decimal points.

For all problems, use $g = 9.8 \text{ m/s}^2$ for the free-fall acceleration due to gravity. Neglect friction and air resistance unless instructed to do otherwise.

• Basic, single-concept problem
•• Intermediate-level problem, may require synthesis of concepts and multiple steps
••• Challenging problem
SSM *Solution is in Student Solutions Manual*

Conceptual Questions

1. •Explain the difference between longitudinal waves and transverse waves and give two examples of each.

2. •Discuss several ways that the human body creates or responds to waves.

3. •When you talk to your friend, are the air molecules that reach his ear the same ones that were in your lungs? Explain your answer. SSM

4. •Draw a sketch of a transverse wave and label the amplitude, wavelength, a crest, and a trough.

5. •If a tree falls in the forest and no humans are there to hear it, was any sound produced?

6. •Are water waves longitudinal or transverse? Explain your answer.

7. •A sound wave passes from air into water. Give a qualitative explanation of how these properties change: (a) wave speed, (b) frequency, and (c) wavelength of the wave. SSM

8. •Two solid rods have the same Young's modulus, but one has larger density than the other. (a) In which rod will the speed of longitudinal waves be greater? (b) Explain your answer making reference to the variables that affect the speed.

9. •Earthquakes produce several types of wave. The most significant are the primary wave (or P wave) and the secondary wave (or S wave). The primary wave is a longitudinal wave that can travel through liquids and solids. The secondary wave is a transverse wave that can only travel through solids. The speed of a P wave usually falls between 1000 m/s and 8000 m/s; the speed of an S wave is around 60–70% of the P wave speeds. For any given seismic event, discuss the damage that might be due to P waves and how that would differ when compared to the damage due to S waves.

10. •A mathematical representation of a wave is $y(x, t) = A \sin k(x - v_p t)$. (a) Explain the meaning of each term and give the units for each variable. (b) Explain the derivation of the angular velocity ω from the basic formula.

11. •Explain the differences and similarities between the concepts of frequency f and angular frequency ω. SSM

12. •A common way to estimate your distance from a lightning strike involves counting off the number of seconds between seeing the flash and hearing the thunder; the strike is 1 km away for every 3 s. One nautical league is equal to 5556 m. How would you estimate the distance from a lightning strike in nautical leagues?

13. •(a) What is a transverse standing wave? (b) For a string stretched between two fixed points, describe how a disturbance on the string might lead to a standing wave.

14. •Search the Internet for the words *rarefaction*, *phonon*, and *compression* in the context of longitudinal waves. Describe how a longitudinal wave is made up of phonons and explain the connection between rarefaction and compression in such a wave.

15. •If you stand beside a railroad track as a train sounding its whistle moves past, you will experience the Doppler effect. Describe any changes in the perceived sound that a person riding on the train will hear.

16. •(a) Describe in words the nature of a sonic boom. (b) Now, referring to the formula for the Doppler shift, explain the phenomenon.

17. •**Calc** Explain the difference between wave speed and the time derivative of the displacement from equilibrium of a transverse wave (dy/dt). SSM

18. •**Calc** The wave equation describes how a disturbance from equilibrium propagates spatially and temporally through a medium. Show that the dimensions of the basic wave equation are consistent:

$$\frac{\partial^2 y}{\partial x^2} = \frac{1}{v^2}\frac{\partial^2 y}{\partial t^2}$$

19. •Explain the concept of phase difference, φ, and predict the outcome of two identical waves interfering when $\varphi = 0°$, $\varphi = 90°$, $\varphi = 180°$, $\varphi = 270°$, and $\varphi = 300°$.

20. •Two waves interfere. (a) How does the concept of path difference affect the outcome? (b) Explain why a path difference of $n\lambda$ does *not* always lead to constructive interference for two wave sources ($n = 1, 2, 3, ...$).

21. •How does the length of an organ pipe determine the fundamental frequency?

22. •Why do the sounds emitted by organ pipes that are closed on one end and open on the other *not* have even harmonics? Include in your explanation a sketch of the resonant waves that are formed in the pipe.

23. •(a) Explain the differences and similarities among the concepts of intensity, sound level, loudness, and power. (b) What happens to intensity as the source of sound moves closer to the observer? (c) What happens to sound level? (d) What happens to power? SSM

24. •Explain how the phenomenon of beats can be used to tune a guitar.

25. •A car radio is tuned to receive a signal from a particular radio station. While the car slows to a stop at a traffic signal, the reception of the radio seems to fade in and out. Use the concept of interference to explain the phenomenon. *Hint:* In broadcast technology, the phenomenon is known as multipathing.

26. •Two pianists sit down to play two identical pianos. However, a string is out of tune on Elaine's piano. The $G_3^\#$ key (208 Hz) appears to be the problem. When George plays the note on his piano and Elaine plays hers, a beat frequency of 6 Hz is heard. Luckily, a piano tuner is present and she is ready to correct the problem. However, in all the confusion, she inadvertently increases the tension in George's $G_3^\#$ string by a factor of 1.058. Now, both pianos are out of tune! Yet oddly, when Elaine plays her $G_3^\#$ note and George plays his $G_3^\#$, there is no beat frequency. Explain what happened.

Multiple-Choice Questions

27. •A visible disturbance propagates around a crowded soccer stadium when fans, section by section, jump up and then sit back down. What type of wave is this?
A. longitudinal wave
B. transverse wave
C. polarized wave
D. spherical wave
E. polarized spherical wave

28. •Two point sources produce waves of the same wavelength that are in phase. At a point midway between the sources, you would expect to observe
A. constructive interference.
B. destructive interference.
C. alternating constructive and destructive interference.
D. constructive or destructive interference depending on the wavelength.
E. no interference.

29. •Standing waves are set up on a string that is fixed at both ends so that the ends are nodes. How many nodes are there in the fourth mode? SSM
A. 2
B. 3
C. 4
D. 5
E. 6

30. •Which of the following frequencies are higher harmonics of a string with fundamental frequency of 80 Hz?
A. 80 Hz
B. 120 Hz
C. 160 Hz
D. 200 Hz
E. 220 Hz

31. •A trombone has a variable length. When a musician blows air into the mouthpiece and causes air in the tube of the horn to vibrate, the waves set up by the vibrations reflect back and forth in the horn to create standing waves. As the length of horn is made shorter, what happens to the frequency?
A. The frequency remains the same.
B. The frequency will increase.
C. The frequency will decrease.
D. The frequency will increase or decrease depending on how hard the horn player blows.
E. The frequency will increase or decrease depending on the diameter of the horn.

This page is intentionally left blank.

For complete end of chapter problem sets, please go to
www.whfreeman.com/kestentauck

23,000 Hz. If the speed of a P wave from an earthquake is 4000 m/s, calculate the minimum wavelength of the seismic waves if they are heard by a dog but not by a human.

13-2: Mathematical Description of a Wave

47. •A wave on a string propagates at 22 m/s. If the frequency is 24 Hz, calculate the wavelength and angular wave number. SSM

48. •The period of a sound wave is 0.01 s. Calculate the frequency f and the angular frequency ω.

49. •Show that the dimensions of speed (m/s) are consistent with both versions of the expression for propagation speed of a wave:

$$v = \frac{\omega}{k} \quad \text{and} \quad v = \lambda f$$

50. ••A transverse wave on a string has an amplitude of 20 cm, a wavelength of 35 cm, and a frequency of 2 Hz. Write the mathematical description of the displacement from equilibrium for the wave if (a) at $t = 0$, $x = 0$ and $y = 0$; (b) at $t = 0$, $x = 0$ and $y = +20$ cm; (c) at $t = 0$, $x = 0$ and $y = -20$ cm; and (d) at $t = 0$, $x = 0$ and $y = 12$ cm.

51. •A wave on a string is described by the equation $y = 0.05 \sin(x - 10t)$ (SI units). What are (a) the frequency, (b) wavelength, and (c) speed of the wave? SSM

52. ••Write the wave equation for a periodic transverse wave traveling in the x direction at a speed of 20 m/s if it has a frequency of 10 Hz and an amplitude in the y direction of 0.10 m.

53. ••The equation for a particular wave is $y(x,t) = 0.10 \sin(kx - \omega t)$ (SI units). If the frequency of the wave is 2.0 Hz, what is the value of y at $x = 0$ when $t = 4.0$ s?

54. ••The pressure wave that travels along the inside of an organ pipe is given in terms of distance and time by the following function (SI units):

$$p(x,t) = (1 \text{ atm}) - (1 \text{ atm})\cos(6x - 4t)$$

(a) What is the pressure amplitude of the wave? (b) What is the wave number of the wave? (c) What is the frequency of the wave? (d) What is the speed of the wave? (e) What is the corresponding spatial displacement, $s(x,t)$, from equilibrium for the pressure wave, assuming the maximum displacement is 2.0 cm?

55. •Write a mathematical description of the wave using the graphs in **Figure 13-36**. SSM

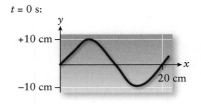

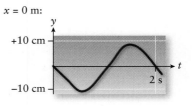

Figure 13-36 Problem 55

56. •Using the graph in **Figure 13-37**, write the mathematical description of the wave if the period of the motion is 4 s and the wave moves to the right (toward the positive x direction).

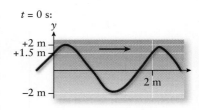

Figure 13-37 Problem 56

13-3: Wave Speed

57. •A string that has a mass of 5.0 g and a length of 2.2 m is pulled taut with a tension of 78 N. Calculate the speed of transverse waves on the string. SSM

58. •A long rope is shaken up and down by a rodeo contestant. The transverse waves travel 12.8 m in a time of 2.1 s. If the tension in the rope is 80 N, calculate the mass per unit length for the rope.

59. ••The violin is a four-stringed instrument tuned so that the ratio of the frequencies of adjacent strings is 3 to 2. (This is the ratio when taken as high frequency to lower frequency.) If the diameter of the E string (the highest frequency) on a violin is 0.25 mm, find the diameters of the remaining strings (A, D, and G), assuming they are tuned to intervals of a perfect fifth, they are made of the same material, and they all have the same tension.

60. •At room temperature the bulk modulus of glycerine is about 4.35×10^9 N/m^2 and the density of glycerine is about 1260 kg/m^3. Calculate the speed of sound in glycerin.

61. •The Young's modulus of water is 2.2×10^9 N/m^2. The density of water is 1000 kg/m^3. Calculate the speed of sound in water. SSM

62. •When sound travels through the ocean, where the Young's modulus is 2.34×10^9 N/m^2, the wavelength associated with 1000 Hz waves is 1.51 m. Calculate the density of seawater.

63. •What is the speed of sound in gasoline? The Young's modulus for gasoline is $1.3 \times 10^9 \, \text{N/m}^2$. The density of gasoline is 0.74 kg/L.

64. •If the Young's modulus for liquid A is twice that of liquid B, and the density of liquid A is one-half of the density of liquid B, what does the ratio of the speeds of sound in the two liquids (v_A/v_B) equal?

65. •The speed of waves in a solid depends on whether it is a longitudinal wave or a transverse wave

$$v_{\text{longitudinal}} = \sqrt{\frac{Y}{\rho}} \qquad v_{\text{transverse}} = \sqrt{\frac{G}{\rho}}$$

where Y is the Young's modulus and G is the shear modulus. Calculate the speed of longitudinal waves in steel versus transverse waves in steel (its density is $7800 \, \text{kg/m}^3$). The Young's modulus for steel is $210 \times 10^9 \, \text{N/m}^2$ and the shear modulus for steel is $84 \times 10^9 \, \text{N/m}^2$.

66. ••Starting with the formula for the speed of sound in air ($v = \sqrt{\frac{B}{\rho}}$), determine how the speed varies with temperature. You will need to use the relationship between bulk modulus and pressure ($B = \gamma P$; γ is the adiabatic constant and is 1.4 for air), the ideal gas law [$PV = nRT$; R equals $8.314 \, \text{J/(mol·K)}$], and fundamental knowledge about air (molar mass of air is 0.02895 kg/mol) to determine the relationship. Don't forget that $T_K = T_C + 273$.

13-4: Superposition and Interference

67. •In Figure 13-38, the square waveforms are approaching each other. Use the ideas of interference to predict the superposed wave that results when the two waves are coincident. SSM

68. •Two waves interfere at the point "X" in Figure 13-39. The resultant wave is shown. Draw the three possible shapes for the two waves that interfere to produce this outcome. In each case one wave should head toward the right and one wave should head toward the left.

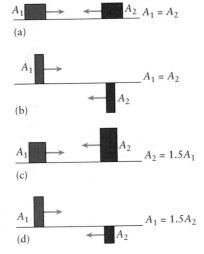

Figure 13-38 Problem 67

Figure 13-39 Problem 68

69. •Construct the resultant wave that is formed when the two waves shown in each case occupy the same space and interfere (**Figure 13-40**).

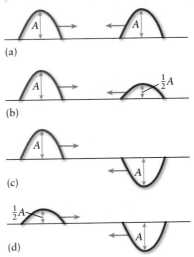

Figure 13-40 Problem 69

70. •Using the concept of interference, draw in the resultant wave for the two waves that coincide together at point X (**Figure 13-41**).

Figure 13-41 Problem 70

71. ••Two identical speakers (1 and 2) are playing a tone with a frequency of 171.5 Hz, in phase (**Figure 13-42**). The speakers are located 6 m apart. Determine what points (A, B, C, D, or E, all separated by 1 m) will experience constructive interference along the line that is 6 m in front of the speakers. Point A is directly in front of speaker 1. The speed of sound is 343 m/s for this problem. SSM

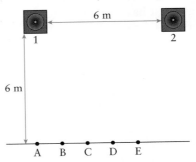

Figure 13-42 Problem 71

72. •Two identical waves are moving at 3 m/s with wavelength 2 m (one to the left, one to the right). The amplitude of each wave is 15 cm, but the wave traveling to the left is out of phase by $+\pi/6$. (a) Find the resultant wave that is formed when the two waves interfere. (b) What is the resultant displacement at $x = 2$ m, when $t = 3$ s?

13-5: Transverse Standing Waves

73. •A string is fixed on both ends with a standing wave vibrating in the fourth harmonic. Draw the shape of the wave and label the location of antinodes (A) and nodes (N).

74. •A string is tied at both ends and a standing wave is established. The length of the string is 2 m and it vibrates in the fundamental mode ($n = 1$). If the speed of waves on the string is 60 m/s, calculate the frequency and wavelength of the waves.

75. •A string that is 1.25 m long has a mass of 0.0548 kg and a tension of 200 N. The string is tied at both ends and vibrated at various frequencies. (a) What frequencies would you need to apply for the first four harmonics? (b) Make a sketch of the first four harmonics for these standing waves. SSM

76. •A 2.35-m-long string is tied at both ends and it vibrates with a fundamental frequency of 24 Hz. Find the frequencies and make a sketch of the next four harmonics. What is the speed of waves on the string?

77. ••A string of length L is tied at both ends and a harmonic mode is created with a frequency of 40 Hz. If the next successive harmonic is at 48 Hz and the speed of transverse waves on the string is 56 m/s, find the length of the string. Note that the fundamental frequency is not necessarily 40 Hz.

78. ••A string vibrates at a frequency of 170 Hz in the nth harmonic. The next successive frequency is 204 Hz (for $n + 1$). Determine whether the string is tied at one end or at both ends. If the speed of transverse waves on the string is 218 m/s, find the length of the string.

79. ••An object of mass M is used to provide tension in a 4.5-m-long string that has a mass of 0.252 kg, as shown in **Figure 13-43**. A standing wave that has a wavelength equal to 1.5 m is produced by a source that vibrates at 30 Hz. Calculate the value of M.

Figure 13-43 Problem 79

80. ••A guitar string has a mass per unit length of 2.35 g/m. If the string is vibrating between points that are 60 cm apart, find the tension when the string is designed to play a note of 440 Hz (A_4).

81. ••An E string on a violin has a diameter of 0.25 mm and is made from steel (density is 7800 kg/m^3). The string is designed to sound a fundamental note of 660 Hz and its unstretched length is 32.5 cm. Calculate the tension in the string. SSM

13-6: Longitudinal Standing Waves

82. •An organ pipe of length L sounds its fundamental tone at 40 Hz. The pipe is open on both ends and the speed of sound in air is 343 m/s. Calculate (a) the length of the pipe and (b) the frequency of the first three overtones.

83. •An organ pipe sounds two successive tones at 228.6 Hz and 274.3 Hz. Determine whether the pipe is open at both ends or open at one end and closed at the other. SSM

84. •A narrow glass tube is 0.40 m long and sealed on the bottom end. It is held under a loudspeaker that sounds a tone at 220 Hz, causing the tube to radically resonate in its first harmonic. Find the speed of sound in the room.

85. •The second overtone of an organ pipe that is open at both ends and is 2.25 m long excites the third overtone in another organ pipe. Determine the length of the other pipe and whether it is open at both ends or open at one end and closed at the other. Assume the speed of sound is 340 m/s in air.

86. •Biology If the human ear canal with a typical length of about 2.8 cm is regarded as a tube open at one end and closed at the eardrum, what is the fundamental frequency that we should hear best? Assume that the speed of sound is 343 m/s in air.

87. •Biology A male alligator emits a subsonic mating call that has a frequency of 18 Hz by taking a large breath into his chest cavity and then releasing it. If the hollow chest cavity of an alligator behaves approximately as a pipe open at only one end, estimate its length. Is your answer consistent with the typical size of an alligator?

88. •Biology The trunk of a very large elephant may extend up to 3 m! It acts much like an organ pipe only open at one end when the elephant blows air through it. (a) Calculate the fundamental frequency of the sound the elephant can create with its trunk. (b) If the elephant blows even harder, the next harmonic may be sounded. What is the frequency of the first overtone?

89. •The longest pipe in the Mormon Tabernacle Organ in Salt Lake City has a "speaking length" of 9.75 m; the smallest pipe is 1.91 cm long. Assuming that the ambient temperature is 20 °C, determine the range of frequencies that the organ can produce if both open–open and open–closed pipes can be used with these lengths. SSM

13-7: Beats

90. •The sound from a tuning fork of 440 Hz produces beats against the unknown emissions from a vibrating string. If beats are heard at a frequency of 4 Hz, what is the vibrational frequency of the string?

91. •A guitar string is "in tune" at 440 Hz. When a standardized tuning fork rated at 440 Hz is simultaneously sounded with the guitar string, a beat frequency of 5 Hz is heard. (a) How far out of tune is the string? (b) What should you do to correct this?

92. •Two tuning forks are both rated at 256 Hz, but when they are struck at the same time, a beat frequency of 4 Hz is created. If you know that one of the tuning forks is in tune (but you are not sure which one), what are the possible values of the "out-of-tune" fork?

93. •A guitar string has a tension of 100 N and is supposed to have a frequency of 110 Hz. When a standard tone of that value is sounded while the string is plucked, a beat frequency of 2 Hz is heard. The peg holding the string is loosened (decreasing the tension) and the beat frequency increases. What should the tension in the string be in order to achieve perfect pitch? SSM

94. ••Two identical guitar strings under 200 N of tension produce sound with frequencies of 290 Hz. If the tension in one string drops to 196 N, what will be the frequency of beats when the two strings are plucked at the same time?

13-8: Volume, Intensity, and Sound Level

95. •By what factor should you move away from a point source of sound waves in order to lower the intensity by (a) a factor of 10? (b) A factor of 3? (c) A factor of 2? (d) Discuss how this problem would change if it were a point source that was radiating in only one-half of the entire spherically isotropic realm.

96. •Calculate the ratio of the acoustic power generated by a blue whale (190 dB) compared to the sound energy generated by a jackhammer (105 dB). Assume that the receiver of the sound is the same distance from the two sources of sound.

97. ••The steady drone of rush hour traffic persists on a stretch of freeway for 4 h each day. A nearby resident undertakes a plan to harness the wasted sound energy and use it for her home. She mounts a 1-m² "microphone" that absorbs 30% of the sound that hits it. If the ambient sound level is 100 dB, calculate the amount of sound energy that is collected. Do you think her plan is "sound"? How could you improve it? SSM

98. •A longitudinal wave has a measured sound level of 85 dB at a distance of 3 m from the speaker that created it. (a) Calculate the intensity of the sound at that point. (b) If the speaker were a point source of sound energy, find its power output.

99. •Biology A single goose sounds a loud warning when an intruder enters the farmyard. Some distance from the goose, you measure the sound level of the warning to be

88 dB. If a gaggle of 30 identical geese is present, and they are all approximately the same distance from you, what will the collective sound level be if they all sound off simultaneously? Neglect any interference effects.

100. •Two sound levels, β_1 and β_2, can be compared relative to each other (rather than to some standardized intensity threshold). Find a formula for this by calculating $\Delta\beta = \beta_2 - \beta_1$. You will need to employ your knowledge of basic logarithm operations to derive the formula.

101. •By what factor must you increase the intensity of a sound in order to hear (a) a 1-dB rise in the sound level? (b) What about a 20-dB rise? SSM

102. •Biology The intensity of a certain sound at your eardrum is 0.003 W/m². (a) Calculate the rate at which sound energy hits your eardrum. Assume that the area of your eardrum is about 55 mm². (b) What power output is required from a point source that is 2 m away in order to create that intensity?

103. •Biology The area of a typical eardrum is 5.0×10^{-5} m². Find the sound power incident on an eardrum at the threshold of pain.

13-9: Moving Sources and Observers of Waves

104. •A fire engine's siren is 1600 Hz when at rest. What frequency do you detect if you move with a speed of 28 m/s (a) toward the fire engine, and (b) away from it? Assume that the speed of sound is 343 m/s in air.

105. •Medical An ultrasound machine can measure blood flow speeds. Assume the machine emits acoustic energy at a frequency of 2.0 MHz, and the speed of the wave in human tissue is taken to be 1500 m/s. What is the beat frequency between the transmitted and reflected waves if blood is flowing in large arteries at 0.020 m/s directly away from the sound source?

106. ••A car sounding its horn (rated by the manufacturer at 600 Hz) is moving at 20 m/s toward the east. A stationary observer is standing due east of the oncoming car. (a) What frequency will he hear assuming that the speed of sound is 343 m/s? (b) What if the observer is standing due west of the car as it drives away?

107. ••A bicyclist is moving toward a sheer wall while holding a tuning fork rated at 484 Hz. If the bicyclist detects a beat frequency of 6 Hz (between the waves coming directly from the tuning fork and the echo waves coming from the sheer wall), calculate the speed of the bicycle. Assume the speed of sound is 343 m/s. SSM

108. ••Medical An ultrasonic scan uses the echo waves coming from something moving (such as the beating heart of a fetus) inside the body and the waves that are directly received from the transmitter to form a measurable beat frequency. This allows the speed of the internal structure to be isolated and analyzed. What is the beat frequency detected when waves with a frequency of 5 MHz are used to scan a fetal heartbeat (moving at a speed of ± 10 cm/s)? The speed of the ultrasound waves in tissue is about 1540 m/s.

109. ••Biology A bat emits a high-pitched squeal at 50,000 Hz as it approaches an insect at 10 m/s. The insect flies away from the bat and the reflected wave that echoes off the insect returns to the bat at a frequency of 50,050 Hz. Calculate the speed of the insect as it tries to avoid being the bat's next meal. (A bat can eat over 3000 mosquitoes in one night!) SSM

General Problems

110. ••Calc A wave is described by the following function:

$$y(x,t) = (20 \text{ cm}) \cos(5x - 4t + \pi/4)$$

(a) What is the value of y when $t = 0$ s and $x = 0$ m? (b) What is the value of y when $t = 1$ s and $x = 1$ m? (c) What is the transverse velocity of any point along the disturbance of the wave in terms of x and t, that is, what is $v(x,t)$? (d) What is the maximum transverse velocity of any point along the wave's disturbance? (e) What is the magnitude of the transverse velocity for the times and locations in parts (a) and (b)?

111. ••A large volcanic eruption triggers a tsunami. At a seismic station 250 km away, the instruments record that the time *difference* between the arrival of the tidal wave and the arrival of the sound of the explosion is 9.25 min. Tsunamis typically travel at approximately 800 km/h. (a) Which sound arrives first, the sound in the air or in the water? Prove your answer numerically. (b) How long after the explosion does it take for the first sound wave to reach the seismic station? (c) How long after the explosion does it take for the tsunami to reach the seismic station?

112. ••Calc A transverse wave propagates according to the following (SI units):

$$y(x,t) = 0.1 \sin(1.25x - 3.5t)$$

(a) Calculate the value of y when $x = 1$ m and $t = 2$ s. (b) If $x = 3$ m, at what time(s) does $y = 0.08$ m? (c) What is the maximum value of the transverse velocity of a point on the wave? (d) What is the maximum value of the transverse acceleration of a point on the wave? (e) What is the value of the transverse velocity when $x = 2$ m and $t = 4$ s? (f) What is the value of the transverse acceleration when $x = 2$ m and $t = 4$ s?

113. ••A transverse wave is propagating according to the following wave function:

$$y(x,t) = (1.25 \text{ m}) \cos(5x - 4t)$$

(a) Plot a graph of y versus x when $t = 0$ s. (b) Repeat when $t = 1$ s.

114. •••Calc The temperature in our atmosphere increases about 1 °C for every decrease in elevation of about 150 m. The speed of sound in air depends on the temperature according to the following formula:

$$v(T) = \sqrt{109,700 + 402T} \quad [T] = °C$$

If a sound is produced at a high elevation and directed straight down, it accelerates toward the surface of Earth. Neglecting any variations in the density of the air with elevation, calculate the acceleration of a sound wave as a function of velocity, $a(v)$. *Hint:* Use the chain rule.

115. ••You may have seen a demonstration in which a person inhales some helium and suddenly speaks in a high pitched voice. Let's investigate the reason for the change in pitch. At 0 °C, the density of air is 1.40 kg/m³, the density of helium is 0.1786 kg/m³, the speed of sound in helium is 972 m/s, and the speed of sound in air is 331 m/s. (a) What is the bulk modulus of helium at 0 °C? (b) If a person produces a sound of frequency 0.500 kHz, while speaking, what frequency sound will that person produce if his respiratory tract is filled with helium instead of air? (c) Use your result to explain why the person sounds strange when he breathes in helium.

116. •Biology When an insect ventures onto a spider-web, a slight vibration is set up alerting the spider. The density of spider silk is approximately 1.3 g/cm³, and its diameter varies considerably depending on the type of spider, but 3.0 μm is typical. If the web is under a tension of 0.50 N when a small beetle crawls onto it 25 cm from the spider, how long will it take for the spider to receive the first waves from the beetle?

117. ••If two musical notes are an *octave* apart, the frequency of the higher note is twice that of the lower note. The note concert A usually has a frequency of 440 Hz (although there is some variation). (a) What is the frequency of a note that is two octaves above (higher than) A in pitch? (b) If a certain string on a viola is tuned to middle C by adjusting its tension to T, what should be the tension (in terms of T) of the string so that it plays a note one octave below middle C? SSM

118. •••In Western music, the octave is divided into 12 notes as follows: C, C#/D^b, D, D#/E^b, E, F, F#/G^b, G, G#/A^b, A, A#/B^b, C'. Note that some of the notes are the same, such as C# and D^b. Each of the 12 notes is called a *semitone*. In the ideal *tempered* scale, the ratio of the frequency of any semitone to the frequency of the note below it is the same for all pairs of adjacent notes. So, for example, $f_D/f_{C\#}$ is the same as $f_{A\#}/f_A$. (a) Show that the ratio of the frequency of any semitone to the frequency of the note just below it is $2^{1/12}$. (b) If A is 440 Hz, what is the frequency of F# in the tempered scale? (c) If you want to tune a string from B^b to B by changing only its tension, by what ratio should you change the tension? Should you increase or decrease the tension?

119. •Find the temperature in an organ loft in Vancouver, British Columbia if the 5th *overtone* associated with the pipe that is resonating corresponds to 1500 Hz. The pipe is 0.7 m long and its type (open–open or open–closed) is not specified. The speed of sound in air depends on the centigrade temperature (T) according to the following: $v(T) = \sqrt{109{,}700 + 402T}$.

120. •**Biology** An adult female ring-necked duck is typically 16 in long and the length of her bill plus neck is about 5.0 cm long. (a) Calculate the expected fundamental frequency of the quack of the duck. For a rough but reasonable approximation, assume that the sound is only produced in the neck and bill. (b) An adult male ring-necked duck is typically 18 in long. If its other linear dimensions are scaled up in the same ratio from those of the female, what would be the fundamental frequency of its quack? (c) Which would produce a higher pitch quack, the male or female?

121. •When a worker puts on ear plugs, the sound level of a jackhammer decreases from 105 dB to 75 dB. Then the worker moves twice as far from the sound. Determine the sound level at the new location with the ear plugs.

122. ••**Medical** High-intensity focused ultrasound (HIFU) is one treatment for certain types of cancer. During the procedure, a narrow beam of high-intensity ultrasound is focused on the tumor, raising its temperature to nearly 90 °C and killing it. A range of frequencies and intensities can be used, but in one treatment a beam of frequency 4.0 MHz produced an intensity of 1500 W/cm². The energy was delivered in short pulses for a total time of 2.5 s over an area measuring 1.4 mm by 5.6 mm. The speed of sound in the soft tissue was 1540 m/s and the density of that tissue was 1058 kg/m³. (a) What was the wavelength of the ultrasound beam? (b) How much energy was delivered to the tissue during the 2.5-s treatment? (c) What was the maximum displacement of the molecules in the tissue as the beam passed through?

123. •Many natural phenomena produce very high energy, but inaudible, sound waves at frequencies below 20 Hz (*infrasound*). During the 2003 eruption of the Fuego volcano in Guatemala, sound waves of frequency 10 Hz (and even less) with a sound level of 120 dB were recorded. (a) What was the maximum displacement of the air molecules produced by the waves? (b) How much energy would such a wave deliver to a 2.0 m by 3.0 m wall in 1.0 min? Assume the density of air is 1.2 kg/m³. SSM

124. •**Medical** A diagnostic *sonogram* produces a picture of interior organs by passing ultrasound through the tissue. In one application, it is used to find the size, location, and shape of the prostate in preparation for surgery or other treatment. The speed of sound in the prostate is 1540 m/s, a diagnostic sonogram uses ultrasound of frequency 2.0 MHz. The density of the prostate is

1060 kg/m³. (a) What is the wavelength of the sonogram ultrasound? (b) What is Young's modulus for the prostate gland?

125. •A jogger hears a car alarm and decides to investigate. While running toward the car, she hears an alarm frequency of 869.5 Hz. After passing the car, she hears the alarm at a frequency of 854.5 Hz. If the speed of sound is 343 m/s, calculate the speed of the jogger.

126. ••Two identical 375-g speakers are mounted on parallel springs, each having a spring constant of 50.0 N/cm. Both speakers face in the same direction and produce a steady tone of 1.00 kHz. Both sounds have an amplitude of 35.0 cm, but they oscillate 180° out of phase with each other. What is the highest frequency of the beat that a person will hear if she stands in front of the speakers?

127. •••A rescuer in an all-terrain vehicle (ATV) is tracking two injured hikers in the desert, each of whom has an emergency locator transmitter (ELT) stored in his backpack (**Figure 13-44**). The beacons give off radio signals at 121.5 MHz, in phase, and there is a receiver in the ATV that is tuned to that frequency. The speed of the radio waves is 300,000,000 m/s. The ATV is traveling due east, 200 m north of the hikers, and the hikers are 100 m apart. If the ATV moves at a constant speed of 15 m/s, how many times per second will the driver detect constructive interference between the two signals?

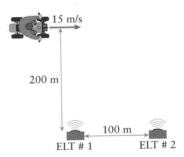

Figure 13-44 Problem 127

128. •••A string that has a mass M, a length L, and a tension T is set into its fundamental harmonic of vibration. The length of the string is changed to $L + \Delta L$. (a) Determine the change in the frequency at which it vibrates $[\Delta f = f(L + \Delta L) - f(L) = ?]$. (b) Convert your answer to part (a) to a function of $\Delta L/L$. Using the binomial expansion, make an approximation for Δf accurate to the first power of $\Delta L/L$ [drop all terms of $(\Delta L/L)^2$ and higher]. (c) If M equals 0.004 kg, L equals 0.80 m, T equals 120 N, and ΔL equals 0.01 m, find the new frequency exactly. Then use your approximation from part (b) and see how well your prediction fits the data.

129. •••Speaker A is sounding a single tone in phase with speaker B (**Figure 13-45**). The two speakers are separated by 8 m. What intensity level will you hear if you are located 6 m in front of speaker A and the frequency of the tone is 857.5 Hz? Assume that the 40-W speakers can be approximated as point sources and the speed of sound is 343 m/s. SSM

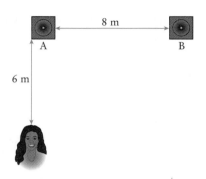

Figure 13-45 Problem 129

MCAT® Section

The section that follows includes material from previously administered MCAT® items and is reprinted with permission of the Association of American Medical Colleges (AAMC).

Passage 1 (Questions 1–6)

Ultrasonic waves are used in many applications by analyzing their reflections and refractions from interfaces between different media. Different intensities of reflections of a scanned object can provide a picture or pattern that can be electronically constructed and displayed on a monitor. The ultrasonic wavelength used for investigation should be equal to or smaller than the size of the object being scanned.

Ultrasonic waves are used for diagnostic purposes in medicine. Waves are directed into a body, and reflections occur at interfaces between different tissues or fluids. A practical depth of penetration of this technique for diagnostic scans is about 200 times the wavelength of the incident ultrasonic waves.

Another use of ultrasonic waves is an application of the Doppler effect. Doctors can detect and monitor blood flow speed as blood moves directly toward the incident ultrasonic waves. Calculations of flow toward the incident waves are made by using the relationship $(f_o/f_s)/(v/v - 2v_s)$, where f_o is the observed frequency, f_s is the incident frequency, v is the speed of ultrasonic waves in the medium, and v_s is the speed of the moving blood.

Ultrasonic waves that are reflected from a moving object are shifted in frequency. If the incident and reflected waves are mixed, this change in frequency causes beats to occur at a frequency that is equal to the absolute value of the difference between the incident and the reflected frequency. This phenomenon can also occur

when waves that have 2 separate frequencies and that have been emitted from 2 stationary sources are mixed.

1. The speed of sound in tissue for an ultrasonic scan is 1,500 m/s. From the information given in the passage, what maximum frequency could be used for a scan when a depth of investigation of 0.20 m is required?
  A. 1.2×10^3 Hz
  B. 6.0×10^3 Hz
  C. 1.2×10^6 Hz
  D. 1.5×10^6 Hz

2. When ultrasonic waves are detected by a transducer that feeds a signal into a display monitor, which of the following is the best description of the energy transformation that occurs at the transducer?
  A. mechanical to electrical
  B. kinetic to mechanical
  C. potential to kinetic
  D. potential to electrical

3. When a trumpet is tuned by comparison with a 512-Hz note from a piano, a beat frequency of 4 Hz is produced. The trumpet could have produced which of the following pairs of frequencies?
  A. 504 Hz and 508 Hz
  B. 508 Hz and 512 Hz
  C. 508 Hz and 516 Hz
  D. 512 Hz and 516 Hz

4. An ultrasonic wave enters a body with a speed of 1,500 m/s, and a reflection is noted at the same position on the surface 4.0×10^{-5} s later. What is the distance between the surface and the reflecting interface?
  A. 1.3×10^{-8} m
  B. 2.7×10^{-8} m
  C. 3.0×10^{-2} m
  D. 6.0×10^{-2} m

5. According to the passage, what is the lowest wave frequency that should be used to provide an image of an vobject that is 10^{-3} m on each side? (*Note:* Assume that the wave travels at 1,500 m/s.)
  A. 1.5 Hz
  B. 1.5×10^6 Hz
  C. 6.7×10^6 Hz
  D. 6.7×10^7 Hz

6. Which of the following best explains why ultrasonic waves are reflected at boundaries within a human body?
  A. Reflections occur at the boundary between objects with different shapes.
  B. Reflections occur at the boundary between objects with different densities.
  C. Reflections occur when the frequency of the ultrasonic wave decreases.
  D. Reflections occur when the frequency of the ultrasonic wave increases.

Passage 2 (Questions 7–13)

A student performed 2 experiments to investigate the behavior of springs.

Experiment 1

The student investigated the stretching of a horizontal spring on a smooth table. One end of the spring was fastened to a vertical wall, and the spring was stretched by a string attached over a pulley to a hanging mass (M) as shown in **Figure 1**. The experimental results showing how the total spring length (L) changed with variations to M are given in **Table 1**.

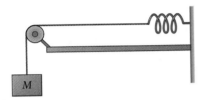

Figure 1

Table 1

Trial	M (kg)	L (m)
1	0	0.400
2	0.1	0.405
3	0.4	0.420
4	2.0	0.500

Experiment 2

The student slid a mass (m) with speed (v) along a horizontal tabletop with which it had negligible friction. The mass collided with a magnetic bumper plate that was attached to a horizontal spring fastened to a wall, as shown in **Figure 2**. When the spring was compressed a distance (x), the spring had a potential energy of $1/2kx^2$, where k is the spring constant. Maximum potential energy occurred at maximum recoil distance, x_{max}. Values of mass and speed used in Experiment 2 are given in **Table 2**.

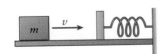

Figure 2

Table 2

Trial	m (kg)	v (m/s)
1	1	5
2	2	4
3	3	3
4	4	2
5	5	1

7. What is the value of k for the spring used in Experiment 1?
   - A. 4 N/m
   - B. 25 N/m
   - C. 100 N/m
   - D. 200 N/m

8. For which of the following trials of Experiment 2 is the recoil distance the greatest?
   - A. Trial 1
   - B. Trial 2
   - C. Trial 3
   - D. Trial 4

9. In Experiment 2, the mass clings to the magnetic bumper plate and vibrates back and forth. Which of the following best compares f_1, the vibration frequency of Trial 1 to f_4, the vibration frequency of Trial 4?
   (*Note*: The period of vibration of this system is $T = 2\sqrt{m/k}$.)
   - A. $f_1 = f_4/4$
   - B. $f_1 = f_4/2$
   - C. $f_1 = f_4$
   - D. $f_1 = 2f_4$

10. In Experiment 2, what minimum force must exist between the mass and the bumper plate so that the mass can vibrate back and forth?
    - A. $k(x_{max})^2/2$
    - B. $k(x_{max})^2/4$
    - C. kx_{max}
    - D. $k(x_{max})^2$

11. Which of the following graphs best describes the potential energy (PE) of the spring as it is compressed to x_{max} in Experiment 2?

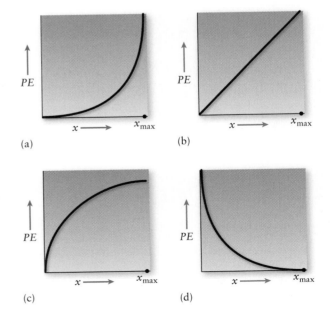

12. Two identical springs are hung side by side and are attached to the same mass. If each spring has a spring constant of 12 N/m, what is the spring constant of the 2-spring system?

    A. 6 N/m

    B. 12 N/m

    C. 18 N/m

    D. 24 N/m

13. After Trial 3 of Experiment 1, an additional mass of 3.0 kg was added to the 0.4-kg mass that was already attached to the system. How much farther did the mass descend when the 3.0-kg mass was added?

    A. 0.15 m

    B. 0.17 m

    C. 0.55 m

    D. 0.57 m

14 Thermodynamics I

(Courtesy H. Schmitz, Bonn University.)

Fire beetles (*Melanophila acuminata*) converge on burning forests from distances of up to 10 km to lay their eggs in trees weakened by fires. How do they detect a fire from so far away? The insects are equipped with special chambers that serve as infrared receptors. When heated even by a small amount, the fluid inside the chambers expands at a higher rate than the chamber walls, causing the pressure inside the chamber to increase; sensory neurons respond to the pressure change. The ability of the fire beetle to detect very small temperature changes is not only good for the beetles; it is also being used in the development of biosensors (which combine sensing chambers from beetles with electronic components) to serve as early warning detectors for forest fires.

We realized when studying fluids that it would be too complicated to describe the properties of liquids or gases by considering each molecule separately. Instead, we found statistical descriptions that represent the fluid as an average of the properties of each particle. It also makes sense to use statistical descriptions when studying *thermodynamic properties*. Thermodynamics, broadly defined, is the branch of physics that deals with relationships between properties of substances (such as temperature, pressure, and volume) and the energy (and flow of energy) associated with such properties. In the next two chapters, we will apply thermodynamics to a broad range of topics, for example, what the temperature of an object is, how a refrigerator works, what the universe and a falling egg have in common, and why arctic hares have rounder bodies and larger ears than rabbits that live in warmer climates.

14-1 Temperature

You likely learned at an early age that normal human body temperature is around 37.0 °C (98.6 °F), and that our bodies do a good job of regulating our internal temperatures even when ambient temperatures vary over a wide range. When you are hot, blood vessels in your skin dilate to dissipate *heat* through the body's surface; when you're cold, blood vessels constrict to reduce blood flow to the skin and thereby conserve your body's heat. In everyday language, temperature is something we read on a *thermometer,* and heat is a sensation associated with high temperatures and spicy foods. But surely as physicists we can be more precise!

Particles in any substance are always in motion. In gases, particles move with a range of speeds in every direction. The same is true for particles of a liquid; however, on average these particles don't move as far as gas particles do in a given time. And although particles in a solid exhibit no net motion, they nevertheless constantly wiggle back and forth. **Temperature** is a measure of the energy associated with the motion of particles. In particular, when two objects that have different temperatures are in thermal contact, so that energy can flow from one to the other, energy flows until both objects reach the same temperature. At that point, the objects are in **thermal equilibrium**.

Various properties of substances change as temperature changes. Nearly all liquids, for example, increase in volume as temperature increases. Some thermometers, devices that measure temperature, take advantage of such changes. Alcohol sealed in a tube can serve as a thermometer. When the tube comes into close contact with another object, the volume of alcohol expands or contracts as the alcohol, the tube, and the object approach thermal equilibrium. When the volume of alcohol stops changing, the alcohol, tube, and object have come into thermal equilibrium and we can say that all are at the same temperature. We can also define temperature as the property of two or more objects that is the same when the objects are in thermal equilibrium.

For a thermometer to be useful, it is important that only a relatively small amount of energy is transferred as the thermometer comes to thermal equilibrium with the object to be measured. For example, when a poor thermometer is placed in a cup of hot tea, the temperature of the tea might decrease considerably as the thermometer warmed up, so that by the time the tea and thermometer are in thermal equilibrium, the tea would be cool. In this case, the reading on the thermometer would not reflect the temperature of the tea when we intended to measure it. When measured with an accurate thermometer, the temperature of the tea hardly changes before the thermometer and tea are in thermal equilibrium.

An accurate thermometer can be used to determine whether two objects are at the same temperature. Imagine we put a thermometer in thermal contact with object A. When they reach thermal equilibrium, we know the temperature of object A. We now put the thermometer in thermal contact with object B and allow them to come to thermal equilibrium. If both A and B are separately in thermal equilibrium with the thermometer and the readings on the thermometer are the same for both, A and B are in thermal equilibrium with each other and therefore at the same temperature. This conclusion is known as the **zeroth law of thermodynamics**, sometimes called the law of thermal equilibrium:

> If two objects are each in thermal equilibrium with a third object, they are also in thermal equilibrium with each other.

It might seem like the zeroth law is too obvious to be of value, but from it we can draw the important conclusion that *temperature is a property of objects.* Indeed, as we noted in Section 1-2, temperature is a physical property used to classify objects and quantify physical processes.

The SI unit of temperature is the **kelvin** (K). Note that the units are kelvin not degrees kelvin or kelvins. At the standard pressure of 1 atm, water freezes at 273.160 K and boils at 373.13 K. Why these seemingly odd values?

The answer requires us to first understand the Celsius scale, which predated the Kelvin scale. In the 18th century, the Swedish astronomer Anders Celsius defined the freezing point of water as 0 °C and the boiling point as 100 °C. Both values were set at a pressure of 1 atm. The Kelvin scale, proposed by 19th century British physicist and engineer William Thomson (known by his title Baron Kelvin of Largs or Lord Kelvin), uses the same interval between integral values of kelvin as between integral values of degrees Celsius. However, zero on the Kelvin scale is based on a special property of gases. So, a change in temperature of 1 °C is the same as a change of 1 K, but the value assigned to any specific temperature is different on the two scales.

The zero point of the Kelvin scale arises from the relationship between pressure and temperature of gases. The pressure in a sealed volume of gas decreases as temperature decreases. Although the rate at which pressure changes with temperature depends on the gas, for idealized gases, the temperature at which the pressure becomes zero is the same for all gases, as shown in **Figure 14-1**. That temperature is called **absolute zero** because a lower temperature is not physically possible.

Using the Kelvin scale, the normal freezing point of water is about 273 K above absolute zero at 1 atm; the normal boiling point of water is about 373 K above absolute zero at 1 atm. The precise values of the differences depend on the **triple point** of water, which is the temperature and pressure at which ice, water, and steam (water vapor) can coexist in thermal equilibrium. (In general, the triple point is defined as the pressure and temperature at which all three common phases of a substance can exist simultaneously.) The triple point temperature and pressure of water are 273.160 K and 0.006030 atm, respectively. (This pressure is clearly quite low compared to 1 atm, the atmospheric pressure at sea level.) Based on the triple point, the Celsius scale was redefined in the 1950s so that at 1 atm of pressure water freezes at 0.010000 °C and boils at 99.9839 °C. To six significant figures these values correspond to 273.160 K and 373.134 K, respectively. So, the difference between a temperature in kelvin and a temperature in degrees Celsius is 273.15:

$$T_C = T_K - 273.15 \tag{14-1}$$

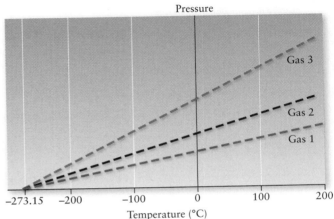

Figure 14-1 The rate at which pressure in a sealed volume of gas decreases as temperature decreases depends on the gas, but pressure becomes zero for all gases at the same temperature. That temperature, −273.15 °C, is termed absolute zero.

Finally, on the Fahrenheit scale (after German physicist Daniel Fahrenheit), water freezes at 32 °F and boils at 212 °F. These curious values are in part due to the use of the normal human body temperature in determining the scale; one of the points Fahrenheit used to fix his scale was determined by the reading of a thermometer when placed in a person's mouth. The range of temperature between freezing and boiling (212 °F − 32 °F = 180 °F) corresponds almost exactly to a difference of 100 °C on the Celsius scale. So a change of 1 °C is nearly equivalent to 100/180 °F or 5/9 °F. To approximate the conversion between Fahrenheit and Celsius, apply the scale factor (1 °C = 5/9 °F), while also accounting for the fact that 0 °C corresponds to 32 °F:

$$T_C = \frac{5}{9}(T_F - 32) \tag{14-2}$$

Estimate It! 14-1 Cold, Hot, and In-between

You've probably grown up using either the Celsius or Fahrenheit temperature scale, and may be uncomfortable with the other scale. What temperatures would you consider to be cold, normal for indoors, and hot? Express all three temperatures in K, °C, and °F.

SET UP

Because both authors grew up using the Fahrenheit scale, we'll start with temperatures in °F and convert to °C. Equation 14-2 provides the conversion from Fahrenheit to Celsius:

$$T_C = \frac{5}{9}(T_F - 32)$$

 Go to Picture It 14-1 for more practice dealing with temperature scales

SOLVE

On a very cold day the temperature might be 5 °F. So

$$T_{C,cold} = \frac{5}{9}(5 - 32) = -15\ ^\circ C$$

A comfortable temperature is 68 °F, widely accepted as "room temperature," so

$$T_{C,RT} = \frac{5}{9}(68 - 32) = 20\ ^\circ C$$

On a hot day the temperature might be 95 °F, so

$$T_{C,hot} = \frac{5}{9}(95 - 32) = 35\ ^\circ C$$

Each of these can be converted to the Kelvin scale by adding approximately 273 to the Celsius value (Equation 14-1): $T_{K,cold} = 258$ K, $T_{K,RT} = 293$ K, and $T_{K,hot} = 308$ K.

REFLECT

It's good to have some references on these scales handy, so that you can get a sense of the temperature even when values are given using a scale with which you are not so familiar. You can now remember these temperatures:

> On a cold day, the temperature might be 258 K, −15 °C, or 5 °F.
> Room temperature is 293 K, 20 °C, or 68 °F.
> On a hot day, the temperature might be 308 K, 35 °C, or 95 °F.

Example 14-1 Which Is Colder?

Which temperature is colder, −40 °C or −40 °F?

SET UP

To compare the two temperatures, we need to convert one so that both are expressed using the same units. We choose to convert −40 °F to degrees Celsius.

SOLVE

Equation 14-2 provides the conversion from Fahrenheit to Celsius; for $T_F = -40$ °F, T_C is

$$T_C = \frac{5}{9}(-40 - 32) = \frac{5}{9}(-72) = -40\ ^\circ C$$

Neither temperature is colder; −40 degrees is the same temperature on the Fahrenheit and Celsius scales.

REFLECT

A one-degree interval on the Celsius and Fahrenheit scales represents a different quantity of energy transfer for any specific object. For example, if a gas is held at fixed pressure, the average kinetic energy per molecule versus temperature has a different slope when plotted against Fahrenheit temperatures rather than Celsius temperatures, as shown in **Figure 14-2**. Notice that the two lines in the figure cross at −40, so indeed −40 °F is the same as −40 °C.

Practice Problem 14-1 Rank these temperatures from coldest to warmest: 280 K, −10 °C, −10 °F.

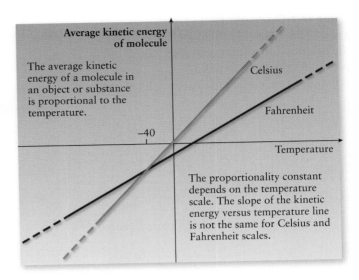

Figure 14-2 The temperature −40 °F is the same as −40 °C.

✳ What's Important 14-1

Temperature, a property of objects, is a measure of the energy associated with the motion of particles in any substance. When two objects that have different temperatures are in thermal contact, so that energy can flow from one to the other, energy flows until both objects reach the same temperature. Objects in contact and at the same temperature are in thermal equilibrium.

14-2 Ideal Gas Law

In the last section, we suggested that the temperature of an object is related to the kinetic energy of its individual particles. Although the conclusion seems reasonable, let's take a closer look at it to see the connection more directly.

In defining absolute zero, the lowest possible temperature, we relied on the fact that the temperature in a sealed container of gas is proportional to its pressure:

$$T \propto P \tag{14-3}$$

Consider a gas confined to a cylinder sealed with a piston on one end, as in **Figure 14-3**. As we push the piston inward, the volume of the gas decreases as the pressure inside the cylinder increases. When the gas is held at constant temperature, the volume is (almost exactly) inversely proportional to the pressure:

$$V \propto \frac{1}{P} \tag{14-4}$$

This relationship is known as *Boyle's law*. In addition, if the pressure in the piston is maintained at a constant value, the volume increases with temperature, so

$$V \propto T \tag{14-5}$$

This equation is known as *Charles' law*. The three equations can be combined a number of ways; for example, because pressure multiplied by volume is constant (Equation 14-4) and volume divided by temperature is constant,

$$P\frac{V}{T} = \text{constant}$$

Figure 14-3 One end of a gas-filled cylinder is closed and the other is sealed by a movable piston. As the volume of gas decreases, pressure increases.

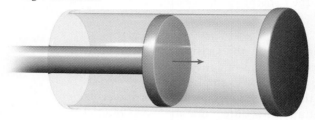

As the piston is pushed in and the volume of the gas decreases...

...the pressure of the gas increases. The pressure is inversely proportional to the volume:

$$P \propto 1/V$$

or

$$PV \propto T$$

This expression is approximately accurate for all gases over a wide range of pressures, volumes, and temperatures. To turn the proportionality into an equality, we need only include the appropriate physical constant. By taking the number of particles (atoms or molecules) N of the gas into account, the value of the proportionality constant k is the same for all gases:

$$PV = NkT \tag{14-6}$$

The **Boltzmann constant k** has the value $k = 1.381 \times 10^{-23}$ J/K. Equation 14-6 is called the **ideal gas law**. Real gases do not obey the equation exactly, especially at high pressures or at temperatures close to the point at which the gases become liquids. At relatively low gas densities, however, the ideal gas law does a good job of representing the relationship between pressure, volume, and temperature for real gases.

？ Got the Concept 14-1
PV

The right side of Equation 14-6 has dimensions of energy, because N has no dimension, k has dimensions of energy over temperature, and of course, T has dimensions of temperature. The inverse temperature in k cancels with T, leaving only energy. Show that the product of pressure and volume—the left side of Equation 14-6—also has dimensions of energy.

Equation 14-6 can be rearranged to give us

$$\frac{PV}{T} = Nk$$

N, the number of particles, is constant for a given sample of ideal gas, so PV/T is constant as well. The product Nk can be written in terms of moles, a standard unit of counting particles of substances such as atoms and molecules. One **mole**, abbreviated mol, contains 6.022×10^{23} particles; this value defines *Avogadro's number* N_A:

$$N_A = 6.022 \times 10^{23} \text{ particles/mol}$$

When the number of particles of a gas is expressed in moles, the product Nk becomes nR, where n is the number of moles and R is called the **universal** or **ideal gas constant**. The value of R is 8.314 J/(mol·K). (N is divided by N_A to obtain the number of moles; therefore, k is multiplied by N_A to get R. To the number of significant digits presented, the value of R is given by the product of N_A and k.)

Using n and R, the ideal gas law is written

$$PV = nRT \qquad\qquad (14\text{-}7)$$

The ideal gas law is important because it is an *equation of state*, that is, it enables us to determine the state of a gas under a broad range of conditions. The quantities pressure, volume, and temperature are examples of **state variables** that not only characterize the current state of a system, but also contain enough information about the system to determine its future behavior. Measurements made on thermodynamic systems like a volume of gas depend on the temperature and pressure at which the measurements were made. For this reason, we define a standard temperature and pressure (STP) to allow comparison between experimental measurements. Standard temperature and pressure is 293 K (20 °C) and 1 atm.

When a volume contains a mixture of gases, the **partial pressure** of each gas is the pressure the gas would have if no other gases were in the volume. The total pressure is the sum of the partial pressures of each individual gas; this conclusion is called the *law of partial pressures*. Mountain climbers, for example, can experience dangerous hypoxia—low levels of oxygen in the blood—when the partial pressure of oxygen in the air they breathe is too low (**Figure 14-4**). Although the atmosphere on the highest mountaintops has essentially the same fraction of oxygen as at sea level, the decrease in pressure at high altitudes results in a smaller number of oxygen molecules per breath and therefore a lower level of oxygen in the blood, depending on the altitude and on the mountain climber's level of acclimatization. Because the pressures we encounter in the physiology of respiration are relatively low, the ideal gas law can be applied to the study of gases in the lungs.

The partial pressure of oxygen $p(O_2)$ in a volume V of air at temperature T is given by the ideal gas law (Equation 14-6):

$$p(O_2) = \frac{N(O_2)kT}{V}$$

where $N(O_2)$ is the number of oxygen molecules in the volume. Similarly, the total pressure P is

$$P = \frac{NkT}{V}$$

where N is the total number of gas molecules in the volume. We can therefore express the partial pressure of oxygen in an air mixture as

$$\frac{p(O_2)}{P} = \frac{N(O_2)}{N}$$

or

$$p(O_2) = P\frac{N(O_2)}{N}$$

The ratio $N(O_2)/N$ is the fraction of oxygen in Earth's atmosphere, which is about 21% and is relatively constant up to an altitude of about 12,000 m. (Most of the rest of the atmosphere is nitrogen.) From the ideal gas law, the partial pressure of oxygen in air is given by the product of the total pressure and the fraction of oxygen in the mixture. This applies to any atmospheric gas.

Figure 14-4 The atmosphere at higher altitudes has essentially the same fraction of oxygen as at sea level. However, the lower pressure results in a smaller number of oxygen molecules per breath and therefore a lower level of oxygen in the blood. (*Courtesy of Prof. Christopher Kulp.*)

Example 14-2 Loss of Cabin Pressure

The average person begins to suffer from the effects of hypoxia at an altitude of around 4 km, where the $p(O_2)$ in the lungs falls below about 0.11 atm (about 80 mmHg). Although commercial jets routinely fly at altitudes of 9 km, where atmospheric pressure is about 0.31 atm, for the safety and comfort of passengers cabin pressure is maintained at 0.75 atm or more. If a jet loses cabin pressure at an altitude of 9 km, what would the $p(O_2)$ in passengers' lungs be? Should passengers breathe from emergency oxygen masks to avoid symptoms of hypoxia? [Note that although the passengers breathe pure oxygen in an emergency, other gases remain in their lungs. For example, the lungs normally retain relatively high levels of carbon dioxide, $p(CO_2) = 0.053 \times 10^5$ Pa (about 40 mmHg), and because the inside surfaces of the respiratory system are moist, air in the lungs is fully saturated with water vapor, $p(H_2O) = 0.062 \times 10^5$ Pa (about 47 mmHg).]

SET UP

The total pressure in the lungs is equal to the sum of the partial pressures of each individual gas and vapor:

$$P_{total} = p(N_2) + p(O_2) + p(CO_2) + p(H_2O)$$

The sum of the partial pressures of nitrogen and oxygen in the lungs is therefore

$$p(N_2) + p(O_2) = P_{total} - p(CO_2) - p(H_2O)$$

Oxygen makes up about 21% of air, so

$$p(O_2) = 0.21 \times [P_{total} - p(CO_2) - p(H_2O)]$$

SOLVE

At an altitude of 9 km, atmospheric pressure is about 0.31×10^5 Pa. The partial pressures of CO_2 and H_2O vapor are $p(CO_2) = 0.053 \times 10^5$ Pa and $p(H_2O) = 0.062 \times 10^5$ Pa, respectively. The partial pressure of oxygen is then

$$p(O_2) = 0.21 \times [0.31 \times 10^5 \, Pa - 0.053 \times 10^5 \, Pa - 0.062 \times 10^5 \, Pa]$$

$$= 0.041 \times 10^5 \, Pa$$

The partial pressure of oxygen in the lungs is 0.041×10^5 Pa or 31 mmHg. Given that symptoms of hypoxia may become pronounced below a $p(O_2)$ of 80 mmHg, emergency oxygen masks are warranted.

REFLECT

This value of $p(O_2)$ is well below the level at which symptoms of hypoxia appear. At this low level of oxygen, hemoglobin picks up about half as much oxygen as it does at sea level. Passengers—and pilots!—should wear oxygen masks. In the United States, federal law requires that supplemental oxygen be available to passengers at altitudes over 15,000 ft (4572 m).

Let's consider a sample of ideal gas in a sealed container in order to study the relationship between the parameters that appear in the ideal gas law. We start from the perspective of pressure, which arises due to the force imparted by individual particles colliding with the walls of the container.

Figure 14-5 shows a cubic container that has sides of length L and contains N atoms of an ideal monatomic gas. Each atom has a mass equal to m. When the gas is in equilibrium, the atoms as a whole exhibit no net motion. One representative atom is shown having just collided with and been reflected from the left

The gas atom shown has just collided with and been reflected from the wall on the left.

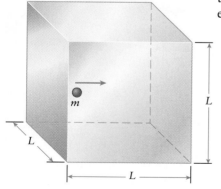

The atom, moving in the x direction, travels a distance L, collides with the right wall, and travels a distance L again before returning to the same position. This trip takes a time Δt.

Figure 14-5 A sealed cube contains gas. One representative atom is shown having just collided with and been reflected from the left wall.

wall. We can study one collision by following the atom for the time Δt that it takes for it to return to the same position after colliding with the wall on the right side of the container. For simplicity, we initially consider the atom moving only in the x direction. Also for simplicity we take the walls of the container to be rigid and the gas atoms to be hard, so that the collisions are elastic, that is, no energy is lost in the process.

Pressure is defined as the force imparted in the collision divided by the area of the wall; for the cube the area of the wall is $A = L^2$:

$$P = \frac{F}{L^2}$$

According to Equation 7-24 the force is given by the change in momentum of the atom divided by the time interval

$$F = \frac{\Delta p}{\Delta t} = \frac{p_i - p_f}{\Delta t} = \frac{mv_x - (-mv_x)}{\Delta t} = \frac{2mv_x}{\Delta t}$$

We have added a subscript "x" to the velocity to remind us that the atom is moving only in the x direction. We also allow the collision to occur with no loss of energy, so that the magnitude of the atom's momentum remains the same after the collision. The momentum after the collision is then just mv_x in the negative x direction.

The last two equations can be combined to eliminate F, which does not enter into the ideal gas equation:

$$P = \frac{F}{L^2} = \frac{2mv_x}{\Delta t L^2}$$

We'd also like to eliminate Δt. Because the atom does not accelerate except instantaneously to reverse direction, its velocity is given by the distance traveled divided by the time interval, so Δt is

$$\Delta t = \frac{2L}{v_x}$$

Combining the previous expressions yields

$$P = \frac{2mv_x}{\Delta t L^2} = \frac{2mv_x}{L^2} \frac{v_x}{2L} = \frac{mv_x^2}{L^3} \tag{14-8}$$

Remembering our original goal in this section (to see more clearly the relationship between temperature and the kinetic energy of atoms of the gas), notice that the term mv_x^2 on the right side is related to translational kinetic energy. We're close.

Equation 14-8 describes the pressure due to the one gas atom we have been considering. The total pressure is found by adding up the pressure that results from all atoms. We rename the pressure in Equation 14-8 P_i, so that we can write the total pressure P as the sum over all N atoms in the gas. Considering only the x direction,

$$P = \sum_{i=1}^{N} P_i = \sum_{i=1}^{N} \frac{m(v_x^2)_i}{L^3} \tag{14-9}$$

All the atoms have the same mass and L is a constant, so both m and L^3 can be removed from the sum:

$$P = \frac{m}{L^3} \sum_{i=1}^{N} (v_x^2)_i \tag{14-10}$$

We can simplify the expression further by noting that the average of a group of N numbers with values a_i is determined by adding them together and dividing by N. We can then define the average of v_x^2 for all of the atoms in the gas as

$$\left(v_x^2\right)_{\text{avg}} = \frac{1}{N}\sum_{i=1}^{N}\left(v_x^2\right)_i$$

This expression allows us to write the sum of the v_x^2 values in Equation 14-10 as

$$\sum_{i=1}^{N}\left(v_x^2\right)_i = N\left(v_x^2\right)_{\text{avg}}$$

The total pressure, Equation 14-10, can then be written in terms of the average velocity of the N gas atoms

$$P = \frac{m}{L^3}N\left(v_x^2\right)_{\text{avg}}$$

In general, the atoms of gas will have velocity in the y and z directions as well as x. The average of the square of the total velocity is the sum of the averages in each direction. Because statistically the average velocity is the same in each direction,

$$\left(v^2\right)_{\text{avg}} = \left(v_x^2\right)_{\text{avg}} + \left(v_y^2\right)_{\text{avg}} + \left(v_z^2\right)_{\text{avg}} = 3\left(v_x^2\right)_{\text{avg}} \tag{14-11}$$

The relationship between pressure and velocity of monatomic gas particles is

$$P = \frac{m}{L^3}\frac{N\left(v^2\right)_{\text{avg}}}{3}$$

To find the relationship between pressure and temperature, we need to connect the expression for pressure above to the pressure term in the ideal gas law, in which T appears explicitly. L^3 is the volume of the container, so we can write

$$P = \frac{N}{V}\frac{m\left(v^2\right)_{\text{avg}}}{3}$$

or

$$PV = N\frac{m\left(v^2\right)_{\text{avg}}}{3} \tag{14-12}$$

We have recovered the left side of the ideal gas law, $PV = NkT$ (Equation 14-6). Therefore, the right side of Equation 14-12 can be set equal to NkT:

$$NkT = N\frac{m\left(v^2\right)_{\text{avg}}}{3}$$

If we include a factor of $\frac{1}{2}$ on both sides of the equation, the result is an alternative expression for Equation 6-10, the average linear kinetic energy $\left[K_{\text{avg}} = \frac{1}{2}m\left(v^2\right)_{\text{avg}}\right]$,

$$\frac{1}{2}kT = \frac{1}{2}\frac{m\left(v^2\right)_{\text{avg}}}{3} = \frac{1}{3}K_{\text{avg}}$$

or

$$K_{\text{avg}} = \frac{3}{2}kT \tag{14-13}$$

The average translational kinetic energy of the atoms in the gas is directly related to the temperature! Equation 14-13 tells us something else, as well. Because k is a constant, the average translational kinetic energy of atoms of an ideal monatomic gas depends *only* on temperature and not on other quantities, such as the mass of the atoms.

Example 14-3 Helium Balloon

Find the average speed of a helium atom in a balloon at room temperature (20.0 °C). The mass of a helium atom is about 6.65×10^{-27} kg.

SET UP
Temperature and velocity are related by kinetic energy. In obtaining Equation 14-13, we uncovered the average translational kinetic energy of an atom of a monatomic gas to be

$$K_{avg} = \frac{1}{2}m(v^2)_{avg}$$

Equation 14-13 provides a relationship between the average kinetic energy and temperature:

$$K_{avg} = \frac{3}{2}kT$$

The relationship between $(v^2)_{avg}$ of a helium atom and its temperature is then

$$\frac{1}{2}m(v^2)_{avg} = \frac{3}{2}kT$$

SOLVE

$$(v^2)_{avg} = \frac{3kT}{m}$$

or

$$v = \sqrt{\frac{3kT}{m}}$$

Don't forget to convert room temperature to kelvin. All calculations must be done using SI units. The temperature 20.0 °C is 293 K, so

$$v = \sqrt{\frac{3(1.381 \times 10^{-23}\,\text{J/K})(293\,\text{K})}{6.65 \times 10^{-27}\,\text{kg}}} = 1.35 \times 10^3\,\text{m/s}$$

REFLECT
In the last step of our solution, we took the square root of the average of velocity squared. The result is not exactly the average velocity, but rather a special average known as the **root mean square (rms)**, so called because to find it we take the square root of the mean (average) velocity squared. The root mean square (rms) value of a quantity has particular significance when the values can be positive or negative. In this case, one helium atom could be moving left to right at $+v_0$ while another right to left at $-v_0$; the average of the two is zero, so the nonzero value of v_{rms} is often more useful. Our answer, then, is that the root mean square speed v_{rms} equals 1.35×10^3 m/s.

Practice Problem 14-3 The rms speed of a gas atom in a container at room temperature (20 °C) is found to be 2.0×10^3 m/s. Find the rms speed of an atom in the gas when the temperature is raised to 40 °C.

Why were we careful to consider specifically a *monatomic* gas in looking for a relationship between temperature and kinetic energy for an ideal gas? To answer the question, we first notice that the factor of 3 in that relationship ($K_{avg} = \frac{3}{2}kT$, Equation 14-13) arose because we first examined the average velocity only in the x direction and later added contributions from the y and z directions. Adding all three contributions together resulted in an average velocity three times the average in only the x direction (Equation 14-11), which is the source of the factor of 3 in Equation 14-13. The general conclusion is that for a gas in equilibrium, the translational kinetic energy is shared equally between energies associated with the motions in each of the three possible directions. Each direction of motion available to the atom contributes average translation kinetic energy $\frac{1}{2}kT$ to the total. We will refer to each of these possible directions as a **degree of freedom** of the system. Strictly speaking, the degrees of freedom of a physical system also include each component of the position of the system.

The number of degrees of freedom of a gas can be more than simply the number of orthogonal directions in which particles of the gas can translate. Imagine a molecular gas, such as ammonia (NH_3), in which the gas particles are composed of more than one atom. An ammonia molecule can certainly have translational kinetic energy, but unlike a single atom, it can also have energy of motion even if its translational velocity is zero. As suggested in **Figure 14-6**, an ammonia molecule can rotate. It can also vibrate in two ways; the bonds between the nitrogen and hydrogen atoms can be stretched and then oscillate, and the hydrogen atoms can be disturbed so that the angles of the bonds oscillate. Each motion represents a possible degree of freedom of the gas molecule and each can contribute energy $\frac{1}{2}kT$ to the total energy. Thus, the average kinetic energy is likely to be more than $\frac{3}{2}kT$ for a gas that is composed of molecules, not single atoms.

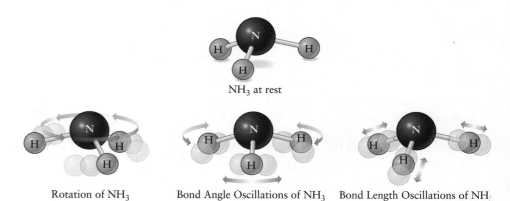

NH₃ at rest

Rotation of NH₃   Bond Angle Oscillations of NH₃   Bond Length Oscillations of NH₃

Figure 14-6 Unlike a monatomic gas, a molecular gas can have energy of motion even if the translational velocity of the molecule is zero because not only can the molecule rotate, but both the bond angle and the bond length can oscillate.

★ What's Important 14-2

Every degree of freedom of the atoms or molecules of a gas can contribute energy $\frac{1}{2}kT$ to the total energy. For a gas composed of single atoms (a monatomic gas), the atoms can move in any of three orthogonal directions; the three degrees of freedom result in an average kinetic energy of $\frac{3}{2}kT$. The quantities pressure, volume, and temperature are examples of state variables that not only characterize the current state of a system, but also contain enough information about the system to determine its future behavior.

14-3 Mean Free Path

A number of the properties of gases, including temperature, depend in part on the collisions that molecules of the gas make, either with the walls of a container or with each other. To better understand the properties of gases, it's helpful to know how long or how far a gas molecule travels between collisions. These are average quantities, of course. The **mean free time** τ is the average time a typical molecule travels between collisions. The **mean free path** λ is the average distance a typical molecule travels between collisions.

To get an idea of the processes at work, imagine you are trying to get to the other side of a terminal at the airport. If you're the only one there, you can walk straight to your destination. The more people present, however, and the faster their average speed, the more potential collisions you'll encounter. Worse, if you're forced to wear a blindfold, which is somewhat artificial but would make you move more like a gas molecule, your path would become more and more random as you bumped against more and more fellow travelers. It is reasonable to treat the gas molecules as if they don't interact with each other unless they collide, so like you walking alone in the airport terminal, an isolated gas molecule would follow a straight path. Collisions between molecules cause them to scatter in random directions; the number of collisions governs the randomness of the net motion. We quantify the amount of scattering by mean free time τ and mean free path λ. More collisions result in a shorter τ as well as a shorter λ.

The molecules of a gas confined to move through a tube experience collisions with the walls of the tube as well as with each other. When the diameter of the tube is large compared to the mean free path of the gas molecules, the rate at which the gas diffuses through the tube, that is, the rate at which the molecules of the gas spread from a region of higher concentration to one of lower concentration through random motion, depends primarily on τ. But when the tube diameter is about the same as the mean free path, the rate of collisions with the walls compared to molecule-on-molecule collisions is large enough to dramatically resist diffusion. For example, the pores in leaves (through which a plant absorbs CO_2 and through which water vapor and oxygen exit) must open during the day to facilitate photosynthesis. As pore diameter increases beyond the mean free path of the various gas molecules, the resistance to diffusion can fall by a factor of 10 or more.

Imagine a spherical gas molecule of radius r traveling in a straight line, as in Figure 14-7. It will collide with another molecule when the center-to-center distance between the two is no more than $2r$. The volume swept out by the molecule as it travels a distance λ is $\pi(2r)^2\lambda$. If the volume contains only the two molecules, then λ is the distance the first molecule travels before a collision, that is, λ is the mean free path. The number of molecules in any volume is given by the product of that volume and the ratio N/V, where N is the total number of molecules in the gas and V is the total volume of the gas. So the mean free path is defined by

$$1 = \pi(2r)^2\lambda\frac{N}{V}$$

or

$$\lambda = \frac{1}{4\pi r^2(N/V)}$$

In our model we treated the first molecule as moving but the other as stationary. If all of the gas molecules are moving, as of course they would be, the average distance between collisions is slightly smaller:

$$\lambda = \frac{1}{4\sqrt{2}\pi r^2(N/V)} \tag{14-14}$$

Figure 14-7 Because only two molecules occupy the cylindrical volume, the mean path length λ is the distance the first molecule travels before colliding with the second, stationary molecule.

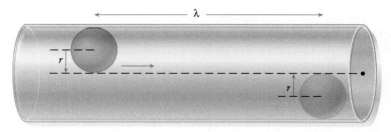

The mean free path λ is defined by the volume swept out by one molecule that includes exactly one other molecule.

In this approximation, one molecule moves at the average speed of the gas molecules...

...while the other is assumed to be stationary. The volume is then $\pi(2r)^2\lambda$. To ensure that exactly one other molecule occupies this volume, we demand that

$$\frac{\text{total number of gas molecules}}{\text{total volume of gas}} \, (\text{this volume}) = 1$$

or

$$\frac{N}{V}\,\pi(2r)^2\lambda = 1$$

So

$$\lambda = \frac{1}{(N/V)\pi(2r)^2}$$

Equation 14-14 gives the mean free path of a molecule that has a radius r in a gas of N molecules occupying a volume V.

We can get a sense of the relationship between the mean free path λ and the pressure in a gas by considering mean free path in an ideal gas. From Equation 14-6 for an ideal gas,

$$\frac{N}{V} = \frac{P}{kT}$$

We can substitute this into Equation 14-14 to obtain

$$\lambda = \frac{kT}{4\sqrt{2}\pi r^2 P} \tag{14-15}$$

So for an ideal gas at fixed temperature, the mean free path is inversely proportional to the pressure, as well as inversely proportional to the size (cross-sectional area) of the molecules.

Estimate It! 14-1 Mean Free Path

Estimate the mean free path of a molecule in a volume of CO_2 gas at STP. Although CO_2 is a linear molecule, treat it as spherical for this estimation. A CO_2 molecule has length of approximately 2.4×10^{-10} m.

SET UP

We will approximate the CO_2 as an ideal gas, so according to Equation 14-15 the mean free path is

$$\lambda = \frac{kT}{4\sqrt{2}\pi r^2 P}$$

We'll treat the CO_2 molecule as spherical, with radius equal to half the length of the molecule.

SOLVE

Taking r to be $\frac{1}{2}(2.4 \times 10^{-10} \text{ m})$ and noting that standard temperature and pressure are 293 K (room temperature) and 1.01×10^5 Pa (1 atm), we find

$$\lambda = \frac{(1.381 \times 10^{-23} \text{ J/K})(293 \text{ K})}{4\sqrt{2}\pi(1.2 \times 10^{-10} \text{ m})^2(1.01 \times 10^5 \text{ Pa})} = 1.6 \times 10^{-7} \text{ m}$$

REFLECT

The mean free path of CO_2 in a gas at STP is approximately 10^{-7} m. So, for example, the pores in the leaves of plants need to open to 10^{-7} m or greater to allow efficient diffusion of CO_2. Also, notice that the units of our answer are $\text{J/m}^2 \cdot \text{Pa}$. However 1 J is equivalent to $1 \text{ kg} \cdot \text{m}^2/\text{s}^2$, and 1 Pa is equivalent to $1 \text{ kg/m} \cdot \text{s}^2$; substituting these into $\text{J/m}^2 \cdot \text{Pa}$ immediately yields the correct units of meters.

★ What's Important 14-3

The mean free time τ is the average time a typical molecule travels between collisions. The mean free path λ is the average distance a typical molecule travels between collisions.

14-4 Thermal Expansion

That objects expand and contract as temperature changes has been known for thousands of years. During ancient times, structures weakened and eventually crumbled because thermal changes caused cracks to form in most building materials. However, many structures built by the Romans, such as the Pont du Gard aqueduct (**Figure 14-8**), survive today because a particular volcanic ash added to the building cement significantly reduces the magnitude of changes in volume due to temperature changes.

A more modern example of objects contracting and expanding due to temperature changes is the SR-71 (**Figure 14-9**), a military reconnaissance aircraft in service until 1998 that was capable of flying at more than three times the speed of sound. As it flew, all pieces of the plane's structure expanded due to heating caused by air drag. To compensate for the expansion, the joints that held the fuel tanks together were designed to be loose until thermal expansion forced them together. So until the SR-71 had been in the air for a while, the fuel tanks leaked. The photograph in Figure 14-9 was taken not long after takeoff, when the plane was still relatively cool; you can see fuel leaking from the joints at the front of the wings!

Figure 14-8 This ancient structure hasn't crumbled because volcanic ash added to the cement reduces the magnitude of thermal expansion and contraction. (*Cristina Fumi Photography /Alamy.*)

Figure 14-9 Heating of the surfaces of the SR-71 strategic reconnaissance aircraft due to extreme drag forces required the seams of the fuel tanks to be loose until thermal expansion forced them together. Until the plane was airborne long enough for the joints to seal, the fuel tanks leaked, as can be seen on the wings of the SR-71 in this photo. *(Courtesy of Lockheed Martin.)*

Nearly all substances and objects expand when heated and contract when cooled, a phenomenon known as **thermal expansion**. The change in each dimension of an object is proportional to the change in temperature. Length, for example, undergoes a change ΔL:

$$\Delta L \propto \Delta T$$

In Chapter 9, we saw that the change in length when an object is stretched depends on its initial length, or $\Delta L = (F/YA)L_0$ (from Equation 9-1). In a similar way, the change in length due to thermal expansion is proportional to the initial length:

$$\Delta L = \alpha L_0 \Delta T \qquad (14\text{-}16)$$

where the coefficient of linear expansion α is the proportionality constant. Equation 14-16 applies to each linear dimension (length, width, and height) of an object.

The coefficient of linear expansion depends on the substance. It can also depend on temperature, but for most substances in the range of temperatures in typical human experiences, α can be treated as a constant. Table 14-1 lists the coefficient of linear expansion for a number of substances. Notice in Equation 14-16 that α has dimensions of inverse temperature ($1/K$ or K^{-1}), so that the dimensions of $\alpha L_0 \Delta T$ on the right side of the equation match those of ΔL on the right side.

! Watch Out
The units of α can be °C^{-1} as well as K^{-1}.

As we have seen, the size of 1 °C and 1 K is the same. That's because a temperature difference has the same magnitude regardless of whether the temperatures are given in degrees Celsius or kelvin. For example, water freezes at 0.01 °C and boils at 99.984 °C when the pressure is 1 atm; the difference between the two is 99.974 °C. Using the Kelvin scale, water freezes at 273.160 K and boils at 373.134 K, a difference of 99.974 K. ΔT has the same value regardless of whether the scale is Celsius or Kelvin, so α for a material must have the same value regardless of whether the units are degrees Celsius or kelvin.

 Go to Picture It 14-2 for more practice dealing with linear expansion

Example 14-4 Buckling Bridge

Workers position two 10-m-long steel beams nearly end to end in the construction of a bridge on a day when the temperature is 18 °C. They leave an expansion gap of length d between the beams (Figure 14-10a) so that even if the temperature were to hit 45 °C on an extraordinarily hot day, the two ends would just barely touch. How wide should the gap be so that the structure does not buckle when the temperature is 45 °C?

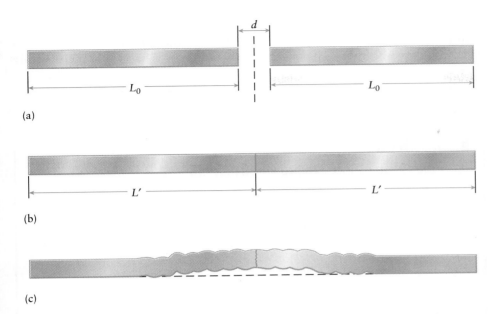

(a)

(b)

(c)

Figure 14-10 (a) A gap of length d separates two beams of a bridge when the temperature is relatively low. (b) At 45 °C, the beams expand enough to fill the gap between them. (c) The bridge would buckle on a day when the temperature is higher than 45 °C if no gap had been built in between the beams.

Table 14-1 Coefficients of Linear Expansion (α) at 1 atm	
Substance	α **(K^{-1} × 10^{-6})**
aluminum	22.2
antimony	10.4
beryllium	11.5
brass	18.7
brick	5.5
bronze	18.0
carbon, diamond	1.2
cement	10.0
concrete	14.5
copper	16.5
glass, hard	5.9
glass, plate	9.0
glass, pyrex	4.0
gold	14.2
graphite	7.9
iron, cast	10.4
iron, forged	11.3
iron, pure	12.0
lead	28.0
marble	12
plaster	25
platinum	9.0
porcelain	3.0
quartz, fused	0.59
rubber	77
silver	19.5
solder	24.0
steel	13.0
wood, oak parallel to grain	4.9

SET UP

As you can see in **Figure 14-10b**, each beam expands so that half of the increase in length fills the gap between the beams. From Equation 14-16, the change in each beam is given by $\Delta L = \alpha L_0 \Delta T$, so the gap is just filled when

$$d = 2\left(\frac{\Delta L}{2}\right) = 2\left(\frac{\alpha L_0 \Delta T}{2}\right) = \alpha L_0 \Delta T$$

Each beam undergoes half of its anticipated thermal expansion in the region of the gap, so the gap must be as wide as the anticipated largest thermal expansion of a single beam.

SOLVE

Using Table 14-1, the coefficient of linear expansion of steel is 13.0×10^{-6} K^{-1}. The beams have a length $L_0 = 10$ m when the temperature is 18 °C, and we are asked to ensure that the beams don't buckle when the temperature is 45 °C. The temperature range is $\Delta T = 45$ °C $-$ 18 °C $= 27$ °C $= 27$ K. The required gap is then

$$d = (13.0 \times 10^{-6} \text{ K}^{-1})(10 \text{ m})(27 \text{ K}) = 3.5 \times 10^{-3} \text{ m}$$

The gap must be 3.5 mm wide to accommodate the thermal expansion that occurs due to a temperature change from 18 °C to 45 °C.

REFLECT

Bridges, buildings, and other structures are built with *expansion joints* that allow for the thermal expansion and contraction of separate, adjacent elements. Without a sufficiently large expansion joint, the ends of the beams in this problem would push against each other and eventually warp or buckle, as suggested in **Figure 14-10c**.

Practice Problem 14-4 Workers position two 5.5-m-long steel beams nearly end to end in the construction of a bridge on a day when the temperature is 18 °C. They leave an expansion gap of length d between the steel beams (Figure 14-10a) so that even if the temperature were to hit 45 °C on an extraordinarily hot day, the two ends would just barely touch. How wide should the gap be so that the structure does not buckle when the temperature is 45 °C?

 Watch Out
Holes in an object expand when heated.

You might be tempted to think that when an object is heated, the substance of which it is made expands to fill any holes in the object. Not so! Every dimension of the object expands, which stretches out the holes as well as the objects, as shown in **Figure 14-11**.

A block with a hole in it is heated.

Termal Energy

In this case, we have exaggerated the expansion so that every dimension of the block doubles. The size of the hole also doubles.

Figure 14-11 When a block with a hole is heated, both the hole and the block expand.

 Got the Concept 14-2
Size of Hole

How does the size of a hole in an object change as temperature changes?

 Got the Concept 14-3
Opening a Jar

A common trick to open a tightly sealed jar is to run hot water on the lid before trying to twist it. Why does this work?

When the temperature of an object changes, its length, width, and height change according to the linear thermal expansion relationship of Equation 14-16. Let's consider a block that has sides of length L_0, width W_0, and height H_0 and is made of a material which has a coefficient of linear expansion α. When the temperature is increased by ΔT, each side expands to new dimensions L_1, W_1, and H_1. The volume also expands, from the initial volume $V_0 = L_0 W_0 H_0$ to a new volume V_1. What is the relationship between the initial and final volumes? Or to match the way we have described linear thermal expansion, what is the relationship between ΔV and V_0? We will assume that the change in lengths and volume is small to take advantage of a useful approximation.

According to Equation 14-16, the new volume is

$$V_1 = (L_0 + \alpha L_0 \Delta T)(W_0 + \alpha W_0 \Delta T)(H_0 + \alpha H_0 \Delta T)$$

or

$$V_1 = L_0 W_0 H_0 (1 + \alpha \Delta T)^3 = L_0 W_0 H_0 (1 + 3\alpha \Delta T + 3(\alpha \Delta T)^2 + (\alpha \Delta T)^3)$$

Because we have assumed that the change in lengths and volume are small, $\alpha \Delta T$ must be small compared to 1 ($\alpha \Delta T \ll 1$). So $(\alpha \Delta T)^2$ and $(\alpha \Delta T)^3$ are really small; we can treat the last two terms as negligible, so

$$V_1 \approx L_0 W_0 H_0 (1 + 3\alpha \Delta T) = V_0 (1 + 3\alpha \Delta T)$$

The change in volume ΔV is the difference between the final and initial volumes:

$$\Delta V = V_1 - V_0 \approx V_0 (1 + 3\alpha \Delta T) - V_0 = V_0 3\alpha \Delta T$$

Writing the result in the same form as Equation 14-16, we obtain

$$\Delta V = 3\alpha V_0 \Delta T \qquad (14\text{-}17)$$

The relationship applies for small changes, so we replaced the approximately equal sign with an equal sign. From Equation 14-17, we see that volume thermal expansion follows the same general relationship as linear thermal expansion, except that the coefficient of expansion is three times as large. When considering the thermal expansion of liquids, it isn't correct, strictly speaking, to define a linear coefficient of thermal expansion, so we can write

$$\Delta V = \beta V_0 \Delta T$$

where β is the coefficient of volume expansion. The coefficient of volume expansion is listed for some liquids in Table 14-2. The units of β, like α, are K^{-1}.

Table 14-2
Coefficients of Volume Expansion (β) at 1 atm

Substance	β ($K^{-1} \times 10^{-6}$)
gasoline	905
ethanol	750
water	207

Math Box 14-1 Volume Change from Length Change

The relationship between volume change and temperature change (Equation 14-17) can be obtained more easily by employing the power of calculus. A cube's volume V and length of a side L are related by

$$V = L^3$$

As the temperature changes, so does the length of one side, so the change in volume with respect to L is

$$\frac{dV}{dL} = 3L^2$$

or

$$dV = 3L^2 \, dL$$

We can express dL using Equation 14-16:

$$dL = \alpha L \, dT$$

so

$$dV = 3L^2 (\alpha L \, dT) = 3\alpha L^3 \, dT$$

Substituting V for L^3 gives

$$dV = 3\alpha V \, dT$$

This equation is equivalent to Equation 14-17.

 Got the Concept 14-4
Coolant Level

Most cars today are equipped with a radiator overflow tank; coolant fluid, which is derived from ethanol, can flow freely back and forth between the radiator and the overflow tank. Before going for a drive in her car, your friend notices that the level of coolant in the overflow tank is low. When she arrives at her destination an hour later, the level of coolant in the tank is high. After the car has been sitting with the engine off for several hours, the coolant is low again. Explain why the coolant level changes.

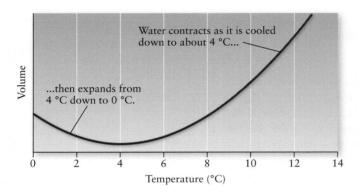

Figure 14-12 Although most substances expand when heated and contract when cooled, water is different. As water is cooled, it stops contracting around 4 °C and expands as the temperature drops to the freezing point.

Nearly all substances expand when heated and contract when cooled. In other words, substances get less dense when heated and more dense when cooled. Curiously, water behaves this way only above 4 °C. As you can see from Figure 14-12, water stops contracting and begins to expand when cooled below about 4 °C (the precise value is closer to 3.98 °C). For this reason, water is less dense at 0 °C than at slightly warmer temperatures. So, as the water in rivers and lakes gets cold during the winter, the coldest water, closest to the freezing point, floats to the surface as the denser 4 °C water sinks to the bottom. The surface water freezes first, so rivers and lakes freeze from the surface down. Were water like most substances, the coldest water and also the ice would sink, causing bodies of water to freeze from the bottom up. Rivers and lakes could freeze solid during extended periods of cold temperatures, possibly destroying aquatic life. Who knows if life on Earth would have even evolved if water did not exhibit the unusual property of slightly expanding as it cools toward the freezing point?

 What's Important 14-4

Nearly all substances and objects expand when heated and contract when cooled, a phenomenon generally known as thermal expansion. The change in each dimension of an object is proportional to the change in temperature. Holes in an object expand when the object is heated.

14-5 Heat

The average kinetic energy of an object's particles determines the object's temperature. Objects of different temperatures reach thermal equilibrium with each other through the transfer of kinetic energy from hotter objects to cooler ones. The energy that flows from one object to another as a result of a temperature difference is termed **heat**.

Take care with the concept of heat; it doesn't have quite the same meaning in the language of physics as in everyday speech. Heat does not imply high temperature and it does not describe the ambient temperature. Heat is not the energy contained in a system, nor is it simply energy. In physics, the word heat refers to the flow of energy from one place to another. Specifically, heat is an energy flow which results due to a temperature difference; in order to emphasize this aspect of heat it is sometimes called *heat transfer*.

When energy flows from one object to another the temperature of the two can change under certain circumstances. When this occurs, the change in temperature ΔT is proportional to the heat, commonly represented by the variable Q,

$$\Delta T \propto Q$$

Heat Q has the SI units of joules, the units of energy. We would intuitively expect that the more massive an object is, the less of an effect a given quantity of energy would have on the object's temperature. The temperature change an object undergoes is inversely proportional to the object's mass, or

$$\Delta T \propto \frac{1}{m} Q$$

Adding a certain amount of energy to a chunk of copper, however, results in a much larger change in temperature than that produced by adding the same amount of energy to a book of the same mass. That is to say, the proportionality constant between heat Q and the change in temperature ΔT depends on the substance. We call that constant the heat capacity. (Don't let the name mislead you; it does not imply that an object contains heat.) However, because two objects made from the same material but of different masses have different heat capacities, a more useful quantity is the **specific heat** c. Specific heat is the heat capacity per unit mass, that is, the ratio of the energy added to an object and the temperature change that results, divided by the mass of the object. We choose to write this relationship for the temperature change to make it clear that the larger the specific heat, the smaller the temperature change for a given mass m and heat Q:

$$\Delta T = \frac{1}{mc} Q \tag{14-18}$$

Specific heat varies somewhat with the temperature of a substance; however, over the range of temperatures we typically experience, variations in the value of specific heat c for any given material are small enough that we will ignore them. Table 14-3 lists the specific heat for a range of substances.

Equation 14-18 can be rearranged as an expression for c:

$$c = \frac{Q}{m\Delta T} \tag{14-19}$$

This equation suggests how specific heat is related to the temperature change that results from adding a certain amount of energy to a substance. The units of c can be inferred from Equation 14-19:

$$[c] = \frac{[Q]}{[m][\Delta T]} = \frac{J}{kg \cdot K}$$

Example 14-5 Camping

Every minute, a certain camping stove burns about 1 g of propane, which releases approximately 5.0×10^4 J of energy. Only half of that energy is transferred to 2.0 L

Table 14-3
Specific Heats (c) at 1 atm

Substance	c [J/(kg · K)]
air (50 °C)	1046
aluminum	910
benzene	1750
copper	387
glass	840
gold	130
ice	(−10 °C to 0 °C) 2093
iron/steel	452
lead	128
marble	858
mercury	138
methyl alcohol	2549
silver	236
steam (100 °C)	2009
water (0 °C to 100 °C)	4186
wood	1700

of water in a pot above the flame. If the water starts out at 20 °C, what will its temperature be after 1 min?

SET UP
Adding energy to the water increases its temperature according to Equation 14-18:

$$\Delta T = \frac{1}{mc}Q$$

We are given the amount of energy transferred, $Q = \frac{1}{2}(5.0 \times 10^4\,\text{J})$, and the mass of 2.0 L of water is $m = 2.0$ kg. The specific heat of water, $4186\,\text{J}/(\text{kg}\cdot\text{K})$, is found in Table 14-3.

SOLVE
The temperature change of the water is

$$\Delta T = \frac{1}{mc}Q = \frac{1}{(2.0\,\text{kg})\left(4186\,\dfrac{\text{J}}{\text{kg}\cdot\text{K}}\right)}\frac{5.0 \times 10^4\,\text{J}}{2} = 3.0\,\text{K}$$

The temperature change is equivalent to 3.0 °C, so the temperature of the water, initially 20 °C, will be 23 °C after absorbing heat from the propane for 1 min.

REFLECT
If you've ever used a camping stove, you know that it takes quite a bit longer than 1 min to bring water to a boil. Using the numbers in this problem, it would take roughly 27 min to increase the temperature of 2 L of water from 20 °C to 100 °C. We also note that while half the energy released by the propane goes into heating and boiling the water, the rest goes into heating the pot and the surrounding air.

Practice Problem 14-5 A certain camping stove burns about 1 g of propane per minute, which releases approximately 5.0×10^4 J of energy. If half of that energy is transferred to 2.0 L of water, how much gas (in g) must be burned to raise the temperature of the water from 20 °C to 40 °C?

Two objects placed in contact eventually reach thermal equilibrium, as energy, in the form of heat, flows from one to the other. For the purposes of this discussion we assume that the objects are thermally isolated, that is, neither one can lose energy to or gain energy from the environment. Because energy flows from the object at higher temperature to the one at lower temperature, the temperature of the initially warmer object decreases while the temperature of the initially cooler one increases as they come into thermal equilibrium with each other.

Consider the two objects in **Figure 14-13**. The temperature of the smaller, red block is higher than the temperature of the blue block, so as the two approach thermal equilibrium, energy Q flows from the warmer block into the cooler one. According to Equation 14-18, the energy leaving the warmer block causes its temperature to decrease:

$$\Delta T_\text{H} = T_\text{H,f} - T_\text{H,i} = \frac{1}{m_\text{H}c_\text{H}}Q_\text{H} \tag{14-20}$$

where Q_H is the energy released from the warmer block. We use the subscript "H" to indicate the warmer block. Similarly, Equation 14-18 can be applied to the cooler block:

$$\Delta T_\text{C} = T_\text{C,f} - T_\text{C,i} = \frac{1}{m_\text{C}c_\text{C}}Q_\text{C} \tag{14-21}$$

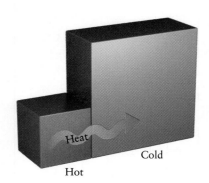

Figure 14-13 Energy in the form of heat (Q) flows from the hot block to the cold one until the two blocks are in thermal equilibrium.

where Q_C is the energy entering cooler block. We use the subscript "C" to indicate the cooler block. In both equations, the initial and final temperatures of each block are labeled as T_i and T_f, the mass of each block is m, and the specific heat of each block is c. Three observations provide valuable insight into the energy exchange process.

First, the final temperature of both blocks must be the same, which is the requirement of thermal equilibrium, so

$$T_{H,f} = T_{C,f} = T_f$$

where T_f is the final temperature of both objects.

Second, the initially warmer object gets cooler during the process while the other, cooler block gets warmer. So $T_{H,i}$ must be greater than T_f and $T_{C,i}$ must be less than T_f. Because $T_f < T_{H,i}$, the change in temperature ΔT_H in Equation 14-20 is negative. Q_H, in turn, is also negative. But ΔT_C in Equation 14-21 is positive, and therefore Q_C is positive. We can conclude that both positive and negative values are allowed for heat. When Q is positive, energy is flowing into an object. When Q is negative, energy is leaving an object.

Finally, because energy is conserved, the energy that leaves the red block enters the blue one. In other words, the absolute magnitude of the heat released from the initially warmer block is equal to the absolute magnitude of the heat absorbed into the initially cooler block:

$$|Q_H| = |Q_C| \tag{14-22}$$

Both Q_H and Q_C are equal in magnitude to the heat Q indicated in Figure 14-13.

The simplicity of Equation 14-22 masks its power. As we have seen in other situations, energy conservation is a powerful tool. Here, conservation of energy enables us to determine the final temperature when two objects come to thermal equilibrium.

To find the final temperature, we rearrange Equation 14-18 as a relationship for Q in terms of m, c, and ΔT:

$$Q = mc\Delta T \tag{14-23}$$

Then, Equations 14-20 and 14-21 can be written as expressions for heat:

$$Q_H = m_H c_H (T_{H,f} - T_{H,i}) \tag{14-24}$$

$$Q_C = m_C c_C (T_{C,f} - T_{C,i}) \tag{14-25}$$

We have seen that Q_H is negative, because we chose T_H to be greater than T_C. However, the absolute magnitudes of Q_H and Q_C are equal (Equation 14-22), so

$$|m_H c_H (T_{H,f} - T_{H,i})| = |m_C c_C (T_{C,f} - T_{C,i})|$$

Another way to express that the magnitudes of Q_H and Q_C are equal is

$$m_H c_H (T_{H,i} - T_{H,f}) = m_C c_C (T_{C,f} - T_{C,i})$$

Again, the final temperatures of both objects are the same value T_f, so

$$m_H c_H (T_{H,i} - T_f) = m_C c_C (T_f - T_{C,i}) \tag{14-26}$$

Conservation of energy, which leads to Equation 14-26, has revealed a way to determine the equilibrium temperature of two objects in thermal contact. We only need a bit of algebra to isolate T_f in terms of the masses and initial temperatures. We'll do the algebra quickly, because the form of the answer is what's important.

$$m_H c_H T_{H,i} - m_H c_H T_f = m_C c_C T_f - m_C c_C T_{C,i}$$

$$m_H c_H T_{H,i} + m_C c_C T_{C,i} = m_H c_H T_f + m_C c_C T_f$$

$$T_f = \frac{m_H c_H T_{H,i} + m_C c_C T_{C,i}}{m_H c_H + m_C c_C} \tag{14-27}$$

Go to Picture It 14-3 for more practice dealing with specific heat

Go to Interactive Exercise 14-1 for more practice dealing with specific heat

The final temperature is a special kind of average of the two initial temperatures, an average in which the values of specific heat and mass of each object determines how much influence the initial temperature of the object has on the final temperature.

Got the Concept 14-5
Final Temperature I

Two objects that are made of the same substance are placed in thermal contact. One object is far more massive than the other. Will the final temperature of the two be closer to the initial temperature of the more massive object, the initial temperature of the less massive object, or about midway between the initial temperatures?

Got the Concept 14-6
Final Temperature II

Two objects that have the same mass and are made from the same substance are placed in thermal contact. One object initially has a temperature $T_{1,i} = 100\ °C$ while the other has a temperature $T_{2,i} = 50\ °C$. Will the final temperature of the two objects be closer to 100 °C, 50 °C, or midway between the two temperatures?

Got the Concept 14-7
Final Temperature III

Two objects that have the same mass are placed in thermal contact. One object is made of a substance that has a much larger specific heat than the substance from which the other object is made. Will the final temperature of the two objects be closer to the initial temperature of the higher specific heat object, the initial temperature of the lower specific heat object, or about midway between the initial temperatures?

The SI unit of heat is the joule, but other units of heat are commonly used. The **calorie (cal)** is defined as the heat required to increase the temperature of 1 g of pure water from 14.5 °C to 15.5 °C. The unit is not, however, the unit associated with food; the food calorie, sometimes written with a capital "C" as in "Calorie" or "Cal," is 1000 calories or 1 kcal. (In many countries other than the United States, the energy content on food labels is given in units of joules, as on the label from a package of flavored sugar produced in Finland, shown in **Figure 14-14.**) The unit of heat in the English system is the **British thermal unit (BTU)**, defined as the heat required to increase the temperature of 1 lb of pure water from 63 °F to 64 °F. The energy flow of air conditioners and heaters is often given in BTUs. To convert values of calories and British thermal units to the SI system, note these relationships:

$$1\ cal = 4.186\ J$$

$$1\ BTU = 1055\ J = 252\ cal$$

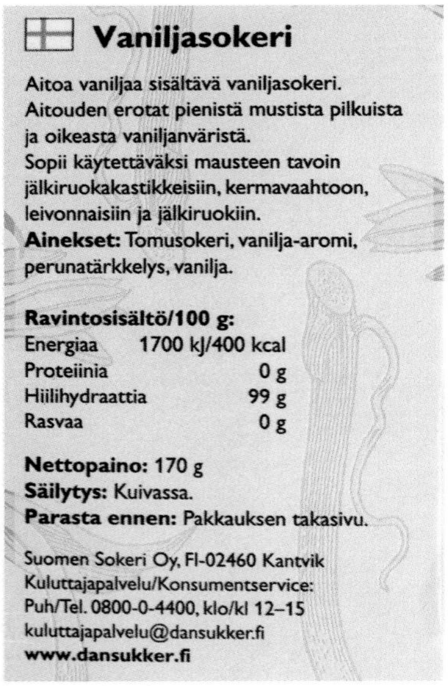

Figure 14-14 The label on a package of flavored sugar from Finland lists the energy content in joules rather than Calories. *(Courtesy of David Tauck.)*

Vaniljasokeri

Aitoa vaniljaa sisältävä vaniljasokeri. Aitouden erotat pienistä mustista pilkuista ja oikeasta vaniljanväristä.
Sopii käytettäväksi mausteen tavoin jälkiruokakastikkeisiin, kermavaahtoon, leivonnaisiin ja jälkiruokiin.
Ainekset: Tomusokeri, vanilja-aromi, perunatärkkelys, vanilja.

Ravintosisältö/100 g:
Energiaa   1700 kJ/400 kcal
Proteiinia   0 g
Hiilihydraattia   99 g
Rasvaa   0 g

Nettopaino: 170 g
Säilytys: Kuivassa.
Parasta ennen: Pakkauksen takasivu.

Suomen Sokeri Oy, FI-02460 Kantvik
Kuluttajapalvelu/Konsumentservice:
Puh/Tel. 0800-0-4400, klo/kl 12–15
kuluttajapalvelu@dansukker.fi
www.dansukker.fi

An average human uses about 2000 kcal (2000 Calories) each day. Estimate the wattage of a light bulb with equivalent power.

SET UP

Power is defined as energy transferred or expended per unit time. To compute power in the SI unit watts (where $1\ W = 1\ J/s$), we need only convert 2000 kcal to joules and 1 day to seconds. The energy used in a day is approximately

$$E = 2000\ \text{Cal}\left(\frac{10^3\ \text{cal}}{1\ \text{Cal}}\right)\left(\frac{4.186\ \text{J}}{1\ \text{cal}}\right) = 8.372 \times 10^6\ \text{J}$$

and 1 day is 86,400 s.

SOLVE

Using the values given, we can calculate an answer

$$P = \frac{E}{t} = \frac{8.372 \times 10^6\ \text{J}}{86,400\ \text{s}} = 96.9\ \text{J/s}$$

We like to use big, round numbers, such as 2000 kcal/day, in our estimations; we are justified in stating the result to one significant digit, or 100 J/s.

REFLECT

The power we consume is equivalent to the energy emitted from a 100-W light bulb every second. Think about how hot a 100-W bulb gets! And also, note that it is not correct to state the power as W *per day*; the SI units of watts include a measure of time.

★ What's Important 14-5

The energy that flows from one object to another as a result of a temperature difference is termed heat. The quantity specific heat is the ratio of the energy added to an object and the temperature change that results, divided by the mass of the object. The larger the specific heat of a substance, the smaller the temperature change for a given mass and energy flow.

14-6 Latent Heat

If you add energy to an object, will its temperature increase? The surprising answer is . . . maybe not! Substances can commonly exist in a solid, liquid, or gaseous phase. For example, we call the phases of H_2O ice, water, and steam (or water vapor). Energy must either be absorbed or released in order for a substance to change from one phase to another, but the temperature of the substance does not change during such a **phase transition**. Energy flow which would otherwise result in a temperature change is completely consumed in rearranging the organization of the molecules to effect the phase transition.

For a particular substance, at any given pressure each phase transition occurs at a specific temperature. In addition, the solid-to-liquid transition occurs at exactly the same temperature as the liquid-to-solid transition. Imagine, for example, an ice cube at the exact temperature at which ice melts. As energy flows into the ice it begins to melt; the resulting water is at the temperature at which water freezes. The temperature at which ice melts and the temperature at which water freezes are the same, so the about-to-melt ice is at the same temperature as the just-melted water.

Adding energy to the ice does not change its temperature but, instead, results in a phase transition.

The process during which a liquid becomes a solid is called *freezing* (or fusion) and the process during which a solid becomes a liquid is called *melting*. Other common phase changes include *boiling* (or vaporization), the transition from liquid to gas, and **condensation**, the transition from the gaseous phase to the liquid phase. Some substances can change from a solid to a gas without an intermediate liquid phase, during a process known as **sublimation**. Carbon dioxide is perhaps the most common example of a substance that undergoes sublimation; at room temperature and atmospheric pressure, CO_2 goes from a solid, commonly known as dry ice, directly to CO_2 gas. Conversely, a transition from gas directly to the solid phase is called **deposition**. Phase transitions can also occur between different arrangements of molecules in the crystalline structures of certain solids. Under high pressure, for example, the crystalline structure of the carbon atoms in graphite can be rearranged to form diamond.

Details of the phase transitions for a particular substance can be summarized in a **phase diagram** such as the generic one shown in **Figure 14-15**. Each of the three lines in the figure delineates a transition between two phases, effectively showing the transition temperature as a function of pressure.

We have defined the triple point as the pressure and temperature at which all three common phases of a substance can exist simultaneously. Three lines extend from the triple point to indicate the pressure and temperature at which the substance changes phase. For example, the line in the upper right shows the conditions under which the substance changes between a liquid and a gas, the boiling point. As the temperature and pressure approach the **critical point**, the properties of the liquid and gas phases approach one another. Beyond the critical point the substance exists in only one phase, called a *supercritical fluid*. Above the critical temperature, unlike below it, no increase in pressure will cause a gas to condense into liquid form.

Notice that in the phase diagram for water (**Figure 14-16a**) the triple point is in the range of the pressures and temperatures we experience on Earth's surface, which means that all three common phases of water are part of our everyday experiences. In contrast, the surface conditions on Titan, Saturn's largest moon, are close to the triple point of methane (CH_4). On Earth methane is normally found only in the gaseous state, but on Titan you might experience methane rain or even methane snow! And on Mars, although there is water present in the thin atmosphere, because the surface pressure is below that of the triple point, no liquid water is present. As the phase diagram suggests, when the temperature of frozen water on Mars increases it sublimates directly to atmospheric water vapor. CO_2 sublimates on Earth's surface for the same reason; typical environmental pressures and temperatures are far below the triple point of CO_2 (**Figure 14-16b**). Under certain conditions, for example, at high altitudes where air pressure is lower, ice or snow can also sublimate on Earth.

The amount of heat per unit mass required to cause a phase transition of a sample of a substance, known as the **latent heat** L, depends on the substance. For example, it takes more heat to melt a 1-kg block of ice than it does to melt a 1-kg block of iron. Yes, the temperature at which iron melts is considerably higher than the temperature at which ice melts, but when both ice and iron are at their respective melting temperatures, the amount of heat that must be added to melt the ice is larger than the heat

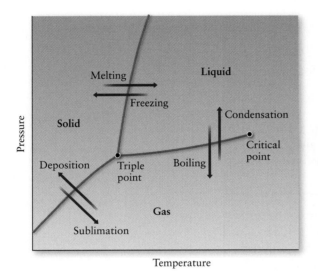

Figure 14-15 Phase diagrams show the relationship between pressure, temperature, and the phase of the substance. Each of the three lines in the figure marks a transition between two phases, effectively showing the transition temperature as a function of pressure. At the triple point, the substance can exist as a gas, liquid, or solid.

required to melt the iron. In addition, the amount of energy needed to cause a phase transition to occur is proportional to the mass of substance present. The more massive the block of ice or iron, for example, the more energy is required to melt it.

We can summarize the relationship between the heat Q associated with a phase transition, mass m of the sample of a substance, and latent heat L as $Q = mL$. Notice that the units of latent heat must be

$$[L] = \frac{[Q]}{[m]} = \frac{J}{kg}$$

Consider a sample of a solid substance. The amount of heat necessary to cause the solid to become a liquid is different than the amount of heat required for the liquid to become a gas. For example, nearly seven times more heat is required to boil 1 kg of water (when the temperature of the water is at its boiling point) than to melt 1 kg of solid water (ice) when the ice is at its melting point. The **latent heat of fusion** L_F is the energy per unit of mass required to change a substance from solid to liquid. In a similar way, the **latent heat of vaporization** L_V is the energy per unit of mass required to change a substance from liquid to gas. Just as energy must be added to a substance to cause it to melt or to boil, energy is released when a substances freezes or liquefies. Thus, the latent heat of fusion is also the energy released per unit mass when a substance changes from a liquid to a solid, and latent heat of vaporization is the energy released per unit mass when a substance changes from gas to liquid.

The heat required to melt a solid into a liquid or released when a liquid freezes is

$$Q = mL_F \qquad (14\text{-}28)$$

The heat required to evaporate a liquid into a gas or released when a gas liquefies is

$$Q = mL_V \qquad (14\text{-}29)$$

Table 14-4 lists the melting and boiling temperatures, and also the latent heats of fusion and the latent heats of vaporization for some common substances.

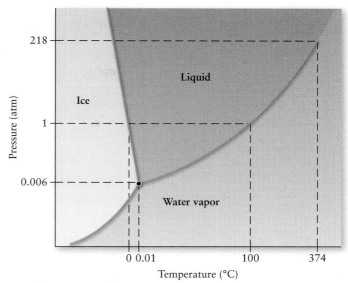

(a) Water

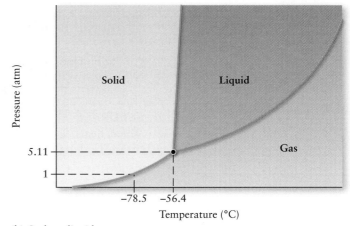

(b) Carbon dioxide

Figure 14-16 (a) Because the triple point of water is in the range of the pressures and temperatures we experience on Earth's surface, water vapor, water, and ice all exist naturally on our planet. (b) The phase diagram for carbon dioxide shows the triple point to be $-56.4\,°C$.

? Got the Concept 14-8
Guess the Liquid

A test tube that appears empty is placed in a dewar containing liquid nitrogen. The open end of the test tube is far above the surface of the liquid and the closed end of the test tube is submersed in the liquid nitrogen. A minute or two later the test tube is removed, revealing that a small amount of a pale blue liquid has collected inside. What is the liquid, and by what process did it collect in the test tube?

Table 14-4 Melting and Boiling Temperatures (T_F and T_V) and Latent Heats of Fusion and Vaporization (L_F and L_V) at 1 atm

Substance	T_F (°C)	T_V (°C)	L_F (kJ/kg)	L_V (kJ/kg)
alcohol (ethyl)	159	351	104	830
aluminum	933	2792	397	10,900
copper	1357	2835	209	4730
gold	1337	3129	63.7	1645
hydrogen (H_2)	14	20	59.5	445
iron	1811	3134	247	6090
lead	600	2022	24.5	866
nitrogen (N_2)	63	77	25.3	199
oxygen (O_2)	54	90	13.7	213
water	273	373	334	2260

Example 14-6 Melting Ice

A 1-kg block of ice initially at a temperature of 0 °C is placed inside an experimental apparatus that releases 250 kJ of energy to the ice. What fraction of the ice melts as a result? What is the final temperature of the melted ice?

SET UP

When heat Q causes a phase transition from solid to liquid, the latent heat of fusion L_F determines the relationship between Q and m, the mass that melts:

$$Q = mL_F \tag{14-28}$$

The mass that melts is then equal to

$$m = \frac{Q}{L_F} \tag{14-30}$$

from which we can determine the fraction of the block of ice that melts.

SOLVE

The fraction of the block that melts is equivalent to the mass of the ice that melts divided by the total mass of the ice:

$$\text{fraction of ice that melts} = \frac{m_{\text{melt}}}{M}$$

Using Equation 14-30,

$$\text{fraction of ice that melts} = \frac{Q/L_F}{M}$$

The latent heat of fusion for water, which can be found in Table 14-4, is $L_F = 334$ kJ/kg. With the addition of 250 kJ of energy the fraction of the ice that melts is

$$\text{fraction of ice that melts} = \frac{250 \times 10^3 \, \text{J}/334 \times 10^3 \, \text{J/kg}}{1 \, \text{kg}} = 0.75$$

Seventy-five percent of the ice block melts. Because all of the added energy goes into melting the ice, the temperature of the resulting water remains at 0 °C.

REFLECT

The amount of energy added to the ice (250 kJ) is less than the energy required to melt the entire 1 kg, so not all of the ice melts. Melting all of the ice would require at least 334 kJ of heat to be transferred to the block.

Practice Problem 14-6 A 3-kg block of ice initially at a temperature of 0 °C is placed inside an experimental apparatus that releases 250 kJ of energy to the ice. What fraction of the ice melts as a result?

Got the Concept 14-9
Adding Energy to Ice

What if more than 334 kJ were added to the 1-kg block of ice in the previous problem?

Energy, in the form of heat, must flow into a substance for a phase change from solid to liquid, or liquid to gas, to occur. Often the source of that energy is another object in thermal contact with the first. For example, as the sweat on our skin evaporates on a hot day, energy flows from our bodies to the beads of perspiration. The process results in the temperatures of our bodies decreasing just a bit. Although other effects can come into play, energy must be conserved, so whatever energy leaves our bodies appears as energy added to the sweat. This balance is expressed in Equation 14-22, which for the body–sweat system takes the form

$$|Q_{body}| = |Q_{sweat}| \tag{14-31}$$

By convention, Q_{sweat}, the heat transferred to the sweat, is positive and Q_{body}, the heat transferred from the body, is negative. As a result, the body rids itself of excess energy; in other words, perspiration helps us cool off.

The same approach can be applied to an ice cube dropped into a glass of water. Heat is transferred into ice in order for the ice to melt (positive quantity), and this heat must equal the magnitude of heat transferred from the water to the ice (negative quantity). The temperature of the water decreases, and if the water loses enough energy, some or all of it could even freeze.

Example 14-7 Cooling Off

A typical person perspires at a rate of about 0.6 L/day or about 7×10^{-6} kg/s. Even though sweat contains low concentrations of ions, for our purposes we can reasonably assume that sweat behaves like water. It is in good thermal contact with the skin, so most of the energy required to vaporize it is drawn from the body. Find the rate in watts at which the body loses energy as sweat evaporates. At human body temperature, sweat's latent heat of vaporization is about 2.430×10^6 J/kg.

SET UP

Energy flows from the body into the sweat. For every mass Δm of water, we need an amount of energy per unit mass of $L_V = 2.430 \times 10^6$ J/kg to cause the water to evaporate. To find the *rate* of cooling, however, we need to include time, which is accounted for in the perspiration rate $\Delta m / \Delta t = 7 \times 10^{-6}$ kg/s. From Equation 14-29 the magnitude of the body's heat loss is

$$|Q_{body}| = \Delta m L_V$$

so the energy flow rate, or power dissipated, is

$$P = \frac{|Q_{\text{body}}|}{\Delta t} = \frac{\Delta m}{\Delta t} L_V$$

SOLVE

Using the given perspiration rate and the latent heat of vaporization of sweat at body temperature, the rate at which energy flows into the sweat to cause it to evaporate is

$$P = \frac{\Delta m}{\Delta t} L_V = (7 \times 10^{-6}\,\text{kg/s})(2.430 \times 10^6\,\text{J/kg}) \approx 20\,\text{J/s}$$

The rate at which energy is dissipated by a typical person due to perspiration is about 20 J/s or 20 W.

REFLECT

This is similar to the power of an incandescent bulb you might find in a small appliance. It's not all that much, but it's enough to keep you cool. But consider that the perspiration rate of 0.6 L per *day* is for an average person. For someone who has spent even just a few weeks in a hot climate and acclimatized to the harsh conditions, the rate can increase to as much as 4 L per *hour*. That's more than 150 times the typical rate, which results in an increase of power dissipated to more than 3000 W—twice that of a typical electric hair dryer.

Example 14-8 Melting Ice, Again

A chunk of ice at $T_{\text{ice}} = -10\,°\text{C}$ is placed in an insulated container that holds 1.0 kg of water at $T_{\text{water}} = 20\,°\text{C}$. When the system comes to thermal equilibrium, all of the ice has melted and the entire system is at a temperature of 0 °C. Find the mass of the ice.

SET UP

Three processes occur as the system reaches thermal equilibrium: The temperature of the ice increases, the ice melts, and the temperature of the water decreases from 20 °C to 0 °C. Energy must be added to the ice in order both for its temperature to increase and for the phase transition to occur. During the third process, the temperature decrease of the water is the result of energy leaving the water. The statement of conservation of energy expressed in Equation 14-31 requires that

$$|Q_{\text{ice}}| = |Q_{\text{water}}|$$

or

$$Q_{\text{ice}} = -Q_{\text{water}}$$

Specifically, using $Q_{\text{ice,temp}}$ to represent the heat required to increase the temperature of the ice from $-10\,°\text{C}$ to 0 °C, $Q_{\text{ice,melt}}$ to represent the heat required to melt the ice, and $Q_{\text{water,temp}}$ to represent the heat (negative quantity) associated with the change in the temperature of the water,

$$Q_{\text{ice,temp}} + Q_{\text{ice,melt}} = -Q_{\text{water,temp}} \tag{14-32}$$

Note that the subscript "water" refers only to the water that was in the liquid state at the start of the problem, *not* the water that results from the melting of the ice. We need to keep track of them separately, because the two quantities of water are only at the same temperature when the system comes to thermal equilibrium. We can apply Equation 14-23 for $Q_{\text{ice,temp}}$ and $Q_{\text{water,temp}}$, and Equation 14-28 for $Q_{\text{ice,melt}}$ to obtain an expression for the mass of the ice.

SOLVE

According to Equation 14-23,

$$Q_{ice,temp} = m_{ice} c_{ice} \Delta T_{ice} = m_{ice} c_{ice} (T_{ice,f} - T_{ice,i})$$

and

$$Q_{water,temp} = m_{water} c_{water} \Delta T_{water} = m_{water} c_{water} (T_{water,f} - T_{water,i})$$

Notice that because $T_{water,i}$ is larger than $T_{water,f}$, $Q_{water,temp}$ is negative. This result is expected, because the water releases energy as it cools.

Using Equation 14-28,

$$Q_{ice,melt} = m_{ice} L_{F,ice}$$

Combining the expressions according to Equation 14-32 gives

$$m_{ice} c_{ice} (T_{ice,f} - T_{ice,i}) + m_{ice} L_{F,ice} = -m_{water} c_{water} (T_{water,f} - T_{water,i})$$

This equation can be solved for m_{ice}, because every other variable has a known value. Rearranging for m_{ice} gives

$$m_{ice} = \frac{-m_{water} c_{water} (T_{water,f} - T_{water,i})}{c_{ice} (T_{ice,f} - T_{ice,i}) + L_{F,ice}}$$

Using the values given in the problem statement:

$$m_{ice} = \frac{-(1.0 \text{ kg})(4186 \text{ J/kg} \cdot {}^\circ\text{C})(0\,{}^\circ\text{C} - 20\,{}^\circ\text{C})}{(2093 \text{ J/kg} \cdot {}^\circ\text{C})[0\,{}^\circ\text{C} - (-10\,{}^\circ\text{C})] + 334 \times 10^3 \text{ J/kg}}$$

or

$$m_{ice} = 0.24 \text{ kg}$$

REFLECT

The mass of ice required to decrease the temperature of 1 kg of water from room temperature to 0 °C while the ice just melts is 0.24 kg. In more familiar terms, 1 kg is the mass of 1 L of water, and an ice cube from a typical ice cube tray is about 0.04 kg (40 g). So it would take about 6 ice cubes to reproduce the process described in this problem, which doesn't seem unreasonable.

Practice Problem 14-8 A chunk of ice at $T_{ice} = -16\,{}^\circ\text{C}$ is placed in an insulated container that holds 1.9 kg of water at $T_{water} = 21\,{}^\circ\text{C}$. When the system comes to thermal equilibrium, all of the ice has melted and the entire system is at a temperature of 0 °C. Find the mass of the ice.

Example 14-9 Ice and Water

Ice that has a mass m_{ice} equal to 0.15 kg and a temperature T_{ice} equal to $-10\,{}^\circ\text{C}$ is placed in an insulated container that holds 1.0 kg of water that has a temperature T_{water} equal to 20 °C. When the system comes to thermal equilibrium, how much ice and how much water are present? What is the temperature?

SET UP

The mass of the ice is small compared to the mass of the water, so we should expect that all of the ice will melt. But the temperature of the water probably won't decrease to 0 °C. What if the mass of the ice is large enough so that the ice doesn't all melt? We could even imagine a case in which there is enough ice relative to water so that the ice doesn't even all get to the melting point. This uncertainty presents a challenge. What change of temperature, for example, should be used in $Q = mc\Delta T$ for the water, and for the ice? Should we include a contribution to the heat associated with some of the ice melting, or some of the water freezing?

The approach we take is similar to having a bank debit card; we put money in and then spend as much as we need as long as we don't overdraw the account. The currency for thermodynamic processes is energy. In this problem, we make "deposits to the account" by considering how much energy the water could release by cooling it down to the freezing point; that's the maximum amount of energy we could "spend" or the maximum amount of energy that can be added to the ice. Recall that the internal energy of a substance is higher in the liquid state than in the solid state but lower than in the gaseous state. Phase changes would also change the spending limit on the account: gas-to-liquid or liquid-to-solid phase changes would increase the amount of energy available to spend. Melting a solid into a liquid or boiling a liquid into a gas would decrease the amount of energy available to spend or use. At the end of the problem, any energy not spent is not used (is kept in the "energy account"). For example, the temperature of the water may not decrease to 0 °C. Perhaps most important, we can spend no more from the energy account than is available, as we cannot violate the conservation of energy.

The energy account method requires us to consider, at each step in a process, whether enough energy is available for the process to continue. It involves a bit more numeric manipulations than we normally prefer but has the advantage of being relatively transparent.

SOLVE

We start by calculating the total energy in our account, that is, the amount of energy that would be released if the temperature of the 1 kg of water were decreased from room temperature to the freezing point. The magnitude is determined from Equation 14-23:

$$|Q_{\text{water,temp}}| = |m_{\text{water}}c_{\text{water}}\Delta T_{\text{water}}| = |m_{\text{water}}c_{\text{water}}(T_{\text{water,f}} - T_{\text{water,i}})|$$

$$= |(1 \text{ kg})[4186 \text{ J/(kg} \cdot {}^\circ\text{C)}](0 \text{ °C} - 20 \text{ °C})|$$

$$= 83{,}720 \text{ J}$$

We can spend as much or as little of this energy as we need. (We didn't put this intermediate result into scientific notation to make it easier to keep track of the running balance in our energy account.) To ensure that we don't introduce rounding errors into the calculations, we'll keep more significant figures than are warranted until we finish the problem.

What is the cost to increase the temperature of the ice to the melting point? Once again, we can find the energy from Equation 14-23:

$$Q_{\text{ice,temp}} = m_{\text{ice}}c_{\text{ice}}\Delta T_{\text{ice}} = m_{\text{ice}}c_{\text{ice}}(T_{\text{ice,f}} - T_{\text{ice,i}})$$

To increase a 0.15-kg sample of ice from −10 °C to 0 °C we need

$$Q_{\text{ice,temp}} = (0.15 \text{ kg})[2093 \text{ J/(kg} \cdot {}^\circ\text{C)}](0 \text{ °C} - [-10 \text{ °C}])$$
$$= 3{,}140 \text{ J}$$

More than enough energy is in the "account" for this process; we therefore "spend" $Q_{\text{ice,temp}} = 3{,}140$ J to increase the temperature of the ice to ice's melting point, leaving 83,720 J − 3,140 J, or 80,580 J, in the energy account. Is the remaining energy enough to melt the ice? Melting the ice requires, from Equation 14-28,

$$Q_{\text{ice,melt}} = m_{\text{ice}}L_{\text{F,ice}} = (0.15 \text{ kg})(334 \times 10^3 \text{ J/kg})$$

$$= 50{,}100 \text{ J}$$

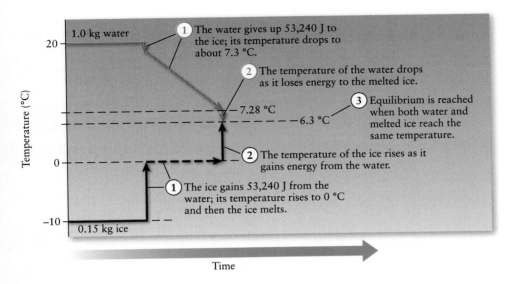

Figure 14-17 As 0.15 kg of ice melts in 1.0 kg of water, the water gives up energy (red arrows) and the ice gains it (blue arrows). Heat transfer stops when the water and the melted ice reach the same temperature.

Yes, enough energy exists in the account to melt all of the ice. We can imagine, then, that all of the ice melts, leaving a mixture of 0.15 kg of water at 0 °C (the result of the ice melting) and 1 kg of water at some temperature lower than the original 20 °C in the insulated container. By how much has the temperature of the water dropped? So far 3,140 J plus 50,100 J, or 53,240 J, of heat has been released from the water. The temperature change of the water is then, from Equation 14-18,

$$\Delta T_{\text{water}} = \frac{1}{m_{\text{water}} c_{\text{water}}} Q_{\text{water}} = \frac{1}{(1 \text{ kg})[4186 \text{ J}/(\text{kg} \cdot {}^{\circ}\text{C})]}(53{,}240 \text{ J})$$

$$= 12.72 \,{}^{\circ}\text{C}$$

The (original) water is now at 20 °C − 12.72 °C, or 7.28 °C. The last step of the problem is then straightforward: find the temperature at thermal equilibrium of 0.15 kg of water at 0 °C combined with 1 kg of water at 7.28 °C. The heat from the melted ice (positive quantity) must be equal in magnitude to the heat from the water (negative quantity); we have previously expressed this as Equation 14-26:

$$m_1 c_{\text{water}}(T_{1,\text{i}} - T_{\text{f}}) = m_2 c_{\text{water}}(T_{\text{f}} - T_{2,\text{i}})$$

where T_{f} is the final temperature of both quantities of water, and subscripts "1" and "2" here represent the water and melted ice, respectively. So,

$$T_{\text{f}} = \frac{m_1 T_{1,\text{i}} + m_2 T_{2,\text{i}}}{m_1 + m_2} = \frac{(1 \text{ kg})(7.28 \,{}^{\circ}\text{C}) + (0.15 \text{ kg})(0 \,{}^{\circ}\text{C})}{1 \text{ kg} + 0.15 \text{ kg}}$$

$$= 6.3 \,{}^{\circ}\text{C}$$

REFLECT

Because there is much more water than ice in the original mixture, we expected all of the ice to melt, as is the case. Thus, the final state of the system is all water, and the final temperature must be above 0 °C. The processes that lead to the equilibrium state are shown graphically in Figure 14-17.

Practice Problem 14-9 Ice that has a mass m_{ice} equal to 0.15 kg and a temperature T_{ice} equal to −10 °C is placed in an insulated container that holds 2.0 kg of water that has a temperature T_{water} equal to 10 °C. When the system comes to thermal equilibrium, how much ice and how much water is present? What is the temperature?

Example 14-10 Ice and Water, Again

Ice that has a mass m_{ice} equal to 150 kg at a temperature T_{ice} equal to −10 °C is placed in an insulated container that holds 1.0 kg of water at a temperature T_{water} equal to 20 °C. When the system comes to thermal equilibrium, how much ice and how much water is present? What is the temperature?

SET UP

We approach the problem in the same way as in Example 14-9. The difference between the two examples is that this time, the mass of the ice is quite large compared to the mass of the water. Because we don't know whether the ice melts or if it does, how much, or whether any or all of the water freezes, we simply apply the energy account approach.

SOLVE

We start by calculating the total energy released by decreasing the temperature of 1 kg of water from room temperature to the freezing point. Again, the magnitude is determined from Equation 14-23:

$$|Q_{water,temp}| = |m_{water}c_{water}\Delta T_{water}| = |m_{water}c_{water}(T_{water,f} - T_{water,i})|$$

$$= |(1\text{ kg})(4186\text{ J/kg}\cdot°\text{C})(0\text{ °C} - 20\text{ °C})|$$

$$= 83{,}720\text{ J}$$

Do we have enough energy to increase the temperature of the ice to the melting point? The energy necessary is

$$Q_{ice,temp} = m_{ice}c_{ice}\Delta T_{ice} = m_{ice}c_{ice}(T_{ice,f} - T_{ice,i}) \qquad (14\text{-}23)$$

So to increase 150 kg of ice from −10 °C to 0 °C, we need

$$Q_{ice,temp} = (150\text{ kg})(2093\text{ J/kg}\cdot°\text{C})(0\text{ °C} - [-10\text{ °C}])$$

$$= 3{,}139{,}500\text{ J}$$

The 83,720 J obtained from cooling the water to 0 °C will increase the temperature of the ice, but it's not nearly enough to increase all of the ice to the melting point. We can use Equation 14-18 to find out the temperature change of the ice:

$$\Delta T_{ice} = \frac{Q_{ice,temp}}{m_{ice}c_{ice}} = \frac{83{,}720\text{ J}}{(150\text{ kg})[2093\text{ J/(kg}\cdot°\text{C})]} = 0.267\text{ °C}$$

Withdrawing the 83,720 J from our energy account leaves us with water at 0 °C and ice at −10 °C plus 0.267 °C, or −9.73 °C. Now what?

Because the temperature of the water is at the freezing point, we can withdraw more energy from the water by letting it freeze. According to Equation 14-28, the energy released is

$$Q = m_{water}L_F = (1\text{ kg})(334 \times 10^3\text{ J/kg}) = 334{,}000\text{ J}$$

This energy can also be spent on increasing the temperature of the ice:

$$\Delta T_{ice} = \frac{Q_{ice,temp}}{m_{ice}c_{ice}} = \frac{334{,}000\text{ J}}{(150\text{ kg})[2093\text{ J/(kg}\cdot°\text{C})]} = 1.06\text{ °C}$$

We now have ice at $-9.73\,°C$ plus $1.06\,°C$, or $-8.67\,°C$, along with the ice which started as water, now at $0\,°C$. All that remains is to determine the temperature at thermal equilibrium of 150 kg of ice at $-8.67\,°C$ combined with 1 kg of ice at $0\,°C$. Once again we employ energy balance in the form of Equation 14-26:

$$m_1 c_{ice}(T_{1,i} - T_f) = m_2 c_{ice}(T_f - T_{2,i})$$

where T_f is the final temperature of both, and subscripts "1" and "2" represent the ice and frozen water, respectively, so

$$T_f = \frac{m_1 T_{1,i} + m_2 T_{2,i}}{m_1 + m_2}$$

$$= \frac{(150\ \text{kg})(-8.67\,°C) + (1\ \text{kg})(0\,°C)}{150\ \text{kg} + 1\ \text{kg}}$$

$$= -8.61\,°C$$

At thermal equilibrium, all of the water has frozen, and the final temperature of the ice is $-8.6\,°C$.

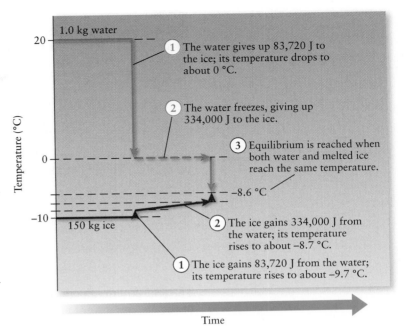

Figure 14-18 As 1.0 kg of water (red arrows) and 150 kg of ice (blue arrows) come into thermal equilibrium, the water freezes and the ice warms up.

REFLECT
Because there is much more ice than water in the original mixture, all of the water freezes, and the final temperature is close to the initial temperature of the ice. The processes that lead to the equilibrium state are shown graphically in Figure 14-18.

Practice Problem 14-10 Ice that has a mass m_{ice} equal to 150 kg at a temperature T_{ice} equal to $-10\,°C$ is placed in an insulated container that holds 2.0 kg of water at a temperature T_{water} equal to $20\,°C$. When the system comes to thermal equilibrium, how much ice and how much water is present? What is the temperature?

Example 14-11 Ice and Water, Yet Again

Ice that has a mass m_{ice} equal to 15 kg at a temperature T_{ice} equal to $-10\,°C$ is placed in an insulated container that holds 1.0 kg of water at a temperature T_{water} equal to $20\,°C$. When the system comes to thermal equilibrium, how much ice and how much water is present? What is the temperature?

SET UP
The only difference between this problem and Examples 14-9 and 14-10 is the mass of the ice. In Example 14-9, the mass of the ice was small relative to that of the water and all of the ice melted. In Example 14-10, the mass of the ice was large relative to the mass of the water and all of the water froze. Here, the masses of ice and water are similar, but we still don't know how much, if any, of the ice melts or how much, if any, of the water freezes. We will apply the energy account approach.

SOLVE

The total energy released by decreasing the temperature of 1 kg of water from room temperature to the freezing point, from Equation 14-23, is

$$|Q_{\text{water,temp}}| = |m_{\text{water}}c_{\text{water}}\Delta T_{\text{water}}| = |m_{\text{water}}c_{\text{water}}(T_{\text{water,f}} - T_{\text{water,i}})|$$

$$= |(1\,\text{kg})[4186\,\text{J/(kg}\cdot\text{°C)}](0\,\text{°C} - 20\,\text{°C})|$$

$$= 83{,}720\,\text{J}$$

Do we have enough energy to increase the temperature of the ice to the melting point? The energy necessary is

$$Q_{\text{ice,temp}} = m_{\text{ice}}c_{\text{ice}}\Delta T_{\text{ice}} = m_{\text{ice}}c_{\text{ice}}(T_{\text{ice,f}} - T_{\text{ice,i}}) \tag{14-23}$$

So to increase 15 kg of ice from -10 °C to 0 °C, we need

$$Q_{\text{ice,temp}} = (15\,\text{kg})[2093\,\text{J/(kg}\cdot\text{°C)}](0\,\text{°C} - [-10\,\text{°C}])$$

$$= 313{,}950\,\text{J}$$

The 83,720 J obtained from cooling the water to 0 °C will increase the temperature of the ice, but it's not enough to increase the temperature of the ice to 0 °C. We can use Equation 14-18 to find out the temperature change of the ice:

$$\Delta T_{\text{ice}} = \frac{Q_{\text{ice,temp}}}{m_{\text{ice}}c_{\text{ice}}} = \frac{83{,}720\,\text{J}}{(15\,\text{kg})(2093\,\text{J/kg}\cdot\text{°C})} = 2.67\,\text{°C}$$

Withdrawing the 83,720 J from our energy account leaves us with water at 0 °C and ice at -10 °C plus 2.67 °C, or -7.33 °C. The process to reach thermal equilibrium is not finished, however, because the ice and water are not at the same temperature. What's next?

Because the temperature of the water is at the freezing point, we can withdraw more energy from the water by letting it freeze. According to Equation 14-28, the energy released if all of the water were to freeze would be

$$Q = m_{\text{water}}L_F = (1\,\text{kg})(334 \times 10^3\,\text{J/kg}) = 334{,}000\,\text{J}$$

This energy can be spent on increasing the temperature of the ice. Is there enough to bring the temperature of the ice from -7.33 °C to 0 °C? We found earlier that 313,950 J is needed to bring the ice from -10.0 °C to 0 °C, of which we've already supplied 83,720 J obtained from cooling the water. Thus, an additional 230,230 J (313,950 J $-$ 83,720 J) is need to bring the ice to 0 °C. More energy would be released were all the water to freeze, which means that some of the water remains in the liquid state. Equation 14-28 can be used again to determine the amount of water that must freeze to release the 230,230 J necessary to increase the ice to 0 °C; from $Q = m_{\text{water}}L_F$,

$$m_{\text{water}} = \frac{Q}{L_F} = \frac{230{,}230\,\text{J}}{334 \times 10^3\,\text{J/kg}} = 0.69\,\text{kg}$$

So 0.69 kg of the water freezes, leaving the rest (0.31 kg) in the liquid phase. We increased the temperature of the original ice to 0 °C but did not melt any of it. Thus, the final state of the system is a mixture of 15 kg plus 0.69 kg of ice at 0 °C, and 0.31 kg of water, also at 0 °C.

REFLECT

In this situation, the final state contains a mixture of two different phases. Notice, however, that the final temperature of the ice and water must be the same, the definition of thermal equilibrium. The processes that lead to the equilibrium state are shown graphically in **Figure 14-19**.

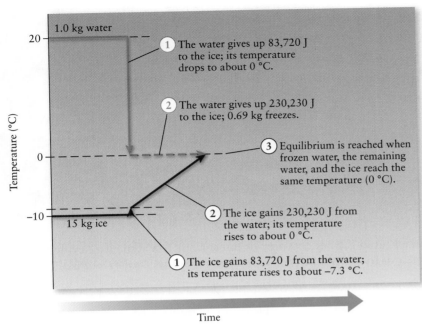

Figure 14-19 As 1.0 kg of water (red arrows) and 15 kg of ice (blue arrows) come into thermal equilibrium, some but not all of the water freezes. The final temperature is 0 °C.

1.0 kg water

20

① The water gives up 83,720 J to the ice; its temperature drops to about 0 °C.

② The water gives up 230,230 J to the ice; 0.69 kg freezes.

③ Equilibrium is reached when frozen water, the remaining water, and the ice reach the same temperature (0 °C).

0

Temperature (°C)

② The ice gains 230,230 J from the water; its temperature rises to about 0 °C.

–10

15 kg ice

① The ice gains 83,720 J from the water; its temperature rises to about –7.3 °C.

Time

Practice Problem 14-11 Ice that has a mass m_{ice} equal to 15 kg at a temperature T_{ice} equal to -10 °C is placed in an insulated container that holds 0.8 kg of water at a temperature T_{water} equal to 20 °C. When the system comes to thermal equilibrium, how much ice and how much water is present? What is the temperature?

✳ What's Important 14-6

Energy, in the form of heat, must either be absorbed or released in order for a substance to change from one phase to another. The latent heat of fusion is the amount of heat per unit mass required to cause a substance to undergo a phase transition between the solid and liquid phases. The latent heat of vaporization is the amount of heat per unit mass required to cause a substance to undergo a phase transition between the liquid and gaseous phases. The triple point is the temperature and pressure at which the solid, liquid, and gaseous phases of a substance can coexist in thermal equilibrium.

14-7 Heat Transfer: Radiation, Convection, Conduction

Animals thrive in a wide variety of harsh environments. Some rabbits, for example, evolved to live in the desert while others adapted to life in the Arctic (Figure 14-20a). The jackrabbit looks scrawny with its thin fur, long and skinny legs, and enormous ears. The large surface area relative to its volume, along with the thin layer of insulation, promotes heat loss and enables the animal to survive in the desert. In contrast, the arctic hare's stubby legs, smaller ears, low surface area-to-volume ratio, and thick layer of insulation help minimize heat loss (Figure 14-20b). Reptiles can't directly control their body temperatures the way mammals such as rabbits can. After a swim in the cold Pacific Ocean, the iguanas in Figure 14-21 have crawled up out of the water to bask on rocks that have been heated by the hot tropical sunshine. In this section, we'll look at how heat is transferred from one place to another, such as from a rabbit to its environment or from a hot rock to a cold iguana.

(a)   (b)

Figure 14-20 (a) Jackrabbits evolved to radiate heat in the hot desert environment. (b) The arctic hare evolved to conserve heat in the cold Arctic environment. *((a) Scott Rheam/U.S. Fish and Wildlife Service. (b) Bonnie Fink/Dreamstime.com.)*

Figure 14-21 Although iguanas are unable to control their body temperature, they warm themselves by basking in the sunshine. They cool themselves by plunging into the cold ocean. *(Michio Hoshino/Minden Pictures.)*

Heat can be transferred from one place or object to another by three processes, **radiation**, **convection**, and **conduction**:

Radiation is the energy transfer by the emission (or absorption) of electromagnetic radiation, for example, light, infrared radiation, and microwaves. (We'll see in Chapter 23 that all are different manifestations of the same physical phenomenon.)

Convection is the energy transfer by the motion of a liquid or gas, for example, by warmer air moving into a cooler region of a system.

Conduction is energy transfer by the collision of particles in one object with the particles in another, or by the collision of more energetic particles in one part of an object with less energetic particles in adjacent parts.

Radiation is the process that warms the rocks and the iguanas as they bask in the sunshine. Convection carries energy away from the arctic hare, as the air warmed by the hare rises away from the animal. And when iguanas spread out on a hot rock, energy is transferred from the rock to the reptiles by conduction.

Radiation

From the definitions of the three heat transfer processes, we can see that radiation is unlike the other two. Both convection and conduction rely on particles bumping against one another. Radiation doesn't; in fact, radiation doesn't involve the kinetic energy of molecules or atoms. The energy emitted by the Sun, for example, a significant fraction of which is in the form of visible or close-to-visible light, reaches Earth after passing through empty space.

As the temperature of an object increases, the amount of energy radiated per second from any region of the object's surface increases dramatically. Experiments show that the radiated power P is proportional to both temperature T and the size, or area A, of the radiating region:

$$P = \sigma \varepsilon A T^4 \tag{14-33}$$

The proportionality constant is $\sigma = 5.6703 \times 10^{-8}\ \text{W}/(\text{m}^2 \cdot \text{K}^4)$, which is the Stefan–Boltzmann constant. Notice that the Stefan–Boltzmann constant carries SI units, which requires that values of T be given in K in order to make computations with Equation 14-33. The additional factor ε is the emissivity of the surface, a factor between 0 and 1 that indicates how well or poorly a surface radiates; a surface with a value of ε close to 1 is a good radiator of thermal energy. The most striking feature of Equation 14-33, however, is that the radiated power increases in proportion to the fourth power of the absolute temperature.

? Got the Concept 14-10
Clay Pot

After a clay pot at room temperature is placed in a kiln, the pot's temperature is twice as high. How much more heat per second is the pot radiating when hot compared to when cool?

? Got the Concept 14-11
Hummingbirds

Hummingbirds have a very large surface area relative to their tiny volume, making them particularly prone to losing heat at night when the ambient temperature decreases well below their body temperature. To conserve energy and extend the time they can survive without food, hummingbirds allow their body temperature to decrease and they enter an altered physiological state called *torpor*. Explain the advantages of torpor to a hummingbird, in the context of heat exchange.

Radiation plays a significant role in heat loss from the human body. Under typical conditions, approximately half of the energy transferred from the body to the environment is in the form of radiation.

Convection

You probably learned long ago that "hot air rises." It does, and the rising air carries energy with it from one region to another in a process called convection. A continuous circulating flow that forms when rising air, say, in a room, forces the air above it to move out of the way and then downward somewhere else, as shown in **Figure 14-22**, is known as a *convection current*.

Nearly all substances expand when heated. As a region of air or other fluid expands it becomes less dense than its surroundings, and so is buoyed up according to Archimedes' principle (Section 11-7). You might like to imagine that hot air floats in cooler air, and that cooler air sinks.

Both local and global climate conditions are driven by convection. For example, sailors and people who live near the coasts of large bodies of water often experience a breeze that blows toward the shore in late afternoon and toward the water

Figure 14-22 A continuous circulating flow forms when warm air rises, forcing the air above it to move out of the way and then downward somewhere else. This is known as a convection current.

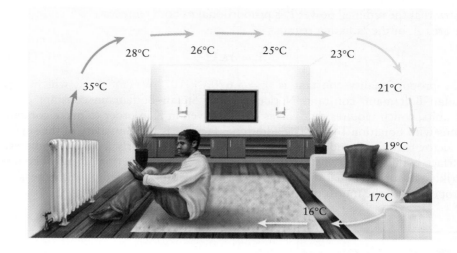

in early morning (Figure 14-23). Because the specific heat of land is much lower than the specific heat of water, the land heats up and cools down more quickly than the water. Especially when a day is hot and the sky is clear, the temperature of the land becomes higher, relative to the temperature of the water, as the day progresses; the air above the land gets warmer and less dense. As the warm air rises, it is replaced by cooler air from above the water, resulting in an onshore wind, or sea breeze. In the early morning, after the land and the air above it has cooled to a temperature below the temperature of the water, air rises above the relatively warmer water, and cooler air from above the land flows to replace it. The process results in an offshore wind, or land breeze.

Onshore and offshore breezes are relatively local phenomena, but because some regions of Earth's surface tend to be warmer or cooler than others, either

Figure 14-23 Convection drives climate. Because the specific heat of land is much lower than the specific heat of water, land heats up and cools down more quickly than the water. This creates offshore breezes during early morning and onshore breezes in late afternoon.

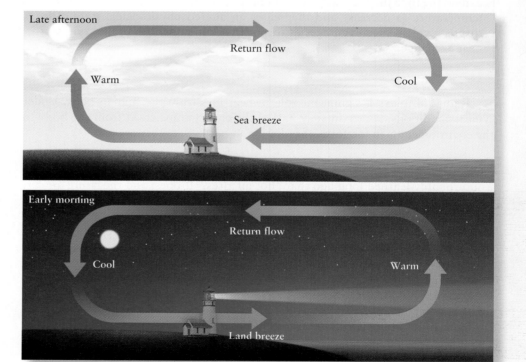

seasonally or for longer periods, convection drives global wind patterns in the atmosphere. Regions near the equator tend to be warm so, for example, winds near the surface in the Indian Ocean (Figure 14-24) tend to blow toward the equator from both north and south.

Conduction

Perhaps you've made the mistake of touching the handle of a metal spoon that has been resting in a pot of soup on the stove. Even though the handle itself is neither touching the stove nor in the soup, it can get hot enough to hurt your fingers if you touch it. Of course, a physical connection to the source of the heat exists. The end you hold is connected to the other end of the spoon, which is in the soup, which is touching the pot, which is touching the stove.

The temperature of an object is directly proportional to the average kinetic energy of its constituent particles. The energetic particles of the hot soup can collide with particles in the end of the spoon in the liquid. As those collisions impart some of the soup's energy to the spoon, that end of the spoon gets hotter. The increased energy gets passed along the spoon, from atom to atom, as collisions between metal atoms occur. Through the process of conduction, the end of the spoon farthest from the soup eventually will also be hot. The rate at which energy is transferred from one region of an object to another is proportional to the difference in temperature between the two.

Porpoise flukes and flippers are thin and relatively uninsulated. Because they have relatively large surface areas compared to their volume, flukes and flippers are potentially major sources of heat loss (Figure 14-25). However, veins carrying cold blood back to a porpoise's body completely surround

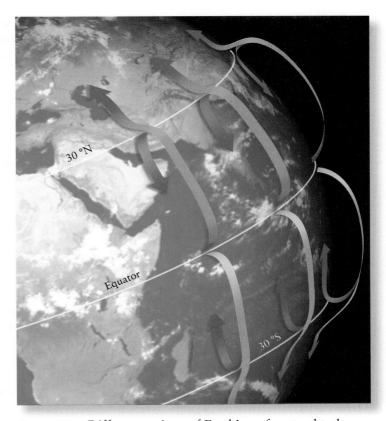

Figure 14-24 Different regions of Earth's surface tend to be warmer or cooler than others. Regions near the equator tend to be warm, for example, so winds in the Indian Ocean tend to blow toward the equator from both north and south.

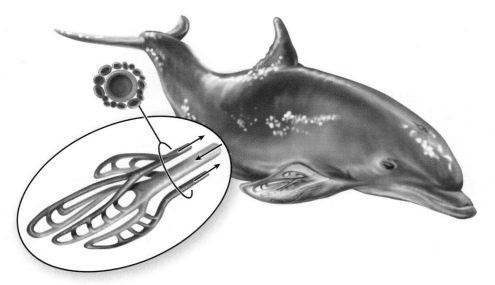

Figure 14-25 Countercurrent heat exchange conserves body heat in many animals, including porpoises. Veins carrying cold blood from the surface of the flipper surround arteries transporting warm blood from the animal's core to the flipper. Heat flows from the arteries to the veins. By the time arterial blood gets to the surface of the flippers, it has already lost much of its heat to venous blood.

The conduction rate, the rate of heat transfer between the hot and cold sides of the cylinder, is proportional to the temperature difference $T_H - T_C$.

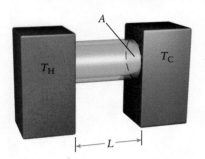

In addition, the rate is higher when the cross-sectional area of the cylinder is larger and the length is smaller, making it easier for energy transfer to occur.

Figure 14-26 The rate of heat exchange between two objects depends on the temperature difference, and on the cross-sectional area and length of the contact region between them.

each artery carrying warm blood into the flipper. The porpoise avoids significant heat loss because as the venous and arterial blood flow past each other, the warm blood heats the cold blood before it leaves the flipper. Moreover, by the time arterial blood reaches tissue near the surface of the flipper, it is already nearly as cold as the seawater, thus minimizing heat loss. This countercurrent heat exchanger depends on good thermal contact between the artery and surrounding veins.

Consider the cylinder in **Figure 14-26**, which is in thermal contact with a hot region, at temperature T_H on the left, and a cold region, at temperature T_C on the right. The rate of heat transfer H will be higher if the cylinder is shorter (when L is smaller) and wider (when A is bigger), thus

$$H = \frac{\Delta Q}{\Delta t} = k\frac{A}{L}(T_H - T_C) \qquad (14\text{-}34)$$

where k, the **thermal conductivity**, is a constant that depends on the material from which the cylinder is made. **Table 14-5** lists the thermal conductivities of various substances. Because energy flow rate $\Delta Q/\Delta t$ (or H) has units of watts, the units of k are

$$[k] = \frac{W}{m \cdot K}$$

A good thermal conductor has a high value of k; the thermal conductivity of aluminum, for example, is about $200\ W/(m \cdot K)$. Materials that are poor thermal conductors—ones that make good thermal insulators—have k values less than $1\ W/(m \cdot K)$.

Notice in **Table 14-5** that the thermal conductivity of air is relatively low. The small value makes air, in particular air trapped between other materials, a good insulator. Most animals rely on air trapped between fur or feathers to slow the rate of heat loss in cold conditions. The phenomenon is even more developed in polar bears. The hair which makes up their fur provides particularly good insulation because it is hollow; the trapped air makes it a good thermal insulator. Clothing, particularly items made from cotton, wool, and other woven cloth, keeps us warm primarily because of the air trapped between the fibers. Fat is also a good insulator. A typical value of thermal conductivity for human fat is 0.2, which is about the same as whale blubber and nearly a third smaller than the value of water (0.58). That's one reason why blubber is so important to marine mammals in frigid polar waters.

> **? Got the Concept 14-12**
> **Wool Sweater**
>
> A thick, wool sweater will keep you warm on a cold day. Will its effectiveness change if you get caught in a rainstorm?

It is important to insulate your house if you live in a cold climate. In the thermograph of a house shown in **Figure 14-27**, the level of heat loss is indicated by a color spectrum from blue (less loss) to red (more loss) to yellow (the most loss). It would appear that while the attic of the house in Figure 14-27 has been adequately insulated, the walls have not. And the bright yellow windows are a sure sign that energy loss through the windows is high.

Figure 14-27 The level of heat loss from a house is indicated by color. The regions where most heat loss occurs are shown in yellow while blue indicates regions of the house that lose less heat. *(Alfred Pasieka/Peter Arnold/Getty Images.)*

Table 14-5
Thermal Conductivities (k) at 1 atm

Material	k [W/(m·k)]
air	0.024
aluminum	235
brick	0.9
copper	401
cotton	0.03
dirt (dry)	1.5
human fat	0.2
fiberglass	0.04
glass (window)	0.96
granite	1.7–4.0
gypsum (plasterboard)	0.17
hydrogen	0.168
nitrogen	0.024
plywood	0.13
sand (dry)	0.35
silver	429
steel	46
styrofoam	0.033
water	0.58
wood (white pine)	0.12
wool	0.04

❓ Got the Concept 14-13
Ice on a plate

The photograph in **Figure 14-28** was taken 30 s after an ice cube was placed on each of two black plates. The plates are the same size and were initially at the same temperature. One of the plates is made of wood, the other of metal. Can you tell which is which?

Figure 14-28 An ice cube was placed on each of two blocks, one made of wood and the other of metal, 30 s before the photo was taken. Can you tell which is which? *(Courtesy of University of Illinois at Urbana-Champaign Physics Department.)*

 *Go to Picture It 14-4 for more practice dealing with thermal conductivity*

Example 14-12 Heat Loss through a Window

Windows are a major source of heat loss from a house because, of the materials typically used in construction, glass is one of the best at conducting heat. Determine the rate of heat loss for a house during a day when the temperatures of the outer and inner surfaces of the windows are 14 °C and 15 °C, respectively. Take the total window area to be 28 m² and the thickness of the windows to be 3.8 mm.

SET UP

The rate of heat exchange H from the inside to the outside of a window is proportional to the temperature difference on the two sides. Notice that although the temperature inside the house is certainly warmer than 15 °C, the inside surface of the window will be much colder.

According to Equation 14-34,

$$H = \frac{\Delta Q}{\Delta t} = k\frac{A}{L}(T_H - T_C)$$

For the windows, A is the total surface area and L is the thickness of the glass. We find $k = 0.96 \text{ W}/(\text{m} \cdot \text{K})$ for glass from Table 14-5. Also, although the SI unit of temperature is the kelvin, we don't need to convert the given temperatures from °C to K because temperature *difference* is the same using either temperature scale.

SOLVE

The temperature difference between the two surfaces of the windows is 1 °C or 1 K, so

$$H = \left(0.96\,\frac{\text{W}}{\text{m} \cdot \text{K}}\right)\left(\frac{28 \text{ m}^2}{0.0038 \text{ m}}\right)(1 \text{ K}) = 7.1 \times 10^3 \text{ W}$$

REFLECT

We found the rate of conduction through the windows to be 7100 W, which is no small amount of power! Compare the answer to the power output of a portable hair dryer, which typically ranges from 1000 to 1500 W. Imagine standing near seven 1000-W hair dryers, and consider your house losing heat at the same rate they are pumping out energy.

Practice Problem 14-12 Determine the rate of heat loss for a house during a day when the temperatures of the outer and inner surfaces of the windows are 14 °C and 20 °C, respectively. Take the total window area to be 28 m² and the thickness of the windows to be 3.8 mm.

? Got the Concept 14-14
Double-pane Windows

Many new homes are built with double-pane windows, windows constructed with two panes of glass that are separated by a thin gap filled with air or an inert gas like argon. What is the advantage of double-pane windows?

> ✳ **What's Important 14-7**
> Heat can be transferred from one place or object to another by three processes: radiation, convection, and conduction. Radiation is the energy transfer by the emission (or absorption) of electromagnetic radiation. Convection is the energy transfer by the motion of a liquid or gas, for example, air that carries warmer air into a cooler region of a system. Conduction is energy transfer by the collisions of particles in one object with the particles in another, or by the collisions of more energetic particles in one part of an object with less energetic particles in adjacent parts.

Answers to Practice Problems

14-1 -10 °F, -10 °C, 280 K

14-3 2.1×10^3 m/s

14-4 1.9×10^{-3} m

14-5   6.7 g

14-6   0.25

14-8   0.45 kg

14-9   all water at 3.4 °C

14-10   all ice -7.2 °C

14-11   15 plus 0.74 kg ice, 0.06 kg water at 0 °C

14-12 42×10^3 W

Answers to Got the Concept Questions

14-1   Pressure is defined as force divided by the area over which the force is applied, or $P = F/A$ (Equation 11-3), so we can write the dimensions of pressure as those of force divided by distance squared:

$$[P] = \frac{[F]}{\text{m}^2}$$

The dimensions of pressure multiplied by volume are then

$$[PV] = \frac{[F]}{\text{distance}^2} \frac{\text{distance}^3}{} = \text{force} \times \text{distance}$$

As we saw in Chapter 6 (Equation 6-1), the product of force and distance gives work, and work is energy.

14-2   Imagine a circular hole of radius r_0 in a thin, flat piece of metal. To understand how the size of the hole changes with temperature, we can use our understanding of linear thermal expansion and apply Equation 14-16 to the circumference c_0 of the hole. (We will not, initially, consider whether the radius expands or contracts, because while the circumference represents a certain length of substance, a hole is the absence of substance so the radius cannot directly be assigned a thermal expansion coefficient.) The circumference is one-dimensional (it has units of length), so the change in circumference Δc is given by

$$\Delta c = \alpha c_0 \Delta T$$

Of course, $c_0 = 2\pi r_0$, and $\Delta c = 2\pi \Delta r$, so

$$2\pi \Delta r = \alpha (2\pi r_0) \Delta T$$

or

$$\Delta r = \alpha r_0 \Delta T$$

Thus, the change in the radius of the hole obeys the linear expansion relationship, and we see that the hole expands as the temperature increases.

If this explanation hasn't convinced you, imagine that the thin flat piece of metal is stretchable. Thermal expansion has the same effect as pulling on the plate in each direction simultaneously. Certainly the hole will expand, not shrink, as you stretch the plate in each direction.

14-3   The coefficient of linear expansion of any common metal is larger than the coefficient of linear expansion of glass. So when the jar and lid expand in the hot water, the lid expands more than the opening to the jar. This will likely loosen the lid enough to allow it to be twisted and removed.

14-4   When the engine runs it gets hot, which in turn increases the temperature of the coolant. The radiator gets hotter, too, but notice from Tables 14-1 and 14-2 that the coefficient of volume expansion for ethanol is much higher than the coefficients of linear expansion for the metals from which a radiator is constructed. Thus, the coolant fluid expands more than the radiator, causing some to spill out. Where does it go? The coolant flows into the overflow tank, causing the level to rise when the engine is hot. Later, when the engine cools down, the coolant in the radiator contracts, drawing fluid back from the overflow tank. As fluid runs back into the radiator, the level in the overflow tank drops.

14-5   When two objects are made of the same substance, intuition tells us the more massive object has more influence on the final temperature. We expect, then, that T_f will be close to the initial temperature of the more massive object. Let's check the conclusion using Equation 14-27. First, if both objects are of the same substance, the specific heat cancels from all four terms in the fraction. Also, by letting object 1 be the more massive object ($m_1 \gg m_2$), we can neglect m_2 in the denominator, so,

$$T_f \approx \frac{m_1 T_{1,i} + m_2 T_{2,i}}{m_1} = T_{1,i} + \frac{m_2}{m_1} T_{2,i}$$

Because $m_1 \gg m_2$, the fraction m_2/m_1 is negligibly small, so $T_f \approx T_{1,i}$. So, as we expect, the final temperature is very close to the initial temperature of the more massive object.

14-6   When two objects are made of the same substance and have the same mass, they each have the same influence on the final temperature. The final temperature should therefore be midway between the initial temperatures. We can check the answer using Equation 14-27. First, if both objects are of the same substance, the specific heat cancels from all four terms in the fraction. Also, if the two masses are the same (that is, $m_1 = m_2 = m$) Equation 14-27 becomes

$$T_f = \frac{m T_{1,i} + m T_{2,i}}{m + m} = \frac{m(T_{1,i} + T_{2,i})}{2m} = \frac{T_{1,i} + T_{2,i}}{2}$$

As we expect, the final temperature is midway between the two initial temperatures.

14-7   We have seen that the larger the specific heat, the smaller the change in temperature for any given amount of energy that flows. So we expect that the temperature of the object with higher specific heat won't change much as the two objects come to thermal equilibrium. T_f will therefore be closer to the initial temperature of the object that has the higher specific heat. Let's check the conclusion by using Equation 14-27. First, because both objects have the same mass, m cancels from all four terms in the fraction. Also, by letting c_1 be much greater than c_2 ($c_1 \gg c_2$) we can neglect c_2 in the denominator, so,

$$T_f \approx \frac{c_1 T_{1,i} + c_2 T_{2,i}}{c_1} = T_{1,i} + \frac{c_2}{c_1} T_{2,i}$$

Because $c_1 \gg c_2$, the fraction c_2/c_1 is negligibly small. As we expect, T_f is approximately equal to $T_{1,i}$; the final

temperature is very close to the initial temperature of the object that has the higher specific heat.

14-8   Primarily liquid oxygen (O_2) has collected in the test tube. Notice in Table 14-4 that the boiling point of nitrogen (N_2) is 77 K, which is 13 K colder than the temperature at which O_2 turns from gas to liquid. So as the temperature of the air inside the test tube (which is not really empty!) decreases to about 77 K, the temperature of the gaseous O_2 in the tube decreases below the temperature at which oxygen undergoes a phase transition to its liquid phase. Liquid O_2 condenses inside the test tube and collects at the bottom. (The temperature of the air in the tube will be just a bit warmer than the liquid nitrogen in the surrounding dewar, so little or none of the N_2 in the air will liquefy.)

14-9   First, 334 kJ would be used to melt the ice. Any additional energy would increase the temperature of the water that results from the ice melting.

14-10   The power radiated by a hot object, and therefore the heat per second due to radiation, is proportional to the fourth power of temperature. Because the temperature of the hot pot is twice that when it was cool, it is radiating 2^4 or 16 times more energy per second.

14-11   Hummingbirds cannot store enough metabolic energy to maintain their normally high core temperature all night long. Moreover, their relatively large surface area to volume ratio means they radiate a significant amount of energy per unit time. By decreasing their body temperature, they reduce the temperature difference between their bodies and the environment, thus decreasing the rate at which heat is radiated and minimizing the amount of energy they need to survive the night.

14-12   Wet clothes are a poor insulator. A thick, wool sweater is an effective insulator primarily because the spaces between the fibers are filled with air. Water has a thermal conductivity more than 25 times larger than air (Table 14-5), so when water replaces the air, the sweater does a better job of conducting energy away from your body to the air. A better conductor is a worse insulator.

14-13   Metals have a higher thermal conductivity than wood. The metal plate will therefore allow energy to flow more quickly from it to the ice cube. The metal plate is the one on which the ice cube has melted more quickly, that is, the one with the relatively larger puddle of water.

14-14   The air or gas trapped between two layers of glass provides much better insulation than a window with a single piece of glass. Although they typically cost a bit more than single-pane windows, double-pane windows can reduce heat loss due to windows by as much as a factor of 3.

SUMMARY

Topic	Summary	Equation or Symbol
absolute zero	The temperature at which the pressure of all gases becomes zero is called absolute zero, because a lower temperature is not physically possible.	
Boltzmann constant	The Boltzmann constant k defines the relationship between the pressure, volume, and temperature of a gas.	$k = 1.381 \times 10^{-23}\,\text{J/K}$
British thermal unit	The unit of heat in the English system is the British thermal unit (BTU), defined as the heat required to increase the temperature of 1 lb of pure water from 63 °F to 64 °F.	
calorie	The calorie (cal) is defined as the heat required to increase the temperature of 1 g of pure water from 14.5 °C to 15.5 °C.	
condensation	Condensation occurs when a substance makes the transition from the gaseous to the liquid phase.	
conduction	Conduction is energy transfer by the collision of particles in one object with the particles in another, or by the collision of more energetic particles in one part of an object with less energetic particles in adjacent parts. The rate H at which energy flows by conduction from a relatively hot region of temperature T_H to a relatively cold region of temperature T_C depends on the cross-sectional area A and length L of the thermal connection between these regions, as well as the thermal conductivity k of the thermal connection.	$H = \dfrac{\Delta Q}{\Delta t} = k\dfrac{A}{L}(T_H - T_C)$ (14-34)
convection	Convection is energy transfer by the motion of a liquid or gas. Convection currents of air, for example, carry warmer air up into cooler regions of a room, or the atmosphere.	
critical point	As temperature and pressure approach the critical point, the properties of the liquid and gas phases of a substance approach one another.	
degree of freedom	Any independent directions of motions through which a physical system can store energy is a degree of freedom. For a gas, the degrees of freedom include the independent directions that the atoms or molecules can move. For a gas composed of molecules, they could also include rotations, springlike motions of the bonds that hold the molecule together, and vibrations that change one or more angles between bonds.	
deposition	A transition from the gaseous phase directly to the solid phase is called deposition.	
heat	The energy that flows from one object to another as a result of a temperature difference is termed heat.	Q
ideal gas constant	The ideal (or universal) gas constant R is the proportionality constant in the fundamental relationship, known as the ideal gas law, between the pressure P, volume V, and temperature T of n moles of an ideal gas.	$R = 8.314\,\text{J}/(\text{mol}\cdot\text{K})$

ideal gas law	The ideal gas law expresses a simple, powerful relationship between the pressure P, the volume V, and the temperature T of a gas. The relationship is typically written either in terms of N (the number of number of atoms or molecules in the gas) and k (the Boltzmann constant), or n (the number of moles of gas) and R (the universal gas constant).	$PV = NkT$ (14-6) $PV = nRT$ (14-7)
kelvin	The SI unit of temperature is the kelvin. Note that the units are kelvin not degrees kelvin or kelvins. At the standard pressure of 1 atmosphere (atm), water freezes at 273.16 K and boils at 373.13 K.	K
latent heat	Latent heat is the amount of heat per unit mass required to cause a phase transition of a sample of a substance.	L
latent heat of fusion	Latent heat of fusion L_F is the energy per unit of mass required to change a substance from solid to liquid. The amount of energy Q required to change the phase of a substance that has a mass m [PE] is therefore the product of the mass m and the latent heat of fusion.	$Q = mL_F$ (14-28)
latent heat of vaporization	The latent heat of vaporization L_V is the energy per unit of mass required to change a substance from liquid to gas. The amount of energy Q required to change the phase of a substance that has a mass m is therefore the product of the mass m and the latent heat of vaporization.	$Q = mL_V$ (14-29)
mean free path	The mean free path λ is the average distance a typical molecule of radius r in a gas of N molecules and volume V travels between collisions.	$\lambda = \dfrac{1}{4\sqrt{2}\pi r^2 (N/V)}$ (14-14)
mean free time	The mean free time is the average time a typical molecule travels between collisions.	τ
mole	One mole contains 6.022×10^{23} particles; this value defines Avogadro's number N_A.	$N_A = 6.022 \times 10^{23}$ particles/mol
partial pressure	When a volume contains a mixture of gases, the partial pressure of each gas is the pressure the gas would have if no other gases were in the volume.	
phase diagram	Details of the phase transitions for a particular substance can be summarized in a phase diagram, which delineates transitions between two phases, effectively showing the transition temperature as a function of pressure.	
phase transition	A substance undergoes a phase transition when it changes from one phase, for example, solid, liquid, or gaseous, to another. Energy must be either absorbed or released in order for a substance to undergo a phase transition, but the temperature of the substance does not change during the process.	
radiation	Radiation is the transfer of energy by the emission (or absorption) of electromagnetic radiation.	

root mean square	The root mean square (rms) is a special average found by taking the square root of the average (mean) value of the square of a quantity. The rms value has particular significance when the quantity can be positive or negative, which could result in the normal average being equal to zero.	
specific heat	Specific heat c is the ratio of the energy added to an object and the temperature change that results, divided by the mass of the object. We choose to write the relationship for the temperature change ΔT, to make it clear that the larger the specific heat, the smaller the temperature change for a given mass m and heat Q.	$\Delta T = \dfrac{1}{mc}Q$ (14-18)
state variables	The quantities pressure, volume, and temperature are examples of state variables that not only characterize the current state of a system, but also contain enough information about the system to determine its future behavior.	
sublimation	Some substances can change from a solid to a gas without an intermediate liquid phase, through a phenomenon known as sublimation.	
temperature	Temperature is a measure of the energy associated with the motion of the constituent particles of a system or object. It is also the property of two or more objects that is the same when the objects are in thermal equilibrium.	T
thermal conductivity	Thermal conductivity determines how well or poorly a substance allows the flow of thermal energy. A good thermal conductor has a high thermal conductivity. Materials that are poor thermal conductors—ones that make good thermal insulators—have thermal conductivity values of less than $1\,\text{W}/(\text{m}\cdot\text{K})$.	k
thermal equilibrium	When two objects that have different temperatures are in thermal contact, energy flows from one to the other until both objects have the same temperature. At that point, the bodies are in thermal equilibrium.	
thermal expansion	Nearly all substances and objects expand when heated and contract when cooled, a phenomenon generally known as thermal expansion. The change in each dimension of an object is proportional to the change in temperature. Specifically, the change in length ΔL is proportional to the change in temperature ΔT and also the original length L_0 of an object; the proportionality constant α is the coefficient of linear expansion. The change in volume ΔV is proportional to the change in temperature ΔT and also the original volume V_0 of an object.	$\Delta L = \alpha L_0 \Delta T$ (14-16) $\Delta V = 3\alpha V_0 \Delta T$ (14-16)
triple point	The triple point is the pressure and temperature at which all three common phases of a substance can exist simultaneously.	
zeroth law of thermodynamics	The zeroth law describes thermal equilibrium. If two objects are each in thermal equilibrium with a third object, they are also in thermal equilibrium with each other.	

QUESTIONS AND PROBLEMS

In a few problems, you are given more data than you actually need; in a few other problems, you are required to supply data from your general knowledge, outside sources, or informed estimate.

Interpret as significant all digits in numerical values that have trailing zeros and no decimal points.

For all problems, use $g = 9.8 \text{ m/s}^2$ for the free-fall acceleration due to gravity. Neglect friction and air resistance unless instructed to do otherwise.

- • Basic, single-concept problem
- •• Intermediate-level problem, may require synthesis of concepts and multiple steps
- ••• Challenging problem

SSM *Solution is in Student Solutions Manual*

Conceptual Questions

1. •The phrase "temperature measures kinetic energy" is somewhat difficult to understand and can be seen as contradictory. Give a few reasons why the explanation actually makes sense and a few reasons why it seems contradictory.

2. •If ideal gas A is thermally in contact with ideal gas B for a significant time, what can you say (if anything) about the following variables: P_A versus P_B, V_A versus V_B, and T_A versus T_B? (P, V, and T stand for pressure, volume, and temperature, respectively.)

3. •A skunk is threatened by a great horned owl and emits its foul-smelling fluid in order to get away. Will the odor be more easily detected on a day when it is cooler or when it is warmer? Explain your answer. **SSM**

4. •A careful physics student is reading about the concept of temperature and has what she thinks is a bright idea. While reading the text, she discovers that the Celsius temperature scale and the Kelvin temperature scale both have the *same* increments for equal temperature differences. However, the Fahrenheit scale was described as having *larger* increments for the same temperature differences. From this information, the student determines that Celsius (or kelvin) thermometers will always be *shorter* than Fahrenheit thermometers. Explain the parts of her idea that are valid and the parts that are not.

5. •A physics student has decided that reading the textbook is very time-consuming and has decided that she will simply attempt the problems in the book without any prior background reading. During the completion of one problem that involves a temperature conversion, the student concludes that the answer is −508 °F. Discuss the validity of the answer.

6. •Search the Internet for *temperature scales*. How many different temperature scales can you find?

7. •The average normal human body temperature falls within the range of 37 ± 0.5 °C or 98.6 ± 0.9 °F. Discuss (a) the difficulties in determining an exact value for normal body temperature and (b) why there is a range of values.

8. •You have probably heard the old saying: "It ain't the heat, it's the humidity that's so unbearable." Discuss what this *really* means and try to include some reference to the definitions of heat and temperature that we have established in this chapter.

9. •(a) Which is hotter: a kilogram of boiling water or a kilogram of steam? (b) Which one can cause a more severe burn? Why? Assume the steam is at the lowest temperature possible for the corresponding pressure. **SSM**

10. •Describe some possible uses for the following thermometers whose temperature ranges are provided in kelvin.

Thermometer A	200–270 K
Thermometer B	230–370 K
Thermometer C	300–550 K
Thermometer D	300–315 K

11. • "Temperature is the physical quantity that is measured with a thermometer." Discuss the limitations of this working definition.

12. •A very old mercury thermometer is discovered in a physics lab. All the markings on the glass have worn away. How could you recalibrate the thermometer.

13. •Using the concepts of heat, temperature changes, and heat capacity, give a simplistic explanation of the global demographic factoid that 90% of the world's population lives within 100 km from a coastline.

14. ••In Section 14-2, it is shown that $K_{avg} = \frac{3}{2}kT$. The result is valid for a three-dimensional collection of atoms. (a) Discuss any changes in the formula that would correspond to a one-dimensional system. (b) What change is necessary for a ten-dimensional space?

15. •In warm regions, highway repair work is often completed in the summer months. Describe why this is thermodynamically sound.

16. •Why are some sidewalks often formed in small segments, separated by a small gap between the segments?

17. •Older mechanical thermostats use a bimetallic strip to open and close a mercury switch that turns the heat on or off. A bimetallic strip is made of two different metals fastened together. One side is made of one metal, and the other side is made of a different metal. Describe how a bimetallic strip functions in such a thermostat. **SSM**

18. •There are several different units besides the joule used to measure thermal energy. (a) List the various units associated with thermal energy and (b) indicate in which area of science or technology they are used.

19. •The ideal gas law can be written as $PV = NkT$ or $PV = nRT$. (a) Explain the different contexts in which you might use one or the other, and (b) define all the variables and constants.

20. •Explain the term latent as it applies to phase changes.

21. •Describe the sequence of thermodynamic steps that a very cold block of ice (below 0 °C) will undergo as it transforms into steam at a temperature above 100 °C.

22. •Write out a brief definition of the three different modes of heat transfer: radiation, convection, and conduction.

23. •Describe how convection causes hot air to rise in a room.

24. •Explain how a light material, such as fiberglass insulation, makes an effective barrier to keep a home warm in the winter and cool in the summer.

25. •State the zeroth law of thermodynamics and explain how it is used in physics.

Multiple-Choice Questions

26. •Two objects that have different sizes, masses, and temperatures come into close contact with each other. Thermal energy is transferred
   A. from the larger object to the smaller object.
   B. from the object that has more mass to the one that has less mass.
   C. from the object that has the higher temperature to the object that has the lower temperature.
   D. from the object that has the lower temperature to the object that has the higher temperature.
   E. back and forth between the two objects until they come to equilibrium.

27. •If you halve the value of the square root of the mean velocity v_{rms} of an ideal gas, the absolute temperature must be
   A. reduced to one-half its original value.
   B. reduced to one-quarter its original value.
   C. unchanged.
   D. increased to twice its original value.
   E. increased to four times its original value. SSM

28. •Two gases each have the same number of molecules, same volume, and same atomic radius, but the atomic mass of gas B is twice that of gas A. Compare the mean free paths.
   A. The mean free path of gas A is four times larger than gas B.
   B. The mean free path of gas A is four times smaller than gas B.
   C. The mean free path of gas A is the same as the gas B.
   D. The mean free path of gas A is two times larger than gas B.
   E. The mean free path of gas A is two times smaller than gas B.

29. •If you heat a thin, circular ring so its temperature is twice what it was originally, the ring's hole
   A. becomes larger.
   B. becomes smaller by an unknown amount.
   C. remains the same size.
   D. becomes four times smaller.
   E. becomes two times smaller.

30. •If you add heat to water at 0 °C, the water will decrease in volume until it reaches
   A) 1 °C
   B) 2 °C
   C) 3 °C
   D) 4 °C
   E) 100 °C

31. •Two objects that are not initially in thermal equilibrium are placed in close contact. After a while,
   A. the specific heats of both objects will be equal.
   B. the thermal conductivity of each object will be the same.
   C. the temperature of the cooler object will rise the same amount that the hotter one drops.
   D. the temperature of each object will be the same.
   E. the temperature of the cooler object will rise twice as much as the temperature of the hotter one drops. SSM

32. •When a substance goes directly from a solid state to a gaseous form, the process is known as
   A. vaporization.
   B. fusion.
   C. melting.
   D. condensation.
   E. sublimation.

33. •Which heat transfer process(es) is (are) important in the transfer of energy from the Sun to Earth?
   A. radiation
   B. convection
   C. conduction
   D. conduction and radiation
   E. conduction, radiation, and convection

34. •A clay pot at room temperature is placed in a kiln, and the pot's temperature doubles. How much more heat per second is the pot radiating when hot compared to when cool?
   A. 2 times
   B. 4 times
   C. 8 times
   D. 16 times
   E. 32 times

This page is intentionally left blank.

For complete end of chapter problem sets, please go to
www.whfreeman.com/kestentauck

53. •(a) Explain the physical significance of the value −273.15 °C.

54. •Starting from $T_C = \frac{5}{9}(T_F - 32)$ (Equation 14-2) 450 m/s, derive a formula for converting from °C to °F. Why is it more common for the multiplicative factor in Equation 14-2 to be written as a fraction rather than a decimal?

55. •The highest temperature ever recorded on Earth is 56.7 °C, in Death Valley, CA in 1913. The lowest temperature on record is −89.2 °C, measured at Vostok Research Station in Antarctica in 1983. Convert these extreme temperatures to °F and kelvin.

56. •In adults the normal range for oral (under the tongue) temperature is approximately 36.7 to 37.0 °C. Calculate the range and differences in °F and in kelvin.

57. •Explain why there must exist a numerical value that is the same on the Celsius scale as on the Fahrenheit scale. Show the calculation that yields the special value. SSM

58. •A thermally isolated system has a temperature of T_A. The temperature of a second isolated system is T_B. When the two systems are placed in thermal contact with each other, they come to an equilibrium temperature of T_C. Describe all the possibilities regarding the relative magnitudes of T_A and T_B when (a) $T_C < T_A$, and (b) $T_C > T_A$.

14-2: Ideal Gas Law

59. •One mole of an ideal gas is at a pressure of 1 atm and occupies a volume of 1 L. (a) What is the temperature of the gas? (b) Convert the temperature to °C and °F.

60. •Calculate the energy of a sample of 1 mol of ideal oxygen (O_2) gas molecules at a temperature of 300 K. Assume that the molecules are free to rotate, vibrate, and move in three dimensions.

61. •A 55-g sample of a certain gas occupies 4.13 L at 20 °C and 10 atm pressure. What is the gas? SSM

62. •The boiling point of water at 1 atm is 373 K. What is the volume occupied by water gas due to evaporation of 10 g of liquid water at 1 atm and 373 K?

63. •An ideal gas is confined to a container at a temperature of 300 K. What is the average kinetic energy of an atom of the gas?

64. •Calculate v_{rms} for a helium atom if 1 mol of the gas is confined to a 1-L container at a pressure of 10 atm.

65. •Two gases present in the atmosphere are water vapor (H_2O) and oxygen (O_2). What is the ratio of their rms speeds? SSM

66. •What is the *average* speed of the oxygen molecules in a 1-L sample of 0.009 mol of air that has a pressure of 1.05 atm? Assume that the air is made up of approximately 78% nitrogen, 21% oxygen, and 1% argon. Also, assume that the oxygen molecule has five degrees of freedom.

67. •What is the temperature of 1 mol of ideal oxygen gas molecules (O_2) if the average speed is 450 m/s? Assume the molecules are free to rotate, vibrate, and move in three dimensions.

68. •On a day when the atmospheric pressure is 0.97 atm and there is 0% humidity (no water vapor), what is the partial pressure of each component of a sample of air? The components of air (as a percent of the volume) are as follows: 78.09% nitrogen, 20.95% oxygen, 0.03% carbon dioxide, and 0.93% argon.

69. •The components of air (as a percent of the volume) are as follows: 78.09% nitrogen, 20.95% oxygen, 0.03% carbon dioxide, and 0.93% argon. Express the components of air as a percentage of mass rather than volume.

14-3: Mean Free Path

70. •State-of-the-art vacuum equipment can attain pressures as low as 7.0×10^{-11} Pa. Suppose that a chamber contains helium at that pressure and at room temperature (300 K). Estimate the mean free path and the collision time for helium in the chamber. Assume the diameter of a helium atom is 1.0×10^{-10} m.

71. •The mean free path for O_2 molecules at a temperature of 300 K and at 1.00 atm pressure is 7.10×10^{-8} m. Use the data to estimate the size of an O_2 molecule. SSM

14-4: Thermal Expansion

72. •Calculate the temperature change needed for a cylinder of gold to increase in length by 0.1%.

73. •Calculate the coefficient of linear expansion for a 10-m-long metal bar that shortens by 0.5 cm when the temperature drops from 25 °C to 10 °C.

74. •By how much would a 1.00-m-long aluminum rod increase in length if its temperature were raised 8.00 °C?

75. •A silver pin is exactly 5.00 cm long when its temperature is 180 °C. How long is the pin when it cools to 28 °C? SSM

76. •A sheet of lead has an 8.00-cm-diameter hole drilled through it while at a temperature of 8.00 °C. What will be the diameter of the hole if the sheet is heated to 208.00 °C?

77. •A thin sheet of copper 80.00 cm by 100.00 cm at 28 °C is heated to 228 °C. What will be the new area of the sheet?

78. ••A sheet of copper at a temperature of 0 °C has dimensions of 20 cm by 30 cm. (a) Calculate the dimensions of the sheet when the temperature rises to 45 °C. (b) By what percent does the area of the sheet of copper change?

79. ••A 5-m-long cylinder of solid aluminum has a radius of 2 cm. (a) If the cylinder is at a temperature of 5 °C, how much will the length change when the temperature rises to 30 °C? (b) Due to the temperature increase, by how much would the density of the aluminum cylinder change? (c) By what percentage does the volume of the cylinder increase? SSM

80. ••A cube of iron is heated uniformly to 100 °C. At that temperature, the volume of the iron is 20 cm³. Find the dimensions of the cube (a) at 100 °C and (b) at 20 °C.

81. ••A sphere of gold has an initial radius of 1 cm when the temperature is 20 °C. (a) If the temperature is raised to 80 °C, calculate the new radius of the sphere. (b) What is the percent change in the volume of the sphere?

14-5: Heat

82. •What is the specific heat of a 500-g metal sample that rises 4.8 °C when 307 J of heat is added to it?

83. •You wish to heat 250 g of water to make a hot cup of coffee. If the water starts at 20 °C and you want your coffee to be 95 °C, calculate the minimum amount of heat required. SSM

84. •In a thermodynamically sealed container, 20 g of 15 °C water is mixed with 40 g of 60 °C water. Calculate the final equilibrium temperature of the water.

85. •How much heat is transferred to the environment when the temperature of 2.00 kg of water drops from 88 °C to 42 °C?

86. •A copper pot has a mass of 1.0 kg and is at 100 °C. How much heat must be removed from it to decrease its temperature to precisely 0 °C? The specific heat of copper is 387 J/(kg·K).

87. •A lake has a specific heat of 4186 J/(kg·K). If we transferred 1.7×10^{14} J of heat to the lake and warmed the water from 10 °C to 15 °C, what is the mass of the water in the lake? Neglect heat released to the surroundings. SSM

88. •Calculate the temperature increase in a 1-kg sample of water that results from the conversion of gravitational potential energy directly to heat energy in the world's tallest waterfall, the 807-m tall *Salto Angel* in Canaima National Park, Venezuela.

89. •A superheated iron bar that has a mass of 5.00 kg absorbs 2.5×10^6 J of heat from a blacksmith's fire. (a) Calculate the temperature increase of the iron. (b) Blacksmiths use a rule when working with iron and steel: "Work the metal when it's heated yellow, put it back in the fire when it's heated red." Should the bar be put back into the fire? Explain your answer.

90. • A 200-g block of ice is at −10 °C. How much heat must be removed to lower its temperature to −40 °C?

91. ••A 250-g sample of copper is heated to 100 °C and placed into a cup containing 300 g of water initially at 30 °C. Ignoring the container holding the water, find the final equilibrium temperature of the mixture in the following cases: (a) The system is completely insulated and no heat is lost or gained to the environment. (b) The copper loses 5% of its initial heat when it is placed into the water (and there are no losses or gains after that).

92. ••A 50-g calorimeter cup made from aluminum contains 100 g of water. Both the aluminum and the water are at 25 °C. A 300-g cube of some unknown metal is heated to 150 °C and placed into the calorimeter; the final equilibrium temperature for the water, aluminum and metal sample is 41 °C. Calculate the specific heat of the unknown metal and make a guess as to its composition.

93. •A hacksaw is used to cut a 20-g steel bolt. Each stroke of the saw supplies 30 J of energy. How many strokes of the saw will it take to raise the temperature of the bolt from 20 °C to 80 °C? Assume none of the energy goes into heating the surroundings. Of course, it will take significantly more than this to also cut the metal! SSM

14-6: Latent Heat

94. •Calculate the amount of heat required to change 25 g of ice at 0 °C to 25 g of water at 0 °C.

95. •How much heat is required to change 25 g of ice at −40 °C to 25 g of steam at 140 °C?

96. •A sealed container (with negligible heat capacity) holds 30 g of 120 °C steam. Describe the final state if 100,000 J of heat is removed from the steam.

97. •How much heat is required to melt a 400-g sample of copper that starts at 20 °C? (L_F equals 205,350 J/kg for copper.) SSM

98. ••Suppose 20 g of ice at −10 °C is placed into 300-g of water in a 200-g copper calorimeter. If the final temperature of the water and copper calorimeter is 18 °C, what was the initial common temperature of the water and copper?

99. ••What mass of ice at −20 °C must be added to 50 g of steam at 120 °C to end up with water at 40 °C?

100. •A 60-kg ice hockey player is moving at 8 m/s when he skids to a stop. If 40% of his kinetic energy goes into melting ice, how much water is created as he comes to a stop? Assume that the surface layer of the ice in the hockey rink has a temperature of 0 °C.

101. •You have 50 g of iron at 120 °C, 60 g of copper at 150 °C, and 30 g of water at 40 °C. Which of those would melt the most ice, starting at −5 °C? How much ice does each melt? SSM

14-7: Heat Transfer: Radiation, Convection, Conduction

102. •A heated bar of gold radiates at a temperature of 300 °C. (a) By what factor does the radiated heat increase if the temperature is increased to 600 °C? (b) What about 900 °C? (c) If the surface area of the gold bar is also doubled, how will the answers be affected?

103. •**Astronomy** An astrophysicist determines the surface temperature of a distant star is 12,000 K. The surface temperature of the Sun is about 5800 K. If the surface temperature of the Sun were to suddenly increase to 12,000 K, by how much would the radiated heat increase?

104. •**Astronomy** A star radiates 3.75 times less heat than our own Sun. What is the ratio of the temperature of the star to the temperature of our Sun?

105. •**Astronomy** A distant star radiates 1000 times more energy than our own Sun even though the temperature of the star is only 70% of the Sun's. If both stars are perfect emitters, estimate the radius of the distant star. Recall, the radius of the Sun is 6.96×10^8 m. SSM

106. ••**Biology** The skin temperature of a nude person is 34 °C and the surroundings are at 20 °C. The emissivity of skin is 0.900 and the surface area of the person is 1.50 m². (a) What is the rate at which energy radiates from the person? (b) What is the net energy loss from the body in 1 min by radiation?

107. •Calculate the heat through a glass window that is 30 cm × 150 cm in area and 1.2 mm thick. Assume the temperature on the inside of the window is 25 °C while the outside temperature is 8 °C. SSM

General Problems

108. •What is the average kinetic energy for a two-dimensional ideal gas made up of diatomic molecules when the temperature is 75 °F?

109. ••**Astronomy** Is the Sun likely to lose its atomic hydrogen? (a) Calculate the escape velocity for hydrogen atoms from the surface (the *photosphere*) of our Sun. (b) If the root-mean-square speed of the hydrogen atoms were equal to the speed you found in part (a), what would be the temperature of the Sun's photosphere? The mass of a hydrogen atom is 1.68×10^{-27} kg. (c) Given that the actual temperature of the photosphere is 5800 K, is the Sun likely to lose its atomic hydrogen?

110. •Consider an ideal gas that has a constant bulk modulus B. If you decrease the volume by a factor of $\frac{1}{2}$, how much will the pressure change?

111. ••**Astronomy** In 1975 and 1976, the *Viking 1* and *Viking 2* probes landed on Mars and radioed direct atmospheric measurements back to Earth. The atmo-sphere on Mars contains 95.4% carbon dioxide, 2.7% nitrogen, 1.6% argon, 0.13% oxygen and 0.07% carbon monoxide, plus small traces of other gases (water, neon, krypton, ozone, and xenon). If the atmospheric pressure of Mars is 0.675 kPa (about 1/150 of Earth's), find the partial pressures of CO_2, N_2, Ar, O_2, and CO. SSM

112. ••**Astronomy** The atmosphere of Jupiter is made up of 89% H_2 and 11% He. If the partial pressure of hydrogen from a sample of Jovian atmosphere is 0.45 atm, calculate (a) the total atmospheric pressure and (b) the partial pressure of helium on Jupiter.

113. ••A company in Hawaii advertises guided, under-water adventures that take scuba divers to a reef 25 m below the surface. Experienced divers can arrange to visit parts of the reef at a depth of 40 m. Divers can experi-ence nitrogen narcosis (the "bends") when the partial pressure of nitrogen in the air they breathe exceeds 3.5×10^5 Pa. Air is approximately 79% nitrogen. (a) Calculate the partial pressure of nitrogen in the lungs when a diver descends to depths of 25 m and 40 m. (b) Why do you think excursions to the deeper reef are limited to experi-enced divers?

114. •••**Biology** With each breath, a person at rest breathes in about 0.50 L of air, 20.9% of which is O_2, and exhales the same volume of air containing 16.3% O_2. In the lungs, oxygen diffuses into the blood, and is then transported throughout the body. Severe illness (altitude sickness) and even death can result if the amount of oxygen is too low. At sea level, atmospheric pressure is 1.00 atm, but at 3048 m (10,000 ft) it is reduced to 0.695 atm; the percentage of oxygen remains the same in both cases. Suppose that the temperature is 20 °C at both altitudes. What is the net number of oxygen molecules in each complete breath (a) at sea level and (b) at an altitude of 3048 m? (c) Use the results above to explain why peo-ple feel "out of breath" and must breathe more rapidly and deeply at high altitudes.

115. •••Derive a formula for the average speed of a component (x) of a sample of gas that is made up of η different monatomic gases. Assume the overall pressure of the gas is P, the total volume is V, and there are n mol of the gas. The percent of the xth component is F_x. Thus, $P = P_1 + P_2 + \cdots + P_x + \cdots + P_\eta$ and $F_1 + F_2 + \cdots + F_x + \cdots + F_\eta = 100\%$.

116. •**Astronomy** Titan, a satellite of Saturn, has a nitro-gen atmosphere with a surface temperature of −179 °C and pressure of 1.5 atm. The mass of a nitrogen molecule is 4.7×10^{-26} kg, and we can model it as a sphere of diameter 2.4×10^{-10} m. The average temperature of Earth's atmosphere is about 10 °C. (a) What is the den-sity of particles in the atmosphere of Titan? (b) Which has a denser atmosphere, Titan or Earth? Justify your answer by calculating the ratio of the particle density on Titan to the particle density on Earth. (c) What is the average distance that a nitrogen molecule travels between collisions on Titan? How does this result compare with the distance for carbon dioxide calculated in Section 14-3? Is your result reasonable?

117. •A brick wall is composed of 19.0-cm-long bricks ($\alpha = 5.5 \times 10^{-6}\,°C^{-1}$) and 1.00-cm-long sections of mortar ($\alpha = 8.0 \times 10^{-6}\,°C^{-1}$) in between the bricks. Describe the expansion effects on a 20-m-long section of wall that undergoes a temperature change of 25 °C. SSM

118. •Derive an approximate formula for the area expansion (ΔA) that a sheet of material undergoes as the temperature changes from T_i to T_f. Assume the linear coefficient of expansion is α for the material.

119. •A 1-cm-diameter sphere of copper is placed concentrically over a 0.99-cm-diameter hole in a sheet of aluminum (**Figure 14-29**). Both the copper and the aluminum start at a temperature of 15 °C. Describe one set of conditions (if any) in which the copper sphere will pass through the hole in the aluminum.

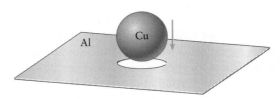

Figure 14-29 Problem 119

120. •A 20.0-m-long bar of steel expands due to a temperature increase. A 10.0-m-long bar of copper also gets longer due to the same temperature rise. If the two bars were originally separated by a gap of 1.5 cm, calculate the change in temperature and the distances that the steel and copper stretch if the gap is exactly "closed" by the expanding bars. Assume the steel and copper bars are fixed on the ends, as shown in **Figure 14-30**.

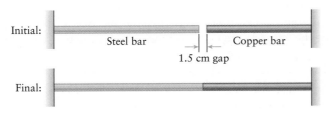

Figure 14-30 Problem 120

121. •Suppose a person who lives in a house next to a busy urban freeway attempts to "harness" the sound energy from the nonstop traffic to heat the water in his home. He places a transducer on his roof, "catches" the sound waves, and converts the sound waves into an electrical signal that warms a cistern of water. However, after running the system for 7 days, the 5 kg of water increases in temperature by only 0.01 °C! Assuming 100% transfer efficiency, calculate the acoustic power "caught" by the transducer. SSM

122. ••Calc The specific heat for a sealed system is not constant, rather, it depends on the temperature as follows: $c(T) = c_0 + c_1 T$ (SI units used throughout; c_0 and c_1 are constants). The mass of the system is m. (a) Write an expression for the heat added to the system when the temperature rises from T_0 to T_f. (b) If $c_0 = 2000\,J/(kg \cdot °C)$ and $c_1 = 40\,J/(kg \cdot °C^2)$, find the heat added for $T_0 = 20\,°C$ and $T_f = 60\,°C$. Take the mass of the system to be 100 g and assume no phase changes occur.

123. ••Calc The specific heat for a sealed system of mass 1 kg, initially at 0 °C, is given by the following: $c(T) = 3000 + 9T^2$ (SI units). Calculate the final temperature of the system if 15 kJ of heat is added, assuming no phase changes occur.

124. •Biology Connie ordinarily eats 2000 kcal of food per day. If her mass is 60 kg and her height is 1.7 m, her surface area is probably close to 1.7 m². Although the body is only about two-thirds water, we will model it as being all water. (a) Typically, 80% of the calories we consume is converted to heat. If Connie's body has no way of getting rid of the heat produced, by how many degrees Celsius would her body temperature rise in a day? (b) Would this be a noticeable increase? (c) How does her body prevent the increase from happening? (See Table 14.3 as needed.)

125. •Helium condenses at −268.93 °C and has a latent heat of vaporization of 21,000 J/kg. If you start with 5 g of helium gas at 30 °C, calculate the amount of heat required to change the sample to liquid helium. The specific heat of helium is 5193 J/(kg·°C) at 300 K. SSM

126. •How much heat is required to convert 1 kg of dry ice (solid CO_2) at −78.5 °C into gaseous carbon dioxide at 20 °C? The heat of sublimation for CO_2 is 573,700 J/kg. The specific heat for CO_2 is not constant, the equation that describes it is a function of the absolute temperature: $c(T_K) = 0.001112 T_K + 0.5128$.

127. •The Arctic perennial sea ice does not melt during the summer and thus lasts all year. NASA found that the perennial sea ice decreased by 14% between 2004 and 2005. The melted ice covered an area of 720,000 km² (the size of Texas!) and was 3.0 m thick on average. The ice is pure water (not salt water) and is 92% as dense as liquid water. Assume that the ice was initially at −10.0 °C and see Tables 14-3 and 14-4 as needed. (a) How much heat was required to melt the ice? (b) Given that 1.0 gal of gasoline releases 1.3×10^8 J of energy when burned, how many gallons of gasoline contain as much energy as in part (a)? (c) A ton of coal releases around 21.5 GJ of energy when burned. How many tons of coal would need to be burned to produce the energy to melt the ice?

128. ••**Biology** Jane's surface area is approximately 1.5 m^2. How much heat is released from her body when the temperature difference across the skin is 1 °C? Assume the average thickness of the skin is 1 mm.

129. ••Building insulation is rated in R-value, the resistance to the transfer of thermal energy. The calculation that defines the R-value for a given material usually uses nonmetric units including degrees Fahrenheit (°F) and the British thermal unit (BTU). Hence, home owners faced with metric units often need to convert quantities to U.S. or English units. Consider the wall of a home that has an area of 25 m^2 (**Figure 14-31**). The wall consists of a 9-cm-thick brick, a 1-cm-thick sheet of plywood, and a layer of R-12 insulation. If the temperature differential between inside and outside is 5 °C, calculate the rate of heat loss from the warm interior to the cooler exterior. Start by calculating the effective R-value for the three materials.

R values for insulation, plywood, brick:
$R_I = 12$ ft$^2 \times$ h $\times$ °F/BTU
$R_P = 0.6$ ft$^2 \times$ h $\times$ °F/BTU
$R_B = 0.6$ ft$^2 \times$ h $\times$ °F/BTU

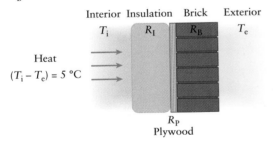

Figure 14-31 Problem 129

130. ••**Sports, Biology** A person can generate about 300 W of power on a treadmill. If the treadmill is inclined at 3% and a 70-kg man runs at 3 m/s for 45 min, (a) calculate the percentage of the power output that goes into heating up his body and the percentage that keeps him moving on the treadmill. (b) How much water would that heat evaporate?

131. ••**Biology** A 1.88 m (6 ft 2 in.) man has a mass of 80 kg, a body surface area of 2.1 m^2 and a skin temperature of 30 °C. Normally 80% of the food calories he consumes go to heat, the rest going to mechanical energy. To keep his body's temperature constant, how many food

calories should he eat per day if he is in a room at 20 °C and he loses heat only through radiation? Does the answer seem reasonable? His emissivity ε is 1 because his body radiates almost entirely nonvisible infrared energy, which is not affected by skin pigment. (*Careful!* His body at 30 °C radiates into the air at 20 °C, but the air also radiates back into his body. The *net* rate of radiation is $P_{net} = P_{body} - P_{air}$.) SSM

132. •**Astronomy** Find the total power radiated by our Sun. Assume it is a perfect emitter of radiation ($\varepsilon = 1$) with a radius of 6.96×10^8 m and a temperature of 5800 K.

133. ••**Biology** You may have noticed that small mammals (such as mice) seem to be constantly eating, whereas some large mammals (such as lions) eat much less frequently. Let us investigate the phenomenon. For simplicity we can model an animal as a sphere. (a) Show that the heat energy stored by an animal is proportional to the cube of its radius, but the rate at which the animal radiates energy away is proportional to the square of its radius. (b) Show that the fraction of the animal's stored energy that it radiates away per second is inversely proportional to the animal's radius. (c) Use the result in part (b) to explain why small animals must eat much more per gram of body weight than very large animals.

134. ••**Astronomy** About 65 million years ago an asteroid struck Earth in the area of the Yucatan Peninsula and wiped out the dinosaurs and many other life forms. Judging from the size of the crater and the effects on Earth, the asteroid was about 10 km in diameter (assumed spherical) and probably had a density of 2.0 g/cm^3 (typical of asteroids). Its speed was at least 11 km/s. (a) What is the maximum amount of ocean water (originally at 20 °C) that it could have evaporated? Express your answer in kilograms and treat the ocean as though it were freshwater. (b) If the water were formed into a cube, how high would it be? (See Tables 14-3 and 14-4 as needed.)

135. •••**Astronomy, Chemistry** We can model chemical reactions as occurring when molecules collide. Therefore the reaction rate r should depend on the mean free time τ between collisions. (a) Would you expect that r is directly proportional to τ or inversely proportional to τ? Explain your answer. (b) Show that a reasonable estimate for the mean free time is $\tau = \lambda/v_{rms}$. (c) On Mars the atmospheric pressure is 650 N/m^2 and the average summer maximum temperature is 0 °C. If a chemical reaction occurs at a rate r on Earth when the temperature is 5 °C, use your results from above to estimate the reaction rate (in terms of r) on Mars during the warmest part of an average summer day.

136. ••A spherical container is constructed from steel and has a radius of 2 m at 15 °C. The container sits in the Sun all day and its temperature rises to 38 °C. The container is initially filled completely with water, but it is not sealed. Describe what will happen to the water after the temperature increase.

137. ••**Astronomy** One method for detecting the presence of a planet orbiting around a distant star is to monitor its brightness. If a planet passes in front of a star and blocks some of its light, the star will appear less bright. We can assume that the planet radiates essentially no visible light because its surface is so much cooler than that of the star. (a) Show that when the planet is in front of the star, the intensity of the star is reduced by the factor $(r/R)^2$, where r is the radius of the planet and R is the radius of the star. (b) If astronomers in an alien civilization were watching our Sun, what percent would its intensity be reduced by (i) Jupiter and (ii) Earth? (c) Would either of the reductions in part (b) be easy to measure? SSM

138. ••A sheet of aluminum has an octagon cut out of it (Figure 14-32). If the temperature of the aluminum increases by 20 °C, calculate the new area of the octagon and compare it to the original area.

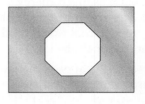

Originally, the octagon has sides of length $s_o = 10$ cm.

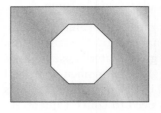

After a temperature increase of $\Delta T = 20$ °C, the new length of the sides is s_f.

Figure 14-32 Problem 138

15 Thermodynamics II

(Dr. Thomas Eisner/Visuals Unlimited.)

The bombardier beetle defends itself by expelling a foul-smelling mixture of hot liquid and gas at attackers like ants and spiders. (As you can see, it can even aim the blast to better deter predators.) To initiate its defense mechanism, the beetle releases certain chemicals into a reaction chamber where they combine with enzymes secreted by the cells that line the chamber's walls. The resulting exothermic reaction releases enough energy to vaporize some of the liquid. The pressure inside the chamber increases rapidly, expelling the hot chemical mixture through openings at the tip of the beetle's abdomen. This process involves pressure, temperature, work—all the stuff of thermodynamics!

Energy used by living organisms ultimately comes from the Sun. Plants harvest solar energy and use it to synthesize the biomolecules that enable them to grow and reproduce. Animals derive their energy from eating plants or other animals. However, all thermodynamic processes are inefficient and waste some energy at each step. You're already intimately familiar with the phenomenon; every time you exercise hard you become hot and sweaty. Like all physical and biological machines, muscles use only some of their energy to do work; the rest of the energy heats the body.

The inefficiency of thermodynamic processes and the resulting tendency of systems to become less ordered through the loss of usable energy define a direction of time. All plants and animals require energy to live, but as time marches forward all living things die and decay into dust. From the tiniest algae to the largest whales, highly complex living things all eventually become more randomly distributed atoms. As we'll see in this chapter, the progression from highly ordered states to random, disordered ones also arises from the laws of thermodynamics. We'll also consider how life can maintain highly complex and ordered structures and functions in the face of a progression toward randomness.

15-1 The First Law of Thermodynamics

As the left ventricle of a human heart pumps blood into the aorta and out to the rest of the body, it repeats a process that involves four distinct phases defined by the relationship between the pressure and volume in the muscular chamber. Graphing pressure versus volume on a pressure–volume or **PV diagram** neatly summarizes the phases, as shown in **Figure 15-1**. At the point on the cycle marked 1, the valve between the left atrium and the left ventricle opens, and blood flows into the ventricle. As the volume in the ventricle increases, the pressure changes very little because the ventricle remains relaxed in this phase. At point 2, the ventricle begins to contract and the valve between the atrium and ventricle closes again. Because the ventricle is now a closed chamber with contracting walls, the pressure increases dramatically while the volume remains the same, resulting in the vertical line between points 2 and 3 on the *PV* diagram. At point 3, the pressure in the ventricle exceeds that in the aorta; blood now pushes open the valve between the ventricle and the aorta, allowing oxygenated blood to leave the heart. Between points 3 and 4, blood volume in the ventricle decreases. Finally, as the ventricle begins to relax, at point 4 the pressure in the ventricle drops below that in the aorta. As blood tries to flow backward from the aorta into the ventricle, the valve between them closes. With valves at both openings of the ventricle closed again, no blood enters or leaves, so volume remains the same—a vertical line on the *PV* diagram—and as the muscle relaxes, the pressure drops. When the pressure in the ventricle drops below that in the atrium at point 1, blood forces the valve open and begins to fill the ventricle again. The cycle repeats with every beat of the heart. We will use *PV* diagrams like this one to understand a range of cycles and processes.

In a *PV* diagram, vertical lines like the two in Figure 15-1 represent processes in which pressure changes but volume does not. Similarly, a horizontal line represents a constant-pressure process for which volume is either increasing or decreasing. Because body temperature is constant from one heartbeat to the next, we did not include temperature in our discussion. In other systems, however, the relationship between pressure and volume depends on temperature. Consider the relationship between the pressure, volume, and temperature of an ideal gas:

$$PV = nRT \tag{14-7}$$

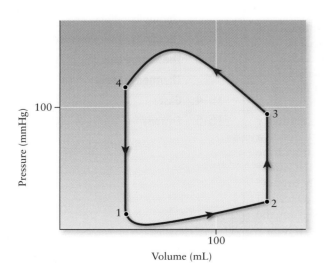

Figure 15-1 The relationship between pressure and volume during a single heartbeat is represented on a *PV* diagram as a cyclic thermodynamic process. At point 1, blood begins to enter the left ventricle from the left atrium. At point 2, the ventricle begins to contract and the valve between the atrium and the ventricle is pushed closed. Between points 2 and 3, as the muscle contracts the volume of blood in the chamber remains constant but the pressure increases. At point 3, the pressure in the ventricle exceeds the pressure in the aorta, so blood pushes open the valve and flows out of the heart. At point 4, the contraction stops. As pressure begins to drop, the valve between the ventricle and aorta closes again, so pressure in the ventricle decreases while volume remains constant. At point 1, the pressure in the atrium exceeds that in the ventricle, so blood pushes open the valve and the process begins again. The area in yellow represents the work done by the heart during a single heartbeat.

Although the variations of pressure and volume of an ideal gas depend on temperature, when temperature is constant, the relationship can be written

$$PV = \text{constant}$$

because n and R are constants. At constant temperature, as the volume of an ideal gas increases, the pressure decreases. This trend can be plotted as a curve called an **isotherm**; three different temperatures are represented by the isotherms shown in **Figure 15-2**. Notice that for any given volume, such as the one shown by the dashed red line, the corresponding pressure of an ideal gas is greater at higher temperatures than at lower ones.

In Got the Concept 14-1 we saw that the quantity PV has dimensions of energy. In particular, the area under a curve on a PV diagram is equal to the work done by the system on its environment as it goes from the initial state to the final state. For example, if we slowly add energy to gas trapped in a cylinder sealed with a piston (**Figure 15-3**), the gas expands, exerting a force on the piston. As the gas pushes the piston, it does work on it.

Can we find a quantity that describes the changes a system undergoes as it evolves from its initial state to some final state? To be of use, we will want a quantity that provides a description uniquely defined by the initial and final states. Notice that a thermodynamic process can be represented as a line between two

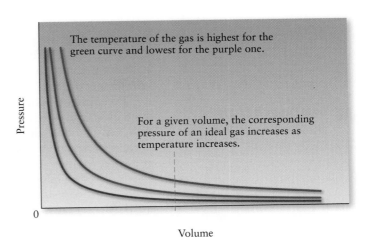

The temperature of the gas is highest for the green curve and lowest for the purple one.

For a given volume, the corresponding pressure of an ideal gas increases as temperature increases.

Pressure

Volume

Figure 15-2 The pressure of an ideal gas decreases as the volume increases while temperature remains constant. Three isotherms, curves of constant temperature, are shown.

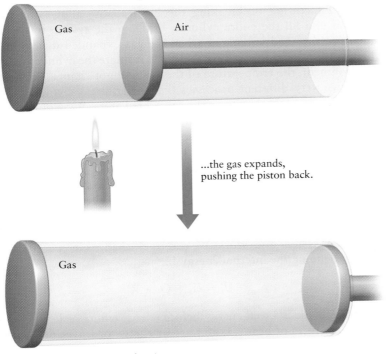

When energyis added slowly to gas trapped in a cylinder with a movable piston...

Gas   Air

...the gas expands, pushing the piston back.

Gas

The gas has done work on the piston.

Figure 15-3 The gas in a cylinder does work on a moveable piston as heat is added, causing the gas to expand.

points—an initial state and a final state—on a PV diagram; any number of such paths, however, are possible between two specific states. We want a **state function**, a quantity that characterizes the state of a system and that does not depend on the path from some previous state through intermediate values of pressure, volume, and temperature.

Perhaps, for example, the work done by a system can serve to describe the final state given the system's initial state. **Figure 15-4a** shows a process that takes a system, say, the ideal gas trapped in the cylinder in Figure 15-3, from an initial pressure, volume, and temperature, P_i, V_i, T_i, respectively, to a final state in which the pressure, volume, and temperature are P_f, V_f, and T_f. The work done in this process equals the area under the curve, shaded yellow in the figure. Let's now consider another process that takes the system from the same initial state to the same final state, but which goes through different intermediate pressure–volume–temperature states. One such process is shown in **Figure 15-4b**. The shaded area again gives the work required for the process; although the initial and final states of the system are identical to those described by the PV diagram in Figure 15-4a, the work here is certainly not the same as the work required for the first process. We must conclude that the final state of a system does not uniquely depend on the work done, because the initial and final states of the two processes are the same but the work done is not.

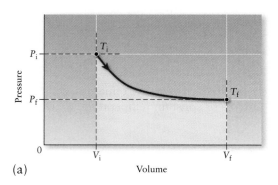

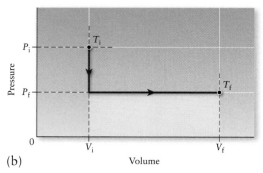

Figure 15-4 The area under a curve plotted on a PV diagram, shown in yellow, represents the work done in a process. (**a**) A process starts at an initial pressure, volume, and temperature P_i, V_i, and T_i, respectively, and progresses to a final state in which the pressure, volume, and temperature are P_f, V_f, and T_f. The work done in the process equals the area under the curve, shaded yellow in the figure. (**b**) Another process takes the same system from the same initial state to the same final state, but goes through different intermediate pressure–volume–temperature states. Although the initial and final states of the system are identical to those described by the PV diagram in (a), the work here is certainly not the same as the work required for the first process.

What about the heat we might add to a system—if we know the initial state, does the amount of heat added uniquely determine the final state? No. Let's return to the ideal gas trapped in a cylinder sealed by a piston, Figure 15-3. By adding heat slowly, the gas expands but the temperature remains constant. (Remember, heat transfer in a system does not necessarily mean that the temperature of the system increases; adding heat only means that energy is added to or removed from the system.) However, imagine that a thin barrier of lightweight material confines the ideal gas to its initial volume and a vacuum exists between the barrier and the other end of the cylinder (**Figure 15-5**). Puncturing or breaking the barrier now allows the gas to expand and fill the entire cylinder, bringing the system to the same final pressure and volume as when the gas expanded due to heat in Figure 15-3. In addition, if we thermally insulate the cylinder from the environment, the temperature of the gas will not change. So, here we have two very different processes that nevertheless change a system from the same initial state to the same final state. We would say that the two processes follow different thermodynamic *paths* to the same ending state. We added heat in one case but not in the other. Therefore, the amount of heat added to a system, by itself, cannot be used to describe the change in state of the system.

The quantity called the *internal energy* of a system can be used to describe the state of a system. Internal energy is the sum of the kinetic energy and potential energy of every atom and molecule in a system. Adding heat increases the energy of a system. Energy leaves a system when that system does work. We can track changes in energy in a system by defining the **internal energy** U:

$$U_f = U_i + Q - W$$

where U_i is the internal energy before energy enters or leaves the system in the form of heat (Q) and before energy or leaves the system in

√x **Go to Picture It 15-1** for more practice dealing with PV diagrams

A thin barrier of lightweight material separates a gas from a vacuum.

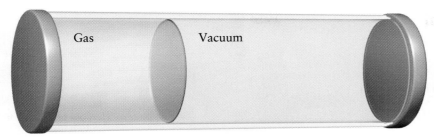

When the barrier ruptures, the gas expands to fill the entire cylinder. No heat was added to the gas in this case, but the final pressure and volume happen to be the same as when heat caused the gas to expand at constant temperature, as in Figure 15-3.

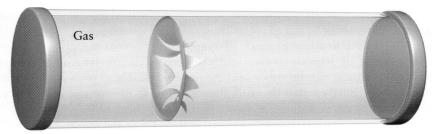

Figure 15-5 A thin barrier of lightweight material confines the ideal gas to its initial volume and a vacuum exists between the barrier and the other end of the cylinder. Puncturing the barrier allows the gas to expand and fill the entire cylinder, bringing the system to the same final pressure and volume as when the gas expanded due to heat in Figure 15-3.

the form of work (W); U_f is the internal energy afterward. The change in internal energy $\Delta U = U_f - U_i$ is then

$$\Delta U = Q - W \qquad (15\text{-}1)$$

This seemingly simple relationship, known as the **first law of thermodynamics,** is fundamentally a statement of the conservation of energy; the net change in the energy of the system equals the difference between how much energy enters and how much energy leaves the system. The change in internal energy described by the first law of thermodynamics provides a way to characterize the change in a system that does not depend on the thermodynamic path from one state to another; the change in internal energy depends uniquely on the initial and final states, not on the path through intermediate pressure, volume, and temperature values that lead from one state to another.

Heat can be transferred into or out of a system, so the quantity Q in the first law of thermodynamics (Equation 15-1) can be either positive or negative. Note that heat released from a system is waste, not work. Similarly, a system can either do work on its environment or have work done on it. We have defined work done *by* a system as positive, which makes sense in the context of the first law of thermodynamics: a positive value of W results in a decrease in the internal energy. Newton's third law demands that the work done *on* a system is negative, so from Equation 15-1:

$$\Delta U = Q - (-W) = Q + W$$

Figure 15-6 Both heat and work can be positive, negative, or zero. The sign conventions for heat and work are shown here.

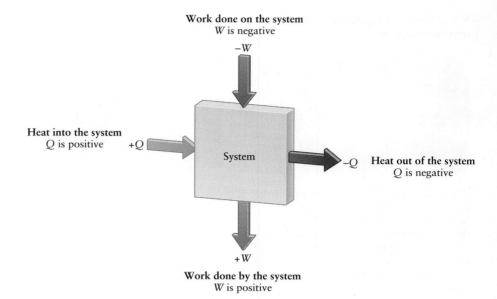

Work done on the system
W is negative
−W

Heat into the system
Q is positive   +Q

System

−Q   Heat out of the system
Q is negative

+W
Work done by the system
W is positive

Not surprisingly then, when work is done on a system, the internal energy increases. The sign conventions for heat and work are shown in **Figure 15-6**.

According to the first law of thermodynamics, the internal energy of a system can either increase or decrease as the system undergoes thermodynamic processes, because both Q and W can be either positive or negative. However, when the heat added to a system exactly equals the work done by the system—for example, if work done to expand the gas in Figure 15-3 exactly equals the heat added—then according to the first law of thermodynamics, the internal energy of the system remains the same. Again, the internal energy doesn't depend on what path the system takes through intermediate states of pressure, volume, and temperature.

> **? Got the Concept 15-1**
> **Ideal Gas in a Cylinder**
> An ideal gas trapped in a thermally isolated cylinder expands slowly by pushing back a piston. Why does the temperature of the gas decrease in the process?

Example 15-1 Climbing It Off

A 62-kg college student intends to climb the 910-m, nearly vertical face of El Capitan (**Figure 15-7**) in Yosemite National Park. How many 220-Calorie energy bars will he "burn off" during the climb? Assume that 20% of the energy he derives

Figure 15-7 El Capitan in California's Yosemite National Park is the world's largest granite monolith; it rises 910 m (over 3000 ft) above the valley floor and attracts climbers from around the world. (*Mike Murphy.*)

from the energy bar results in work, and neglect his resting metabolism, that is, neglect the fact that he could burn off the caloric content of the energy bars simply by sitting at the base of the rock formation.

SET UP

While burning off the energy bars, the student has no net change to his internal energy. When $\Delta U = 0$, $0 = Q - W$ or $Q = W$ according to the first law of thermodynamics. Only 20% of the energy derived from the energy bars goes into work. (Basal metabolism uses the other 80% to drive the chemical reactions required to maintain life; the energy lost as heat in each of those chemical reactions raises his body temperature above the ambient temperature.) So with Q representing the caloric content of N energy bars, each of caloric content Q_0 (= 220 Cal), the energy that goes into the work done to climb is

$$W_{climb} = 0.20NQ_0$$

The work he does is also equal to his change in gravitational potential energy

$$W_{climb} = mgh$$

Setting these two equal yields

$$mgh = 0.20NQ_0$$

SOLVE

The number of energy bars required is

$$N = \frac{mgh}{0.20Q_0}$$

To get a numeric answer, we need to convert the energy of the energy bar from Calories to joules. Noting that 1 Cal is equivalent to 1000 cal and 1 Cal equals 4186 J, we arrive at a solution by substituting in known values from the problem statement

$$N = \frac{(62\,\text{kg})(9.8\,\text{m/s})(910\,\text{m})}{(0.20)(220\,\text{Cal})(4186\,\text{J/Cal})} = 3.0$$

In addition to the energy required to sustain life, the student would require the caloric intake of three energy bars in a climb to the top of El Capitan.

REFLECT

Human metabolism is much more complicated than suggested by this simple example.

Practice Problem 15-1 A runner does 5.6×10^5 J of work and gives off 4.0×10^5 J of energy as heat. What is the change in her internal energy?

★ **What's Important 15-1**

On a pressure–volume (PV) diagram, in which pressure is plotted as a function of volume, the area under the curve represents the work done. The internal energy of a system (the sum of the kinetic energy and potential energy of every atom and molecule in the system) describes the state of a system in a way that does not depend on the specific thermodynamic processes that lead to it. Adding heat increases the internal energy of a system, and energy leaves a system when that system does work; this is known as the first law of thermodynamics.

15-2 Thermodynamic Processes

Figure 15-2 shows a series of curves on a PV diagram, each representing a basic thermodynamic process that occurs in an ideal gas at a different, constant temperature. Although a system can undergo processes in which any number of the thermodynamic variables can remain constant or change simultaneously, considering

processes in which at least one variable remains constant makes it easier to understand the underlying principles. We'll consider only systems in which the total
amount of material remains constant. Such systems could be air sealed in a container or water that boils resulting in an equal number of steam molecules. We'll
also require that changes to a system mostly occur slowly, so that even if temperature and pressure change, they remain uniform throughout the system. Using the
first law of thermodynamics, we will analyze the following processes:

- **Isothermal:** temperature remains constant
- **Adiabatic:** no heat transfers in or out of a system
- **Isobaric:** pressure remains constant
- **Isochoric:** volume remains constant

Isothermal Processes

When a substance freezes or boils, the phase transition occurs with no change in
temperature. Although heating water from some lower temperature up to its boiling
point requires energy, as additional heat is transferred into the system at the boiling
point, the liquid becomes vapor but the temperature remains constant. The first law
of thermodynamics ($\Delta U = Q - W$, Equation 15-1) describes, for example, what
happens when we add energy to a pot of 100°C water. In this case, Q represents the
energy added to the water and W is the work that the expanding water vapor does
on the surrounding air.

As we saw in the last chapter, the kinetic energy K and therefore the internal
energy of a gas is directly proportional to its temperature T; for example, for a
monatomic ideal gas, each atom, on average, contributes

$$K_{avg} = \frac{3}{2}kT \tag{14-13}$$

to the total kinetic energy. Recall that k is the Boltzmann constant, with value
1.381×10^{-23} J/K. For an isothermal process, then, because temperature remains
constant, the kinetic energy is constant. Therefore, as long as the potential energy
remains constant, the internal energy of the gas particles also remains constant.
This conclusion doesn't imply that no work is done or that no heat is transferred,
but the first law of thermodynamics does allow us to relate work and heat. When
$\Delta U = 0$, Equation 15-1 becomes

$$0 = Q - W$$

or

$$Q = W$$

So for an isothermal process in an ideal gas, whatever energy is added to a system
appears as work done by the system.

On a PV diagram, in which pressure is plotted as a function of volume, the area
under the curve represents the work done. The integral of a function gives the area
under the curve that represents it, so the work done in a thermodynamic process is

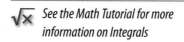
√x̄ *See the Math Tutorial for more*
information on Integrals

$$W = \int_{V_i}^{V_f} PdV \tag{15-2}$$

where V_i and V_f are the system's initial and final volumes during the process. We
know from the ideal gas law (Equation 14-7) that the pressure P is a function of
volume and temperature:

$$P = \frac{nRT}{V} \tag{15-3}$$

So from Equation 15-2 the work done by an ideal gas in a thermodynamic process is

$$W = \int_{V_i}^{V_f} \frac{nRT}{V} dV$$

as the volume of the gas changes from V_i to V_f. For an isothermal process, the constants n, R, and T can be removed from the integral:

$$W = nRT \int_{V_i}^{V_f} \frac{dV}{V}$$

The form of this integral is listed in the Math Tutorial as $\int At^{-1} dt = A \ln|t|$; so

$$W = nRT(\ln V \,|_{V_i}^{V_f}) = nRT(\ln V_f - \ln V_i)$$

or

$$W = nRT \ln \left(\frac{V_f}{V_i} \right) \qquad (15\text{-}4)$$

The work done in an isothermal expansion of an ideal gas therefore depends on the ratio of the final to the initial volume of the gas.

Example 15-2   Isothermal Expansion

A cylinder sealed with a piston contains 0.1 mol of an ideal gas at atmospheric pressure and a temperature of 300 K. Heat is transferred slowly into the gas so that it expands isothermally from an initial volume of 0.3×10^{-3} m^3 to 1.0×10^{-3} m^3 (Figure 15-8). How much heat is transferred into the gas during the process?

SET UP
The change in internal energy is zero for an ideal gas undergoing an isothermal process, so as we have seen, $Q = W$. All of the energy we add to the system causes the gas to expand. The area under the isothermal curve at that temperature gives

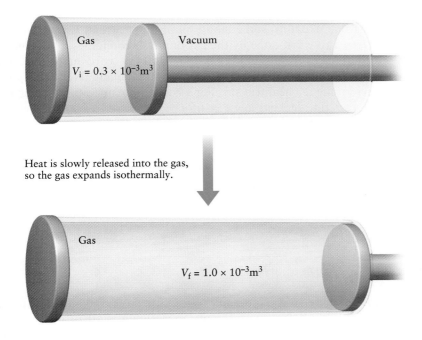

Heat is slowly released into the gas, so the gas expands isothermally.

Figure 15-8 Heat is released slowly into the gas so it expands isothermally.

Figure 15-9 The area shaded yellow under the isotherm represents the work done by the gas as it expands.

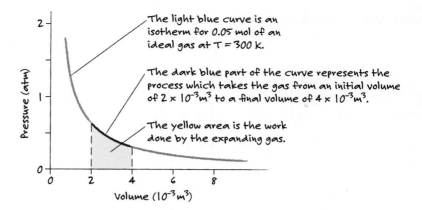

the work done. Note that for any specific temperature, there is only one isotherm for an ideal gas (Figure 15-9), so

$$W = nRT \ln\left(\frac{V_f}{V_i}\right) \tag{15-4}$$

gives a unique solution for any given value of T.

SOLVE

The heat added to the expanding gas is given by

$$Q = nRT \ln\left(\frac{V_f}{V_i}\right)$$

Using the values given in the problem statement,

$$Q = (0.1 \text{ mol})\left[8.314 \text{ J/(mol·K)}\right](300 \text{ K})\ln\left(\frac{1.0 \times 10^{-3} \text{ m}^3}{0.3 \times 10^{-3} \text{ m}^3}\right)$$

or

$$Q = 300 \text{ J}$$

REFLECT

For an ideal gas that expands isothermally, all of the heat that is transferred into the gas is converted to work done by the gas on its surroundings. Here, the 300 J of energy added to the gas results in the gas doing 300 J of work on the surrounding air as it expands. Also, note that we didn't need to use the initial (or final) pressure of the gas to determine the heat transfer. For a constant temperature (which occurs during this process), the ideal gas law $P = nRT/V$ (Equation 15-3) uniquely determines pressure given any value of volume. So, only the values of initial and final volumes are needed to determine Q.

Practice Problem 15-2 An ideal gas (0.37 mol) at 290 K is made to expand isothermally so that its volume triples. How much energy was added to the gas in the process?

Example 15-3 Isothermal Piston

A cylinder sealed at one end by a moveable piston contains 0.010 mol of an ideal gas at a temperature of 280 K. You push the piston in slowly, so that the temperature of the gas remains unchanged, until the final volume of the gas is one-half the original volume. How much work do you do in the process?

SET UP

To push the piston, you must overcome the force that the gas exerts on it. The work you do, therefore, is the negative of the work done by the gas. According to Equation 15-4, the work you do is

$$W = -nRT \ln \left(\frac{V_f}{V_i} \right)$$

SOLVE

Using the values given in the problem statement,

$$W = -(0.010 \, \text{mol}) \left[8.314 \, \text{J} / (\text{mol} \cdot \text{K}) \right] (280 \, \text{K}) \ln \left(\frac{1}{2} \right)$$

or

$$W = 16 \, \text{J}$$

REFLECT

We don't need to know the specific initial and final values of the volume to compute the work done in a process like this one; we only need to know the ratio of the initial and final volumes.

Practice Problem 15-3 A cylinder sealed at one end by a moveable piston contains 0.37 mol of an ideal gas at 290 K. You push the piston in slowly, so that the temperature of the gas remains unchanged, until the final volume of the gas is one-third the original volume. How much work do you do in the process?

Adiabatic Processes

An adiabatic process is one in which no heat transfers into or out of a system. Many animals stay warm in cold environments through a (nearly) adiabatic process that involves trapping air, a relatively poor thermal conductor, next to their bodies in a layer of fur or feathers. Little heat transfers into or out of a mass of air as the air experiences pressure and volume changes; the thermodynamic processes that the mass of air undergoes therefore tend to be nearly adiabatic. As we saw in Chapter 9, air in a scuba diver's lungs expands as he ascends to regions of lower pressure, and the exhaled air bubbles continue to expand as they rise to the surface (Example 9-2). In a similar way, when winds carry air up and over high mountains, the lower pressure of higher altitudes causes the air mass to expand. The expansion occurs nearly adiabatically, so we let Q be zero. From the first law of thermodynamics (Equation 15-1),

$$\Delta U = 0 - W$$

The expanding air does (positive) work on the surrounding air, resulting in a decrease in internal energy and a corresponding decrease in temperature of as much as 10°C for every 1 km of altitude. As temperature decreases, the amount of water vapor air can hold decreases as well. Because the actual amount of water vapor remains the same, however, the air eventually becomes saturated, that is, it can hold no more water vapor. Further cooling results in condensation—clouds form and rain falls. A satellite image of Molokai Island in Hawaii (**Figure 15-10**) presents a dramatic contrast caused by the process of adiabatic cooling. As the winds blow up and

Figure 15-10 Winds blow up and over the volcano that formed the northeastern part of Molokai Island in Hawaii. In the process, nearly all of the moisture is wrung out of the air, creating the lush rainforest visible as bright green in the satellite photograph. With its moisture content depleted, dry air blows down the other side of the volcano. The dry air results in the arid, tan-colored plain on the southern and western parts of the island. *(NASA.)*

Wind direction

over the volcano that formed the northeastern part of the island, nearly all of the moisture is wrung out of the air, creating a lush rainforest. With its moisture content depleted, dry air blows down the other side of the volcano. The dry air contributes to the arid, desertlike conditions on the southern and western parts of the island.

A second thermodynamic effect also comes into play as the air rises and falls while crossing mountains. Adding heat changes the temperature of saturated air more slowly than unsaturated air because heat is released as water vapor condenses. So, the temperature of the air rising on the windward side of the volcano decreases more slowly than the temperature of the unsaturated air flowing back down the other side increases. As a result, the air over the dry, downwind slope is much warmer than the wind blowing off the ocean. The vegetation on Molokai is evidence of these thermodynamic effects.

 Got the Concept 15-2
PV Diagram

Figure 15-11 shows three processes on a *PV* diagram. All three curves reach the same high volume state at the same pressure, so the temperatures of all three processes are the same in that state. The three processes start the same high pressure value at different volumes, however, so each high pressure state must have a different temperature. The orange curve is an isotherm. Which of the other two curves represents an adiabatic process?

Figure 15-11 The orange curve on this *PV* diagram is an isotherm. Which curve represents an adiabatic process?

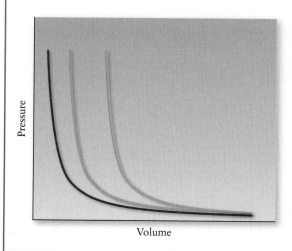

Pressure

Volume

The piston reaches its equilibrium position when the pressure of the gas in the cylinder equals atmospheric pressure.

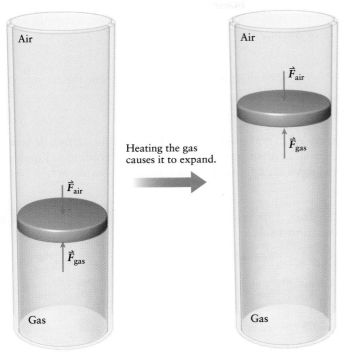

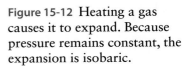

Figure 15-12 Heating a gas causes it to expand. Because pressure remains constant, the expansion is isobaric.

Heating the gas causes it to expand.

Even after the gas expands, the gas pressure must equal atmospheric pressure. Because neither gas pressure nor atmospheric pressure has changed, the expansion is isobaric.

Isobaric Processes

Isobaric processes occur with no change in pressure. Imagine a cylinder containing an ideal gas and sealed with a piston of negligible mass, as in Figure 15-12. At equilibrium the upward force on the piston due the pressure of the gas equals the downward force due to atmospheric pressure. This condition doesn't change when the gas is heated and expands; because atmospheric pressure doesn't change, the pressure in the gas must remain constant as well. The process of adding heat to the gas is therefore isobaric. The heat results in increased volume—the gas does work on the piston—but the pressure of the gas remains constant. This process is represented by a horizontal line on a PV diagram (Figure 15-13).

As always, the area under the curve on a PV diagram (Equation 15-2) represents the work done by a system undergoing a thermodynamic process. In the case of an isobaric change, the shape of the area, as you can verify from Figure 15-13, is rectangular; so the work done is the area of a rectangle of height P_0 and base $V_f - V_i$:

$$W = P_0(V_f - V_i) \qquad (15\text{-}5)$$

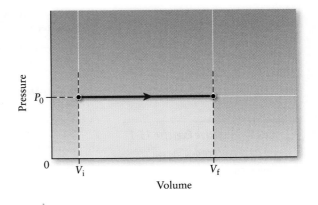

Figure 15-13 On a PV diagram, an isobaric process is represented as a horizontal line, and the work done is computed as the area of a rectangle.

Example 15-4 Isobaric Piston

Boiling 1.000 g (1.000 cm³) of water at 1.000 atm results in 1671 cm³ of water vapor. Imagine 1.000 g of water at the boiling point in an insulated cylinder sealed with a piston. Determine the change in internal energy of the water when just enough heat is added to boil all of the liquid.

SET UP

We will apply the first law of thermodynamics to determine the change in internal energy:

$$\Delta U = Q - W \tag{15-1}$$

Here Q is the heat added to make the water undergo a phase transition to vapor and W is the work the gas does on the piston as it expands. According to Equation 14-29, the energy required to cause a liquid-to-vapor phase transition is

$$Q = mL_V \tag{14-29}$$

where L_V is the latent heat of vaporization. The work done by the gas is negative and is given by Equation 15-5:

$$W = -P_0(V_f - V_i)$$

The process takes place at $P_0 = 1.000$ atm $= 1.013 \times 10^5$ Pa.

SOLVE

The change in internal energy is then

$$\Delta U = Q - W = mL_V + P_0(V_f - V_i)$$

The latent heat of vaporization of water L_V, from Table 14-4, is 2260 kJ/kg, so

$$\Delta U = (0.0010 \text{ kg})(2260 \times 10^3 \text{ J/kg})$$
$$+ (1.013 \times 10^5 \text{ Pa})(1671 \times 10^{-6} \text{ m}^3 - 1 \times 10^{-6} \text{ m}^3)$$
$$= 2260 \text{ J} + 169 \text{ J}$$
$$= 2429 \text{ J}$$

REFLECT

We were required to add 2260 J of energy to boil the water. Most of this energy goes into increasing the internal energy of the system. Because of conservation of energy, we know that the rest of the energy is not lost; it leaves the system in the form of work done on the piston.

Isochoric Processes

Go to Interactive Exercise 15-1 *for more practice dealing with isobaric and isochoric processes*

A gas is sealed inside a container of fixed volume. Adding heat will increase the pressure and temperature, but the volume of the gas cannot change. This kind of thermodynamic process is called isochoric. Volume remains constant in an isochoric process.

Systems undergoing an isochoric change do no work. Recall that the definition of work relies on a force being applied over a certain distance. Although the gas pressure results in forces on the wall of the container, constant volume means that the relative positions of the walls remain the same; no net displacement means no work is done. From the first law of thermodynamics, then, $\Delta U = Q$ for an isochoric process.

Example 15-5 Checking the Tire Pressure

You measure the air pressure in your tires to be 33.0 lb/in.^{2} (pounds per square inch, or psi) on a day when the temperature is 15.0°C. After a long drive you stop and measure the pressure again. Knowing that the temperature of the air in a car's tires can increase 30.0°C above ambient temperature when the car is in motion, what pressure would you expect?

SET UP

The change in the volume of a car's tires is relatively small when pressure and temperature change, so we can treat this process as isochoric. By treating the air in the tires as an ideal gas we can use the ideal gas law (Equation 14-7):

$$\frac{P}{T} = \frac{nR}{V}$$

Because the process is isochoric, the right-hand side of the equation is a constant. So, $P/T =$ constant, or

$$\frac{P_1}{T_1} = \frac{P_2}{T_2} \qquad (15\text{-}6)$$

for any two pressure and temperature states of a system.

SOLVE

We are given the pressure in the tires at a certain temperature, and asked for the pressure at some other temperature. We therefore rearrange Equation 15-6 for P_2:

$$P_2 = P_1 \frac{T_2}{T_1}$$

To get a numeric solution, temperature must first be expressed in SI units of kelvin; $T_1 = 15.0°C = 288$ K and $T_2 = 15.0°C + 30.0°C = 45.0°C = 318$ K. Also, note that "tire pressure" is gauge pressure; that is, it does not include the 14.7 psi of pressure that surrounds us and "fills" even tires that are not pressurized. From the problem statement we know that $P_1 = 33.0$ psi + 14.7 psi = 47.7 psi, so

$$P_2 = (47.7 \text{ psi}) \frac{318 \text{ K}}{288 \text{ K}} = 52.7 \text{ psi}$$

The pressure will be 52.7 psi, so a reading on a tire pressure gauge will be 52.7 − 14.7 = 38.0 psi.

REFLECT

After driving the car and causing the air in the tires to get hot, you measure a tire pressure of 38.0 psi which is 5.0 psi or 15% higher than when the tires are cold. If you conclude that the pressure in your tires is too high and let out some air, you might be driving on underinflated tires once the tires cool, which could be unsafe. For this reason, always check the air pressure in the tires when a car has been stationary for a while.

Practice Problem 15-5 After a long drive the air pressure in your tires is 39.0 psi. If the temperature of the air in the tires at that moment is 50.0°C, what tire pressure would you expect to measure when the air in the tires has cooled to 20.0°C?

Example 15-6 *PV* Diagram and the First Law

The *PV* diagram in **Figure 15-14** shows two thermodynamic processes that occur in an ideal gas that is initially at pressure $P_i = 1.1 \times 10^5$ Pa and volume $V_i = 3.0 \times 10^{-3}$ m³, point *A* on the diagram. Energy is removed from the gas in such a way that its pressure decreases to $P_f = 0.7 \times 10^5$ Pa while its volume remains the same, taking the system from *A* to *B*. The gas is then allowed to expand adiabatically to a volume $V_f = 4.7 \times 10^{-3}$ m³ (*B* to *C*) so that it returns to its initial temperature. How much energy flows in or out of the gas during these processes?

SET UP

The first law of thermodynamics guides the solution to this problem. The temperature changes from A to B and from B to C, but because both A and C are on the same isotherm, drawn in light blue in the figure, we know that the temperature at C is the same as at A. There is therefore no net change in temperature as a result of the two processes. Because internal energy is directly related to temperature, there is no net change in internal energy from A to C, so

$$\Delta U_{AB} = \Delta U_{BC} \qquad (15\text{-}7)$$

Now consider the two processes separately. Because the process from A to B occurs at constant volume, no net work is done. With $W = 0$, the equation for first law of thermodynamics (Equation 15-1) becomes

$$\Delta U_{AB} = Q_{AB} - W = Q_{AB}$$

By again applying the first law of thermodynamics, we find the change in internal energy in the process from B to C

$$\Delta U_{BC} = Q_{BC} - W_{BC}$$

The process from B to C occurs adiabatically, so $Q_{BC} = 0$. The work for this process is not zero, however, because the process occurs at constant pressure. We can find the work done from B to C directly from Equation 15-5:

$$W_{BC} = P_f(V_f - V_i)$$

so, by substitution we find the change in internal energy of the system between points B and C

$$\Delta U_{BC} = -P_f(V_f - V_i)$$

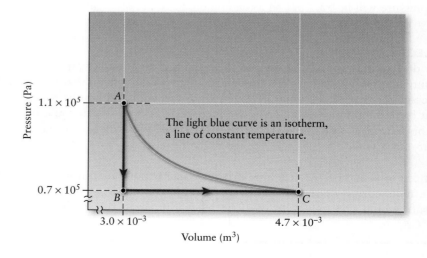

Figure 15-14 Between points *A* and *B*, energy is removed from the gas in such a way that its pressure decreases but its volume remains the same, so no work is done during the process. The gas expands adiabatically between points *B* and *C* so that it returns to its initial temperature.

SOLVE

We can now set the changes in internal energy equal according to Equation 15-7:

$$\Delta U_{AB} = \Delta U_{BC}$$

or

$$Q_{AB} = -P_f(V_f - V_i)$$

Using the values given in the problem statement,

$$Q_{AB} = -(0.7 \times 10^5\,\text{Pa})(4.7 \times 10^{-3}\,\text{m}^3 - 3.0 \times 10^{-3}\,\text{m}^3) = -119\,\text{J}$$

Because $Q_{BC} = 0$, the heat transfer for the process from A to C is Q_{AB} or -119 J.

REFLECT

As the system goes from state A to state B the heat transfer is negative because energy leaves the system as the pressure of the gas decreases.

Practice Problem 15-6 The *PV* diagram in Figure 15-14 shows the initial and final pressures and volumes of two thermodynamic processes that occur in an ideal gas. (a) If the initial pressure is $P_i = 2.2 \times 10^5$ Pa, determine how much energy flows into or out of the gas during the cycle from A to B to C. (b) If the final pressure is 1.4×10^5 Pa, determine how much energy flows into or out of the gas during the cycle from A to B to C.

> ★ **What's Important 15-2**
> Temperature remains constant in isothermal processes. No heat transfers into or out of a system during an adiabatic process. Isobaric processes occur with no change in pressure. Isochoric processes occur with no change in volume.

15-3 The Second Law of Thermodynamics

When an organism dies, its body decomposes into the inorganic components that make it up. The atoms that once made up a long-dead dinosaur don't reassemble themselves into a living, breathing *Tyrannosaurus rex*. Why not? Figure 15-15 shows two frames from a video in which someone inflates a balloon until it pops. How do you know that the upper frame happens before the lower one? None of the physical laws that we have studied so far preclude a set of balloon fragments from flying toward each other and reassembling as an intact balloon. Think of the fundamental laws we have uncovered, that total energy is always conserved, for example, and that momentum is conserved in an isolated system. None would be violated were the balloon fragments to reassemble. What about the pictures of an ice cube melting on a piece of granite in Figure 15-16? Clearly, each image was taken before the one below it, but why? Here again, the reverse process would be permitted by the fundamental laws we know. For example, the total energy of the system would be conserved if a puddle of water were placed on the granite, and the granite then drew enough energy from the water to cause it to freeze. This is because the amount of energy released when water freezes is the same as that required to melt a chunk of ice of equal mass. Yet we have declared that heat transfer is always from a hotter to a cooler object. Why?

For both the balloon and the ice cube, an understanding of the advance of time helps us know which event comes first and which happens later. We can think of

Figure 15-15 Both images were taken from a video of someone blowing up a balloon until it pops. How do you know that the upper frame happens before the lower one? *(Ted Kinsman/ Photo Researchers.)*

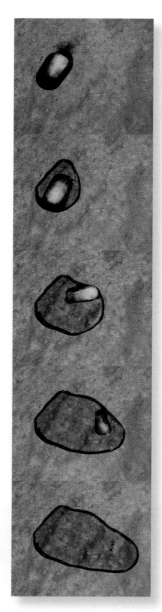

Figure 15-16 An ice cube melts. Each image was taken before the one below it, but how do you know? *(Courtesy of David Tauck.)*

time as an arrow pointing from the past to the future. The direction of the arrow is directly related to the level of organization or order in a closed system. Throw a fresh deck of cards into the air, and the cards will land on the floor scattered in a random way. You would be stunned if you threw a well-shuffled deck in the air and the cards came out in perfect order. The arrow of time takes us from ordered to random. This observation can be generalized to the **second law of thermodynamics**, which tells us that randomness in systems tends to increase:

> The amount of order in an isolated system either always decreases or, if the system is in equilibrium, stays the same.

Any system described by the second law must be isolated to ensure that nothing outside the system can cause its state to change in nonnatural ways. For example, we could manually sort out the fragments of the balloon in Figure 15-15 and carefully reassemble them. The balloon system would experience increasing order, but only as a result of our external intervention; we could not treat the balloon as isolated. The most general example of an isolated system is the entire universe; thus the second law of thermodynamics predicts that the amount of randomness in the universe always increases until it reaches equilibrium.

The second law of thermodynamics predicts that it is likely that a highly ordered system such as a chicken egg could evolve into a state of random bits of shell and scrambled egg but is unlikely that a disordered system like a scrambled egg will ever spontaneously arrange itself back into a highly ordered whole egg. Similarly, because the energy added to an ice cube increases the random motion of its molecules, the liquid water is in a less ordered state than the ice. The direction of time described by the second law of thermodynamics makes it highly unlikely that the molecules of water will give energy back to the environment and reform themselves into an ice cube. The egg, the ice cube and the universe have the second law of thermodynamics—and direction of time—in common.

The second law of thermodynamics allows us to analyze a wide range of systems that convert heat to work or work to heat, including mechanical systems such as the engine in a car (heat to work) and a kitchen refrigerator (work to heat), as well as living systems such as the human body, in which energy derived from food powers muscle contraction (heat to work) and produces sweat which in turn cools the body down (work to heat). Remember that heat in physics is energy *flow*; physics heat can be used to cool down an object, such as a refrigerator or a person!

A system or device that converts heat to work is a **heat engine** or simply an engine. Heat engines are cyclic: Some part of the system absorbs energy, work is done, and the system returns to its original state in order for the cycle to begin again. Although the term "engine" might call to mind the complex device that powers an automobile, in thermodynamics a heat engine can be as simple as gas in a piston that expands as heat is added and contracts as the gas cools. You can also imagine tiny, molecular engines in living cells. Though much smaller than a piston or the engine in a car, such engines perform the same function, transforming energy into motion. In the chloroplasts of all plant cells, as we noted in Section 6-6, hydrogen ions flow down their concentration gradient through ATP synthase, a curious enzyme that has a structure not unlike a waterwheel. The hydrogen ions cause the stalk of the ATP synthase to spin, which in turn pushes phosphate ions and adenosine diphosphate molecules together to form adenosine triphosphate (ATP). ATP synthase is in many ways analogous to an engine that converts the kinetic energy of hydrogen ions into a high-energy chemical bond that stores energy in a form that can be used by all cells. Enzymes such as ATP synthase follow a cycle, using energy to do work and then returning to their original state so that the cycle can start again.

Consider the generic heat engine shown in **Figure 15-17**. It is thermally connected to a *reservoir* of higher (hotter) temperature (T_H) and to a reservoir of lower (cooler) temperature (T_C). Energy Q_H flows from the hot reservoir; during this process some of the energy is used in doing work (W). The second law of thermodynamics casts its shadow over such a cycle; in order for the heat engine system to move toward increasing disorder, not all of the energy that flows from the hot reservoir can be converted to work. Some of this energy, Q_C, flows into the cold reservoir and cannot be recovered. Because this energy does not result in useful work, it is considered a waste by-product of the process. The same thing happens in the human body; at best less than a third of the energy in food is actually used by your cells to do work. The rest of the energy generates body heat.

Note that we define a **reservoir** as a part of a system with a heat capacity large enough either to absorb or supply heat without a change in temperature. In an old-time steam engine, for example, the furnace serves as the hot reservoir and the surrounding atmosphere acts as the cold reservoir.

It is not possible to create a perfect heat engine, or more precisely,

No process is possible in which heat is absorbed from a reservoir and converted completely into work.

This statement is often referred to as the *Kelvin–Planck form of the second law of thermodynamics*. It is impossible to extract an amount of heat Q_H from a hot reservoir and use it all to do work W; some amount of heat Q_C must be exhausted to a cold reservoir. This statement is described mathematically by

$$W = Q_H - |Q_C| \qquad (15\text{-}8)$$

taking into account the sign convention we have defined, in which heat into a system is positive and heat out of a system is negative. The **efficiency** of a process is defined as the useful work expressed as a percentage of the amount of energy required to do it. In many systems, the heat Q_H comes from burning fuel, for example, coal or oil. Remember that heat is energy flow (not high temperature as is the common meaning in everyday language), so burning fuel supplies energy to the engine. Said loosely then, efficiency tells you how much you get "out" for what you put "in." We write the efficiency e as

$$e = \frac{W}{Q_H} \qquad (15\text{-}9)$$

Efficiency would be 100%, or $e = 1$, if all of the input energy were converted to work; but, the Kelvin-Plank form of the second law of thermodynamics tells us that this is impossible. The best combustion engines, such as the ones found in automobiles, operate at an efficiency of about 30% ($e \approx 0.30$). Similarly, when human muscles contract, only about 25% of the input energy does work; the rest is lost as heat. Using Equation 15-8, the efficiency of a heat engine (Equation 15-9) can be expressed in terms of the input heat and the wasted heat:

$$e = \frac{W}{Q_H} = \frac{Q_H - |Q_C|}{Q_H} = 1 - \frac{|Q_C|}{Q_H}$$

For the efficiency equation, we adopt the standard convention of writing the exhaust or waste energy as a positive value (think of it as heat *lost*), so that the equation becomes

$$e = 1 - \frac{Q_C}{Q_H} \qquad (15\text{-}10)$$

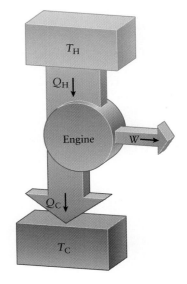

Figure 15-17 A generic heat engine, represented by a blue circle, is thermally connected to a reservoir of higher temperature (T_H) and to a reservoir of lower temperature (T_C). Energy Q_H flows from the hot reservoir; during the process some of the energy is used in doing work W.

Watch Out

When computing efficiency, Q_C is always positive.

Entering a negative value for Q_C into the equation for efficiency would result in e greater than 1, a nonsensical value.

Got the Concept 15-3
Cycle Forever

A charlatan claims to have invented a device which will cycle forever without the need for fuel. The device has two pistons; as the gas in one expands, the gas in the other is compressed. By thermally insulating the device, she claims that the cycle of compression and expansion will repeat endlessly without the need to add fuel. Do you believe her claim?

Example 15-7 Efficiency of an Engine

The combustion of gasoline in a lawnmower engine releases 22.0 J per cycle. Of that, 15.2 J is lost to warming the body of the engine and the surrounding air. What is the efficiency of the lawnmower engine?

SET UP

Some of the energy Q_H input to the engine is used to do work W. The rest, Q_C, is wasted. The engine's efficiency, as given in Equation 15-9, is the work done expressed as a fraction of the energy input. For the lawnmower, the 22.0 J of energy released by the combustion of gasoline is Q_H, and the 15.7 J lost to the environment is Q_C. Also, from Equation 15-8, the work done per cycle is

$$W = Q_H - |Q_C| = 22.0\,\text{J} - 15.7\,\text{J} = 6.30\,\text{J}$$

SOLVE

From Equation 15-9, the efficiency is then

$$e = \frac{6.30\,\text{J}}{22.0\,\text{J}} = 0.286$$

or 28.6%. Note that we could use Equation 15-10 to obtain the same result:

$$e = 1 - \frac{Q_C}{Q_H} = 1 - \frac{15.7\,\text{J}}{22.0\,\text{J}} = 0.286$$

REFLECT

An efficiency of around 30% is typical for an internal combustion engine.

Practice Problem 15-7 An engine that has a measured efficiency of 23.0% releases 19.5 J per cycle. How much of the energy is lost per cycle?

Go to Picture It 15-2 for more practice dealing with the Carnot heat engine

Go to Interactive Exercise 15-2 for more practice dealing with efficiency

The process that underlies the combustion engine found in automobiles is shown as a *PV* diagram in Figure 15-18. Beginning at the state marked 1, a mixture of fuel and air is compressed in a piston. At state 2 a spark ignites the fuel resulting in a positive heat transfer. The associated increase in pressure and temperature

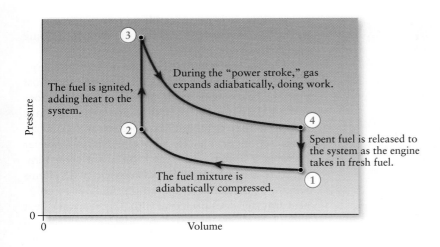

The fuel is ignited, adding heat to the system.

During the "power stroke," gas expands adiabatically, doing work.

Spent fuel is released to the system as the engine takes in fresh fuel.

The fuel mixture is adiabatically compressed.

Figure 15-18 A *PV* diagram shows the process that underlies the combustion engine found in automobiles. A mixture of fuel and air is compressed in a piston (point 1) followed by a spark that ignites the fuel (point 2). The increase in pressure and temperature (from point 2 to point 3) associated with the resulting positive heat transfer occurs so rapidly that the volume of the gas in the piston remains nearly constant. The high pressure causes the gas to expand (point 3), and in that process, going from point 3 to point 4, the engine does work. The work causes the wheels to turn.

from state 2 to state 3 occurs so rapidly that the volume of the gas in the piston remains nearly constant. At state 3, the high pressure causes the gas to expand. The expansion is adiabatic; because $Q = 0$ in an adiabatic process, the first law of thermodynamics ($\Delta U = Q - W$) tells us that all of the change in internal energy results in work by the gas. The work done by the engine during the process from state 3 to state 4 is the output of the engine; this work causes the wheels to turn. As the system returns to its initial state (the process from state 4 to state 1) exhaust is expelled and new fuel is injected into the piston.

Of the many cyclic heat engine processes, the **Carnot cycle**, first proposed in the nineteenth century by French engineer Sadi Carnot, is of particular importance because it provides the highest efficiency of any thermodynamic cycle. Carnot based his design on a requirement that all of the processes that comprise the cycle be **reversible**.

A system experiences a reversible thermodynamic process when the system and its surroundings can be returned to their initial states through another thermodynamic process and without any permanent change to the system or surroundings. Freezing water is reversible. Extract a certain amount of energy from the water and it becomes ice; allow that same amount of energy to be added back to the ice and it melts into water. Both the water and the environment will be in the same thermodynamic state as before the processes started. Sliding a block along a tabletop is an irreversible process. The block slows and stops as the surfaces lose energy due to frictional heating; there is no way to harness that energy and convert it back into translational kinetic energy of the block. If it isn't possible to discern the direction of time during a process as described by the second law of thermodynamics, then the process is reversible. For example, if it isn't possible to tell whether a video of a process is being played forward or backward, that process is reversible. (Note that no real process can be completely reversible, so for us, reversibility is an idealization.)

A system undergoes a reversible process when it stays close to equilibrium that any change in state of the system can be reversed by only an infinitesimally small change. When each intermediate state of a system is close to equilibrium, the process can go either forward or backward due to only a small change. Often a **quasi-static process**, one that proceeds very slowly, satisfies this requirement and is reversible. For example, a slow, adiabatic expansion of a gas can be reversed by a small increase in pressure, which results in a slow, adiabatic compression of the gas back to the initial pressure and volume. The work done by the expansion is equal to the work required to increase the pressure and return the system to the initial state, and no loss of energy occurs. However, when a gas under pressure suddenly expands to a larger volume (as in Figure 15-5), the free expansion of the gas is

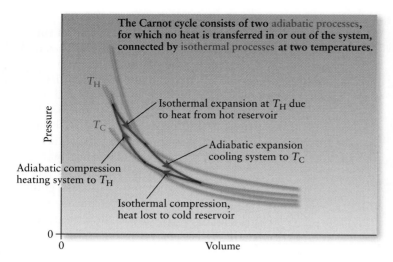

The Carnot cycle consists of two adiabatic processes, for which no heat is transferred in or out of the system, connected by isothermal processes at two temperatures.

T_H

T_C

Isothermal expansion at T_H due to heat from hot reservoir

Adiabatic expansion cooling system to T_C

Adiabatic compression heating system to T_H

Isothermal compression, heat lost to cold reservoir

Pressure

Volume

Figure 15-19 Carnot's heat engine employs an adiabatic expansion and an adiabatic compression connected by an isothermal expansion and an isothermal compression. Because all four processes are reversible, at least as an idealization, the Carnot heat engine has the most efficient heat engine cycle possible.

irreversible. It is not possible to return the system to the initial pressure and volume without adding energy by doing work.

Carnot realized that because no loss of energy occurs in a perfectly reversible process, the most efficient heat engine cycle possible must therefore involve only (nearly) reversible processes. We have just seen that adiabatic expansions and contractions are reversible. In addition, energy transfer between a system and a reservoir at the same temperature is also reversible. Because the two systems remain at the same temperature, the "hot-to-cold" heat transfer does not come into play. Carnot's heat engine therefore employs an adiabatic expansion and an adiabatic compression connected by an isothermal expansion and an isothermal compression, shown in **Figure 15-19**. Because all four processes are reversible, the Carnot heat engine has the most efficient heat engine cycle possible. Note, however, that the Carnot cycle is an idealization; it is not possible to build a perfect Carnot heat engine.

The efficiency of a heat engine obeys Equation 15-10:

$$e = 1 - \frac{Q_C}{Q_H}$$

According to the first law of thermodynamics, for an isothermal process in which no change in internal energy occurs, heat transfer equals work done:

$$Q = W$$

To find the theoretical limit of efficiency for the Carnot engine we use an ideal gas, so that the work done by the system during the isothermal processes is given by Equation 15-4:

$$W = nRT \ln\left(\frac{V_f}{V_i}\right)$$

During the isothermal expansion,

$$Q_H = W_{3\rightarrow4} = nRT_H \ln\left(\frac{V_4}{V_3}\right)$$

and during the isothermal compression,

$$Q_C = W_{1\rightarrow2} = nRT_C \ln\left(\frac{V_2}{V_1}\right)$$

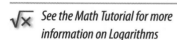

See the Math Tutorial for more information on Logarithms

Q_C is negative; we defined work done on a system to be negative (in addition, the logarithm of a number less than 1 is negative). However, recall that for the purpose of computing efficiency, we must treat Q_C as positive in order to ensure that $e < 1$. Because $-\ln a = \ln(1/a)$, changing the sign of Q_C involves inverting the fraction V_2/V_1, giving

$$Q_C = nRT_C \ln\left(\frac{V_1}{V_2}\right)$$

Equation 15-10 then becomes

$$e = 1 - \left[\frac{nRT_C \ln\left(\frac{V_1}{V_2}\right)}{nRT_H \ln\left(\frac{V_4}{V_3}\right)}\right]$$

For the two adiabatic processes in the Carnot cycle, the ratios of the volumes are the same, that is,

$$\frac{V_1}{V_2} = \frac{V_4}{V_3}$$

so,

$$e = 1 - \frac{T_C}{T_H} \qquad (15\text{-}11)$$

The efficiency of the ideal Carnot heat engine depends only on the high and low temperatures between which the engine operates. Again, this efficiency is the theoretical limit for the efficiency of a heat engine.

In theory, the efficiency of an ideal Carnot engine can be closer and closer to 100% by decreasing T_C and increasing T_H. In order for the efficiency to be exactly 1, T_C must equal 0 K. It is not possible, however, to attain a temperature of absolute zero. This statement is referred to as the **third law of thermodynamics:**

> It is possible for a system to asymptotically approach but never reach a temperature of absolute zero.

A conclusion of the third law of thermodynamics is that it is impossible to make a heat engine that is 100% efficient.

Example 15-8 Efficiency of a Car Engine

The ignition temperature of regular grade gasoline is about 430°C. Consider the cold reservoir for the system to be the liquid coolant, which has a temperature of 95°C. What is the theoretical maximum efficiency of such an engine?

SET UP
The theoretical maximum efficiency would be achieved by the engine operating according to an ideal Carnot cycle. The efficiency of such an engine depends only on the temperatures of the hot and cold reservoirs:

$$e = 1 - \frac{T_C}{T_H} \qquad (15\text{-}11)$$

SOLVE
We have been given $T_H = 430°C$ and $T_C = 95°C$. To correctly compute the efficiency, note that the temperatures must be in SI units, so

$$e = 1 - \frac{95 + 273.15 \text{ K}}{430 + 273.15 \text{ K}} = 0.48$$

In theory, as much as 48% of the energy input to the engine can be used to do work. The remaining 52% would be wasted.

REFLECT
The maximum efficiency for a real combustion engine is approximately 37%, and typical engines run at 18 to 20%. No real engine can be as efficient as the Carnot engine prediction, because no real process can be perfectly reversible.

Practice Problem 15-8 The temperature inside the boiler of a certain steam engine is 215°C. The temperature of the exhaust is 22.0°C. What is the theoretical maximum efficiency of the engine?

Example 15-9 Theoretical and Actual Efficiencies

A Carnot engine is designed to operate between 290 K and 450 K. The engine produces 1.5×10^2 J of mechanical energy for every 6.0×10^2 J of heat absorbed from the combustion of fuel. Compare the theoretical efficiency with the actual efficiency of the engine.

SET UP

We want to compare the actual efficiency (that is, the ratio of work done to energy received and described by Equation 15-9) and the efficiency of a Carnot engine (Equation 15-11).

SOLVE

The actual efficiency of the engine is then

$$e = \frac{W}{Q_H} = \frac{1.5 \times 10^2 \, \text{J}}{6.0 \times 10^2 \, \text{J}} = 0.25$$

The theoretical maximum predicted by the Carnot engine is

$$e = 1 - \frac{T_C}{T_H} = 1 - \frac{290 \, \text{K}}{450 \, \text{K}} = 0.36$$

REFLECT

The actual efficiency of this engine, 0.25, is about 70% of its theoretical maximum, 0.36.

Practice Problem 15-9 A Carnot engine produces 1.2×10^2 J of mechanical energy for every 8.0×10^2 J of heat absorbed from the combustion of fuel. Determine the efficiency of the engine.

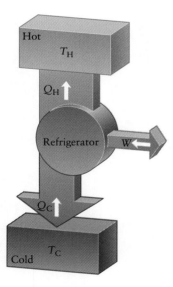

Figure 15-20 The generic refrigerator shown as a blue circle looks similar to the heat engine in Figure 15-17, but notice the important difference: All of the arrows are in the opposite direction. Work is an *input* to the device; as a result heat Q_C is made to go from the colder region to the hotter one.

A **refrigerator** is a device that converts work or mechanical energy to heat. In some sense, a refrigerator is an engine running backward. Like heat engines, refrigerators are cyclic, and like heat engines, refrigerators are constrained as described by the second law of thermodynamics.

> No process is possible in which heat is absorbed from a cold reservoir and transferred completely to a hot reservoir.

In other words, it isn't possible to make a perfect refrigerator. This statement is referred to as the *Clausius form of the second law of thermodynamics,* after the 19th-century German physicist Robert Clausius.

The generic refrigerator shown in **Figure 15-20** looks similar to the heat engine in Figure 15-17, but notice the important difference: All of the arrows are in the opposite direction. Work is an *input* to the device; as a result heat Q_C is made to go from the colder region to the hotter one. Such transfer is, of course, exactly what you'd want in order to cool down a place to store food (a refrigerator) or a building (an air conditioner). A **heat pump**, a device used to warm up a region, is identical to a refrigerator except that we're interested in how much energy can be added to the hot reservoir rather than how much can be extracted from the cold reservoir.

As a refrigerator runs, work is done in order to continually return the system to its initial pressure and volume. In a kitchen refrigerator, for example, the cycle shown in **Figure 15-21** involves changes in the pressure, volume, and temperature of CH_2FCF_3, an inert gas also known as 1,1,1,2-Tetrafluoroethane. CH_2FCF_3 vapor at a temperature below the boiling point (about $-26°C$) is forced through a tube

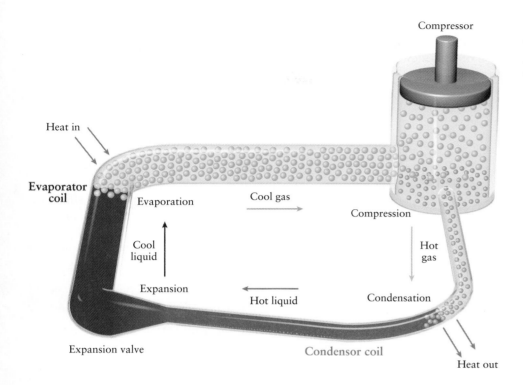

Figure 15-21 The cyclic process that cools a kitchen refrigerator involves compressing and expanding CH_2FCF_3 vapor (or some other refrigerant).

(the evaporator coil) that passes through the freezer and refrigerator compartments. The CH_2FCF_3 absorbs energy from the air, cooling the air and warming the CH_2FCF_3. A compressor pump then pressurizes the CH_2FCF_3, which not only drives the gas around the system again, but also increases its temperature even more. As the hot gas passes through the long, narrow condenser coil tube, heat is transferred from the gas to the cooler air in the kitchen. This part of the process cools the gas enough to cause it to liquefy, and the liquid CH_2FCF_3 is further cooled by letting it expand rapidly as it passes through a valve from a narrow tube to a much wider one. The cold liquid then makes its way back to the evaporator coil to start the process again. In this way, energy is removed from inside the refrigerator and delivered to the environment that surrounds it.

The less work required to extract heat Q_C from the cold reservoir, the more efficient the refrigerator. (A perfect refrigerator would require no work to be done in order to cause heat to be transferred from the cold reservoir to the hot reservoir, but such a process isn't possible according to the second law of thermodynamics.) The efficiency of a refrigerator is given by the **coefficient of performance CP**:

$$CP = \frac{Q_C}{W} \tag{15-12}$$

As with efficiency, CP is essentially what you get out in terms of what you put in. The smaller W is, the larger CP will be; a typical value for the coefficient of performance for a kitchen refrigerator or a home air conditioner is 3.

To ensure that energy is conserved, the first law of thermodynamics requires that the total energy that goes into the refrigerator in Figure 15-20 comes out:

$$Q_C + W = Q_H$$

Equation 15-12 then becomes

$$CP = \frac{Q_C}{Q_H - Q_C} \tag{15-13}$$

We found that the theoretical maximum efficiency of a heat engine is obtained when the engine runs on the Carnot cycle. For the Carnot engine, the general definition of efficiency

$$e = 1 - \frac{Q_C}{Q_H} \qquad (15\text{-}10)$$

becomes

$$e = 1 - \frac{T_C}{T_H} \qquad (15\text{-}11)$$

The maximum coefficient of performance of a refrigerator is obtained when the system is based on the Carnot cycle; in an analogous way to heat engine efficiency, the coefficient of performance of a refrigerator based on the Carnot cycle is

$$CP = \frac{T_C}{T_H - T_C} \qquad \textbf{(15-14)}$$

Example 15-10 Kitchen Refrigerator

A kitchen refrigerator has a CP of 3.3. When the refrigerator pump is running, it removes 760 J per second from inside the refrigerator. How much work must be done per second to extract this energy? How much energy is dumped into the kitchen in the process? If the refrigerator operates between 2°C and 25°C, what is the theoretical maximum CP?

SET UP
The coefficient of performance

$$CP = \frac{Q_C}{W} \qquad (15\text{-}12)$$

is a measure of how much cooling a refrigerator provides (Q_C) relative to the work W required to produce that cooling. The work is then given by

$$W = \frac{Q_C}{CP}$$

Note that the energy dumped into the kitchen per second is the 760 J plus the work the refrigerator does—that is, the energy added to the system in order to remove the 760 J.

SOLVE
The work that the refrigerator does is

$$W = \frac{760\,\text{J}}{3.3} = 230\,\text{J}$$

The energy dumped into the kitchen is then 760 J + 230 J = 990 J.

The maximum coefficient of performance is based on the Carnot cycle and is given by:

$$CP = \frac{T_C}{T_H - T_C} \qquad \textbf{(15-14)}$$

We need to convert $T_C = 2°C$ to a Kelvin temperature, so $T_C = 275.15$ K. Notice that the temperature difference $T_H - T_C$ is the same ($25 - 2 = 23$ K) whether or not the values are converted to kelvin, so,

$$CP = \frac{275.15\,\text{K}}{23\,\text{K}} = 12$$

REFLECT

Notice that the higher the actual value of CP, the less work must be done to achieve the same amount of cooling inside the refrigerator. Less work means less waste, and as a result, less heating of the kitchen.

Practice Problem 15-10 A kitchen refrigerator has a CP of 3.0. When the refrigerator pump is running, it dumps 920 J per second into the kitchen. How much work does the refrigerator pump per second to accomplish this?

★ What's Important 15-3

Randomness in systems tends to increase over time; this is embodied in the second law of thermodynamics. A system or device that converts heat to work is a heat engine; it is not possible to create an engine or process in which heat is absorbed from a reservoir and converted completely into work. The temperature of a system can asymptotically approach but never reach absolute zero; this is the third law of thermodynamics.

15-4 Gases

Removing energy from a liquid such as water, or from a solid such as a chunk of copper, causes a decrease in the temperature of the substance, or a phase change of the substance, or both. Unlike liquids and solids, gases are far more easily compressed, so it is also possible that the volume decreases as energy leaves the gas. In general, for gases we must consider the relationship between heat and changes in volume, pressure, and temperature. You may likely have experienced this physics directly, for example, if you've ever noticed how warm a bicycle pump gets when you add air to a tire. If you're a diver, you may have noticed that scuba tanks get warmer as they are filled.

Figure 15-22 shows two possible ways that a gas in a cylinder with a movable piston can be slowly heated from a lower temperature to a higher one. The process represented by the red arrow takes place at constant pressure, and the process represented by the orange arrow takes place at constant volume. Although the change in temperature, and therefore the change in internal energy, is the same for both, there is a significant difference between the two. Because work done depends on a change in volume (as in Equation 15-2), no work is done by the gas when it undergoes the constant volume process. At constant volume, all of the energy added as heat results in increased internal energy. At constant pressure, however, only some of the heat increases the internal energy, with the rest going into work done by the gas to move the piston. To achieve the same change in internal energy in both cases, then, more heat must be added in the constant pressure process.

Let Q_P and Q_V represent the heat required for a given change in internal energy in constant pressure and constant volume processes, respectively. The first law of thermodynamics (Equation 15-1) then gives

$$\Delta U = Q_P - W$$

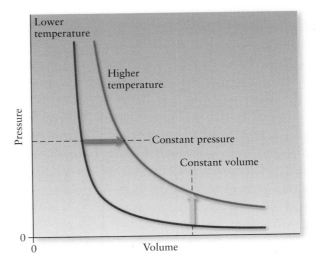

Figure 15-22 Shown are two possible ways that gas in a cylinder with a movable piston can be slowly heated from a lower temperature to a higher one. The process represented by the red arrow takes place at constant pressure, and the process represented by the orange arrow takes place at constant volume. To achieve the same change in internal energy in both cases, the constant pressure process requires the input of more heat because the work of moving the piston uses some energy.

and, because no work is done when the process occurs at constant volume,

$$\Delta U = Q_V \qquad (15\text{-}15)$$

Setting the two equal confirms mathematically that more heat must be added in the constant pressure process to achieve the same change in internal energy:

$$Q_P - W = Q_V$$

or

$$Q_P = Q_V + W \qquad (15\text{-}16)$$

The work done by the expanding gas on the piston is positive, so clearly Q_P is larger than Q_V.

We can draw a second, important conclusion from the application of the first law of thermodynamics to the constant pressure and constant volume heating of an ideal gas. When the *same* amount of heat is added to the gas in both cases, the change in temperature of the gas will be lower in the constant pressure case than in the constant volume one. For gases, then, we must revise the relationship that we developed in the last chapter between heat and temperature change. Whereas the specific heat (Equation 14-23)

$$Q = mc\Delta T$$

applies to solids, for gases we must define one specific heat for constant pressure processes and another for constant volume processes. When heat is added to a gas and either pressure or volume is held constant, the relationship between heat and change in temperature ΔT is

$$Q_P = nC_P\Delta T \qquad (15\text{-}17)$$

$$Q_V = nC_V\Delta T \qquad (15\text{-}18)$$

We have chosen to write Equations 15-17 and 15-18 in terms of the number of moles n in the gas. C_P is the molar specific heat of a gas at constant pressure. C_V is the molar specific heat of a gas at constant volume. Table 15-1 lists the molar specific heats for some gases.

Table 15-1	Molar Specific Heats of Gases at 300 K and 1 atm [J/mol·K)]				
	Gas	C_P	C_V	$C_P - C_V$	$\gamma = C_P/C_V$
Monatomic	Ar	20.8	12.5	8.28	1.66
	He	20.8	12.5	8.28	1.66
	Ne	20.8	12.7	8.08	1.63
Diatomic	H_2	28.8	20.4	8.37	1.41
	N_2	29.1	20.8	8.33	1.40
	O_2	29.4	21.1	8.33	1.39
	CO	29.3	21.0	8.28	1.39
Triatomic	CO_2	37.0	28.5	8.49	1.30
	SO_2	40.4	31.4	9.00	1.29
	H_2O (373 K)	34.3	25.9	8.37	1.32

Equation 15-16 tells us that more heat must be added to a gas at constant pressure than at constant volume in order to produce the same change in temperature. Considering the relationships between Q, ΔT, and C_P and C_V (Equations 15-17 and 15-18), it must also be true that C_P is greater than C_V. To confirm this, combine Equations 15-17 and 15-18 with Equation 15-16:

$$nC_P\Delta T = nC_V\Delta T + W$$

From the definition of work (Equation 15-2), for an ideal gas the work done by the gas is $P\Delta V$ (from Equation 14-7),

$$P\Delta V = nR\Delta T$$

so

$$nC_P\Delta T = nC_V\Delta T + nR\Delta T$$

or

$$C_P = C_V + R \qquad (15\text{-}19)$$

Not only is C_P greater than C_V, it is also larger by R, the universal gas constant.

Equation 15-19 is a theoretical relationship for an ideal gas; you can compare the experimentally determined values for $C_P - C_V$ (Table 15-1) with the value of R, 8.314 J/(mol·K).

Notice that the values of both C_V and C_P are essentially the same for all gases with the same number of atoms. We can quantify this by considering the change in internal energy as heat is added. At constant volume, from Equations 15-15 and 15-17

$$\Delta U = Q_V = nC_V\Delta T \qquad (15\text{-}20)$$

A gas molecule has energy $\frac{1}{2}kT$ for each degree of freedom. In a monatomic gas each atom has three degrees of freedom (it can move in the x, y, or z directions) so the total internal energy of N atoms of gas is $\frac{3}{2}NkT$. Recall that for an ideal gas $PV = NkT$ (Equation 14-6) and, for n moles of gas, $PV = nRT$ (Equation 14-7). So nR is equivalent to Nk; for n moles of a monatomic gas, then, the internal energy is

$$U = \frac{3}{2}nRT$$

If the temperature undergoes a change ΔT, the corresponding change in internal energy ΔU is

$$\Delta U = \frac{3}{2}nR\Delta T$$

Combining this with Equation 15-20 gives

$$nC_V\Delta T = \frac{3}{2}nR\Delta T$$

or

$$C_V = \frac{3}{2}R$$

For all monatomic ideal gases, the molar specific heat at constant volume C_V is the same. Because C_P and C_V differ only by a constant for a given gas (Equation 15-19), the same is true for the molar specific heat at constant pressure. The experimentally determined values of C_V for the monatomic gases listed in Table 15-1 are all close to the value of $\frac{3}{2}R$.

Notice that the values of C_V for the diatomic gases listed in Table 15-1 are greater than $\frac{3}{2}R$. For N_2, for example, C_V equals 20.8 J/(mol·K) or about $\frac{5}{2}R$. At temperatures in the range of room temperature, a diatomic molecule can rotate around two different axes in addition to being able to move in the x, y, or z directions. A diatomic gas molecule therefore has five degrees of freedom, so its total internal energy is $\frac{5}{2}nRT$. Following the analysis above, we see that $C_V = \frac{5}{2}R$ for a diatomic gas at around room temperature. At higher temperatures the atoms in a diatomic molecule can vibrate (much like objects attached by a spring), and for molecules made up of more than two atoms, additional degrees of freedom result in larger values of C_V and C_P.

The ratio of C_P to C_V is often used to directly identify the number of degrees of freedom of the molecules in a gas, a useful number when studying the characteristics of gases. By convention the ratio is γ:

$$\gamma = \frac{C_P}{C_V}$$

Using DOF to represent the degrees of freedom,

$$C_V = \frac{\text{DOF}}{2}R$$

and from Equation 15-19

$$C_P = C_V + R = \frac{\text{DOF}}{2}R + R = \frac{\text{DOF} + 2}{2}R$$

so

$$\gamma = \frac{C_P}{C_V} = \frac{\text{DOF} + 2}{2}R\frac{2}{\text{DOF}}\frac{1}{R} = \frac{\text{DOF} + 2}{\text{DOF}}$$

For a monatomic ideal gas, for example, which has three degrees of freedom,

$$\gamma = \frac{3 + 2}{3} = 1.67$$

Experimental measurements support this, as evidenced by the values of γ listed in Table 15-1.

? Got the Concept 15-4

Helium in a Cylinder

A quantity of He gas at room temperature is placed in a cylinder with a movable piston. When heat Q_0 is added so that the volume of gas remains the same, the temperature of the gas increases by ΔT_0. If the process were repeated so that the pressure in the gas, rather than the volume of it, remains the same, will the temperature change of the gas be lower than, the same as, or higher than ΔT_0?

? Got the Concept 15-5
Helium Versus Nitrogen

Quantities of He and N_2 gas are placed in separate cylinders with movable pistons. Each is at room temperature. Heat Q_0 can be added so that either the volume or the pressure of each gas remains the same. Which gas will experience the greatest difference in final temperature between the constant pressure and constant volume processes?

Because transfer of heat is a relatively slow process, thermodynamic processes that are carried out quickly often involve relatively little energy flow. We can consider such processes to be nearly adiabatic. In addition, because gases such as air are poor conductors of heat, relatively little energy transfer occurs when a gas expands or is compressed. Especially because air is all around us, it is therefore useful to be able to characterize adiabatic expansion and compression of a gas.

Consider an ideal gas that is allowed to expand slowly in a thermally isolated cylinder fitted with a movable piston. By ensuring the cylinder is thermally isolated, such an expansion proceeds without heat transfer and is adiabatic. For an ideal gas that undergoes such an adiabatic expansion

$$PV^\gamma = \text{constant} \tag{15-21}$$

where γ is the ratio of C_P to C_V. As long as the thermodynamic changes that take place in an ideal gas are adiabatic, the product of pressure and temperature taken to the power of γ remains the same. As volume goes down, pressure goes up.

Equation 15-21 notwithstanding, the ideal gas law still holds. So, for example, from Equation 14-7,

$$P = \frac{nRT}{V}$$

Combining this with Equation 15-21 gives

$$\frac{nRT}{V}V^\gamma = \text{constant}$$

or because n and R are constants

$$TV^{\gamma-1} = \text{constant} \tag{15-22}$$

As volume decreases, temperature increases.

To arrive at an expression that relates the pressure and temperature in an ideal gas that experiences an adiabatic expansion, write the ideal gas law in terms of volume:

$$V = \frac{nRT}{P}$$

Then Equation 15-21 becomes

$$P\left(\frac{nRT}{P}\right)^\gamma = \text{constant}$$

or

$$P^{1-\gamma}T^\gamma = \text{constant} \tag{15-23}$$

Notice that the exponent $1 - \gamma$ is always a negative number, which means that as pressure goes up, so does temperature. Using a pump to increase the pressure in a bicycle tire or a scuba tank causes the temperature to rise.

Air is a poor heat conductor, so when a mass of air rises and expands, the process is nearly adiabatic. From Equation 15-21, as the volume of the air mass grows the pressure drops, and from either Equation 15-22 or Equation 15-23, as the volume increases and the pressure decreases, the temperature must decrease. That's why it's cold at the top of tall mountains!

Example 15-11 Compression in an Engine Cylinder

In a combustion engine, a mixture of air and fuel is injected into a cylinder, compressed, and then ignited by a spark. If the air–fuel mixture is injected at room temperature (293 K) and atmospheric pressure and then compressed from 800 cm³ to 80 cm³, what is the temperature of the mixture immediately before the spark ignites it? Assume that the mixture is an ideal, diatomic gas and that the process is adiabatic.

SET UP
We take the mixture to be a diatomic gas, not unreasonable because air is composed primarily of N_2 and O_2. The number of degrees of freedom (DOF) of the mixture is five and

$$\gamma = \frac{\text{DOF} + 2}{\text{DOF}} = \frac{7}{5} = 1.4$$

We can then apply

$$TV^{\gamma - 1} = \text{constant} \tag{15-22}$$

at a time immediately before and after the compression occurs.

SOLVE
Applying Equation 15-22 right before and right after the compression gives

$$T_{\text{before}} V_{\text{before}}^{\gamma - 1} = T_{\text{after}} V_{\text{after}}^{\gamma - 1}$$

so

$$T_{\text{after}} = T_{\text{before}} \left(\frac{V_{\text{before}}}{V_{\text{after}}} \right)^{\gamma - 1}$$

Using the equation above and the values from the problem statement, the temperature is

$$T_{\text{after}} = 293 \text{ K} \left(\frac{800 \text{ cm}^3}{80 \text{ cm}^3} \right)^{1.4 - 1} = 736 \text{ K}$$

REFLECT
During the compression stage of the combustion engine cycle, the temperature of the air–fuel mixture rises from 20°C to well over 400°C. The high temperature improves the rate and efficiency of combustion. In some engines, for example, diesel engines, the compression of the air–fuel mixture is large enough that the temperature exceeds the ignition point, and no spark is required to cause the fuel to burn.

Estimate It! 15-1 Hot Nozzles

Estimate the temperature of the nozzle on a bicycle pump after it has been used to pressurize a tire to about 2 atm gauge pressure.

SET UP

The air pressure increases from 1 atm to 3 atm (2 atm gauge pressure is 3 atm absolute pressure). Air can be considered diatomic, so γ is estimated to be 1.4. To determine the temperature of the nozzle, use

$$T_{\text{before}}^{\gamma} P_{\text{before}}^{\gamma-1} = T_{\text{after}}^{\gamma} P_{\text{after}}^{\gamma-1}$$

SOLVE

Rearrange the equation so that the final temperature appears alone on the left side:

$$T_{\text{after}} = T_{\text{before}} \left(\frac{P_{\text{before}}}{P_{\text{after}}} \right)^{(\gamma-1)/\gamma}$$

We can take as a big, round—but reasonable—value for initial temperature of the nozzle to be room temperature or so, or around 300 K. The temperature of the nozzle after using the pump is then

$$T_{\text{after}} = (300 \text{ K}) \left(\frac{1}{3} \right)^{(1-1.4)/1.4}$$

Note there is no need to convert atmospheres to pascals because pressure units cancel. So

$$T_{\text{after}} \cong 400 \text{ K}$$

REFLECT

We find the temperature to be about 400 K, or about 130°C. This result is no doubt higher than you would experience in real life. Approximating the gas as ideal is not unreasonable, and the compression process is nearly adiabatic because on a short time scale relatively little energy will leave the system as heat. However, all parts of the pump will be heated by conduction, reducing the temperature increase in any one part of the system. Nevertheless, the estimation correctly suggests that the gas as well as the pump nozzle will get hot to the touch.

✳ What's Important 15-4

Because energy flow (heat) is a relatively slow process, thermodynamic processes that are carried out quickly often involve relatively little energy flow. Because gases are poor conductors of heat, relatively little energy transfer occurs when a gas expands or is compressed. More heat must be added to a gas at constant pressure than at constant volume in order to produce the same change in temperature.

15-5 Entropy

Imagine that your sock drawer is an unorganized jumble of an equal number of blue socks and red socks. If you grab four socks without looking at their colors, what distribution are you most likely to have? The solution requires us to pay attention to the level of randomness in the distribution of the four selected socks. For example, four red socks and no blue socks is a highly ordered grouping, while the colors in a group of two blue socks and two red socks are less ordered, or more randomly distributed. It is far less likely to pull four red socks from the drawer than a mixture of red and blue. Although much simpler than the kind of order addressed

Table 15-2		Combinations of Four Red and Blue Socks	
Blue	Red	Combinations	Number
0	4	(●●●●)	1
1	3	(●●●●), (●●●●), (●●●●), (●●●●)	4
2	2	(●●●●), (●●●●), (●●●●), (●●●●), (●●●●), (●●●●)	6
3	1	(●●●●), (●●●●), (●●●●), (●●●●)	4
4	0	(●●●●)	1

by the second law of thermodynamics and seen in, say, heat engines and ATP synthase, this example allows us to see the connection between order (and, of course, randomness) and the probability that a system will evolve from some particular initial state to a specific final state.

There is only one way to end up with four red socks in our final state: Each separate selection must be a red sock. There are four ways to have three red socks and one blue one, however. One way is to get a red sock during each of the first three selections and then get one blue one. Another way is for the first sock chosen to be blue and the last three red. We can summarize the possible ways to get three red and one blue sock this way: (●●●●), (●●●●), (●●●●), and (●●●●). Table 15-2 summarizes all possible four-sock combinations. You can verify from the table that the most probable grouping of four socks is two red and two blue, which is also the most randomly distributed four-sock state. This observation, although based on a simple demonstration, can nonetheless be generalized. The most probable final or equilibrium state of a system is the one that is the least ordered. So, for example, two objects put into thermal contact come to thermal equilibrium not because no other possibility is allowed, but because the state is the least ordered and therefore the most probable. Energy would still be conserved if the hotter object got hotter while the colder one got colder, but this would be highly improbable.

We have seen that the thermodynamic state of a system is described by such quantities as temperature, pressure, volume, and internal energy. To represent order, we will add to those state variables the quantity **entropy**. Entropy S is defined so that the smaller its value, the less randomness (or more order) is present in a system. The second law of thermodynamics therefore predicts that the entropy of an isolated system will either remain constant or increase. The second law of thermodynamics also predicts that the entropy of the universe, in other words, all things and all processes, will increase until equilibrium is reached. Entropy addresses the time evolution of thermodynamic processes.

On what does entropy depend? Recall that the Carnot cycle, the most efficient cyclic thermodynamic process, is reversible. The efficiency of the Carnot cycle depends only on the temperatures of the hot and cold reservoirs:

$$e = 1 - \frac{T_C}{T_H} \tag{15-11}$$

In general, the efficiency of a cycle depends on the heat transfers from the hot reservoir and to the cold reservoir:

$$e = 1 - \frac{Q_C}{Q_H} \tag{15-10}$$

For the Carnot cycle, then,

$$\frac{T_C}{T_H} = \frac{Q_C}{Q_H}$$

or

$$\frac{Q_C}{T_C} = \frac{Q_H}{T_H}$$

In other words, the ratio of Q to T is the same for both the cold and hot reservoirs. It was this realization that prompted the physicist Clausius to associate entropy with this ratio, specifically he equated Q/T with the change in S

$$\Delta S = \frac{Q}{T} \qquad (15\text{-}24)$$

The SI units of ΔS are joules per kelvin (J/K).

A natural, spontaneous process either increases entropy or leaves it constant. Consider a process that increases the entropy of a system. To reverse the process would require that entropy of the system decreases, so the reverse process could not happen spontaneously. Some external intervention is required, for example, an external agent that does work on the system. A process can only be reversed without heat or work if the system was in equilibrium both before and after the process occurred. We saw in Section 15-3 that reversible processes have the highest efficiency. Now we see that reversible processes are those processes for which the entropy remains constant. Entropy increases when an irreversible process occurs.

We obtained the relationship $\Delta S = Q/T$ (Equation 15-24) by considering the isothermal expansion of an ideal gas. As we saw in the discussion of the Carnot cycle (Section 15-3), the isothermal expansion of an ideal gas is a reversible process. Strictly speaking, the relationship $\Delta S = Q/T$ applies only to processes that are reversible. Because reversible processes are an idealization, the change in entropy predicted by Equation 15-24 is also an idealization, the smallest change possible for a real, irreversible process.

Watch Out

The entropy of a system can go down, but only through a process which does not occur spontaneously or for a system which is not isolated.

Plant cells synthesize sugar by photosynthesis:

$$\text{light energy} + 6\,CO_2 + 6\,H_2O \rightarrow C_6H_{12}O_6 + 6\,O_2$$

The sugar molecules $C_6H_{12}O_6$ exhibit far more order than the raw materials, so the change in entropy in this process is negative. Indeed, photosynthesis is not spontaneous; it requires energy in the form of light and the action of enzymes to occur.

Animal and plant cells break down sugar in order to convert the energy stored in its chemical bonds into a usable form, a process called *cellular respiration*. The change in entropy is positive during respiration; energy is eventually released as complex—highly ordered—sugar molecules break down and combine with oxygen to form carbon dioxide and water molecules. Both carbon dioxide and water are simpler and therefore exhibit less order than the initial sugar molecules so the change in entropy is positive.

? Got the Concept 15-6
Two Pistons

The same quantity of heat is added to two identical pistons containing ideal gases so that each gas expands isothermally (**Figure 15-23**). The gas in the piston on the left is at a higher temperature than the gas in the piston on the right. How does the change in entropy compare for the two processes? Is the change in entropy ΔS the same for both processes, and if not, does the process of adding heat to the piston containing the hotter gas have a smaller or a greater ΔS than the other process?

The same quantity of an ideal gas is placed in two pistons. The temperature of the gas on the left is higher than the temperature of the gas on the right.

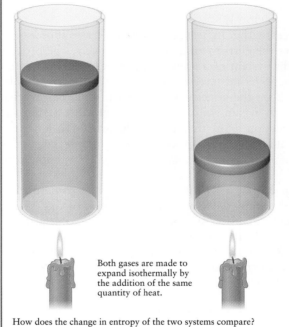

Both gases are made to expand isothermally by the addition of the same quantity of heat.

How does the change in entropy of the two systems compare?

Figure 15-23 The same quantity of heat is added to two identical pistons containing ideal gases, so that each gas expands isothermally. The gas in the piston on the left is at a higher temperature than in the gas in the piston on the right.

? Got the Concept 15-7
Entropy at the South Pole

A scientist places a tray of water outside the sealed habitat of the Amundsen–Scott South Pole Station (**Figure 15-24**). It quickly freezes. Has the entropy of the universe gone down, stayed the same, or gone up?

Figure 15-24 A tray of water quickly freezes when placed outside the sealed habitat of the Amundsen–Scott South Pole Station, where the annual mean temperature is –49°C. *(Courtesy: National Science Foundation.)*

Example 15-12 Boiling Water

What is the minimum change of entropy that occurs in 0.380 kg of water at 100°C when enough energy is added so that it boils?

SET UP
The change in entropy of the water must be equal to or larger than the minimum predicted by Equation 15-24 for a reversible process:

$$\Delta S = \frac{Q}{T}$$

To apply this relationship we need to know Q, the amount of energy required to boil the water, which (from Equation 14-29) depends on the latent heat of vaporization of water L_V as well as the mass of the water m:

$$Q = mL_V$$

so

$$\Delta S = \frac{mL_V}{T}$$

From Table 14-4, the latent heat of fusion of water is 2260 kJ/kg.

SOLVE
The temperature must be converted to kelvin before we can determine a numerical answer, so $T = 100°C = 373.15$ K. Now we substitute in the values given in the problem statement

$$\Delta S = \frac{(0.380 \text{ kg})(2260 \times 10^3 \text{ J/kg})}{373.15 \text{ K}} = 2.30 \times 10^3 \text{ J/K}$$

REFLECT
When the liquid water turns to vapor, the positions of the molecules become less localized, so we should expect a positive change in entropy. What would ΔS be if 0.380 kg of steam condensed to water? The magnitude of the change would be the same, but because the water state is more ordered, entropy would decrease. So for a steam-to-water transition the change in entropy would be -2.30×10^3 J/K.

Practice Problem 15-12 What is the minimum change of entropy that occurs in 3.25 kg of water at 100°C when enough energy is added so that it boils?

 Go to Interactive Exercise 15-3
for more practice dealing with entropy

Example 15-13 Entropy in a Carnot Engine

The Carnot engine of Example 15-9 is designed to operate between 290 K and 450 K. We computed the theoretical limit of the engine's efficiency to be 0.36. Now let's determine the change of entropy of the hot reservoir and the cold reservoir separately, and then for the entire engine, based on this theoretical limit. Assume that 6.0×10^2 J of heat Q is drawn from the hot reservoir.

SET UP
We can apply Equation 15-24 to each reservoir separately, taking care to use the appropriate value of heat for each. For the hot reservoir, the magnitude of Q_H is the 6.0×10^2 J of energy given in the problem statement. In addition, Q_H is negative because energy is leaving the hot reservoir. The heat that is absorbed by the cold

reservoir Q_C is the waste or the heat from Q_H that is not converted to work by the engine.

SOLVE

For the hot reservoir, we determine the change in entropy using Equation 15-24

$$\Delta S_H = \frac{Q_H}{T_H} = \frac{-6.0 \times 10^2 \, J}{450 \, K} = -1.3 \frac{J}{K}$$

The engine has an efficiency of 36%, which means that 36% of Q_H is converted to work and the remaining 64% is wasted and therefore absorbed by the cold reservoir

$$Q_C = 0.64 (6 \times 10^2 \, J) = 384 \, J$$

We arrive at a value for the change in entropy by substituting known values into Equation 15-24

$$\Delta S_C = \frac{Q_C}{T_C} = \frac{384 \, J}{290 \, K} = 1.3 \frac{J}{K}$$

The net change in entropy for the entire engine is the sum of the change for each reservoir, that is,

$$\Delta S = \Delta S_H + \Delta S_C$$

You can see that the net change in entropy is 0:

$$\Delta S = -1.3 \frac{J}{K} + 1.3 \frac{J}{K} = 0 \, J/K$$

REFLECT

The Carnot cycle provides the theoretically most efficient cyclic thermodynamic process. Each step is reversible, and entropy remains constant in reversible processes. We should therefore expect the net change in entropy to be zero for the engine described in this problem. If the engine were not operating at its theoretical maximum this would not be the case. For example, in Example 15-9 we found the actual efficiency of the engine to be 25%. The change in entropy for the hot reservoir of $-1.3 \, J/K$ is the same change in either case, but because the engine would waste 75% of Q_H, the change in entropy for the cold reservoir is different:

$$Q_C = 0.75 (6 \times 10^2 \, J) = 450 \, J$$

so,

$$\Delta S_C = \frac{Q_C}{T_C} = \frac{450 \, J}{290 \, K} = 1.6 \frac{J}{K}$$

Thus, the net change in entropy of the system—and the universe—is

$$\Delta S = -1.3 \frac{J}{K} + 1.6 \frac{J}{K} = 0.3 \frac{J}{K}$$

The net change in entropy for the real engine is positive, as it must be.

Practice Problem 15-13 The Carnot engine of Example 15-9 is designed to operate between 290 K and 450 K. We computed the theoretical limit of the engine's efficiency to be 0.36. Determine the change of entropy of the hot reservoir and the cold reservoir separately, and then for the entire engine based on the theoretical limit. Assume that the process draws $12.0 \times 10^2 \, J$ of heat Q from the hot reservoir.

Earth's biosphere has evolved over the past 3 billion years, from a time when only the simplest organisms existed to today when millions of complex living things inhabit the planet. Does life violate the second law of thermodynamics? Life moves in the direction of more complexity and therefore more order. Moreover, humans and all other complex organisms start out as single cells and use energy to grow into highly ordered populations of cells. (Of course, even the simplest bacterium or virus is incredibly complex and represents a much more ordered state than the organic molecules from which it formed.) Complex organisms are more efficient, require less energy, and are thereby better able to survive in an environment of limited resources. Life necessarily introduces more order and therefore lowers entropy, but does it violate the second law of thermodynamics?

No, life does not violate the second law of thermodynamics! This fundamental tenet of thermodynamics predicts that an isolated system, say, the universe as a whole, tends to greater disorder. Individual components of the system may not. For example, a living organism can at least in a general way be thought of as a heat engine; it consumes and metabolizes food, uses some of the energy to do work, for example, to grow, but wastes some of that energy, too. It is this inefficiency that causes the *net* effect to be an increase in the entropy of the universe. When we consider the thermodynamics of life, we must remember that the organism is not an isolated system and include its environment in our deliberations.

> ## ✳ What's Important 15-5
> The state variable entropy describes the amount of order in a system; systems with a low value of entropy are more ordered. Entropy remains constant when a reversible cycle occurs but increases when an irreversible cycle occurs. The most probable final or equilibrium state of a system is the one that has higher entropy, that is, the one that is the least ordered. The entropy of the universe increases until equilibrium is reached.

Answers to Practice Problems

15-1 -9.6×10^5 J

15-2   980 J

15-3   980 J

15-5   34.0 psi

15-6   (a) -119 J, (b) -238 J

15-7   15.0 J

15-8   39.5%

15-9   15%

15-10 230 J

15-12 2.15×10^4 J/K

15-13 0 J/K

Answers to Got the Concept Questions

15-1 The gas in the cylinder does work on the piston as it expands. Because the cylinder is thermally isolated, Q must be zero: No heat can transfer into or out of the gas. According to the first law of thermodynamics (Equation 15-1), the change in internal energy is $\Delta U = -W$, if $Q = 0$. The internal energy of the gas therefore decreases, and a decrease in internal energy corresponds to a decrease in temperature because temperature is directly related to the kinetic energy of the constituent atoms and molecules of a system (Section 14-2). Note that if the cylinder were not thermally isolated, we could make the gas expand without a change in temperature as energy is drawn in from the environment through the walls of the cylinder.

15-2 All three processes come to the same final temperature, as evidenced by the fact that all three curves in Figure 15-11 come to the same pressure and volume. However, $Q = 0$ for an adiabatic process, so according to the first law of thermodynamics (Equation 15-1), $\Delta U = -W$. The internal energy and therefore the temperature of the system must therefore decrease. Comparing Figure 15-11 to Figure 15-2, you can see that the initial temperature of the light blue curve in Figure 15-11 is higher than that of the orange isotherm, while the initial temperature of the dark blue curve is lower. Of the two (blue) nonisothermal processes, then, the light blue one exhibits a decrease in temperature and therefore represents an adiabatic process. As always, note that the transfer of heat does not necessarily require a change in temperature. Heat in physics, the flow of energy, should not be confused with the commonly used term that implies that an object is warm!

15-3 Don't believe it! People, including reputable scientists and engineers, have been inventing such "perpetual motion machines" for almost a thousand years. Some have been based on trickery, employing a hidden motor of some kind. But most fail simply in accordance with the second law of thermodynamics; over time, friction or some other dissipative force drains the device of usable energy.

15-4 The temperature change will be smaller. For a given amount of heat the temperature change ΔT is smaller for larger values of molar specific heat for example, $\Delta T = \frac{1}{nC_V}Q_V$ from Equation 15-18. The ratio of C_P to C_V is 1.67 for a monatomic gas such as He. Because C_P is larger than C_V, the change in temperature at constant pressure is smaller than the change in temperature at constant volume. All of the energy added as heat when volume is held constant goes into increasing the internal energy. When the gas can expand, some of the energy goes into the work the gas does on the piston. Notice that we can draw the same conclusion from Equation 15-16, $Q_P = Q_V + W$.

15-5 The difference in final temperature between the constant pressure and constant volume process will be larger for N_2. As we saw above, the ratio of C_P to C_V is 1.67 for a monatomic gas such as He. For a diatomic gas such as N_2 which has five degrees of freedom (DOF) at room temperature, the ratio is

$$\gamma = \frac{C_P}{C_V} = \frac{\text{DOF} + 2}{\text{DOF}} = \frac{5 + 2}{5} = \frac{7}{5} = 1.4$$

So C_P is larger, relative to C_V, for He than for N_2. The larger the C_P, assuming all other quantities are the same, the smaller the change in temperature from the initial to final state.

15-6 According to the definition of entropy S in Equation 15-24, for a specific value of heat Q, the higher the temperature T the smaller the change in entropy ΔS. For this reason, the piston on the left, at higher temperature, experiences a smaller change in entropy. The initial state matters when determining the change in entropy.

15-7 The motion of the molecules is less random in the ice than in the water, so even if the temperature of the water remains the same, its entropy decreases as a result of the phase change to ice. However, heat from the water goes into the atmosphere, so ΔS of the atmosphere is positive. In addition, the second law of thermodynamics requires that work must be done to draw energy from the water, and that process can never be 100% efficient. So the positive change in entropy of the air is greater than the magnitude of the negative change in entropy of the water. The entropy of the universe increases.

SUMMARY

Topic	Summary	Equation or Symbol
adiabatic process	No heat transfers in or out of a system during an adiabatic process.	
Carnot cycle	The Carnot cycle is a thermodynamic process that takes a system from an initial pressure, volume, temperature state through three intermediate states and back to the initial state. The Carnot cycle provides the highest efficiency of any thermodynamic cycle.	
coefficient of performance	The efficiency of a refrigerator is quantified by the coefficient of performance (CP). CP is defined as the heat Q_C extracted from the cold reservoir divided by the work done to extract it—what you get out relative to what you put in.	$CP = \dfrac{Q_C}{W}$ (15-12)

efficiency	The efficiency e of a process is defined as the useful work W expressed as a fraction of the amount of energy Q_H required to do it—what you get out relative to what you put in.	$e = \dfrac{W}{Q_H}$ (15-9)
entropy	Entropy is defined so that the smaller its value, the less randomness (or more order) is present in a system. The most probable final or equilibrium state of a system is the one that has higher entropy, that is, the one that is the least ordered. Entropy therefore addresses the time evolution of thermodynamic processes; systems tend to become less ordered over time. Entropy S can be defined in terms of the energy flow Q into or out of a system and the temperature T of the system.	$\Delta S = \dfrac{Q}{T}$ (15-24)
first law of thermodynamics	The first law of thermodynamics is a statement of the conservation of energy; the net change in the internal energy U of a system equals the difference between how much energy Q enters the system as heat and how much energy W leaves the system as work done by the system.	$\Delta U = Q - W$ (15-1)
heat engine	A heat engine is a general term for a thermodynamic device or system that converts heat to work. Heat engines are cyclic. Some part of the system absorbs energy, work is done, and the system returns to its original state in order for the cycle to begin again.	
heat pump	A heat pump, a device used to warm up a region, does work in order to transfer energy to a hot reservoir. A heat pump is identical to a refrigerator except that the quantity of interest is the energy that can be added to the hot reservoir rather than the energy that can be extracted from the cold reservoir.	
internal energy	Internal energy is the sum of the kinetic energy and potential energy of every atom and molecule in a system. The internal energy of a system does not depend on the specific thermodynamic process(es) that lead(s) to a particular state of the system.	U
isobaric processes	Isobaric processes occur with no change in pressure.	
isochoric processes	Isochoric processes occur with no change in volume.	
isotherm	An isotherm describes a series of states of a system that, regardless of pressure and volume, all have the same temperature.	
isothermal processes	Temperature remains constant in isothermal processes.	
***PV* diagram**	The pressure P of a system is plotted as a function of volume V on a PV diagram. The area under a curve on a PV diagram represents the work W done in that process.	$W = \displaystyle\int_{V_i}^{V_f} P\,dV$ (15-2)
quasistatic processes	A quasistatic process proceeds slowly and is therefore reversible.	
refrigerator	A refrigerator is a device that converts work or mechanical energy to heat. In some sense, a refrigerator is a heat engine running backward.	

reservoir	A reservoir is a part of a system with a heat capacity large enough either to absorb or supply heat without a change in temperature.
reversible	A system experiences a reversible thermodynamic process when the system and its surroundings can be returned to their initial states through another thermodynamic process and without any permanent change to the system or surroundings.
second law of thermodynamics	According to the second law of thermodynamics, the amount of order in an isolated system either always decreases or, if the system is in equilibrium, stays the same.
state function	A state function is a quantity that characterizes the state of a system that does not depend on pressure, volume, or temperature.
third law of thermodynamics	The third law of thermodynamics precludes a system from reaching a temperature of absolute zero.

QUESTIONS AND PROBLEMS

In a few problems, you are given more data than you actually need; in a few other problems, you are required to supply data from your general knowledge, outside sources, or informed estimate.

Interpret as significant all digits in numerical values that have trailing zeros and no decimal points.

For all problems, use $g = 9.8 \text{ m/s}^2$ for the free-fall acceleration due to gravity. Neglect friction and air resistance unless instructed to do otherwise.

• Basic, single-concept problem

•• Intermediate-level problem, may require synthesis of concepts and multiple steps

••• Challenging problem

SSM *Solution is in Student Solutions Manual*

Conceptual Questions

1. •Clearly define and give an example of each of the following thermodynamic processes: (a) isothermal, (b) adiabatic, (c) isobaric, and (d) isochoric. SSM

2. •Can a system absorb heat without increasing its internal energy?

3. •Why does the temperature of a gas increase when it is quickly compressed?

4. •Why is it possible for the temperature of a system to remain constant even though heat is released or absorbed by the system?

5. •When we say "engine," we think of something mechanical with moving parts. In such an engine, friction always reduces the engine's efficiency. Why? SSM

6. •There are people who try to keep cool on a hot summer day by leaving the refrigerator door open, but you can't cool your kitchen this way! Why not?

7. •If the coefficient of performance is greater than 1, do we get more energy out than we put in, violating conservation of energy? Why or why not?

8. •How does the time required to freeze water vary with each of the following parameters: mass of water, power of the refrigerator, and temperature of the outside air?

9. •By how much is the entropy of the universe changed when heat is released from a hotter object to a colder one? In what sense does this correspond to energy becoming unavailable for doing work? SSM

10. •In a slow, steady isothermal expansion of an ideal gas against a piston, the work done is equal to the heat input. Is this consistent with the first law of thermodynamics?

11. •If a gas expands freely into a larger volume in an insulated container so that no heat is added to the gas, its entropy increases. Using the definition of ΔS, explain this statement.

12. •Why do engineers designing a steam-electric generating plant always try to design for as high a feed-steam temperature as possible?

13. •Conduction across a temperature difference is an irreversible process, but the object that lost heat can always be rewarmed, and the one that gained heat can be recooled. An object sliding across a rough table and slows down and warms up as mechanical energy dissipates. This process is irreversible, but the object can be cooled and set moving again at its original speed. So in just what sense are these processes "irreversible"? SSM

14. •The frictional drag of the atmosphere causes an orbiting satellite to move closer to Earth and to gain kinetic energy. In what way does energy become unavailable for doing work in this irreversible process?

15. •Is a process necessarily reversible if there is no exchange of heat between the system in which the process takes place and its surroundings?

16. •Is the operation of an automobile engine reversible?

17. •In discussing the Carnot cycle, we say that extracting heat from a reservoir isothermally does not change the entropy of the universe. In a real process, this is a limiting situation that can never quite be reached. Why not? What is the effect on the entropy of the universe? SSM

18. •A pot full of hot water is placed in a cold room, and the pot gradually cools. How does the entropy of the water change?

19. •If you drop a glass cup on the floor, it will shatter into fragments. If you then drop the fragments on the floor, why will they not become a glass cup?

20. •Why is the entropy of 1 kg of liquid iron greater than that of 1 kg of solid iron? Explain your answer.

Multiple-Choice Questions

21. •An ideal gas trapped inside a thermally isolated cylinder expands slowly by pushing back against a piston. The temperature of the gas
    A. increases.
    B. decreases.
    C. remains the same.
    D. increases if the process occurs quickly.
    E. remains the same if the process occurs quickly. SSM

22. •A gas is compressed adiabatically by a force of 800 N acting over a distance of 5.0 cm. The net change in its internal energy is
    A. +800 J
    B. +40 J
    C. −800 J
    D. −40 J
    E. 0

23. •An ideal gas is contained in a cylinder of fixed length and diameter. Eighty joules of heat is added while the piston is held in place. The work done by the gas on the walls of the cylinder is
    A. 80 J
    B. 0 J
    C. less than 80 J
    D. more than 80 J
    E. not specified by the information given

24. •In an isothermal process, there is no change in
    A. pressure.
    B. temperature.
    C. volume.
    D. internal energy.
    E. internal energy *and* pressure.

25. •In an isobaric process, there is no change in
    A. pressure.
    B. temperature.
    C. volume.
    D. internal energy.
    E. internal energy *and* pressure. SSM

26. •In an isochoric process, there is no change in
    A. pressure.
    B. temperature.
    C. volume.
    D. internal energy.
    E. internal energy *and* pressure.

27. •A gas quickly expands in an isolated environment. During the process, the gas exchanges no heat with its surroundings. The process is
    A. isothermal.
    B. isobaric.
    C. isochoric.
    D. adiabatic.
    E. isotonic.

28. •The statement that no process is possible in which heat is absorbed from a cold reservoir and transferred completely to a hot reservoir is
    A. not always true.
    B. only true for isothermal processes.
    C. the first law of thermodynamics.
    D. the second law of thermodynamics.
    E. the zeroth law of thermodynamics.

29. •Carnot's heat engine employs
    A. two adiabatic processes and two isothermal processes.
    B. two adiabatic processes and two isobaric processes.
    C. two adiabatic processes and two isochoric processes.
    D. two isothermal processes and two isochoric processes.
    E. two isothermal processes and two isobaric processes. SSM

30. •Compare two methods to improve the theoretical efficiency of a heat engine: lower T_C by 10 K or raise T_H by 10 K. Which one is better?
    A. Lower T_C by 10 K.
    B. Raise T_H by 10 K.
    C. Both changes would give the same result.
    D. The best method would depend on the difference between T_C and T_H.

This page is intentionally left blank.

For complete end of chapter problem sets, please go to
www.whfreeman.com/kestentauck

43. •Calculate the amount of work done on a gas that undergoes a change of state described by the *PV* diagram shown in **Figure 15-26.**

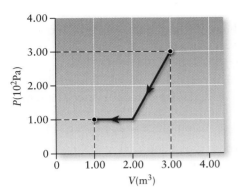

Figure 15-26 Problem 43

44. •A gas is heated and is allowed to expand such that it follows a horizontal line path on a *PV* diagram from its initial state (1.0×10^5 Pa, 1.0 m³) to its final state (1.0×10^5 Pa, 2.0 m³). Calculate the work done by the gas on its surroundings.

45. •A gas is heated such that it follows a vertical line path on a *PV* diagram from its initial state (1.0×10^5 Pa, 3.0 m³) to its final state (2.0×10^5 Pa, 3.0 m³). Calculate the work done by the gas on its surroundings. SSM

46. •If 800 J of heat is added to a system that does no external work, how much does the internal energy of the system increase?

47. •Five hundred joules of heat is absorbed by a system that does 200 J of work on its surroundings. What is the change in the internal energy of the system?

15-2: Thermodynamic Processes

48. •A sealed cylinder has a piston and contains 8000 cm³ of an ideal gas at a pressure of 8.0 atm. Heat is slowly introduced and the gas isothermally expands to 16,000 cm³. How much work does the gas do on the piston?

49. •An ideal gas expands isothermally, performing 8.8 kJ of work in the process. Calculate the heat absorbed during the expansion. SSM

50. •Heat is added to 8.0 m³ of helium gas in an expandable chamber that increases its volume by 2.0 m³. If in the isothermal expansion process 2.0 kJ of work is done by the gas, what was its original pressure?

51. •A cylinder that has a piston contains 2.00 mol of an idea gas and undergoes a reversible isothermal expansion at 400 K from an initial pressure of 12 atm down to 3 atm. Determine the amount of work done by the gas.

52. •A gas contained in a cylinder that has a piston is kept at a constant pressure of 2.8×10^5 Pa. The gas

expands from 0.5 m³ to 1.5 m³ when 300 kJ of heat is added to the cylinder. What is the change in internal energy of the gas?

53. •The pressure in an ideal gas is slowly reduced to 1/4 its initial value, while being kept in a container with rigid walls. In the process, 800 kJ of heat leaves the gas. What is the change in internal energy of the gas during this process? SSM

54. •An ideal gas is compressed adiabatically to half its volume. In doing so, 1888 J of work is done on the gas. What is the change in internal energy of the gas?

55. •Two moles of an ideal monatomic gas expand adiabatically, performing 8.0 kJ of work in the process. What is the change in temperature of the gas during the expansion?

56. ••Calc A sample of monatomic gas undergoes a thermodynamic process where the volume and pressure are both tripled, as shown in **Figure 15-27.** Find the work done by the gas in terms of V_0 and P_0.

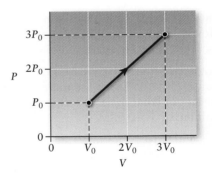

Figure 15-27 Problem 56

15-3: Second Law of Thermodynamics

57. •An engine doing work takes in 10 kJ and exhausts 6 kJ. What is the efficiency of the engine? SSM

58. •What is the theoretical maximum efficiency of an engine operating between 100°C and 500°C?

59. •A heat engine operating between 473 K and 373 K runs at about 70% of its theoretical maximum efficiency. What is its efficiency?

60. •An engine operates between 10°C and 200°C. At the very best, how much heat should we be prepared to supply in order to output 1000 J of work?

61. •A furnace supplies 28 kW of thermal power at 300°C to an engine and exhausts waste energy at 20°C. At the very best, how much work could we expect to get out of the system per second? SSM

62. •A kitchen refrigerator extracts 75 kJ of energy from a cool chamber while exhausting 100 kJ per second to the room. What is its coefficient of performance?

63. •What is the coefficient of performance of a Carnot refrigerator operating between 0°C and 80°C?

64. •An electric refrigerator removes 13.0 MJ of heat from its interior for each kilowatt-hour of electric energy used. What is its coefficient of performance?

65. •A certain refrigerator requires 35 J of work to remove 190 J of heat from its interior. (a) What is its coefficient of performance? (b) How much heat is ejected to the surroundings at 22°C? (c) If the refrigerator cycle is reversible, what is the temperature inside the refrigerator? SSM

66. •A refrigerator is rated at 370 W. Its interior is at 0°C and its surroundings are at 20°C. If the second law efficiency of its cycle is 66%, how much heat can it remove from its interior in 1 min?

15-4: Gases

67. •A container holds 32 g of oxygen at a pressure of 8.0 atm. How much heat is required to increase the temperature by 100°C at constant pressure?

68. •A container holds 32 g of oxygen at a pressure of 8.0 atm. How much heat is required to increase the temperature by 100°C at constant volume?

69. •The temperature of 4 g of helium is increased at constant volume by 1°C. Using the same amount of heat, the temperature of what mass of oxygen will increase at constant volume by 1°C? SSM

70. •Heat is added to 1 mol of air at constant pressure, resulting in a temperature increase of 100°C. If the same amount of heat is instead added at constant volume, what is the temperature increase? The molar specific heat ratio $\gamma = C_P/C_V$ for the air is 1.4.

71. ••The volume of a gas is halved during an adiabatic compression that increases the pressure by a factor of 2.6. What is the molar specific heat ratio $\gamma = C_P/C_V$?

72. ••The volume of a gas is halved during an adiabatic compression that increases the pressure by a factor of 2.5. By what factor does the temperature increase?

73. ••What ratio of initial volume to final volume, V_i/V_f, will raise the temperature of air from 27°C to 857°C in an adiabatic process? The molar specific heat ratio $\gamma = C_P/C_V$ for the air is 1.4. SSM

74. ••A monatomic ideal gas at a pressure of 1 atm expands adiabatically from an initial volume of 1.5 m³ to a final volume of 3.0 m³. What is the new pressure?

75. ••One mole of an ideal monatomic gas ($\gamma = 1.66$), initially at a temperature of 0.00°C, undergoes an adiabatic expansion from a pressure of 10 atm to a pressure of 2 atm. Find the work done on the gas.

15-5: Entropy

76. •A reservoir at a temperature of 400 K gains 100 J of heat from another reservoir. What is its entropy change?

77. •What is the minimum change of entropy that occurs in 0.200 kg of ice at 273 K when 6.68×10^4 J of heat is added so that it melts to water? SSM

78. •If, in a reversible process, enough heat is added to change a 500-g block of ice to water at a temperature of 273 K, what is the change in the entropy of the system? The heat of fusion of ice is 334 kJ/kg.

79. •A room is at a constant 295 K maintained by an air conditioner that pumps heat out. What is the entropy change for each 5.0 kJ of heat removed?

80. •One mole of ideal gas expands isothermally from 1.0 m³ to 2.0 m³. What is the entropy change for the gas?

81. •An 1800-kg car traveling at 80 km/h crashes into a concrete wall. If the temperature of the air is 27°C, find the entropy change of the universe. SSM

82. •A 1000-kg rock at 20°C falls 100 m into a large lake, also at 20°C. Assuming that all of the rock's kinetic energy on entering the lake converts to thermal energy absorbed by the lake, what is the change in entropy of the lake?

83. •**Astro** The surface of the Sun is about 5700 K, and the temperature of Earth's surface is about 293 K. What entropy change occurs when 8000 J of energy is transferred by heat from the Sun to Earth?

84. •When 1 kg of water is frozen under standard conditions, by how much does its entropy change?

85. •In a vacuum bottle, 350 g of water and 150 g of ice are initially in equilibrium at 0°C. The bottle is not a perfect insulator. Over time, its contents come to thermal equilibrium with the outside air at 25°C. How much does the entropy of universe increase in the process? SSM

86. ••You have a cup containing 220 g of freshly brewed coffee which, at 75°C, is too warm to drink. If you cool the coffee a little by pouring 60 g of tap water at 26°C into the cup, estimate how much the entropy of the universe increases.

General Problems

87. •A certain refrigerator has a power rating of 88.0 W. Consider it to be a reversible refrigerator. If the temperature of the room is 26.0°C, how long will it take to freeze 2.50 kg of water that is put into the refrigerator at 0°C?

88. •••A vertical metal cylinder contains an ideal gas. The top of the cylinder is closed off by a piston of mass m that is free to move up and down with no appreciable

friction. The piston is a height h above the bottom of the cylinder when the gas alone supports it. Sand is now very slowly poured onto the piston until the weight of the sand is equal to the weight of the piston. Find the new height of the piston (in terms of h) if the ideal gas in the cylinder is (a) oxygen O_2, (b) helium He, or (c) hydrogen H_2.

89. •A Carnot engine on a ship extracts heat from seawater at 18.0°C and exhausts the heat to evaporating dry ice at −78.0°C. If the ship's engines are to run at 8000 horsepower, what is the minimum amount of dry ice the ship must carry for the ship to run for a single day? SSM

90. •A Carnot engine removes 1200 J of heat from a high-temperature source and dumps 600 J to the atmosphere at 20°C. (a) What is the efficiency of the engine? (b) What is the temperature of the hot reservoir?

91. ••A certain engine has a second-law efficiency of 85.0%. During each cycle, it absorbs 480 J of heat from a reservoir at 300°C and dumps 300 J of heat to a cold-temperature reservoir. (a) What is the temperature of the cold reservoir? (b) How much more work could be done by a Carnot engine working between the same two reservoirs and extracting the same 480 J of heat in each cycle?

92. ••The interior of a refrigerator's freezing compartment is at a temperature of 10°F. The temperature of the kitchen is 78°F. Suppose that heat leaks through the walls into the freezing compartment at a rate of 70.0 cal/min. (a) In 1 h, how much has the entropy of the universe been increased by the heat leakage? (b) How much energy becomes unavailable for doing work when the heat leaks into the freezer compartment?

93. ••Medical During a high fever, a 60-kg-patient's normal metabolism is increased by 10%. This results in an increase of 10% in the heat given off by the person. When the person slowly walks up five flights of stairs (20 m), she normally releases 100 kJ of heat. Compare her efficiency when she has a fever to when her temperature is normal. SSM

94. ••A rigid 5.50-L pressure cooker contains steam initially at 100°C under a pressure of 1.00 atm. Consult Table 15-1 as needed and assume that the values given there remain constant. The mass of a water molecule is 2.99×10^{-26} kg. (a) To what temperature (in °C) would you have to heat the steam so that its pressure was 1.25 atm? (b) How much heat would you need in part (a)? (c) Calculate the specific heat of the steam in part (a) in units of $J/(kg \cdot K)$ and in $cal/(g \cdot K)$.

95. ••Calc A sample of 0.2 mol of an ideal gas at 320 K undergoes an isothermal expansion from 2 L to 8 L. (a) Draw the PV diagram (with appropriate units) for the process. (b) Calculate the work done for the 6-L change in volume. (c) Calculate the heat lost or gained by the gas. (d) Calculate the change in internal energy of the gas.

96. ••A certain electric generating plant produces electricity by using steam that enters its turbine at a temperature of 320°C and leaves it at 40°C. Over the course of a year, the plant consumes 4.4×10^{16} J of heat and produces an average electric power output of 600 MW. What is its second-law efficiency?

97. ••As we drill down into the rocks of Earth's crust, the temperature typically increases by 3.0°C for every 100 m of depth. Oil wells are commonly drilled to depths of 1830 m. If water is pumped into the shaft of the well, it will be heated by the hot rock at the bottom and the resulting heated steam can be used as a heat engine. Assume that the surface temperature is 20°C. (a) Using such a 1830-m well as a heat engine, what is the maximum efficiency possible? (b) If a combination of such wells is to produce a 2.5-MW power plant, how much energy will it absorb from the interior of Earth each day? SSM

98. •••The energy efficiency ratio (or rating)—the EER—for air conditioners, refrigerators and freezers is defined as the ratio of the input rate of heat (Q_C/t, in BTU/h) to the output rate of work (W/t, in W):
$$EER = \frac{Q_C/t \ (BTU/hr)}{W/t \ (W)}.$$ (a) Show that the EER can be expressed as $EER = \dfrac{Q_C \ (BTU)}{W \ (W \cdot h)}$ and is therefore nothing more than the coefficient of performance CP expressed in mixed units. (b) Show that the EER is related to the coefficient of performance CP by the equation $CP = EER/3.412$. (c) Typical home freezers have EER ratings of about 5.1 and operate between an interior freezer temperature of 0°F and an outside kitchen temperature of about 70°F. What is the coefficient of performance for such a freezer, and how does it compare to the coefficient of performance of the best possible freezer operating between those temperatures? (d) What is the EER of the best possible freezer in part (c)?

99. ••Your energy efficient home freezer has an EER of 6.50 (see problem 15-98). In preparation for a picnic, you put 1.50 L of water at 20°C into the freezer to make ice at 0°C for your ice chest. (See Tables 11-1, 14-3, and 14-4 as needed.) (a) How much electrical energy (which runs the freezer) is required to make the ice? Express your answer in J and kWh. (b) How much heat is ejected into your kitchen, which is at 22°C? (c) How much does making the ice change the entropy of your kitchen?

100. ••Biology The volume of air taken in during a typical breath is 0.5 L. The inhaled air is heated to 37°C (the internal body temperature) as it enters the lungs. Because air is about 80% nitrogen N_2, we can model it as

an ideal gas. Suppose that the outside air is at room temperature (20°C) and that you take two breaths every 3.0 s. Assume that the pressure does not change during the process. (a) How many joules of heat does it take to warm the air in a single breath? (b) How many food calories (kcal) are used up per day in heating the air you breathe? Is this a significant amount of typical daily caloric intake?

101. ••A heat engine works in a cycle between reservoirs at 273 K and 490 K. In each cycle, the engine absorbs 1250 cal of heat from the high-temperature reservoir and does 475 J of work. (a) What is its efficiency? (b) By how much is the entropy of the universe changed when the engine goes through one full cycle? (c) How much energy becomes unavailable for doing work when the engine goes through one full cycle? SSM

102. •••You have a cabin on the plains of central Saskatchewan. It is built on a 8.50-m by 12.5-m rectangular foundation with walls 3.00 m tall. The wooden walls and flat roof are made of white pine that is 9.00 cm thick. To conserve heat, the windows are negligibly small. The floor is well insulated so you lose negligible heat through it. The cabin is heated by an electrically powered heat pump operating on the Carnot cycle between the inside and outside air. When the outside temperature is a frigid −10°F, how much electrical energy does the heat pump consume per second to keep the interior temperature a steady and toasty 70°F? Assume that the surfaces of the walls and roof are at the same temperature as the air with which they are in contact and neglect radiation. (Consult Table 14-5 as needed.)

103. •**Sports** In an international diving competition, divers fall from a platform 10.0 m above the surface of the water in a very large pool. The diver leaves the platform with negligible initial speed. What is the maximum change in the entropy of the water in the pool at 25°C when a 75.0-kg diver executes his dive? Does the pool's entropy increase or decrease?

104. •Eighty kilograms of pure water at 313 K is mixed with 80 kg of pure water at 305 K. How much does the entropy of the universe change during the mixing?

105. ••**Biology** A 68-kg person typically eats about 2250 kcal per day, 20% of which goes to mechanical energy and the rest to heat. If she spends most of her time in her apartment at 22°C, how much does the entropy of

her apartment change in one day? Does entropy increase or decrease? SSM

106. ••A heat engine works in a cycle between reservoirs at 273 K and 490 K. In each cycle the engine absorbs 1250 J of heat from the high temperature reservoir and does 475 J of work. (a) What is its efficiency? (b) What is the change in entropy of the universe when the engine goes through one complete cycle? (c) How much energy becomes unavailable for doing work when the engine goes through one complete cycle?

107. •••Consider an engine in which the working substance is 1.23 mol of an ideal gas for which $\gamma = 1.41$. The engine runs reversibly in the cycle shown on the PV diagram (**Figure 15-28**). The cycle consists of an isobaric (constant pressure) expansion a at a pressure of 15.0 atm, during which the temperature of the gas increases from 300 K to 600 K, followed by an isothermal expansion b until its pressure becomes 3.00 atm. Next is an isobaric compression c at a pressure of 3.00 atm, during which the temperature decreases from 600 K to 300 K, followed by an isothermal compression d until its pressure returns to 15 atm. Find the work done by the gas, the heat absorbed by the gas, the internal energy change, and the entropy change of the gas, first for each part of the cycle and then for the complete cycle.

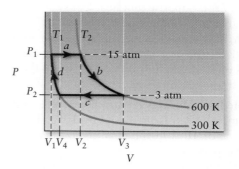

Figure 15-28 Problem 107

108. •••**Calc** Starting with the differential form of the first law of thermodynamics, $dU = dQ + P\, dV$, derive the condition for the adiabatic expansion of an ideal gas, $PV^\gamma = $ constant. Note that $\gamma = C_P/C_V$.

APPENDIX A
SI Units and Conversion Factors

Base Units*

Length	The *meter* (m) is the distance traveled by light in a vacuum in 1/299,792,458 s.
Time	The *second* (s) is the duration of 9,192,631,770 periods of the radiation corresponding to the transition between the two hyperfine levels of the ground state of the ^{133}Cs atom.
Mass	The *kilogram* (kg) is the mass of the international standard body preserved at Sèvres, France.
Mole	The *mole* (mol) is the amount of substance of a system which contains as many elementary entities as there are atoms in 0.012 kg of carbon-12.
Current	The *ampere* (A) is that constant current which, if maintained in two straight parallel conductors of infinite length, of negligible circular cross section, and placed 1 m apart in vacuum, would produce between the conductors a force equal to 2×10^{-7} N/m of length.
Temperature	The *kelvin* (K) is 1/273.16 of the thermodynamic temperature of the triple point of water.
Luminous intensity	The *candela* (cd) is the luminous intensity in a given direction, of a source that emits monochromatic radiation of frequency 540×10^{12} Hz and that has a radiant intensity, in that direction of 1/683 W/steradian.

*These definitions are found on the Internet at http://physics.nist.gov/cuu/Units/current.html.

Derived Units

Force	newton (N)	$1\ N = 1\ kg \cdot m/s^2$
Work, energy	joule (J)	$1\ J = 1\ N \cdot m$
Power	watt (W)	$1\ W = 1\ J/s$
Frequency	hertz (Hz)	$1\ Hz = cy/s$
Charge	coulomb (C)	$1\ C = 1\ A \cdot s$
Potential	volt (V)	$1\ V = 1\ J/C$
Resistance	ohm (Ω)	$1\ \Omega = 1\ V/A$
Capacitance	farad (F)	$1\ F = 1\ C/V$
Magnetic field	tesla (T)	$1\ T = 1\ N/(A \cdot m)$
Magnetic flux	weber (Wb)	$1\ Wb = 1\ T \cdot m^2$
Inductance	henry (H)	$1\ H = 1\ J/A^2$

Conversion Factors

Conversion factors are written as equations for simplicity;
relations marked with an asterisk are exact.

Length

1 km = 0.6214 mi

1 mi = 1.609 km

1 m = 1.0936 yard = 3.281 ft = 39.37 in.

*1 in. = 2.54 cm

*1 ft = 12 in. = 30.48 cm

*1 yard = 3 ft = 91.44 cm

1 light-year = 1 $c \cdot y$ = 9.461 × 10^{15} m

*1 Å = 0.1 nm

Area

*1 m^2 = 10^4 cm^2

1 km^2 = 0.3861 mi^2 = 247.1 acres

*1 in.^{2} = 6.4516 cm^2

1 ft^2 = 9.29 × 10^{-2} m^2

1 m^2 = 10.76 ft^2

*1 acre = 43 560 ft^2

1 mi^2 = 640 acres = 2.590 km^2

Volume

*1 m^3 = 10^6 cm^3

*1 L = 1000 cm^3 = 10^{-3} m^3

1 gal = 3.785 L

1 gal = 4 qt = 8 pt = 128 oz = 231 in.^{3}

1 in.^{3} = 16.39 cm^3

1 ft^3 = 1728 in.^{3} = 28.32 L
      = 2.832 × 10^4 cm^3

Time

*1 h = 60 min = 3.6 ks

*1 d = 24 h = 1440 min = 86.4 ks

1 y = 365.25 day = 3.156 × 10^7 s

Speed

*1 m/s = 3.6 km/h

1 km/h = 0.2778 m/s = 0.6214 mi/h

1 mi/h = 0.4470 m/s = 1.609 km/h

1 mi/h = 1.467 ft/s

Angle and Angular Speed

*π rad = 180°

1 rad = 57.30°

1° = 1.745 × 10^{-2} rad

1 rev/min = 0.1047 rad/s

1 rad/s = 9.549 rev/min

Mass

*1 kg = 1000 g

*1 tonne = 1000 kg = 1 Mg

1 u = 1.6605 × 10^{-27} kg
       931.49 MeV/c^2

1 kg = 6.022 × 10^{26} u

1 slug = 14.59 kg

1 kg = 6.852 × 10^{-2} slug

Density

*1 g/cm^3 = 1000 kg/m^3 = 1 kg/L

(1 g/cm^3)g = 62.4 lb/ft^3

Force

1 N = 0.2248 lb = 10^5 dyn

*1 lb = 4.448222 N

(1 kg)g = 2.2046 lb

Pressure

*1 Pa = 1 N/m^2

*1 atm = 101.325 kPa = 1.01325 bar

1 atm = 14.7 lb/in.^{2} = 760 mmHg
       = 29.9 in.Hg = 33.9 ftH$_2$O

1 lb/in.^{2} = 6.895 kPa

1 torr = 1 mmHg = 133.32 Pa

1 bar = 100 kPa

Energy

*1 kW · h = 3.6 MJ

*1 cal = 4.186 J

1 ft · lb = 1.356 J = 1.286 × 10^{-3} BTU

*1 L · atm = 101.325 J

1 L · atm = 24.217 cal

1 BTU = 778 ft · lb = 252 cal = 1054.35 J

1 eV = 1.602 × 10^{-19} J

1 u · c^2 = 931.49 MeV

*1 erg = 10^{-7} J

Power

1 horsepower = 550 ft · lb/s = 745.7 W

1 BTU/h = 2.931 × 10^{-4} kW

1 W = 1.341 × 10^{-3} horsepower
     = 0.7376 ft · lb/s

Magnetic Field

*1 T = 10^4 G

Thermal Conductivity

1 W/(m · K) = 6.938 BTU · in./(h · ft^2 · °F)

1 BTU · in./(h · ft^2 · °F) = 0.1441 W/(m · K)

APPENDIX B
Numerical Data

Terrestrial Data

Free-fall acceleration g	
Standard value (at sea level at 45° latitude)*	$9.806\ 65$ m/s^2; 32.1740 ft/s^2
At equator*	9.7804 m/s^2
At poles*	9.8322 m/s^2
Mass of Earth M_E	5.98×10^{24} kg
Radius of Earth R_E, mean	6.38×10^6 m; 3960 mi
Escape speed	1.12×10^4 m/s; 6.95 mi/s
Solar constant†	1.37 kW/m^2
Standard temperature and pressure (STP):	
Temperature	293.15 K
Pressure	101.3 kPa (1.00 atm)
Molar mass of air	28.97 g/mol
Density of air (273.15 K, 101.3 kPa), ρ_{air}	1.217 kg/m^3
Speed of sound (273.15 K, 101.3 kPa)	331 m/s
Latent heat of fusion of H_2O (0°C, 1 atm)	334 kJ/kg
Latent heat of vaporization of H_2O (100°C, 1 atm)	2.26 MJ/kg

* Measured relative to Earth's surface.
† Average power incident normally on 1 m^2 outside Earth's atmosphere at the mean distance from Earth to the Sun.

Astronomical Data*

Earth	
Distance to the Moon, mean†	3.844×10^8 m; 2.389×10^5 mi
Distance to the Sun, mean†	1.496×10^{11} m; 9.30×10^7 mi; 1.00 AU
Orbital speed, mean	2.98×10^4 m/s
Moon	
Mass	7.35×10^{22} kg
Radius	1.737×10^6 m
Period	27.32 day
Acceleration of gravity at surface	1.62 m/s^2
Sun	
Mass	1.99×10^{30} kg
Radius	6.96×10^8 m

* Additional solar system data are available from NASA at http://nssdc.gsfc.nasa.gov/planetary/planetfact.html.
† Center to center.

Physical Constants*

Universal constant of gravitation	G	$6.673\ 84(80) \times 10^{-11}\ \text{N} \cdot \text{m}^2/\text{kg}^2$
Speed of light	c	$2.997\ 924\ 58 \times 10^8\ \text{m/s}$
Fundamental charge	e	$1.602\ 176\ 565(35) \times 10^{-19}\ \text{C}$
Avogadro's constant	N_A	$6.022\ 141\ 29(27) \times 10^{23}\ \text{particles/mol}$
Gas constant	R	$8.314\ 462\ 1(75)\ \text{J/(mol} \cdot \text{K)}$
		$1.987\ 206\ 5(36)\ \text{cal/(mol} \cdot \text{K)}$
		$8.205\ 746(15) \times 10^{-2}\ \text{L} \cdot \text{atm/(mol} \cdot \text{K)}$
Boltzmann constant	$k = R/N_A$	$1.380\ 648\ 8(13) \times 10^{-23}\ \text{J/K}$
		$8.617\ 332\ 4(78) \times 10^{-5}\ \text{eV/K}$
Stefan-Boltzmann constant	$\sigma = (\pi^2/60)\text{K}^4/(\hbar^3 c^2)$	$5.670\ 373(21) \times 10^{-8}\ \text{W/(m}^2 \cdot \text{K}^4)$
Atomic mass constant	$m_u = 1/12m(^{12}\text{C})$	$1.660\ 538\ 921(73) \times 10^{-27}\ \text{kg} = 1\ \text{u}$
Permeability of free space	μ_0	$4\pi \times 10^{-7}\ \text{N/A}^2$
		$1.256\ 637 \ldots \times 10^{-6}\ \text{N/A}^2$
Permittivity of free space	$\varepsilon_0 = 1/(\mu_0 \text{C}^2)$	$8.854\ 187\ 817 \ldots \times 10^{-12}\ \text{C}^2/(\text{N} \cdot \text{m}^2)$
Coulomb constant	$k = 1/(4\pi\varepsilon_0)$	$8.987\ 551\ 787 \ldots \times 10^9\ \text{N} \cdot \text{m}^2/\text{C}^2$
Planck's constant	h	$6.626\ 069\ 57(29) \times 10^{-34}\ \text{J} \cdot \text{s}$
		$4.135\ 667\ 516(91) \times 10^{-15}\ \text{eV} \cdot \text{s}$
	$\hbar = h/2\pi$	$1.054\ 571\ 726(47) \times 10^{-34}\ \text{J} \cdot \text{s}$
		$6.582\ 119\ 28(15) \times 10^{-16}\ \text{eV} \cdot \text{s}$
Mass of electron	m_e	$9.109\ 382\ 91(40) \times 10^{-31}\ \text{kg}$
		$0.510\ 998\ 928(11)\ \text{MeV}/c^2$
Mass of proton	m_p	$1.672\ 621\ 777(74) \times 10^{-27}\ \text{kg}$
		$938.272\ 046(21)\ \text{MeV}/c^2$
Mass of neutron	m_n	$1.674\ 927\ 351(74) \times 10^{-27}\ \text{kg}$
		$939.565\ 379(21)\ \text{MeV}/c^2$
Bohr magneton	$m_B = eh/2m_e$	$9.274\ 009\ 68(20) \times 10^{-24}\ \text{J/T}$
		$5.788\ 381\ 806\ 6(38) \times 10^{-5}\ \text{eV/T}$
Nuclear magneton	$m_n = eh/2m_p$	$5.050\ 783\ 53(11) \times 10^{-27}\ \text{J/T}$
		$3.152\ 451\ 260\ 5(22) \times 10^{-8}\ \text{eV/T}$
Magnetic flux quantum	$\phi_0 = h/2e$	$2.067\ 833\ 758(46) \times 10^{-15}\ \text{T} \cdot \text{m}^2$
Quantized Hall resistance	$R_K = h/e^2$	$2.581\ 280\ 744\ 34(84) \times 10^4\ \Omega$
Rydberg constant	R_H	$1.097\ 373\ 156\ 853\ 9(55) \times 10^7\ \text{m}^{-1}$
Josephson frequency–voltage quotient	$K_J = 2e/h$	$4.835\ 978\ 70(11) \times 10^{14}\ \text{Hz/V}$
Compton wavelength	$\lambda_C = h/m_e c$	$2.426\ 310\ 238\ 9(16) \times 10^{-12}\ \text{m}$

* The values for these and other constants may be found on the Internet at http://physics.nist.gov/cuu/Constants/index.html. The numbers in parentheses represent the uncertainties in the last two digits. (For example, 2.044 43(13) stands for 2.044 43 ± 0.000 13.) Values without uncertainties are exact, including those values with ellipses (such as the value of pi is exactly 3.1415…).

APPENDIX C
Periodic Table of Elements*

1																	18
1 H	**2**											13	14	15	16	17	2 He
3 Li	4 Be											5 B	6 C	7 N	8 O	9 F	10 Ne
11 Na	12 Mg	**3**	**4**	**5**	**6**	**7**	**8**	**9**	**10**	**11**	**12**	13 Al	14 Si	15 P	16 S	17 Cl	18 Ar
19 K	20 Ca	21 Sc	22 Ti	23 V	24 Cr	25 Mn	26 Fe	27 Co	28 Ni	29 Cu	30 Zn	31 Ga	32 Ge	33 As	34 Se	35 Br	36 Kr
37 Rb	38 Sr	39 Y	40 Zr	41 Nb	42 Mo	43 Tc	44 Ru	45 Rh	46 Pd	47 Ag	48 Cd	49 In	50 Sn	51 Sb	52 Te	53 I	54 Xe
55 Cs	56 Ba	57–71 Lanthanoids	72 Hf	73 Ta	74 W	75 Re	76 Os	77 Ir	78 Pt	79 Au	80 Hg	81 Tl	82 Pb	83 Bi	84 Po	85 At	86 Rn
87 Fr	88 Ra	89–103 Actinoids	104 Rf	105 Db	106 Sg	107 Bh	108 Hs	109 Mt	110 Ds	111 Rg	112 Cn						

Lanthanoids	57 La	58 Ce	59 Pr	60 Nd	61 Pm	62 Sm	63 Eu	64 Gd	65 Tb	66 Dy	67 Ho	68 Er	69 Tm	70 Yb	71 Lu
Actinoids	89 Ac	90 Th	91 Pa	92 U	93 Np	94 Pu	95 Am	96 Cm	97 Bk	98 Cf	99 Es	100 Fm	101 Md	102 No	103 Lr

* From http://old.iupac.org/reports/periodic_table/IUPAC_Periodic_Table-21Jan11.pdf.

Atomic Numbers and Atomic Weights*

Atomic Number	Name	Symbol	Weight	Atomic Number	Name	Symbol	Weight
1	Hydrogen	H	[1.007 84; 1.008 11]	57	Lanthanum	La	138.90547(7)
2	Helium	He	4.002602(2)	58	Cerium	Ce	140.116(1)
3	Lithium	Li	[6.938; 6.997]	59	Praseodymium	Pr	140.90765(2)
4	Beryllium	Be	9.012182(3)	60	Neodymium	Nd	144.242(3)
5	Boron	B	[10.806; 10.821]	61	Promethium	Pm	
6	Carbon	C	[12.009 6; 12.011 6]	62	Samarium	Sm	150.36(2)
7	Nitrogen	N	[14.006 43; 14.007 28]	63	Europium	Eu	151.964(1)
8	Oxygen	O	[15.999 03; 15.999 77]	64	Gadolinium	Gd	157.25(3)
9	Fluorine	F	18.9984032(5)	65	Terbium	Tb	158.92535(2)
10	Neon	Ne	20.1797(6)	66	Dysprosium	Dy	162.500(1)
11	Sodium	Na	22.98976928(2)	67	Holmium	Ho	164.93032(2)
12	Magnesium	Mg	24.3050(6)	68	Erbium	Er	167.259(3)
13	Aluminum	Al	26.9815386(8)	69	Thulium	Tm	168.93421(2)
14	Silicon	Si	[28.084; 28.086]	70	Ytterbium	Yb	173.054(5)
15	Phosphorus	P	30.973762(2)	71	Lutetium	Lu	174.966 8(1)
16	Sulfur	S	[32.059; 32.076]	72	Hafnium	Hf	178.49(2)
17	Chlorine	Cl	[35.446; 35.457]	73	Tantalum	Ta	180.94788(2)
18	Argon	Ar	39.948(1)	74	Tungsten	W	183.84(1)
19	Potassium	K	39.0983(1)	75	Rhenium	Re	186.207(1)
20	Calcium	Ca	40.078(4)	76	Osmium	Os	190.23(3)
21	Scandium	Sc	44.955912(6)	77	Iridium	Ir	192.217(3)
22	Titanium	Ti	47.867(1)	78	Platinum	Pt	195.084(9)
23	Vanadium	V	50.9415(1)	79	Gold	Au	196.966569(4)
24	Chromium	Cr	51.9961(6)	80	Mercury	Hg	200.59(2)
25	Manganese	Mn	54.938045(5)	81	Thallium	Tl	[204.382; 204.385]
26	Iron	Fe	55.845(2)	82	Lead	Pb	207.2(1)
27	Cobalt	Co	58.933195(5)	83	Bismuth	Bi	208.98040(1)
28	Nickel	Ni	58.6934(2)	84	Polonium	Po	
29	Copper	Cu	63.546(3)	85	Astatine	At	
30	Zinc	Zn	65.38 (2)	86	Radon	Rn	
31	Gallium	Ga	69.723(1)	87	Francium	Fr	
32	Germanium	Ge	72.63 (1)	88	Radium	Ra	
33	Arsenic	As	74.92160(2)	89	Actinium	Ac	
34	Selenium	Se	78.96(3)	90	Thorium	Th	232.03806(2)
35	Bromine	Br	79.904(1)	91	Protactinium	Pa	231.03588(2)
36	Krypton	Kr	83.798(2)	92	Uranium	U	238.02891(3)
37	Rubidium	Rb	85.4678(3)	93	Neptunium	Np	
38	Strontium	Sr	87.62(1)	94	Plutonium	Pu	
39	Yttrium	Y	88.90585(2)	95	Americium	Am	
40	Zirconium	Zr	91.224(2)	96	Curium	Cm	
41	Niobium	Nb	92.90638(2)	97	Berkelium	Bk	
42	Molybdenum	Mo	95.96 (2)	98	Californium	Cf	
43	Technetium	Tc		99	Einsteinium	Es	
44	Ruthenium	Ru	101.07(2)	100	Fermiun	Fm	
45	Rhodium	Rh	102.90550(2)	101	Mendelevium	Md	
46	Palladium	Pd	106.42(1)	102	Nobelium	No	
47	Silver	Ag	107.8682(2)	103	Lawrencium	Lr	
48	Cadmium	Cd	112.411(8)	104	Rutherfordium	Rf	
49	Indium	In	114.818(3)	105	Dubnium	Db	
50	Tin	Sn	118.710(7)	106	Seaborgium	Sg	
51	Antimony	Sb	121.760(1)	107	Bohrium	Bh	
52	Tellurium	Te	127.60(3)	108	Hassium	Hs	
53	Iodine	I	126.90447(3)	109	Meitnerium	Mt	
54	Xenon	Xe	131.293(6)	110	Darmstadtium	Ds	
55	Cesium	Cs	132.9054519(2)	111	Roentgenium	Rg	
56	Barium	Ba	137.327(7)	112	Copernicium	Cn	

*Some weights are listed as intervals ([a; b]; a ≤ atomic weight ≤ b) because these weights are not constant but depend on the physical, chemical, and nuclear histories of the samples used. Atomic weights are not listed for some elements because these elements do not have stable isotopes. Exceptions are thorium, protactinium, and uranium. Elements 113 to 118 are not listed in this table although they have been reported (IUPAC has not named them). From *Atomic Weights of the Elements 2009 (IUPAC Technical Report)*, Pure Appl. Chem., Vol. 83(2), pp. 359–396, 2011.

Table of Atomic Masses

Element	Symbol	Mass number (*indicates radioactive)	Atomic mass	Percent abundance	Half-life and decay mode (if unstable)	
(Neutron)	*n*	1*	1.008665		10.4 m	β⁻
Hydrogen	H	1	1.007825	99.985		
Deuterium	D	2	2.014102	0.015		
Tritium	T	3*	3.016049		12.33 y	β⁻
Helium	He	3	3.016029	0.00014		
		4	4.002602	99.99986		
		6*	6.018886		0.81 s	β⁻
		8*	8.033922		0.12 s	β⁻
Lithium	Li	6	6.015121	7.5		
		7	7.016003	92.5		
		8*	8.022486		0.84 s	β⁻
		9*	9.026789		0.18 s	β⁻
		11*	11.043897		8.7 ms	β⁻
Beryllium	Be	7*	7.016928		53.3 d	ec
		9	9.012174	100		
		10*	10.013534		1.5×10^6 y	β⁻
		11*	11.021657		13.8 s	β⁻
		12*	12.026921		23.6 ms	β⁻
		14*	14.042866		4.3 ms	β⁻
Boron	B	8*	8.024605		0.77 s	β⁺
		10	10.012936	19.9		
		11	11.009305	80.1		
		12*	12.014352		0.0202 s	β⁻
		13*	13.017780		17.4 ms	β⁻
		14*	14.025404		13.8 ms	β⁻
		15*	15.031100		10.3 ms	β⁻
Carbon	C	9*	9.031030		0.13 s	β⁺
		10*	10.016854		19.3 s	β⁺
		11*	11.011433		20.4 m	β⁺
		12	12.000000	98.90		
		13	13.003355	1.10		
		14*	14.003242		5730 y	β⁻
		15*	15.010599		2.45 s	β⁻
		16*	16.014701		0.75 s	β⁻
		17*	17.022582		0.20 s	β⁻

(Continued)

Element	Symbol	Mass number (*indicates radioactive)	Atomic mass	Percent abundance	Half-life and decay mode (if unstable)	
Nitrogen	N	12*	12.018613		0.0110 s	β⁺
		13*	13.005738		9.96 m	β⁺
		14	14.003074	99.63		
		15	15.000108	0.37		
		16*	16.006100		7.13 s	β⁻
		17*	17.008450		4.17 s	β⁻
		18*	18.014082		0.62 s	β⁻
		19*	19.017038		0.24 s	β⁻
Oxygen	O	13*	13.024813		8.6 ms	β⁺
		14*	14.008595		70.6 s	β⁺
		15*	15.003065		122 s	β⁺
		16	15.994915	99.71		
		17	16.999132	0.039		
		18	17.999160	0.20		
		19*	19.003577		26.9 s	β⁻
		20*	20.004076		13.6 s	β⁻
		21*	21.008595		3.4 s	β⁻
Fluorine	F	17*	17.002094		64.5 s	β⁺
		18*	18.000937		109.8 m	β⁺
		19	18.998404	100		
		20*	19.999982		11.0 s	β⁻
		21*	20.999950		4.2 s	β⁻
		22*	22.003036		4.2 s	β⁻
		23*	23.003564		2.2 s	β⁻
Neon	Ne	18*	18.005710		1.67 s	β⁺
		19*	19.001880		17.2 s	β⁺
		20	19.992435	90.48		
		21	20.993841	0.27		
		22	21.991383	9.25		
		23*	22.994465		37.2 s	β⁻
		24*	23.993999		3.38 m	β⁻
		25*	24.997789		0.60 s	β⁻
Sodium	Na	21*	20.997650		22.5 s	β⁺
		22*	21.994434		2.61 y	β⁺
		23	22.989767	100		
		24*	23.990961		14.96 h	β⁻
		25*	24.989951		59.1 s	β⁻
		26*	25.992588		1.07 s	β⁻
Magnesium	Mg	23*	22.994124		11.3 s	β⁺
		24	23.985042	78.99		
		25	24.985838	10.00		
		26	25.982594	11.01		
		27*	26.984341		9.46 m	β⁻
		28*	27.983876		20.9 h	β⁻
		29*	28.375346		1.30 s	β⁻

Element	Symbol	Mass number (*indicates radioactive)	Atomic mass	Percent abundance	Half-life and decay mode (if unstable)	
Aluminum	Al	25*	24.990429		7.18 s	β^+
		26*	25.986892		7.4×10^5 y	β^+
		27	26.981538	100		
		28*	27.981910		2.24 m	β^-
		29*	28.980445		6.56 m	β^-
		30*	29.982965		3.60 s	β^-
Silicon	Si	27*	26.986704		4.16 s	β^+
		28	27.976927	92.23		
		29	28.976495	4.67		
		30	28.973770	3.10		
		31*	30.975362		2.62 h	β^-
		32*	31.974148		172 y	β^-
		33*	32.977928		6.13 s	β^-
Phosphorus	P	30*	29.978307		2.50 m	β^+
		31	30.973762	100		
		32*	31.973762		14.26 d	β^-
		33*	32.971725		25.3 d	β^-
		34*	33.973636		12.43 s	β^-
Sulfur	S	31*	30.979554		2.57 s	β^+
		32	31.972071	95.02		
		33	32.971459	0.75		
		34	33.967867	4.21		
		35*	34.969033		87.5 d	β^-
		36	35.967081	0.02		
Chlorine	Cl	34*	33.973763		32.2 m	β^+
		35	34.968853	75.77		
		36*	35.968307		3.0×10^5 y	β^-
		37	36.965903	24.23		
		38*	37.968010		37.3 m	β^-
Argon	Ar	36	35.967547	0.337		
		37*	36.966776		35.04 d	ec
		38	37.962732	0.063		
		39*	38.964314		269 y	β^-
		40	39.962384	99.600		
		42*	41.963049		33 y	β^-
Potassium	K	39	38.963708	93.2581		
		40*	39.964000	0.0117	1.28×10^9 y	β^+, ec, β^-
		41	40.961827	6.7302		
		42*	41.962404		12.4 h	β^-
		43*	42.960716		22.3 h	β^-

(Continued)

Element	Symbol	Mass number (*indicates radioactive)	Atomic mass	Percent abundance	Half-life and decay mode (if unstable)	
Calcium	Ca	40	39.962591	96.941		
		41*	40.962279		1.0×10^5 y	ec
		42	41.958618	0.647		
		43	42.958767	0.135		
		44	43.955481	2.086		
		46	45.953687	0.004		
		48	47.952534	0.187		
Scandium	Sc	41*	40.969250		0.596 s	β^+
		43*	42.961151		3.89 h	β^+
		45	44.955911	100		
		46*	45.955170		83.8 d	β^-
Titanium	Ti	44*	43.959691		49 y	ec
		46	45.952630	8.0		
		47	46.951765	7.3		
		48	47.947947	73.8		
		49	48.947871	5.5		
		50	49.944792	5.4		
Vanadium	V	48*	47.952255			
		50*	49.947161	0.25	15.97 d	β^+
		51	50.943962	99.75	1.5×10^{17} y	β^+
Chromium	Cr	48*	47.954033		21.6 h	ec
		50	49.946047	4.345		
		52	51.940511	83.79		
		53	52.940652	9.50		
		54	53.938883	2.365		
Manganese	Mn	53*	52.941292		3.74×10^6 y	ec
		54*	53.940361		312.1 d	ec
		55	54.938048	100		
		56*	55.938908		2.58 h	β^-
Iron	Fe	54	53.939613	5.9		
		55*	54.938297		2.7 y	ec
		56	55.934940	91.72		
		57	56.935396	2.1		
		58	57.933278	0.28		
		60*	59.934078		1.5×10^6 y	β^-
Cobalt	Co	57*	56.936294		271.8 d	ec
		58*	57.935755		70.9 d	ec, β^+
		59	58.933198	100		
		60*	59.933820		5.27 y	β^-
		61*	60.932478		1.65 h	β^-
Nickel	Ni	58	57.935346	68.077		
		59*	58.934350		7.5×10^4 y	ec, β^+
		60	59.930789	26.223		
		61	60.931058	1.140		
		62	61.928346	3.634		
		63*	62.929670		100 y	β^-
		64	63.927967	0.926		

Element	Symbol	Mass number (*indicates radioactive)	Atomic mass	Percent abundance	Half-life and decay mode (if unstable)
Copper	Cu	63	62.929599	69.17	
		64*	63.929765		12.7 h ec
		65	64.927791	30.83	
		66*	65.928871		5.1 m β^-
Zinc	Zn	64	63.929144	48.6	
		66	65.926035	27.9	
		67	66.927129	4.1	
		68	67.924845	18.8	
		70	69.925323	0.6	
Gallium	Ga	69	68.925580	60.108	
		70*	69.926027		21.1 m β^-
		71	70.924703	39.892	
		72*	71.926367		14.1 h β^-
Germanium	Ge	69*	68.927969		39.1 h ec, β^+
		70	69.924250	21.23	
		72	71.922079	27.66	
		73	72.923462	7.73	
		74	73.921177	35.94	
		76	75.921402	7.44	
		77*	76.923547		11.3 h β^-
Arsenic	As	73*	72.923827		80.3 d ec
		74*	73.923928		17.8 d ec, β^+
		75	74.921594	100	
		76*	75.922393		1.1 d β^-
		77*	76.920645		38.8 h β^-
Selenium	Se	74	73.922474	0.89	
		76	75.919212	9.36	
		77	76.919913	7.63	
		78	77.917307	23.78	
		79*	78.918497		$\leq 6.5 \times 10^4$ y β^-
		80	79.916519	49.61	
		82*	81.916697	8.73	1.4×10^{20} y $2\beta^-$
Bromine	Br	79	78.918336	50.69	
		80*	79.918528		17.7 m β^+
		81	80.916287	49.31	
		82*	81.916802		35.3 h β^-
Krypton	Kr	78	77.920400	0.35	
		80	79.916377	2.25	
		81*	80.916589		2.11×10^5 y ec
		82	81.913481	11.6	
		83	82.914136	11.5	
		84	83.911508	57.0	
		85*	84.912531		10.76 y β^-
		86	85.910615	17.3	

(Continued)

Element	Symbol	Mass number (*indicates radioactive)	Atomic mass	Percent abundance	Half-life and decay mode (if unstable)
Rubidium	Rb	85	84.911793	72.17	
		86*	85.911171		18.6 d β^-
		87*	86.909186	27.83	4.75×10^{10} y β^-
		88*	87.911325		17.8 m β^-
Strontium	Sr	84	83.913428	0.56	
		86	85.909266	9.86	
		87	86.908883	7.00	
		88	87.905618	82.58	
		90*	89.907737		29.1 y β^-
Yttrium	Y	88*	87.909507		106.6 d ec, β^+
		89	88.905847	100	
		90*	89.914811		2.67 d β^-
Zirconium	Zr	90	89.904702	51.45	
		91	90.905643	11.22	
		92	91.905038	17.15	
		93*	92.906473		1.5×10^6 y β^-
		94	93.906314	17.38	
		96	95.908274	2.80	
Niobium	Nb	91*	90.906988		6.8×10^2 y ec
		92*	91.907191		3.5×10^7 y ec
		93	92.906376	100	
		94*	93.907280		2×10^4 y β^-
Molybdenum	Mo	92	91.906807	14.84	
		93*	92.906811		3.5×10^3 y ec
		94	93.905085	9.25	
		95	94.905841	15.92	
		96	95.904678	16.68	
		97	96.906020	9.55	
		98	97.905407	24.13	
		100	99.907476	9.63	
Technetium	Tc	97*	96.906363		2.6×10^6 y ec
		98*	97.907215		4.2×10^6 y β^-
		99*	98.906254		2.1×10^5 y β^-
Ruthenium	Ru	96	95.907597	5.54	
		98	97.905287	1.86	
		99	98.905939	12.7	
		100	99.904219	12.6	
		101	100.905558	17.1	
		102	101.904348	31.6	
		104	103.905428	18.6	
Rhodium	Rh	102*	101.906794		207 d ec
		103	102.905502	100	
		104*	103.906654		42 s β^-

Element	Symbol	Mass number (*indicates radioactive)	Atomic mass	Percent abundance	Half-life and decay mode (if unstable)
Palladium	Pd	102	101.905616	1.02	
		104	103.904033	11.14	
		105	104.905082	22.33	
		106	105.903481	27.33	
		107*	106.905126		6.5×10^6 y β^-
		108	107.903893	26.46	
		110	109.905158	11.72	
Silver	Ag	107	106.905091	51.84	
		108*	107.905953		2.39 m ec, β^+, β^-
		109	108.904754	48.16	
		110*	109.906110		24.6 s β^-
Cadmium	Cd	106	105.906457	1.25	
		108	107.904183	0.89	
		109*	108.904984		462 d ec
		110	109.903004	12.49	
		111	110.904182	12.80	
		112	111.902760	24.13	
		113*	112.904401	12.22	9.3×10^{15} y β^-
		114	113.903359	28.73	
		116	115.904755	7.49	
Indium	In	113	112.904060	4.3	
		114*	113.904916		1.2 m β^-
		115*	114.903876	95.7	4.4×10^{14} y β^-
		116*	115.905258		54.4 m β^-
Tin	Sn	112	111.904822	0.97	
		114	113.902780	0.65	
		115	114.903345	0.36	
		116	115.901743	14.53	
		117	116.902953	7.68	
		118	117.901605	24.22	
		119	118.903308	8.58	
		120	119.902197	32.59	
		121*	120.904237		55 y β^-
		122	121.903439	4.63	
		124	123.905274	5.79	
Antimony	Sb	121	120.903820	57.36	
		123	122.904215	42.64	
		125*	124.905251		2.7 y β^-
Tellurium	Te	120	119.904040	0.095	
		122	121.903052	2.59	
		123*	122.904271	0.905	1.3×10^{13} y ec
		124	123.902817	4.79	
		125	124.904429	7.12	
		126	125.903309	18.93	
		128*	127.904463	31.70	$> 8 \times 10^{24}$ y $2\beta^-$
		130*	129.906228	33.87	1.2×10^{21} y $2\beta^-$

(*Continued*)

Element	Symbol	Mass number (*indicates radioactive)	Atomic mass	Percent abundance	Half-life and decay mode (if unstable)
Iodine	I	126*	125.905619		13 d ec, β^+, β^-
		127	126.904474	100	
		128*	127.905812		25 m β^-, ec, β^+ β^-
		129*	128.904984		1.6×10^7 y
Xenon	Xe	124	123.905894	0.10	
		126	125.904268	0.09	
		128	127.903531	1.91	
		129	128.904779	26.4	
		130	129.903509	4.1	
		131	130.905069	21.2	
		132	131.904141	26.9	
		134	133.905394	10.4	
		136	135.907215	8.9	
Cesium	Cs	133	132.905436	100	
		134*	133.906703		2.1 y β^-
		135*	134.905891		2×10^6 y β^-
		137*	136.907078		30 y β^-
Barium	Ba	130	129.906289	0.106	
		132	131.905048	0.101	
		133*	132.905990		10.5 y ec
		134	133.904492	2.42	
		135	134.905671	6.593	
		136	135.904559	7.85	
		137	136.905816	11.23	
		138	137.905236	71.70	
Lanthanum	La	137*	136.906462		6×10^4 y ec
		138*	137.907105	0.0902	1.05×10^{11} y ec, β^+
		139	138.906346	99.9098	
Cerium	Ce	136	135.907139	0.19	
		138	137.905986	0.25	
		140	139.905434	88.43	
		142	141.909241	11.13	
Praseodymium	Pr	140*	139.909071		3.39 m ec, β^+
		141	140.907647	100	
		142*	141.910040		25.0 m β^-
Neodymium	Nd	142	141.907718	27.13	
		143	142.909809	12.18	
		144*	143.910082	23.80	2.3×10^{15} y α
		145	144.912568	8.30	
		146	145.913113	17.19	
		148	147.916888	5.76	
		150	149.920887	5.64	
Promethium	Pm	143*	142.910928		265 d ec
		145*	144.912745		17.7 y ec
		146*	145.914698		5.5 y ec
		147*	146.915134		2.623 y β^-

Element	Symbol	Mass number (*indicates radioactive)	Atomic mass	Percent abundance	Half-life and decay mode (if unstable)
Samarium	Sm	144	143.911996	3.1	
		146*	145.913043		1.0×10^8 y α
		147*	146.914894	15.0	1.06×10^{11} y α
		148*	147.914819	11.3	7×10^{15} y α
		149	148.917180	13.8	
		150	149.917273	7.4	
		151*	150.919928		90 y β^-
		152	151.919728	26.7	
		154	153.922206	22.7	
Europium	Eu	151	150.919846	47.8	
		152*	151.921740		13.5 y ec, β^+
		153	152.921226	52.2	
		154*	153.922975		8.59 y β^-
		155*	154.922888		4.7 y β^-
Gadolinium	Gd	148*	147.918112		75 y α
		150*	149.918657		1.8×10^6 y α
		152*	151.919787	0.20	1.1×10^{14} y α
		154	153.920862	2.18	
		155	154.922618	14.80	
		156	155.922119	20.47	
		157	156.923957	15.65	
		158	157.924099	24.84	
		160	159.927050	21.86	
Terbium	Tb	158*	157.925411		180 y ec, β^+, β^-
		159	158.925345	100	
		160*	159.927551		72.3 d β^-
Dysprosium	Dy	156	155.924277	0.06	
		158	157.924403	0.10	
		160	159.925193	2.34	
		161	160.926930	18.9	
		162	161.926796	25.5	
		163	162.928729	24.9	
		164	163.929172	28.2	
Holmium	Ho	165	164.930316	100	
		166*	165.932282		1.2×10^3 y β^-
Erbium	Er	162	161.928775	0.14	
		164	163.929198	1.61	
		166	165.930292	33.6	
		167	166.932047	22.95	
		168	167.932369	27.8	
		170	169.935462	14.9	
Thulium	Tm	169	168.934213	100	
		171*	170.936428		1.92 y β^-

(Continued)

Element	Symbol	Mass number (*indicates radioactive)	Atomic mass	Percent abundance	Half-life and decay mode (if unstable)
Ytterbium	Yb	168	167.933897	0.13	
		170	169.934761	3.05	
		171	170.936324	14.3	
		172	171.936380	21.9	
		173	172.938209	16.12	
		174	173.938861	31.8	
		176	175.942564	12.7	
Lutetium	Lu	173*	172.938930		1.37 y ec
		175	174.940772	97.41	
		176*	175.942679	2.59	3.8×10^{10} y β^-
Hafnium	Hf	174*	173.940042	0.162	2.0×10^{15} y α
		176	175.941404	5.206	
		177	176.943218	18.606	
		178	177.943697	27.297	
		179	178.945813	13.629	
		180	179.946547	35.100	
Tantalum	Ta	180	179.947542	0.012	
		181	180.947993	99.988	
Tungsten (Wolfram)	W	180	179.946702	0.12	
		182	181.948202	26.3	
		183	182.950221	14.28	
		184	183.950929	30.7	
		186	185.954358	28.6	
Rhenium	Re	185	184.952951	37.40	
		187*	186.955746	62.60	4.4×10^{10} y β^-
Osmium	Os	184	183.952486	0.02	
		186*	185.953834	1.58	2.0×10^{15} y α
		187	186.955744	1.6	
		188	187.955744	13.3	
		189	188.958139	16.1	
		190	189.958439	26.4	
		192	191.961468	41.0	
		194*	193.965172		6.0 y β^-
Iridium	Ir	191	190.960585	37.3	
		193	192.962916	62.7	
Platinum	Pt	190*	189.959926	0.01	6.5×10^{11} y α
		192	191.961027	0.79	
		194	193.962655	32.9	
		195	194.964765	33.8	
		196	195.964926	25.3	
		198	197.967867	7.2	
Gold	Au	197	196.966543	100	
		198*	197.968217		2.70 d β^-
		199*	198.968740		3.14 d β^-

Element	Symbol	Mass number (*indicates radioactive)	Atomic mass	Percent abundance	Half-life and decay mode (if unstable)
Mercury	Hg	196	195.965806	0.15	
		198	197.966743	9.97	
		199	198.968253	16.87	
		200	199.968299	23.10	
		201	200.970276	13.10	
		202	201.970617	29.86	
		204	203.973466	6.87	
Thallium	Tl	203	202.972320	29.524	
		204*	203.973839		3.78 y β^-
		205	204.974400	70.476	
	(Ra E″)	206*	205.976084		4.2 m β^-
	(Ac C″)	207*	206.977403		4.77 m β^-
	(Th C″)	208*	207.981992		3.053 m β^-
	(Ra C″)	210*	209.990057		1.30 m β^-
Lead	Pb	202*	201.972134		5×10^4 y ec
		204	203.973020	1.4	
		205*	204.974457		1.5×10^7 y ec
		206	205.974440	24.1	
		207	206.975871	22.1	
		208	207.976627	52.4	
	(Ra D)	210*	209.984163		22.3 y β^-
	(Ac B)	211*	210.988734		36.1 m β^-
	(Th B)	212*	211.991872		10.64 h β^-
	(Ra B)	214*	213.999798		26.8 m β^-
Bismuth	Bi	207*	206.978444		32.2 y ec, β^+
		208*	207.979717		3.7×10^5 y ec
		209	208.980374	100	
	(Ra E)	210*	209.984096		5.01 d α, β^-
	(Th C)	211*	210.987254		2.14 m α
	(Ra C)	212*	211.991259		60.6 m α, β^-
		214*	213.998692		19.9 m β^-
		215*	215.001836		7.4 m β^-
Polonium	Po	209*	208.982405		102 y α
	(Ra F)	210*	209.982848		138.38 d α
	(Ac C′)	211*	210.986627		0.52 s α
	(Th C′)	212*	211.988842		0.30 μs α
	(Ra C′)	214*	213.995177		164 μs α
	(Ac A)	215*	214.999418		0.0018 s α
	(Th A)	216*	216.001889		0.145 s α
	(Ra A)	218*	218.008965		3.10 m α
Astatine	At	215*	214.998638		$\approx$100 μs α
		218*	218.008685		1.6 s α
		219*	219.011297		0.9 m α
Radon	Rn				
	(An)	219*	219.009477		3.96 s α
	(Tn)	220*	220.011369		55.6 s α
	(Rn)	222*	222.017571		3.823 d α

(Continued)

Element	Symbol	Mass number (*indicates radioactive)	Atomic mass	Percent abundance	Half-life and decay mode (if unstable)
Francium		221*	221.01425		4.18 m α
	Fr	222*	222.017585		14.2 m β^-
	(Ac K)	223*	223.019733		22 m β^-
Radium	Ra	221*	221.01391		29 s α
	(Ac X)	223*	223.018499		11.43 d α
	(Th X)	224*	224.020187		3.66 d α
		225*			14.9 d β^-
	(Ra)	226*	226.025402		1600 y α
	(MsTh$_1$)	228*	228.031064		5.75 y β^-
Actinium	Ac	225*			10 d α
	(Ms Th$_2$)	227*	227.027749		21.77 y β^-
		228*	228.031015		6.15 h β^-
		229*			1.04 h β^-
Thorium	Th				
	(Rd Ac)	227*	227.027701		18.72 d α
	(Rd Th)	228*	228.028716		1.913 y α
		229*	229.031757		7300 y α
	(Io)	230*	230.033127		75,000 y α, sf
	(UY)	231*	231.036299	100	25.52 h β^-
	(Th)	232*	232.038051		1.40×10^{10} y α
	(UX$_1$)	234*	234.043593		24.1 d β^-
Protactinium	Pa	231*	231.035880		32,760 y α
	(UZ)	234*	234.043300		6.7 h β^-
Uranium	U	231*	231.036264		4.2 d β^+
		232*	232.037131		69 y α
		233*	233.039630		1.59×10^5 y α
	(UII)	234*	234.040946	0.0055	2.45×10^5 y α
	(Ac U)	235*	235.043924	0.720	7.04×10^8 y α
	(UI)	236*	236.045562		2.34×10^7 y α
		238*	238.050784	99.2745	4.47×10^9 y α
		239*	239.054290		23.5 m β^-
Neptunium	Np	235*	235.044057		396 d α
		236*	236.046559		1.54×10^5 y ec
		237*	237.048168		2.14×10^6 y α
Plutonium	Pu	236*	236.046033		2.87 y α, sf
		238*	238.049555		87.7 y α, sf
		239*	239.052157		24,120 y α, sf
		240*	240.053808		6560 y α, sf
		241*	241.056846		14.4 y β^-
		242*	242.058737		3.7×10^5 y α, sf
		244*	244.064200		8.1×10^7 y α, sf
Americium	Am	240*	240.055285		2.12 d ec
		241*	241.056824		432 y α, sf
Curium	Cm	247*	247.070347		1.56×10^7 y α
		248*	248.072344		3.4×10^5 y α, sf

Element	Symbol	Mass number (*indicates radioactive)	Atomic mass	Percent abundance	Half-life and decay mode (if unstable)
Berkelium	Bk	247*	247.070300		1380 y α
		249*	249.074979		327 d β^-
Californium	Cm	250*	250.076400		13.1 y α, sf
		251*	251.079580		898 y α
Einsteinium	Es	252*	252.082974		1.29 y α
		253*	253.084817		2.02 d α, sf
Fermium	Fm	253*	253.085173		3.00 d ec
		254*	254.086849		3.24 h α, sf
Mendelevium	Md	256*	256.093988		75.6 m ec, β^+
		258*	258.098594		55 d α
Nobelium	No	257*	257.096855		25 s α
		259*	259.100932		58 m α, sf
Lawrencium	Lr	259*	259.102888		6.14 s α, sf
		260*	260.105346		3.0 m α, sf
Rutherfordium	Rf	260*	260.160302		24 ms sf
		261*	261.108588		65 s α, sf
Dubnium	Db	261*	261.111830		1.8 s α
		262*	262.113763		35 s α
Seaborgium	Sg	263*	263.118310		0.78 s α, sf
Bohrium	Bh	262*	262.123081		0.10 s α, sf
Hassium	Hs	265*	265.129984		1.8 ms α
		267*	267.131770		60 ms α
Meitnerium	Mt	266*	266.137789		3.4 ms α, sf
		268*	268.138820		70 ms α
Darmstadtium	Ds	269*	269.145140		0.17 ms α
		271*	271.146080		1.1 ms α
		273*	272.153480		8.6 ms α
Roentgenium	Rg	272*	272.153480		1.5 ms α
Copernicium	Cn	277*	?		0.2 ms α
Ununtrium	Unt	284*	?		? α
Ununquadium	Unq	289*	?		? α
Ununpentium	Unp	288*	?		? α
Ununhexium	Unh	292*	?		? α
Ununseptium	Uus				
Ununoctium	Uno	294*	?		? α

Math Tutorial

In this tutorial, we review some of the basic results of algebra, geometry, trigonometry, and calculus. In many cases, we merely state results without proof. Table M-1 lists some mathematical symbols.

M-1 Significant Figures

Many numbers we work with in science are the result of measurement and are therefore known only within a degree of uncertainty. This uncertainty should be reflected in the number of digits used. For example, if you have a 1-meter-long rule with scale spacing of 1 cm, you know that you can measure the height of a box to within a fifth of a centimeter or so. Using this rule, you might find that the box height is 27.0 cm. If there is a scale with a spacing of 1 mm on your rule, you might perhaps measure the box height to be 27.03 cm. However, if there is a scale with a spacing of 1 mm on your rule, you might not be able to measure the height more accurately than 27.03 cm because the height might vary by 0.01 cm or so, depending on which part of the box you measure the height at. When you write down that the height of the box is 27.03 cm, you are stating that your best estimate of the height is 27.03 cm, but you are not claiming that it is exactly 27.030000 . . . cm high. The four digits in 27.03 cm are called **significant figures**. Your measured length, 2.703 m, has four significant digits. Significant figures are also called significant digits.

The number of significant digits in an answer to a calculation will depend on the number of significant digits in the given data. When you work with numbers that have uncertainties, you should be careful not to include more digits than the certainty of measurement warrants. *Approximate* calculations (order-of-magnitude estimates) always result in answers that have only one significant digit or none. When you multiply, divide, add, or subtract numbers, you must consider the accuracy of the results. Listed below are some rules that will help you determine the number of significant digits of your results.

1. When multiplying or dividing quantities, the number of significant digits in the final answer is no greater than that in the quantity with the fewest significant digits.
2. When adding or subtracting quantities, the number of decimal places in the answer should match that of the term with the smallest number of decimal places.
3. Exact values have an unlimited number of significant digits. For example, a value determined by counting, such as 2 tables, has no uncertainty and is an exact value. In addition, the conversion factor 0.0254000 . . . m/in. is an exact value because 1.000 . . . inches is exactly equal to 0.0254000 . . .

Table M-1	Mathematical Symbols
$=$	is equal to
$\neq$	is not equal to
$\approx$	is approximately equal to
$\sim$	is of the order of
$\propto$	is proportional to
$>$	is greater than
$\geq$	is greater than or equal to
$\gg$	is much greater than
$<$	is less than
$\leq$	is less than or equal to
$\ll$	is much less than
Δx	change in x
$\|x\|$	absolute value of x
$n!$	$n(n-1)(n-2)\ldots 1$
Σ	sum
lim	limit
$\Delta t \rightarrow 0$	Δt approaches zero
dx/dt	derivative of x with respect to t
$\partial x/\partial t$	partial derivative of x with respect to t
$\int$	integral

meters. (The yard is, by definition, equal to exactly 0.9144 m, and 0.9144 divided by 36 is exactly equal to 0.0254.)

4. Sometimes zeros are significant and sometimes they are not. If a zero is before a leading nonzero digit, then the zero is not significant. For example, the number 0.00890 has three significant digits. The first three zeroes are not significant digits but are merely markers to locate the decimal point. Note that the zero after the nine is significant.

5. Zeros that are between nonzero digits are significant. For example, 5603 has four significant digits.

6. The number of significant digits in numbers with trailing zeros and no decimal point is ambiguous. For example 31,000 could have as many as five significant digits or as few as two significant digits. To prevent ambiguity, you should report numbers by using scientific notation or by using a decimal point.

Example M-1   Finding the Average of Three Numbers

Find the average of 19.90, −7.524, and −11.8179.

SETUP

You will be adding 3 numbers and then dividing the result by 3. The first number has three significant digits, the second number has four, and the third number has six.

SOLVE

1. Sum the three numbers.

$$19.90 + (-7.524) + (-11.8179) = 0.5581$$

2. If the problem only asked for the sum of the three numbers, we would round the answer to the least number of decimal places among all the numbers being added. However, we must divide this intermediate result by 3, so we use the intermediate answer with the two extra digits (italicized and red).

$$\frac{0.5581}{3} = 0.1860333\ldots$$

3. Only two of the digits in the intermediate answer, 0.1860333 . . . , are significant digits, so we must round this number to get our final answer. The number 3 in the denominator is a whole number and has an unlimited number of significant digits. Thus, the final answer has the same number of significant digits as the numerator, which is 2.

The final answer is $\boxed{0.19.}$

REFLECT

The sum in step 1 has two significant digits following the decimal point, the same as the number being summed with the least number of significant digits after the decimal point.

Practice Problems

1. $\dfrac{5.3 \text{ mol}}{22.4 \text{ mol/L}}$

2. $57.8 \text{ m/s} - 26.24 \text{ m/s}$

M-2 Equations

An **equation** is a statement written using numbers and symbols to indicate that two quantities, written on either side of an equal sign (=), are equal. The quantity on either side of the equal sign may consist of a single term, or of a sum or difference of two or more **terms**. For example, the equation $x = 1 - (ay + b)/(cx - d)$ contains three terms, x, 1 and $(ay + b)/(cx - d)$.

You can perform the following operations on equations:

1. The same quantity can be added to or subtracted from each side of an equation.
2. Each side of an equation can be multiplied or divided by the same quantity.
3. Each side of an equation can be raised to the same power.

These operations are meant to be applied to each *side* of the equation rather than each term in the equation. (Because multiplication is distributive over addition, operation 2—and only operation 2—of the preceding operations also applies term by term.)

Caution: Division by zero is forbidden at any *stage in solving an equation; results (if any) would be invalid.*

Adding or Subtracting Equal Amounts

To find x when $x - 3 = 7$, add 3 to both sides of the equation: $(x - 3) + 3 = 7 + 3$; thus, $x = 10$.

Multiplying or Dividing by Equal Amounts

If $3x = 7$, solve x for by dividing both sides of the equation by 3; thus, $x = \frac{17}{3}$, or 5.7.

Example M-2 Simplifying Reciprocals in an Equation

Solve the following equation for x:

$$\frac{1}{x} + \frac{1}{4} = \frac{1}{3}$$

Equations containing reciprocals of unknowns occur in geometric optics and in electric circuit analysis—for example, in finding the net resistance of parallel resistors.

SETUP

In this equation, the term containing is on the same side of the equation as a term not containing Furthermore, is found in the denominator of a fraction.

SOLVE

1. Subtract $\frac{1}{4}$ from each side:

$$\frac{1}{x} = \frac{1}{3} - \frac{1}{4}$$

2. Simplify the right side of the equation by using the lowest common denominator:

$$\frac{1}{x} = \frac{1}{3} - \frac{1}{4} = \frac{4}{12} - \frac{3}{12}$$

$$= \frac{4 - 3}{12} = \frac{1}{12} \quad \text{so} \quad \frac{1}{x} = \frac{1}{12}$$

3. Multiply both sides of the equation by to determine the value of x:

$$12x\frac{1}{x} = 12x\frac{1}{12}$$

$$\boxed{12} = x$$

REFLECT

Substitute 12 for x in the left side of original equation.

$$\frac{1}{x} + \frac{1}{4} = \frac{1}{12} + \frac{3}{12} = \frac{4}{12} = \frac{1}{3}$$

Practice Problems Solve each of the following for

3. $(7.0 \text{ cm}^3)x = 18 \text{ kg} + (4.0 \text{ cm}^3)x$

4. $\dfrac{4}{x} + \dfrac{1}{3} = \dfrac{3}{x}$

M-3 Direct and Inverse Proportions

When we say variable quantities and are **directly proportional**, we mean that as x and y change, the ratio x/y is constant. To say that two quantities are proportional is to say that they are directly proportional. When we say variable quantities x and y are **inversely proportional**, we mean that as x and y change, the ratio xy is constant.

Relationships of direct and inverse proportion are common in physics. Objects moving at the same velocity have momenta directly proportional to their masses. The ideal gas law ($PV = nRT$) states that pressure P is directly proportional to (absolute) temperature T, when volume V remains constant, and is inversely proportional to volume, when temperature remains constant. Ohm's law ($V = IR$) states that the voltage V across a resistor is directly proportional to the electric current in the resistor when the resistance remains constant.

Constant of Proportionality

When two quantities are directly proportional, the two quantities are related by a *constant of proportionality*. If you are paid for working at a regular rate R in dollars per day, for example, the money m you earn is directly proportional to the time t you work; the rate R is the constant of proportionality that relates the money earned in dollars to the time worked t in days:

$$\frac{m}{t} = R \quad \text{or} \quad m = Rt$$

If you earn $400 in 5 days, the value of R is $400/(5 \text{ days}) = $80/\text{day}$. To find the amount you earn in 8 days, you could perform the calculation

$$m = ($80/\text{day})(8 \text{ days}) = $640$$

Sometimes the constant of proportionality can be ignored in proportion problems. Because the amount you earn in 8 days is $\frac{8}{5}$ times what you earn in 5 days, this amount is

$$m = \frac{8}{5}($400) = $640$$

Example M-3 Painting Cubes

You need 15.4 mL of paint to cover one side of a cube. The area of one side of the cube is 426 cm². What is the relation between the volume of paint needed and the area to be covered? How much paint do you need to paint one side of a cube in which the one side has an area of 503 cm²?

SETUP
To determine the amount of paint for the side whose area is you will need to set up a proportion.

SOLVE

1. The volume V of paint needed increases in proportion to the area A to be covered.

   > V and A are directly proportional.

   That is, $\dfrac{V}{A} = k$ or $V = kA$

   where k is the proportionality constant

2. Determine the value of the proportionality constant using the given values $V_1 = 15.4$ mL and $A_1 = 426$ cm²:

   $k = \dfrac{V_1}{A_1} = \dfrac{15.4 \text{ mL}}{426 \text{ cm}^2} = 0.0361 \text{ mL/cm}^3$

3. Determine the volume of paint needed to paint a side of a cube whose area is 503 cm² using the proportionality constant in step 1:

   $V_2 = kA_2 = (0.0361 \text{ mL/cm}^2)(503 \text{ cm}^2)$
   $= \boxed{18.2 \text{ mL}}$

REFLECT
Our value for V_2 is greater than the value for V_1, as expected. The amount of paint needed to cover an area equal 503 cm² to should be greater than the amount of paint needed to cover an area of 426 cm² because 503 cm² is larger than 426 cm².

Practice Problems

5. A cylindrical container holds 0.384 L of water when full. How much water would the container hold if its radius were doubled and its height remained unchanged?

   Hint: The volume of a right circular cylinder is given by $V = \pi r^2 h$, where r is its radius and h is its height. Thus, V is directly proportional to r^2 when h remains constant.

6. For the container in Practice Problem 5, how much water would the container hold if both its height and its radius were doubled?

   Hint: The volume V of a right circular cylinder is given by $V = \pi r^2 h$, where r is its radius and h is its height.

M-4 Linear Equations

A **linear equation** is an equation of the form $x + 2y - 4z = 3$. That is, an equation is linear if each term either is constant or is the product of a constant and a variable raised to the first power. Such equations are said to be linear because the plots of these equations form straight lines or planes. The equations of direct proportion between two variables are linear equations.

Graph of a Straight Line

A linear equation relating and can always be put into the standard form

$$y = mx + b \qquad \text{M-1}$$

where m and b are constants that may be either positive or negative. Figure M-1 shows a graph of the values of x and y that satisfy Equation M-1. The constant b, called the **y intercept**, is the value of y at $x = 0$. The constant m is the **slope** of the

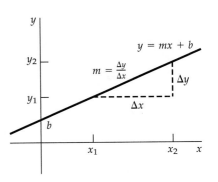

Figure M-1 Graph of the linear equation $y = mx + b$, where b is the y intercept and $m = \Delta y / \Delta x$ is the slope.

line, which equals the ratio of the change in y to the corresponding change in x. In the figure, we have indicated two points on the line, (x_1, y_1) and (x_2, y_2), and the changes $\Delta x = x_2 - x_1$ and $\Delta y = y_2 - y_1$. The slope m is then

$$m = \frac{y_2 - y_1}{x_2 - x_1} = \frac{\Delta y}{\Delta x}$$

If x and y are both unknown in the equation $y = mx + b$, there are no unique values of x and y that are solutions to the equation. Any pair of values (x_1, y_1) on the line in Figure M-1 will satisfy the equation. If we have two equations, each with the same two unknowns x and y, the equations can be solved simultaneously for the unknowns. Example M-4 shows how simultaneous linear equations can be solved.

Example M-4 Using Two Equations to Solve for Two Unknowns

Find any and all values of x and y that simultaneously satisfy

$$3x - 2y = 8 \qquad\qquad \text{M-2}$$

and

$$y - x = 2 \qquad\qquad \text{M-3}$$

SETUP

Figure M-2 shows a graph of the two equations. At the point where the lines intersect, the values of x and y satisfy both equations. We can solve two simultaneous equations by first solving either equation for one variable in terms of the other variable and then substituting the result into the other equation.

SOLVE

1. Solve Equation M-3 for y: $\qquad y = x + 2$

2. Substitute this value for y $\qquad 3x - 2(x + 2) = 8$
into Equation M-2:

3. Simplify the equation and $\qquad 3x - 2x - 4) = 8$
solve for x: $\qquad\qquad\qquad (x - 4) = 8$
$\qquad\qquad\qquad\qquad\qquad x = \boxed{12}$

4. Use your solution for x and $\qquad y - x = 2$, where $x = 12$
one of the given equations $\qquad y - 12 = 2$
to find the value of y: $\qquad y = 2 + 12 = \boxed{14}$

REFLECT

An alternative method is to multiply one equation by a constant such that one of the unknown terms is eliminated when the equations are added or subtracted. We can multiply through Equation M-3 by 2

$$2(y - x) = 2(2)$$
$$2y - 2x = 4$$

and add the result to Equation M-2 and solve for x:

$$2y - 2x = 4$$
$$\underline{3x - 2y = 8}$$
$$3x - 2x = 12 \Rightarrow x = 12$$

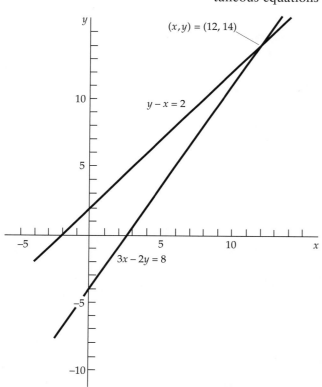

Figure M-2 Graph of Equations M-2 and M-3. At the point where the lines intersect, the values of x and y satisfy both equations.

Substitute into Equation M-3 and solve for y:

$$y - 12 = 2 \Rightarrow y = 14$$

Practice Problems

7. True or false: $xy = 4$ is a linear equation.
8. At time $t = 0.0$ s, the position of a particle moving along the x axis at a constant velocity is $x = 3.0$ m. At $t = 2.0$ s, the position is $x = 12.0$ m. Write a linear equation showing the relation of x to t.
9. Solve the following pair of simultaneous equations for x and y:

$$\frac{5}{4}x + \frac{1}{3}y = 30$$

$$y - 5x = 20$$

M-5 Quadratic Equations and Factoring

A **quadratic equation** is an equation of the form $ax^2 + bxy + cy^2 + ex + fy + g = 0$, where x and y are variables and and a, b, c, e, f, and g are constants. In each term of the equation the powers of the variables are integers that sum to 2, 1, or 0. The designation *quadratic equation* usually applies to an equation of one variable that can be written in the standard form

$$ax^2 + bx + c = 0 \qquad\qquad \textbf{M-4}$$

where a, b, and c are constants. The quadratic equation has two solutions or **roots**—values of x for which the equation is true.

Factoring

We can solve some quadratic equations by **factoring**. Very often terms of an equation can be grouped or organized into other terms. When we factor terms, we look for multipliers and multiplicands—which we now call **factors**—that will yield two or more new terms as a product. For example, we can find the roots of the quadratic equation $x^2 - 3x + 2 = 0$ by factoring the left side, to get $(x - 2)(x - 1) = 0$. The roots are $x = 2$ and $x = 1$.

Factoring is useful for simplifying equations and for understanding the relationships between quantities. You should be familiar with the multiplication of the factors $(ax + by)(cx + dy) = acx^2 + (ad + bc)xy + bdy^2$.

You should readily recognize some typical factorable combinations:

1. Common factor: $2ax + 3ay = a(2x + 3y)$
2. Perfect square: $x^2 - 2xy + y^2 = (x - y)^2$ (If the expression on the left side of a quadratic equation in standard form is a perfect square, the two roots will be equal.)
3. Difference of squares: $x^2 - y^2 = (x + y)(x - y)$

Also, look for factors that are prime numbers (2, 5, 7, etc.) because these factors can help you factor and simplify terms quickly. For example, the equation $98x^2 - 140 = 0$ can be simplified because 98 and 140 share the common factor 2. That is, $98x^2 - 140 = 0$ becomes $2(49x^2 - 70) = 0$, so we have $49x^2 - 70 = 0$.

This result can be further simplified because 49 and 70 share the common factor 7. Thus, $49x^2 - 70 = 0$ becomes $7(7x^2 - 10) = 0$, so we have $7x^2 - 10 = 0$.

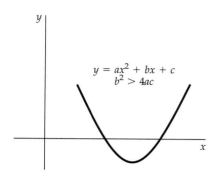

Figure M-3 Graph of y versus x when $y = ax^2 + bx + c$ for the case $b^2 > 4ac$. The two values of x for which $y = 0$ satisfy the quadratic equation (Equation M-4).

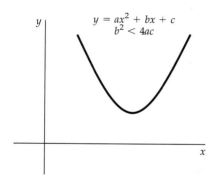

Figure M-4 Graph of y versus x when $y = ax^2 + bx + c$ for the case $b^2 < 4ac$. In this case, there are no real values of x for which $y = 0$.

The Quadratic Formula

Not all quadratic equations can be solved by factoring. However, *any* quadratic equation in the standard form $ax^2 + bx + c = 0$ can be solved by the **quadratic formula**,

$$x = \frac{-b \pm \sqrt{b^2 - 4ac}}{2a} = -\frac{b}{2a} \pm \frac{1}{2a}\sqrt{b^2 - 4ac} \qquad \text{M-5}$$

When b^2 is greater than $4ac$, there are two solutions corresponding to the $+$ and $-$ signs. Figure M-3 shows a graph of y versus x where $y = ax^2 + bx + c$. The curve, a **parabola**, crosses the x axis twice. (The simplest representation of a parabola in (x, y) coordinates is an equation of the form $y = ax^2 + bx + c$.) The two roots of this equation are the values for which $y = 0$; that is, they are the x *intercepts*.

When b^2 is less than $4ac$, the graph of y versus x does not intersect the x axis, as is shown in Figure M-4; there are still two roots, but they are not real numbers (see the discussion of complex numbers beginning on page M-19). When $b^2 = 4ac$, the graph of y versus x is tangent to the x axis at the point $x = -b/2a$; the two roots are each equal to $-b/2a$.

Example M-5   Factoring a Second-Degree Polynomial

Factor the expression $6x^2 + 19xy + 10y^2$.

SETUP

We examine the coefficients of the terms to see whether the expression can be factored without resorting to more advanced methods. Remember that the multiplication $(ax + by)(cx + dy) = acx^2 + (ad + bc)xy + bdy^2$.

SOLVE

1. The coefficient of x^2 is 6 which can be factored two ways:

   $ac = 6$
   $3 \cdot 2 = 6 \quad$ or $\quad 6 \cdot 1 = 6$

2. The coefficient of y^2 is 10 which can also be factored two ways:

   $bd = 10$
   $5 \cdot 2 = 10 \qquad 10 \cdot 1 = 10$

3. List the possibilities for a, b, c, and d in a table. Include a column for $ad + bc$.
   If $a = 3$, then $c = 2$, and vice versa. In addition, if $a = b$, then $c = 1$, and vice versa. For each value of a there are four values for b.

a	b	c	d	$ad + bc$
3	5	2	2	16
3	2	2	5	19
3	10	2	1	23
3	1	2	10	32
2	5	3	2	19
2	2	3	5	16
2	10	3	1	32
2	1	3	10	23
6	5	1	2	17
6	2	1	5	32
6	10	1	1	16
6	1	1	10	61
1	5	6	2	32
1	2	6	5	17
1	10	6	1	61
1	1	6	10	16

4. Find a combination such that $ad + bc = 19$. As you can see from the table there are two such combinations, and each gives the same results:

$ad + bc = 19$
$3 \cdot 5 + 2 \cdot 2 = 19$ and
$2 \cdot 2 + 5 \cdot 3 = 19$

5. Use the combination in the second row of the table to factor the expression in question:

$6x^2 + 19xy + 10y^2$
$= (3x + 2y)(2x + 5y)$

REFLECT

As a check, expand $(3x + 2y)(2x + 5y)$.

$$(3x + 2y)(2x + 5y) = 6x^2 + 15xy + 4xy + 10y^2 = 6x^2 + 19xy + 10y^2$$

The combination in the fifth row of the table also gives the step-4 result.

Practice Problems

10. Show that the combination in the fifth row of the table also gives the step-4 result.
11. Factor $2x^2 - 4xy + 2y^2$.
12. Factor $2x^2 + 10x^3 + 12x^2$.

M-6 Exponents and Logarithms

Exponents

The notation x^n stands for the quantity obtained by multiplying x times itself n times. For example, $x^2 = x \cdot x$ and $x^3 = x \cdot x \cdot x$. The quantity n is called the **power**, or the **exponent**, of x (the **base**). Listed below are some rules that will help you simplify terms that have exponents.

1. When two powers of x are multiplied, the exponents are added:

    $$(x^m)(x^n) = x^{m+n}$$ **M-6**

    Example: $x^2 x^3 = x^{2+3} = (x \cdot x)(x \cdot x \cdot x) = x^5$.

2. Any number (except 0) raised to the 0 power is defined to be 1:

    $$x^0 = 1$$ **M-7**

3. Based on rule 2,

    $$x^n x^{-n} = x^0 = 1$$

    $$x^{-n} = \frac{1}{x^n}$$ **M-8**

4. When two powers are divided, the exponents are subtracted:

    $$\frac{x^n}{x^m} = x^n x^{-m} = x^{n-m}$$ **M-9**

5. When a power is raised to another power, the exponents are multiplied:

    $$(x^n)^m = x^{nm}$$ **M-10**

6. When exponents are written as fractions, they represent the roots of the base. For example,

    $$x^{1/2} \cdot x^{1/2} = x$$

    so

    $$x^{1/2} = \sqrt{x} \quad (x > 0)$$

Example M-6 Simplifying a Quantity That Has Exponents

Simplify $\dfrac{x^4x^7}{x^8}$.

SET UP
According to rule 1, when two powers of x are multiplied, the exponents are added. Rule 4 states that when two powers are divided, the exponents are subtracted.

SOLVE

1. Simplify the numerator x^4x^7 using rule 1. $x^4x^7 = x^{4+7} = x^{11}$

2. Simplify $\dfrac{x^{11}}{x^8}$ using rule 4: $\dfrac{x^{11}}{x^8} = x^{11}x^{-8} = x^{11-8} = x^3$

REFLECT
Use the value $x = 2$ to determine if your answer is correct.

$$\frac{2^4 2^7}{2^8} = 2^3 = 8$$

$$\frac{2^4 2^7}{2^8} = \frac{(16)(128)}{256} = \frac{2048}{256} = 8$$

Practice Problems
13. $(x^{1/18})^9 =$
14. $x^6x^0 =$

Logarithms

Any positive number can be expressed as some power of any other positive number except one. If y is related to x by $y = a^x$, then the number x is said to be the **logarithm** of y to the **base** a, and the relation is written

$$x = \log_a y$$

Thus, logarithms are *exponents*, and the rules for working with logarithms correspond to similar laws for exponents. Listed below are some rules that will help you simplify terms that have logarithms.

1. If $y_1 = a^n$ and $y_2 = a^m$, then

$$y_1 y_2 = a^n a^m = a^{n+m}$$

   Correspondingly,

   $\log_a y_1 y_2 = \log_a a^{n+m} = n + m = \log_a a^n + \log_a a^m = \log_a y_1 + \log_a y_2$ **M-11**

   It then follows that

$$\log_a y^n = n \log_a y \qquad\qquad \text{M-12}$$

2. Because $a^1 = a$ and $a^0 = 1$,

$$\log_a a = 1 \qquad\qquad \text{M-13}$$

   and

$$\log_a 1 = 0 \qquad\qquad \text{M-14}$$

There are two bases in common use: logarithms to base 10 are called **common logarithms**, and logarithms to base e (where $e = 2.718 \ldots$) are called **natural logarithms**.

In this text, the symbol ln is used for natural logarithms and the symbol log, without a subscript, is used for common logarithms. Thus,

$$\log_e x = \ln x \quad \text{and} \quad \log_{10} x = \log x \qquad \text{M-15}$$

and $y = \ln x$ implies

$$x = e^y \qquad \text{M-16}$$

Logarithms can be changed from one base to another. Suppose that

$$z = \log x \qquad \text{M-17}$$

Then

$$10^z = 10^{\log x} = x \qquad \text{M-18}$$

Taking the natural logarithm of both sides of Equation M-18, we obtain

$$z \ln 10 = \ln x$$

Substituting $\log x$ for z (see Equation M-17) gives

$$\ln x = (\ln 10)\log x \qquad \text{M-19}$$

Example M-7 Converting between Common Logarithms and Natural Logarithms

The steps leading to Equation M-19 show that, in general, $\log_b x = (\log_b a)\log_e x$, and thus that conversion of logarithms from one base to another requires only multiplication by a constant. Describe the mathematical relation between the constant for converting common logarithms to natural logarithms and the constant for converting natural logarithms to common logarithms.

SET UP
We have a general mathematical formula for converting logarithms from one base to another. We look for the mathematical relation by exchanging a for b and vice versa in the formula.

SOLVE
1. You have a formula for converting logarithms from base a to base b:

$$\log_b x = (\log_b a)\log_a x$$

2. To convert from base b to base a, exchange all a for b and vice versa:

$$\log_a x = (\log_a b)\log_b x$$

3. Divide both sides of the equation in step 1 by $\log_a x$:

$$\frac{\log_b x}{\log_a x} = \log_b a$$

4. Divide both sides of the equation in step 2 by $(\log_a b)\log_a x$:

$$\frac{1}{\log_a b} = \frac{\log_b x}{\log_a x}$$

5. The results show that the conversion factors $\log_b a$ and $\log_a b$ are reciprocals of one other:

$$\frac{1}{\log_a b} = \log_b a$$

REFLECT
For the value of $\log_{10} e$, your calculator will give 0.43429. For ln 10, your calculator will give 2.3026. Multiply 0.43429 by 2.3026; you will get 1.0000.

Practice Problems
15. Evaluate $\log_{10} 1000$.
16. Evaluate $\log_2 5$.

Area of a circle $A = \pi r^2$

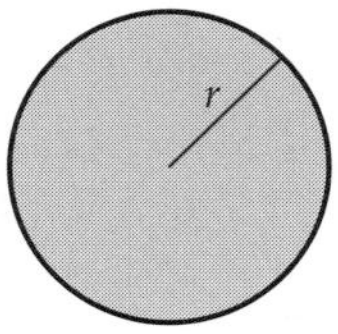

Figure M-5 Area of a circle.

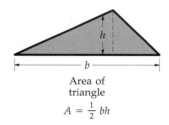

Area of parallelogram
$A = bh$

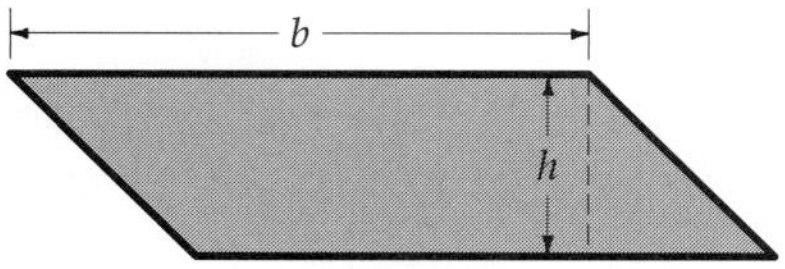

Figure M-6 Area of a parallelogram.

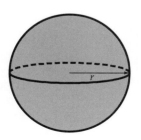

Area of triangle
$A = \frac{1}{2}bh$

Figure M-7 Area of a triangle.

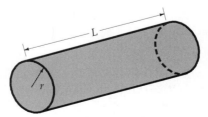

Spherical surface area
$A = 4\pi r^2$
Spherical volume
$V = \frac{4}{3}\pi r^3$

Figure M-8 Surface area and volume of a sphere.

M-7 Geometry

The properties of the most common **geometric figures**—bounded shapes in two or three dimensions whose lengths, areas, or volumes are governed by specific ratios—are a basic analytical tool in physics. For example, the characteristic ratios within triangles give us the laws of *trigonometry* (see Section M-8), which in turn give us the theory of vectors, essential in analyzing motion in two or more dimensions. Circles and spheres are essential for understanding, among other concepts, angular momentum and the probability densities of quantum mechanics.

Basic Formulas in Geometry

Circle The ratio of the circumference of a circle to its diameter is a number π, which has the approximate value

$$\pi = 3.141\ 592$$

The circumference C of a circle is thus related to its diameter d and its radius r by

$$C = \pi d = 2\pi r \quad \text{circumference of circle} \qquad \textbf{M-20}$$

The area of a circle is (**Figure M-5**)

$$A = \pi r^2 \quad \text{area of circle} \qquad \textbf{M-21}$$

Parallelogram The area of a parallelogram is the base b multiplied by the height h (**Figure M-6**):

$$A = bh$$

Triangle The area of a triangle is one-half the base multiplied by the height (**Figure M-7**)

$$A = \frac{1}{2}bh$$

Sphere A sphere of radius r (**Figure M-8**) has a surface area given by

$$A = 4\pi r^2 \quad \text{surface area of sphere} \qquad \textbf{M-22}$$

and a volume given by

$$V = \frac{4}{3}\pi r^3 \quad \text{volume of sphere} \qquad \textbf{M-23}$$

Cylinder A cylinder of radius r and length L (**Figure M-9**) has a surface area (not including the end faces) of

$$A = 2\pi r L \quad \text{surface of cylinder} \qquad \textbf{M-24}$$

and volume of

$$V = \pi r^2 L \quad \text{volume of cylinder} \qquad \textbf{M-25}$$

Cylindrical surface area
$A = 2\pi r L$
Cylindrical volume
$V = \pi r^2 L$

Figure M-9 Surface area (not including the end faces) and the volume of a cylinder.

Example M-8 Calculating the Mass of a Spherical Shell

An aluminum spherical shell has an outer diameter of 40.0 cm and an inner diameter of 39.0 cm. Find the volume of the aluminum in this shell.

SET UP

The volume of the aluminum in the spherical shell is the volume that remains when we subtract the volume of the inner sphere having $d_i = 2r_i = 39.0$ from the volume of the outer sphere having $d_o = 2r_o = 40.0$ cm.

SOLVE

1. Subtract the volume of the sphere of radius r_i from the volume of the sphere of radius r_o:

$$V = \tfrac{4}{3}\pi r_o^3 - \tfrac{4}{3}\pi r_i^3 = \tfrac{4}{3}\pi (r_o^3 - r_i^3)$$

2. Substitute 20.0 cm for r_o and 19.5 cm for r_i:

$$V = \tfrac{4}{3}\pi \left[(20.0 \text{ cm})^3 - (19.5 \text{ cm})^3 \right]$$
$$= \boxed{2.45 \times 10^3 \text{ cm}^3}$$

REFLECT

The volume of the shell is expected to be the same order of magnitude as the volume of a hollow cube with an outside edge length of 40.0 cm and an inside edge length of 39.0 cm. The volume of such a hollow cube is $(40.0 \text{ cm})^3 - (39.0 \text{ cm})^3 = 4.68 \times 10^3 \text{ cm}^3$. The result meets the expectation that the volume of the shell is the same order of magnitude as the volume of the hollow cube.

Practice Problems

17. Find the ratio between the volume V and the surface A of a sphere of radius r.
18. What is the area of a cylinder that has a radius that is $1/3$ its length?

M-8 Trigonometry

Trigonometry, which gets its name from Greek roots meaning "triangle" and "measure," is the study of some important mathematical functions, called **trigonometric functions**. These functions are most simply defined as ratios of the sides of right triangles. However, these right-triangle definitions are of limited use because they are valid only for angles between zero and 90°. However, the validity of the right-triangle definitions can be extended by defining the trigonometric functions in terms of the ratio of the coordinates of points on a circle of unit radius drawn centered at the origin of the xy plane.

In physics, we first encounter trigonometric functions when we use vectors to analyze motion in two dimensions. Trigonometric functions are also essential in the analysis of any kind of periodic behavior, such as circular motion, oscillatory motion, and wave mechanics.

Angles and Their Measure: Degrees and Radians

The size of an angle formed by two intersecting straight lines is known as its **measure**. The standard way of finding the measure of an angle is to place the angle so that its **vertex**, or point of intersection of the two lines that form the angle, is at the center of a circle located at the origin of a graph that has Cartesian coordinates and one of the lines extends rightward on the positive x axis. The distance traveled *counterclockwise* on the circumference from the positive x axis to reach the intersection of the circumference with the other line defines the measure of the angle. (Traveling clockwise to the second line would simply give us a negative measure; to

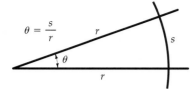

Figure M-10 The angle θ in radians is defined to be the ratio s/r, where s is the arc length intercepted on a circle of radius r.

illustrate basic concepts, we position the angle so that the smaller rotation will be in the counterclockwise direction.)

The most familiar unit for expressing the measure of an angle is the **degree**, which equals $1/360$ of the full distance around the circumference of the circle. For greater precision, or for smaller angles, we either show degrees plus minutes (′) and seconds (″), with $1' = 1°/60$ and $1'' = 1'/60 = 1°/3600$; or show degrees as an ordinary decimal number.

For scientific work, a more useful measure of an angle is the **radian** (rad). Again, place the angle with its vertex at the center of a circle and measure counterclockwise rotation around the circumference. The measure of the angle in radians is then defined as the length of the circular arc from one line to the other divided by the radius of the circle (**Figure M-10**). If s is the arc length and r is the radius of the circle, the angle θ measured in radians is

$$\theta = \frac{s}{r} \qquad \textbf{M-26}$$

Because the angle measured in radians is the ratio of two lengths, it is dimensionless. The relation between radians and degrees is

$$360° = 2\pi \text{ rad}$$

or

$$1 \text{ rad} = \frac{360°}{2\pi} = 57.3°$$

Figure M-11 shows some useful relations for angles.

The Trigonometric Functions

Figure M-12 shows a right triangle formed by drawing the line BC perpendicular to AC. The lengths of the sides are labeled a, b, and c. The right-triangle definitions of the trigonometric functions $\sin\theta$ (the **sine**), $\cos\theta$ (the **cosine**), and $\tan\theta$ (the **tangent**) for an acute angle θ are

$$\sin\theta = \frac{a}{c} = \frac{\text{opposite side}}{\text{hypotenuse}} \qquad \textbf{M-27}$$

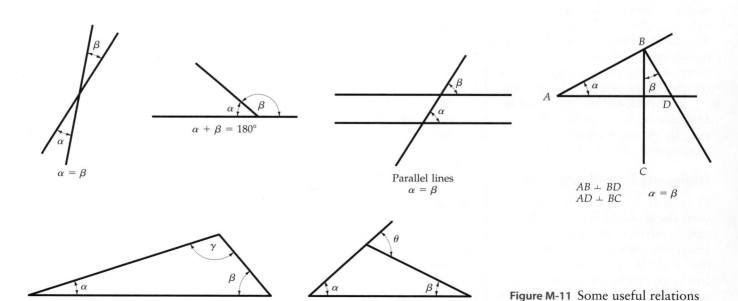

Figure M-11 Some useful relations for angles.

$$\cos \theta = \frac{b}{c} = \frac{\text{adjacent side}}{\text{hypotenuse}} \qquad \text{M-28}$$

$$\tan \theta = \frac{a}{b} = \frac{\text{opposite side}}{\text{adjacent side}} = \frac{\sin \theta}{\cos \theta} \qquad \text{M-29}$$

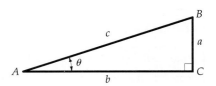

Figure M-12 A right triangle with sides of length a and b and a hypotenuse of length c.

(**Acute angles** are angles whose positive rotation around the circumference of a circle measures less than 90° or $\pi/2$.) Three other trigonometric functions—the **secant** (sec), the **cosecant** (csc), and the **cotangent** (cot), defined as the reciprocals of these functions—are

$$\sec \theta = \frac{c}{b} = \frac{1}{\cos \theta} \qquad \text{M-30}$$

$$\csc \theta = \frac{c}{a} = \frac{1}{\sin \theta} \qquad \text{M-31}$$

$$\cot \theta = \frac{b}{a} = \frac{1}{\tan \theta} = \frac{\cos \theta}{\sin \theta} \qquad \text{M-32}$$

The angle θ, whose sine is x, is called the arcsine of x, and is written $\sin^{-1} x$. That is, if

$$\sin \theta = x$$

then

$$\theta = \arcsin x = \sin^{-1} x \qquad \text{M-33}$$

The arcsine is the inverse of the sine. The inverse of the cosine and tangent are defined similarly. The angle whose cosine is y is the arccosine of y. That is, if

$$\cos \theta = y$$

then

$$\theta = \arccos y = \cos^{-1} y \qquad \text{M-34}$$

The angle whose tangent is z is the arctangent of z. That is, if

$$\tan \theta = z$$

then

$$\theta = \arctan z = \tan^{-1} z \qquad \text{M-35}$$

Trigonometric Identities

We can derive several useful formulas, called **trigonometric identities,** by examining relationships between the trigonometric functions. Equations M-30 through M-32 list three of the most obvious identities, formulas expressing some trigonometric functions as reciprocals of others. Almost as easy to discern are identities derived from the **Pythagorean theorem,**

$$a^2 + b^2 = c^2 \qquad \text{M-36}$$

(Figure M-13 illustrates a graphic proof of the theorem.) Simple algebraic manipulation of Equation M-36 gives us three more identities. First, if we divide each term in Equation M-36 by c^2, we obtain

$$\frac{a^2}{c^2} + \frac{b^2}{c^2} = 1$$

or, from the definitions of $\sin \theta$ (which is a/c) and $\cos \theta$ (which is b/c)

$$\sin^2 \theta + \cos^2 \theta = 1 \qquad \text{M-37}$$

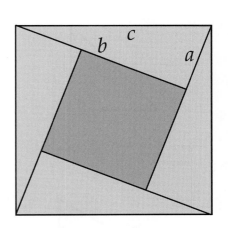

Figure M-13 When this figure was first published, the letters were absent and it was accompanied by the single word "Behold!" Using the drawing, establish the Pythagorean theorem ($a^2 + b^2 = c^2$).

Table M-2   Trigonometric Identities

$\sin(A \pm B) = \sin A \cos B \pm \cos A \sin B$

$\cos(A \pm B) = \cos A \cos B \mp \sin A \sin B$

$\tan(A \pm B) = \dfrac{\tan A \pm \tan B}{1 \mp \tan A \tan B}$

$\sin A \pm \sin B = 2 \sin\left[\dfrac{1}{2}(A \pm B)\right]\cos\left[\dfrac{1}{2}(A \mp B)\right]$

$\cos A + \cos B = 2 \cos\left[\dfrac{1}{2}(A + B)\right]\cos\left[\dfrac{1}{2}(A - B)\right]$

$\cos A - \cos B = 2 \sin\left[\dfrac{1}{2}(A + B)\right]\sin\left[\dfrac{1}{2}(B - A)\right]$

$\tan A \pm \tan B = \dfrac{\sin(A \pm B)}{\cos A \cos B}$

$\sin^2\theta + \cos^2\theta = 1; \sec^2\theta - \tan^2\theta = 1; \csc^2\theta - \cot^2\theta = 1$

$\sin 2\theta = 2 \sin\theta \cos\theta$

$\cos 2\theta = \cos^2\theta - \sin^2\theta = 2\cos^2\theta - 1 = 1 - 2\sin^2\theta$

$\tan 2\theta = \dfrac{2\tan\theta}{1 - \tan^2\theta}$

$\sin\dfrac{1}{2}\theta = \pm\sqrt{\dfrac{1 - \cos\theta}{2}}; \cos\dfrac{1}{2}\theta = \pm\sqrt{\dfrac{1 + \cos\theta}{2}};$

$\tan\dfrac{1}{2}\theta = \pm\sqrt{\dfrac{1 - \cos\theta}{1 + \cos\theta}}$

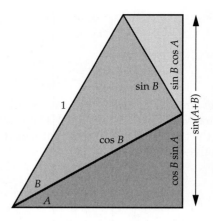

Figure M-14 Using this drawing, establish the identity $\sin(A + B) = \sin A \cos B + \cos A \sin B$. You can also use it to establish the identity $\cos(A + B) = \cos A \cos B - \sin A \sin B$. Try it.

Similarly, we can divide each term in Equation M-36 by a^2 or b^2 and obtain

$$1 + \cot^2\theta = \csc^2\theta \qquad\qquad \textbf{M-38}$$

and

$$1 + \tan^2\theta = \sec^2\theta \qquad\qquad \textbf{M-39}$$

Table M-2 lists these last three and many more trigonometric identities. Notice that they fall into four categories: functions of sums or differences of angles, sums or differences of squared functions, functions of double angles (2θ), and functions of half angles ($\frac{1}{2}\theta$). Notice that some of the formulas contain paired alternatives, expressed with the signs $\pm$ and $\mp$; in such formulas, remember to always apply the formula with either all the upper or all the lower alternatives. **Figure M-14** shows a graphic proof of the first two sum-of-angle identities.

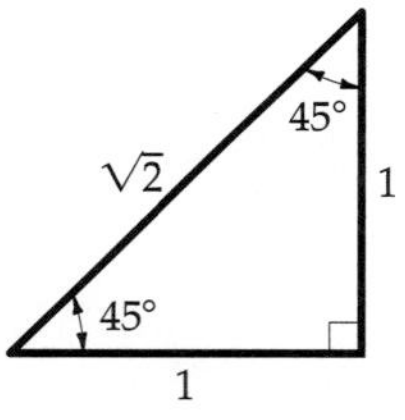

Some Important Values of the Functions

Figure M-15 is a diagram of an *isosceles* right triangle (an isosceles triangle is a triangle with two equal sides), from which we can find the sine, cosine, and tangent of 45°. The two acute angles of this triangle are equal. Because the sum of the three angles in a triangle must equal 180° and the right angle is 90°, each acute angle must be 45°. For convenience, let us assume that the equal sides each have a length of 1 unit. The Pythagorean theorem gives us a value for the hypotenuse of

$$c = \sqrt{a^2 + b^2} = \sqrt{1^2 + 1^2} = \sqrt{2} \text{ units}$$

Figure M-15 An isosceles right triangle.

We calculate the values of the functions as follows:

$$\sin 45° = \frac{a}{c} = \frac{1}{\sqrt{2}} = 0.707 \quad \cos 45° = \frac{b}{c} = \frac{1}{\sqrt{2}} = 0.707 \quad \tan 45° = \frac{a}{b} = \frac{1}{1} = 1$$

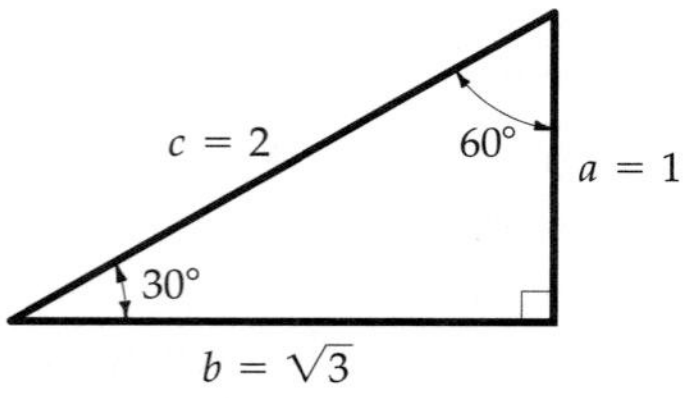

Another common triangle, a 30°–60° right triangle, is shown in **Figure M-16**. Because this particular right triangle is in effect half of an *equilateral triangle* (a 60°–60°–60° triangle or a triangle having three equal sides and three equal angles),

Figure M-16 A 30°–60° right triangle.

we can see that the sine of 30° must be exactly 0.5 (**Figure M-17**). The equilateral triangle must have all sides equal to c, the hypotenuse of the 30°–60° right triangle. Thus, side a is one-half the length of the hypotenuse, and so

$$\sin 30° = \frac{1}{2}$$

To find the other ratios within the 30°–60° right triangle, let us assign a value of 1 to the side opposite the 30° angle. Then

$$c = \frac{1}{0.5} = 2 \qquad\qquad b = \sqrt{c^2 - a^2} = \sqrt{2^2 - 1^2} = \sqrt{3}$$

$$\cos 30° = \frac{b}{c} = \frac{\sqrt{3}}{2} = 0.866 \qquad \tan 30° = \frac{a}{b} = \frac{1}{\sqrt{3}} = 0.577$$

$$\sin 60° = \cos 30° = 0.866 \qquad \cos 60° = \frac{a}{c} = \sin 30° = \frac{1}{2}$$

$$\tan 60° = \frac{b}{a} = \frac{\sqrt{3}}{1} = 1.732$$

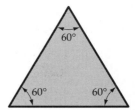 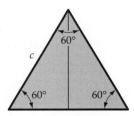

Figure M-17 (a) An equilateral triangle. (b) An equilateral triangle that has been bisected to form two 30°–60° right triangles.

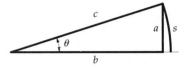

Figure M-18 For small angles, $\sin\theta = a/c$, $\tan\theta = a/b$, and the angle $\theta = s/c$ are all approximately equal.

Small-Angle Approximation

For small angles, the length a is nearly equal to the arc length s, as can be seen in **Figure M-18**. The angle $\theta = s/c$ is therefore nearly equal to $\sin\theta = a/c$:

$$\sin\theta \approx \theta \quad \text{for small values of } \theta \qquad\qquad \textbf{M-40}$$

Similarly, the lengths c and b are nearly equal, so $\tan\theta = a/b$ is nearly equal to both θ and $\sin\theta$ for small values of θ:

$$\tan\theta \approx \sin\theta \approx \theta \quad \text{for small values of } \theta \qquad\qquad \textbf{M-41}$$

Equations M-40 and M-41 hold only if θ is measured in radians. Because $\cos\theta = b/c$, and because these lengths are nearly equal for small values of θ, we have

$$\cos\theta \approx 1 \quad \text{for small values of } \theta \qquad\qquad \textbf{M-42}$$

Figure M-19 shows graphs of θ, $\sin\theta$, and $\tan\theta$ versus θ for small values of θ. If accuracy of a few percent is needed, small-angle approximations can be used only for angles of about a quarter of a radian (or about 15°) or less. Below this value, as the angle becomes smaller, the approximation $\theta \approx \sin\theta \approx \tan\theta$ is even more accurate.

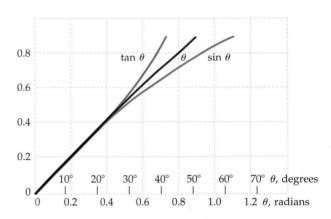

Figure M-19 Graphs of $\tan\theta$, θ, and $\sin\theta$ versus θ for small values of θ.

Trigonometric Functions as Functions of Real Numbers

So far we have illustrated the trigonometric functions as properties of angles. **Figure M-20** shows an *obtuse* angle with its vertex at the origin and one side along the x axis. The trigonometric functions for a "general" angle such as this are defined by

$$\sin\theta = \frac{y}{c} \qquad\qquad \textbf{M-43}$$

$$\cos\theta = \frac{x}{c} \qquad\qquad \textbf{M-44}$$

$$\tan\theta = \frac{y}{x} \qquad\qquad \textbf{M-45}$$

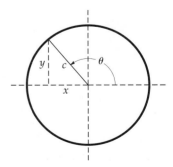

Figure M-20 Diagram for defining the trigonometric functions for an obtuse angle.

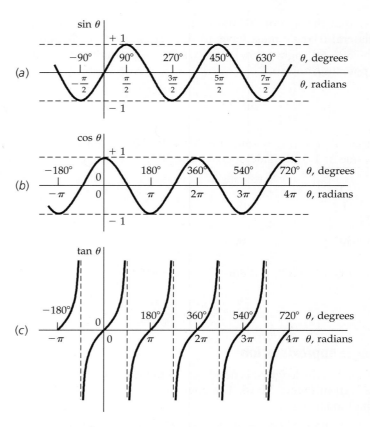

Figure M-21 The trigonometric functions sin θ, cos θ, and tan θ versus θ.

It is important to remember that values of x to the left of the vertical axis and values of y below the horizontal axis are negative; c in the figure is always regarded as positive. **Figure M-21** shows plots of the general sine, cosine, and tangent functions versus θ. The sine function has a period of 2π rad. Thus, for any value of θ, $\sin(\theta + 2\pi) = \sin\theta$, and so forth. That is, when an angle changes by 2π rad, the function returns to its original value. The tangent function has a period of π rad. Thus, $\tan(\theta + \pi) = \tan\theta$, and so forth. Some other useful relations are

$$\sin(\pi - \theta) = \sin\theta \qquad\qquad \text{M-46}$$

$$\cos(\pi - \theta) = -\cos\theta \qquad\qquad \text{M-47}$$

$$\sin(\tfrac{1}{2}\pi - \theta) = \cos\theta \qquad\qquad \text{M-48}$$

$$\cos(\tfrac{1}{2}\pi - \theta) = \sin\theta \qquad\qquad \text{M-49}$$

Because the radian is dimensionless, it is not hard to see from the plots in Figure M-21 that the trigonometric functions are functions of all real numbers. The functions can also be expressed as power series in θ. The series for sin θ and cos θ are

$$\sin\theta = \theta - \frac{\theta^3}{3!} + \frac{\theta^5}{5!} - \frac{\theta^7}{7!} + \cdots \qquad\qquad \text{M-50}$$

$$\cos\theta = 1 - \frac{\theta^2}{2!} + \frac{\theta^4}{4!} - \frac{\theta^6}{6!} + \cdots \qquad\qquad \text{M-51}$$

When θ is small, good approximations are obtained using only the first few terms in the series.

Example M-9 Cosine of a Sum

Using the suitable trigonometric identity from Table M-2, find $\cos(135° + 22°)$. Give your answer in four significant figures.

SET UP
As long as all angles are given in degrees, there is no need to convert to radians, because all operations are numerical values of the functions. Be sure, however, that your calculator is in degree mode. The suitable identity is $\cos(A \pm B) = \cos A \cos B \mp \sin A \sin B$, where the upper signs are appropriate.

SOLVE
1. Write the trigonometric identity for the cosine of a sum, with $A = 135°$ and $B = 22°$:
    $$\cos(135° + 22°) = (\cos 135°)(\cos 22°) - (\sin 135°)(\sin 22°)$$

2. Using a calculator, find $\cos 135°$, $\sin 135°$, $\cos 22°$, and $\sin 22°$:
    $$\cos 135° = -0.7071$$
    $$\cos 22° = 0.9272$$
    $$\sin 135° = 0.7071$$
    $$\sin 22° = 0.3746$$

3. Enter the values in the formula and calculate the answer:
    $$\cos(135° + 22°) = (-0.7071)(0.9272) - (0.7071)(0.3746) = -0.9205$$

REFLECT
The calculator shows that the $\cos(135° + 22°) = \cos(157°) = -0.9205$.

Practice Problems
19. Find $\sin \theta$ and $\cos \theta$ for the right triangle shown in Figure M-12 in which $a = 4$ cm and $b = 7$ cm. What is the value for θ?
20. Find $\sin \theta$ where $\theta = 8.2°$. Is your answer consistent with the small-angle approximation?

M-9 The Binomial Expansion

A **binomial** is an expression consisting of two terms joined by a plus sign or a minus sign. The **binomial theorem** states that a binomial raised to a power can be written, or *expanded*, as a series of terms. If we raise the binomial $(1 + x)$ to a power n, the binomial theorem takes the form

$$(1 + x)^n = 1 + nx + \frac{n(n - 1)}{2!}x^2 + \frac{n(n - 1)(n - 2)}{3!}x^3 + \cdots \qquad \text{M-52}$$

The series is valid for any value of n if $|x|$ is less than 1. The binomial expansion is very useful for approximating algebraic expressions, because when $|x| < 1$, the higher-order terms in the sum are small. (The order of a term is the power of x in the term. Thus, the terms explicitly shown in Equation M-52 are of order 0, 1, 2, and 3.) The series is particularly useful in situations where $|x|$ is small compared with 1; then each term is *much* smaller than the previous term and we can drop all but the first two or three terms in the expansion. If $|x|$ is much less than 1, we have

$$(1 + x)^n \approx 1 + nx, \qquad |x| \ll 1 \qquad \text{M-53}$$

The binomial expansion is used in deriving many formulas of calculus that are important in physics. A well-known use in physics of the approximation in Equation

M-53 is the proof that relativistic kinetic energy reduces to the classic formula when the velocity of a particle is very small compared with the velocity of light c.

Example M-10 Using the Binomial Expansion to Find a Power of a Number

---

Use Equation M-53 to find an approximate value for the square root of 101.

SET UP

The number 101 readily suggests a binomial, namely, $(100 + 1)$. To approximate the answer using the binomial expansion, we must manipulate the expression to get a binomial consisting of 1 and a term less than 1.

SOLVE

1. Write $(101)^{1/2}$ to give an expression $(1 + x)^n$ in which x is much less than 1:

$$(101)^{1/2} = (100 + 1)^{1/2} = (100)^{1/2}(1 + 0.01)^{1/2} = 10(1 + 0.01)^{1/2}$$

2. Use Equation M-53 with $n = \frac{1}{2}$ and $x = 0.01$ to expand $(1 + 0.01)^{1/2}$:

$$(1 + 0.01)^{1/2} = 1 + \tfrac{1}{2}(0.01) + \frac{\frac{1}{2}(-\frac{1}{2})}{2}(0.01)^2 + \cdots$$

3. Because $|x| \ll 1$, we expect the magnitude of terms of order 2 and higher to be significantly smaller than the magnitude of the first-order term. Approximate the binomial (1) by keeping only the zeroth and first-order terms, and (2) by keeping only the first 3 terms:

Keeping only the zeroth and first-order terms gives

$$(1 + 0.01)^{1/2} \approx 1 + \tfrac{1}{2}(0.01) = 1 + 0.005\,000\,0 = 1.005\,000\,0$$

Keeping only the zeroth, first-, and second-order terms gives

$$(1 + 0.01)^{1/2} \approx 1 + \tfrac{1}{2}(0.01) + \frac{\frac{1}{2}(-\frac{1}{2})}{2}(0.01)^2$$
$$\approx 1 + 0.005\,000\,0 - 0.000\,012\,5$$
$$= 1.004\,987\,5$$

4. Substitute these results into the equation in step 1:

Keeping only the zeroth and first-order terms gives
$$(101)^{1/2} = 10(1 + 0.01)^{1/2} \approx \boxed{10.050\,000}$$

Keeping only the zeroth, first-, and second-order terms gives
$$(101)^{1/2} = 10(1 + 0.01)^{1/2} \approx \boxed{10.049\,875}$$

REFLECT

We therefore expect our answer to be correct to within about 0.001%. The value of $(101)^{1/2}$, to eight figures, is 10.049 876. This differs from 10.050 000 by 0.000 124, or about one part in 10^5, and differs from 10.049 875 by about one part in 10^7.

Practice Problems For the following, calculate the answer keeping the zeroth and first-order terms in the binomial series (Equation M-53), find the answer using your calculator, and show the percentage discrepancy between the two values:

21. $(1 + 0.001)^{-4}$
22. $(1 - 0.001)^{40}$

M-10 Complex Numbers

Real numbers are all numbers, from $-\infty$ to $+\infty$, that can be *ordered*. We know that, given two real numbers, one is always equal to, greater than, or less than the other. For example, $3 > 2$, $1.4 < \sqrt{2} < 1.5$, and $3.14 < \pi < 3.15$. A number that *cannot* be ordered is $\sqrt{-1}$; we cannot measure the size of this number, and so it makes no sense to say, for example, that $3 \times \sqrt{-1}$ is greater than or less than $2 \times \sqrt{-1}$. The earliest mathematicians who dealt with numbers containing $\sqrt{-1}$ referred to these numbers as *imaginary* numbers because they could not be used to measure or count something. In mathematics the symbol i is used to represent $\sqrt{-1}$.

Equation M-5, the quadratic formula, applies to equations of the form

$$ax^2 + bx + c = 0$$

The formula shows that there are no real roots when $b^2 < 4ac$. There are, however, still two roots. Each root is a number containing two terms: a real number and a multiple of $i = \sqrt{-1}$. The multiple of i is called an **imaginary number**, and i is called the **unit imaginary**.

A general **complex number** z can be written

$$z = a + bi \qquad \textbf{M-54}$$

where a and b are real numbers. The quantity a is called the real part of z or Re(z), and the quantity b is called the imaginary part of z or Im(z). We can represent a complex number z as a point in a plane, called the complex plane, as shown in **Figure M-22**, where the x axis is the **real axis** and the y axis is the **imaginary axis**. We can also use the relations $a = r\cos\theta$ and $b = r\sin\theta$ from Figure M-22 to write the complex number z in **polar coordinates** (a system in which a point is designated by the counterclockwise angle of rotation θ and the distance r in the direction of θ):

$$z = r\cos\theta + ir\sin\theta \qquad \textbf{M-55}$$

where $r = \sqrt{a^2 + b^2}$ is called the **magnitude** of z.

When complex numbers are added or subtracted, the real and imaginary parts are added or subtracted separately:

$$z_1 + z_2 = (a_1 + ib_1) + (a_2 + ib_2) = (a_1 + a_2) + i(b_1 + b_2) \qquad \textbf{M-56}$$

However, when two complex numbers are multiplied, each part of one number is multiplied by each part of the other number:

$$z_1 z_2 = (a_1 + ib_1)(a_2 + ib_2) = a_1 a_2 + i^2 b_1 b_2 + i(a_1 b_2 + a_2 b_1)$$
$$= a_1 a_2 - b_1 b_2 + i(a_1 b_2 + a_2 b_1) \qquad \textbf{M-57}$$

where we have used $i^2 = -1$.

The **complex conjugate** z^* of the complex number z is that number obtained by replacing i with $-i$ when writing z. If $z = a + ib$, then

$$z^* = (a + ib)^* = a - ib \qquad \textbf{M-58}$$

(When a quadratic equation has complex roots, the roots are **conjugate complex numbers**, in the form $a \pm bi$.) The product of a complex number and its complex conjugate equals the square of the magnitude of the number:

$$zz^* = (a + ib)(a - ib) = a^2 + b^2 = r^2 \qquad \textbf{M-59}$$

A particularly useful function of a complex number is the exponential $e^{i\theta}$. Using an expansion for e^x, we have

$$e^{i\theta} = 1 + i\theta + \frac{(i\theta)^2}{2!} + \frac{(i\theta)^3}{3!} + \frac{(i\theta)^4}{4!} + \cdots$$

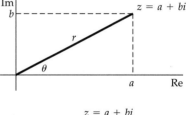

$$z = a + bi$$
$$= r\cos\theta + (r\sin\theta)i$$
$$= r(\cos\theta + i\sin\theta)$$

Figure M-22 Representation of a complex number in a plane. The real part of the complex number is plotted along the horizontal axis, and the imaginary part is plotted along the vertical axis.

Using $i^2 = -1$, $i^3 = -i$, $i^4 = +1$, and so forth, and separating the real parts from the imaginary parts, this expansion can be written

$$e^{i\theta} = \left(1 - \frac{\theta^2}{2!} + \frac{\theta^4}{4!} - \cdots\right) + i\left(\theta - \frac{\theta^3}{3!} + \cdots\right)$$

Comparing this result with Equations M-50 and M-51, we can see that

$$e^{i\theta} = \cos\theta + i\sin\theta \qquad\qquad \textbf{M-60}$$

Using this result, we can express a general complex number as an exponential:

$$z = a + ib = r\cos\theta + ir\sin\theta = re^{i\theta} \qquad\qquad \textbf{M-61}$$

If $z = x + iy$, where x and y are real variables, then z is called a **complex variable**.

Complex Variables in Physics

Complex variables are often used in formulas describing AC circuits: the impedance of a capacitor or an inductor includes a real part (the resistance) and an imaginary part (the reactance). (There are alternative ways, however, of analyzing AC circuits—such as rotating vectors called *phasors*—that do not require assigning imaginary values.) Complex variables are also important in the study of harmonic waves through Fourier analysis and synthesis. The time-dependent Schrödinger equation contains a complex-valued function of position and time.

Example M-11 Finding a Power of a Complex Number

Calculate $(1 + 3i)^4$ by using the binomial expansion.

SET UP
The expression is of the form $(1 + x)^n$. Because n is a positive integer, the expansion is valid for any value of x, and all terms, other than those of order n or lower must equal zero.

SOLVE

1. Write out the expansion of $(1 + 3i)^4$ to show the terms up through the fouth-order term:

$$1 + 4\cdot 3i + \frac{4(3)}{2!}(3i)^2 + \frac{4(3)(2)}{3!}(3i)^3$$
$$+ \frac{4(3)(2)(1)}{4!}(3i)^4$$

2. Evaluate each term, remembering that $i^2 = -1$, $i^3 = -1$ and $i^4 = +1$:

$$1 + 12i - 54 - 108i + 81$$

3. Show the result in the form $a + bi$:

$$(1 + 3i)^4 = \boxed{28 - 96i}$$

REFLECT
We can solve the problem algebraically to show that the answer is correct. We first square $(1 + 3i)$ and then square the result, to get $(1 + 3i)^4$:

$$(1 + 3i)^2 = 1\cdot 1 + 2\cdot 1\cdot 3i + (3i)^2 = 1 + 6i - 9 = -8 + 6i$$

$$(-8 + 6i)^2 = (-8)(-8) + 2(-8)(6i) + (6i^2)$$
$$= 64 - 96i - 36 = 28 - 96i$$

Practice Problems Express in the form $a + bi$:
23. $e^{i\pi}$
24. $e^{i\pi/2}$

M-11 Differential Calculus

Calculus is the branch of mathematics that allows us to deal with instantaneous rates of change of functions and variables. From the equation of a function—say, x as a function of t—we can always find x for a particular t, but with the methods of calculus you can go much further. You can know where x will have certain properties, such as a maximum or a minimum value, without having to try endless values of t. With calculus, if given the proper data, you can find, for example, the location of maximum stress on a beam, or the velocity or position of a falling object at a time t, or the energy a falling object has acquired at the time of impact. The principles of calculus are derived from examining functions at the infinitesimal level—analyzing how, say, x will change when the change in t becomes vanishingly small. We start with **differential calculus**, in which we determine the *limit* of the rate of change of x with respect to t as the change in t becomes closer and closer to zero.

Figure M-23 is a graph of x versus t for a typical function $x(t)$. At a particular value $t = t_1$, x has the value of x_1, as indicated. At another value t_2, x has the value x_2. The change in t, $t_2 - t_1$, is written $\Delta t = t_2 - t_1$; and the corresponding change in x is written $\Delta x = x_2 - x_1$. The ratio $\Delta x / \Delta t$ is the slope of the straight line connecting (x_1, t_1) and (x_2, t_2). If we take the limit as t_2 approaches t_1 (as Δt approaches zero) the slope of the line connecting (x_1, t_1) and (x_2, t_2) approaches the slope of the line that is tangent to the curve at the point (x_1, t_1). The slope of this tangent line is equal to the **derivative** of x with respect to t and is written dx/dt:

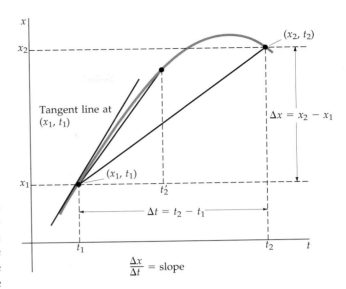

Figure M-23 Graph of a typical function $x(t)$. The points (x_1, t_1) and (x_2, t_2) are connected by a straight line. The slope of this line is $\Delta x / \Delta t$. As the time interval beginning at t_1 is decreased, the slope for that interval approaches the slope of the line tangent to the curve at time t_1, which is the derivative of x with respect to t.

$$\frac{dx}{dt} = \lim_{\Delta t \to 0} \frac{\Delta x}{\Delta t} \qquad\qquad \text{M-62}$$

(When we find the derivative of a function, we say that we are **differentiating** the function; and the very small dx and dt elements are called **differentials** of x and t, respectively.) The derivative of a function of t is another function of t. If x is a constant and does not change, the graph of x versus t is a horizontal line with zero slope. The derivative of a constant is thus zero. In Figure M-24, x is not constant but is proportional to t:

$$x = Ct$$

This function has a constant slope equal to C. Thus the derivative of Ct is C. Table M-3 lists some properties of derivatives and the derivatives of some particular functions that occur often in physics. It is followed by comments aimed at making these properties and rules clearer. More detailed discussion can be found in most calculus textbooks.

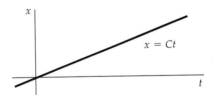

Figure M-24 Graph of the linear function $x = Ct$. This function has a constant slope C.

Comments On Rules 1 Through 5

Rules 1 and 2 follow from the fact that the limiting process is linear. We can understand rule 3, the chain rule, by multiplying $\Delta f / \Delta t$ by $\Delta x / \Delta x$ and noting that as Δt approaches zero, Δx also approaches zero. That is,

$$\lim_{\Delta t \to 0} \frac{\Delta f}{\Delta t} = \lim_{\Delta t \to 0} \left(\frac{\Delta f}{\Delta t} \frac{\Delta x}{\Delta x} \right) = \lim_{\Delta t \to 0} \left(\frac{\Delta f}{\Delta x} \frac{\Delta x}{\Delta t} \right) = \left(\lim_{\Delta t \to 0} \frac{\Delta f}{\Delta x} \right) \left(\lim_{\Delta t \to 0} \frac{\Delta x}{\Delta t} \right) = \frac{df}{dx} \frac{dx}{dt}$$

where we have used that the limit of the product is equal to product of the limits.

Table M-3 Properties of Derivatives and Derivatives of Particular Functions

Linearity

1. The derivative of a constant C multiplied by a function $f(t)$ equals the constant multiplied by the derivative of the function:

$$\frac{d}{dt}[Cf(t)] = C\frac{df(t)}{dt}$$

2. The derivative of a sum of functions equals the sum of the derivatives of the functions:

$$\frac{d}{dt}[f(t) + g(t)] = \frac{df(t)}{dt} + \frac{dg(t)}{dt}$$

Chain rule

3. If f is a function of x and x is in turn a function of t, the derivative of f with respect to t equals the product of the derivative of f with respect to x and the derivative of x with respect to t:

$$\frac{d}{dt}f(x(t)) = \frac{df}{dx}\frac{dx}{dt}$$

Derivative of a product

4. The derivative of a product of functions $f(t)g(t)$ equals the first function multiplied by the derivative of the second plus the second function multiplied by the derivative of the first:

$$\frac{d}{dt}[f(t)g(t)] = f(t)\frac{dg(t)}{dt} + g(t)\frac{df(t)}{dt}$$

Reciprocal derivative

5. The derivative of t with respect to x is the reciprocal of the derivative of x with respect to t, assuming that neither derivative is zero:

$$\frac{dt}{dx} = \left(\frac{dx}{dt}\right)^{-1} \quad \text{if } \frac{dt}{dx} \neq 0 \quad \text{and} \quad \frac{dx}{dt} \neq 0$$

Derivatives of particular functions

6. If C is a constant, then $dC/dt = 0$.

7. $\dfrac{d(t^n)}{dt} = nt^{n-1}$ If n is constant.

8. $\dfrac{d}{dt}\sin \omega t = \omega \cos \omega t$ If ω is constant.

9. $\dfrac{d}{dt}\cos \omega t = -\omega \sin \omega t$ If ω is constant.

10. $\dfrac{d}{dt}\tan \omega t = \omega \sin^2 \omega t$ If ω is constant.

11. $\dfrac{d}{dt}e^{bt} = be^{bt}$ If b is constant.

12. $\dfrac{d}{dt}\ln bt = \dfrac{1}{t}$ If b is constant.

Rule 4 is not immediately apparent. The derivative of a product of functions is the limit of the ratio

$$\frac{f(t + \Delta t)g(t + \Delta t) - f(t)g(t)}{\Delta t}$$

If we add and subtract the quantity $f(t + \Delta t)g(t)$ in the numerator, we can write this ratio as

$$\frac{f(t + \Delta t)g(t + \Delta t) - f(t + \Delta t)g(t) + f(t + \Delta t)g(t) - f(t)g(t)}{\Delta t}$$

$$= f(t + \Delta t)\left[\frac{g(t + \Delta t) - g(t)}{\Delta t}\right] + g(t)\left[\frac{f(t + \Delta t) - f(t)}{\Delta t}\right]$$

As Δt approaches zero, the terms in square brackets become $dg(t)/dt$ and $df(t)/dt$, respectively, and the limit of the expression is

$$f(t)\frac{dg(t)}{dt} + g(t)\frac{df(t)}{dt}$$

Rule 5 follows directly from the definition:

$$\frac{dx}{dt} = \lim_{\Delta t \to 0}\frac{\Delta x}{\Delta t} = \lim_{\Delta x \to 0}\left(\frac{\Delta t}{\Delta x}\right)^{-1} = \left(\frac{dt}{dx}\right)^{-1}$$

Comments on Rule 7

We can obtain this important result using the binomial expansion. We have

$$f(t) = t^n$$

$$f(t + \Delta t) = (t + \Delta t)^n = t^n\left(1 + \frac{\Delta t}{t}\right)^n$$

$$= t^n\left[1 + n\frac{\Delta t}{t} + \frac{n(n-1)}{2!}\left(\frac{\Delta t}{t}\right)^2 + \frac{n(n-1)(n-2)}{3!}\left(\frac{\Delta t}{t}\right)^3 + \cdots\right]$$

Then

$$f(t + \Delta t) - f(t) = t^n\left[n\frac{\Delta t}{t} + \frac{n(n-1)}{2!}\left(\frac{\Delta t}{t}\right)^2 + \cdots\right]$$

and

$$\frac{f(t + \Delta t) - f(t)}{\Delta t} = nt^{n-1} + \frac{n(n-1)}{2!}t^{n-2}\Delta t + \cdots$$

The next term omitted from the last sum is proportional to $(\Delta t)^2$, the following to $(\Delta t)^3$, and so on. Each term except the first approaches zero as Δt approaches zero. Thus

$$\frac{df}{dt} = \lim_{\Delta x \to 0}\frac{f(t + \Delta t) - f(t)}{\Delta t} = nt^{n-1}$$

Comments on Rules 8 to 10

We first write $\sin \omega t = \sin \theta$ with $\theta = \omega t$ and use the chain rule,

$$\frac{d\sin\theta}{dt} = \frac{d\sin\theta}{d\theta}\frac{d\theta}{dt} = \omega\frac{d\sin\theta}{d\theta}$$

We then use the trigonometric formula for the sine of the sum of two angles θ and $\Delta\theta$:

$$\sin(\theta + \Delta\theta) = \sin\Delta\theta\cos\theta + \cos\Delta\theta\sin\theta$$

Because $\Delta\theta$ is to approach zero, we can use the small-angle approximations

$$\sin\Delta\theta \approx \Delta\theta \quad\text{and}\quad \cos\Delta\theta \approx 1$$

Then

$$\sin(\theta + \Delta\theta) \approx \Delta\theta \cos\theta + \sin\theta$$

and

$$\frac{\sin(\theta + \Delta\theta) - \sin\theta}{\Delta\theta} \approx \cos\theta$$

Similar reasoning can be applied to the cosine function to obtain rule 9.

Rule 10 is obtained by writing $\tan\theta = \sin\theta/\cos\theta$ and applying rule 4 along with rules 8 and 9:

$$\frac{d}{dt}(\tan\theta) = \frac{d}{dt}(\sin\theta)(\cos\theta)^{-1} = \sin\theta\frac{d}{dt}(\cos\theta)^{-1} + \frac{d(\sin\theta)}{dt}(\cos\theta)^{-1}$$

$$= \sin\theta(-1)(\cos\theta)^{-2}(-\sin\theta) + (\cos\theta)(\cos\theta)^{-1}$$

$$= \frac{\sin^2\theta}{\cos^2\theta} + 1 = \tan^2\theta + 1 = \sec^2\theta$$

To obtain rule 10, let $\theta = \omega t$ and use the chain rule.

Comments on Rule 11

Again we use the chain rule

$$\frac{de^\theta}{dt} = \frac{b\,de^\theta}{b\,dt} = b\frac{de^\theta}{d(bt)} = b\frac{de^\theta}{d\theta} \quad \text{with} \quad \theta = bt$$

and the series expansion for the exponential function:

$$e^{\theta + \Delta\theta} = e^\theta e^{\Delta\theta} = e^\theta\left[1 + \Delta\theta + \frac{(\Delta\theta)^2}{2!} + \frac{(\Delta\theta)^3}{3!} + \cdots\right]$$

Then

$$\frac{e^{\theta + \Delta\theta} - e^\theta}{\Delta\theta} = e^\theta + e^\theta\frac{\Delta\theta}{2!} + e^\theta\frac{(\Delta\theta)^2}{3!} + \cdots$$

As $\Delta\theta$ approaches zero, the right side of this equation approaches e^θ.

Comments on Rule 12

Let

$$y = \ln bt$$

Then

$$e^y = bt \Rightarrow t = \frac{1}{b}e^y$$

Then, using rule 11, we obtain

$$\frac{dt}{dy} = \frac{1}{b}e^y \therefore \frac{dt}{dy} = t$$

Then, using rule 5, we obtain

$$\frac{dy}{dt} = \left(\frac{dt}{dy}\right)^{-1} = \frac{1}{t}$$

Second- and Higher-Order Derivatives; Dimensional Analysis

Once we have differentiated a function, we can differentiate the resulting derivative as long as terms remain to differentiate. A function such as $x = e^{bt}$ can be differentiated indefinitely: $dx/dt = be^{bt}$ (this function differentiates to give $b^2 e^{bt}$, and so on).

Consider velocity and acceleration. We can define velocity as the rate of change of position of a particle or dx/dt, and acceleration as the rate of change of velocity, or the *second* derivative of x with respect to t, written dx^2/dt^2. If a particle moves at a constant velocity, then dx/dt will equal a constant. The acceleration, however, will be zero: Having constant velocity is the same as having no acceleration, and the derivative of a constant is zero. Now consider a falling object, subject to the constant acceleration of gravity: The velocity itself will be time-dependent, so the *second* derivative, dx^2/dt^2 will be a constant.

The *physical dimensions* of a derivative with respect to a variable are those that would result if the original function of the variable were divided by a value of the variable. For example, the dimension of an equation in which one term is x (for position) is that of length (L); the dimensions of the derivative of x with respect to time t are those of velocity (L/T), and the dimensions of dx^2/dt^2 are those of acceleration (L/T^2).

Example M-12 Position, Velocity, and Acceleration

Find the first and the second derivatives of $x = \frac{1}{2}at^2 + bt + c$ where a, b, and c are constants. The function gives the position (in m) of a particle in one dimension, where t is the time (in s), a is acceleration (in m/s^2), b is velocity (in m/s) at a time $t = 0$, and c is the position (in m) of the particle at $t = 0$.

SET UP
Both the first and the second derivatives are sums of terms; for each differentiation we take the derivative of each term separately and add the results.

SOLVE

1. To find the first derivative, first compute the derivative of the first term:

$$\frac{d\left(\frac{1}{2}at^2\right)}{dt} = \left(\frac{1}{2}a\right)2t^1 = at$$

2. Compute the first derivative of the second and third terms:

$$\frac{d(bt)}{dt} = b, \quad \frac{d(c)}{dt} = 0$$

3. Add these results:

$$\frac{dx}{dt} = at + b$$

4. To compute the second derivative, repeat the process for the result in step 3:

$$\frac{d^2x}{dt^2} = at + 0 = a$$

REFLECT
The physical dimensions show that the answer is plausible. The original function is an equation for position; all terms are in meters—the units of t^2 and t cancel the units of s^2 and s in the constants a and b, respectively. In the function for dx/dt, all terms are similarly in m/s: the constant c has differentiated to zero, and the unit for t cancels one of the units for s in the constant a. In the function for dx^2/dt^2, only the acceleration constant remains; as expected, its dimensions are L/T^2.

Practice Problems
25. Find dy/dx for $y = \frac{5}{8}x^3 - 24x - \frac{5}{8}$.
26. Find dy/dt for $y = ate^{bt}$, where a and b are constants.

Solving Differential Equations Using Complex Numbers

A **differential equation** is an equation in which the derivatives of a function appear as variables. It is an equation in which the variables are related to each other through their derivatives. Consider an equation of the form

$$a\frac{d^2x}{dt^2} + b\frac{dx}{dt} + cx = A\cos\omega t \qquad \text{M-63}$$

that represents a physical process, such as a damped harmonic oscillator driven by a sinusoidal force, or a series RLC combination being driven by a sinusoidal potential drop. Although each of the parameters in Equation M-63 is a real number, the time-dependent cosine term suggests that we might find the steady-state solution to this equation by introducing complex numbers. We first construct the "parallel" equation

$$a\frac{d^2y}{dt^2} + b\frac{dy}{dt} + cy = A\sin\omega t \qquad \text{M-64}$$

Equation M-64 has no physical meaning of its own, and we have no interest in solving it. However, it is of use in solving Equation M-63. After multiplying through Equation M-64 by the unit imaginary i, we add Equation M-64 and Equation M-63 to obtain

$$\left(a\frac{d^2x}{dt^2} + ai\frac{d^2y}{dt^2}\right) + \left(b\frac{dx}{dt} + bi\frac{dy}{dt}\right) + (cx + ciy) = A\cos\omega t + Ai\sin\omega t$$

We next combine terms to get

$$a\frac{d^2(x + iy)}{dt^2} + b\frac{d(x + iy)}{dt} + c(x + iy) = A(\cos\omega t + i\sin\omega t) \qquad \text{M-65}$$

which is valid because the derivative of a sum is equal to the sum of the derivatives. We simplify our result by defining $z = x + iy$ and by using the identity $e^{i\omega t} = \cos\omega t + i\sin\omega t$. Substituting these into Equation M-65, we obtain

$$a\frac{d^2z}{dt^2} + b\frac{dz}{dt} + cz = Ae^{i\omega t} \qquad \text{M-66}$$

which we now solve for z. Once z is obtained, we can solve for x using $x = \text{Re}(z)$.

Because we are looking only for the steady-state solution for Equation M-65, we can assume its solution is of the form $x = x_0\cos(\omega t - \varphi)$, where φ is a constant. This is equivalent to assuming that the solution to Equation M-66 is of the form $z = \eta e^{i\omega t}$, where η, pronounced eta (like beta without the b), is a constant complex number. Then $dz/dt = i\omega z$, $d^2z/dt^2 = -\omega^2 z$, and $e^{i\omega t} = z/\eta$. Substituting these into Equation M-65 gives

$$-a\omega^2 z + i\omega bz + cz = A\frac{z}{\eta}$$

Dividing both sides of this equation by z and solving for η gives

$$\eta = \frac{A}{-a\omega^2 + i\omega b + c}$$

Expressing the denominator in polar form gives

$$(-a\omega^2 + c) + i\omega b = \sqrt{(-a\omega^2 + c)^2 + \omega^2 b^2}\,e^{i\varphi}$$

where $\tan\varphi = \omega^2 b^2/(-a\omega^2 + c)$. Thus,

$$\eta = \frac{A}{\sqrt{(-a\omega^2 + c)^2 + \omega^2 b^2}}\,e^{-i\varphi}$$

so

$$z = \eta e^{i\omega t} = \frac{A}{\sqrt{(-a\omega^2 + c)^2 + \omega^2 b^2}} e^{i(\omega t - \varphi)}$$

$$= \frac{A}{\sqrt{(-a\omega^2 + c)^2 + \omega^2 b^2}} [\cos(\omega t - \varphi) + i\sin(\omega t - \varphi)] \qquad \textbf{M-67}$$

It follows that

$$x = \text{Re}(z) = \frac{A}{\sqrt{(-a\omega^2 + c)^2 + \omega^2 b^2}} \cos(\omega t - \varphi) \qquad \textbf{M-68}$$

The Exponential Function

An **exponential function** is a function of the form a^{bx}, where $a > 0$ and b are constants. The function is usually written as e^{cx}, where c is constant.

When the rate of change of a quantity is proportional to the quantity itself, the quantity increases or decreases exponentially, depending on the sign of the proportionality constant. An example of an *exponentially* decreasing function is nuclear decay. If N is the number of radioactive nuclei at some time, then the change dN in some very small time interval dt will be proportional to N and to dt:

$$dN = -\lambda N\,dt$$

where λ is the *decay constant* (not to be confused with the decay rate dN/dt, which decreases exponentially). The function N satisfying this equation is

$$N = N_0 e^{-\lambda t} \qquad \textbf{M-69}$$

where N_0 is the value of N at time $t = 0$. **Figure M-25** shows N versus t. A characteristic of exponential decay is that N decreases by a constant factor in a given time interval. The time interval for N to decrease to half its original value is its *half-life* $t_{1/2}$. The half-life is obtained from Equation M-69 by setting $N = \frac{1}{2}N_0$ and solving for the time. This gives

$$t_{1/2} = \frac{\ln 2}{\lambda} = \frac{0.693}{\lambda} \qquad \textbf{M-70}$$

An example of *exponential increase* is population growth. If the number of organisms is N, the change in N after a very small time interval dt is given by

$$dN = +\lambda N\,dt$$

where λ is now the *growth constant*. The function N satisfying this equation is

$$N = N_0 e^{\lambda t} \qquad \textbf{M-71}$$

(Note the change of sign in the exponent.) A graph of this function is shown in **Figure M-26**. An exponential increase can be characterized by a doubling time T_2, which is related to λ by

$$T_2 = \frac{\ln 2}{\lambda} = \frac{0.693}{\lambda} \qquad \textbf{M-72}$$

Very often, we know population growth as an annual percentage increase and wish to calculate the doubling time. In this case, we find T_2 (in years) from the equation

$$T_2 = \frac{69.3}{r} \qquad \textbf{M-73}$$

where r is the percent per year. For example, if the population increases by 2 percent per year, the population will double every $69.3/2 \approx 35$ years. **Table M-4** lists some useful relations for exponential and logarithmic functions.

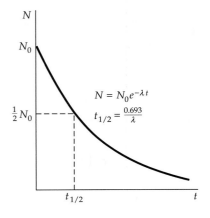

Figure M-25 Graph of N versus t when N decreases exponentially. The time $t_{1/2}$ is the time it takes for N to decrease by one-half.

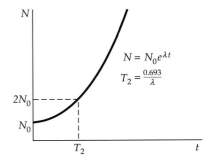

Figure M-26 Graph of N versus t when N increases exponentially. The time T_2 is the time it takes for N to double.

Table M-4 Exponential and Logarithmic Functions

$e = 2.718\ 28\ldots$

$e^0 = 1$

If $y = e^x$, then $x = \ln y$

$e^{\ln x} = x$

$e^x e^y = e^{(x+y)}$

$(e^x)^y = e^{xy} = (e^y)^x$

$\ln e = 1; \ln 1 = 0$

$\ln xy = \ln x + \ln y$

$\ln \dfrac{x}{y} = \ln x - \ln y$

$\ln e^x = x; \ln a^x = x \ln a$

$\ln x = (\ln 10) \log x$

$\quad = 2.30\ 26 \log x$

$\log x = (\log e) \ln x = 0.43429 \ln x$

$e^x = 1 + x + \dfrac{x^2}{2!} + \dfrac{x^3}{3!} + \cdots$

$\ln(1 + x)$

$\quad = x - \dfrac{x^2}{2} + \dfrac{x^3}{3} - \dfrac{x^4}{4} + \cdots$

Example M-13 Radioactive Decay of Cobalt-60

The half-life of cobalt-60 (^{60}Co) is 5.27 y. At $t = 0$ you have a sample of ^{60}Co that has a mass equal to 1.20 mg. At what time t (in years) will 0.400 mg of the sample of ^{60}Co have decayed?

SET UP
When we derived the half-life in exponential decay, we set $N/N_0 = 1/2$. In this example, we are to find the time at which two-thirds of a sample remains, and so the ratio N/N_0 will be 0.667.

SOLVE

1. Express the ratio N/N_0 as an exponential function:

$$\frac{N}{N_0} = 0.667 = e^{-\lambda t}$$

2. Take the reciprocal of both sides:

$$\frac{N_0}{N} = 1.50 = e^{\lambda t}$$

3. Solve for t:

$$t = \frac{\ln 1.50}{\lambda} = \frac{0.405}{\lambda}$$

4. The decay constant is related to the half-life by $\lambda = (\ln 2)/t_{1/2}$ (Equation M-70). Substitute $(\ln 2)/t_{1/2}$ for λ and evaluate the time:

$$t = \frac{\ln 1.5}{\ln 2} t_{1/2} = \frac{\ln 1.5}{\ln 2} \times 5.27 \text{ y} = 3.08 \text{ y}$$

REFLECT
It takes 5.27 y for the mass of a sample of ^{60}Co to decrease to 50% of its initial mass. Thus, we expect it to take less than 5.27 y for the sample to lose 33.3% of its mass. Our step-4 result of 3.08 y is less than 5.27 y, as expected.

Practice Problems

27. The discharge time constant τ of a capacitor in an RC circuit is the time in which the capacitor discharges to e^{-1} (or 0.368) times its charge at $t = 0$. If $\tau = 1$ s for a capacitor, at what time t (in seconds) will it have discharged to 50.0% of its initial charge?

28. If the coyote population in your state is increasing at a rate of 8.0% a decade and continues increasing at the same rate indefinitely, in how many years will it reach 1.5 times its current level?

M-12 Integral Calculus

Integration can be considered the inverse of differentiation. If a function $f(t)$ is *integrated*, a function $F(t)$ is found for which $f(t)$ is the derivative of $F(t)$ with respect to t.

The Integral as an Area Under a Curve; Dimensional Analysis

The process of finding the area under a curve on a graph illustrates integration. **Figure M-27** shows a function $f(t)$. The area of the shaded element is approximately $f_i \Delta t_i$ where f_i is evaluated anywhere in the interval Δt_i. This approximation is highly accurate if Δt_i is very small. The total area under some stretch of the curve is found by summing all the area elements it covers and taking the limit as

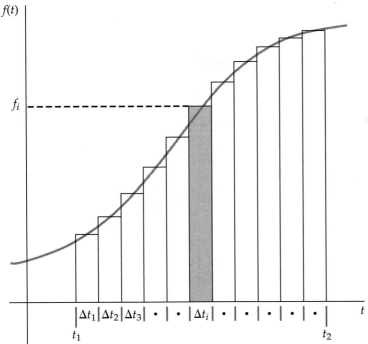

Figure M-27 A general function $f(t)$. The area of the shaded element is approximately $f_i \Delta t_i$, where f_i is evaluated anywhere in the interval.

each Δt_i approaches zero. This limit is called the **integral** of f over t and is written

$$\int f \, dt = \text{area}_i = \lim_{\Delta t \to 0} \sum_i f_i \, \Delta t_i \qquad \textbf{M-74}$$

The *physical dimensions* of an integral of a function $f(t)$ are found by multiplying the dimensions of the *integrand* (the function being integrated) and the dimensions of the integration variable t. For example, if the integrand is a velocity function $v(t)$ (dimensions L/T) and the integration variable is time t, the dimension of the integral is L = (L/T) × T. That is, the dimensions of the integral are those of velocity times time.

Let

$$y = \int_{t_1}^{t} f \, dt \qquad \textbf{M-75}$$

The function y is the area under the f versus t curve from t_1 to a general value t. For a small interval Δt, the change in the area Δy is approximately $f \, \Delta t$:

$$\Delta y \approx f \, \Delta t$$

$$f \approx \frac{\Delta y}{\Delta t}$$

If we take the limit as Δt approaches 0, we can see that f is the derivative of y:

$$f = \frac{dy}{dt} \qquad \textbf{M-76}$$

Indefinite Integrals And Definite Integrals

When we write

$$y = \int f \, dt \qquad \textbf{M-77}$$

Table M-5 Integration Formulas[†]
1. $\int A \, dt = At$
2. $\int At \, dt = \dfrac{1}{2} At^2$
3. $\int At^n \, dt = A\dfrac{t^{n+1}}{n+1}, \; n \neq -1$
4. $\int At^{-1} \, dt = A \ln\lvert t \rvert$
5. $\int e^{bt} \, dt = \dfrac{1}{b} e^{bt}$
6. $\int \cos \omega t \, dt = \dfrac{1}{\omega} \sin \omega t$
7. $\int \sin \omega t \, dt = -\dfrac{1}{\omega} \cos \omega t$
8. $\displaystyle\int_0^\infty e^{-ax} \, dx = \dfrac{1}{a}$
9. $\displaystyle\int_0^\infty e^{-ax^2} \, dx = \dfrac{1}{2} \sqrt{\dfrac{\pi}{a}}$
10. $\displaystyle\int_0^\infty xe^{-ax^2} \, dx = \dfrac{2}{a}$
11. $\displaystyle\int_0^\infty x^2 e^{-ax^2} \, dx = \dfrac{1}{4} \sqrt{\dfrac{\pi}{a^3}}$
12. $\displaystyle\int_0^\infty x^3 e^{-ax^2} \, dx = \dfrac{4}{a^2}$
13. $\displaystyle\int_0^\infty x^4 e^{-ax^2} \, dx = \dfrac{3}{8} \sqrt{\dfrac{\pi}{a^5}}$

[†]In these formulas, A, b, and ω are constants. In formulas 1 through 7, an arbitrary constant C can be added to the right side of each equation. The constant a is greater than zero.

we are showing y as an **indefinite integral** of f over t. To evaluate an indefinite integral, we find the function y whose derivative is f. Because that function could contain a constant term that differentiated to zero, we include as our final term a **constant of integration** C. If we are integrating the function over a known segment—such as t_1 to t_2 in Figure M-27—we can find a **definite integral**, eliminating the unknown constant C:

$$\int_{t_1}^{t_2} f\, dt = y(t_2) - y(t_1) \qquad \text{M-78}$$

Table M-5 lists some important integration formulas. More extensive lists of integration formulas can be found in any calculus textbook or by searching for "table of integrals" on the Internet.

Example M-14 Integrating Equations of Motion

A particle is moving at a constant acceleration a. Write a formula for position x at time t given that the position and velocity are x_0 and v_0 at time $t = 0$.

SET UP
Velocity v is the derivative of x with respect to time t, and acceleration is the derivative of v with respect to t. We should be able to write a function $x(t)$ by performing two integrations.

SOLVE

1. Integrate a with respect to t to find the v as a function of t. The a can be factored from the integrand because a is constant:

$$v = \int a\, dt = a \int dt$$
$$v = at + C_1$$

where C_1 represents a multiplied by the constant of integration.

2. The velocity $v = v_0$ when $t = 0$:

$$v_0 = 0 + C_1 \Rightarrow C_1 = v_0$$
$$\text{so } v = v_0 + at$$

3. Integrate v with respect to t to find x as a function of t:

$$x = \int v\, dt = \int (v_0 + at)dt = \int v_0 dt + \int at\, dt$$
$$x = v_0 \int dt + a \int t\, dt = v_0 t + \tfrac{1}{2}at^2 + C_2$$

where C_2 represents the combined constants of integration.

4. The position $x = x_0$ when $t = 0$ is

$$x_0 = 0 + 0 + C_2$$
$$\text{so } x = x_0 + v_0 t + \tfrac{1}{2}at^2$$

REFLECT
Differentiate the step-4 result twice to get the acceleration

$$v = \frac{dx}{dt} = \frac{d}{dt}(x_0 + v_0 t + \tfrac{1}{2}at^2) = 0 + v_0 + at$$

$$a = \frac{dv}{dt} = \frac{d}{dt}(v_0 + at) = a$$

Practice Problems

29. $\displaystyle\int_{3}^{6} 3\, dx =$

30. $\displaystyle V = \int_{5}^{8} \pi r^2 dL =$

Answers to Practice Problems

1. 0.24 L
2. 31.6 m/s
3. 6.0 kg/cm^3
4. -3
5. 1.54 L
6. 3.07 L
7. False
8. $x = (4.5\ \text{m/s})t + 3.0\ \text{m}$
9. $x = 8, y = 60$
11. $2(x - y)^2$
12. $x^2(2x + 4)(x + 3)$
13. $x^{1/2}$
14. x^6
15. 3
16. ~2.322

17. $V/A = \frac{1}{3}r$
18. $A = \dfrac{2}{3}\pi L^2$
19. $\sin\theta = 0.496, \cos\theta = 0.868, \theta = 29.7°$
20. $\sin 8.2° = 0.1426, 8.2° = 0.1431$ rad
21. 0.996, 0.996 00, close to 0%
22. 0.96, 0.960 77, $\ll 1\%$
23. $-1 + 0i = -1$
24. $0 + i = i$
25. $dy/dx = \frac{15}{8}x^2 - 24$
26. $dy/dt = ae^{bt}(bt + 1)$
27. 0.693 s
28. 51 y
29. 9
30. $3\pi r^2$

This page is intentionally left blank.
In the first edition this page will provide
answers to odd-numbered problems.

This page is intentionally left blank.
In the first edition this page will provide
answers to odd-numbered problems.

This page is intentionally left blank.
In the first edition this page will provide
answers to odd-numbered problems.

This page is intentionally left blank.
In the first edition this page will provide
answers to odd-numbered problems.

This page is intentionally left blank.
In the first edition this page will provide
answers to odd-numbered problems.

This page is intentionally left blank.
In the first edition this page will provide
answers to odd-numbered problems.

This page is intentionally left blank.
In the first edition this page will provide
answers to odd-numbered problems.

This page is intentionally left blank.
In the first edition this page will provide
answers to odd-numbered problems.

This page is intentionally left blank.
In the first edition this page will provide
answers to odd-numbered problems.

This page is intentionally left blank.
In the first edition this page will provide
answers to odd-numbered problems.

This page is intentionally left blank.
In the first edition this page will provide
answers to odd-numbered problems.

This page is intentionally left blank.
In the first edition this page will provide
answers to odd-numbered problems.

This page is intentionally left blank.
In the first edition this page will provide
answers to odd-numbered problems.

This page is intentionally left blank.
In the first edition this page will provide
answers to odd-numbered problems.

This page is intentionally left blank.
In the first edition this page will provide
answers to odd-numbered problems.

This page is intentionally left blank.
In the first edition this page will provide
answers to odd-numbered problems.

This page is intentionally left blank.
In the first edition this page will provide
answers to odd-numbered problems.

This page is intentionally left blank.
In the first edition this page will provide
answers to odd-numbered problems.

This page is intentionally left blank.
In the first edition this page will provide
answers to odd-numbered problems.

This page is intentionally left blank.
In the first edition this page will provide
answers to odd-numbered problems.

This page is intentionally left blank.
In the first edition this page will provide
answers to odd-numbered problems.

This page is intentionally left blank.
In the first edition this page will provide
answers to odd-numbered problems.

This page is intentionally left blank.
In the first edition this page will provide
answers to odd-numbered problems.

This page is intentionally left blank.
In the first edition this page will provide
answers to odd-numbered problems.

INDEX